MANUFACTURING TECHNOLOGY

Volume I

Foundry, Forming and Welding

Fifth Edition

About The Author

P N Rao received his BE degree in Mechanical Engineering from the Government Engineering College, Anantapur (Sri Venkateswara University, Tirupati), in 1970, ME degree from Birla Institute of Technology and Science, Pilani, in 1973 and PhD from the Indian Institute of Technology, Delhi, in 1981.

Presently, he is working as a Professor in the University of Northern Iowa, Cedar Falls, USA, in the Department of Technology, since Fall of 2001. Earlier he worked in the Indian Institute of Technology, New Delhi, from 1975 till 1997, the final position being as a Professor in the Mechanical Engineering Department between 1990 and 1997. He has also taught at the MARA University of Technology, Shah Alam, Malaysia, between 1995 and 2001. Prior to joining IIT Delhi, he worked in the Birla Institute of Technology and Science, Pilani, for a period of two years. He has conducted post-doctoral research in UMIST, Manchester, UK (1980–1981), and Loughborough University of Technology, UK (1989).

His active areas of research are Manufacturing Engineering, Design Engineering, Sustainable Manufacturing, CAD/CAM,CIM, Tool design and MEMS, and Nano education. He is the author and co-author of more than 250 papers, published and presented in national and international journals and conferences, respectively.

He also has wide interaction with the industry through the process of consultancy work. He conducts continuing education programmes on various aspects related to modern manufacturing. He has received a number of grants from various bodies to support his research activities.

He has authored three books, all published by McGraw-Hill Education (India).

- Manufacturing *Technology, Volume 2: Metal Cutting and Machine Tools,* 4e
- *Computer Aided Manufacturing,* 1e (with *Kundra* and *Tewari* as co-authors)
- *CAD/CAM: Principles and Applications*, 3e

He has also worked as a co-editor of *Emerging Trends in Manufacturing* (Proceedings of the 12th All India Machine Tool Design and Research Conference, 1986) published by McGraw-Hill Education (India).

Dr P N Rao's biographical information has appeared in *Marquis Who's Who in The World, Marquis Who's Who in Finance and Industry, Marquis Who's Who in Science and Engineering, Marquis Who's Who in America, Marquis Who's Who in American Education, and Who's Who in Asia and the Pacific Nations.*

MANUFACTURING TECHNOLOGY

Volume I

Foundry, Forming and Welding

Fifth Edition

P N Rao
Professor
Department of Technology
University of Northern Iowa
Cedar Falls
USA

McGraw Hill Education (India) Private Limited

Published by McGraw Hill Education (India) Private Limited,
Registered Office: Anjana Complex No: 5/90 A, Butt Road, St. Thomas Mount, Chennai – 600016, Tamil Nadu, India.

Manufacturing Technology—Foundry, Forming and Welding, 5e (Volume 1)

Tenth reprint 2026

Print Edition:
ISBN-13: 978-93-5316-050-0
ISBN-10: 93-5316-050-2

E-Book Edition:
ISBN-13: 978-93-5316-051-7
ISBN-10: 93-5316-051-0

11 12 13 14 15 RAJ 30 29 28 27 26

Typeset at the Composers, 260, C.A Apt., Paschim Vihar, New Delhi 110063, and Printed and bound in India at Rajkamal Electric Press, Plot No. 2, Phase-IV, Kundli, Haryana

Cover Designer: The Computer Minds
Cover Image Source: Panksvatouny/Shutterstock

Visit us at: www.mheducation.co.in
Toll free in India: 1800 103 5875
Write to us at: support.india@mheducation.com
CIN: U80302TN2010PTC111532

DEDICATED TO

My Parents, Suramma and Kondala Rayudu

Contents

Preface to the Fifth Edition

I am happy to note that the book is getting continuous support from the technical educational community as a textbook for manufacturing technology among most of the universities in India. As the technology is changing at a rapid rate, some of these technologies will be percolating into manufacturing at various levels. The universities therefore make changes to the curricula to reflect these advances. This edition is therefore an attempt to fulfil all those requirements.

Content Overview

Whenever a process or equipment is described, the practical information, such as specifications, operating parameters and designing for the process, have all been highlighted. Each process is supplemented with simple illustrations, numerical calculations for the design process and a discussion of the results so obtained. A large number of well-labelled illustrations are provided to give the necessary insight into the process and its design.

Chapter 1 introduces the different manufacturing processes and the importance of manufacturing for humankind. *Chapter 2* discusses engineering materials. *Chapter 3* is on metal-casting processes. *Chapter 4* describes gating systems for casting. *Chapter 5* discusses melting and casting quality. *Chapter 6* is on special casting processes. *Chapter 7* explains the different types of metal-forming processes, while *Chapter 8* describes some sheet-metal operations. *Chapter 9* deals with welding processes, and *Chapter 10* is on other fabrication processes. *Chapter 11* describes powder metallurgy. *Chapter 12* deals with plastic processing. *Chapter 13* deals with ceramics and glass materials processing. Finally, *Chapter 14* deals with composite material processing.

New Features

All the chapters were thoroughly checked to see that written material is in line with the current practice such that some of the obsolete details are removed. In addition to that, some of the major additions are detailed as follows:

Chapter 2 – Shot peening
Chapter 3 – Additive manufacturing (rapid prototyping) for pattern making
Chapter 5 – Non-destructive testing
Chapter 6 – Thixocasting
Chapter 8 – Lancing
Chapter 9 – Non-destructive testing of weldments
Chapter 11 – Metal injection moulding

Two new chapters are also added:

Chapter 13 – Ceramics and Glasses
Chapter 14 – Composite Materials

A few case studies have been added to provide further discussion on the topics covered, from the practical viewpoint. The instructors can discuss these case studies in more detail in the class so that it will generate interest in the students towards the subject matter. For GATE aspirants, a dedicated section at book end is provided for thorough practice.

With these additions and changes, it is hoped that the current edition will be able to fully satisfy the curricula of most of the universities and thus serve the intended purpose with which I started in the beginning.

Online Learning Centre

The text is supported with online supplements for instructors and students. Instructor resources are available through *https://connect.mheducation.com* while student resources are available through *https://www.mheducation.co.in* (see the back cover for more details).

Instructors will have access to Solution Manual and PowerPoint Lecture Slides. Chapter-wise GATE Previous Years Questions are given for students.

Acknowledgements

Special thanks are due to my colleague, Dr Srikanth Revuru, who has helped me with the development of all the case studies presented in the book. I wish to express my sincere thanks to University of Northern Iowa, for providing excellent environment and facilities so that I could carry out the updating of this book. Special thanks are extended to my colleagues, Dr Scott Geise and Mr Sairam Ravi, of the Metal Casting Center of our department, who helped me with the material for the casting simulation using Magma software. I am delighted to record my appreciation for the editorial staff at McGraw Hill (India) who really nudged me to complete this task as fast as possible such that the book is produced with the highest quality within the shortest time so that the book is available in time for the new academic session.

I would like to thank the reviewers who have gone through the revised content of the book and have given their valuable suggestions. In particular, I would like to thank:

Pravin Kumar	Delhi Technological University, Delhi
Dhananjay A. Jolhe	Visvesvaraya National Institute of Technology (VNIT), Nagpur
Mukund Dutt Sharma	National Institute of Technology (NIT), Hamirpur, Himachal Pradesh
Harmesh Kumar	University Institute of Engineering and Technology, Panjab University, Chandigarh
Shiv Lal	Government Engineering College, Banswara, Rajasthan Technical University (RTU), Kota

Feedback

Despite the utmost care taken, it is not uncommon that some errors will be left in the book uncorrected, and I would request the readers to communicate any such errors and omissions so that I will be able to correct them at the earliest possible opportunity.

P. N. RAO

Preface to the First Edition

This new book on manufacturing technology relates to its practice with as much of scientific aspects as possible. The study of manufacturing processes forms a core subject area for a majority of engineering students. In particular, this is an essential subject for all mechanical engineering students. But it is my experience that more often the subject is taught with greater emphasis on the descriptive aspect rather than from a scientific and practical viewpoint. As a result, a fresh engineering graduate, when he enters a manufacturing unit, would be at a loss as to the means of correlating what he learned and what is required in practice. With this in mind, an attempt has been made to bring in as much of practice as possible into this book to make it more useful for engineering students.

This book is the outgrowth of material used by me for teaching two undergraduate courses relating to manufacturing processes. Whenever a process is described, the practical information, such as specifications, operating parameters and designing for the process, have all been highlighted. Each process is supplemented with simple illustrations, numerical calculations for the design process and a discussion of the results so obtained. A large number of well-labelled illustrations are provided to give the necessary insight into the process and its design.

After an introductory chapter, a short chapter is given to provide the necessary details of engineering properties as related to manufacturing processes. Three chapters have been devoted to the essential details of ferrous and non-ferrous materials along with their heat treatment methods.

Another 11 chapters have been devoted to the technology of casting processes, wherein the sand casting process has been extensively dealt with from the technological viewpoint. In a concise form the available knowledge relating to the pattern, gating, risering and product design has been presented in these chapters. It is expected that an engineering student should be able to do a complete sand casting design from the information provided in this book.

Six chapters are devoted to the technological aspects related to metal forming processes, such as rolling, forging, extrusion and sheet metal operations. A good number of design examples as well as design exercises are provided wherever necessary.

Lastly, the welding and allied processes are covered in eight chapters, again giving a good amount of practical information. The emphasis is more on the commercial welding processes such as arc welding as well as the modern developments in welding processes.

SI units have been used throughout the book without exception. Countries all over the world have adopted these units because of the unambiguous representation of the various quantities. Hence it is imperative that engineering education should be done in SI units so that change over takes place at the earliest and with least resistance. A brief introduction to SI units as related to the measurements used in this book has been provided in the appendix along with the essential data related to conversions from British or fps units to SI units.

References were provided at the end of each chapter which should be useful for those interested in studying further into the specific aspects. Also, a list of Indian standards that are relevant for the subject under discussion has been provided at the end of each chapter. This information together with the numerous tables that are provided should help the practising engineer as well in the proper utilisation of the manufacturing processes.

I wish to express my sincere gratitude to Prof. U.R.K.Rao, Prof. N.K.Tewari and Mr. S. Swaminathan who have provided constant encouragement to undertake this type of productive activity over a period of time. I am particularly thankful to Mr S.Swaminathan and Prof. U.R.K.Rao for having taken the pains to review major portions of the manuscript and for giving useful suggestions for improvements. It is a pleasure to express heartfelt gratitude to my family members who have borne long hours of inconvenience during the preparation of the manuscript. I am indebted to the authorities of Indian Institute of Technology, New Delhi, for having allowed me to undertake this activity and provided the necessary facilities. Lastly, I wish to acknowledge the contributions made by many of my students over the past years during the teaching of the courses which enriched the book in many forms.

I am open to criticism and suggestions regarding the coverage in the book, and would welcome any helpful suggestions for improvement in future editions.

P. N. RAO

Visual Walkthrough

Historical Perspective

Provides a brief perspective of historical developments related to the processes discussed in the chapter

11.1 INTRODUCTION

Powder metallurgy is the name given to a process in which metallic powders are heated below their melting temperatures to achieve the bonding. In reality the powder metallurgy (P/M) process involves compacting of metal or alloy powders into the desired shape after blending and then heated in a controlled atmosphere at a temperature below the melting point in order to achieve the bonding of the particles to get the desired properties. The powder metallurgy process enables to produce parts in their final shape eliminating the need for any additional machining. Raw material is not wasted during the processing while unusual materials or mixtures can be utilized. It is possible to get parts with unique properties not possible by any other manufacturing process. Most of the powder metallurgy parts are in the size range of less than 2 kg, though parts as large as 20 kg were made. Large parts require very expensive tooling and as such are not widely made by powder metallurgy.

The limit to the size of a PM component is based on the final density and the available press size. Typical values are the projected area (area perpendicular to the pressing direction) usually between 4 and 16,000 mm^2, with length between 0.8 to 150 mm, although 75 mm is the practical maximum. There are parts weighing above 10 kg in production today.

Though powder metallurgy was appeared to have been used by Egyptians around 3000 BC, the modern developments have started in the mid or late nineteenth century. The early interest was in the self lubricated bearings. Later the invention of incandescent light required the filament to be made by P/M process. The invention of tungsten carbide in 1920's utilized the P/M process. Later the automobile industry had utilized the P/M process to a great extent and even today it accounts for a large volume of its usage. Many of the enhancements in various powder metallurgy technologies have taken place in the last 70 years.

Chapter Objectives

Each chapter begins with a clearly defined set of objectives that provide a quick reference to the chapter's key aspects. These help students better anticipate what they will be studying and help instructors measure student's understanding.

Objectives

Sand casting processes described so far are not suitable and economical in many applications. In such situations special casting processes would be more appropriate. After completing this chapter, the reader will be able to

- Understand the shell moulding as a process for higher quality castings with thin surface details
- Know precision investment casting used for complex shapes
- Use permanent mould casting for mass production of relatively simple shapes
- Apply die casting process for complex parts
- Understand the advantages of using centrifugal casting process
- Learn the advantages and applications of various other special casting process

Well-labelled Illustrations

Neat illustrations within the chapters provide a complete description of the object in question, labelling the various parts describing the function.

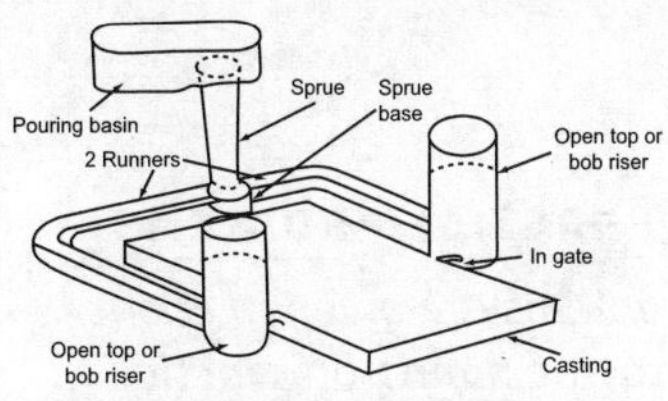

Fig. 4.1 *Typical gating system*

Example 4.1

Calculate the optimum pouring time for a casting whose mass is 20 kg and having an average section thickness of 15 mm. The materials of the casting are grey cast iron and steel. Take the fluidity of iron as 28 inches.

Grey cast iron

Pouring time, $t = K\left(1.41 + \frac{T}{14.59}\right)\sqrt{W}$ s

$= \frac{28}{40}\left(1.41 + \frac{15}{14.59}\right)\sqrt{20} = 7.632$ s

Steel

Pouring time, $t = (2.4335 - 0.3953 \log W)\sqrt{W}$ s

$= (2.4335 - 0.3953 \log 20)\sqrt{20} = 8.5825$ s

Example 4.2

Calculate the optimum pouring time for a casting whose mass is 100 kg and a thickness of 25 mm. Fluidity of iron is 32 inches. Calculate both for cast iron and steel.

Grey cast iron

Pouring time, $t = \frac{32}{40}\left(1.41 + \frac{25}{14.59}\right)\sqrt{100} = 24.988$ s

Steel

Pouring time, $t = (2.4335 - 0.3953 \log 100)\sqrt{100} = 16.429$ s

Solved Examples

Throughout each chapter, various solved examples are given that help readers understand and apply the concept learnt in the chapter.

Summary

A detailed chapter-end summary is provided for a quick review of the important concepts discussed in the chapter.

SUMMARY

Gating and risering systems are important for the final quality of the casting achieved. To this extent, there are a number of elements that are present in the path of the molten metal.

- Pouring basin, sprue, sprue-base well, runner and runner extension serve the purpose of allowing clean molten metal to enter the mould cavity.
- Parting gate is the most widely used gate while the top and bottom gates are sometimes used for specific applications that favour them.
- Fluid mechanics laws together with empirical relations are applied to design the optimum gating system.
- It is important to make sure that slag entering the gating system be removed completely before the metal enters the mould cavity.
- Risers compensate for the volumetric shrinkage taking place during the solidification of the molten metal inside the mould cavity.
- Caine's method provides a simple procedure to size the risers in sandcastings, while the modulus method is more elaborate and provides a better design.
- Sometimes chills may need to be added to reduce porosity at isolated sections that are not fed by risers.
- To reduce the size of risers, feeding aids such as riser sleeves are often used.

Questions

7.1 List the various processes of metal working.

7.2 What are the advantages of hot working over cold working of metals?

7.3 What are specific merits of cold working over hot working?

7.4 For hot working it is often necessary to heat the work piece in a furnace and there are scale losses and other problems. Why is hot working sometimes preferred to cold working in spite of such disadvantages?

7.5 What is the significance of recrystallisation temperature in metal working?

7.6 To what temperature should a given plain-carbon steel be heated for hot working?

7.7 State two parameters which determine the extent of deformation possible in a given material in a metal-working process.

7.8 What is meant by 'grain flow' in the case of forged or rolled components?

7.9 Give a line diagram of the various processes involved in the manufacture of the steel sheet, starting from steel ingots.

Revision Questions

A set of review questions has been carefully constructed to help students review their understanding of the concepts.

Practice Problems

Practice problems are given in each chapter to provide hands-on practice to students in solving problems related to real-life situations.

Problems

9.1 In a given arc welding operation, the power source is at 20 V and current at 300 A. If the electrode travel speed is 6 mm/s, calculate the cross-sectional area of the joint. The heat transfer efficiency is taken as 0.80 and melting efficiency as 0.30. Heat required to melt the steel is 10 J/mm^3.

9.2 In a resistance welding of a lap joint or two mild steel sheets of 1.5 mm thick, a current of 10 000 A is passed for a period of 0.1 seconds. The effective resistance of the joint is 120 micro ohms. Density of steel is 0.00786 g/mm^3 and heat required to melt is 1381 J/g. The joint can be considered as a cylinder of 5 mm diameter and 2.25 mm in height. Calculate the percentage of heat distributed to the surroundings.

9.3 A projection welding of 3 spots is obtained with a current of 35 000 A for a period of 0.01 second. If the effective resistance of the joint is taken as 120 micro ohms. If the joint can be considered as a cylinder of 6 mm diameter and 2.5 mm in height, calculate the heat distributed to the surroundings.

9.4 The arc welding of 6 mm steel plates in butt welding is being carried out at a welding speed of 9 mm/s. The welding transformer is set at 25 V and the current flowing is 300 A. Taking the arc efficiency as 0.85, calculate the temperature to which the plates should be preheated such that the critical cooling rate is kept below 6 °C/s at a temperature of 550°C for satisfactory metallurgical quality.

Multiple Choice Questions

11.1 Powder metallurgy process involves the following operations in sequence
(a) Powder mixing, sintering, compacting, and finishing
(b) Powder mixing, sintering and finishing
(c) Powder mixing, compacting, sintering and finishing
(d) Powder mixing, compacting, and sintering

11.2 Metal powder particle size is reduced during atomisation by
(a) Decreasing gas velocity
(b) Increasing gas velocity
(c) Decreasing gas pressure
(d) Decreasing metal volume

11.3 Identify the correct statement among the following:
(a) Blending and mixing refer to the same process of mixing different types of metal or alloy powders of different size distributions
(b) Blending is mixing the different metal or alloy powders of different size distributions while mixing is mixing same types
(c) Any type (single or multiple) of punch will be able to achieve good compacting of the powder metallurgy parts
(d) Punching with variable pressure will be able to achieve good compacting of the powder metallurgy parts

11.5 Green density of the powder metallurgy part will be increased by
(a) Increasing the sintering temperature
(b) Decreasing the sintering temperature
(c) Increasing the compacting pressure
(d) Decreasing the compacting pressure

11.6 Final strength of a powder metallurgy part can be increased by
(a) Increasing the sintering temperature
(b) Decreasing the sintering temperature
(c) Increasing the lubricant in the powder mix
(d) Decreasing the compacting pressure

11.7 To improve the mechanical properties of a powder metallurgy part, the following finishing operation is used
(a) Repressing (b) Sizing
(c) Coining (d) Impregnation

Multiple Choice Questions

These help the students to have a quick recap of the important terms and concepts learnt in the chapter.

Case Studies

Provides a practical view point of the subject to increase interest and understanding.

Case Study

ENERGY CONSUMPTION IN BULK FORMING

Due to the increasing population of the world, the energy demand is severe. Further, energy generation leads to a significant carbon footprint that affects the environment. Nearly about 20% of the energy is wasted in the industrial sector in the United States. Hence, it is the need of the hour to choose the processes and improve them to make them more sustainable. In this case study, the energy consumption of bulk forming was compared with additive and subtractive manufacturing. Bulk forming processes for plastics include injection moulding, thermoforming and compression moulding, where as in metals, the methods include die casting, sand casting, investment casting and forging.

In the present case study, injection moulding, milling (subtractive) and fused deposition modeling (FDM-Additive) are considered. A model of 30 × 30 × 5 mm with two holes and slots was selected. Specific Energy Consumption (SEC) is used for the purpose of comparing the different manufacturing processes. The energy consumed in the production of a component part is defined as SEC. It is defined as 1 joule per unit volume of material for subtractive processes and 1 joule per unit mass of material for additive processes.

Injection Moulding

The mould was machined using a CNC machine. The prepared aluminium mould was used for injection moulding. The warmup of the moulds took 27 minutes. The average power consumption was 534 W. The total energy consumption for cutting the mould and the injection moulding process was 832.4 Wh/part. However, the specific energy consumption (SEC) drastically decreased with the number of parts as the actual processing energy consumption was only 37 Wh/part.

Milling

In this study a CNC machine was used to manufacture the part. The process consisted of milling drilling and slotting. The total processing time was 1123 s including 15 min preparation time. The total energy consumption was 40.8 Wh.

GATE Previous Years Questions

Chapter 2

2.1 The process of reheating the martensitic steel to reduce its brittleness without any significant loss in its hardness is
(a) Normalizing (b) Annealing
(c) Quenching (d) Tempering
(GATE-2014-ME-SET-1, 1-Mark)

2.2 During normalizing process of steel, the specimen is heated
(a) Between the upper and lower critical temperature and cooled in still air
(b) Above the upper critical temperature and cooled in furnace

Process	Characteristics
P. Tempering	1. Austenite is converted into bainite
Q. Austempering	2. Austenite is converted into martensite
R. Martempering	3. Cementite is converted into globular structure
	4. Both hardness and brittleness are reduced
	5. Carbon is absorbed into the metal

(a) P-3, Q-1, R-5 (b) P-4, Q-3, R-2
(c) P-4, Q-1, R-2 (d) P-1, Q-5, R-4
(GATE-2004-ME-2-Marks)

GATE Section

Chapter-wise previous years GATE questions given for the students to have hands on practice.

CHAPTER 1

Introduction

Objectives

After completing this chapter, the reader will be able to

- Understand the importance of manufacturing for the mankind
- Know the different types of manufacturing processes
- Evaluate the feasibility of a process by carrying out break-even analysis

1.1 MANUFACTURING PROCESSES

The benefits of civilisation which we enjoy today are essentially due to the improved quality of products available to us. The improvement in the quality of the goods can be achieved with proper design that takes into consideration the functional requirement and its manufacturing aspects. The design process that would take proper care of the manufacturing process as well would be the ideal one. This would ensure a better product is being made available at an economical cost.

Manufacturing involved in turning raw materials to finished products is to be used for same purpose. In the present age there have been increasing demands on the product performance by way of desirable exotic properties such as resistance to high temperature, higher operating speed and extra load. These in turn would require a variety of new materials and its associated processing. Also, demanding working conditions that are desired in the modern industrial operations make large demands on the manufacturing industry.

Further, the economics of the manufacturing operations is a very important consideration. To be viable in the modern environment, a product has to be competitively priced besides having the functional and aesthetic appeal. Therefore it is necessary for an engineer to give a proper thought to various aspects of manufacturing.

Manufacturing processes is a very fundamental subject since it is of interest not only to mechanical engineers but also to those from practically every discipline of engineering. It is so because engineering as a whole is meant for providing various materials for human consumption. For various products such as plant machinery required for chemical, civil, electrical, electronic, textile, etc., the manufacturing process forms a vital ingredient.

A detailed understanding of the manufacturing processes is thus essential for every engineer. This helps them appreciate the capabilities, advantages and also the limitations of the various manufacturing processes. This in turn helps in the proper design of any product required by them. Firstly, they would be able to assess the manufacturing feasibility of their designs. Secondly, they may also find out that there is more than one process available for manufacturing a particular product. Keeping this in mind, they can make a proper choice of the process which would require the lowest manufacturing cost and would deliver the product of desired quality. This may result in slight modification in their design to suit the particular manufacturing process they choose.

Manufacturing, as defined by Merriam Webster online dictionary, is to make into a product suitable to use. Manufacturing remained as a craft from the first industrial revolution till the end of 18th century with low volumes and single piece production. It required highly skilled craftsmen to individually produce the pieces and fit them when the assembly was required. This was a slow and expensive process but in the absence of any machine that was the only thing possible.

The availability of steam engine and the coal fired furnaces have made the invention of a number of steam power driven machinery to greatly increase the manufacturing capacity. A large number of inventions related to machine tools have taken place during this period and continued into the 19th century. Towards the end of 19th century, with the invention of electricity and better engineering materials, manufacturing operations became more productive. The developments in the automobiles at the beginning of the 20th century are instrumental in the growth of a variety of manufacturing methods and practices.

There are a large number of processes available for manufacture to the engineer. These processes can be broadly classified into four categories:

(a) Casting Processes
(b) Forming Processes
(c) Fabrication Processes
(d) Material Removal Processes

1.1.1 Casting Processes

Casting is also one of the oldest known manufacturing processes. These are the only processes where liquid metal is used. It requires preparation of a cavity usually in a refractory material to resemble closely the final object to be made. Molten metal is poured into this refractory mould cavity and is allowed to solidify. The object after solidification is removed from the mould. Casting processes are universally used for manufacture of a wide variety of products. The principal process among these is sand casting where sand is used as the refractory material. The process is equally suitable for the production of a very small batch as well as on a very large scale.

Some of the other casting processes for specialised needs are as follows:

- Shell Mould Casting
- Precision Investment Casting
- Plaster Mould Casting
- Permanent Mould Casting
- Die Casting
- Centrifugal Casting

1.1.2 Forming Processes

These are solid state manufacturing processes involving minimum amount of material wastage and faster production. In a forming process, metal may be heated to a temperature which is slightly below the solidus temperature and then a large force is applied such that the material flows and takes the desired shape. The desired shape is controlled by means of a set of tools called dies which may be completely or partially closed during manufacturing. These processes are normally used for large-scale production rates. These are generally economical and in many cases improve the mechanical properties too.

Some of the metal forming processes are as follows:

- Rolling
- Drop Forging
- Press Forging
- Upset Forging
- Extrusion
- Wire Drawing
- Sheet Metal Operations

1.1.3 Fabrication Processes

These are secondary manufacturing processes where the starting raw materials are processed by any of the previous manufacturing processes described. It essentially involves joining pieces either permanently or temporarily to perform the necessary function. The joining can be achieved by either or both of heat and pressure and/or a joining material. Many of the steel structural constructions we see are first rolled and then joined together by a fabrication process.

Some of the processes of interest in this category are as follows:

- Gas Welding
- Electric Arc Welding
- Electric Resistance Welding
- Thermit Welding
- cold Welding
- Brazing
- Soldering

1.1.4 Material Removal Processes

These are also the secondary manufacturing processes where the additional unwanted material is removed in form of chips from the blank material by a harder tool to obtain the final desired shape. Material removal is normally the most expensive manufacturing process because more energy is consumed and also a lot of waste material is generated in the process. Still this is widely used because it delivers very good dimensional accuracy and good surface finish. It also generates accurate contours. Material removal processes are also called machining processes.

Various processes in this category are as follows:

- Turning
- Drilling
- Shaping and Planning
- Milling
- Grinding
- Broaching
- Sawing

All these manufacturing processes have been continuously developed to obtain better products at a reduced cost. Development of computers and their effect on the manufacturing processes has been of particular interest. The arrival of computers has made a remarkable difference to most of the above manufacturing processes. They have contributed greatly to both automation and designing of the process.

1.2 BREAK-EVEN ANALYSIS

An important responsibility of the engineer is to choose a manufacturing process which makes the required quality of a product to the specifications and at the lowest cost possible. To fulfil both the conditions, one would have to do a break-even analysis of the various processes suitable for the production of the given object.

In the break-even analysis, two types of costs are considered:

(a) Fixed cost relates to the initial investment on the equipment and tools required for the process. This cost would be constant and does not vary with quantity produced.

(b) Variable cost on the other hand varies with the actual number of objects made. This takes into account the raw material required, energy consumed, labour cost, cost of special toolings, cost of tools used and other administrative overheads. These requirements are directly proportional to the quantity produced.

The total cost of manufacture is thus the sum of both fixed cost and variable cost. This may be written in equation form as

$$TC = FC + VC \times Q$$

Where, TC = total cost

FC = fixed cost

VC = variable cost per piece

Q = production quantity

The same is shown in graphical form in Fig. 1.1. Normally as the quantity of production increases, the cost per unit produced decreases. This is due to the fact that the fixed cost associated with the production remains constant.

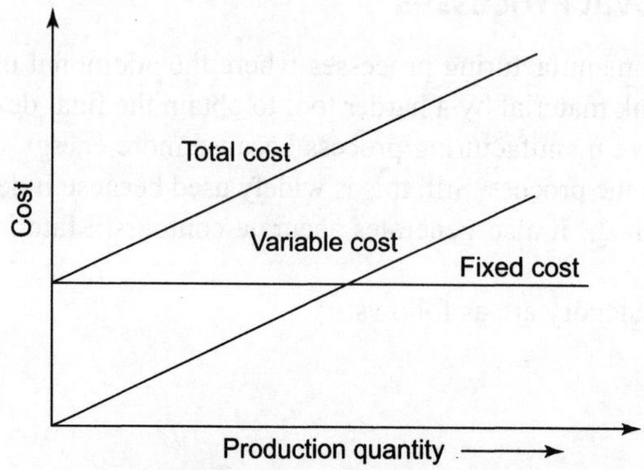

FIG. 1.1 *Typical cost curves*

If an engineer has to make a choice between two different processes for manufacturing the same component, he or she may first estimate the fixed costs and variable costs for both the processes and then plot the total cost curves for both as in Figure 1.2. The point where both these curves intersect is called the break-even point. It implies that the first process will be economical if the quantity of production is less than that of the break-even point, that while beyond it the second process is economical.

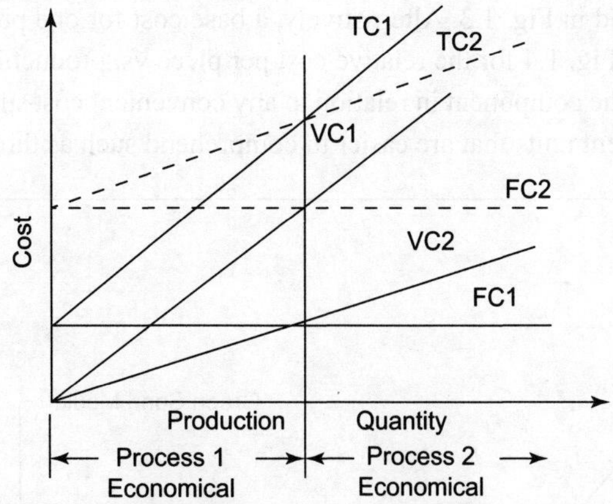

FIG. 1.2 *Break-even curve*

The break-even quantity '*n*' can also be obtained by equating the total costs in both the processes. That is

$$FC_1 + n \times VC_1 = FC_2 + n \times VC_2$$

$$n = \frac{FC_2 - FC_1}{VC_1 - VC_2}$$

This would give a positive value when fixed cost of a process is greater and variable cost lesser than those of the second. If both fixed and variable costs are lower than the other process, then the latter process is always uneconomical what ever may be the production quantity.

Another way of plotting a cost comparison curve is to calculate the average cost of production for different production quantity. The average cost is given by

$$\frac{\text{Total cost}}{\text{Production quantity}}$$

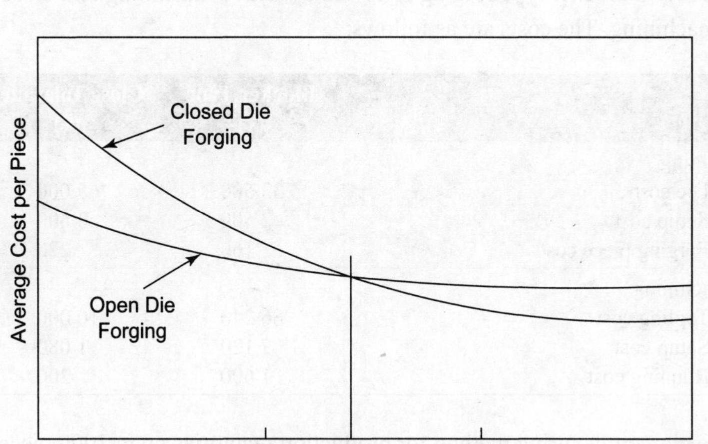

FIG. 1.3 *Break-even chart*

A typical curve is presented in Fig. 1.3. Alternatively, a base cost for one process could be taken and the curves could be plotted as in Fig. 1.4 for the relative cost per piece vs. production quantity. The relative cost can be defined as the cost of the component in relation to any convenient cost such that the abscissa would be represented by more convenient units that are easier to comprehend such as those shown in Fig. 1.4.

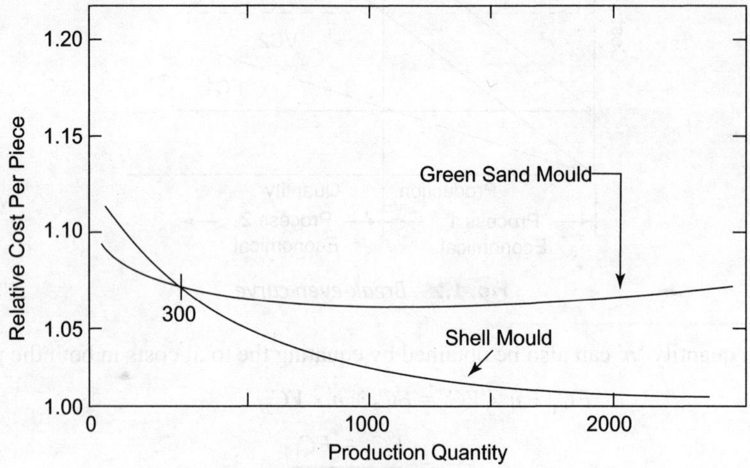

FIG. 1.4 *Break-even chart*

The manufacturing of a component may involve more than one process or more variants of the same process. In the first case it is necessary to consider the costs of all the processes while computing the total cost for each of the processes.

Example 1.1

In the manufacture of a machine fitting involving drop forging and machining there are two options available. First option is to produce a blocker type forging followed by heavy machining or a close toleranced forging followed by finish machining. The costs are as follows:

	Blocker Type (₹)	Close Tolerance (₹)
Forging		
Die cost	33 500	65 000
Setup cost	2 500	3 600
Forging piece cost	160	220
Machining		
Tooling cost	86 240	10 000
Setup cost	7 120	1 080
Running cost	1 600	200

Find the economical manufacturing method for production quantities up to 1000 pieces.

Fixed costs are (a) Forging die, (b) Forging die setup, (c) Machining tooling, and (d) Machining setup.

Variable costs are (a) Forging piece cost, and (b) Machining running cost.

	Blocker Type (₹)	Close Tolerance (₹)
Fixed cost	33 500 2 500 86 240 7 120	65 000 3 600 10 000 1 080
Total	129 360	79 680
Variable cost	160 1 600	220 200
Total	1 760	420

Since in this particular case, both fixed cost and variable cost for close precision forging happen to be lower than the other option, it is economical for any production quantity.

Example 1.2

In the production of flap hinges (of titanium) the comparison is to be made between open die forging and blocker type closed die forging. Both these are to be followed by machining. The costs are

	Open Die (₹)	Blocker (₹)
Forging		
Die cost		93 900
Setup		3 375
Forging piece	35 820	18 150
Machining		
Tooling	93 000	93 000
Running cost	53 625	33 450

Prepare the break-even analysis for the above two variants.

	Open Die (₹)	Blocker (₹)
Fixed cost	93 000	93 900 3 375 93 000
Total	93 000	190 275
Variable cost	35 820 53 625	18 150 33 450
	89 445	51 600

Production	Production Cost per Piece (₹)	
Quantity	Open Die	Blocker
1	182 445	241 875
2	135 945	146 738
3	120 445	115 025
4	112 695	99 169
5	108 045	89 655
10	98 745	70 628
50	91 305	55 406
100	90 375	53 503

$$\text{Break-even quantity} = \frac{FC_2 - FC_1}{VC_1 - VC_2} = \frac{190\,275 - 93\,000}{89\,445 - 51\,600} = 2.57 \text{ pieces}$$

SUMMARY

Knowledge of manufacturing processes is important to all disciplines in engineering to help in developing better products.

- The manufacturing processes have developed over the years with need for large volumes and variety of consumer requirements.
- To cater to the varying requirements, a number of processes under the broad categories of casting, forming, welding and machining processes have been developed.
- For any given application there can be more than one feasible manufacturing solution.
- Break-even analysis helps in deciding the most economical solution for a given application.

Questions

1.1 Explain why the understanding of manufacturing processes is essential for better products?

1.2 Why is it necessary for all engineers to be familiar with manufacturing processes?

1.3 What are the broad classifications of the manufacturing processes?

1.4 Define break-even analysis.

1.5 Describe various ways in which a break-even analysis plot could be made.

Problems

1.1 To produce a steel eye the two processes suitable are green sand and shell moulding. The various associated costs are

	Shell (₹)	Green Sand (₹)
Equipment cost	100 000	35 000
Cost per casting		
Metal cost	3.07	9.00
Moulding cost	6.07	4.85
Cleaning and inspection	2.26	3.83
Overheads	4.04	4.76

Compare the costs of above two processes with the production quantity ranging from 1 to 10 000.

[Ans. Green sand is cheaper compared to shell moulding throughout the production range]

1.2 An aluminium canopy can be obtained by either permanent mould casting or die casting process. The costs in rupees in either case are

	Permanent Mould Casting (₹)	Die Casting (₹)
Tooling	112 500	210 000
Setup cost	202	540
Labour cost	14	10
Material cost	14	7

Find out the break-even quantity of production from 1000 to 15,000 pieces. [Ans. 8894]

Multiple Choice Questions

1.1 Break-even analysis classifies the various costs as
(a) Fixed costs and indirect costs
(b) Fixed costs and variable costs
(c) Variable costs and indirect costs
(d) Variable costs and direct costs

1.2 For a typical manufacturing operation, the variable cost may include the following costs
(a) Cost of the die
(b) Tooling cost
(c) Part material cost
(d) Setup cost

1.3 For a typical manufacturing operation, the fixed cost may include the following costs
(a) Overhead cost
(b) Tooling cost
(c) Part material cost
(d) Running cost

Answers to MCQs

1.1 (b) 1.2 (c) 1.3 (b)

CHAPTER 2

Engineering Materials

Objectives

After completing this chapter, the reader will be able to

- Understand the importance of the properties of engineering materials such as strength, hardness, ductility and toughness
- Learn about the fundamentals of structure of engineering materials and how they control their properties
- Get the various details of the plain carbon steels and the variables that control their properties
- Understand the importance of different alloying elements in promoting the properties in alloy steels
- Learn different non-ferrous materials from the engineering standpoint
- Understand the heat treatment process and learn about the various heat treatment methods, and their application

2.1 ENGINEERING PROPERTIES

Manufacturing of a component is normally influenced by the mechanical and thermal properties of the work material. Also the mechanical properties are affected by the manufacturing process employed. Either way the knowledge of mechanical properties of engineering materials is important to a manufacturing engineer.

2.1.1 Strength

The resistance offered by a material on application of external force is called strength. Depending on the type of load applied, the strength could be tensile, compressive or shear. By application of load the material is elastically deformed which is called strain. It can be defined as

$$\text{Strain} = \frac{\text{Change in dimension}}{\text{original dimension}}$$

The resistance offered by the material is also referred to as stress which can then be defined as

$$\text{Stress} = \frac{\text{applied load}}{\text{area of cross section opposing the load}}$$

The deformation caused in a material is of two types, elastic and plastic. Elastic deformation is that part of the deformed material which when the applied load is removed, would spring back to its normal shape. Plastic deformation is, on the other hand, permanently set in a material and cannot be regained.

Tensile strength is measured by a tensile test carried out on a universal testing machine. This involves the preparation of a test specimen as per standard shown in Fig. 2.1. The standard specimen can be either rectangular or cylindrical. Rectangular is generally used for plate type specimens. Care has to be taken to see that the specimen prepared is such that the surface is smooth and without any deep cracks or nicks.

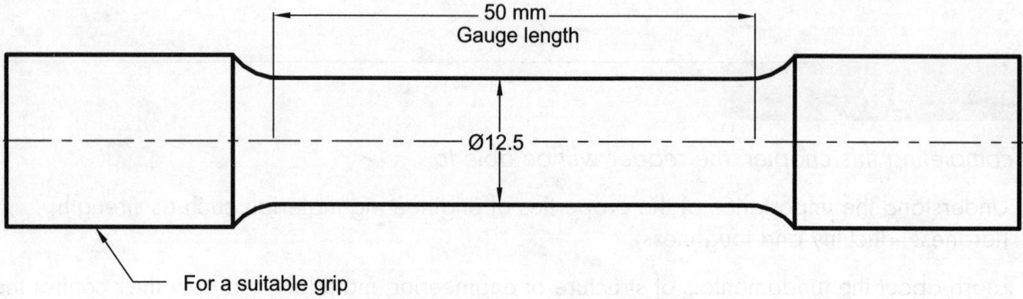

Fig. 2.1 *Standard test specimen for tensile test*

The tensile testing machine consists of a heavy test frame with a lower fixed beam and an upper crosshead which is the moving beam used to apply the tensile force on the specimen as shown in Fig. 2.2. The specimen is mounted between two grips, one attached to the fixed beam and the other attached to the crosshead. An extensometer is attached to the specimen to measure the elongation of the test specimen as the force is being applied. Then a uniformly increasing tensile load is applied on the specimen. The movement of the crosshead relative to the fixed beam generates strain within the specimen and a corresponding load. As the load increases the specimen initially gets elastically elongated. As can be noted from Fig. 2.1, the central portion is smaller compared to the gripping section. On further elongation, the specimen starts necking at some point (within the gauge marks) when the material goes beyond the elastic range. The reduced width

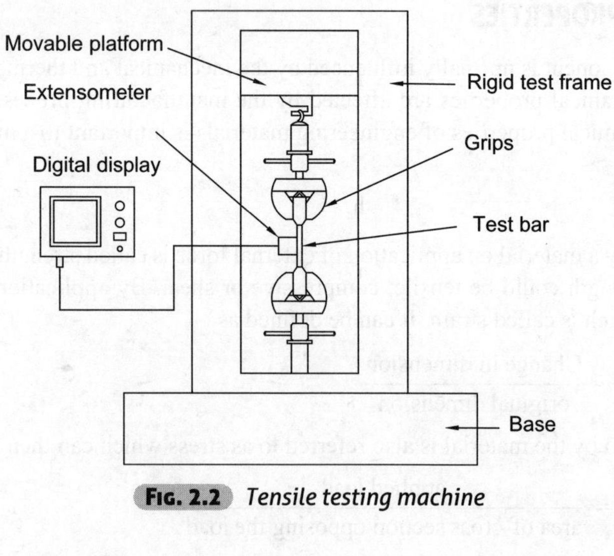

Fig. 2.2 *Tensile testing machine*

of the specimen would further be reduced under the force of the load and finally develop fractures when the test is completed. If the necking does not happen within the gauge marks, the test needs to be repeated with a different specimen. During the test, a record is maintained of the load and the corresponding elongation. The stresses and strains are calculated from the data and plotted in a diagram as shown in Fig. 2.3.

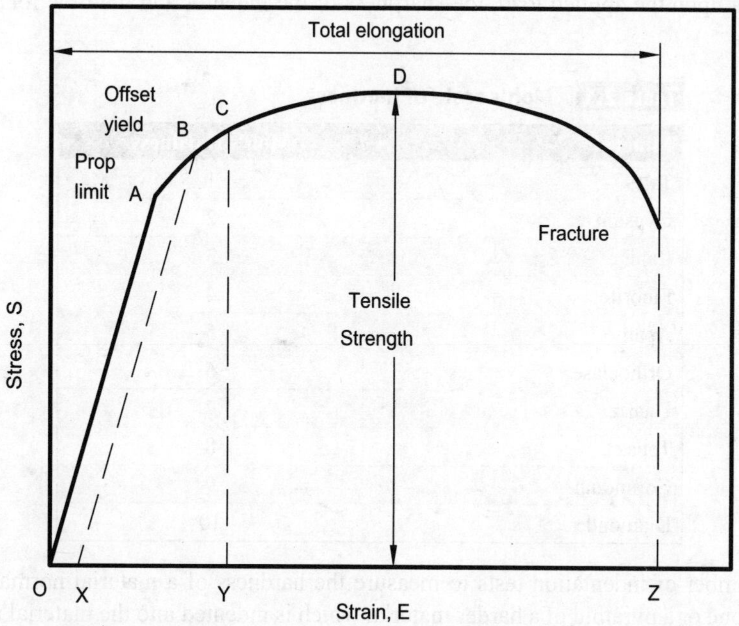

Fig. 2.3 *Engineering stress - strain curve (not drawn to scale)*

It can be observed that there is a limit up to which the applied stress is directly proportional to the induced strain as represented by the linear portion of the curve up to A which is termed as the proportional limit. The end of this linear portion is the yield point of the material above which the material starts plastically deforming. In the plastic region, there is a non-linear relationship between the stress and strain as evidenced by the bow shaped portion of the curve, Finally the specimen breaks when the force of the applied load goes beyond the limit that can be borne by the material.

The proportional limit is often difficult to calculate, therefore, two practical measurements, offset yield strength (OYS) and yield by extension under load (EUL) were developed to approximate the proportional limit. Point B represents the offset yield strength and is found by constructing a line X-B parallel to the curve in the elastic region. Line X-B is offset a strain amount O-X that is typically 0.2% of the gage length. Point C represents the yield strength by extension under load (EUL) and is found by constructing a vertical line Y-C. Line Y-C is offset a strain amount O-Y that is typically 0.5% of gage length. The ultimate tensile strength or peak stress is represented by point D. Total elongation, which includes both elastic and plastic deformation, is the amount of uniaxial strain at fracture and is depicted as strain at point Z. Percent elongation at break is determined by removing the fractured specimen from the grips, fitting the broken ends together and measuring the distance between gage marks. Percent elongation at break reports the amount of plastic deformation only. The gage length used for measurement is reported with the result.

Similar tests can also be conducted for measuring the compression and shear strength.

2.1.2 Hardness

Hardness is a very important property since the manufacturing depends on it to a great extent. Hardness is the resistance offered by a material to indentation. Moh's scale of hardness is based on ten standard minerals as shown in Table 2.1. It can generally be measured by the indentation made by a harder material. The indentation made depends upon the applied load, the sharpness of the indentor and the time for which the applied load is maintained.

TABLE 2.1 Moh's scale of hardness

Mineral	Moh's Hardness
Talc	1
Gypsum	2
Calcite	3
Fluorite	4
Apatite	5
Orthoclase	6
Quartz	7
Topaz	8
Corundum	9
Diamond	10

There are a number of indentation tests to measure the hardness of a material normally. These usually involve a ball, a cone or a pyramid of a harder material which is indented into the material under the test with a specified load. The permanent indentation thus made is measured to give an indication of the hardness on the given scale for the tests.

The most commonly used tests are:

Brinell hardness test where a sphere (usually of diameter 10 + 0.01 mm) made of steel or tungsten carbide is indented with a gradually applied load at right angles to the specimen surface and the indentation diameter made on the specimen measured. Then the Brinell hardness number, BHN, is given by

$$\text{BHN} = \frac{2P}{\pi D[D - \sqrt{D^2 - d^2}]}$$

Where, P is the applied load in kg

D is the diameter of the ball in mm

d is the diameter of the indentation in mm

The applied load, P, is based on the hardness of the specimen material being tested.

Some precautions that need to be taken during the use of Brinell tests are that the surface of the material whose hardness is being tested should be smooth and even, and there is enough material underneath the indentation and on all sides to give a proper value for the indentation.

The Brinell hardness of steel is roughly related to the tensile strength of the material utilizing the following empirical relationship for plain carbon steels.

Tensile strength (in MPa) = 3.55 × BHN when BHN ≤ 175

Tensile strength (in MPa) = 3.38 × BHN when BHN > 175

It may be noted that these conversions are only approximate. Also, the actual relationship depends upon the chemical composition of the alloy as well as the heat treatment process. Comparative values for these are given in Table 2.2.

TABLE 2.2 **Comparison of different hardness scales for steels**

Brinell Hardness Number	Vickers Hardness Number	Rockwell Hardness Number			Scleroscope Hardness	Tensile Strength (Approximate) Equivalent MPa
		A Scale	B Scale	C Scale		
578	633	80	—	57	77	2096
555	613	79	—	56	75	2027
555	595	79	—	55	74	1979
534	577	78	—	54	72	1924
514	560	77	—	53	71	1855
495	544	77	—	52	69	1800
495	528	76	—	51	68	1751
477	513	76	—	50	67	1689
461	498	75	—	49	65	1641
444	484	75	—	48	64	1600
444	471	74	—	47	63	1551
432	458	73	—	46	62	1510
415	446	73	—	45	61	1455
415	434	73	—	44	59	1420
401	423	72	—	43	58	1393
388	412	72	—	42	56	1365
388	402	71	—	41	55	1317
375	392	70	—	40	54	1276
363	382	70	—	39	53	1248
352	372	69	—	38	51	1213
341	363	69	109	37	50	1179
331	354	68	109	36	49	1158
331	345	68	108	35	48	1124
321	336	67	108	34	46	1096
311	327	67	107	33	45	1062
302	318	66	106	32	44	1034
293	310	66	106	31	43	1007
285	302	65	105	30	42	979
277	294	65	104	29	41	951
269	286	64	103	28	40	924

(Contd.)

262	279	64	103	27	39	903
255	272	63	102	26	38	869
255	266	63	101	25	37	855
248	260	62	100	24	37	841
241	254	62	99	23	36	814
235	248	62	99	22	35	800
229	243	61	98	21	35	779
23	238	61	97	20	34	765
217	230	—	95	18	33	738
212	222	—	94	16	32	703
203	213	—	92	14	31	676
192	204	—	90	12	29	634
187	195	—	89	10	28	621
179	187	—	87	8	27	600
170	180	—	85	6	26	572
166	173	—	84	4	25	545
156	166	—	82	2	25	531
156	160	—	81	—	25	510
149	156	—	79	—	23	503
143	150	—	77	—	22	483
137	143	—	74	—	21	462
131	137	—	72	—	—	448
126	132	—	70	—	20	427
121	127	—	68	—	19	414
116	122	—	65	—	18	400
112	117	—	—	—	15	386

Rockwell tests utilise the principle that the depth of penetration of the indentor is proportional to the material hardness. Thus the hardness measurement is faster compared to other methods. In Rockwell test, a sphero-conical diamond cone of 120° angle and a spherical apex of radius 0.2 mm is used to make the indentation and the depth of the indentation, t, is used as a criterion to calculate the hardness number. The Rockwell hardness number (R) is given by

$$R = 100 - 500\,t$$

Depending on the load used for indentation there are a number of scales A, B, C, etc., available in Rockwell tests. These are used for materials with different hardnesses. In the Rockwell B test, a steel ball of 0.0625 inch diameter is used with a load of 100 kg. This test is normally used for low and medium carbon steels. It should not be used for materials whose hardness is above RB 100. For materials with hardness above this value, Rockwell C test (30 RB corresponds to 0 RC) is more generally used. This is the fastest way of measuring hardness because the hardness can directly be read from the scale on the Rockwell tester.

Besides, there are other tests available such as Vickers and Knoop's hardness tests to measure hardness over a small area. In Vickers hardness test, a square base pyramid diamond indentor with 136° angle between the opposite faces is used. The Vickers hardness number (VHN) is calculated by

$$\text{VHN} = \frac{1.854L}{D^2}$$

Where, L is the applied load in kg which is normally about 30 kg though provision up to 120 kg would be available on the testing machine for harder materials, and

D is the measured average diagonal of the indentation in mm

The Scleroscope uses a different principle to measure the hardness which is based on the rebound height of a ball from the specimen. Harder the material, higher is the height to which the ball rebounds.

Because of these hardness tests being conducted under arbitrary conditions, it is difficult to have any proper correlation between the different scales. An approximate relationship of these hardness values is presented in Table 2.2, merely as a guideline.

The hardness values are normally represented as 200 BHN, 32 RC, etc. Representative values of some commonly used materials are presented in Table 2.3.

TABLE 2.3 Hardness values of some commonly used materials

Material	Brinell Hardness Number	Rockwell, B
Aluminium	20	
Chromium	125	37
Copper		
Magnesium	46	
Nickel	70	
Tungsten	350	
Iron	70	
C10 steel	105	
C25 steel	126	
C35 steel	163	
C45 steel	179	
C90 steel	197	
30 Mn2 steel	179 – 235	
32 Mo25 steel	174 – 217	
60 Si2 steel	184 – 235	

2.1.3 Ductility

It is the measure of the amount of plastic deformation a material can undergo under tensile forces without fracture. In quantitative terms it is normally measured as the ratio of elongation of the material at fracture during the tensile test to the original length, expressed as a percentage. The final value of elongation obtained during the tensile test immediately after the fracture could be taken as the ductility. Since the elongation is dependent upon the gauge length chosen for the tensile test, the length needs to be specified along with the elongation values. Alternatively, it may also be expressed as the ratio of reduction in cross-sectional area in the fractured specimen to the original cross-section area. This is independent of the gauge length and hence is a more convenient measure for ductility. It is also termed as the ability of a material to be drawn into wires since only ductile materials can be drawn into continuous wires without breaking in-between. Typical values of ductilities are given in Table 2.4 for engineering materials.

Besides the tensile tests, special forms of ductility testing based on the desired application can be done using the bend test and cupping tests for the purpose of sheet metal forming applications. More details of these tests can be found in literature.

Brittleness is the property opposite to that of ductility.

TABLE 2.4 Ductility values

Material	Ductility, % Reduction in Area
Cast iron	0
Structural steel	70 – 40
Cold rolled steel, SAE 1112	45
Cold rolled carbon steels	
C10 steel	57
C20 steel	52
C30 steel	48
C40 steel	40
Steel SAE 1300, annealed	70
Quenched, drawn 700°C	65
Drawn 540°C	60
Drawn 370°C	45
Drawn 200°C	30
Steel SAE 4340, annealed	70
Quenched, drawn 700°C	60
Drawn 540°C	50
Drawn 370°C	48
Drawn 200°C	44
Stainless steel	75 – 65
Aluminium alloys	35 – 39
Copper annealed	73
Monel, Ni-Cu alloy	75

2.1.4 Toughness

This is the property which signifies the amount of energy absorbed by a material at the time of fracture under impact loading. In short it is the capacity to take impact load. It can be considered as the total area under the stress strain curve since it is an indication of the amount of work done on the material without causing fracture. Thus toughness can be considered as a parameter consisting of both strength and ductility.

Toughness of a material is measured by means of impact tests, where a notched bar prepared as per the standards from the test material is held in a vice and a weight is allowed to swing from a known height in such a way that it hits the notched bar in its path and breaks it. Since the material has absorbed some amount of energy during its fracture, the swinging mass loses part of its energy and therefore will not be able to reach the same height from where it started, as shown in Fig. 2.4. The loss in height (h) multiplied by the weight represents the energy absorbed by the specimen during fracture which can be directly measured from the indicator on the tester.

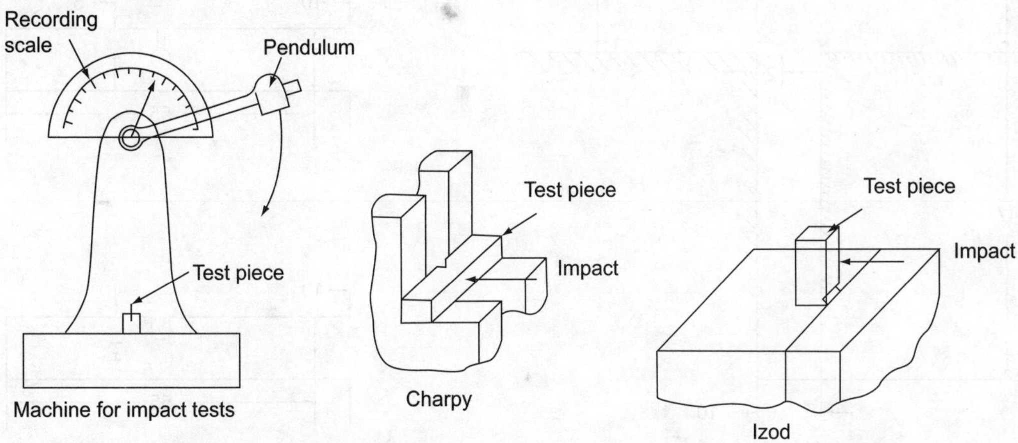

Fig. 2.4 *Impact testing machine variations*

In Charpy impact test, the specimen is held between two grips whereas in Izod impact test the specimen is held at one end like a cantilever as represented in Fig. 2.4. The standard test specimens used in the Izod and Charpy impact tests are made of bars with a square cross section 10 mm, the details of which are given in Fig. 2.5. The impact resistance is dependent upon the material composition as well as the heat treatment process given to it. The annealed materials normally would have better toughness than the corresponding normalised or quenched specimens. Coarse grained structures would tend to have higher ductility compared to fine grain structure and consequently better toughness. Typical values of impact resistance measured from these tests are given in Table 2.5 below.

Table 2.5 **Toughness values**

Material	Toughness, Izod Impact Value, N m
Steel 20 Mn 2	47
Steel 37 Mn 2	47
Steel 35 Mn 2 Mo 28	54
Steel 40 Cr 1	54
Steel 40 Cr 1 Mo 28	54
Steel 40 Ni 2 Cr 1 Mo 28	54

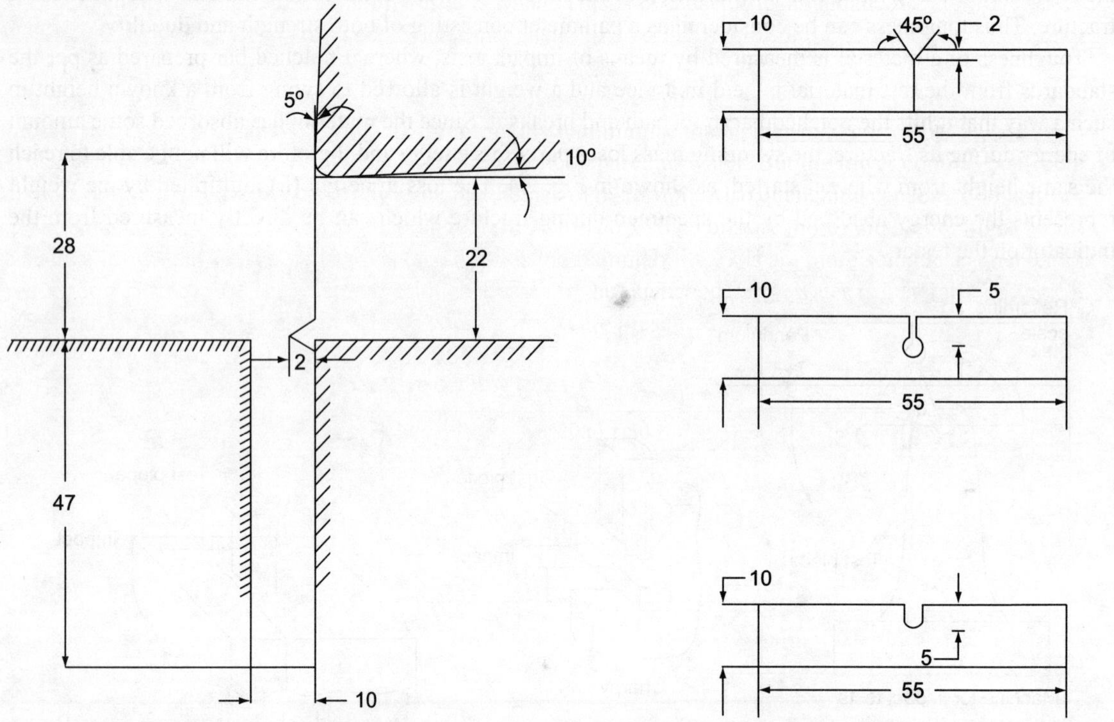

(a) Specimen mounting for Izod impact test (b) Standard specimens for Charpy impact test

FIG. 2.5 *Standard test specimens for impact tests*

2.2 FERROUS MATERIALS

We use a variety of metals, non-metals and their compounds in our daily life. For example, a typical automobile may contain the following materials:

Steel	1530 kg
Cast Iron	350 kg
Rubber	60 kg
Plastic	55 kg
Glass	52 kg
Aluminium	30 kg
Zinc	26 kg
Copper	16 kg
Lead	15 kg
Wood, ceramics, etc., in smaller quantities	

Based on their origin and composition, these materials may be broadly classified as shown in Fig. 2.6.

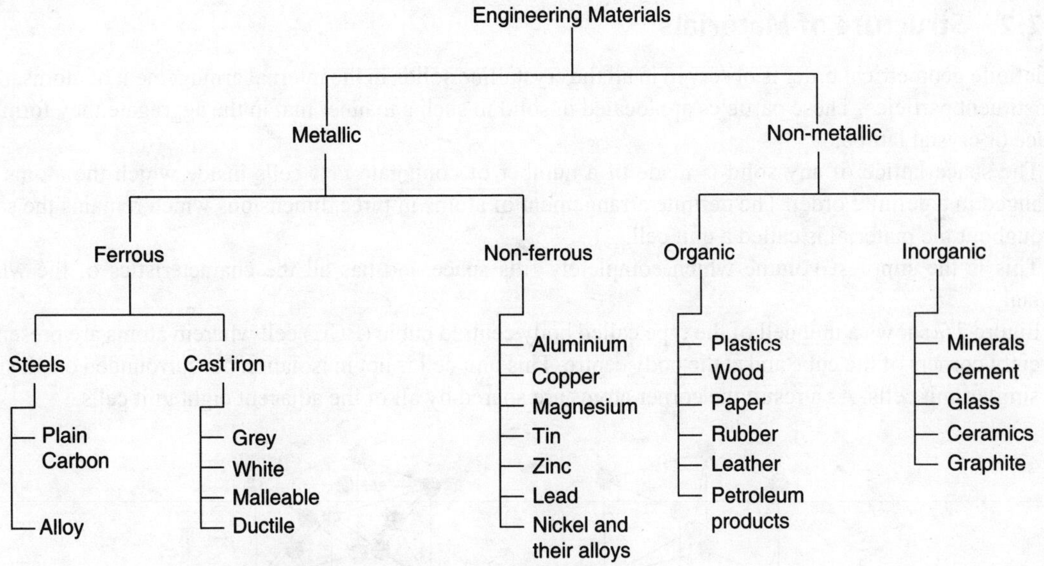

FIG. 2.6 *The classification of engineering materials*

Among these, ferrous materials are by far used most extensively because of their better and varied mechanical properties, and lower costs.

2.2.1 Iron

The basic source of all iron and steels is iron ore, which is an oxide of iron mixed with alumina, silica, phosphorous, manganese, sulphur and other materials. Major iron ores are haematite and magnetite which contain about 55% iron.

Pig iron, the principal base material for all steel furnaces, is a product of the blast furnace. Pig iron contains about 4% carbon, 1% silicon, 1% manganese and smaller percentages of phosphorous and sulphur. Pig iron is hard and brittle. It lacks the great strength, ductility and resistance to shock what steel possesses.

Absolute pure iron is very difficult to obtain. In this state, it is a soft and highly plastic metal of a light grey colour having a specific weight of 7.86. The mechanical properties of commercial grade iron containing 0.1 to 0.2% impurities are:

Hardness	60 to 80 BHN
Tensile strength	180 to 310 MPa
Yield point	200 MPa
Reduction in area	75%

The only application that can be found for pure iron is in making of magnets in view of its high permeability. Otherwise, the extensive use of iron is in the form of its large number of alloys. Iron can be alloyed with many elements. Alloys of iron and carbon are most widely used in engineering. They contain certain amounts of silicon, manganese, chromium, nickel and other elements.

Before proceeding to study the properties of ferrous alloys, a brief survey of the structure of materials would be beneficial for better understanding of the subject.

2.2.2 Structure of Materials

A definite geometrical order is observed in all the crystalline solids in the internal arrangement of atoms, their constituent particles. These particles are located in solid in such a manner that in the aggregate they form the space or crystal lattice.

The space lattice of any solid is made of a number of conjugate unit cells inside which the atoms are arranged in a definite order. The definite arrangement of atoms in three dimensions which remains the same throughout the material is called a unit cell.

This is the simplest volume which completely fills space and has all the characteristics of the whole crystal.

Figure 2.7 shows a unit cell of the type called body centred cubic (BCC) cell wherein atoms are present at all eight corners of the cube and at the body centre. This unit cell is not in isolation but surrounded on all sides by similar unit cells. As a result the corner atoms are shared by all of the adjacent eight unit cells.

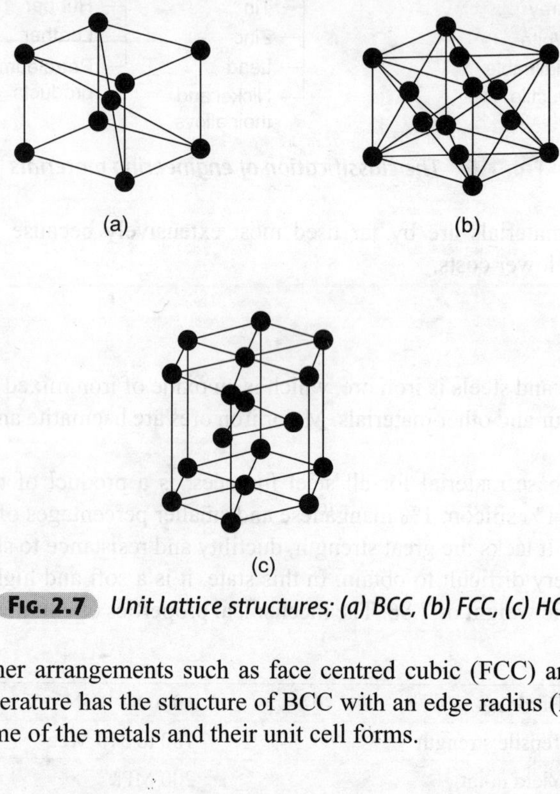

Fig. 2.7 *Unit lattice structures; (a) BCC, (b) FCC, (c) HCP*

Similarly, there are other arrangements such as face centred cubic (FCC) and hexagonal close packed (HCP). Iron at room temperature has the structure of BCC with an edge radius (lattice constant) of 0.24824 nm. The following are some of the metals and their unit cell forms.

Face centered cubic:

Ca, Ni, Cu, Ag, Pt, Au, Pb, Al

Body centered cubic:

Na, K, V, Mo, Ta, W

Hexagonal close packed:

Be, Mg, Zn, Cd, Te

Some of the materials exist in more than one form depending upon the temperature. The following is a list of such allotropic forms:

Cobalt

Hexagonal close packed	below 420°C
Face centered cubic	420°C to 1495°C

Chromium

Hexagonal close packed β	below 20°C
Body centered cubic α	20°C to 1799°C

Pure metals are seldom used as compared to the alloys which are commonly used. Alloys are metallic solids, complex in composition and are formed as a result of solidification of molten liquid of two or more elements. An alloy normally has properties which are a compromise between those of its constituents and sometimes has properties superior to and quite different from those of its constituents.

Alloys are formed by combining different metals in a number of ways. One predominant form is solid solution. A solution is formed when solute atoms are dissolved in solvent atoms. Similar to familiar liquid solutions, it is also feasible to have a solid solution. If a solution is allowed to freeze without separating the constituents, a solid solution would result. In a solid solution the materials are present only as a mixture but not as chemical compounds. The solid solutions are the essential parts of an alloy.

There are generally two types of solid solutions; the interstitial and the substitutional. In an interstitial solid solution, the solute atom would be positioned in the interstitial sites (empty space between the adjacent atoms) formed by the solvent atoms. It is possible only when the solvent atom is much larger compared to the solute atom. Also the extent of solubility depends on the difference in the atomic sizes. For example, carbon would form an interstitial solid solution with iron.

$$\text{Carbon atomic radius} = 0.0750 \text{ nm}$$
$$\text{Iron atomic radius} = 0.1241 \text{ nm}$$

Iron changes to an FCC structure above 720°C. If a carbon atom is placed in the FCC structure of iron at 730°C, the iron atoms are to be displaced by a distance of 0.0430 nm. In BCC structure at room temperature, if a carbon atom is to be located, the iron atoms are to be displaced by a distance of about 0.111 nm. Because of this distortion, iron at room temperature would only dissolve a maximum of 0.006% carbon. But above 730°C, up to 2% carbon can be dissolved.

In the substitutional solid solutions, the solute atoms would replace the solvent atoms. This is only possible if both the atoms are similar in size and also in nature.

Diffusion is the process of movement of atoms from one location of higher concentration to another of lower concentration or to a vacant place. Diffusion of atoms would be faster at high temperatures and in liquid phase. It is also a time dependent phenomenon as the atoms have to physically travel from one site to the other.

In an alloy system the component metals may combine within a certain temperature range to form two homogeneous coexisting portions. Each of these portions may have different compositions and consequently different properties. These homogeneous physically different portions of the alloy systems are termed as phase. A phase may be defined as any part of a chemical system that possesses distinctive physical characteristics. An alloy may consist of one phase or a combination of different phases. In liquid state, alloys are entirely homogeneous and from the physical point of view, constitute a single phase. Non-homogeneity may appear when an alloy is transformed from liquid to solid state, i.e., several solid phases may be formed.

Sometimes when the amount of alloyed metal that can be dissolved in a solid solution is exceeded, the 'parent' metal and the alloyed metal will together form an intermetallic compound. Though the intermetallic compounds are shown with a chemical symbol such as Fe_3C they are not like the chemical compounds such as CO_2 which are actually formed by a chemical reaction. In the present case it is only an atomic arrangement. Intermetallic compounds are usually very hard. Even if only a small amount of intermetallic compound is present, the alloy will combine the toughness of a solid solution with the hardness of the intermetallic compound.

Eutectic Composition

In certain alloy systems, alloying causes a lowering of the melting points and at a certain composition, called the eutectic composition for the system, the melting point is the lowest. For example, the alloy of 62% tin and 38% lead has a melting point of 183°C whereas lead melts at 327°C and tin melts at 232°C as in Fig. 2.8. The low melting point of this alloy called solder enables delicate parts of metal to be soldered without damage by heat.

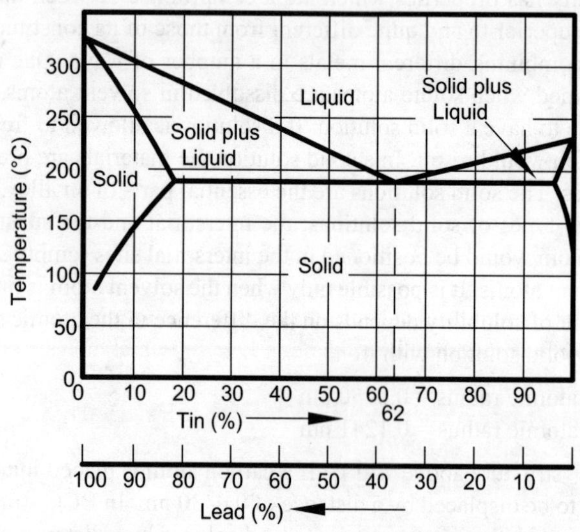

FIG. 2.8 *Equilibrium diagram of lead-tin system*

Equilibrium Diagrams

A cooling curve is drawn with temperature against time for a given metal or alloy showing the phases present in it. But for an alloy system containing various compositions of the constituent elements, it is necessary to draw a phase diagram or equilibrium diagram wherein the phases found at various temperatures and compositions are plotted in a single chart. The equilibrium diagrams are thus the generalised results of investigations into the solidification process and the structural and phase transformation of the given alloy system.

These diagrams enable to determine the temperature levels at the beginning and end of melting for alloys of various compositions, the structure of alloys for various temperatures under equilibrium conditions and also the transformation the alloys are subjected to during cooling and heating. Equilibrium conditions mean that sufficient time would be available for the changes of phases to take place.

An equilibrium diagram is plotted by laying off the percentage concentrations of the two components along the abscissa and the temperature along the ordinate. Any point on the diagram refers to a definite composition of the alloy at a particular temperature, as shown in Fig. 2.8.

Nucleation and Grain Growth

When the free energy of a parent phase is reduced by means of temperature or pressure then there is a driving force leading to crystallisation. For example, at the melting point, the thermal fluctuations result in the formation of tiny particles (containing only a few atoms) of the product phase within the parent volume. Such a tiny particle has an interface that separates it from the parent matrix. It grows by transfer of atoms across its interface.

- The process of formation of the first stable tiny particles is called nucleation.
- The process of increase in the sizes of these particles is called grain growth.

The grain size in the product phase depends on the relative rates of nucleation and growth. Each nucleating particle becomes a grain in the final product. So a high nucleation rate means a larger number of grains. Also when this is combined with a low growth rate, more time is available for further nucleation to take place in the parent phase that lies between slowly growing particles. So a combination of a high nucleation rate and a low growth rate yields a fine grain size. On the other hand, a low nucleation rate combined with a high growth rate yields a coarse grain size.

The temperature of maximum rate of nucleation is lower than the temperature of maximum growth rate. An increase in cooling rate lowers the effective transformation temperature and results in the combination of a high nucleation rate and a relatively slow growth rate ultimately yielding a fine grain size. A representation of grains is presented in the Fig. 2.9. In this figure, each square represents a unit cell.

Coarse grain steels are not as strong and have a higher tendency towards distortion than those having fine grain, although they offer better machinability and greater depth hardening qualities. Fine grained steels are tougher, more ductile and do not distort or crack during heat treatment.

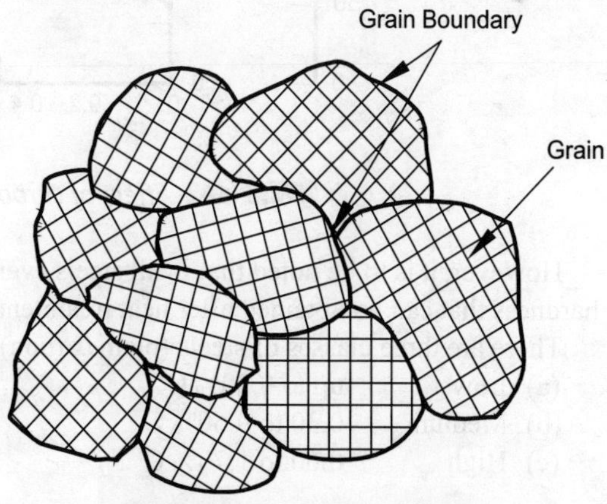

Fig. 2.9 *Grain representation*

2.2.3 Plain Carbon Steels

As has been mentioned earlier, iron in its purest form is not used as an engineering material because it lacks tensile strength and hardness. But when alloyed with other elements, the properties can be controlled greatly. Out of the various alloying elements, carbon is the most important because it is found in all the alloys of iron.

The maximum amount of carbon that can be alloyed with iron is 6.67%. Alloys containing up to 2% carbon are termed steels and above 2% are called cast irons. Besides carbon, these alloys also contain small amounts of manganese, sulphur, phosphorous and silicon. These are generally considered as impurities and hence need to be controlled. However, in certain conditions some of these such as silicon are treated as alloying elements.

The properties of steel are influenced significantly by an increase in carbon content (Fig. 2.10) with a result that:

- tensile strength is increased
- greater hardness is obtained
- ductility is decreased
- weldability is decreased

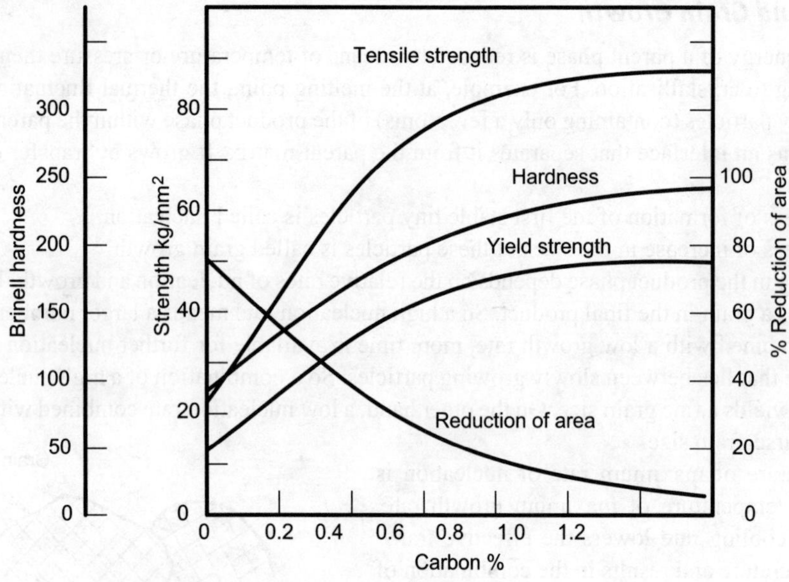

Fig. 2.10 *Effect of carbon on mechanical properties of steels*

However, it is to be noted that in all steels even when alloyed with other alloying elements, the maximum hardness that can be obtained after heat treatment processes is basically a function of the carbon content.

There are three classes of steels (plain carbon)

(a) Low up to 0.30% C
(b) Medium 0.30 to 0.60% C
(c) High 0.60 to 1.70% C

Low Carbon Steel

This is generally known as 'soft' or mild steel. It is used where ductility and softness are important, and high tensile strength is not required. These are tough but not resistant to wear. Due to softness, these can be very easily formed and later can be carburised to increase the hardness and wear resistance.

Low carbon steels are used for operations such as spinning, cold bending, riveting, swaging, etc. These are not responsive to normal heat treatment but for case hardening. They form the largest percentage of steel produced because being the cheapest engineering material. Products such as screws, nails, nuts, bolts, washers, wire fences, light and heavy structural members, machine parts and forged parts can be made from low carbon steel. It is also used for tin plate and automobile body sheet. It is available in the form of sheets, squares, rounds, plates and wires.

Medium Carbon Steel

They are less ductile but harder and have greater tensile strength than low carbon steels. They also have better machining qualities and are more responsive to heat treatment.

These are widely used in the industry. Medium carbon steels are used for making shafts, connecting rods, spindles, rail axles, gears, turbine bucket wheels, steering arms and other machine parts requiring medium strength and wear resisting surfaces.

High Carbon Steel

They have higher tensile strength and are harder than the other plain carbon steels. They also readily respond to heat treatment. These are used for making hand tools such as wrenches, chisels, punches, files, cutting tools such as drills, wood working tools, rail road wheels, rails, bars for reinforcing of concrete, etc.

Effect of Small Quantities of Other Elements

In addition to carbon the plain carbon steels contain small quantities of other elements more as impurities. They affect the properties in the following way.

Sulphur

Iron forms iron sulphide (FeS) with sulphur, which solidifies along the grain boundaries making the steel brittle and lower hot working properties such as ductility. If equal amount of manganese is present in the steel then manganese sulphide, MnS, forms and the harmful effects of sulphur are reduced. It is generally recommended that manganese should at least be 3 times that of sulphur. However, very small quantities (0.075 to 0.15%) that are generally present contribute to the better machinability.

Phosphorous

Phosphorous in small amounts increases the strength and hardness of steels. Most of the steels contain a very small percentage of about 0.05% phosphorous.

Silicon

Silicon in very small amounts of the order of less than 0.2% does not have any effect. When it is between 0.2 and 0.4%, it raises the elastic limit and ultimate strength of the steel without greatly reducing the ductility. More than this percentage it reduces the ductility.

The main limitations of plain carbon steels are:

- Low hardenability
- Loss of hardness during tempering
- Low strength at elevated temperature
- Lower resistance to corrosion and oxidation

Alloy steels have been specifically developed to overcome these limitations and are discussed later in this chapter.

2.2.4 Iron Carbon Equilibrium Diagram

The structural form of pure iron at room temperature is called ferrite or α-iron. Ferrite is soft and ductile. Since ferrite has a body centered cubic structure, the inter-atomic spaces are small and pronouncedly oblate and cannot readily accommodate even a small carbon atom. Therefore, solubility of carbon in ferrite is very low, of the order of 0.006% at room temperature. The maximum carbon content in ferrite is 0.05% at 723°C. In addition to carbon, a certain amount of silicon, manganese and phosphorous may be found in ferrite.

The face centred modification of iron is called austenite or γ-iron. It is the stable form of pure iron at temperatures between 910°C and 1400°C. At its stable temperature austenite is soft and ductile and consequently, is well suited for manufacturing processes. The face centered cubic structure of iron has larger inter atomic spacing than in ferrite. Even so, in FCC structure the interstices are barely large enough to accommodate carbon atoms and lattice strains are produced. As a result, not all the interstitial sites can be filled at any one time. The maximum solubility is only 2% of carbon at 1130°C.

Above 1400°C, austenite is no longer the most stable form of iron and the crystal structure changes back to a body centred cubic phase called δ-iron. This is the same phase as the α-iron except for its temperature range. The solubility of carbon in δ-ferrite is small, but it is appreciably larger than in α-ferrite because of higher temperature. The maximum solubility of carbon in δ-iron is 0.1% at 1490°C.

In iron-carbon alloys, carbon in excess of the solubility limit must form a second phase, which is called iron carbide or cementite. Iron carbide has the chemical composition of Fe_3C. This does not mean that iron carbide forms molecules of Fe_3C but simply that the crystal lattice contains iron and carbon atoms in a 3:1 ratio. The compound Fe_3C has an orthorhombic unit cell with twelve iron atoms and four carbon atoms per cell, and thus has a carbon content of 6.67%.

As compared to austenite and ferrite, cementite being an inter-metallic compound is very hard and brittle. The presence of iron carbide with ferrite in steel greatly increases the strength of steel.

The iron carbon equilibrium diagram is shown in Fig. 2.11. The solidification of the liquid iron and carbon melt begins along the liquidus denoted in the figure by ABCD. Above the liquidus, the alloy is in a liquid state and is a homogeneous system. Along the liquidus, AB the crystals of the solid solution of carbon in γ-iron are separated from the liquid.

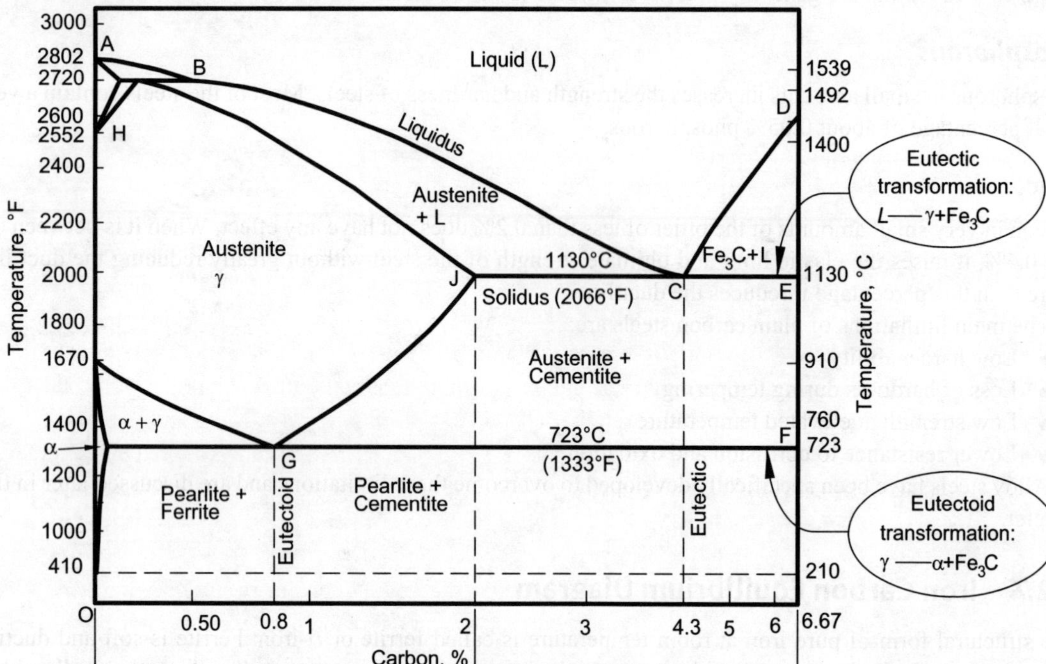

FIG. 2.11 *Iron carbon equilibrium diagram*

Crystals of austenite are separated from the liquid along the line BC with the compositions ranging from 0.18 to 2.0%. The complete solidification of these alloys proceeds along the solidus line HJCE. Alloys containing 0.18 to 2.0% carbon become solid along the solidus HJE. Those with 2.0 to 4.3% carbon are completely solidified on line EC. The solidification of the last portion of the liquid phase, enriched in carbon to 4.3% takes along this line. All this liquid is completely solidified at 1130°C and at the same time crystals of austenite containing 2% carbon and cementite containing 6.67% carbon get separated from it. The solidification of alloys containing 4.3 to 6.67% carbon begins along line CD with the separation of primary cementite from the melt.

At the lower temperatures, the eutectoid reaction, i.e., formation of two solids from a single solid occurs at a temperature of 723°C. This is called the eutectoid temperature and the composition at which this reaction occurs (0.80% C) is called the eutectoid composition. The reaction may be shown as follows:

$$\text{solid 1} \underset{\text{heating}}{\overset{\text{cooling}}{\longleftrightarrow}} \text{solid 2} + \text{solid 3}$$

$$\text{Austenite (0.77\% C)} \underset{\text{heating}}{\overset{\text{cooling}}{\longleftrightarrow}} \text{Ferrite (0.02\% C)} + \text{Cementite (6.67\% C)}$$

In the reaction, the simultaneous formation of ferrite and cementite from austenite results at the temperature of 723°C and composition of 0.80% carbon. There are nearly 12% of iron carbide and slightly more than 88% of ferrite in the resulting mixture. Since the ferrite and cementite are formed simultaneously, they are intimately mixed. Characteristically, the mixture is lamellar, i.e., it is composed of alternate layers of ferrite and cementite. This micro-structure is called pearlite which is very important in iron and steel technology because it can be formed in almost all steels by means of suitable heat treatments.

The alloy containing 0.80% of carbon is called the eutectoid steel. Upon cooling the eutectoid steel below 723°C, all of the austenite is transformed into pearlite. Alloys with less than 0.80% C are called hypo-eutectoid steels and those with higher composition are called hyper-eutectoid steels.

When hyper-eutectoid steels are cooled below line SE, the austenite decomposes with the separation of cementite. Since this cementite contains 6.67% C, the carbon concentration in the remaining austenite changes continuously along the line SE until the eutectoid composition of 0.80% C.

2.2.5 Cast Irons

The ferrous alloys which have carbon contents of more than 2% are called cast irons. Though cast irons can have any carbon percentage between 2 to 6.67, the practical limit is normally between 2 and 4%. These are important mainly because of their excellent casting qualities.

From the iron carbon equilibrium diagram (Fig. 2.11), it can be observed that cast irons have essentially cementite and ferrite. Because of the larger percentage of carbon, the amount of cementite is high resulting in very high hardness and brittleness qualities for cast iron.

When cast iron is slowly cooled, the cementite decomposes into iron and carbon in form of graphite which is called graphitisation. Cast irons where a large percentage of cementite is decomposed by graphitisation are called grey cast irons. Cast iron in which graphitisation has not taken place, i.e., all the carbon is in the combined form, is called white cast iron. The graphitisation process requires time and therefore, when liquid cast iron is cooled rapidly, white cast iron would result. White cast iron is comparable in properties to that of high carbon steels. However it is highly brittle and as such is not used for structural parts. It is useful for parts where abrasive wear is present. Tensile strength varies between 170 to 345 MPa and is usually about 240 MPa. The hardness ranges from 350 to 500 BHN. In view of the very high hardness the machinability is poor and is commonly finished by grinding.

In the presence of graphitising elements such as Nickel and Silicon, iron carbide decomposes as follows:

$$Fe_3C \rightarrow 3Fe + C$$

The decomposition is controlled by the graphitising agents as well as the cooling rate.

The dissociated carbon is in the form of graphite which is very soft and without any strength thereby reducing the hardness and increasing the machinability of cast iron. The shape of graphite present in cast irons would greatly affect its strength. When it is in a flake like shape as in grey cast iron, the graphite breaks up continuity of iron and greatly weakens it. But it also helps in absorbing vibrational energy as a result of which grey cast iron is normally used for the beds of machine tools. Grey cast iron is easily machinable and is the cheapest form of cast iron. It is extensively used in casting process because of its low melting temperature, higher fluidity and negligible shrinkage on cooling.

The other form of cast iron is known as malleable iron in which free carbon is present in the form of nodules in the matrix of cementite and ferrite. This is achieved by first chilling the casting so that all white cast iron is formed, followed by a controlled heat treatment process so that some of the cementite is transformed to ferrite and nodules of free carbon. This material is more ductile than grey cast iron. This form is suitable only for components with very small section thicknesses since all white cast iron is used to form the starting point for malleable iron.

When graphite is present as small, round, and well distributed particles, its weakening effect is small and such cast irons would have higher ductility. This type of cast iron is called ductile or nodular iron or spheroidal graphite or simply SG iron. This form of graphite can be achieved by adding elemental magnesium or cerium or a combination of the two elements to molten cast iron. Magnesium is added in quantities of 0.07 to 0.10% followed by the addition of ferro-silicon to promote graphitisation. During solidification, magnesium helps in the distribution of graphite throughout the metal.

Ductile iron has better strength to weight ratio, better machinability and higher impact value. More over the ductile iron components are produced by casting process where in better control of component shape can be achieved compared to drop forging. Thus many a components such as crank shafts and connecting rods manufactured usually by drop forging is increasingly being replaced by ductile iron castings. Typical comparison of different casting materials is given in Table 2.6.

TABLE 2.6 Comparative properties of cast irons

Property	Grey Cast Iron	Malleable Iron	Ductile Iron	C30 Steel
Melt temperature, °C	1 175	1 200	1 150	1 450
Specific gravity, kg/m^3	6 920	6 920	6 920	7 750
Vibration damping	Excellent	Good	Good	Poor
Modulus of elasticity, MPa	126 174	175 126	173 745	210 290
Modulus of rigidity, MPa	48 955	70 329	66 190	78 600

2.2.6 Other Alloying Elements

Steel is an alloy of iron. Normally, ferrous alloys containing only carbon as the alloying element are called plain carbon steels or simply steels while those containing, besides carbon, some other alloying elements such as chromium are termed alloy steels. In fact, the definition given by American Iron and Steel Institute (AISI) is as follows:

"Steel is considered to be alloy steel when the maximum of the range given for the content of alloying elements exceeds one or more of the following limits: manganese, 1.65%; silicon, 0.60%; copper, 0.60%; or in which a definite range or a definite minimum quantity of any of the following elements is specified or required within the limits of the recognised field of constructional alloy steels: aluminium, boron, chromium up to 3.99%, cobalt, columbium, molybdenum, nickel, titanium, tungsten, vanadium, zirconium, or any other alloying element added to obtain a desired alloying effect."

The alloy steels are normally required when additional properties such as strength, ductility, toughness or corrosion resistance are desirable in large measures. The various ways in which special alloying elements are used to improve the properties of steels are:

1. To improve hardenability
2. To improve mechanical properties at low or elevated temperatures
3. To improve the corrosion and oxidation resistance
4. To increase the machinability
5. To increase the electrical and magnetic properties
6. To increase resistance to softening on tempering
7. To increase abrasion resistance
8. To increase hardness of steels that cannot be quenched

Since the micro-structure essentially consists of ferrite and cementite, the mechanical properties can be controlled by changing either the properties of carbide and ferrite phases by the alloying elements or by controlled dispersion of carbide in the ferrite matrix. The carbide phase present in alloy steels is not pure iron carbide but a complex combination of iron and alloy carbides. Some of the alloying elements act as austenite stabilisers. The austenite stabilisers lower the eutectoid temperature thereby expanding the temperature range in which austenite is stable.

The effect of alloying elements can also be described by means of the following empirical relationships which show the critical temperatures in the iron-carbon equilibrium diagrams. Ac_1 refer to the boundary between austenite and pearlite, while Ac_3 refers to the temperature separating austenite with austenite and cementite.

$$Ac_1 = 723 - 10.7\,Mn - 16.9\,Ni + 29.1\,Si + 16.9\,Cr + 290\,As + 6.38\,W$$

$$Ac_3 = 910 - 203\sqrt{C} - 15.2\,Ni + 44.7\,Si + 104\,V + 31.5\,Mo + 13.1\,W$$

Out of the various alloying elements, nickel, silicon and aluminium do not form any carbide whereas manganese, chromium, tungsten, molybdenum, vanadium, titanium and niobium have increasing carbide stability in that order. When nitrogen is present, many of these carbide formers form carbo-nitrides or nitrides which are highly abrasion resistant. The following is a detailed account of the effect of individual alloying elements on the mechanical properties of alloy steels.

Manganese

This is the most common alloying element in all steels. It decreases the critical temperatures appreciably and thus lets the steel oil harden. Also it is a cheap way of increasing the hardenability of steels. It forms carbide Mn_3C but its carbide forming tendency is the lowest of all alloying elements. It counteracts the brittleness caused by sulphur in steels. In steel, equal amount of manganese and sulphur forms manganese disulphide readily which is evenly distributed in it. This greatly improves the hot working characteristics and also the lubrication in machining ensuring good surface finish. Manganese in amounts 2 to 10% imparts brittleness to steel. The particular composition of 11 to 14% manganese and 1 to 1.4% carbon has too high a resistance to wear and abrasion even under high impact stresses. With such large manganese content the critical temperature is reduced and martensite could be obtained even with slow cooling during quenching. Very rapid quenching allows for retained austenite thus giving it high ductility and toughness. This is quite useful for service involving continuous impact loads but not suitable for abrasive loads.

Chromium

It is a strong carbide former and forms complex series of carbide compounds of chromium and iron. It raises the critical temperature appreciably and thus resists tempering. It increases hardenability, wear resistance, corrosion and oxidation resistance. The primary use of chromium in steels, thus, is because of its high hardness and corrosion resistance.

Nickel

Nickel is not a carbide former but strengthens and toughens the ferrite phase. It reduces the critical cooling rate required for quenching and therefore they can be very easily heat treated. It increases the tensile strength without appreciable decrease in elongation and decrease in area. In many ways its effect on properties is similar to manganese. In combination with chromium, it provides high elastic ratios, greater hardenability, higher impact and fatigue resistance.

Tungsten

It is a very strong carbide former and forms abrasive resistant particles in tool steels. At larger percentages, it improves hot hardness and hot strength, and as such is useful in cutting of hot working tools. It is not softened by tempering. In tungsten steels much higher tempering temperatures may be employed with less loss in hardness with reduction in internal strains compared to plain carbon steels.

Molybdenum

It greatly increases the hardenability. It is also a strong carbide former. It increases the hot hardness and hot strength when used in combination with chromium and vanadium. Since molybdenum is cheaper than tungsten, it is replacing tungsten to a great extent in tool steels. It raises the critical temperature for tempering and so does not soften even at a higher temperature. The typical amounts present are 0.15 to 0.50% molybdenum in general tool steels. The main problem with molybdenum is the promotion of skin decarburization which is to be properly accounted for during heat treatment.

Vanadium

It is a strong carbide former. It increases the hardenability and also the secondary hardening effect upon tempering. Grain growth tendency at heat treating temperatures is minimised. Vanadium steels have a much finer structure than steels without vanadium. It increases hardness at elevated temperatures.

Silicon

It slightly increases the hardenability. It sustains hardness during tempering. One of the important uses of silicon is as a deoxidiser in molten steel and for its ability to resist oxidation in steel. In large quantities above 2.5%, it increases the strength of ferrite phase without affecting its ductility. But at this percentage it develops poor machinability and is susceptible to decarburization. Silicon increases the electrical resistivity of iron thus reducing eddy current effects with alternating current. Thus silicon steels are used extensively for electrical applications. Silicon steels can be easily magnetized in a direction parallel to their crystallographic cubic edge. By a careful combination of rolling and annealing, it is possible to align the grains in the desired direction thus silicon steels are also used for magnetic applications.

Aluminium

It is primarily used as a deoxidiser in steels. It is most effective in inhibiting grain growth. In steels which are to be nitrided, aluminium provides an extremely high hardness of the nitrided case due to the formation of hard and stable aluminium nitride compound.

Titanium

It has the highest carbide forming tendency of all the alloying elements. It has no effect on hardenability. It is a good deoxidiser and inhibits grain growth. Because of its strong carbide forming tendencies, medium carbon steels do not quench harden.

Niobium

Also called 'columbium', it reduces the hardenability and increases the ductility slightly which results in a marked increase in impact strength. It imparts a fine grain structure to steels and retards softening during tempering.

Cobalt

It decreases the hardenability. Cobalt strengthens ferrite when dissolved in it and resists softening under elevated temperatures. It promotes skin decarburization.

The comparative effect of the alloying elements on the hardness of alloy steels is presented in Fig. 2.12. It can be seen from this figure that the effect of each alloying elements is different and the effect is also pronounced by an increase in the composition percentage.

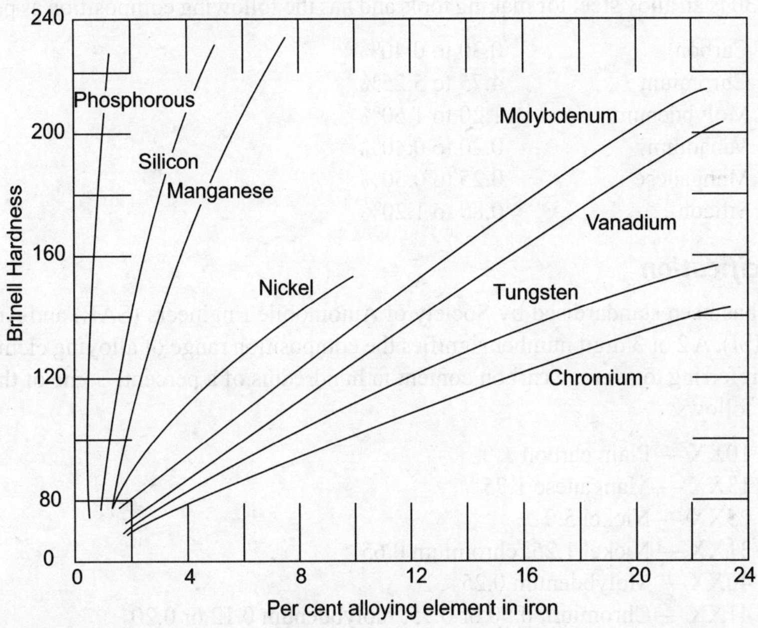

FIG. 2.12 *Effect of alloying elements on the probable hardness of alloy steels*

2.2.7 Steel Designation

To refer materials, there are a number of ways based on standard specifications. The following are some of the most commonly referred to designations.

Indian Standard Designation

Among plain carbon steels, the structural steels where no heat treatment is required, are designated by its tensile strength in kgf/mm^2, for example, st 42 means a structural steel with a minimum tensile strength of 42 kgf/mm^2.

Other plain carbon steels with carbon as the chief alloying element, are referred to by its chemical symbol C followed by a number representing its percentage in hundredths. For example, C20 refers to plain carbon steel with an average carbon percentage of 0.20. If sulphur and phosphorous are to be denoted as limited, a suffix K would follow like C25K.

Tool steels would be designated by the letter T instead of C. For example, T103 is a tool steel with an average carbon percentage of 1.03 (0.95 to 1.10).

Alloy steels are designated by the chemical symbols of the significant elements followed by their percentages rounded up to the nearest whole number or if the percentage is less than one then the hundredths of a percent with an underline.

For example, 35Mn2Mo45 represents the following average composition.

Carbon	0.35%
Manganese	2.00%
Molybdenum	0.45%

T35Cr5Mo1V30 is an alloy steel for making tools and has the following composition as per IS: 3748-1966.

Carbon	0.30 to 0.40%
Chromium	4.75 to 5.25%
Molybdenum	1.20 to 1.60%
Vanadium	0.20 to 0.40%
Manganese	0.25 to 0.50%
Silicon	0.80 to 1.20%

American Specification

This designation has been standardised by Society of Automobile Engineers (SAE) and American Iron and Steel Institute (AISI). A 2 or 3 digit number signifies the composition range of alloying elements followed by a 2-digit number referring to average carbon content in hundredths of a percent. Some of the most generally used steels are as follows:

10XX — Plain carbon
13XX — Manganese 1.75
25XX — Nickel 5.0
31XX — Nickel 1.25; chromium 0.65
40XX — Molybdenum 0.25
41XX — Chromium 0.50 or 0.95; molybdenum 0.12 or 0.20
43XX — Nickel 1.80; chromium 0.50 or 0.80; molybdenum 0.25
44XX — Manganese 0.80; molybdenum 0.40
46XX — Nickel 1.85; molybdenum 0.25
51XX — Chromium 0.80, 0.88, 0.93, 0.95 or 1.00
61XX — Chromium 0.60, 0.80 or 0.95; vanadium 0.12 or 0.10 min or 0.15 min
81XX — Nickel 0.30; chromium 0.40; molybdenum 0.12
92XX — Manganese 0.85; silicon 2.00
93XX — Nickel 3.25; chromium 1.20; molybdenum 0.12

In the following tables comparative designations of steels which are used for manufacture are given.

TABLE 2.7 General steels with comparative standard designations

Usual Description	IS Designation	British (1955) Designation	American Designation	German Designation
Mild steel	C20	En 3A	AISI 1020	C22
Medium carbon steel	C35	En 5	AISI 1035	C35
	C45	En 8D	AISI 1045	Ck45
	C55	En 9	AISI 1055	C55
	C60	En 43D	AISI 1060	C60
	C75	En 42	AISI 1078	C75
High carbon steel	C100	En 44	AISI 1095	
Low alloy steel	40Cr90Mo15	En 19c	AISI 4142	42CrMo4
	40Ni2Cr1Mo25	En 24	AISI 4340	36CrNiMo4
	30Ni4Cr1	En 30A	—	35NiCr18
	55Si2Mn90	En 45	AISI 9260	
	50Cr1V23	En 47	AISI 6150	50CrV4
	17Mn1Cr95	En 207	AISI 5115	16MnCr5
	13Ni3Cr80	En 36	AISI 9310	14NiCr14
		En 111	AISI 3120	
		En 18A	AISI 5140	41Cr4

TABLE 2.8 Tool and die steels used for hot working

IS Designation	British (1971) Designation	American Designation
T33W9Cr3V38	BH21	H21
T35Cr5Mo1V30	BH11	H11
T35Cr5MoV1	BH13	H13
T35Cr5MoW1V30	BH12	H12
T55W14Cr3V45	BH26	H24

TABLE 2.9 Tool and die steels used for cold working

IS Designation	American Designation
T70Mn65	AISI A6
T118Cr45	AISI O5
T55Ni2Cr65Mo30	AISI A9
T105W2Cr60V25	AISI A5
T110W2Cr1	AISI A4
T90Mn2W50Cr45	ASTM O2
T55Cr70V15	AISI 6155

2.3 NON-FERROUS MATERIALS

Ferrous materials are extensively used in engineering industry because of their superior range of mechanical properties and lower costs. Still non-ferrous materials are also used in various applications for their specific properties compared to ferrous alloys in spite of their generally high cost. Desired mechanical properties can be obtained in these alloys by work hardening, age hardening, etc. but not through normal heat treatment processes used for ferrous alloys. Some of the principal non-ferrous materials of interest are:

- Aluminium
- Copper
- Zinc
- Magnesium

2.3.1 Aluminium

Of all non-ferrous alloys, aluminium and its alloys are the most important because of their excellent properties. Some of the properties of pure aluminium for which it is used in engineering industry are:

1. Excellent thermal conductivity (0.53 cal/cm/°C)
2. Excellent electrical conductivity (376 600 ohm/cm)
3. Low mass density (2.7 g/cm^3)
4. Low melting point (658°C)
5. Excellent corrosion resistance. Aluminium in fact has greater affinity towards oxygen. As a result, when aluminium is exposed to air, the outer surface readily gets oxidised forming aluminium oxide. This oxide skin has good bond with the parent metal and thus protects it from further oxidation.
6. It is non-toxic.
7. It has got one of the highest reflectivity (85 to 95%) and very low emissivity (4 to 5%).
8. It is very soft and ductile as a result of which it has got very good manufacturing properties.

Some of the applications where pure aluminium is generally used are in electrical conductors, radiator fin material, air conditioning units, optical and light reflectors, foil and packaging material.

In spite of the above useful applications, pure aluminium is not widely used because of the following problems:

1. It has got low tensile strength (65 MPa) and hardness (20 BHN).
2. It is very difficult to weld or solder.

The mechanical properties of aluminium can be substantially improved by alloying. The principal alloying elements used are copper, manganese, silicon, nickel and zinc.

Aluminium and copper form the chemical compound Cu Al_2. Above a temperature of 548°C, it is dissolved completely in liquid aluminium. When this is quenched and artificially aged (prolonged holding at 100 – 150°C), a hardened alloy is obtained. The Cu Al_2 which is not aged does not have time to precipitate from the solid solution of aluminium and copper and thus is in an unstable position (super saturated at room temperature). The ageing process precipitates very fine particles of Cu Al_2 which causes the strengthening of the alloy. This process is called solution hardening.

The other alloying elements used are magnesium up to 7%, manganese up to 1.5%, silicon up to 13 %, nickel up to 2 %, zinc up to 5 % and iron up to 1.5 %. Besides these, titanium, chromium and columbium may also be added in small percentages. The composition of some typical aluminium alloys used in permanent moulding and die casting is given in Table 2.10 with their applications. The mechanical properties expected of these materials after these are cast using permanent moulds or pressure die casting is shown in Table 2.11.

TABLE 2.10 Some aluminium alloys

Unified Designation	Aluminium Association	ASTM	Cu	Fe	Mg	Mn	Ni	Si	Zn	Applications
A03550	AA355.0	B26 (355.0)	1.0–1.5	0.6	0.4–0.6	0.5	Ti = 0.25 Cr = 0.25	4.5–5.5	0.35	Timing gears, impellers, aircraft and missile parts of high strength
A03600	AA360.0	B85 (360.0)	0.6	2.0	0.4–0.6	0.35	0.5	9.0–10.0	0.5	Improved corrosion resistance, superior strength
A03800	AA380.0	B85 (380.0)	3.0–4.0	2.0	0.1	0.5	0.5	7.5–9.5	3.0	Lawn mower housings, electronics chassis, engine components, home appliances, tools
A03830	AA383.0	B85 (383.0)	2.0–3.0	1.3	0.1	0.5	0.3	9.5–11.5	3.0	Intricate components
A03840	AA384.0	B85 (384.0)	3.0–4.5	1.3	0.1	0.5	0.5	10.5–12.0	3.0	Intricate components
A03900	AA390.0	B85 (SC174A)	4.0–5.0	1.3	0.45–0.65	0.1	Ti = 0.2	16.0–18.0	0.1	Wear resistance
A04130	AA413.0	B85 (413.0)	0.6	2.0	0.1	0.35	0.5	11.0–13.0	0.5	Intricate details, excellent pressure tightness
A04430	AA443.0	B26 (443.0)	0.6	0.6	0.05	0.5	Ti = 0.25 Cr = 0.25	4.5–6.0	0.5	Greatest ductility
A05180	AA518.0	B85 (68A)	0.25	1.8	7.5–8.5	0.35	0.15	0.35	0.15	Used for marine and aircraft hardware

TABLE 2.11 Properties of cast aluminium alloys

Unified	Aluminium Association	Hardness BHN	Ultimate Tensile Strength, MPa	Yield Strength, MPa	Modulus of Elasticity, GPa	Fatigue Strength, MPa
A03550	AA355.0	75.0–105	255	185	70.3	69.0
A03600	AA360.0	75.0	300	170	71.0	138.0
A03800	AA380.0	80.0	317	159	71.0	138.0
A03830	AA383.0	75.0	310	152	—	145.0
A03840	AA384.0	85.0	331	165	—	140.0
A03900	AA390.0	120.0	280	240	81.2	140.0
A04130	AA413.0	80.0	296	145	71.0	130.0
A04430	AA443.0	30.0–60.0	145	48.3	71.0	—
A05180	AA518.0	80	310	193	69.0	160.0

2.3.2 Copper

Similar to aluminium, pure copper also finds wide application because of its following properties:

1. The electrical conductivity of pure copper is high (5.8×10^5 ohm/cm) in its purest form. Any small impurity brings down the conductivity drastically. For example, 0.1% phosphorous reduces the conductivity by 40 %.
2. It has a very high thermal conductivity (0.92 cal/cm/°C).
3. It is a heavy metal (specific gravity 8.93).
4. It can readily be joined together by brazing.
5. It resists corrosion.
6. It has got a pleasing colour.

Pure copper is used in manufacture of electrical wire, bus bars, transmission cables, refrigerator tubing and piping. The mechanical properties of copper in its purest state are not very good. It is soft and relatively weak. It can be alloyed profitably to improve the mechanical properties. The main alloying elements used are zinc, tin, lead and phosphorous.

The alloys of copper and zinc are called brasses. With zinc content up to 39%, copper forms a single phase (α-phase) structure. Such alloys have high ductility. The colour of the alloy remains red up to a zinc content of 20% but beyond that it becomes yellow. A second structural component called β-phase appears between 39 to 46 % of zinc. It is actually the inter-metallic compound CuZn which is responsible for the increased hardness. The strength of brass gets further increased when small amounts of manganese and nickel are added.

The alloys of copper with tin are called bronzes. Increase in tin content increases the hardness and strength of bronze. The ductility is also reduced with the increase in tin percentage above 5. When aluminium is also added (4 to 11%) the resulting alloy is termed as aluminium bronze which has a considerably higher corrosion resistance. Bronzes are comparatively costly compared to brasses due to the presence of tin which is an expensive metal.

Some of the copper alloys with their compositions and applications are presented in Table 2.12.

TABLE 2.12 Some copper alloys

Alloy	Copper	Tin	Zinc	Others	Applications
Gun metal	88	10	2	—	Bearings, Bushes.
Bronze	89	11	—	—	Bearings
Phosphor bronze	89.5	10	—	P = 0.5	Bearings, Springs
Bell bronze	75–80	20–25	—	—	Bells
Admiralty brass	70	1	29	—	Heat exchangers
Gilding metal	85	—	15	—	Coins
Naval brass	60	1	39	—	Marine applications
Cartridge brass	70	—	30	—	Deep drawing
Clock brass	58–60	—	38–40	Pb = 1.5 – 2.5	Clock parts and engravings
Aluminium brass	76	—	22	Al = 2	Marine applications

2.3.3 Other Materials

Zinc

Zinc is principally used in engineering because of its low melting temperature (419.4°C) and higher corrosion resistance which increases with the purity of zinc. The corrosion resistance is caused by the formation of a protective oxide coating on the surface. Principal applications of zinc are in galvanising to protect steel from corrosion, in printing industry and for die casting.

The disadvantages of zinc are the strong anisotropy exhibited under deformed conditions, lack of dimensional stability under ageing conditions, a reduction in impact strength at lower temperatures and the susceptibility to inter-granular corrosion. It cannot be used for service above a temperature of 95°C because it will cause substantial reduction in tensile strength and hardness.

Its widespread use in diecastings is because it requires lower pressure which results in higher die life compared to other diecasting alloys. Further, it has very good machinability. The finish obtained by zinc diecasting is often adequate to warrant any further processing, except for the removal of the flash present in the parting plane.

Magnesium

Because of their light weight and good mechanical strength, magnesium alloys are used in applications where the weight is important, for example, in aerospace industries and in applications involving very high speeds. For the same stiffness, magnesium alloys require only 37.2% of the weight of C25 steel thus saving in weight. The two principal alloying elements used are aluminium and zinc. Magnesium alloys can be sand cast, permanent mould cast or diecast. The properties of sand cast magnesium alloy components are comparable with those of the permanent mould cast or diecast components. The die casting alloys generally have high copper content allowing them to be made from the secondary metals to reduce the costs. They are used for making automobile wheels, crank cases, etc. Higher the content, higher is the mechanical strength of magnesium wrought alloys such as rolled and forged components. Magnesium alloys can be readily welded by most of the traditional welding processes. A very useful property of magnesium alloys is their high machinability. They only require about 15% of power for machining compared to low carbon steel.

Titanium

Titanium (Ti) is a light weight, strong, corrosion resistant and refractory (Melting Point- 1668°C) metal. Alloys containing titanium are known for their high strength, light weight, and exceptional corrosion resistance. Since it is 40% lighter and as strong as steel, it is the preferred material for aerospace applications. Titanium forms a thin layer of titanium dioxide (TiO_2) on its surface which is extremely difficult for water and chemicals to penetrate thus providing exceptional corrosion resistance. Titanium has a low modulus of elasticity making it less flexible. It returns to its original shape after bending, resulting in its importance to shape memory alloys. Also titanium being non-magnetic and biocompatible (non-toxic, non-allergenic) has led to its increasing use in the medical field. Alloys of titanium are critical to the aerospace industry but due to their numerous unique properties, they are also used in medical, chemical and military applications, as well as in sporting goods.

Major application of titanium is as titanium dioxide (TiO_2), a whitening pigment used in paints, foods, medicines, and cosmetics. The average titanium content in Boeing's commercial airlines is approximately 15% of body weight. Titanium is used in aerospace engine applications such as rotors, compressor blades, hydraulic system components and nacelles. Titanium alloys are used in a variety of components in aircraft like critical structural parts, fire walls, landing gear, exhaust ducts (helicopters), hydraulic systems, armour plating, naval ships, spacecraft, and missiles. Titanium is also used in different medical applications such as dental implants, surgical implements and other implants like hip balls and sockets (joint replacement) that can stay in place for up to 20 years.

2.4 HEAT TREATMENT OF METALS

Earlier a detailed description of the iron carbon equilibrium diagram has been presented where all transformations that occur are under equilibrium conditions allowing sufficient time for the reaction to take place. Therefore, time would never be represented in that diagram. By controlling the cooling rate, it is possible to alter the properties of steels. The process of affecting properties by controlled heating and cooling rates is called heat treatment.

2.4.1 Transformation Curves

The basis for heat treatment is the time-temperature-transformation curves or TTT curves where in a single diagram all the three parameters are plotted. Because of the shape of the curves, they are also sometimes called C-curves or S-curves.

To plot TTT curves, the particular steel is held at a given temperature and the structure is examined at predetermined intervals to record the amount of transformation that taken place. From Fig. 2.11, it is known that the eutectoid steel (C80) under equilibrium conditions contains all austenite above 723°C, whereas below it is pearlite. To form pearlite, the carbon atoms should diffuse to form cementite. The diffusion being a rate process would require sufficient time for complete transformation of austenite to pearlite. From different samples, it is possible to note the amount of the transformation taking place at any temperature. These points are then plotted on a graph with time and temperature as the axes. Through these points, transformation curves can be plotted as shown in Fig. 2.13 for eutectoid steel. The curve at extreme left represents the time required for the transformation of austenite to pearlite to start at any given temperature. Similarly, the extreme right curve represents the time required for completing the transformation. In between the two curves are the points representing partial transformation.

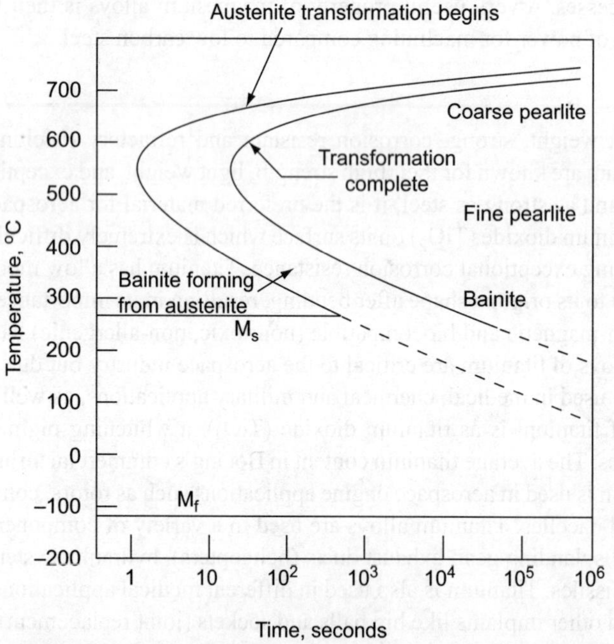

Fig. 2.13 *Isothermal decomposition diagram of C80 steel*

The horizontal lines Ms and Mf represents the start and finish of martensitic transformation. Since austenite needs time for transformation to pearlite, super cooled austenite would not have sufficient time for all of its carbon atoms to properly diffuse and form cementite. As a result, the carbon atoms would be trapped in the unit cell of iron. Since iron does not have enough interstitial space, the unit cell gets distorted with the extra carbon atoms. The distorted lattice structure is a body centered tetragonal and is called martensite. The degree of distortion depends on the number of carbon atoms trapped in the cell. It is formed in steels when cooled rapidly at a high rate which is of the order of 500°C per second for plain carbon steels.

Martensite has a needle like structure. It is extremely hard and brittle. As shown in Fig. 2.14, the maximum hardness of steel varies depending on the amount of carbon in steels. It is not possible to get any hardness in low carbon steels (less than 0.3% C). The maximum attainable hardness is reached by a carbon percentage of 0.7 to 0.8. The strength of martensite is derived from the fact that the movement of dislocations is effectively blocked by the distorted lattice.

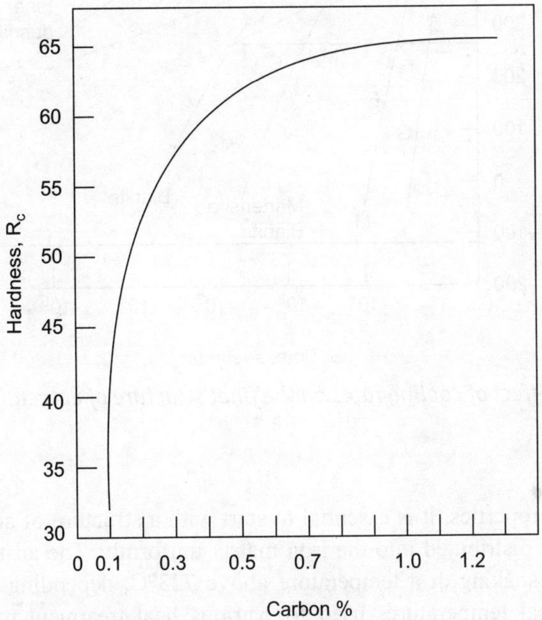

FIG. 2.14 *Effect of carbon percentage on the hardness of martensite produced*

Cooling rates employed to lower the austenitic temperature to room temperature control the final micro-structure of steel. The final microstructure could be complete martensite, martensite with pearlite or complete pearlite. Also the final grain size of pearlite produced could vary from very fine to coarse thus giving rise to wide varieties of properties that could be expected of them. Schematically, the cooling curves have been superimposed on the TTT curve in Fig. 2.15.

As seen from the TTT diagram, all the retained austenite may not be transformed unless the temperature reaches the Mf temperature. This is generally below the room temperature. The retained austenite (without transformation) can cause loss of strength or hardness, dimensional instability or cracking. In alloy steels since the alloying elements decrease the martensite transformation starting and ending temperatures, it may be necessary to cool the material to a lower temperature by the use of liquid nitrogen to get the full hardness.

The curve A represents extremely slow cooling rate which gives rise to conditions conducive to grain growth thus producing very coarse grains of pearlite in steels. The curve B which shows faster cooling rate than A would give rise to medium sized pearlitic grains. Similarly, the curves C and D which represents faster rates than B but not so fast as to miss the complete transformation curve produce complete pearlite which is extremely fine and called bainite. The bainitic structure is somewhat between pearlite and martensite. The cooling rates are faster in curves E, F and G than in the rest, not converting all the austenite into pearlite. The retained austenite below the Ms line is all converted into martensite and the final microstructure would be martensite and bainite. The amount of martensite formed increases as the cooling rate rises.

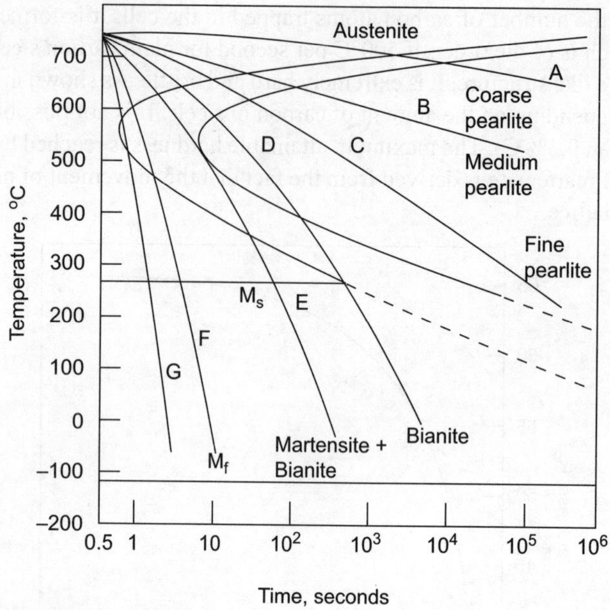

FIG. 2.15 *Effect of cooling rates on the final structure of eutectoid (C80) steel*

2.4.2 Hardening

To control the mechanical properties, it is essential to start with a structure of austenite since it is the state where the carbon atoms are distributed into the iron matrix uniformly. The austenitic structure in steels is achieved by heating it and soaking at a temperature above 723°C, depending on the carbon percentage. Table 2.13 shows the typical temperatures used for various heat treatment processes of general steels. Relevant portion of the iron-carbon equilibrium diagram is presented in Fig. 2.16, with the temperatures marked for the various carbon compositions. Up to the eutectoid composition, complete austenite is achieved by heating above the line KL but in hyper-eutectoid steels not all cementite is dissolved in austenite by heating just above the line KM (Fig. 2.16).

TABLE 2.13 Comparative heat treatment temperatures for steels

Steel	Quenching, °C	Normalizing, °C	Annealing, °C
C10 steel	900–925	900–955	540–730
C20 steel	870–910	900–955	540–730
C30 steel	855–900	870–915	675–745
C40 steel	815–855	855–900	845–885
C50 steel	800–845	845–900	830–870
C60 steel	790–845	830–885	815–855

Hardening or quenching process involves, first, bringing the steels into austenitic range by proper soaking at a temperature which depends on the thickness of the section. This steel is then rapidly cooled using any of the quenching media such as salt solution, water or oil. The heated material is directly dipped into the quenching bath as required for the particular steel. The critical cooling rate required for getting the complete martensitic structure in steels depends on the carbon percentage as shown in Fig. 2.17. Higher the carbon, lower is the cooling rate required.

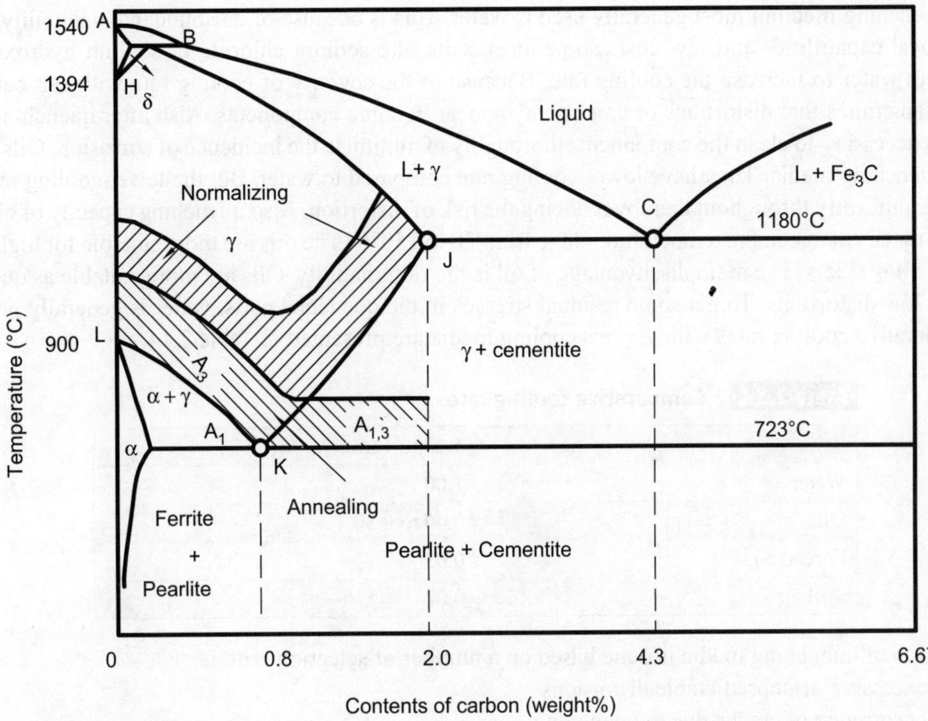

FIG. 2.16 *Part of iron carbon equilibrium diagram relevant for heat treatment of steels*

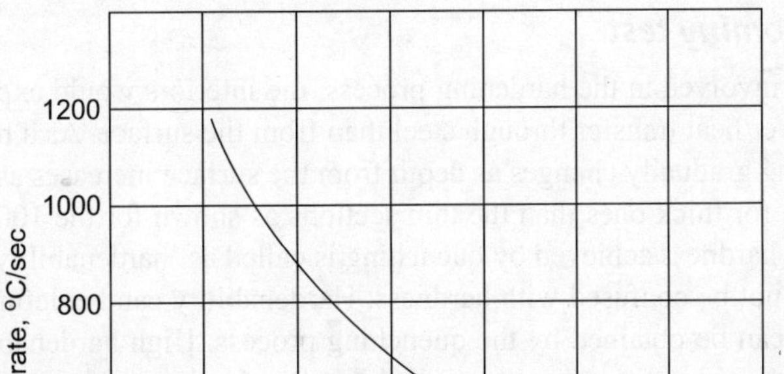

FIG. 2.17 *Influence of carbon content on critical cooling rate for missing the transformation curve*

The quenching medium most generally used is water. This is because of its abundant availability, greater heat removal capabilities and low cost. Sometimes, salts like sodium chloride or sodium hydroxide may be added to water to increase the cooling rate. Because of the severity of cooling rate with salt baths, it is possible sometimes that distortions or cracks may appear in some components. Also after quenching in salt baths, it is necessary to clean the components thoroughly to minimise the incidence of corrosion. Oils are also used as quenching media. They have lower cooling rate compared to water. But the lower cooling rate cools steel more uniformly throughout thereby reducing the risk of distortion. Also quenching capacity of oils is not subject to much change in the temperature range from 20 to 150°C. The oils are more suitable for high carbon steels and alloy steels. The main disadvantage of oil is the flammability. Oils are more suitable as quenching media for low distortions. To get some residual stresses in the quenched parts, water is generally preferred. The comparative cooling rates with various cooling media are presented in Table 2.14.

TABLE 2.14 Comparative cooling rates

Brine	1.20 to 1.30
Water	1.00
Oil	0.40 to 0.50
Forced air	0.03
Still air	0.02

The selection of quenching media is done based on a number of selection criteria.

- No excessive or unpredictable distortions
- No occurrence of cracks due to quenching
- To obtain the correct micro structure
- To produce the correct residual stress pattern

Hardenability and Jominy test

When thick sections are involved in the hardening process, the interiors would experience somewhat lower cooling rates due to slower heat transfer through steel than from the surface. As a result, the hardness of the material during quenching gradually changes as depth from the surface increases as shown in Fig. 2.18. The variation would be more for thick ones than the thin sections as shown for the 100 mm and 15 mm bars in Fig. 2.18. The depth and hardness achieved by quenching is called as 'hardenability'.

Hardenability should not be confused with hardness. Hardenability can be defined as the depth to which a certain hardness level can be obtained by the quenching process. High hardenability steel would be able to be thoroughly hardened without too severe a quenching rate. Under standard conditions, it is possible to determine the hardenability characteristics of different materials by Jominy's end quench hardenability test.

In this test, a standard test specimen (25.4 mm diameter and 102 mm long) of the material as shown in Fig. 2.19 is prepared. The test piece is heated to transform it into austenite completely and quenched by means of a water jet applied against the end of the specimen. Though the temperature of water and the flow rate is not critical to the hardness achieved, it is desirable to adhere to the standard condition as prescribed in the standards for the sake of comparison. When the specimen reaches the room temperature, a flat is ground to a depth of 0.4 mm on the side taking precaution that no grinding burn is caused. Hardness measured from the end using a micro hardness tester, at regular intervals (1/16 inch or 1.5875 mm), would give the hardenability characteristics of the material under consideration. This test is suitable for all steels except for those with very low or very high hardenability.

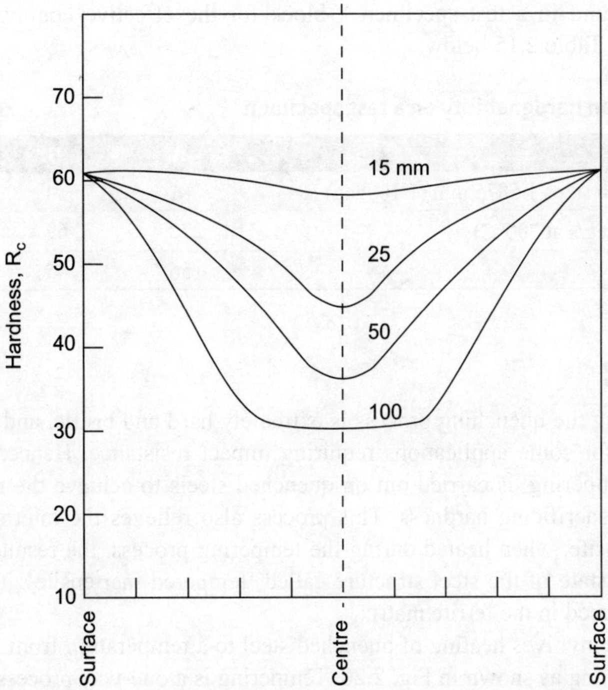

FIG. 2.18 *Variation of hardness across the depth of C50 steel bar after quenching*

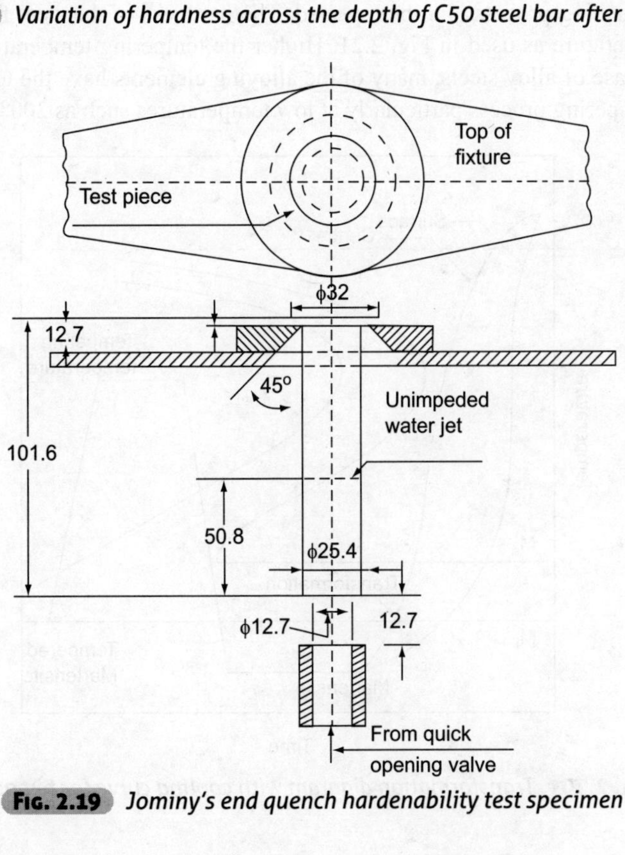

FIG. 2.19 *Jominy's end quench hardenability test specimen*

Typical values as found in a test specimen V-block for the effective cooling rates and the resultant hardnesses are shown in Table 2.15 below.

TABLE 2.15 Test data on hardenability on a test specimen

Location on V Block	A	B	C	D
End quench bar distance in steps 1.5875 mm (1/16 inch)	1	4	4.8	12.1
Equivalent cooling rate (deg/s at 700°C)	272	69	47	9
Block hardness, Rc	56	45	37	27

2.4.3 Tempering

Martensite formed during the quenching process is extremely hard and brittle, and lacks toughness making these steels unsuitable for some applications requiring impact resistance. Hence, a secondary heat treatment process called 'tempering' is carried out on quenched steels to achieve the necessary toughness and ductility by marginally sacrificing hardness. This process also relieves the internal stresses thus improving the ductility. Martensite, when heated during the tempering process, the resultant transformation leads towards an equilibrium state of the steel structure called 'tempered martensite'. This is a structure where cementite is finely dispersed in the ferrite matrix.

Tempering procedure involves heating of quenched steel to a temperature from 200 to 400°C, holding it for a while and then cooling as shown in Fig. 2.20. Tempering is a one-way process. If the hardness is to be regained, then the quenching process is to be repeated. The hardness of the steel finally achieved depends on the tempering temperature as used in Fig. 2.21. Higher the tempering temperature, lower would be the final hardness. In the case of alloy steels, many of the alloying elements have the tendency for reduction in hardness due to the tempering process particularly at low temperatures such as 200°C.

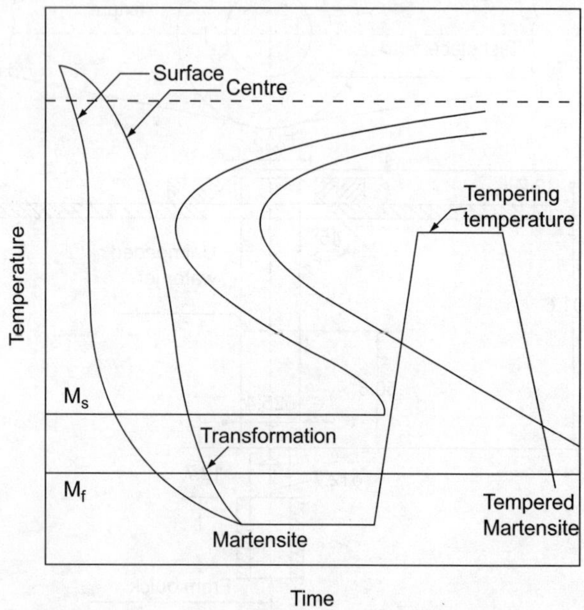

FIG. 2.20 *Transformation diagram with cooling curve for tempering*

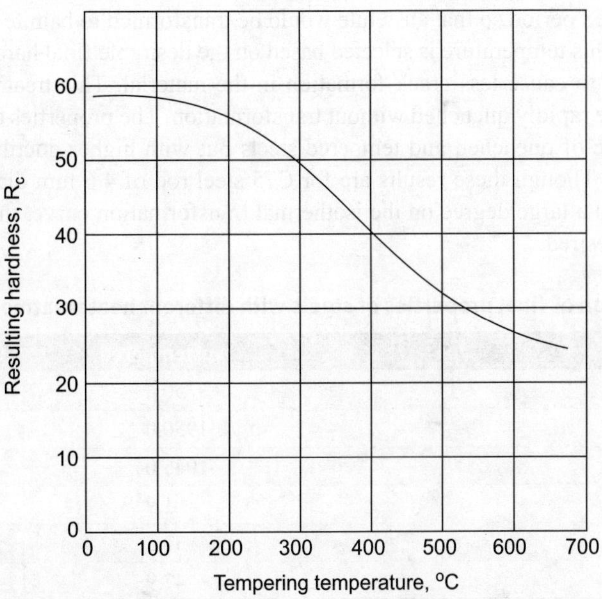

Fig. 2.21 *Effect of tempering temperature on the resulting hardness of the plain carbon*

Austempering

The quenching process being severe, the martensite formed would be extremely crack prone. A variation of the hardening process is austempering wherein the material is quickly quenched a little above the Ms temperature to just miss the nose of the cooling curve, as shown in Fig. 2.22. Then it is maintained at the same

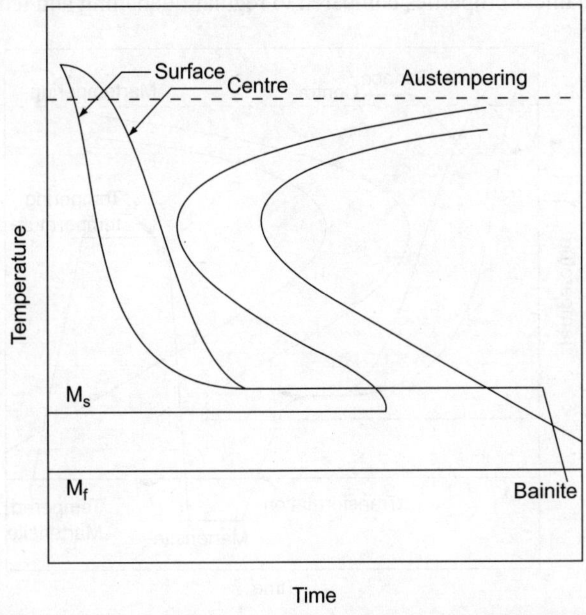

Fig. 2.22 *Transformation diagram with cooling curve for austempering*

temperature for prolonged period so that austenite would be transformed to bainite (very fine pearlite) under isothermal conditions. This temperature is selected based on the desirable final hardness. The bainite formed is stress free and likely to cause less crack formation in the material. This treatment is possible only for those steels which can be rapidly quenched without transformation. The properties obtained by austempering compare well with those of quenched and tempered steels but with higher ductility and impact resistance as shown in Table 2.16. Though these results are for C75 steel rod of 4.6 mm diameter, the application of austempering depends to a large degree on the isothermal transformation curve, the size of the work pieces and the final hardness desired.

TABLE 2.16 Comparison of final properties of steels with different heat treatment processes

	Austempered	Quenched and Tempered
Hardness Rockwell C	50.4	50.2
Ultimate strength, MPa	1950.0	1700.0
Yield strength, MPa	1045.0	840.0
Elongation in 150 mm, %	1.9	0.3
Reduction in area, %	34.5	0.7
Impact energy, N m	47.9	4.0

Martempering

In martempering, which is similar to austempering, the steel is heated to austenitic range, followed by rapid quenching (water bath) to a temperature above the Ms temperature (Fig. 2.23). Thereafter the material is maintained at a constant temperature (oil bath) such that the entire section is brought to a uniform temperature. As soon as the work piece attains uniform temperature, it is taken out of the bath and cooled in air. The air cooling is sufficiently fast so that martensite is formed. This provides relative stress free material without any distortion and better mechanical properties compared to regular quenching and tempering process.

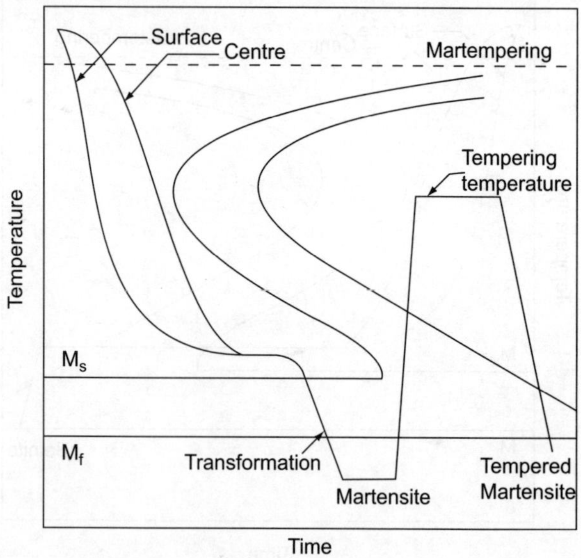

FIG. 2.23 *Transformation diagram with cooling curve for martempering*

2.4.4 Annealing and Normalising

Annealing is a widely used heat treatment process which involves heating the material to the austenitic temperature and subsequently cooling it very slowly in a way similar to that of the cooling line A in Fig. 2.15. Such cooling rate is achieved by leaving the steel inside the heating furnace itself without any further heat input. Because of the extremely slow cooling rate, final structure of the steel would be very coarse pearlite. This structure would eliminate all internal stresses, reduce hardness, increase ductility, enhance machinability, refine grain size and also affect some changes in electrical and magnetic properties.

In normalising, a somewhat faster cooling rate is used as in curve C in Fig. 2.15. This may be achieved by cooling the heated specimen in still air. Because of the faster cooling rate, the pearlitic structure obtained would be of uniformly fine grain size. In the final structure there is better dispersion of cementite. This would result in higher tensile strength and hardness than what are possible by annealing. The mechanical properties achieved through normalising depend to a great extent on the thickness of the section. In case of thicker objects or sections, the outer surface may be normalised but the core would be annealed.

2.4.5 Case Hardening

Heat treatment processes covered so far are meant to be effective for the entire material whereas case hardening processes are essentially meant for improving the hardness of the outer layers only, leaving the core to retain their original softness. In applications where it is desired to have greater hardness on the surface to take care of the abrasive forces and a soft core to provide the necessary toughness for absorbing any impact forces, case hardening would generally be ideal.

Case Carburising

This is carried out for low carbon steels which do not respond readily to quenching process because of the very low carbon content. In this process, the specimen along with a carbonaceous material such as charcoal, is packed in a sealed container and kept in a furnace. The pack is then heated between temperatures from 800 to 950°C and is held for a period of 4 to 20 hours. On heating, the oxygen present in small amount in the sealed container reacts with the charcoal and forms carbon monoxide (CO). The carbon monoxide when comes in contact with the specimen, releases oxygen while the carbon diffuses into the outer skin of the specimen. The released oxygen will again form CO with charcoal and the whole process gets repeated. As the time continues, sufficient amount of carbon gets into the material resulting in a medium to high carbon steel in the skin of the specimen.

The amount of carbon diffused into steel depends on the carburising temperature and time. The case depth achieved with carburising time is shown in Fig. 2.24. The carburising rate is initially high, but as time progresses, the rate decreases slightly. The case depth achieved is generally between 1 to 2 mm. Also the carbon content varies with very high amounts on the surface but decreasing gradually as depth increases. The process is called pack carburising since the specimen is packed with charcoal or any other carbonaceous material. The case depth achieved is a function of the quality of the coal and as such there is little uncertainty.

The other type which is more generally used is the gas carburising where carbon in gaseous form such as natural gas, propane or methane is used instead of charcoal. The process is very similar to pack carburising. The main advantage of gas carburising is to have a better control of surface carbon and case depth than possible in pack carburising.

In case carburising because of the prolonged heating grain size may be affected. The case carburised steel is to be further hardened by a secondary heat treatment process of quenching and tempering depending on the desired application. Some normalisation may also be done to refine the grain structure.

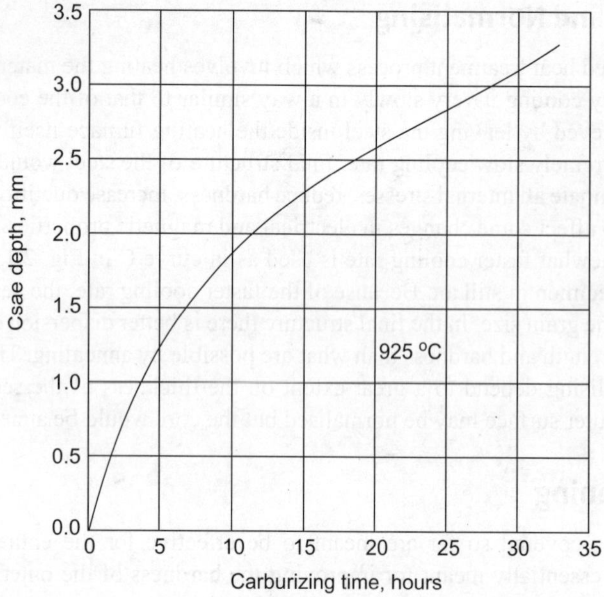

Fig. 2.24 *Effect of pack carburising on the case depth produced in 15Ni1Cr65*

The characteristics of the process are:

- Hard and highly wear resistant surface (medium case depths)
- Excellent capacity for contact load
- Good bending fatigue strength
- Good resistance to seizure
- Excellent freedom from quench cracks
- High capital investment

Cyaniding

Also called 'liquid carbonitriding', the cyaniding process involves addition of carbon and nitrogen to the cases of carbon steels and alloy steels. This is achieved by heating the steel in contact with a molten bath of cyanide. The cyanide bath may consist of sodium cyanide, potassium cyanide or potassium ferro-cyanide with inert salts such as sodium chloride and sodium carbonate which provide the necessary fluidity to the cyanide bath. The bath is maintained at a temperature from 750 to 850°C and the contact time is between 30 and 60 minutes. This produces a reasonable case depth of nearly 0.15 mm. Longer contact times can produce a larger case depth. In view of the large heat transfer coefficient of the liquid bath, the distortion of work piece is minimised. After the cyaniding process, the steel is quenched either in oil or in water based quenching medium to produce the desired case hardness. If necessary it may be followed with a low temperature tempering.

The process is faster than the carburising process but achieves shallow case-depths. It is generally employed for parts that do not require finishing after hardening. This process is not suitable for parts requiring high impact resistance since nitrogen addition is detrimental for these properties.

The characteristics of the process are:

- Hard and highly wear resistant surface (shallow and uniform case depths)
- Fair capacity for contact load (even at relatively high operating temperatures)
- Good bending fatigue strength

- Good resistance to seizure
- Good dimensional control possible
- Excellent freedom from quench cracks
- Medium capital investment

Nitriding

Alloying elements in steels such as aluminium, chromium, vanadium and molybdenum would form very hard nitrides when they come in contact with nitrogen. This is of use in the process of nitriding where alloy steels are case hardened without any quenching process. The nitriding temperatures are in the range of 500 to 575°C. The steels which are already quenched and tempered are put in a sealed container with ammonia gas and then heated to the nitriding temperature. The process is maintained for a period of 8 to 40 hours (as high as 125 hours in some cases) for getting the necessary case depths. The nitrogen forms complex nitrides with the alloying elements, whose depth depends on the nitriding time.

Because of the lower temperatures employed in nitriding, there are less possibilities of distortion and deformation than any other case hardening process. Also, this process increases the wear resistance and fatigue life greatly. Further, no finish machining is required.

The characteristics of the process are:

- Hard and highly wear resistant surface (shallow case depths)
- Fair capacity for contact load
- Good bending fatigue strength
- Excellent resistance to seizure
- Excellent dimensional control possible
- Excellent freedom from quench cracks
- Medium capital investment

Flame Hardening

This is the simplest form of heat treatment process. The work piece is heated by means of a gas torch (oxy-acetylene flame) followed by a water spray on the heated parts. The heat from the torch penetrates only to a small depth on the surface and consequently the steel in the outer layers gets quenched to martensite and bainite. Case depths up to 3 mm can be achieved by this process. This process can be followed by heating to about 200°C for the purpose of stress relieving. The surface hardness is not appreciably affected by this reheating operation. This process is suitable for any complex shape of the component such a crank shaft, large gears, cams, etc., with carbon percentages ranging from 0.3 to 0.6%. Though high carbon steels can also be flame hardened but greater care is needed to avoid surface cracking.

The characteristics of the process are:

- Hard and highly wear resistant surface (deep case depths)
- Good capacity for contact load
- Good bending fatigue strength
- Fair resistance to seizure
- Fair dimensional control possible
- Fair freedom from quench cracks
- Medium capital investment.

Induction Hardening

This is similar to flame hardening process where the heating of the component surface is achieved by the electromagnetic induction. The work piece such as crank shaft is enclosed in the magnetic field of an alternating (10 kHz to 2 MHz) current conductor to obtain case depths of the order of 0.25 to 1.5 mm. This causes induction heating of the work piece. The heated work piece is then quenched by water spray. The induction heat penetrates only the outer surface of the work piece and as a result only the skin gets hardened by the quenching process. The whole process is very fast (5 seconds to 4 minutes) and results in hard outer surface (50 to 60 RC) which is wear resistant.

2.4.6 Laser Heat Treatment

Hardening of ferrous alloy surfaces using laser radiation relies on the ability of the surface to get austenitized and then get transformed to martensite/bainite/pearlite depending upon the associated cooling rate achieved. A laser beam is scanned across the component that causes surface to heat rapidly. The surrounding material acts as an efficient heat sink, leading to rapid quenching and hardening phase transformations. As a result, hardened surface layer is produced, while the desirable bulk properties, such as toughness and ductility, remain unaffected throughout the material. Components made from hardenable ferrous alloys are particularly suitable for laser hardening. Depending on the material, hardness values up to about 1,000 HV can be achieved to a depth of around 1.5 mm through solid-state transformation without surface melting.

During the process of laser heat treatment the base material microstructure (generally pearlite and ferrite in as-received condition) will be transformed to austenite by the laser heating process. Subsequently, the heat from the processing zone is removed by the bulk of the material through heat conduction, providing sufficient cooling rate for the metallurgical transformations that lead to the martensitic and bainite formation, thus improving the hardness.

In order for the laser heat treatment to be successful the following conditions should be satisfied:

- Temperature for the zone being hardened should reach well into the austenitizing zone.
- Between heating and cooling cycles, the substrate should be maintained at the austenitizing temperature long enough for carbon diffusion.
- There should be enough mass so that the cooling rate by self-quenching is such that it could satisfy the critical quenching rate requirement.

Laser heat treatment process is applied to a variety of components and is credited with the successes that are not possible with other heat treatment processes. A number of benefits have been claimed for the laser heat treatment process. Some of the advantages are:

- High processing speeds since laser radiation will not have any inertia. This allows for rapid starting and stopping, as well as high production rates. This provides great flexibility in the operation.
- Improved surface hardness, strength, lubrication, wear and fatigue properties.
- Surface hardening can be achieved with little or no contamination (i.e., without using chemicals such as fuels and cooling media), allowing for very minimal environmental impact.
- No separate quenching is required since the mass of the part acts as a heat sink with sufficient cooling rate. Some small parts may require quenching with an external medium.
- It is possible to obtain higher case hardness compared to the conventional case hardening processes.
- Can be even applied to the lower carbon steel because of extremely rapid cooling rate that can be achieved by laser heat treatment.
- Greater product design flexibility with respect to material selection and geometry.
- Selective hardening of localized work piece surfaces. Here a selective hard surface arises with a tough centre and uninfluenced areas in the direct neighbourhood.
- Special geometries such as hardening tracks at cutting edges, guide tracks, grooves and free mould surfaces with continuous crossing geometry can be easily hardened using laser.
- It is possible to achieve good hardening of surfaces that are difficult to reach such as craned grooves, flutes, inside fillets and interior surfaces.

- It is also possible to dot harden at naps and in blind holes.
- Hardening can be done for some thin, warp-prone parts by a selective, quick heat input at an extremely short dwell time.
- Use of a few hardened tracks or paths instead of a fully hardened surface.
- Greater flexibility and precise localization of the hardened area making use of computer controlled operation for the positioning of the work piece surface. It is possible to automate the process so that it can be integrated into machining cells.
- Greater precision and lower energy input leading to reduced post-treatment work. The distortion of the components due to laser heating will be very low.

Limitations to the Process

Also there are certain limitations of the process which need to be well understood and taken care of in the process and product design.

- The metallic surfaces generally have reflectivity which will not allow all the laser power to be absorbed. To overcome this difficulty, the surface needs to be painted by black paint (or pickled) so that as much of laser energy as possible can be absorbed. This means produced component has to be cleaned and dried before the paint can be applied. This calls for additional operations and cost. Also after the operation, the leftover residue of the black paint needs to be cleaned.
- Laser hardening process is suitable for single pass operation for a finite surface. However, if a large surface is to be hardened then the laser beam has to traverse the entire surface in a suitable manner to completely cover it. One possibility is to have a helical path for cylindrical surface. During this process there is a possibility of some overlap heating of the surfaces by the laser which will temper part of the surface. This may produce some soft spots in otherwise hard surface produced. It needs to be established whether such soft areas are useful for the operation or detrimental.
- Suitable only for steel and cast iron that are heat treatable.
- Parts without sufficient mass may heat up (ahead of the beam) thus curtailing self-quenching, therefore, may require a separate quenching operation.
- Alloys that require higher soaking times for austenitization, such as steels, containing stable spheroidal carbides and ferritic cast irons are not suitable for laser heat treatment.
- The depth of the hardness produced is very small and generally limited to about a few millimeters.

The laser hardening has been used since the early 1970's for hardening components, particularly in the automotive and machine tool industries. Camshaft lobes, axle bearings, ratchets, cylinder liners, blanking dies, piston rings, gear housings, gear teeth and valve guides are typical examples. In general, if the surface area to be hardened is less than 10% of the total surface area of the component and the material can be hardened through thermally-induced phase transformations, then laser hardening is often the most economic hardening process.

2.4.7 Heat Treatment of Non-Ferrous Alloys

One of the principal mechanisms by which the non-ferrous alloys are heat treated is the precipitation hardening. In this process, a second phase of the alloy is dispersed finely in the matrix of the first phase in solid state thereby increasing the strength. The inter-metallic compound present in a solid solution is generally hard and its presence increases the hardness even in a single phase alloy. For example, in a copper alloy with zinc forming the alpha solid solution the hardness increases by about 39% when the zinc percentage is increased from 0 to 30%.

In the precipitation hardening, the alloy is heated and soaked to form a single phase solid solution. This then is cooled quickly (e.g. quenched in water) so that the single phase structure is retained even at the room temperature. When this material is heated to a certain temperature and retained at that temperature for specified periods, then the precipitation of the second phase at a number of random locations throughout the solid solution takes place, thus increasing the strength of the alloy.

The hardness obtained is a function of precipitation temperature since the size of the precipitating particles and their distribution controls the movement of dislocations. If the particles are too coarse, then they are too dispersed and will not give rise to high strength. Hence, it is important to know the temperature and time to get an optimum size of the particles and their distribution so that best strength can be obtained by the precipitation hardening process. If the precipitation process occurs at the room temperature, it is called age hardening.

2.4.8 Shot Peening

The word peen means to flatten with a small hammer. Shot peening is a carefully controlled cold working process of hitting the surface of the metal at a relatively high velocity with small spherical metallic elements called shot. As the old blacksmiths used to hammer hot forged parts even after it cooled, making it tough and have a long life, today it is used to increase the fatigue life of metal in a way that is not attainable by any other similar production means. It is used successfully on engineering components such as mechanical power transmission and other components for the aeronautical, automotive and industrial fields as well as for many other applications whether constructed of ferrous or non-ferrous materials. Shot peening affects a small layer of metal about 1.5 mm from the surface.

Each shot when it impinges on the surface forms a small crater as shown in Fig. 2.25(b). This thin surface layer of the crater is in tension when the shot is in contact with the surface. When the shot rebounds, the material below attempts forcing the surface layer spring back to its original condition. This puts the complete surface of the crater in a highly compressed cold worked state. When subsequent shots strike the surface where there is no crater, the same phenomenon repeats with a new crater. If it strikes an existing crater, the already cold worked surface prevents any new formation. Where there are no craters, new craters will be formed creating a uniform layer on the surface as shown in Fig. 2.25(c). This action creates a continuous formation of cold worked craters which put the total area into a layer of compressive stress as shown in Fig. 2.25(d).

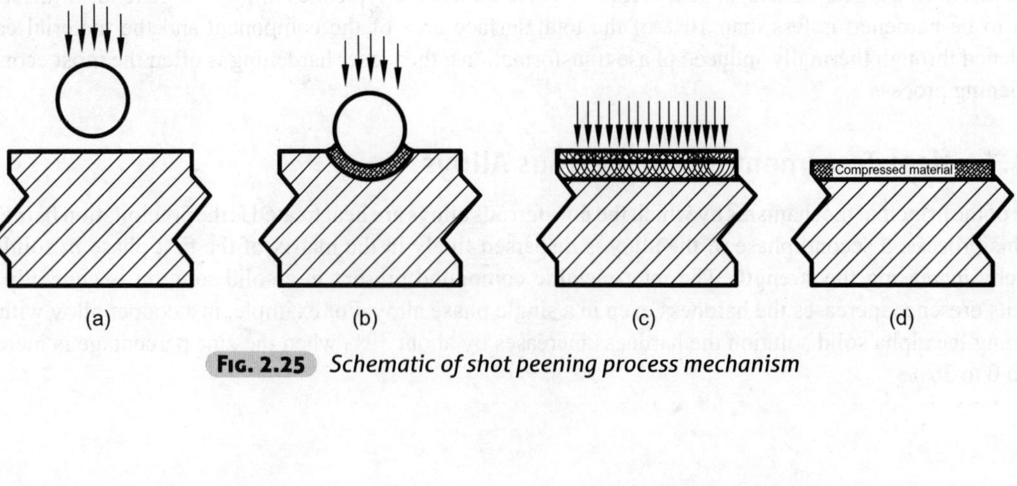

Fig. 2.25 *Schematic of shot peening process mechanism*

The various media that can be used in shot peening are steel spheres, glass beads and ceramic beads. It is important to have uniform size and shape of the shot used in shot peening to expose the surface to uniform residual stresses. The conventional shot peening involves accelerating shot material with compressed air. Shot is introduced into a high velocity air stream that accelerates the shot to speed up to 80 m/s at the time of impinging on the surface. The second method involves accelerating the shot with a wheel. The shot gets dropped onto the middle of the wheel and accelerates to the outer edge where it leaves on a tangential path.

Most failures start at the surface of a part in tension. Since the entire working area of a shot peened part is in compression, the part will experience increased life.

In general, shot peening will increase the life of a part if it is subjected to bending or twisting stress. However, it has little effect on the life of a part that is subject to axial (push-pull) stress since such stresses are reacted by the entire cross-section of the part rather than principally on the outer fibres.

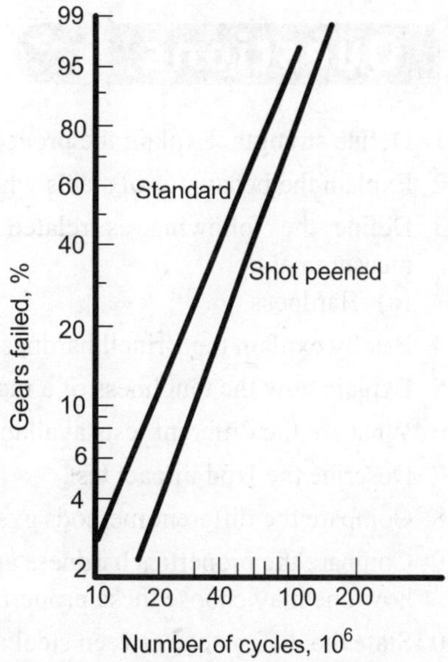

FIG. 2.26 *Shot peened spur gears exhibiting 1.6 times longer surface pitting fatigue life*

SUMMARY

Knowledge of the properties of engineering materials is an essential requirement in understanding the various manufacturing processes and their behaviour.

- There are a number of properties of engineering materials such as strength, ductility, hardness and toughness that are important from the application point of view.
- There are specified tests for each of these properties that provide standardised methods for measurement and understanding their relevance.
- Iron alloyed with carbon plays an important role in providing a large range of properties suitable for different engineering applications. Properties of steels (alloys of iron and carbon) depend on the amount of carbon present as well as the microstructure of the compounds formed by iron and carbon.
- Addition of different alloying elements to steels enhances special class of properties for specific applications, thereby extending the functionality of plain carbon steels.
- Aluminium and copper are the two most widely used non-ferrous material used in engineering industries. Their properties can be manipulated by the addition of alloying elements similar to steels. In addition there are other non-ferrous alloys which are being used for specific applications.
- Specialised heat treatment processes help in improving the properties of ferrous and non-ferrous alloys to suit the specific applications. Quenching is the process used for medium and high carbon steels to improve the hardness by heating above the austenitic range followed by the sudden cooling. Other heat treatment methods reduce the internal stresses and improve ductility. Other class of heat treatment processes increase the surface hardness while retaining the softness of the core.

Questions

2.1 Define strength. Explain the procedure for measuring the tensile strength of steels.

2.2 Explain the behaviour of steels when they are tensile loaded.

2.3 Define the following as related to engineering materials and explain the principles of their measurement.

(a) Hardness (b) Ductility

2.4 Briefly explain the Brinell hardness test.

2.5 Explain how the toughness of a material is measured.

2.6 What are the different tests available for hardness measurement? Compare their individual merits.

2.7 Describe the Izod impact test.

2.8 Compare the different methods available for measuring impact resistance of engineering materials.

2.9 Compare the properties hardness and toughness from application engineering point of view. Mention how one may choose these properties for a given application.

2.10 State the difference between steel and cast iron with respect to their compositions.

2.11 State the reason why white cast iron is more brittle than grey cast iron.

2.12 State how carbon content influences the strength and ductility of plain carbon steels.

2.13 State how the properties of alloy steels are affected by following alloying elements: Manganese, chromium and tungsten.

2.14 Name two alloying elements other than carbon, commonly employed in steel and state how they influence the properties of steel.

2.15 State the composition and application of an alloy steel.

2.16 State the ranges of composition for low, medium and high carbon steels. Give two applications for each range.

2.17 Make a classification of the commonly used engineering materials. Insert the following materials under the appropriate head: Grey cast iron, brass, duralumin, stainless steel, nylon.

2.18 Differentiate between steel and cast iron with reference to their carbon percentage, tensile strength, hardness and toughness.

2.19 State the composition and applications of two of the following: Low carbon steel, Duralumin, Yellow brass.

2.20 What do you understand by the term 'steel'? Explain the effect of each of the alloying element present in an alloy steel.

2.21 State along with an example how the Indian standard designation for structural steels is specified.

2.22 Write the composition of the following steels:

25 Mn 1 S 14 — 15 Mn 75
55 Si 2 Mn 90 — 55 Mn 1
50 Cr 1 V 23 — 35 Mn 2 Mo 45
40 Ni 2 Cr 1 Mo 25 — 31 Ni 3 Cr 65 Mo 55
T 105 Cr 1 Mn 60 — T 50 W 2 Cr 1 V 18
T 75 W 18 Cr 4 V 1 — 16 Ni Cr 2 Mo 20

2.23 List any two commonly used non-ferrous alloys stating their composition and application.

2.24 State the advantages of aluminium alloys over ferrous alloys.

2.25 Give the composition of two copper based alloys and their applications.

2.26 What is meant by solution hardening?

2.27 Explain the effect of copper as an alloying element in aluminium alloys.

2.28 What are the specific advantages of copper which make it a good engineering material compared to steel?

2.29 Describe briefly about brasses.

2.30 Make a short note on bronzes.

2.31 Give the composition of any two aluminium alloys with applications.

2.32 Why is titanium a preferred material for aerospace applications? Give some applications of titanium as used in aerospace components.

2.33 Give the application of titanium and titanium alloys in engineering applications.

2.34 Distinguish between the following:

(a) Quenching and tempering

(b) Normalising and annealing

(c) Cementite and martensite

2.35 Why do not the low carbon steels respond to hardening by quenching process?

2.36 Distinguish clearly between hardness and hardenability. Explain a method of measuring the hardenability of steel.

2.37 Explain why tempering follows the quenching process in the heat treatment of steels.

2.38 Differentiate between normalising and annealing in terms of the process and nature of the product.

2.39 Why is annealing done?

2.40 State the purpose served by quenching and tempering.

2.41 What are the case hardening processes that are normally used?

2.42 Is case carburising done to high carbon steels? State reasons to support your answer.

2.43 Briefly explain the process and application of the following:

(a) Case hardening (b) Annealing

2.44 Briefly describe the following processes:

(a) Carburising (b) Nitriding.

2.45 Explain briefly your understanding about the TTT curves.

2.46 Explain the necessity of heat treatment for steels. Describe the process of quenching.

2.47 What are the methods used for hardening low carbon steels? Explain with reasons.

2.48 Explain the concepts of laser hardening. What are the applications of laser hardening?

2.49 Explain the purpose of shot peening as used in industrial components.

Multiple Choice Questions

2.1 Strength of a material can be defined as the
(a) Resistance offered to indentation
(b) Resistance offered on application of external force
(c) Resistance offered to impact forces
(d) Resistance offered to sudden application of impact forces

2.2 Strain in a given material is the ratio of
(a) Applied load to the area of cross section opposing the load
(b) Change in dimension to original dimension
(c) Applied load to original dimension
(d) Change in dimension to the area of cross section opposing the load

2.3 Brinell hardness test uses an indenter that is a
(a) Sphero-conical diamond
(b) Sphere made of steel or tungsten carbide
(c) Square base pyramid diamond
(d) Rectangular base pyramid diamond

2.4 Vickers hardness test uses an indenter that is a
(a) Sphero-conical diamond
(b) Sphere made of steel or tungsten carbide
(c) Square base pyramid diamond
(d) Rectangular base pyramid diamond

2.5 Ductility of a material can be defined as the
(a) Resistance offered to indentation
(b) Resistance offered on application of external force
(c) Resistance offered to impact forces
(d) None of these

2.6 Pig iron is
(a) Pure iron with no carbon
(b) Iron with 0.40% carbon
(c) Iron with 4.0% carbon
(d) Iron with 2.40% carbon

2.7 In a face centred cubic (FCC) cell the atoms are present
(a) At all eight corners of the cube and at the centre of each face.
(b) At all eight corners of the cube and at the body centre.
(c) At all eight corners of the cube, at the centre of each face and at the body centre.
(d) None of the above

2.8 Fine grain size during the solidification of a metal is achieved by
(a) Lower nucleation rate
(b) Higher nucleation rate with lower growth rate
(c) Higher nucleation rate with higher growth rate
(d) Lower growth rate

2.9 Maximum carbon contained in steels is up to
(a) 0.2% (b) 6.67%
(c) 2% (d) 1.2%

2.10 Austenite is the
(a) BCC form of iron
(b) FCC form of iron
(c) Pure iron without any carbon
(d) Inter metallic compound iron carbide

2.11 Ferrite is the
(a) BCC form of iron
(b) FCC form of iron
(c) Pure iron with very low carbon
(d) Inter metallic compound iron carbide

2.12 Gray cast iron is
(a) Iron with the colour gray
(b) Iron where all the carbon is in combined form
(c) Iron where part of the carbon is in graphite form
(d) Graphite present is in spherical form

2.13 Ductile iron is
(a) Iron with the colour gray
(b) Iron where all the carbon is in combined form
(c) Iron where part of the carbon is in graphite form
(d) Graphite present is in spherical form

2.14 Chromium as an alloying element in alloy steels is used principally to
(a) Improve hardenability
(b) Improve mechanical properties at low or elevated temperatures
(c) Improve the corrosion and oxidation resistance
(d) Increase the machinability

2.15 Tungsten as an alloying element in alloy steels is used principally to
(a) Improve hardenability
(b) Improve mechanical properties at elevated temperatures
(c) Improve the corrosion and oxidation resistance
(d) Increase the machinability

2.16 Aluminium alloys are generally used for their
(a) Low thermal conductivity
(b) Low electrical conductivity
(c) Excellent corrosion resistance
(d) High mass density

2.17 Martensite is
(a) BCC form of iron
(b) Body centred tetragonal form of iron
(c) Pure iron with very low carbon
(d) Inter metallic compound iron carbide

2.18 Tempering process is used to achieve in steels
(a) Lower hardness by transforming austenite to pearlite
(b) Lower hardness by transforming martensite to pearlite
(c) High hardness by transforming austenite to martensite
(d) High hardness by transforming martensite to pearlite

2.19 Normalising process is used to achieve in steels
(a) Improve ductility by transforming austenite to martensite
(b) Improve mechanical properties by transforming austenite to fine grain pearlite
(c) Increase hardness by transforming austenite to martensite
(d) Decrease hardness by transforming martensite to pearlite

2.20 Nitriding process is used to increase surface hardness for
(a) Low carbon steels
(b) Alloy steels
(c) Medium carbon steels
(d) High carbon steels

Answers to MCQs

2.1 (b)	2.2 (b)	2.3 (b)	2.4 (c)	2.5 (d)
2.6 (c)	2.7 (a)	2.8 (b)	2.9 (c)	2.10 (b)
2.11 (c)	2.12 (c)	2.13 (d)	2.14 (c)	2.15 (b)
2.16 (c)	2.17 (b)	2.18 (b)	2.19 (b)	2.20 (b)

CHAPTER 3

Metal Casting Processes

Objectives

Metal casting processes are the primary manufacturing processes that are used to make complex parts utilising liquid metal. After completing this chapter, the reader will be able to

- Understand the history of metal casting process
- Design patterns and cores for metal casting process
- Understand the various moulding materials used in the making of moulds and cores

3.1 INTRODUCTION

Casting is one of the earliest metal shaping methods known to human being. It generally means pouring molten metal into a refractory mould with a cavity of the shape to be made and allowing it to solidify. When solidified, the desired metal object is taken out from the refractory mould either by breaking the mould or taking the mould apart. The solidified object is called casting. This process is also called founding.

3.1.1 History of Casting Process

The discovery of the casting process was done probably around 3500 B.C. in Mesopotamia. In many parts of the world during that period, copper axes and other flat objects were made in open moulds made of stone or baked clay. These moulds are essentially in single piece. But in later periods, when round objects were required to be made, the mould was split into two or more parts to facilitate the withdrawal of round objects.

The Bronze Age (2000 B.C.) brought far more refinement into casting process. For the first time perhaps, core for making hollow sockets in the objects was invented. These cores were made of baked clay. Also the cire perdue or lost wax process was extensively used for making ornaments and fine work.

Casting technology was greatly improved by Chinese from around 1500 B.C. Before that there is no evidence of any casting activity found in China. They do not appear to have been greatly familiar with the *cire perdue* process nor used it extensively but instead specialised in the multi-piece moulds for making highly intricate jobs. They spent a lot of time in perfecting the mould to the last detail so that hardly any finishing work was required on the casting made from the moulds. They had probably made piece moulds containing carefully fitted pieces numbering thirty or more. In fact, many such moulds have been unearthed during the archaeological excavations in various parts of China.

Indus valley civilisation is also known for their extensive use of casting of copper and bronze for ornaments, weapons, tools and utensils. But there was not much of improvement in the technology. From the various objects and figurines that were excavated from the Indus valley sites, they appear to have been familiar with all the known casting methods such as open mould, piece mould and the cire perdue process.

Though India could be credited with the invention of crucible steel, not much of iron founding was evident in India. There is evidence that iron founding had started around 1000 B.C. in Syria and Persia. It appears that iron casting technology in India has been in use from the times of the invasion of Alexander the Great, around 300 B.C.

The famous iron pillar located inside the Qutub Minar complex in Delhi is an example of the metallurgical skills of ancient Indians. It is 7.2 m long and is made of pure malleable iron. This is assumed to be of the period of Chandragupta II (375–413 A.D) of Gupta dynasty. The rate of rusting of outside of this pillar is practically zero and even the buried portion is rusting at extremely slow rate. This must have been first cast and then hammered to the final shape.

3.1.2 Advantages and Limitations

Casting process is extensively used in manufacturing because of its many advantages. Molten material flows into any small section in the mould cavity and as such any intricate shapes, internal or external, can be made with the casting process. It is possible to cast practically any material, be it ferrous or non–ferrous. Further, the necessary tools required for casting moulds are very simple and inexpensive. As a result, for trial production or production of a small lot, it is an ideal method. It is possible in casting process to place the amount of material where exactly required. As a result, weight reduction in design can be achieved. Castings are generally cooled uniformly from all sides and therefore they are expected to have no directional properties. There are certain metals and alloys which can only be processed by the casting and not by any other process like forging because of the metallurgical considerations. Casting of any size and weight, even up to 200 tons, can be made.

However, the dimensional accuracy and surface finish achieved by normal sand casting process would not be adequate for final application in many cases. To take these cases into consideration, some special casting processes such as die casting have been developed, which are discussed in later chapters. Also the sand casting process is labour intensive to some extent and therefore many improvements are aimed at it like machine moulding and foundry mechanisation. With some materials it is often difficult to remove defects arising out of the moisture present in sand castings.

3.1.3 Applications

Typical applications of sand casting process are cylinder blocks, liners, machine tool beds, pistons, piston rings, mill rolls, wheels, housings, water supply pipes and specials, and bells.

3.1.4 Casting Terms

In the following chapters, the details of sand casting which represents the basic process of casting would be seen. Before going into the details of the process defining a number of casting vocabulary words would be appropriate. Reference may please be made to Fig. 3.1.

Flask A moulding flask is one which holds the sand mould intact. Depending upon the position of the flask in the mould structure, it is referred to by various names such as drag, cope and cheek. It is made up of wood for temporary applications or more generally of metal for long term use.

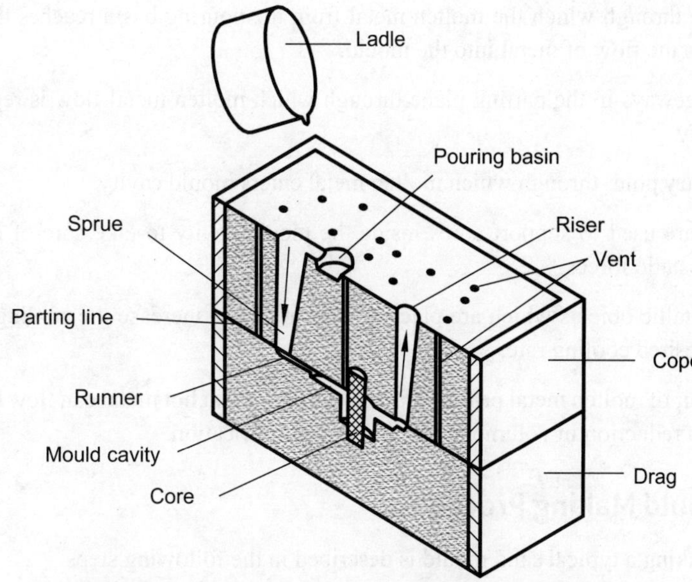

Fig. 3.1 *Cross section of a sand mould ready for pouring*

Drag Lower moulding flask.

Cope Upper moulding flask.

Cheek Intermediate moulding flask used in three piece moulding.

Pattern Pattern is a replica of the final object to be made with some modifications. The mould cavity is made with the help of the pattern.

Parting line This is the dividing line between the two moulding flasks that makes up the sand mould. In split pattern it is also the dividing line between the two halves of the pattern.

Bottom board This is a board which is normally made of wood and is used at the start of the mould making. The pattern is first kept on the bottom board, sand is sprinkled on it and then the ramming is done in the drag.

Facing sand The small amount of carbonaceous material sprinkled on the inner surface of the moulding cavity to give better surface finish to the castings.

Moulding sand It is the freshly prepared refractory material used for making the mould cavity. It is a mixture of silica, clay and moisture in appropriate proportions to get the desired results and it surrounds the pattern while making the mould.

Backing sand It is what constitutes most of the refractory material found in the mould. This is made up of used and burnt sand.

Core It is used for making hollow cavities in castings.

Pouring basin A small funnel shaped cavity at the top of the mould into which the molten metal is poured.

Sprue The passage through which the molten metal from the pouring basin reaches the mould cavity. In many cases it controls the flow of metal into the mould.

Runner The passageways in the parting plane through which molten metal flow is regulated before they reach the mould cavity.

Gate The actual entry point through which molten metal enters mould cavity.

Chaplet Chaplets are used to support cores inside the mould cavity to take care of its own weight and overcome the metallostatic forces.

Chill Chills are metallic objects which are placed in the mould to increase the cooling rate of castings to provide uniform or desired cooling rate.

Riser It is a reservoir of molten metal provided in the casting so that hot metal can flow back into the mould cavity when there is a reduction in volume of metal due to solidification.

3.1.5 Sand Mould Making Procedure

The procedure for making a typical sand mould is described in the following steps.

First a bottom board is placed either on the moulding platform or on the floor, making the surface even. The drag moulding flask is kept upside down on the bottom board along with the drag part of the pattern at the centre of the flask on the board. There should be enough clearance between the pattern and the walls of the flask which should be of the order of 50 to 100 mm. Dry facing sand is sprinkled over the board and pattern to provide a non-sticky layer. Freshly prepared moulding sand of requisite quality is now poured into the drag and on the pattern to a thickness of 30 to 50 mm. Rest of the drag flask is completely filled with the backup sand and uniformly rammed to compact the sand. The ramming of the sand should be done properly, neither to compact it too hard which makes the escape of gases difficult, nor too loose, so that mould would not have enough strength. After the ramming is over, the excess sand in the flask is completely scrapped using a flat bar to the level of the flask edges.

Now, with a vent wire, which is a wire of 1 to 2 mm diameter with a pointed end, vent holes are made in the drag to the full depth of the flask as well as to the pattern to facilitate the removal of gases during casting solidification. This completes the preparation of the drag.

The finished drag flask is now rolled over to the bottom board exposing the pattern as shown in Fig. 3.2(c). Using a slick, the edges of sand around the pattern is repaired and cope half of the pattern is placed over the drag pattern, aligning it with the help of dowel pins. The cope flask on top of the drag is located aligning again with the help of the pins. The dry parting sand is sprinkled all over the drag and on the pattern.

A sprue pin for making the sprue passage is located at a small distance of about 50 mm from the pattern. Also a riser pin, if required, is kept at an appropriate place and freshly prepared moulding sand similar to that of the drag along with the backing sand is sprinkled. The sand is thoroughly rammed, excess sand scraped and vent holes are made all over in the cope as in the drag.

The sprue pin and the riser pin are carefully withdrawn from the flask. Later the pouring basin is cut near the top of the sprue. The cope is separated from the drag and any loose sand on the cope and drag interface of the drag is blown off with the help of bellows. Now the cope and the drag pattern halves are withdrawn by using the draw spikes and rapping the pattern all around to slightly enlarge the mould cavity so that the mould walls are not spoiled by the withdrawing pattern. The runners and the gates are cut in the mould carefully

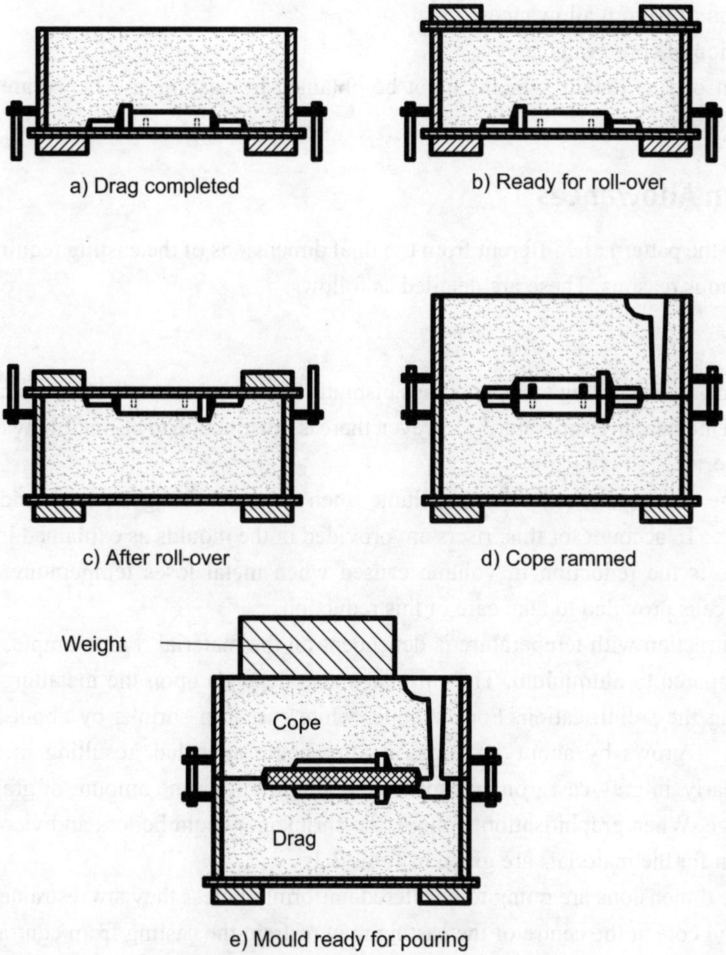

FIG. 3.2 *Sand mould making procedure*

without spoiling the mould. Any excess or loose sand found in the runners and mould cavity is blown away using the bellows. Now the facing sand in the form of a paste is applied all over the mould cavity and the runners which would give the finished casting a good surface finish.

A dry sand core is prepared using a core box. After suitable baking, it is placed in the mould cavity as shown in Fig. 3.2. The cope is replaced on the drag taking care of the alignment of the two by means of the pins. A suitable weight is kept on the cope to take care of the upward metallostatic force during the pouring of molten metal. The mould now, as shown in Fig. 3.2(e), is ready for pouring.

3.2 PATTERNS

As has been defined earlier, a pattern is a replica of the object to be made by the casting process, with some modifications. The main modifications are:

(a) The addition of pattern allowances,
(b) The provision of core prints, and
(c) Elimination of fine details which cannot be obtained by casting and hence are to be obtained by further processing.

3.2.1 Pattern Allowances

The dimensions of the pattern are different from the final dimensions of the casting required. This is required because of the various reasons. These are detailed as follows.

Shrinkage

All the metals shrink when cooling except perhaps bismuth. This is because of inter– atomic vibrations which are amplified by an increase in temperature. However, there is a distinction to be made between liquid shrinkage and solid shrinkage.

Liquid shrinkage refers to the reduction in volume when the metal changes from liquid to solid state at the solidus temperature. To account for this, risers are provided in the moulds as explained in Chapter 4.

Solid shrinkage is the reduction in volume caused when metal loses temperature in solid state. The shrinkage allowance is provided to take care of this reduction.

The rate of contraction with temperature is dependent on the material. For example, steel contracts to a higher degree compared to aluminium. The shrinkage also depends upon the metallurgical transformation taking place during the solidification. For example, white cast iron shrinks by about 21.0 mm/m during casting. However, it grows by about 10.5 mm/m after being annealed, resulting in a net shrinkage of 10.5 mm/m. Similarly, in gray cast iron and spheroidal graphite iron, the amount of graphitisation controls the actual shrinkage. When graphitisation is more, the shrinkage would be less and vice versa. The various rates of contraction for the materials are given in Table 3.1.

As a rule all the dimensions are going to be altered uniformly unless they are restrained in some way. For example, a dry sand core at the centre of the casting may restrain the casting from contracting but the edges are not restrained. Thus, it may be desirable to provide a higher shrinkage allowance for outer dimensions compared to ones which may be restrained. The actual value of shrinkage depends on various factors specific to a particular casting, namely the actual composition of the alloy cast, mould materials used, mould design, complexity of the pattern and the component size. The pattern maker's experience and a little bit of trial are to be used in arriving at the final shrinkages provided on the pattern. The values shown in Table 3.1 are average values and higher values are to be used for smaller dimensions and vice versa.

The shrinkage allowance is always to be added to the linear dimensions. Even in case of internal dimensions (e.g. internal diameters of cylinders), the material has a tendency to contract towards the centre and thus are to be increased. It is also possible to obtain shrink rulers for specific materials such as steels which are nothing but special scales where dimensions shown are actually longer by a measure equal to the shrinkage allowance. Dimensions provided by such a rule can be used at the time of making the pattern. Different shrink rulers are used for different casting materials.

TABLE 3.1 Shrinkage allowances for various metals

Material	Pattern Dimension, mm	Section Thickness, mm	Shrinkage Allowance, mm/m
Grey cast iron	up to 600	–	10.5
	600 to 1200	–	8.5
	over 1200	–	7.0
White cast iron	–	–	16.0 to 23.0
Ductile iron	–	–	8.3 to 10.4
Malleable iron	–	6	11.8
		9	10.5
		12	9.2
		15	7.9
		18	6.6
		22	4.0
		25	2.6
Plain carbon steel	up to 600	–	21.0
	600 to 1800	–	16.0
	over 1800	–	13.0
Chromium steel	–	–	20.0
Manganese steel	–	–	25.0 to 38.0
Aluminium	–	–	13.0
Aluminium bronze	–	–	20.0 to 23.0
Copper	–	–	16.0
Brass	–	–	15.5
Bronze	–	–	15.5 to 22.0
Gunmetal	–	–	10.0 to 16.0
Manganese bronze	–	–	15.6
Silicon bronze	–	–	10.4
Tin bronze	–	–	10.4
Chromium copper	–	–	20.8
Lead	–	–	26.0
Monel	–	–	20.0
Magnesium	–	–	13.0
Magnesium alloys	–	–	16.0
White metal	–	–	6.0
Zinc	–	–	10.0 to 15.0

Example 3.1

The casting shown in Fig. 3.3(a) is to be made in plain carbon steel using a wooden pattern. Assuming only shrinkage allowance, calculate the dimensions of the pattern.

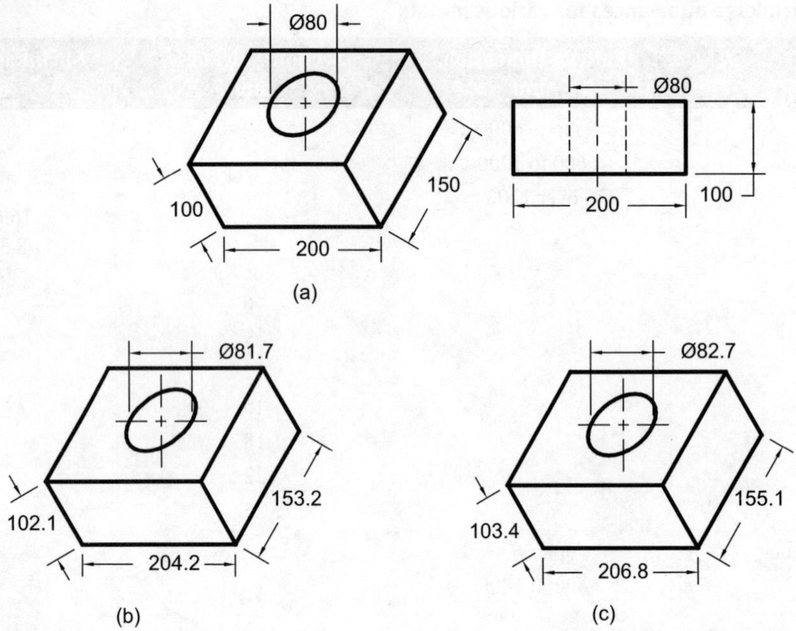

Fig. 3.3 *Provision of allowances for patterns in sand castings (All dimensions in mm)*

From Table 3.1, for steel the shrinkage allowance is 21.0 mm/m.
For dimension 200, allowance is $200 \times 21.0/1000 = 4.20$ mm
For dimension 150, allowance is $150 \times 21.0/1000 = 3.15 \approx 3.20$ mm
For dimension 100, allowance is $100 \times 21.0/1000 = 2.10$ mm
For dimension 80, allowance is $80 \times 21.0/1000 = 1.68 \approx 1.70$ mm
The pattern drawing with required dimensions taking shrinkage into account is shown in Fig. 3.3(b).

Double shrinkage allowance is to be provided on the pattern dimensions, if it is to be used for casting the metallic pattern, which would ultimately be used for moulding to take care of the shrinkage of the actual metal cast as well as the shrinkage of the pattern metal. This is illustrated in the following example.

Example 3.2

For the same example as shown in Fig. 3.3(a), if the master pattern is to be made of aluminium then calculate the dimensions of the wooden pattern which is to be used for making the aluminium pattern.

Figure 3.3(b) shows the dimensions of the aluminium pattern.

From Table 3.1, for aluminium the shrinkage allowance is 13.0 mm/m. For plain carbon steel it is 21.0 mm/m. The total shrinkage = $13.0 + 21.0 = 34.0$ mm/m

For dimension 200, allowance is $200 \times 34.0/1000 = 6.80$ mm
For dimension 150, allowance is $150 \times 34.0/1000 = 5.10$ mm
For dimension 100, allowance is $100 \times 34.0/1000 = 3.40$ mm
For dimension 80, allowance is $80 \times 34.0/1000 = 2.72$ mm

The final dimensions of the wooden pattern for making the aluminium master pattern are shown in Fig. 3.3(c). The pattern allowance applied for this wooden pattern is called double shrinkage allowance.

Finish or Machining Allowance

The finish and accuracy achieved in sand casting are generally poor and therefore when the casting is functionally required to be of good surface finish or dimensionally accurate, it is generally achieved by subsequent machining. Also ferrous materials would have scales on the skin which are to be removed by cleaning. Hence, extra material is to be provided which is to be subsequently removed by machining or cleaning process. This depends on dimensions, the type of casting material and the finish required. This may range from 2 to 20 mm. General guidelines for machining allowances are provided in the Table 3.2. The machining allowance provided ultimately has to be removed by machining. Hence, the cost of providing additional machining allowance should be carefully examined before finalising.

The type of machining allowance provided would depend on the metal cast, the type of moulding used, the class of accuracy required on the surface and the complexity of surface details. One way of reducing the machining allowance is to keep entire casting in the drag flask such that dimensional variation and other defects due to the parting plane are reduced to a minimum.

Example 3.3

In the previous Example 3.2, what will be the pattern dimension if all the surfaces of the casting need to be machined?

From Table 3.2,
Machining allowance for bore = 3 mm
Machining allowance for all surfaces = 3 mm
Machining allowance for cope side = 6 mm

The dimension 80, $80 + 2 \times 3 = 86$ mm
The dimension 100, $100 + 3 + 6 = 109$ mm
The dimension 150, $150 + 3 + 3 = 156$ mm
The dimension 200, $200 + 5.5 + 5.5 = 211$ mm
The final dimensions are shown in Fig. 3.4.

TABLE 3.2 Machining allowances on patterns for sand castings

Dimension, mm	Allowance, mm		
	Bore	Surface	Cope Side
Cast iron			
up to 300	3.0	3.0	5.5
301 to 500	5.0	4.0	6.0
501 to 900	6.0	5.0	6.0
Cast steel			
up to 150	3.0	3.0	6.0
151 to 500	6.0	5.5	7.0
501 to 900	7.0	6.0	9.0
Non-ferrous			
up to 200	2.0	1.5	2.0
201 to 300	2.5	1.5	3.0
301 to 900	3.0	2.5	3.0

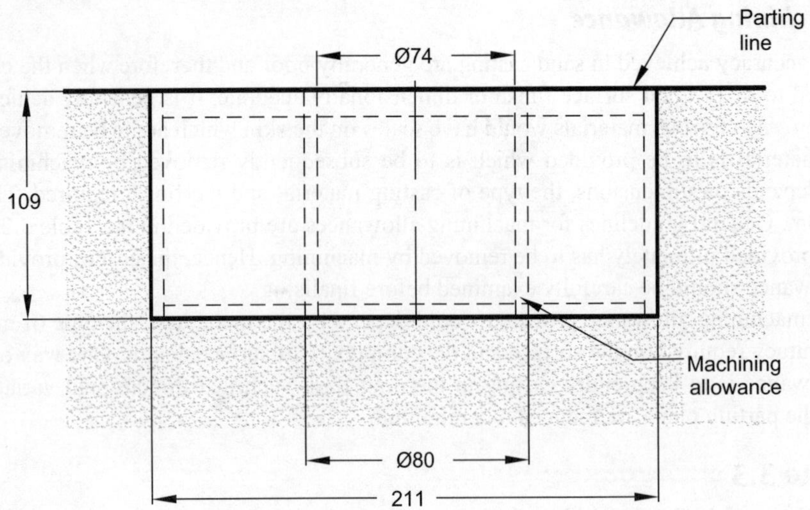

FIG. 3.4 *Pattern after providing machining allowance (All dimensions in mm)*

Draft

At the time of withdrawing the pattern from the sand mould, the vertical faces of the pattern are in continual contact with the sand which may damage the mould cavity, as shown in Fig. 3.5(a). To reduce its chances, the vertical faces of the pattern are always tapered from the parting line (Fig. 3.5(b)). This provision is called draft allowance.

Draft allowance varies with the complexity of the job. But in general, inner details of the pattern require higher draft than outer surfaces. Table 3.3 is a general guide to the provision of drafts. The draft allowance given varies for hand moulding and machine moulding. More draft need to be provided for hand moulding compared to machine moulding. In machine moulding, the actual draft given varies with the condition of the machine (new, rigid, properly aligned, etc., require less draft).

An observation here is that draft is always provided as an extra metal over and above the original casting dimensions as shown in the following example.

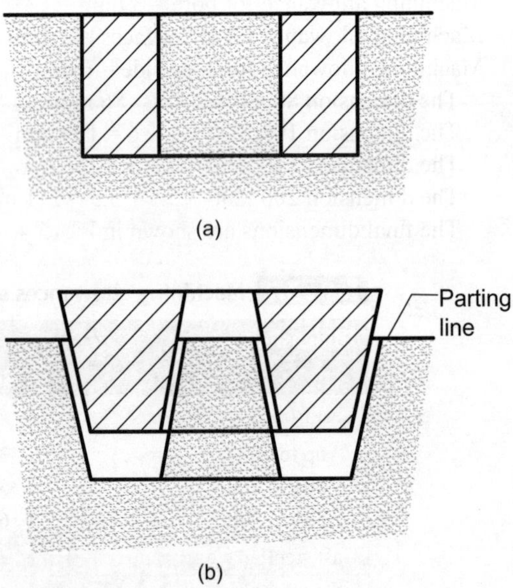

FIG. 3.5 *Effect of draft on pattern withdrawing*

Example 3.4

Provide draft allowance to the pattern shown in Fig. 3.4 after providing the machining allowance.

From Table 3.3, the draft angle is 0.75° for external details and 1° for internal details.

Draft need to be provided only for the dimensions that are perpendicular to the parting line. Hence, for 109 mm size, the taper required is

External = 109 × tan (0.75) = 1.4268 ≈ 1.40 mm

Internal = 109 × tan (1.00) = 1.9026 ≈ 1.90 mm

Based on those dimensions, the sizes are:

The bore dimension = 74 – 2 × 1.90 = 70.20 mm

External dimension = 211 + 2 × 1.40 = 213.80 mm

After providing for this taper, the pattern drawing is as shown in Fig. 3.6(b) compared to Fig. 3.6(a) which has been before providing the draft allowance.

TABLE 3.3 Suggested draft values for patterns

Pattern Material	Height of the Given Surface, mm	Draft Angle of Surfaces, Degrees	
		External Surface	Internal Surface
Wood	20	3.00	3.00
	21 to 50	1.50	2.50
	51 to 100	1.00	1.50
	101 to 200	0.75	1.00
	201 to 300	0.50	1.00
	301 to 800	0.50	0.75
	801 to 2000	0.35	0.50
	over 2000	—	0.25
Metal and plastic	20	1.50	3.00
	21 to 50	1.00	2.00
	51 to 100	0.75	1.00
	101 to 200	0.50	0.75
	201 to 300	0.50	0.75
	301 to 800	0.35	0.50

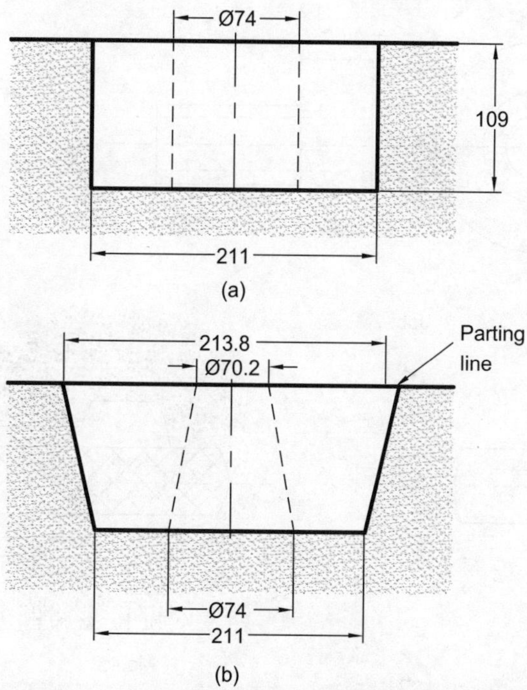

FIG. 3.6 *Example showing the application of draft (All dimensions in mm)*

Shake Allowance

Before withdrawal from the sand mould, the pattern is rapped all around the vertical faces to enlarge the mould cavity slightly which facilitates its removal. Since it enlarges the final casting made, it is desirable that the original pattern dimensions should be reduced to account for this increase. There is no sure way of quantifying this allowance since it is highly dependent on the foundry personnel and practices involved.

It is a negative allowance and is to be applied only to those dimensions which are parallel to the parting plane. One way of reducing this allowance is to increase the draft which can be removed during the subsequent machining.

Distortion Allowance

A metal when has just solidified is very weak and therefore is likely to be distortion prone. This is particularly so for weaker sections such as long flat portions, V, U sections or in a complicated casting which may have thin and long sections connected to thick sections. The foundry practice should be to make extra material provision for reducing the distortion. Alternatively, the shape of pattern itself should be given a distortion of equal amount in the opposite direction of the likely distortion direction. This can be done by trial and error basis to get the distortion amount. Some data about a few test cases may be available from the literature.

3.2.2 Core Prints

For all those castings where coring is required, provision should be made to support the core inside the mould cavity. One of the methods that are universally followed is to provide core prints where possible. Figure 3.7 shows an example of the provision of core prints. The size of the core prints to be provided is estimated based on the specific casting, the details of which are provided later.

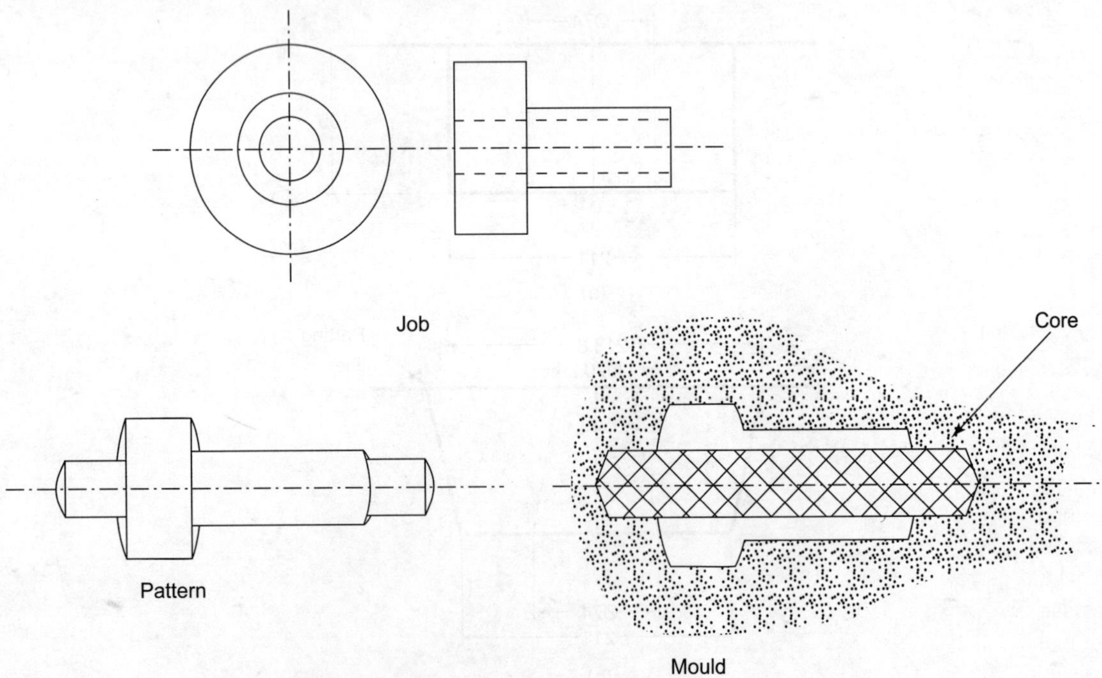

FIG. 3.7 *Typical job, its pattern and the mould cavity*

3.2.3 Elimination of Details

Often it is not possible to get very fine details on the surface or very small holes by sand casting. In such cases, it is desirable to simplify the casting process by eliminating those details and get them during the finishing process employed. One such example is presented in Fig. 3.8. The types of details that one would like to eliminate in the casting depend on the required accuracy, the capability of the chosen casting process and the moulding method employed.

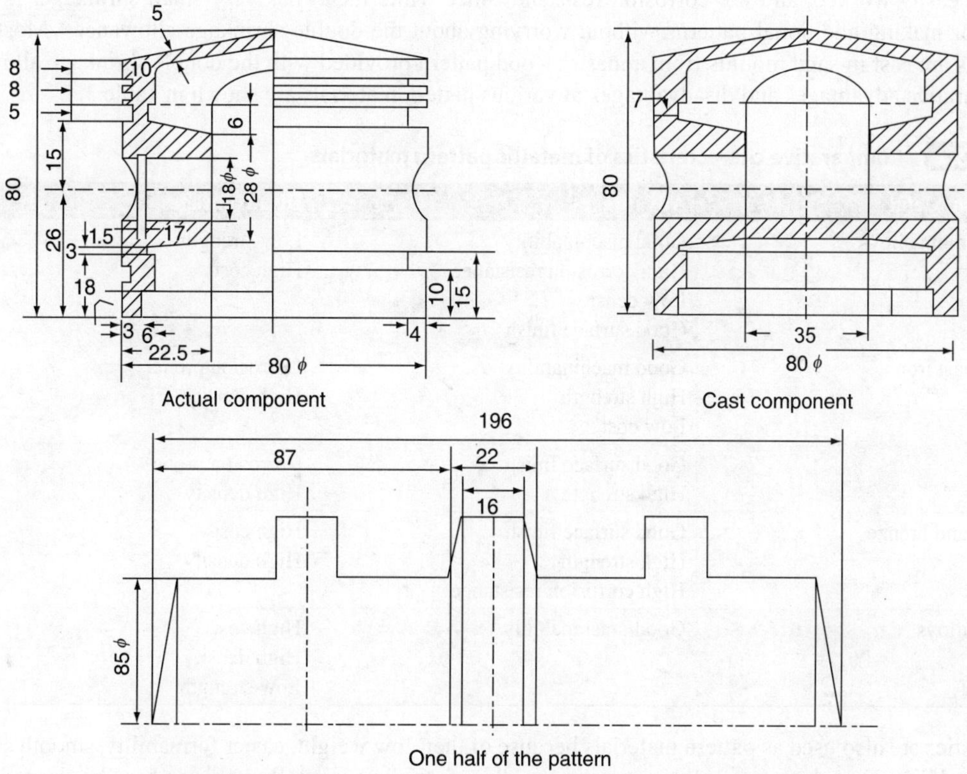

FIG. 3.8 *Elimination of details on a casting to simplify moulding*

3.2.4 Pattern Materials

The usual pattern materials are wood, metal and plastics. The most commonly used pattern material is wood, because of its easy availability, low weight, can be easily shaped and is relatively cheap. But the main disadvantage of wood is its absorption of moisture as a result of which distortions and dimensional changes occur. A good construction may be able to reduce the warpage to some extent. Hence, proper seasoning and upkeep of wood is almost a pre-requisite for large–scale use of wood as a pattern material.

The usual varieties of wood commonly used for making patterns are pine, mahogany, teak, walnut and deodar. Besides the wood, the plywood boards of the veneer type as well as the particle boards are also used for making patterns. Because of their availability in various thicknesses, their higher strength and no need for seasoning are the reasons for their usage. However, they can be used only in patterns which are of flat type (pattern plates) and no three dimensional contours.

Choice of the pattern material depends essentially on the size of the casting, the number of castings to be made from the pattern and the dimensional accuracy required. For very large castings, wood may be the only practical pattern material. Moulding sand being highly abrasive for large–scale production, wood may not be suitable as a pattern material and one may have to opt for metal patterns.

Because of their durability and smooth surface finish, metal patterns are extensively used for large scale casting production and for closer dimensional tolerances. Though many materials such as cast iron, brass, etc., can be used as pattern materials, aluminium and white metal are most commonly used. These are light, can be easily worked, and are corrosion resistant. Since white metal has very small shrinkage, it can be used for making additional patterns without worrying about the double shrinkage allowances. Most metal patterns are cast in sand moulds from a master wood pattern provided with the double shrinkage allowance. Comparative advantages and disadvantages of various pattern materials are shown in Table 3.4.

TABLE 3.4 Comparative characteristics of metallic pattern materials

Pattern Metal	Advantages	Disadvantages
Aluminium alloys	Good machinability High corrosion resistance Low density Good surface finish	Low strength High cost
Grey cast iron	Good machinability High strength Low cost	Corrosion prone High density
Steel	Good surface finish High strength	Corrosion prone High density
Brass and bronze	Good surface finish High strength High corrosion resistance	High cost High density
Lead alloys	Good machinability	High cost High density Low strength

Plastics are also used as pattern materials because of their low weight, easier formability, smooth surfaces and durability. They do not absorb moisture and are therefore, dimensionally stable and can be cleaned easily. The making of a plastic pattern can be done in sand clay moulds or moulds made of plaster of paris. The most generally used plastics are cold setting epoxy resins with suitable fillers. With a proper combination it is possible to obtain no shrink plastic material. In such a case double shrinkage allowances may not be required.

Polyurethane foam is also used as pattern material. It is very light and can be easily formed into any shape required. It can be used for light duty work for small number of castings for the conventional casting and for single casting in the case of full mould process where the pattern is burned inside the mould without withdrawing. This plastic has very low ash content and hence can be burned inside the mould.

The pattern material is to be chosen based on the expected life of the pattern. Table 3.5 gives comparative values of pattern material choices.

3.2.5 Types of Patterns

There are various types of patterns depending upon the complexity of the job, the number of castings required and the moulding procedure adopted.

TABLE 3.5 **Pattern materials based on expected life**

Number of Castings Produced before Pattern Equipment Repair		Pattern Material
Pattern	Core	
Small castings (under 600 mm)		
2 000 6 000 100 000	2 000 6 000 100 000	Hard wood Aluminium, Plastic Cast iron
Medium castings (600–1800 mm)		
1 000 3 000	750 3 000	Hard wood Aluminium, Plastic
Large castings (above 1800 mm)		
200 500	150 500	Soft wood Hard wood metal reinforced

Single Piece Pattern

These are inexpensive and the simplest type of patterns. As the name indicates, they are made of a single piece as shown in Fig. 3.9. This type of pattern is used only in cases where the job is very simple and does not create any withdrawal problems. It is also used for applications in very small–scale production or in prototype development. This pattern is expected to be entirely in the drag. One of the surfaces is expected to be flat which is used as the parting plane. If no such flat surface exists, the moulding may become complicated with the necessity of a follow board as explained later.

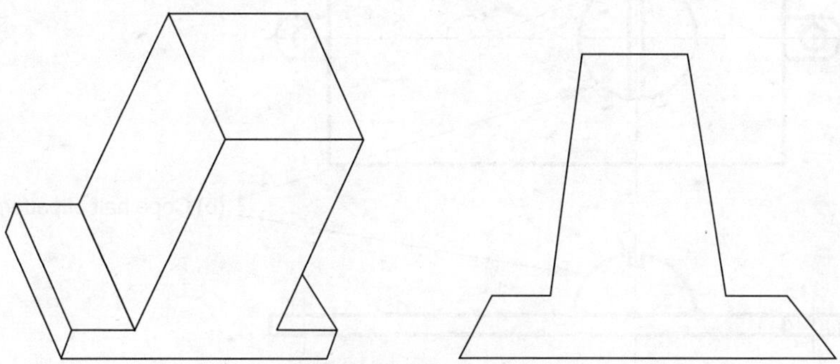

(a) Casting (b) Pattern

FIG. 3.9 *Single piece pattern*

Split Pattern or Two Piece Pattern

This is the most widely used type of pattern for intricate castings. When the contour of the casting makes its withdrawal from the mould difficult or when the depth of the casting is too high, then the pattern is split into two parts so that one part is in the drag and the other in the cope. The split surface of the pattern is same as the parting plane of the mould. The two halves of the pattern should be aligned properly by making use of the dowel pins which are fitted to the cope half. These dowel pins match with the precisely made holes in the drag half of the pattern and thus align the two halves properly as seen in Fig. 3.10.

Gated Pattern

This is an improvement over the simple pattern where the gating and runner system are integral with the pattern. This would eliminate the hand cutting of the runners and gates, and will help in improving the productivity of a moulder.

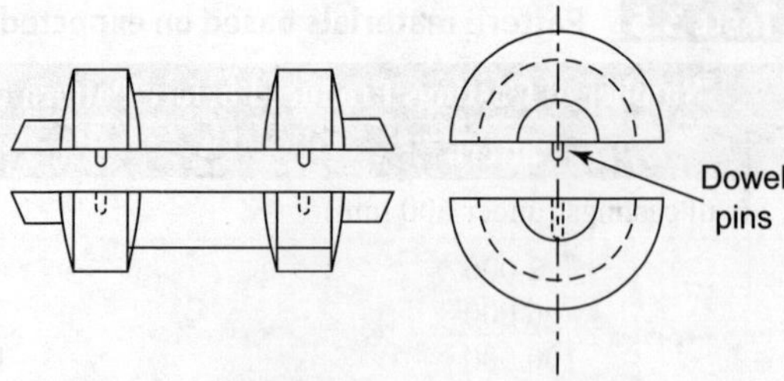

Fig. 3.10 *Split pattern*

Cope and Drag Pattern

These are similar to split patterns. In addition to splitting the pattern, the cope and drag halves of the pattern along with the gating and risering systems are attached separately to the metal or wooden plates along with the alignment pins (Fig. 3.11). They are called the cope and drag patterns. The cope and drag moulds may be produced using these patterns separately by two moulders but they can be assembled to form a complete mould. These types of patterns are used for castings which are heavy and inconvenient for handling as also for continuous production.

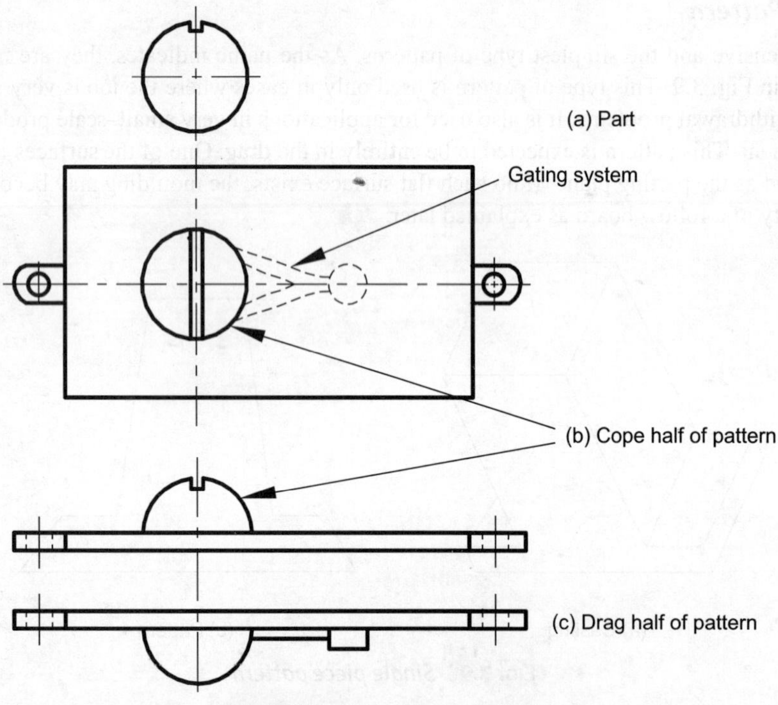

Fig. 3.11 *Cope and drag pattern*

Match Plate Pattern

These are extensions of the previous type. Here the cope and drag patterns along with the gating and the risering are mounted on a single matching metal or wooden plate on either side as shown in Fig. 3.12. On one side of the match plate the cope flask is prepared and on the other, the drag flask. After moulding when the match plate is removed, a complete mould with gating is obtained by joining the cope and the drag together.

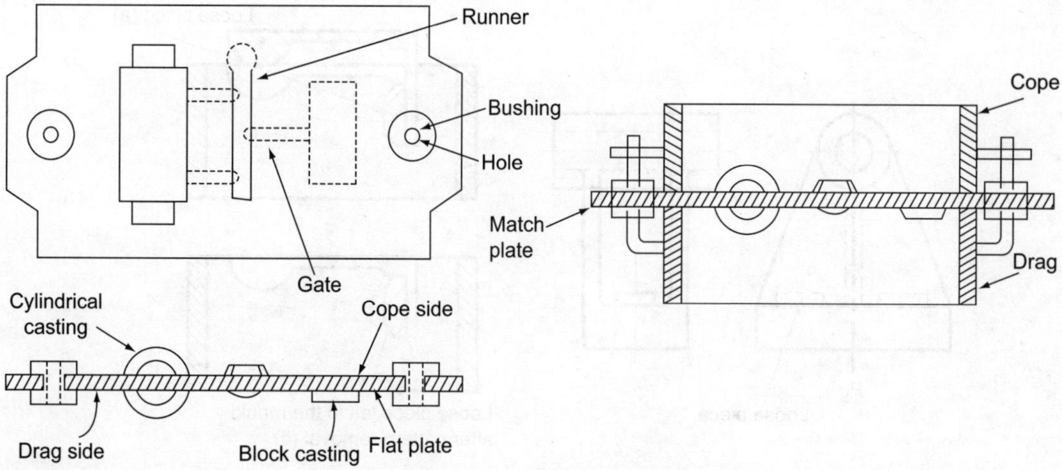

Fig. 3.12 *Match plate pattern (courtesy J.S. Campbell: Principles of Manufacturing Materials and Processes, p 153, McGraw-Hill, New York)*

The complete pattern with match plate is entirely made of metal, usually aluminium for its light weight and machinability. But when dimensions are critical, the match plate may be made of steel with necessary case hardening of the critical wear points. The pattern and gating are either screwed to the match plate in the case of a flat parting plane or are made integral in case of an irregular parting plane. The casting of a match plate pattern is done usually in plaster moulds but sometimes sand moulds are also used. When the cope and the drag patterns are similar, the pattern may be kept on only one side of the plate and is used for making both the drag as well as the cope.

These are generally used for small castings with higher dimensional accuracy and large production. The gating system is already made and attached to the match plate. Several patterns can be fixed to a single match plate, if they are sufficiently small in size. These patterns are used for machine moulding. They are expensive but since they increase productivity, the additional cost is justified.

Loose Piece Pattern

This type of pattern is also used when the contour of the part is such that withdrawing the pattern from the mould is not possible. Hence during moulding the obstructing part of the contour is held as a loose piece by a wire. After moulding is over, first the main pattern is removed and then the loose pieces are recovered through the gap generated by the main pattern (Fig. 3.13). Moulding with loose pieces is a highly skilled job and is generally expensive and therefore, should be avoided where possible.

Follow Board Pattern

This type of pattern is adopted for those castings where there are some portions which are structurally weak and if not supported properly are likely to break under the force of ramming. Hence the bottom board is modified as a follow board to closely fit the contour of the weak pattern and thus support it during the ramming of the drag. During the preparation of the cope, no follow board is necessary because the sand that is already compacted in the drag will support the fragile pattern. An example is shown in Fig. 3.14.

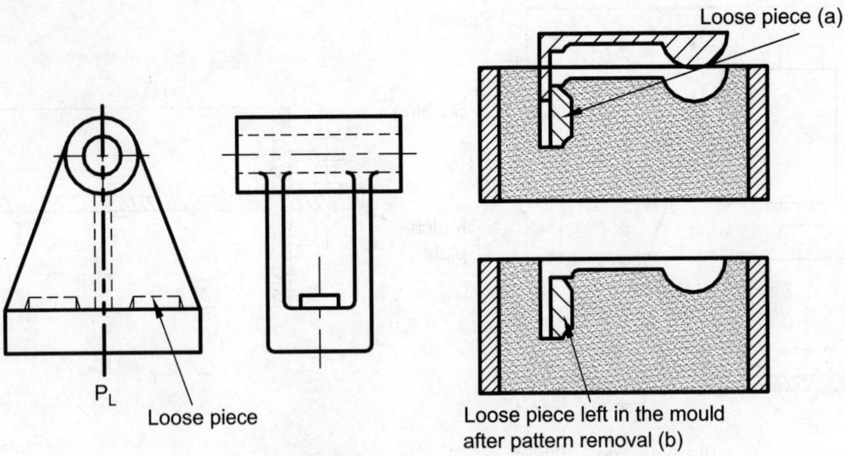

Fig. 3.13 *Loose piece pattern*

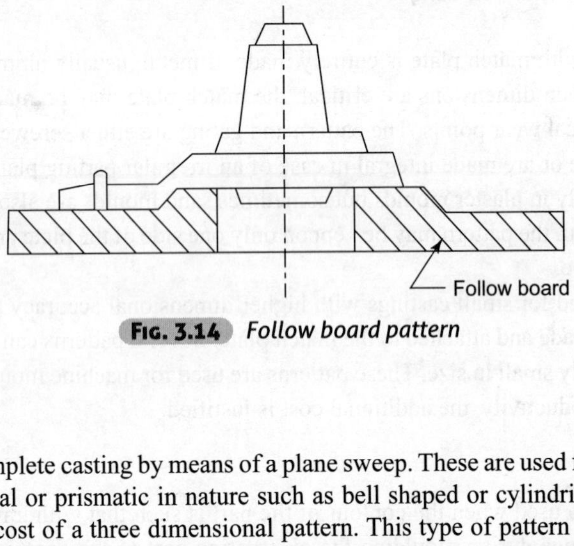

Fig. 3.14 *Follow board pattern*

Sweep Pattern

It is used to sweep the complete casting by means of a plane sweep. These are used for generating large shapes which are axi–symmetrical or prismatic in nature such as bell shaped or cylindrical as shown in Fig. 3.15. This greatly reduces the cost of a three dimensional pattern. This type of pattern is particularly suitable for very large castings such as bells for ornamental purposes used, which are generally cast in pit moulds.

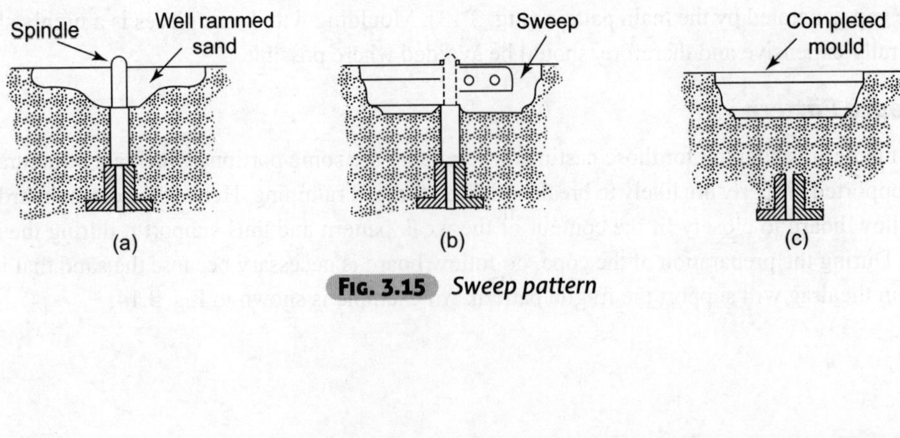

Fig. 3.15 *Sweep pattern*

Skeleton Pattern

A skeleton of the pattern made of strips of wood is used for building the final pattern by packing sand around the skeleton. After packing the sand, the desired form is obtained with the help of a strickle as shown in Fig. 3.16. The type of skeleton to be made is dependent upon the geometry of the work piece. This type of pattern is useful generally for very large castings required in small quantities where large expense on complete wooden pattern is not justified.

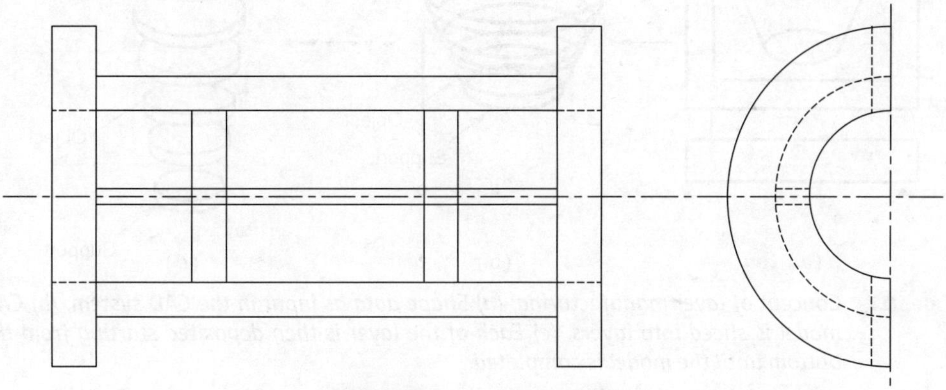

Fig. 3.16 *Skeleton pattern*

3.2.6 Pattern Colour Code

The patterns are normally painted with contrasting colours such that the mould maker would be able to understand the functions clearly. The colour code used is

1. Red or orange on surfaces, not to be finished and left as cast
2. Yellow on surfaces to be machined
3. Black on core prints for unmachined openings
4. Yellow stripes on black on core prints for machined openings
5. Green on seats of and for loose pieces and loose core prints
6. Diagonal black stripes with clear varnish on to strengthen the weak patterns or to shorten a casting

3.2.7 Additive Manufacturing (Rapid Prototyping)

Additive Manufacturing (AM) is a process that uses direct digital 3D model data to build physical parts by depositing layer by layer of the build material. It is also more prominently being called "*3D printing*" as a synonym for Additive Manufacturing. This process is in contrast from traditional material removal processes where a work piece is obtained from solid block by removing the unwanted material. The process was first demonstrated in 1987 by 3D Systems and is originally conceived as a rapid prototyping process to produce prototypes during the product design stage. Since then the process evolved into more prominence by offering its use for making final parts as well as tooling required for the manufacturing operations. Though the first demonstrated process utilized photopolymer as the build material, now a range of different metals, plastics and composite materials are available for additive manufacturing.

The process starts with developing the part model in a 3D CAD system. The CAD model is then translated into STL (Standard Tessellation Language that was originally adopted by the Stereo Lithography process by 3D Systems) format acceptable for AM machines. CAD model of the part is then sliced using a number of horizontal planes that are parallel to each other and separated by a distance equivalent to the resolution of the AM machine such as 0.125 mm as shown in Fig. 3.17. The AM machine will then add support material where required to have a bottoms up build process and all the layers are supported uniformly.

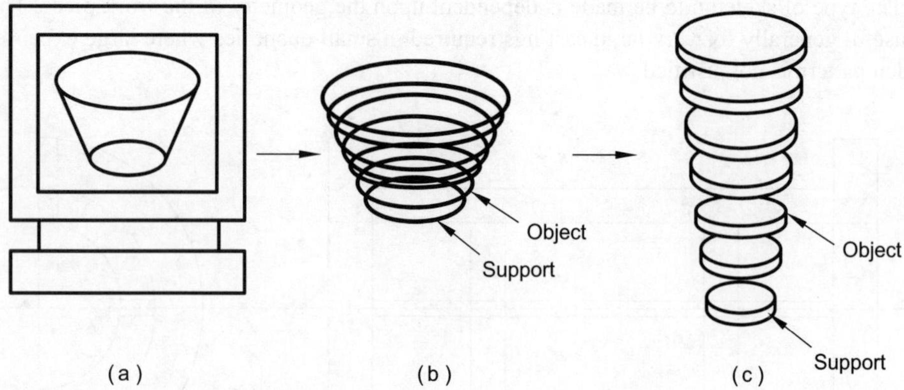

FIG. 3.17 *Concept of layer manufacturing; (a) Shape data as input in the CAD system, (b) CAD model is sliced into layers, (c) Each of the layer is then deposited starting from the bottom until the model is completed.*

The machine then starts creating the bottom most layers. The system then generates trajectories for the material to be added in each layer by the RP machine. The sacrificial supporting layers are also simultaneously generated to keep the unconnected layers in proper position. The resultant separate cross–sectional layers of very small thickness when assembled (glued) together will form the final object required. After completing the building process, the model and supports, if any, are removed from the machine and the surface of the model is then finished and cleaned.

As already mentioned, additive manufacturing was originally developed for prototyping. However, in view of its versatility, it has been used in a number of ways in manufacturing. One of the first uses was for developing tooling required for other manufacturing operations such as patterns for investment casting and permanent moulds. It is normally called as Rapid tooling as shown in Fig. 3.18. The other way to use AM is rapid manufacturing where the actual parts required are directly manufactured using AM. Typical examples are the difficult to get spare parts and complex parts that cannot be manufactured by normal manufacturing operations.

The ASTM Committee F42 on Additive Manufacturing Technologies formed in 2009 looks after the various matters related to AM. The ASTM approved AM process categories (ISO/ASTM52900–15) are:

- **Vat Photopolymerization** – "an additive manufacturing process in which liquid photopolymer in a vat is selectively cured by light–activated polymerization". An example is SLA – Stereo Lithography Apparatus. Stereo lithography was patented in 1986 by Charles W. Hull.
- **Sheet Lamination** – "an additive manufacturing process in which sheets of material are bonded to form an object". An example is LOM – Laminated object manufacturing that utilized adhesive backed paper as the build material.
- **Material Jetting** – "an additive manufacturing process in which droplets of build material are selectively deposited". An example is Poly–Jet that dispenses photopolymer resin through multiple jets that would be solidified by exposure to UV light.

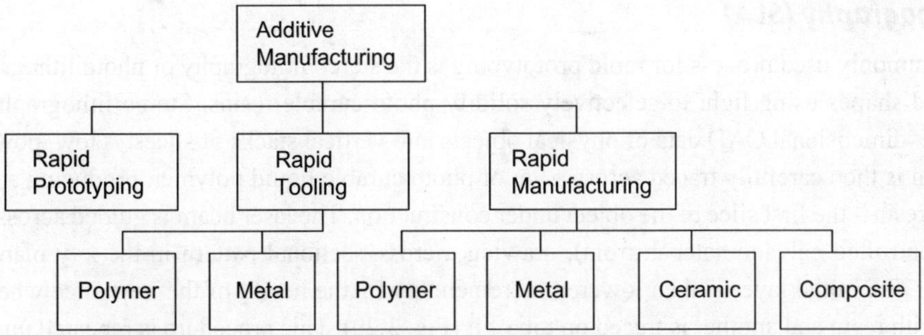

FIG. 3.18 *The three branches of additive manufacturing operations*

- **Powder Bed Fusion** – "an additive manufacturing process in which thermal energy selectively fuses regions of a powder bed". An example is SLS – Selective laser sintering, where the heat from laser fuses the powder particles to form the layers.
- **Binder Jetting** – "an additive manufacturing process in which a liquid bonding agent is selectively deposited to join powder materials". An example is 3DP – 3D printing also known as Inkjet Powder Printing using a technology invented by MIT.
- **Material Extrusion** – "an additive manufacturing process in which material is selectively dispensed through a nozzle or orifice". An example is FDM – Fused deposition modeling where a plastic wire is melted in a nozzle and dispensed to form the solid layer.
- **Directed Energy Deposition** – "an additive manufacturing process in which focused thermal energy is used to fuse materials by melting as the material is being deposited". An example is LENS – Laser engineered net shaping pioneered by Sandia National Laboratories.

These processes use the build material that is either liquid powder or solid and can therefore be classified based on the nature of build material as shown in Fig. 3.19.

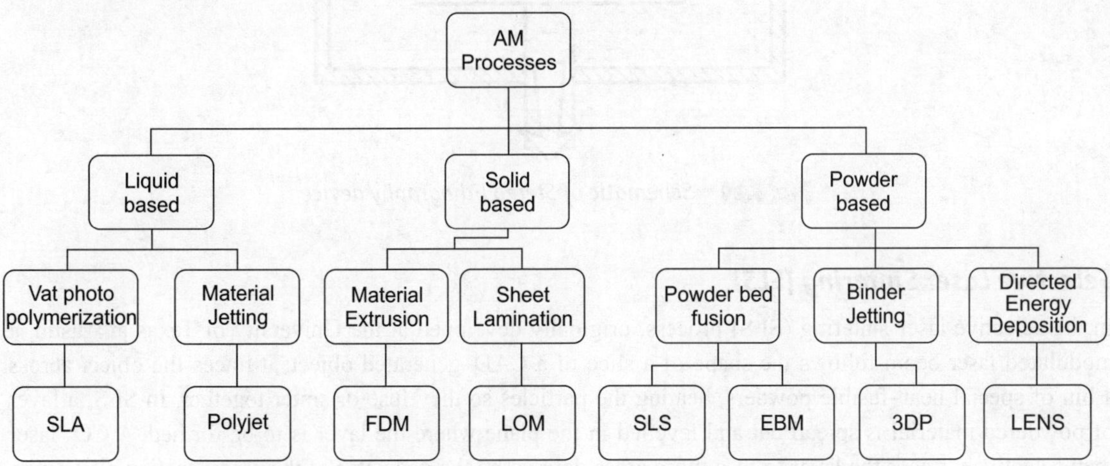

FIG. 3.19 *Classification of the additive manufacturing technologies*

Stereo Lithography (SLA)

The most commonly used process for rapid prototyping is the stereo lithography or photo lithography. These systems build shapes using light to selectively solidify photo curable resins. Stereo lithography machines convert three–dimensional CAD data of physical objects into vertical stacks of slices. A low–power ultraviolet laser beam is then carefully traced across a vat of photo curable liquid polymer, producing a single layer of solidified resin – the first slice of the object under construction. The laser beam is guided across the surface (by servo–controlled galvanometer mirrors), drawing a cross–sectional pattern in the x–y plane to form a solid section. The initial layer is then lowered incrementally by the height of the next slice where the layer is recoated with resin and another is traced on top of it (Fig. 3.20). This procedure is repeated until the entire part is fabricated. After completing the build process, the part is taken out of the SLA machine and cured as the UV laser used in the process does not have sufficient strength to completely cure the polymer. Though this was the first process commercialised, it is expensive and is limited to some of the photo–curable plastic materials only. This process is used to make complex master patterns for all casting processes. Sometimes it may also be used for low volume production of investment casting patterns as well.

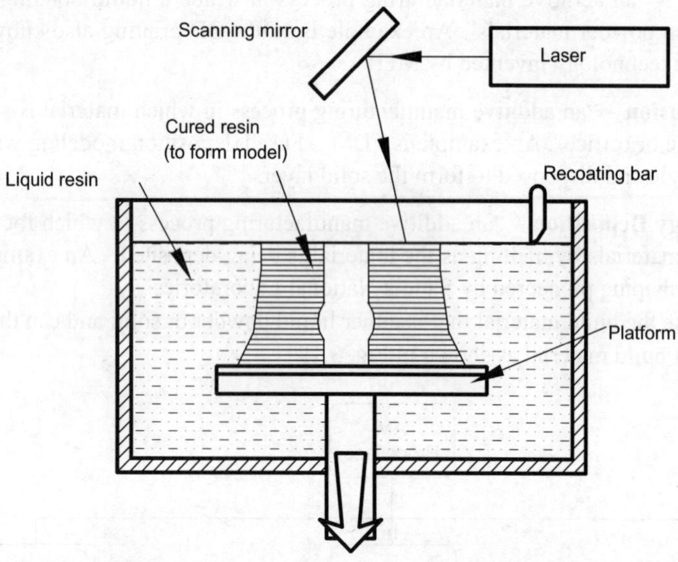

Fig. 3.20 *Schematic of Stereo lithography device*

Selective Laser Sintering (SLS)

In the selective laser sintering (SLS) process, originally developed at the University of Texas at Austin, a modulated laser beam follows the shape of a slice of a CAD–generated object; it traces the object across a bin of special heat–fusible powders, heating the particles so they fuse or sinter together. In SLS, a layer of powdered material is spread out and levelled in the plane where the layer is to be formed. A CO_2 laser then selectively traces the layer to fuse those areas defined by the geometry of the cross–section along with fusing to the bottom layer. The powders can be joined by melting or surface bonding. The unfused material remains in place as the support structure. After the initial layer is formed, powder is reapplied and the laser processes the next layer. Some of the materials used are plastics, waxes and low–melting–temperature metal alloys. Because of the use of metal powders, this process is greatly used in applications such as direct tooling applications for investment and die casting applications discussed later. This process is used for making low volume metallic moulds used for permanent mould casting and die casting. It can also make the dies for making the patterns for investment casting.

3–D Printing (3DP)

3–D printing can be compared to SLS with a difference that instead of laser beam, liquid binder is applied to bond the powder particles. A 3D Printer is operated in the following sequence. The Printer spreads a layer of powder from the feed box to cover the surface of the build platform and then prints binder solution onto the loose powder, forming the first cross–section of the part. Where the binder is printed, the powder's particles are glued together. The remaining powder is loose and supports the part as it is being printed. When the cross–section is complete, the build platform is lowered slightly and a new layer of powder is spread over its surface. The process is repeated until the whole model is completed. The build platform is raised and the loose powder is vacuumed away, revealing the completed part.

Fused Deposition Modelling (FDM)

In this process a plastic filament is unwound from a coil and supplies material to an extrusion nozzle. The nozzle is heated to melt the plastic and has a mechanism which allows the flow of the melted plastic to be turned on and off. The nozzle is mounted to a mechanical platform which can be moved in both horizontal and vertical directions. As the nozzle is moved over the table in the required geometry, it deposits a thin bead of extruded plastic to form each layer. The plastic hardens immediately after being squirted from the nozzle and bonds to the layer below. Several materials are available for the process including investment–casting wax. Some FDM systems utilize two extrusion nozzles: one for deposition of a build material and second for deposition of washable material to make support environment. Several FDM materials are in the engineering–development stage, including polycarbonate, polypropylene, PMMA (polymethylmethacrylate), and other various polyesters.

Laminated Object Manufacturing (LOM)

Laminated Object Manufacturing (LOM) machine works by actually cutting the "slices" of the object out of a sheet of paper foil and then bonding them together. The foil comes off the material supply roll and the laser then cuts around the outline of the layer. It also hatches the foil around the edge so that this can be easily broken away when all of the layers have been bonded together. After the laser has cut out the top layer, a heated roller moves over the top of the foil to bond the layer to the rest of the object. A sensor is used to measure the thickness of the foil as this can vary and the machine will automatically adjust the dimensions of the layer being cut to account for any variation. The result is a part that looks like laminated wood.

A comparison of these technologies is given in Table 3.6. These methods are used for making the patterns and core boxes. Some of the processes are also used for making the metal moulds and dies for large volume production.

TABLE 3.6 Summary of some rapid prototyping technologies

System	Max. Build Size (mm)	Dimensional Accuracy (mm)	Materials	Advantages	Disadvantages
Stereo–lithography	(500×400×200)	(0.1–0.2)	Liquid photosensitive resins	High accuracy, medium range of materials, large build size	High cost process, support structures needed, post cure required
Selective Laser Sintering	(550×550×750)	(0.1–0.2)	Nylon based materials, elastomer, rapid steel, cast form, sand form	Large range of materials, good accuracy, large build size	High cost process, poor surface finish
Fused Deposition Modelling	(600×500×600)	(0.1–0.2)	ABS, elastomer and wax	Good accuracy, functional materials, medium range of materials	Support structures needed
Laminated Object Modelling	(815×560×508)	(0.1–0.2)	Paper	Good accuracy, large build size	Limited range of materials, support removal necessary, poor material properties

3.3 MOULDING MATERIALS

A large variety of moulding materials are used in foundries for manufacturing moulds and cores. They are:

- Moulding sand
- System sand (backing sand)
- Rebonded sand
- Facing sand
- Parting sand
- Core sand

The choice of moulding materials is based on their processing properties. The properties that are generally required in moulding materials are:

Refractoriness It is the ability of the moulding material to withstand the high temperatures of the molten metal so that it does not cause fusion. Some refractory materials are given in Table 3.7.

Green Strength The moulding sand that contains moisture is termed as green sand. The green sand should have enough strength so that the constructed mould retains its shape.

Dry Strength When the moisture in the moulding sand is completely expelled, it is called dry sand. When molten metal is poured into a mould, the sand around the mould cavity is quickly converted into dry sand as the moisture in the sand immediately evaporates due to the heat in the molten metal. At this stage, it should retain the mould cavity and at the same time withstand the metallostatic forces.

Hot Strength After all the moisture is eliminated, the sand would reach a high temperature when the metal in the mould is still in the liquid state. The strength of the sand that is required to hold the shape of the mould cavity is called hot strength.

TABLE 3.7 Properties of some refractory materials

Material	Melting Point, °C	Coefficient of Linear Expansion, × 10^6/°C
Silica (SiO_2)	1710	16.2
Alumina (Al_2O_3)	2020	8.0
Magnesia (MgO)	2800	13.5
Thoria (ThO_2)	3050	9.5
Zirconia (ZrO_2)	2700	6.5
Zircon (ZrO_2. SiO_2)	2650	4.5
Silicon Carbide (SiC)	~2700	3.5
Graphite	~4200	—

Permeability During the solidification of a casting, large amounts of gases are to be expelled from the mould. The gases are those which have been absorbed by the metal in the furnace, air absorbed from the atmosphere, steam and other gases that are generated by the moulding and core sands. If these gases are not allowed to escape from the mould, they would be trapped inside the casting causing defects. The moulding sand should be sufficiently porous so that the gases are allowed to escape from the mould. This gas evolution capability of the moulding sand is termed as permeability.

Besides these specific properties, the moulding sand should also have collapsibility so that during the contraction of the solidified casting, it does not provide any resistance which may result in cracks in the casting. They should be reusable and should have good thermal conductivity so that heat from the casting is quickly transferred.

3.3.1 Moulding Sand Composition

The main ingredients of any moulding sand are:

- Silica grains (SiO_2)
- Clay as binder
- Moisture to activate the clay and provide plasticity

Besides, some other materials are also added to these to enhance the specific properties of moulding sands, the details of which are given in later chapters.

Silica Sand The sand which forms the major portion of the moulding sand (up to 96 %) is essentially silica grains, the rest being the other oxides such as alumina, sodium ($Na_2O + K_2O$) and magnesium oxide (MgO + CaO). These impurities should be minimised to about 2% since they affect the fusion point of the silica sands. The main source is the river sand which is used with or without washing. Ideally the fusion point of sands should be about 1450°C for cast irons and about 1550°C for steels. In the river sand, all sizes and shapes of grains are mixed. The sand grains may vary in size from a few micrometers to a few millimetres. Shape of the grains may be round, sub-angular, angular and very angular. The size and shapes of these sand grains greatly affect the properties of the moulding sands which are explained later in this chapter.

Zircon sand is basically a zirconium silicate ($ZrSiO_4$). The typical composition is ZrO_2 = 66.25%, SiO_2 = 30.96%, Al_2O_3 = 1.92%, Fe_2O_3 = 0.74% and traces of other oxides. It is very expensive. In India it is available in Quilon beach of Kerala. It has a fusion point of about 2400°C and also a low coefficient of thermal expansion. The other advantages are high thermal conductivity, high chilling power and high density. It requires a very small amount of binder (about 3%). It is generally used to manufacture precision steel castings requiring better surface finish and for precision investment casting.

Chromite sand is crushed from the chrome ore whose typical composition is Cr_2O_3 = 44%, Fe_2O_3 = 28%, SiO_2 = 2.5%, CaO = 0.5%, and Al_2O_3 + MgO = 25%. The fusion point is about 1800°C. It also requires a very small amount of binder (about 3%). It is also used to manufacture heavy steel castings requiring better surface finish. It is best suited to austenitic manganese steel castings.

Olivine sand contains the minerals fosterite (Mg_2SiO_4) and fayalite (Fe_2SiO_4). It is very versatile sand and the same mixture can be used for a range of steels.

Comparative properties relevant for moulding of these various base sands are given in Table 3.8.

Clay Clays are the most generally used binding agents mixed with the moulding sands to provide the strength because of their low cost and wider utility. The most popular clay types used are:

Kaolinite or fire clay (Al_2O_3 $2SiO_2$ $2H_2O$), and
Bentonite (Al_2O_3 $4SiO_2$ H_2O nH_2O)

Kaolinite has a melting point of 1750 to 1787°C and Bentonite has melting temperature range of 1250 to 1300°C. Of the two, bentonite can absorb more water which increases its bonding power. The clays besides these basic constituents may also contain some mixtures of lime, alkalies and other oxides which tend to reduce their refractoriness.

There are basically two types of bentonites, one with sodium as adsorbed ion often called western bentonite and the other with calcium ion called southern bentonite. Sodium bentonites produce better swelling properties. Volume increases some 10 to 20 times, high dry strength which lowers the risk of erosion, better tolerance of variations in water content, low green strength and high resistance to burnout, which reduces clay consumption. In contrast the calcium bentonites have low dry strength but higher green strength. It is possible to improve the properties of calcium bentonite by treating it chemically with soda ash (sodium carbonate).

The clay chosen for moulding sand should give it the requisite strength for the given application taking into account the metal being cast and thickness of the casting. Normally, the river sand contains a large amount of clay and therefore can be directly used.

TABLE 3.8 Comparison of Foundry base sand properties

	Silica	Olivine	Chromite	Zircon
Colour	White–light Brown	Greenish Gray	Black	White–brown
Hardness	6.0–7.0	6.5–7.0	5.5–7.0	7.0–7.5
Dry bulk density (lb/ft^3)	85–100	100–125	155–165	160–185
Specific gravity	2.2–2.6	3.2–3.6	4.3–4.5	4.4–4.7
Grain shape	Angular/ Rounded	Angular	Angular	Rounded/ Angular
Thermal expansion (mm/mm/°C)	0.018	0.0083	0.005	0.003
Apparent heat transfer	Average	Low	Very high	High
Fusion point, °C	1427–1760	1538–1760	1760–1982	2038–2204
High temperature reaction	Acid	Basic	Basic	Acid
Wettability with molten metal	Easily	Not generally	Resistant	Resistant
Grain distribution	2–5 screens	3–4 screens	4–5 screens	2–3 screens
AFS Grain fineness number	25–180	40–160	50–90	95–160

Water Clay is activated by water so that it develops the necessary plasticity and strength. The amount of water used should be properly controlled. This is because a part of the water absorbed by clay helps in bonding while the remainder up to a limit helps in improving the plasticity, but more than that it would decrease the strength and formability. The normal percentages of water used are from 2 to 8.

Besides these three main ingredients, many other materials also may be added to enhance the specific properties. For example, cereal binder up to 2% increases the strength; pitch obtained as a by–product in coke making if used in percentages up to 3 would improve the hot strength and saw dust up to 2% may improve the collapsibility by slowly burning and increase the permeability. There are other materials such as sea coal, asphalt, fuel oil, graphite, molasses, iron oxide, etc., which are also used for obtaining specific properties.

Comparative properties of moulding sands with various clays are shown in Table 3.9 below.

TABLE 3.9 Comparison of Foundry sand properties with various clays

	Sodium Bentonite	Fire Clay	Calcium Bentonite
Silica sand, AFS GFN 60, weight %	95	88	95
Clay bond, weight %	5	12	5
Moisture, weight %	2.5	3.0	2.5
Permeability	110	60	108
Green compression strength, kPa	76.53	63.43	97.22
Dry compression strength, kPa	675.70	468.84	372.32

3.3.2 Testing Sand Properties

The moulding sand, after it is prepared, should be properly tested to see that the requisite properties are achieved. There are standard tests to be used which are given in relevant Indian standards and that of other foundry societies, a list of which is given at the end of this chapter. However, a brief review is made here of the general testing methods.

Sample Preparation

Tests are conducted on a sample of the standard sand. The moulding sand should be prepared exactly as is done in the shop on the standard equipment and then carefully enclosed in a closed container to safeguard its moisture content.

Moisture Content

As explained earlier, moisture is an important element of the moulding sand as it affects many properties. To test the moisture of moulding sand a carefully weighed test sample of 50 g is dried at a temperature of 105°C to 110°C for 2 hours, by which all the moisture in the sand would have been evaporated. The sample is then weighed. The weight difference in grams when multiplied by two would give the percentage of moisture contained in the moulding sand.

Alternatively, a 'moisture teller' can also be used for measuring the moisture content. In this test, the sand is dried by suspending the sample on a fine metallic screen and allowing hot air to flow through the sample. This method of drying completes the removal of moisture in a matter of minutes compared to 2 hours as in the earlier method.

Another moisture teller utilises calcium carbide to measure the moisture content. A measured amount of calcium carbide (a little more than actually required for complete reaction) in a container along with a separate cap consisting of measured quantity of moulding sand is kept in the moisture teller. Care has to be taken before closing the apparatus that carbide and sand do not come into contact. The apparatus is then shaken vigorously such that the following reaction takes place:

$$CaC_2 + 2H_2O \rightarrow C_2H_2 + Ca(OH)_2$$

The acetylene (C_2H_2) coming out will be collected in the space above the sand raising the pressure. A pressure gauge connected to the apparatus would give directly the amount of acetylene generated which is proportional to the moisture present. It is possible to calibrate the pressure gauge (or manometer) to directly read the amount of moisture.

Clay Content

The clay content of moulding sand is determined by dissolving or washing it off the sand. To determine the clay percentage a 50 g sample is dried at 105 to 110°C. The dried sample is taken in a 1 litre glass flask and added with 475 ml of distilled water and 25 ml of a 1% solution of caustic soda (NaOH 25 g per litre). This sample is then thoroughly stirred.

After the stirring for a period of 5 minutes, the sample is diluted with fresh water up to a 150 mm graduation mark and the sample is left undisturbed for 10 minutes to settle. The sand settles at the bottom and the clay particles washed from the sand would be floating in the water. 125 mm of this water is siphoned off the flask and it is again topped to the same level and allowed to settle for 5 minutes. The above operation is repeated till the water above the sand becomes clear indicating that all the clay in the moulding sand has been removed. Now, the sand is removed from the flask and dried by heating. The difference in weight of the dried sand and 50 g when multiplied by two gives the clay percentage in the moulding sand.

Sand Grain Size

To find out the sand grain size, a sand sample which is devoid of moisture and clay such as the one obtained after the previous testing is to be used. The dried clay-free sand grains are placed on the top sieve of a sieve shaker which contains a series of sieves one upon the other with gradually decreasing mesh sizes. The mesh sizes are standardised as shown in Table 3.10. The sieves are shaken continuously for a period of 15 min. After this shaking operation, the sieves are taken apart and the sand left over on each of the sieve is carefully weighed.

TABLE 3.10 AFS Sieve numbers and their Indian equivalents with sizes

US Series Equivalent No. (ASTM)	Mesh Opening (mm)	IS Sieve No. Microns	Multiplying Factor
6	3.327	3.35	3
12	1.651	1.70	5
20	0.833	850	10
30	0.589	600	20
40	0.414	425	30
50	0.295	300	40
70	0.208	212	50
100	0.147	150	70
140	0.104	106	100
200	0.074	75	140
270	0.053	53	200
Pan	—	—	300

The sand retained on each of the sieve expressed as a percentage of the total mass can be plotted against the sieve number as in Fig. 3.21 to obtain the grain distribution. But more important is the Grain Fineness Number (GFN) which is a quantitative indication of the grain distribution. To calculate the grain fineness number, each sieve has been given a weightage factor as shown in Table 3.10. The amount retained on each sieve is multiplied by the respective weightage factor, summed up, and then divided by the total mass of the sample, which gives the grain fineness number. The same can be expressed as

$$\text{GFN} = \frac{\Sigma M_i f_i}{\Sigma f_i}$$

Where, M_i = multiplying factor for the i_{th} sieve

f_i = amount of sand retained on the i_{th} sieve

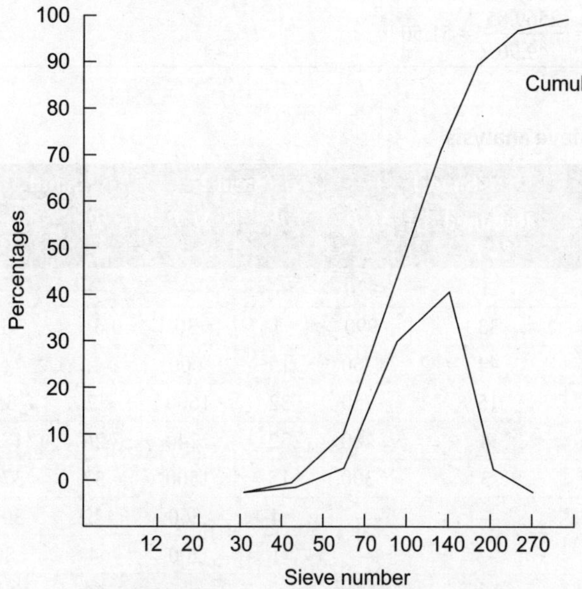

FIG. 3.21 *Sand grain size distribution*

A typical example is presented in Table 3.11 for calculating the grain fineness number. This type of sand is normally used in malleable and gray iron foundries for castings which weigh more than 20 kg.

By the above definition, the grain fineness number is the average grain size and corresponds to a sieve number through which all the sand grains would pass through, if they were all of the same size. This is a very convenient way of describing the grain size and its value can be expected between 40 and 220 for the sands used by most of the foundries. Though the sand properties depend on both the grain size and the grain size distribution, GFN is a very convenient way of finding the sand properties since it takes both into account.

Table 3.12 gives the sieve analysis for the typical sands, which may be used for the purposes mentioned.

Permeability

The rate of flow of air passing through a standard specimen under a standard pressure is termed as permeability number.

Table 3.11 Sieve analysis

Sieve Number	Multiplying Factor, M_i	Retained Sample f_i (g)	Retained Percentage P_i	$M_i \times P_i$	$M_i \times f_i$
40	30	2.495	5	150	74.85
50	40	13.972	28	1120	558.88
70	50	23.952	48	2400	1197.60
100	70	6.986	14	980	489.02
140	100	2.495	5	500	249.50
		49.900	100	5150	2569.85
$GFN = \frac{5150}{100} = 51.50$					
$GFN = \frac{2569.85}{49.90} = 51.50$					

Table 3.12 Typical sand sieve analysis

Sieve Number	Multiplier M_i	Sand 1		Sand 2		Sand 3		Sand 4	
		Retained % P_i	$M_i P_i$	P_i	$M_i P_i$	P_i	$M_i P_i$	P_i	$M_i P_i$
30	20	1	20	—	—	—	—	—	—
40	30	33	990	1	30	0.6	18	1.2	36
50	40	44	1760	15	600	1.4	56	6	240
70	50	15	750	32	1600	7	350	7.2	360
100	70	4	280	32	2240	26	1820	14.4	1008
140	100	3	300	15	1500	37	3700	18	1800
200	140	—	—	4	560	22	3080	16.8	2352
270	200	—	—	1	200	4	800	18	3600
Pan	300	—	—	—	—	2	600	18.7	5610
	Total		4100		6730		10 424		15 006
	GFN		41.00		67.30		104.24		150.06

Sand 1: Steel castings 50 kg and more
Sand 2: Small steel castings
Sand 3: Light cast iron and copper castings
Sand 4: Aluminium alloy castings

The standard permeability test is to measure time taken by a 2000 cub cm of air at a pressure typically of 980 Pa (10 g/cm^2), to pass through a standard sand specimen confined in a specimen tube (Fig. 3.22). The standard specimen size is 50.8 mm in diameter and a length of 50.8 mm. Then, the permeability number, P is obtained by

$$P = \frac{V \cdot H}{p \cdot A \cdot T}$$

Where, V = volume of air = 2000 cm^3

H = height of the sand specimen = 5.08 cm

p = air pressure, g/cm^2

A = cross sectional area of sand specimen = 20.268 cm^2

T = time in minutes for the complete air to pass through

Inserting the above standard values into the expression, we get

$$P = \frac{501.28}{p \cdot T}$$

Specimen Preparation

Since the permeability of sand is dependent to a great extent on the degree of ramming, it is necessary that the specimen be prepared under standard conditions. To get reproducible ramming conditions, a laboratory sand rammer is used along with a specimen tube. The measured amount of sand is filled in the specimen tube and a fixed weight of 6.35 to 7.25 kg is allowed to fall on the sand three times from a height of 50.8 + 0.125 mm. The specimen thus produced should have a height of 50.8 + 0.8 mm. To produce this size of specimen usually sand of 145 to 175 g would be required.

After preparing a test sample of sand as described, 2000 cm^3 of air is passed through the sample and the time taken by it to completely pass through the specimen is noted. Then from the above equation the permeability number can be calculated.

Fig. 3.22 *Permeability meter for measuring the permeability of green sand*

Example 3.5

Calculate the permeability number of sand if it takes 1 min 25 s to pass 2000 cm^3 of air at a pressure of 5 g/cm^2 through the standard sample.

$$p = 5.0 \text{ g/cm}^2$$

$$T = 1 \text{ min } 25 \text{ s} = 1 + \frac{25}{60} = 1.417 \text{ min}$$

$$\text{Permeability number, } p = \frac{501.28}{5 \times 1.417} = 70.75$$

The permeability test is conducted for two types of sands.

(a) Green permeability is the permeability of the green sand.

(b) Dry permeability is the permeability of the moulding sand, dried at 105 to 110°C to remove the moisture completely.

Strength

Measurement of strength of moulding sands can be carried out on the universal sand strength testing machine. The strength can be measured in compression, shear and tension. The sands that could be tested are green sand, dry sand or core sand. The compression and shear test involve the standard cylindrical specimen that was used for the permeability test.

Green Compression Strength

Green compression strength or simply green strength generally refers to the stress required to rupture the sand specimen under compressive loading. The sand specimen taken out of the specimen tube is immediately (any delay causes the drying of the sample which increases the strength) put on the strength testing machine and the force required to cause the compression failure is determined. The green strength of sands is generally in the range of 30 to 160 kPa.

Green Shear Strength

With a sand sample similar to the above test, a different adapter is fitted in the universal machine so that the loading now be made for the shearing of the sand sample. The stress required to shear the specimen along the axis is then represented as the green shear strength. The green shear strengths may vary from 10 to 50 kPa.

Dry Strength

The tests similar to the above can also be carried with the standard specimens dried between 105° to 110°C for 2 hours. Since the strength greatly increase with drying, it may be necessary to apply larger stresses than the previous tests. The range of dry compression strengths found in moulding sands is from 140 to 1800 kPa, depending on the sand sample.

Mould Hardness

The mould hardness is measured by a method similar to the Brinell hardness test. A spring loaded steel ball with a mass of 0.9 kg is indented into the standard sand specimen prepared. The depth of indentation can be directly measured on a scale which shows units 0 to 100. When no penetration occurs, then it is a mould hardness of 100 and when it sinks completely, the reading is zero indicating a very soft mould.

Besides these, there are other tests to determine such properties as deformation, green tensile strength, hot strength, expansion, etc. The details of these testing methods can be found in the references cited at the end of this chapter.

3.3.3 Sand Preparation

One of the important requirements for the preparation of sand is thorough mixing of its various ingredients. This is essential to ensure uniform distribution of the various components in the entire bulk of the sand.

During the mixing process any lump present in sand is broken up and clay is uniformly enveloped around the sand grains and moisture is uniformly distributed. Besides manual mixing, equipment called mueller is normally used in foundries to mix the sands. These are essentially of two types, batch type and continuous.

As shown in Fig. 3.23, a batch Mueller consists of one or two Mueller wheels and equal number of plough blades connected to a single driving source. The Mueller wheels are large and heavy, and continuously roll inside the Mueller bowl. The plough blades ensure that the sand is continuously agitated. The combined action of both these is a sort of kneading action which makes the clay and the moisture uniformly distributed throughout the sand.

A continuous Mueller consists of two bowls with Mueller wheels and ploughs, such that sand, clay and moisture are fed through a hopper into one of the bowls which after getting mulled moves into the second one and then finally out. Thus, well prepared moulding sand is continuously available for use. It is generally used for large–scale production.

3.3.4 Moulding Sand Properties

The properties of moulding sand are dependent to a great extent on a number of variables. The important among them are:

- Sand grain shape and size
- Clay type and amount
- Moisture content
- Method of preparing sand mould

Fig. 3.23 *Batch Mueller*

Sand Grains

The shape and size of the sand grain would greatly affect the various moulding sand properties. The sand grain size could be coarse or fine. Similarly the grain shape could be round or angular.

The coarse grains would have more void space between the grains, which increases the permeability. Similarly, the finer grains would have lower permeability; however they provide better surface finish to the casting produced. The distribution of the grain size also plays an important role. For example widely distributed sand would have higher permeability than the one with the same fineness number but where all the grains have the same size.

Angular sand grains require higher amounts of binder. The round grains would have lower permeability compared to angular grains because of the irregular shape of the latter. The same has been depicted in Fig. 3.24.

The grain size also affects the refractoriness. Higher the grain size, higher would be the refractoriness. The purity of sand grains also improves the refractoriness. For example, sand grains with a GFN of 30 to 45 may have a higher fusion point of the order of 1650°C. But finer grains and impurities in the sand tend to lower the refractoriness by promoting fusion. The impurities that may be found are iron oxide, feldspar and lime stone.

The strength of the moulding sand is also affected by the grain size and shape. Figure 3.25 shows that the green compression strength increases with a decrease in the grain size because the fine grain size provides a large surface area for the binder to act.

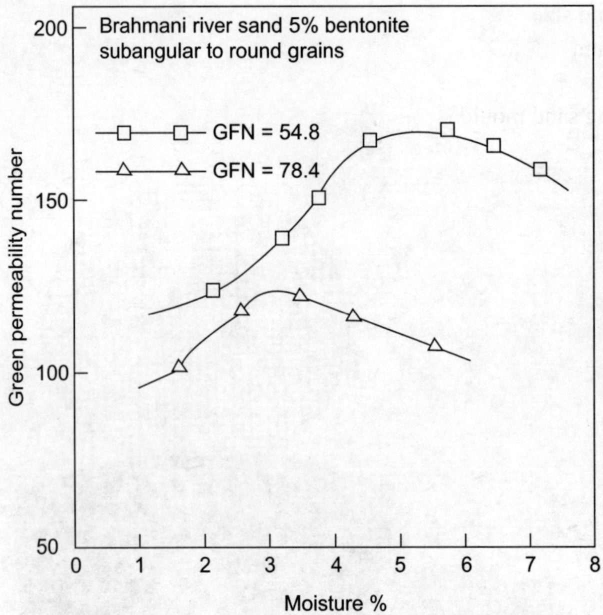

Fig. 3.24 *Variation of permeability with grain size*

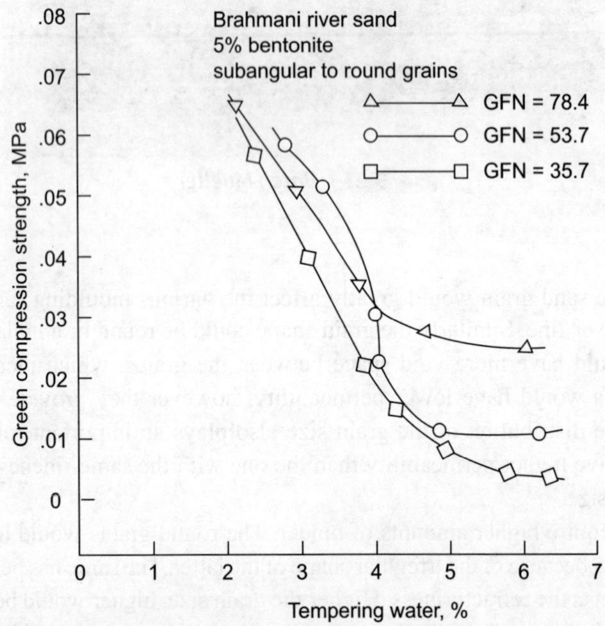

Fig. 3.25 *Variation of green compression strength with sand grain size*

Clay and Water

Besides the sand grains, clay and water have large influence on major properties of the moulding sands. Figure 3.26(a) and (b) shows the relationship between the green compression strength and water content for various clay percentages in the moulding sand.

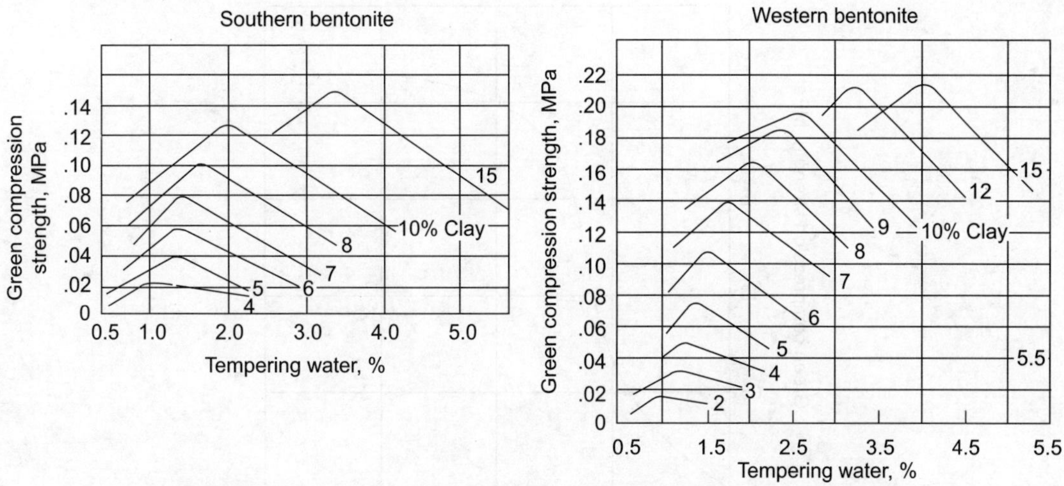

FIG. 3.26 *Variation of green compression strength with clay and water*

As can be seen, there is an optimum amount of water to be used for given clay content to obtain maximum green compression strength. During the sand preparation clay is uniformly coated around the sand grains. The water then reacts with the clay and forms a linkage of silica-water-clay-water- silica (or clay) throughout the moulding sand. For this linkage to develop there is only certain amount of water required which is specified by the type and amount of clay present. Any additional amount of water increases the plasticity and dry strength (Fig. 3.27) but reduces the green compression strength.

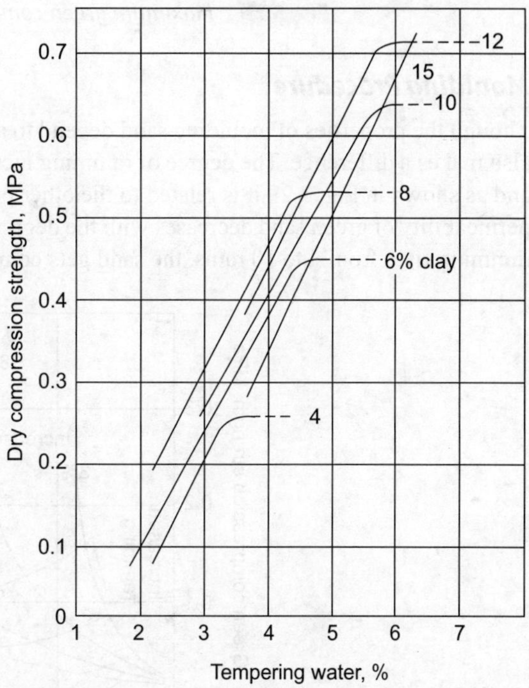

FIG. 3.27 *Variation of dry compression strength with water content*

There is a maximum limit up to which the green compression strength of a moulding sand could be increased which is shown in Fig. 3.28. It is possible to theoretically calculate the water percentage required to achieve this strength, the details of which are given in references at the end of this chapter. These sands are called clay saturated sands and are used to advantage in cast iron and heavy non-ferrous metal foundries. These sands would be able to reduce some of the casting defects such as sand expansion, erosion, cuts and washes which are essentially caused by the moulding sand expansion. Since such sands have higher green compression strength in the range of 100 to 250 kPa, they are to be rammed properly to develop these properties.

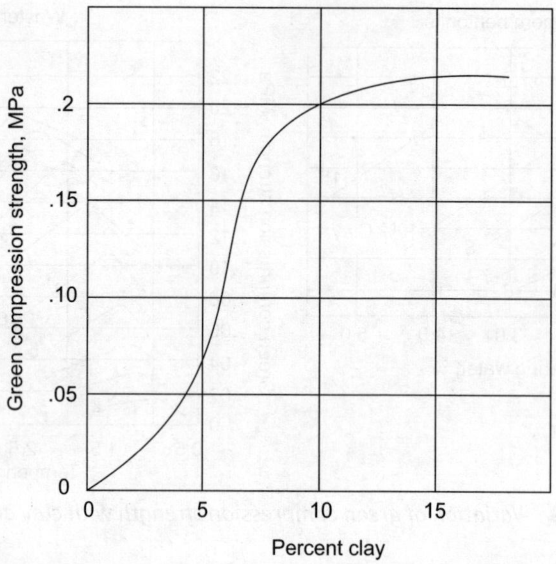

FIG. 3.28 *Maximum green compression strength obtainable with clay*

Moulding Procedure

Though the properties of moulding sand depend to a great extent on its constituents, the moulding procedure also makes a difference. The degree of ramming increases the bulk density or the mould hardness of the sand and as shown in Fig. 3.29 it is related to the other properties. Increased ramming increases the strength. The permeability of green sand decreases with the degree of ramming. Figure 3.30 shows that when the degree of ramming rises from 3 to 10 rams, the sand gets compacted and in the process reduces the voids in the sand.

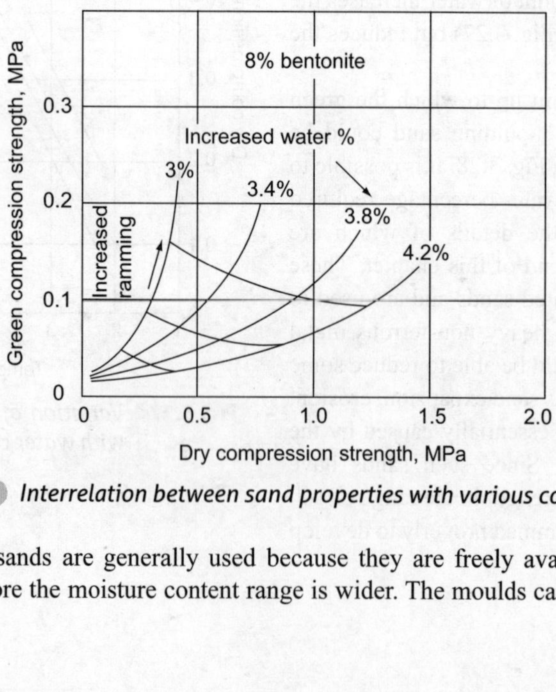

FIG. 3.29 *Interrelation between sand properties with various constituents*

The natural moulding sands are generally used because they are freely available. They contain large amount of clay and therefore the moisture content range is wider. The moulds can be prepared easily as not

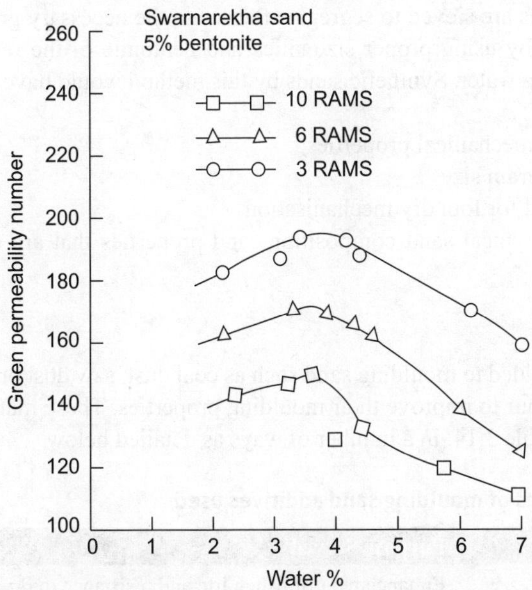

Fig. 3.30 *Change in the permeability of moulding sand with the degree of ramming*

much control is needed in the sand preparation. In spite of these advantages, natural sands lack the necessary refractoriness and other desirable properties.

To achieve the controlled properties for sands as shown in Table 3.13, it is necessary to prepare moulding sands synthetically. The natural sand obtained is thoroughly washed so as to remove as much clay as possible

Table 3.13 **Control sand test data for various types of castings**

Casting Material	Moisture %	Perme-ability	Green Com-pression Strength kPa	Deforma-tion mm	Clay %	Fineness Number GFN	Sintering Tempera-ture °C
Aluminium	6.5–8.5	7–13	46–53	0.45–0.60	12–18	225–160	1300
Brass and bronze	6–8	13–20	49–56	0.35–0.50	12–14	150–140	1300
Copper and Nickel	6–7.5	37–50	46–56	0.35–0.50	12–14	130–120	1325
Grey iron	6.5–8.5	10–15	42–53	0.45–0.55	10–12	200–180	1300
Light grey iron (Squeeze moulds)	6–7.5	18–25	–	0.48–0.55	12–14	120–87	1325
Medium grey iron (floor moulds)	5.5–7	40–60	49–56	0.25–0.35	11–14	86–70	1325
Medium grey iron (synthetic sand)	4–6	50–80	–	0.39–0.43	4–10	75–55	1350
Heavy grey iron	4–6.5	80–120	35–53	0.30–0.40	8–13	61–50	1380
Light malleable	6–8	20–30	46–53	0.43–0.50	8–13	120–92	1380
Heavy malliable	5.5–7.5	40–60	46–53	0.30–0.45	8–13	85–70	1380
Light steel	2–4	125–200	46–53	0.50–0.75	4–10	56–45	1450
Heavy steel	2–4	130–300	46–53	0.50–0.75	4–10	62–38	1500
Steel (dry sand)	4–6	100–200	46–53	0.75–1.00	6–12	60–45	1450

and the resultant silica grains are sieved to segregate them into the necessary grain sizes. Then the requisite moulding sand is prepared by using proper size silica and bentonite of the required type. The mixture is thoroughly mulled by adding water. Synthetic sands by this method would have had

- Higher refractoriness
- Better and controlled mechanical properties
- Uniform or requisite grain size

Further, this is well suited for foundry mechanisation

Table 3.13 shows some typical sand compositions and properties that are desirable in various casting materials.

Additives

A number of materials are added to moulding sand such as coal dust, saw dust or wood flour, starch and dextrin, iron oxide and silica flour to improve their moulding properties. These materials help in improving the casting quality, shown in Table 3.14, in a number of ways as detailed below.

TABLE 3.14 Different types of moulding sand additives used

Additives	Purpose Served
Molasses Cereals Ethylene glycol	Enhancement of bench life and resistance to drying out
Iron oxide Silica flour	Hot strength development
Coal dust Silica flour	Surface finish and resistance to metal penetration
Cereals Saw dust	Collapsibility and resistance to expansion defects

Coal Dust

It is basically used for providing better surface finish to the castings. This when comes into contact with the molten metal would provide a gaseous envelope to keep the molten metal from fusing with the sand thus providing good surface finish.

Saw Dust or Wood Flour

The addition of wood flour (about 1.5% max) to moulding sand widens the range of water that can be added to get proper green strength. It reduces the expansion defects while improving the flowability of the moulding sand and helps maintain the uniform mould density. Too high wood flour makes the moulding sand brittle.

Starch and Dextrin

These are the organic binding materials used with mould and core sands. These increase resistance to deformation, skin hardness and expansion defects such as scab.

Iron Oxide

Iron oxide (up to 3% max) in moulding sand improves surface finish, decreases metal penetration, reduces burn–on, increases the chilling effect of the mould and increases glazing. With enough iron oxide in combination with silica flour, mould washes could be avoided. Iron oxide decreases green strength and permeability while improving the hot strength. It reduces collapsibility and makes the shake out of the mould difficult.

3.3.5 Indian Sands

The normal practice of Indian foundries is to use the locally available river sand. For example, the foundries in Jamshedpur use the local Swarnarekha river sand. The river sands normally are of medium to fine grain and sub–angular in shape. They contain about 80 to 85 % silica and the rest are alumina and other impurities. A brief description of the properties of various sands analysed by National Metallurgical Laboratory is presented in Table 3.15. Typical properties of natural moulding sand as affected by the moisture content is presented in Fig. 3.31 for yadgiri sand.

TABLE 3.15 Indian sands (natural)

Sand Location	Moisture %	Green Permeability Number	Green Compression Strength, kPa	Clay %	Fineness Number GFN	Sintering Temperature °C	Grain Type*	Usage
Londha	3–6	25–48	140–240	16–26	80–130	1100–1300	B	Cast iron
Bhavanagar	2.5–6	6–35	100–250	13–30	90–140	1050–1150	B	Only for non-ferrous
Goriali, Jayadi (Gujarat)	1–4	70–85	40–120	6–11	90	1050–1100	B	Cast iron
Rajkot	2–6	20–125	100–160	19	75	1150–1200	A, B	Only for non-ferrous
Jabalpur	4.5–6	165–205	60–80	6–8	40–70	1250–1350	A, C	Steel
Kanpur	4.8–8	4–6	50–60	40	247	1050–1100	B	Only for non-ferrous
Guntur (A.P.)	5.5–6.8	30–32	70	13	80	1150–1200	B	Cast iron, non-ferrous
Secunderabad	6–7	–	–	14.2	–	–	A, B	Cast iron, non-ferrous
Balamrai (A.P.)	–	–	–	38	78	–	A, B	Cast iron, non-ferrous
Oyaria (W.B.)	3.8–5.6	210–220	70– 90	11.6	47	1250–1300	B, C	Cast iron
Barakar	–	–	–	11.3	72	–	–	Cast iron
Mogra (W.B.)	–	–	–	0.22	–	–	B	Cast iron, non-ferrous

*Where, A = angular grains,
B = sub angular grains, and
C = round grains.

High silica sands which are normally used by steel foundries are available from Rajmahal sand (Bihar), quartzitic sand (Jamshedpur), Allahabad, Jabalpur, Hyderabad, Madras and Vengurla (Maharashtra). Many of these sands contain very small amounts of clay and grains are generally of the requisite grade. The typical properties and uses of these sands are presented in Table 3.15. The properties of Hyderabad sand when washed and mixed with 5% bentonite, are presented in Fig. 3.32.

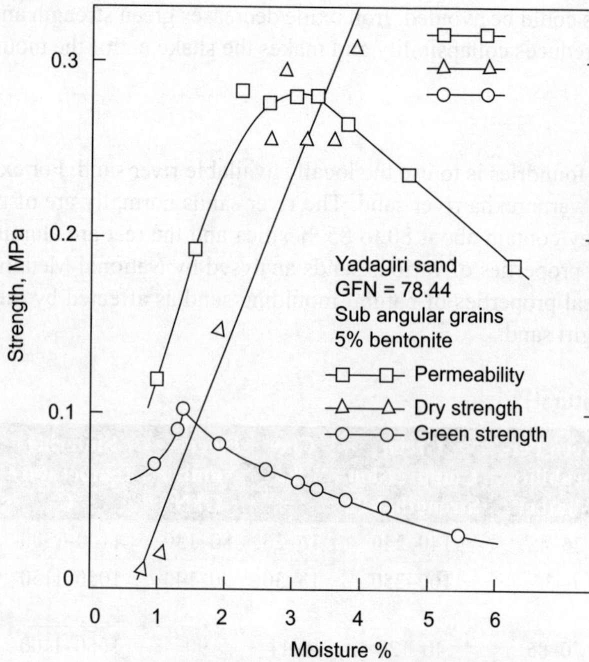

FIG. 3.31 *Properties of Yadgiri sand*

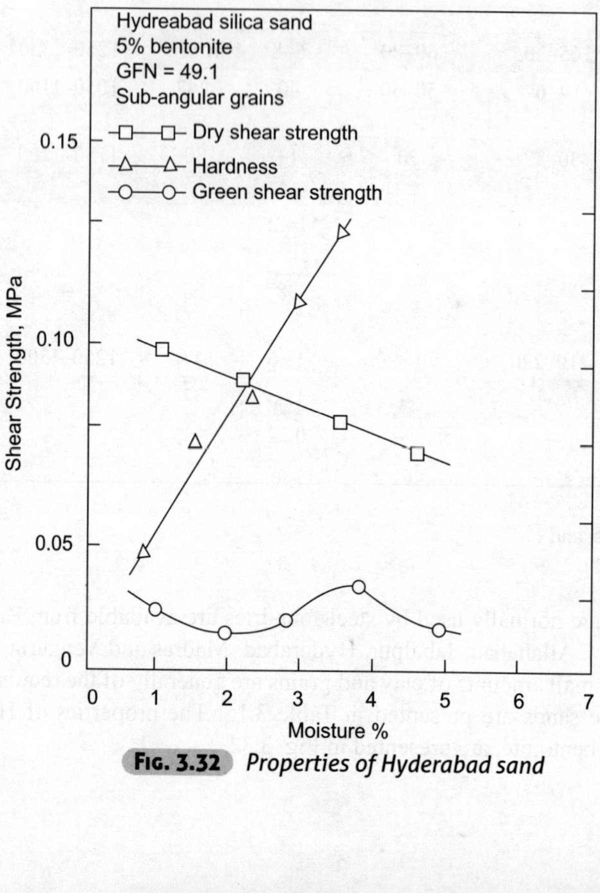

FIG. 3.32 *Properties of Hyderabad sand*

TABLE 3.16 High silica Indian sands (synthetic)

Sand Location	Moisture %	Green Perme-ability Number	Green Compression Strength, kPa	Clay %	Fineness Number GFN	Sintering Temperature °C	Grain Type*	Usage
Rajkot	1.5–3.4	160–190	40–60	3	53	1350	–	Cast iron, steels
Hardwar	3–6	85	30–50	1	57	1250–1300	B, C	Steels
Brahmani river (Orissa)	2.5–4.5	150–500	20–50	0.4–3	28–78	1250–1350	B, C	Cast iron
Gidni (W.B.)	2.4–3.4	175–335	80–120	5.7	27	1150–1200	A, B	Steel with addition
Adjoy river (W.B.)	2.2–4	155	30–40	1.4	48.6	1200–1250	B, C	Cast iron
Yadgiri (A.P.)	1.3–3.3	130–180	40–70	3.2	78	1100–1150	B	Non–ferrous
Rajmahal (Bihar)	3.4	200	60	1.02	52	–	B	Steels
Durgapur	3	250	30	0.55	34.69	> 1400	B, C	Steels
Jabalpur	3.2	130	120	5.14	53	1350–1450	B, C	Steels
Allahabad	3.2	80	35	–	54	1350–1450	B, C	Steels
Madras	2.4	740	50	–	26.4	1450–1550	B, C	Steels
Hyderabad	2 –3.8	160–170	30–50	3	49.1	1400–1450	B	Steels, Cast iron
Vengurla	1.5–3	175–255	25–40	1	39.1	1350–1450	A, B	Steels
Ratnagiri	1.2–2.6	300–375	20–40	0.7	42.19	1450–1500	B, C	Steels with addition

*Where, A = angular grains,
B = sub angular grains, and
C = round grains.

The change in permeability that can be achieved by means of washing is presented in Fig. 3.33 where the permeability of the Rajkot sand is improved by the washing of the clay and adding the 3% bentonite.

The bentonite deposits suitable for foundry use are found in Bihar, Kashmir and Rajasthan. Both sodium bentonite and calcium bentonite are available. These bentonites are light green in colour and primarily with sodium base. Rajasthan bentonite can be used for light and medium steel castings while those found in Bihar can be used for gray iron and steel castings. Comparative properties of Rajasthan and Bihar bentonites as they affect the Rajmahal sand are presented in Fig. 3.34.

For a detailed description of the composition of the various Indian sands, clays and their properties reference may be made to the NML book.

3.3.6 Other Sands

Though moulding sands are the prime mould materials used in a foundry, there are a number of other materials which are also used for a number of specific properties.

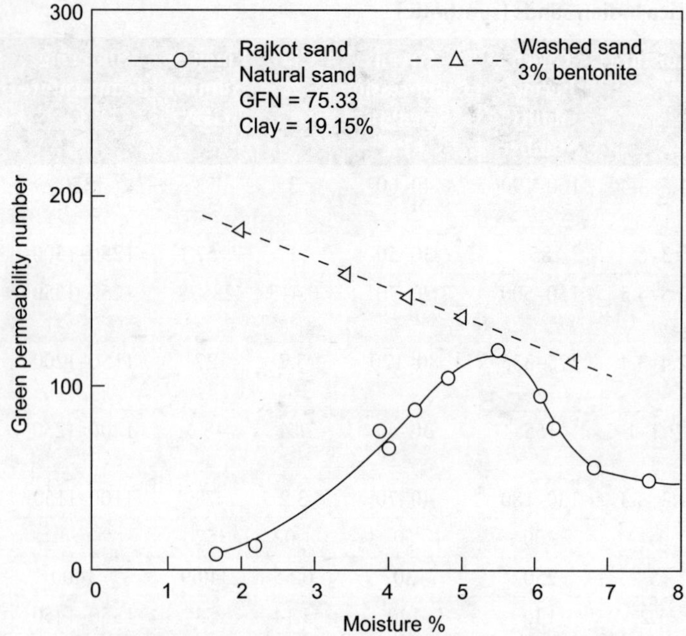

FIG. 3.33 *Permeability of sand as affected by washing*

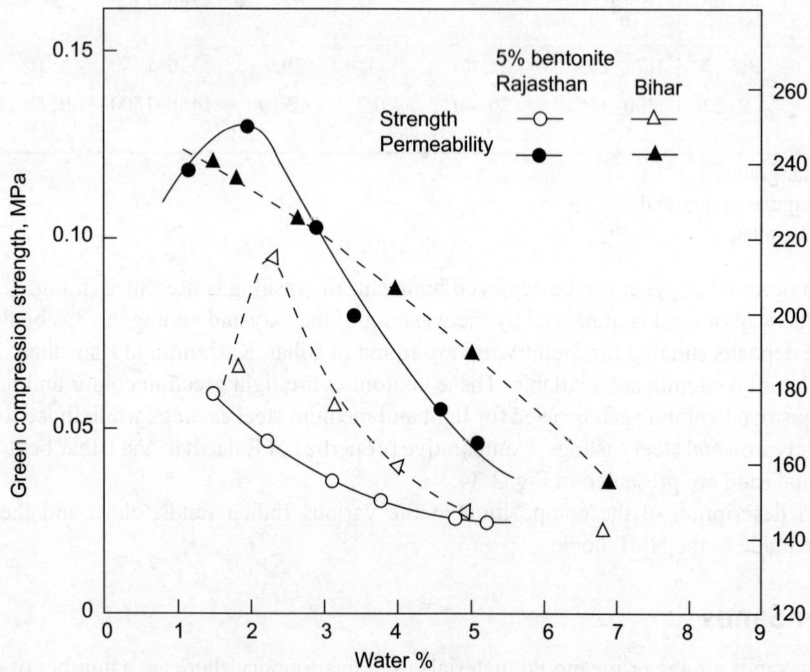

FIG. 3.34 *Moulding properties of Rajmahal sand as affected by Bihar and Rajasthan bentonites*

Facing Sand

This sand is used next to the pattern to obtain cleaner and smoother casting surfaces. Generally, sea coal or coal dust (finely divided bituminous coal of 2 to 8%) is mixed with the system sand to improve the mouldability and surface finish. The sea coal being carbonaceous, will slowly burn due to the heat from the molten metal and give off small amounts of reducing gases. This creates a small gas pressure in the surroundings of the cavity that molten metal is prevented from entering into the silica grains or fuse with them. This helps in generating good casting surface and also lets the moulding sand peel off from the casting during shake out.

Mould Wash

Purely carbonaceous materials such as sea coal, finely powdered graphite or proprietary compounds are also applied on to the mould cavity after the pattern is withdrawn. This is called the mould wash and is done by spraying, swabbing or painting in the form of a wet paste. These are used essentially for the following reasons:

(i) To prevent metal penetration into the sand grains and thus ensuring a good casting finish, and
(ii) To avoid mould-metal interaction and prevent sand fusion.

For depositing the mould wash, either water or alcohol can be used as a carrier. But because of the problem of getting the water out of the mould, alcohol is preferred as a carrier. The proprietary washes are available in powder, paste or liquid form. The powder needs to be first prepared and applied whereas the paste and liquid can be straight away applied.

Backing Sand

This is normally the reconditioned foundry sand and is used for ramming the bulk of the moulding flask. The moulding flask is completely filled with backing sand after the pattern is covered with a thin layer of facing sand. Since the casting is not affected to any great extent by the backing sand, it usually contains the burnt facing sand, moulding sand and clay.

Parting Sand

This is the material which is sprinkled on the pattern and to the parting surfaces of the mould halves before they are prepared to prevent the adherence of the moulding sand. This helps in easy withdrawal of the pattern and easier separation of the cope and drag flasks at parting surface. It is essentially a non–sticky material such as washed silica grains.

3.3.7 Fluidity

The term fluidity is normally used in a foundry, to designate the casting material's ability to fill the mould cavity. The fluidity therefore is a complex property and there are a number of variables affecting it. Fluidity depends on the casting material as well as the mould.

The properties of casting material affecting the fluidity to a great extent are: viscosity of the melt, heat content of the melt, surface tension, freezing range and specific weight of the liquid metal. Lower the coefficient of viscosity of the molten metal, higher would be the fluidity since the melt will be able to flow freely. Since the heat content and superheat of the molten metal decrease the coefficient of viscosity, it would also be responsible for an increase in the fluidity. Similarly lower surface tension which promotes wetting of the mould by the melt would quickly fill the mould, particularly the narrow sections. In general, the alloys with a narrow freezing range have a higher fluidity compared to the wide freezing range ones. In wide freezing range alloys, during the process of solidification the dendrites are spread over a much larger part of the mould and thus reduce the flow of the metal, decreasing the fluidity.

The mould properties that affect the fluidity are thermal characteristics, permeability and the mould cavity surface. The way in which heat is transferred by the mould from the melt affects the ability of the melt to fill the mould cavity. For example, in a green sand mould the fluidity would be lower since more heat is extracted by it than in a dry sand mould. The fluidity of a mould, which is coated with refractory washes, would be high compared to the one with no coatings. Similarly, a higher permeable mould is conducive to better fluidity.

Fluidity Tests

Since fluidity depends on such a large number of diverse variables, it is not feasible theoretically to determine it and therefore, practical testing is resorted to. The most commonly used testing is the spiral fluidity test. The mould of the same composition that is to be used is prepared with the standard spiral pattern with gating as shown in Fig. 3.35. The casting prepared from the mould is measured for the fluidity which is represented by the length of the spiral solidified.

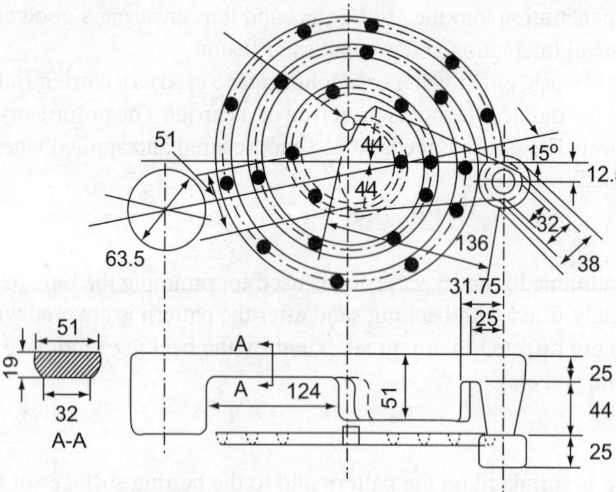

FIG. 3.35 *Fluidity spiral design*

A modified spiral pattern for gray cast iron is presented in Fig. 3.36. Gray cast iron is the most fluid of all the ferrous alloys. The fluidity of gray cast iron is affected by the composition factor (CF), which can be expressed as

$$CF = \%\,C + 0.25 \times \%\,Si + 0.5 \times \%\,P$$

Then the fluidity spiral length in cm can be estimated by means of the following empirical relation:

$$\text{Fluidity, cm} = 37.846 \times CF + 0.228 \times T - 389.6$$

Where, T = pouring temperature, °C

The results obtained by Porter and Rosenthal for the fluidity of gray cast iron as affected by the composition and the melting temperatures are presented in Fig. 3.37.

Since the fluidity is affected by the pouring temperature and to affect the completion of the mould filling in a reasonable time, the pouring temperatures of the moulds are accordingly controlled. The pouring temperature should be a little above the melting temperature of the alloy along with a sufficient enough superheat to account for the cooling of the molten metal from the time it is tapped from the furnace till it is

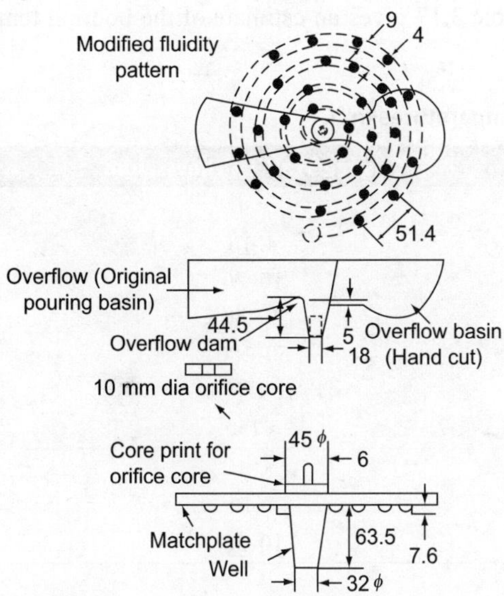

Fig. 3.36 *Fluidity spiral for gray cast iron*

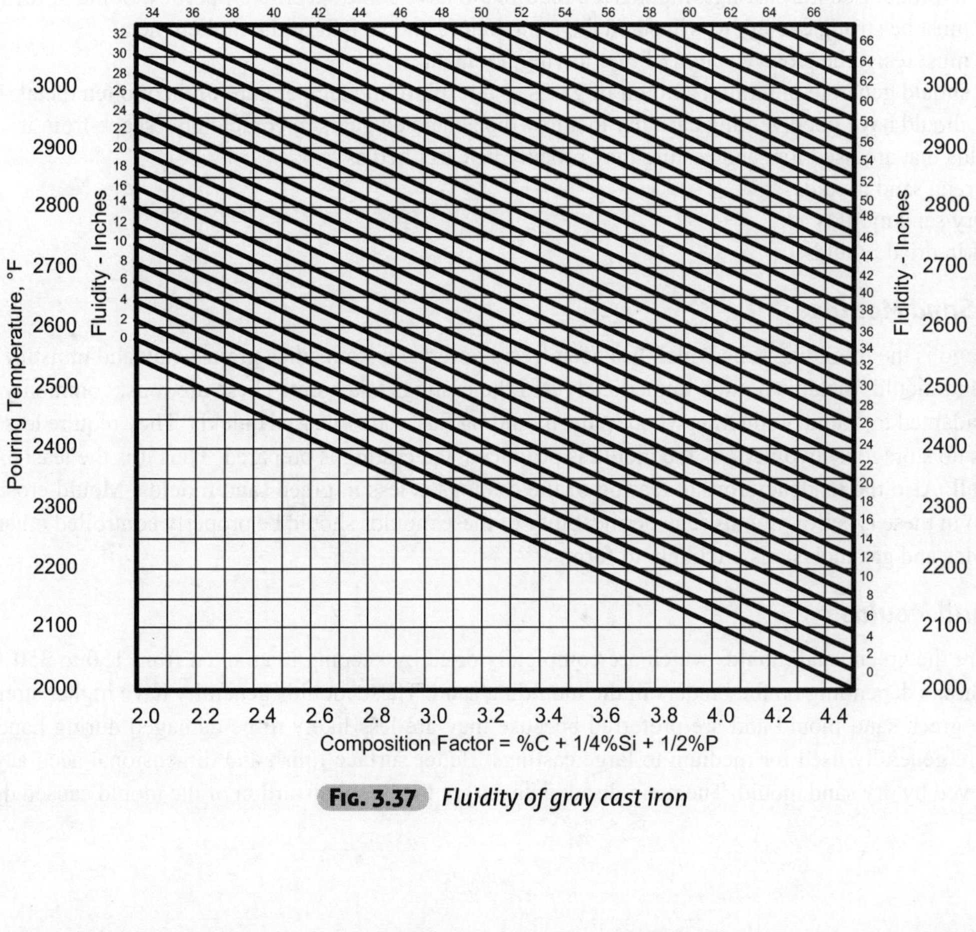

Fig. 3.37 *Fluidity of gray cast iron*

poured into the mould. The Table 3.17 gives an estimate of the pouring temperatures based on the casting material and section thicknesses.

TABLE 3.17 Metal pouring temperatures in °C

	Casting Wall Thickness, mm	Pouring Temperatures, °C
Gray and high strength iron	< 4	1450–1360
	4–10	1430–1340
	10–20	1400–1320
	20–50	1380–1300
	50–100	1340–1250
	100–150	1300–1230
	> 150	1280–1220
Malleable iron	< 4	1480–1450
	4–10	1450–1380
	10–20	1430–1360

3.3.8 Types of Sand Moulds

In order to produce sound castings, moulds are required to have some specific properties. Some of them are:

- It must be strong enough to withstand the temperature and weight of the molten metal.
- It must resist the erosive action of the flowing hot metal.
- It should generate minimum amount of gases as a result of the temperature of the molten metal.
- It should have good venting capacity to allow the generated gases to completely escape from it.

Moulds that are used for sand casting may broadly be classified as

- Green sand moulds
- Dry sand moulds
- Skin dried moulds

Green Sand Moulds

Green sand is the moulding sand which has been freshly prepared from silica grains, clay and moisture. In a green sand mould, metal is poured immediately and the castings taken out. These are most commonly used and are adapted for rapid production whereas the moulding flasks are released quickly. They require less floor space as no storage is involved. As the mould is produced, the casting is prepared. Thus it is the least expensive of all. Also the tendency for hot tearing of the castings is less in green sand moulds. Mould erosion is common in these types of moulds. The permeability of these moulds should be properly controlled otherwise blow holes and gas inclusions are likely to form.

Dry Sand Moulds

These are the green sand moulds which are completely dried by keeping in an oven from 150 to 350°C for 8 to 48 hours depending on the binders in the moulding sand. These moulds generally have higher strengths than the green sand mould and are preferred because they are less likely to be damaged during handling. These are generally used for medium to large castings. Better surface finish and dimensional accuracy can be achieved by dry sand mould. The main disadvantages are the likely distortion of the mould caused during the baking process; susceptibility to hot tearing of castings and longer production cycles. Also this is more expensive than the green sand mould.

Skin Dried Mould

Though the dry sand mould is preferable for large moulds because of the expense involved, a compromise is achieved by drying only the skin of the mould cavity with which the molten metal comes into contact, instead of the full mould. The skin is normally dried to a depth of 15 to 25 mm, using either torches or by simply allowing them to dry in atmosphere. This can also be done in pit moulding. However, pouring of metal should be completed immediately after the drying process such that moisture from the undried portion would not penetrate the dried skin.

3.3.9 Moulding Machines

Hand moulding as described earlier is suitable for small batch production or for large size castings. However, for production work involving large batches of the same type of casting is to be produced, machine moulding is a necessity. In the machine moulding, there are basically three methods used for ramming the sand into the moulding flasks. They are:

- Jolting
- Squeezing
- Sand slinging

In jolt moulding the sand is first filled into the flask generally from an overhead hopper and then it is raised to a certain height before it is allowed to free fall on to a solid bed plate. The resulting impact forces the sand to get compacted uniformly into the mould. This lifting and dropping process continues repeatedly till the required mould hardness is achieved. This type of ramming is suitable for horizontal surfaces. It is very severe on the moulding equipment and also likely to give rise to uneven packing. The sand at the bottom experiences the highest force and consequently is packed well compared to the sand in the top layers.

In the squeeze ramming, a plate slightly smaller than the inside dimensions of the moulding flask is fitted into the flask already filled with the moulding sand as shown in Fig. 3.38(a). A uniform pressure is applied on the plate by either moving it down or by moving the flask upwards. The resulting force compacts the sand uniformly. The sand next to the plate rams hardest while the sand below is progressively less hard. Hardness achieved is also dependent upon the pressure applied on the squeeze plate. Since the hardness achieved follows contour of the squeeze plate which is in contact with the sand, sometimes the plate is provided with contour as shown in Fig. 3.38(b), to match the pattern for achieving better uniform hardness of the mould. This type is suitable for small castings and is generally suitable for shallow flasks. It is also possible to use a diaphragm to provide differential ramming force required for the contour of the pattern as shown in Fig. 3.38(c).

A combination of jolt squeeze machines are more common in view of the more uniform ramming that could be achieved by combining the two methods as described above.

Sand slinging is done by throwing sand into the flask rapidly and with great force. This process develops uniformly high mould hardness. The process is very fast and gives high uniform sand ramming. The sand should be uniformly sprayed into the mould by moving the nozzle through the entire area of the flask. Also it is necessary that the sand be sprayed in layers to achieve uniform hardness. However, the initial cost of the equipment is high compared to the other form of moulding machines described above.

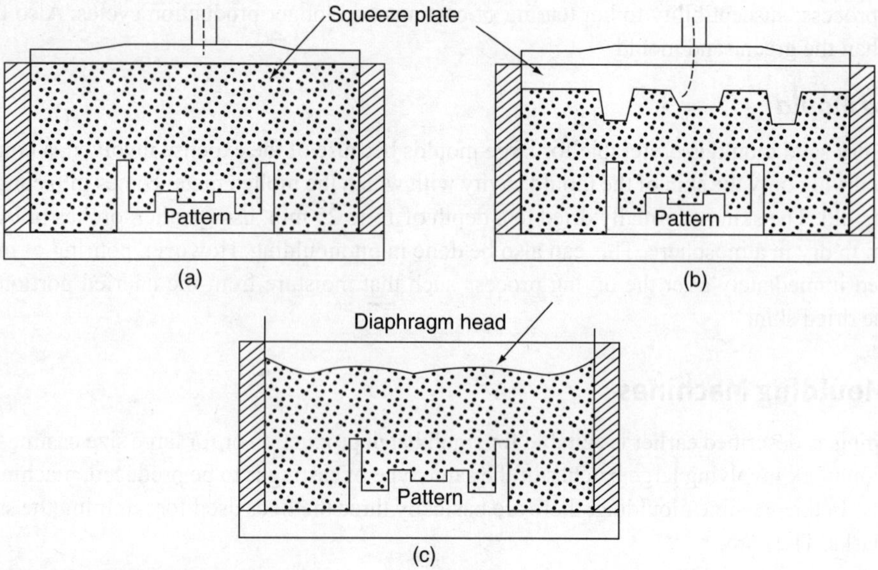

FIG. 3.38 *Types of squeeze heads used for machine moulding, (a) conventional squeeze, (b) profile squeeze head, and (c) diaphragm squeeze*

3.3.10 Nobake Moulding Process

Nobake is a casting process that involves the use of chemical binders (furan or urethane) to bond the moulding sand in place of clay normally used in green sand moulding. Clay free silica sand is conveyed to the mould fill station in preparation for filling of the mould. A mixer is then used to thoroughly blend the sand with the chemical binder and a catalyst. As the sand exits the mixer, the binder begins the chemical process of hardening. The chemical mixed sand is then poured into the moulding flask that is fitted with the cope/drag pattern plate that is complete with the necessary gating and risering system as designed. Each mould half is then compacted to form a strong and dense mould. After a specified period of time (from as little as 10 sec to as long as the foundry requires depending upon mould size), the sand mixture hardens (resembling a brick in strength) to form the mould halves and the cope/drag pattern plates are removed. Cores, if required, can also be made using the same process. After the sand has set, a mould wash may be applied. The cores, if required, are set into the drag and the cope is closed over the cores to complete the mould.

Pouring of molten metal is done normally and the mould is allowed to cool before shake–out.

The broken pieces of moulded sand during shake–out are further broken down until they are close to the sand grain size. The sand can now be either reclaimed for reuse in the casting process or removed for disposal.

Advantages

- It is possible to get good dimensional tolerances (±0.125–0.375 mm) because of the high strength of the mould. Nobake mould can withstand the metallostatic pressure exerted by the molten metal during casting.
- Because of the rigidity and tensile strength of nobake moulds, thin sections about 2.5 mm thick can be routinely produced. Also the draft required is less compared to green sand moulding.
- Nobake mould produces fine surface finishes that can be upgraded further with the mould and core coatings to support special finishing on the cast components such as paint or dressing.

- Less skill and labour are required than in conventional sand moulding.
- Nobake process is compatible with practically all types of pattern materials as well as it imparts minimal tooling wear. If needed inexpensive tooling can be used for casting runs for prototyping.
- It is possible to pour any ferrous/non–ferrous metals in a nobake mould.
- Cores and moulds prepared with nobake process can be stored indefinitely.
- Most of the nobake systems have excellent shake–out properties.
- Nobake moulding reduces the opportunity for gas–related defects as the nitrogen content of most binder systems used minimizes susceptibility to gas porosity.
- Though this process is adaptable to any volume, it is preferable for low to medium volume production capability.

3.3.11 Plaster Moulding

Plaster mould casting, also called rubber plaster moulding (RPM), is a specialized casting process used for producing non–ferrous (aluminium and copper base alloys) castings by pouring liquid metal into moulds made of Plaster of Paris (gypsum).

In this process, Plaster of Paris ($CaSO_4.½H_2O$) is mixed with water to form a slurry with additives such as ceramic talc, fiberglass, clay, silica flour, fly ash, etc., with water and then poured over the pattern. The additives are used to enhance the mould properties such as green and dry strength, permeability, and castability. The slurry is allowed to preset to a rubbery consistency which allows it to be stripped from the pattern but which is sufficiently strong to return to the shape it had when on the pattern. The preset mould is then ignited to burn off the volatile content in the set gel and baked at about 120 to 260°C. This step results in a hard and rigid mould containing microscopic cracks. Cores can also be made using similar process.

Cores, if required, are then placed in the drag half. The cope half is then aligned with the drag half and locating pins on cope going into the matching pin holes in the cope. This assembling process can be done while the moulds are still hot or after they have cooled to room temperature depending on mould complexity. Dried plaster moulds have extremely low permeability of about 1 to 2 AFS. In view of this low permeability, gravity pouring is rarely employed and a vacuum assist is usually required for the pouring of moulds.

Advantages

- Castings produced by plaster moulding have high dimensional accuracy that is comparable to that of the castings made in investment moulds.
- Using plaster moulding a surface finish of the order of 1.25 to 3.25 μm can be readily obtained.
- Complex designs can be obtained by using flexible rubber patterns and utilizing the investment casting techniques.
- Since plaster has low thermal conductivity, castings cool slowly that will allow thin sections to be cast. Wall thicknesses as low as 0.75 to 1.00 mm can be cast at selected locations.
- Enhanced mechanical properties can be obtained by appropriately controlling the thermal gradients while slow cooling minimizes warpage and promotes uniformity of structure.

Disadvantages

- The process is expensive because of the lengthy processing procedures and the one time use of the mould materials.
- Compared to green sand moulding plaster moulding requires more equipment.
- Permeability of plaster moulds is inherently low when compared to the sand casting process.

3.3.12 Pit Moulding

Pit moulding is used for large and heavy castings weighing up to 150 tons where using a moulding flask for the process becomes almost impossible and too expensive for the amount of handling is concerned. The pit is normally formed in the foundry floor with sand. The pit acts as a drag where the cavity is formed. Pattern is placed in the pit and the sand is rammed around it to form the drag cavity. A separate cope can then be used above the pit which will be at the floor level. The cope can be constructed with bricks and normally houses the pouring cup and sprue. Vent pipes will be added into the pit to facilitate the escape of gases. This is a slow process and requires a lot of time to complete the mould.

3.3.13 Loam Moulding

Loam sand contains many ingredients like fine sand particles, finely ground refractories, clay, graphite and fibre reinforcements. Loam soils generally have better drainage than clay soils. They retain water easily. Organic matter such as chopped straw is added to the sand to provide good ventilation. Loam moulds are generally employed for making large castings without using the expensive full patterns and moulding flasks. They use skeleton patterns and sweeps to reduce the cost of patterns. Objects such as large cylinders, chemical pans, large gears, round bottoms, kettles and other machining parts are produced in loam moulding. Big moulds are constructed using brick framework that will be lined with loam sand and dried. Sweeps, etc. are used for getting the requisite profile of the casting.

3.3.14 Additive Manufacturing for Mould Preparation

It is possible to use Additive Manufacturing (AM) to directly produce sand moulds without using a pattern. Most of the systems use binder jetting or 3D Printing (3DP) process. This process can be used to directly fabricate sand moulds as well as cores by selectively jetting binder into a powder bed of foundry sand. ExOne and voxeljet are the two major manufacturers that provide 3D building machines utilizing different technologies and materials.

The moulding material used by ExOne is common silica sand with different binders depending upon the application similar to the nobake process discussed earlier. One of the binders provided by ExOne is furan binder for traditional foundries so that they do not have to make any changes to their moulding procedures. Silicate binder offered by ExOne is environment friendly with low gas emissions. ExOne also offers phenolic binder that is suited for thin walls similar to shell moulding. Voxeljet offers furan binder with silica sand normal for easier integration into existing foundries.

3.4 CORES

Cores are the materials used for making cavities and hollow projections which cannot normally be produced by the pattern alone. Any complicated contour or cavity can be made by means of cores so that really intricate shapes can be easily obtained. These are generally made of sand and are even used in permanent moulds. In general, cores are surrounded on all sides by the molten metal and are therefore subjected to much more severe thermal and mechanical conditions. As a result, the core sand should be of higher strength than the moulding sand.

The normal characteristics desired of a core are:

1. Green strength: A core made of green sand should be strong enough to retain the shape till it goes for baking.

2. Dry strength: It should have adequate dry strength so that when the core is placed in the mould, it should be able to resist the metal pressure acting on it.
3. Refractoriness: Since in most cases the core is surrounded all around, it is desirable that the core material should have higher refractoriness.
4. Permeability: Some of the gases evolving from the molten metal and generated from the mould may have to go through the core to escape out of the mould. Hence cores are required to have higher permeability.
5. Collapsibility: As the casting cools, it shrinks. Unless the core has good collapsibility (ability to decrease in size), it is likely to provide resistance against shrinkage and thus can cause hot tears.
6. Friability: After the casting is completely cooled, the core should be removed from the casting before it is processed further. Hence the friability (the ability to crumble) should also be a very important consideration.
7. Smoothness: The surface of the core should be smooth so as to provide a good finish to the casting.
8. Low gas emission: Because of the high temperatures to which a core is subjected to, it should allow only minimal amount of gases to be evolved such that voids in the castings can be eliminated.

3.4.1 Core Sands

The core sand should contain the sand grains, binders and other additives to provide specific properties.

Sand The silica sand, which is completely devoid of clay, is generally used for making core sands. Coarse silica, because of its higher refractoriness, is used in steel foundries. The finer sands are used for cast irons and non–ferrous alloys.

Binders As explained earlier, core sands need to be stronger than the moulding sand and therefore the clay binder used in moulding sands is not enough but somewhat better binders need to be used. The normal binders are organic in nature because these would be burnt away by the heat of the molten metal and thus make the core collapsible during the cooling of the casting. The amount of binder required depends to a great extent on the fineness of the sand grains. Also, the amount of clay left in the sand increases the consumption of the binder.

The binders generally used are linseed oil, core oil, resins, dextrin, molasses, etc. Core oils are mixtures of linseed, soy, fish, petroleum oils and coal tar.

The general composition of a core sand mixture could be core oil (1%) and water (2.5 to 6%). The actual composition to be used depends on the size and shape of the core and the alloy being cast. The properties as controlled by the composition are shown in Fig. 3.39. These properties are obtained by baking the standard AFS test samples for 1.5 hours at 175°C.

The organic binders develop strength by means of polymerisation and cross linking. To affect this, the cores after preparation need to be baked. A proper combination of baking time is to be chosen so as to optimise the core properties, as represented in Fig. 3.40.

3.4.2 Carbon Dioxide Moulding

The other method which is widely used in making cores and occasionally for moulds is to use sodium silicate (water glass, SiO_2:Na_2O) as a binder. This is essentially a quick process of core or mould preparation. The mould is prepared with a mixture of sodium silicate and sand which is treated with carbon dioxide

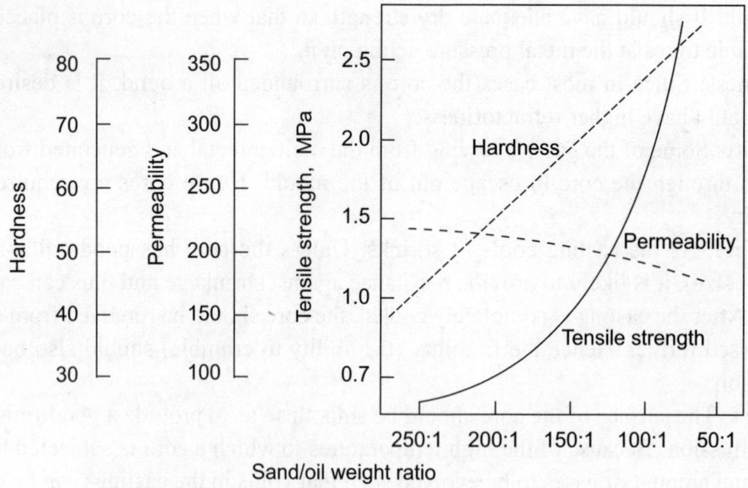

FIG. 3.39 *Core sand properties varied by composition*

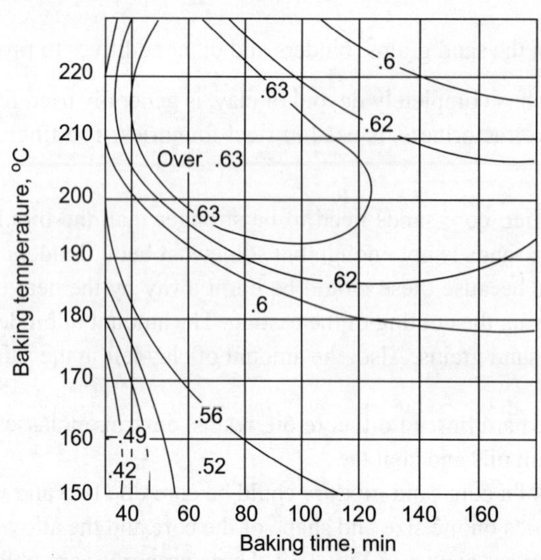

FIG. 3.40 *Core properties as affected by baking time and temperature*

for two to three minutes such that a dry compressive strength of over 1.4 MPa is arrived at. The carbon dioxide is expected to form a weak acid which hydrolyses the sodium silicate resulting in amorphous silica forming the bond. The introduction of CO_2 gas starts the reaction by forming hydrated sodium carbonate ($Na_2CO_3 + H_2O$). This gelling reaction increases the viscosity of the binder till it becomes solid. The compressive strength of the bond increases with standing time due to dehydration.

Because of the high strength of the bond, the core need not be provided with any other reinforcements. The equipment required is fairly simple because the main requirement is carbon dioxide, which is easily available, and no expensive core baking equipment is needed. Also it does not involve any distortions due to baking and also better dimensional accuracies are achieved. But it is a little more expensive process. Also, sand mixture does not have good shelf life and therefore should be used immediately after preparation. The shakeout properties of carbon dioxide moulding are also poor compared to the normal moulding. Moulds and cores get deteriorated from water pick up if they are kept stored for longer periods before use. They develop non-uniform strength due to the difficulty in achieving uniform gassing.

The core sand should be completely free from clay, moisture and any other impurities; otherwise the binder consumption would increase. The GFN for the sand grains may range from 55 to 85. The amount of binder required increases from 3% of sodium silicate (by weight) for a grain size of 55 to 4.5% for the grain size of 85.

In addition to the sand and binder, some more additives are added to improve the specific properties. For example, kaolin clay improves mould stability, aluminium oxide improves hot strength of sand at about 800°C and molasses or inverted sugar are used to retard the strength of the mould after pouring to eliminate shake out problems.

The gassing of carbon dioxide into the moulds and cores is done by a number of ways depending essentially on the size of the core or the mould. For smaller cores, a gas-tight hood, which is connected to a carbon dioxide source, is placed directly on the core box. Medium sized cores are fed directly through the core box or through the pattern in the case of moulds. For very large moulds special probes are inserted into the mould which may be connected to a manifold getting the carbon dioxide supply. The gas pressure should be maintained around 0.14 to 0.28 MPa depending largely on the thickness of the section to be gassed.

3.4.3 Types of Cores

Based on the type of sand used, the cores are essentially of two types.

Green sand cores are those which are obtained by the pattern itself during moulding. Though this is the most economical way of preparing core, the green sand being low in strength cannot be used for fairly deep holes. Also a large amount of draft is to be provided so that the pattern can be withdrawn. This is used only for those types of cavities which permit the withdrawal of the pattern as in Fig. 3.41.

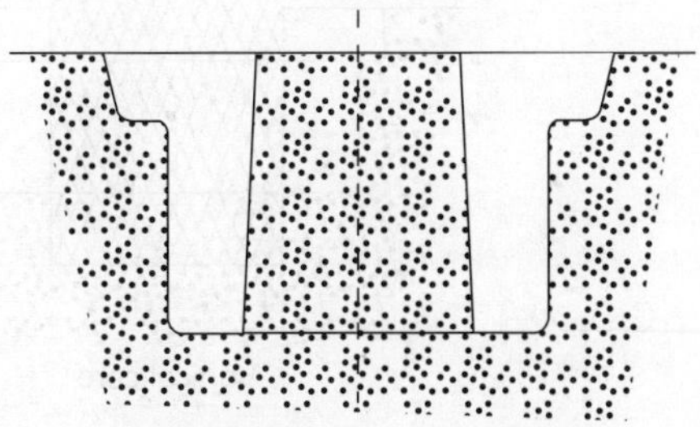

FIG. 3.41 *Green sand core*

Dry sand cores are those which are made by means of special core sands in a separate core box, baked and then placed in the mould before pouring. Some of the types of cores used in various situations are presented in Fig. 3.42.

In designing the core boxes, care should be taken to consider the strength of the core. Before the baking process, core is generally weak and should be well supported. If the core is simple and strong in itself, no special precaution is required. But for slender and complicated cores it may be necessary to place the core in the core box during the baking process. In such a situation the core box should be small. An example of a core box is shown in Fig. 3.43.

If a core is symmetrical, then it can be made into two equal parts and then assembled together by adhesives or fastened by wires. The glues that are used are mixtures of talc, dextrin, flour, molasses and water. Care should be taken while applying the glue to not close the core vent holes made earlier for the gases to escape. When very large cores are to be joined, it may be necessary to use nuts and bolts. The bolt holes are generally covered with a core plug. If the core is to be mounted in a particular orientation, then some specific provision should be made in the core prints so that the core can be placed in the mould in only one position, as in Fig. 3.44.

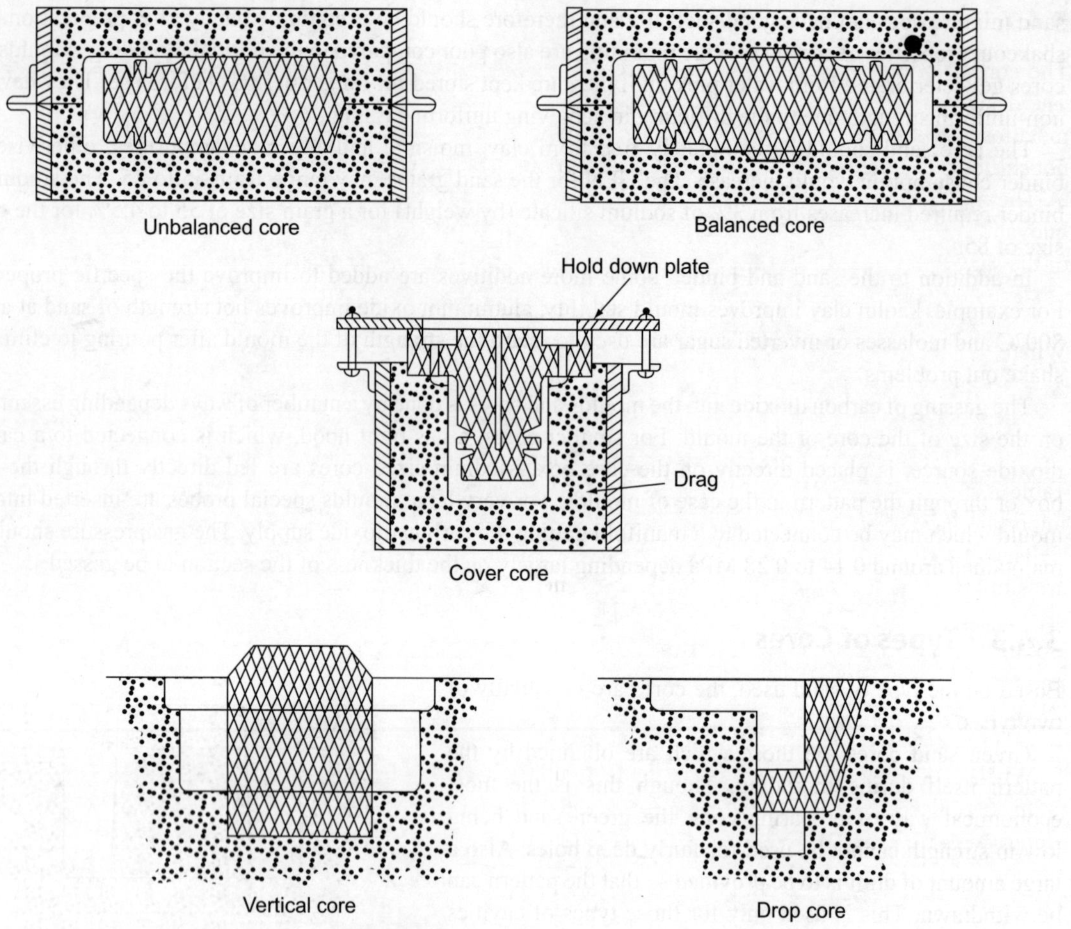

Fig. 3.42 *Dry sand core types*

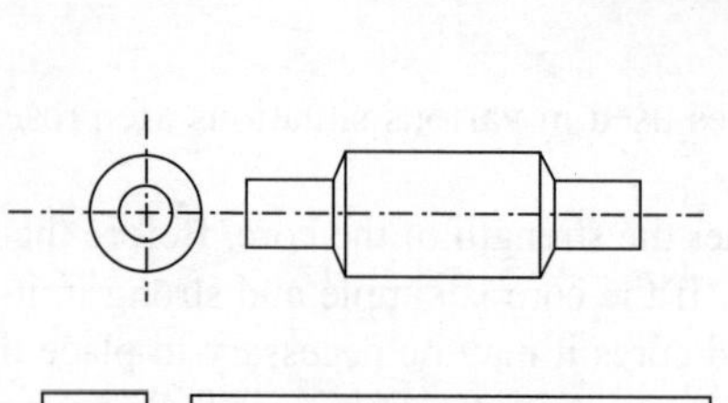

Fig. 3.43 *A typical core box*

Fig. 3.44 *Unsymmetrical core location*

3.4.4 Core Prints

The core prints are provided so that the cores are securely and correctly positioned in the mould cavity. The design of core prints is such as to take care of the weight of the core before pouring and the upward metallostatic pressure of the molten metal after pouring. The core prints should also ensure that the core is not shifted during the entry of the metal into the mould cavity.

The main force acting on the core, when metal is poured into the mould cavity, is due to buoyancy. The buoyant force can be calculated as the difference in the weight of the liquid metal to that of the core material of the same volume as that of the exposed core. It can be written as

$$P = V(\rho - d)$$

Where, P = buoyant force, N
V = volume of the core in the mould cavity, cm^3
ρ = weight density of the liquid metal, N/cm^3
d = weight density of the core material = 0.0165 N/cm^3

Table 3.18 lists the weight densities of some of the important foundry materials.

TABLE 3.18 Liquid densities of foundry materials

Material	Density, N/cm^3
Aluminium	0.0265
Copper	0.0878
Magnesium	0.0171
Zinc	0.0700
Lead	0.1113
Carbon steel	0.0771
Gray cast iron	0.0686 to 0.0735
White cast iron	0.0755
Moulding sand	0.0157

The above equation would be valid for cases that are similar to the one illustrated in Fig. 3.45, which are more common, where V is given by $0.25\,\pi D^2 H$. But for vertical cores as those shown in Fig. 3.46, the buoyant force is given by

$$P = 0.25\pi\,(D_1^2 - D^2)\,H\rho - V\,d$$

Where, V = total volume of the core in the mould.

In order to keep the core in position, it is empirically suggested that core print will be able to support a load of 3.5 N/cm^2 of surface area. Hence to fully support the buoyant force, it is necessary that the following condition is satisfied.

$$P \le 350A$$

Where, A = core print area, mm^2

If the above is not satisfied, then it would be necessary to provide additional support by way of chaplets as described later. From the core print area, necessary core print sizes can be calculated.

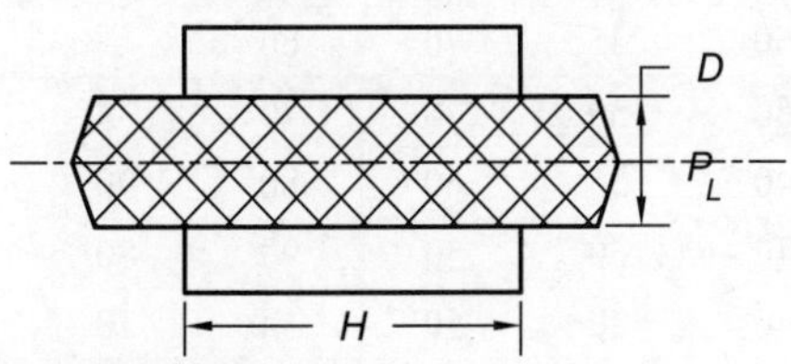

FIG. 3.45 *Horizontal core position*

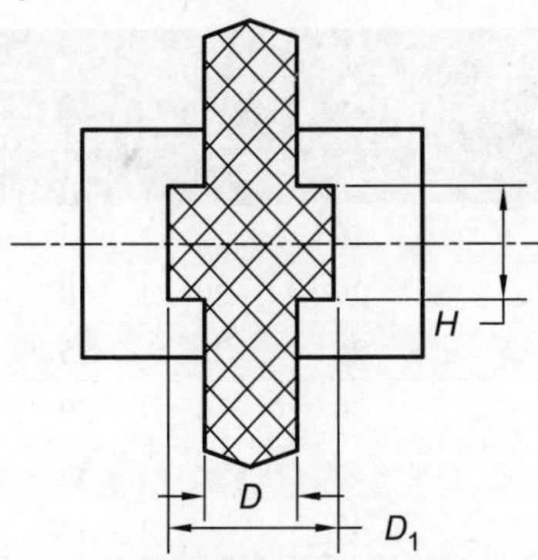

FIG. 3.46 *Vertical core*

Another set of empirical proportions for the core print dimensions which were found suitable in practice are presented in Fig. 3.47.

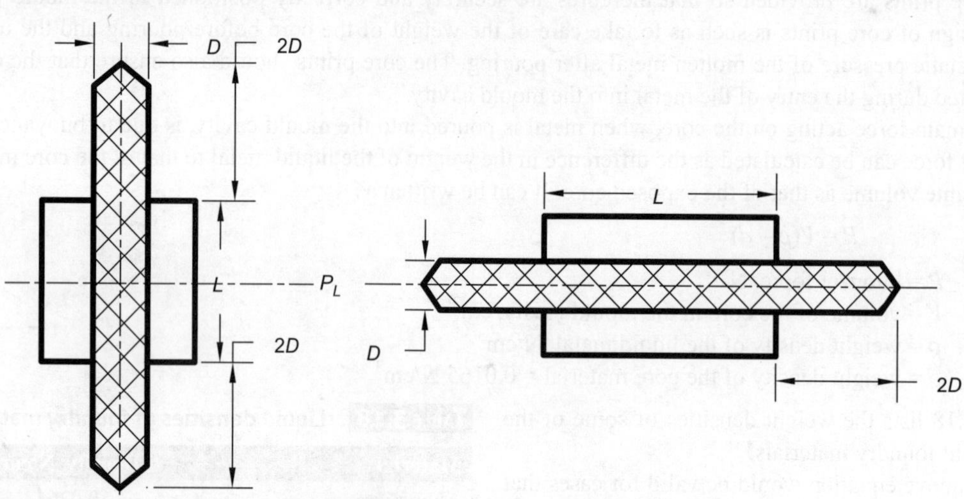

FIG. 3.47 *Core print proportions*

The Russian practice of dimensioning the core prints is to make the pressure acting on the core bearing area (i.e., *A*, the core print surface area) to be less than 50 to 75% of the moulding sand compression strength. Hence

$$A > K\frac{V(\rho - d)}{\sigma}$$

Where, V = total volume of the core including prints, cm^3
K = 0.50 to 0.75
ρ = weight density of the liquid metal, N/cm^3
d = weight density of the core material = 0.0165 N/cm^3
σ = compression strength of the moulding sand, N/cm^2

A suggested guide for core prints from Russian literature is given in Table 3.19 with reference to Fig. 3.48. The draft angles to be used for these core prints are given in Table 3.20 with reference to Fig. 3.48.

TABLE 3.19 **Core print dimensions**

Core Diameter D, mm	Core Length L, mm									
	< 50		51–150		151–300		310–500		500–750	
	v_1	h_1	v_1	h_1	v_1	h_1	v_1	h_1	v_1	h_1
up to 25	20	15	25	25	–	40	–	–	–	–
26–50	20	20	40	35	60	45	70	60	–	–
51–100	25	25	35	40	50	50	70	70	100	90
101–200	30	30	30	50	40	55	60	80	90	100
201–300	35	–	35	–	40	60	50	90	80	110
301–400	40	–	40	–	40	80	50	100	70	120
401–500	40	–	40	–	40	110	50	120	60	130

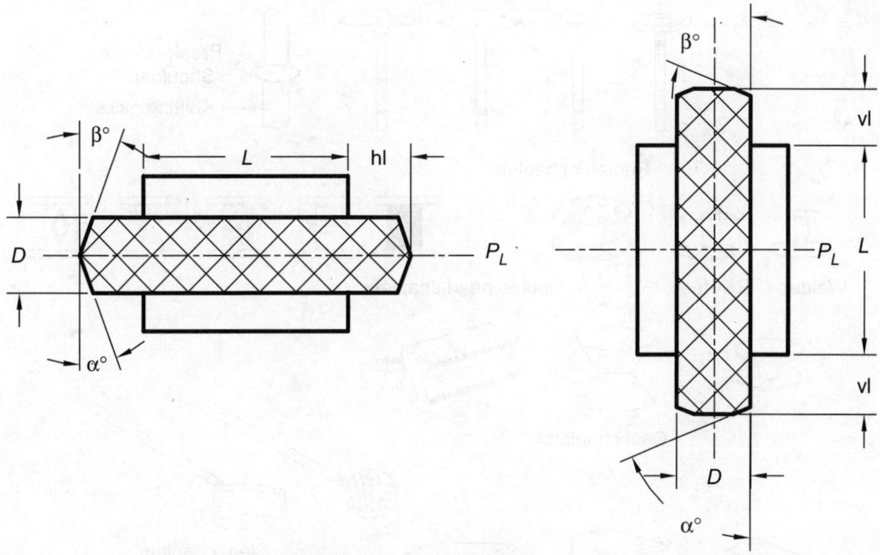

(a) Horizontal core (b) Vertical core

Fig. 3.48 *Core print sizes*

Table 3.20 Draft angles for core prints

Core Print, h_l or v_l, mm	Vertical, Degrees		Horizontal, Degrees	
	α	β	α	β
< 20	10	15	10	15
21 to 50	7	10	7	10
51 to 100	6	8	6	8
101 to 200	5	6	5	

3.4.5 Chaplets

Chaplets are metallic supports often kept inside the mould cavity to support the cores. These are of the same composition as that of the pouring metal so that the molten metal would provide enough heat to completely melt them and thus fuse with it during solidification. Some of the types of chaplets normally used are shown in Fig. 3.49 with an example of chaplet support in Fig. 3.50.

Though the chaplet is supposed to fuse with the parent metal, in practice it is difficult to achieve and normally it forms a weak joint in the casting. The other likely problem encountered in chaplets is the condensation of moisture which finally ends up as blow holes. Generally, the chaplets before they are placed in the mould should be thoroughly cleaned of any dirt, oil or grease. Because of the problems associated with chaplets, it is desirable to redesign the castings, as far as possible.

In order to calculate the required chaplet area Ac, we need to know the unsupported load. Referring back to the discussion on core prints, it was suggested that

$$P \leq 3.5 \text{ Ac}$$

Hence, Unsupported load = $P - 3.5$ Ac

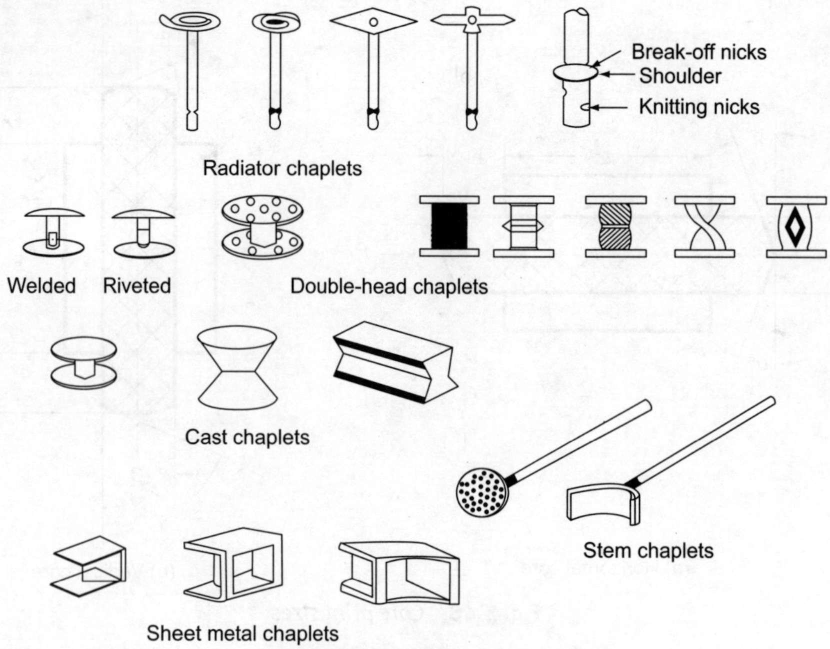

Fig. 3.49 *Types of chaplets*

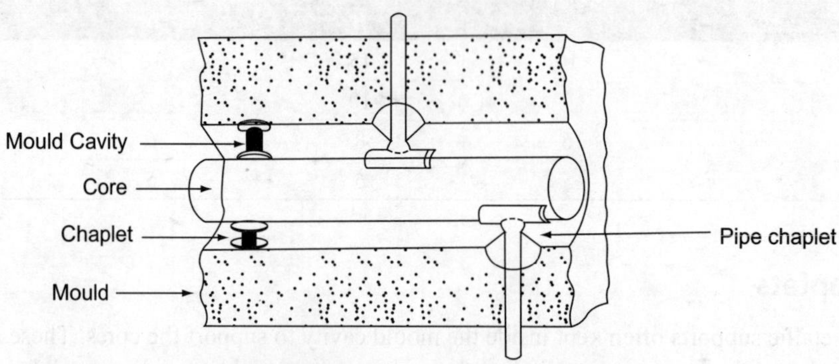

Fig. 3.50 *Core supported by chaplet*

If the unsupported load is less than or equal to zero, no chaplet is required. But if it is greater than zero then the chaplet area required is 29 mm^2 for every Newton of unsupported load.

3.4.6 Forces acting on the Moulding Flasks

Earlier, the buoyancy force exerted by the core has been calculated. This force is transmitted by the core to the cope and would tend to lift the cope away from the drag. In addition to this, there is another force termed as 'metallostatic' force also present inside the moulding cavity. This 'metallostatic' force is exerted by the molten metal in all the directions of the cavity. However, we are interested in the force exerted in the upward direction.

The 'metallostatic' force is coming because of the head with which the metal is entering the mould cavity. This force, Fm, can be estimated by taking the area of cross section of the casting on which this is acting. The projected area, A_p, of the casting in the parting plane is the area on which the metal pressure will be acting. The head of the metal as shown in Fig. 3.51 is given by $(h - c)$. Hence the force, F_m is given by

$$F_m = A_p \rho(h - c)$$

Care has to be taken to see that proper units are taken into account.

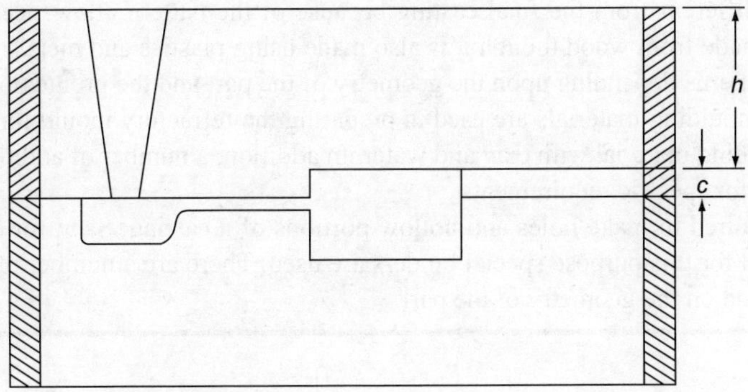

FIG. 3.51 *Head of the metal in sand casting*

Example 3.6

Find the weights that need to be kept to compensate for the forces during the pouring in a sand casting of a cast iron pipe of 12.5 cm OD and 10 cm ID with a length of 180 cm. The metal head is to be about 20 cm while the moulding flask size used for the purpose is 200 × 25 × 20 cm in size. Take the density of the core sand to be 0.0165 N/cm³ and the liquid metal density to be 0.0771 N/cm³.

$$\text{Volume of the core} = \frac{\pi \cdot 10^2 \cdot 180}{4} = 14\,137.2 \text{ cm}^3$$

Buoyancy force = $V(\rho - d)$ = 14137.2(0.0771 – 0.0165) = 856.7 N

Projected area of the casting = 180 × 12.5 = 2250 cm²

Upward metallostatic force = 2250 × 0.0771 × 20 = 3469.5 N

The total upward force = 3469.5 + 856.7 = 4326.7 N

Volume of sand in the cope = 200 × 25 × 20 = 100 000 cm³

Weight of the cope = 100 000 × 0.0165 = 1650 N

Net upward force acting = 4326.7 – 1650 = 2676.7 N

The foundry has to keep a total weight of about 3000 N to compensate for this upward thrust exerted by the core and the metallostatic forces.

In the above example, only the static force was considered. However, when the metal is entering the mould cavity at high flow rate, the loss of momentum would exert much higher pressure than that accounted for the static condition. Consequently more force would be acting on the mould flask than that calculated in the above procedure. It is general practice in the foundries to keep a weight which is about 50% more than the calculated one to account for this dynamic effect.

SUMMARY

Metal casting processes are the primary manufacturing processes that are used to make complex parts utilising liquid metal.

- Metal casting is one of the oldest manufacturing processes invented and practiced over the last 5000 years. However the major developments have taken place in the last 100 years in adopting it as a major primary manufacturing process.
- A pattern is different from the final casting because of the pattern allowances to be added. Most patterns are made from wood though it is also made using plastics and metals. There are a number of types of patterns depending upon the geometry of the part and the production volumes required.
- A variety of moulding materials are used in preparing the refractory mould for casting. Sand is the primary moulding material with clay and water. In addition, a number of additives to sand improve its properties for specific requirements.
- Cores are required to make holes and hollow portions of a casting. Sand for cores requires better properties and for this purpose special binders are used. There are a number of types of cores used in casting based on the geometry of the part.

Questions

3.1 List the main advantages of the casting process.

3.2 State the typical applications of casting process as used in automobile sector.

3.3 What are the major limitations of the sand casting process and how are they overcome?

3.4 What are the points which a designer should consider while going for a casting process?

3.5 Define the following terms as related to casting:
Cope, drag, bottom board

3.6 Give a brief write up on the following casting terms:
Sprue, gate, runner, riser

3.7 Sketch the cross section of a sand mould which is ready for pouring and label the various important parts.

3.8 Briefly explain the procedure to be followed for making a sand mould.

3.9 Distinguish clearly between the following casting terms:
Moulding sand, backing sand and facing sand

3.10 What are the distinguishing features between a casting and a pattern?

3.11 Distinguish between liquid shrinkage and solid shrinkage as related to castings. Explain how these are taken care of in designing sand castings.

3.12 Name the pattern allowances which can be quantitatively specified. Write a brief note on each of them.

3.13 What is draft allowance? How is it provided for patterns?

3.14 Briefly discuss why draft allowance is important for patterns.

3.15 Describe the allowances given on pattern for shaking and distortion.

3.16 What is meant by core prints? Explain how they are to be provided.

3.17 What are the materials that are generally used for preparing patterns?

3.18 What are the specific applications for the following pattern materials?
Wood, metal and plastics.

3.19 What metals are generally used for making patterns? Explain the reasons for their selection.

3.20 What is meant by double shrinkage allowance?

3.21 Name the various patterns that are normally encountered in foundry practice.

3.22 Name any two allowances provided on the pattern for a sand casting and state the reasons why they are provided.

3.23 Explain where skeleton patterns are employed.

3.24 Which is the most widely used type of pattern? Describe its advantages.

3.25 What are the situations in which a single piece pattern is advantageously used?

3.26 Sketch an example showing the cope and the drag type pattern.

3.27 Why is a loose piece pattern used? Give its problems.

3.28 Explain with a neat sketch, the use of a follow board pattern.

3.29 Describe the pattern colour code normally used in foundries.

3.30 What properties are desirable of moulding sand from the stand point of sound castings?

3.31 Explain the importance of permeability of moulding sands.

3.32 State the essential ingredients of a moulding sand.

3.33 Explain the method of determining the moisture content in a moulding sand.

3.34 What is the role played by clay in a moulding sand? What is the method adopted for determining the clay content in a moulding sand?

3.35 What is meant by grain fineness number? Explain the procedure for determining this number for a moulding sand.

3.36 How is a standard sand test specimen prepared for testing the strength or permeability?

3.37 Describe the method used for determining the permeability of any moulding sand.

3.38 Show graphically how the green strength of sand varies with moisture and clay contents. Describe the behaviour.

3.39 State the detriment caused to the mould properties if the moulding sand contains
(a) Too much clay,
(b) Too much moisture.

3.40 "Natural sand is often not suitable for moulding purposes" – comment on this statement giving reasons.

3.41 Describe the method of sand preparation in a small batch foundry.

3.42 Explain how the shape and the size of sand grains affect the permeability and green strength of moulding sand.

3.43 How does the dry compression strength vary with the moisture content? Show graphically.

3.44 Explain the advantages derived by synthetic sands compared to natural sands.

3.45 Write a brief note on the moulding sands available in India.

3.46 What is the relevance of the facing sand towards the casting quality? Describe the facing sands normally used.

3.47 Write briefly about the mould washes and their methods of application.

3.48 Distinguish between green sand moulds and dry sand moulds from the point of view of
(a) Process
(b) Application.

3.49 What is meant by skin dried mould? Explain its applications.

3.50 Explain the desirable characteristics of any core in sand casting.

3.51 What are the normal binders used in core sand?

3.52 Why are organic binders preferred in core making?

3.53 What is meant by carbon dioxide moulding?

3.54 What are the specific advantages of using carbon dioxide in core making?

3.55 Explain the disadvantages of carbon dioxide moulding.

3.56 Differentiate between green sand and dry sand cores with reference to their applications.

3.57 Sketch different types of cores used in foundry practice.

3.58 How is the core boxes designed?

3.59 Describe a method of calculating the core print dimensions.

3.60 Briefly explain the application of chaplets.

3.61 Briefly explain the nobake moulding process.

3.62 Explain the advantages of the nobake moulding process.

3.63 Briefly explain the plaster moulding process.

3.64 Explain the advantages and disadvantages of the plaster moulding process.

3.65 Briefly explain the pit moulding.

3.66 Explain how additive manufacturing can be used for making sand moulds and cores.

3.67 Briefly explain loam moulding.

Problems

3.1 A cast iron cylinder of 450 mm outside diameter, 75 mm inside diameter and 150 mm long is to be obtained by sand casting. Design the requisite pattern assuming that the internal hole is to be finished by machining.

3.2 A master aluminium pattern is to be cast using a wooden pattern for a hollow steel cylindrical piece with a flange of width 50 mm and diameter 250 mm. The cylindrical portion is 200 mm long with an outer diameter 150 mm and an inner diameter of 75 mm. If the face of the flange is to be machined, design the wooden pattern.

3.3 Design the patterns for the following sand cast water pipes and specials of gray cast iron:
(a) DF pipe, Fig. 3.52
(b) DF (double flanged) 1/4 bend, Fig. 3.53
(c) All flanged tee, Fig. 3.54
In each of the above, choose the proper type of pattern and its parting line.

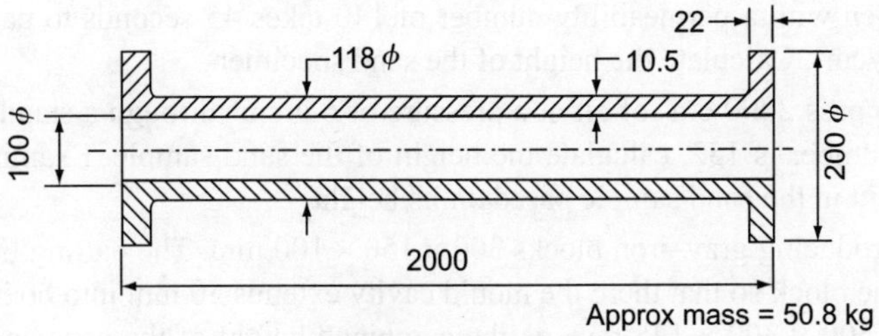

Fig. 3.52 *DF pipe*

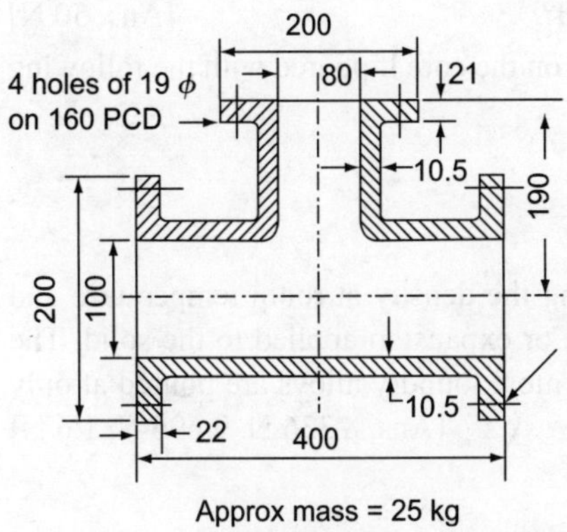

Fig. 3.53 *DF 1/4 bend*

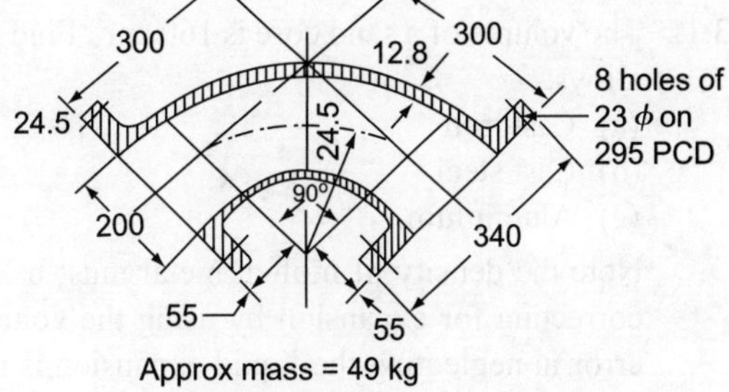

Fig. 3.54 *All flanged tee*

3.4 A sample of Londha (Maharashtra) sand of 50 g is weighed after washing and drying and found to weigh 41.6 g. What would be the clay percentage in this sand sample? [Ans. 16.8%]

3.5 The Bhavanagar (Gujarat) sand sample when sieved through the standard sieves was found to retain the following amounts of sand on the respective sieves.

Sieve no.	12	20	30	40	50	70	100	140	200	270	Pan
Retained weight, g	0.73	1.26	1.03	1.44	2.20	3.57	5.77	14.15	11.13	3.98	4.67

Plot the cumulative grading curve and calculate the grain fineness number. [Ans. 120]

3.6 The zircon sand obtained from Quilon (Kerala) has produced the following results on sieve analysis:

Sieve no.	70	100	140	200	270	Pan
Retained (%)	0.04	1.01	45.21	48.02	5.38	0.36

Calculate the grain fineness number and explain its uses. [Ans. 125]

3.7 The Madras sand has been found to contain the following size grading.

Sieve no.	12	20	30	40	50	70	100	140	200	270	Pan
Retained (%)	0.75	18.44	28.91	29.82	16.91	3.62	0.75	0.20	0.06	0.03	0.11

Calculate the grain fineness number. If the above sand has a fusion temperature of 1710°C, what are the applications in which this sand could be beneficially used? [Ans. 26]

3.8 Calculate the permeability number of a sand specimen if it takes 1 min 15 s to pass 2000 cm^3 of air at a pressure of 6 g/cm^2 through the standard sample. [Ans. 67]

3.9 A sand specimen with a permeability number of 140 takes 45 seconds to pass 2000 cm^3 of air at a pressure of 5 g/cm^2. Calculate the height of the sand specimen. [Ans. 5.32 cm]

3.10 It takes 35 s to pass 2000 cm^3 of air at a pressure of 6 g/cm^2 through a standard sand sample. If the permeability number is 152, calculate the height of the sand sample. Examine the adequacy of the moisture content in the sand sample based on its height. [Ans. 5.391 cm]

3.11 A foundry is producing gray–iron blocks 300 × 150 × 100 mm. The parting line is at the midpoint of the height of the block so that there the mould cavity extends 50 mm into both the cope and the drag. The flasks are 500 × 300 × 125 mm, so the combined height of the cope and drag is 250 mm. The mould is poured with molten gray–iron which weighs 0.0597 N/cm^3 and the compacted sand weighs 0.0157 N/cm^3. What would be the total lifting force in Newtons tending to separate the cope from the drag as a result of the metallostatic pressure within the mould? [Ans. 50 N]

3.12 The volume of a sand core is 160 cm^3. Find the buoyant force on the core if poured with the following alloys:

(a) Cast iron
(b) Cast steel
(c) Aluminium

Note the density of molten metal must be obtained by taking the density at room temperature and correcting for expansion by using the volumetric coefficient or expansion applied to the solid. The error in neglecting the liquid expansion is not large because most foundry alloys are poured at only 50 to 100 degrees superheat. [Ans. 8.336 N, 9.696 N, 1.6 N]

Multiple Choice Questions

3.1 Sand casting is preferred mainly because
(a) Good dimensional accuracy can be achieved
(b) Molten material flows into any small section in the mould cavity and as such any intricate shapes internal or external can be obtained
(c) Better surface finish is achieved
(d) Thin sections can be easily cast.

3.2 The lower moulding flask used in sand casting is called
(a) Cope (b) Cheek
(c) Drag (d) Core

3.3 The upper moulding flask used in sand casting is called
(a) Cope (b) Cheek
(c) Drag (d) Core

3.4 The intermediate moulding flask used in sand casting is called
(a) Cope (b) Cheek
(c) Drag (d) Core

3.5 Solid shrinkage in a metal during solidification is
(a) Reduction in volume when liquid metal temperature is reduced from pouring temperature to liquidus temperature
(b) Reduction in volume when solid metal temperature is reduced from solidus to room temperature
(c) Reduction in volume when solid metal temperature is reduced from liquidus to room temperature
(d) Reduction in volume when the metal changes from liquid to solid state at the solidus temperature

3.6 Liquid shrinkage in a metal during solidification is
(a) Reduction in volume when liquid metal temperature is reduced from pouring temperature to liquidus temperature
(b) Reduction in volume when solid metal temperature is reduced from solidus to room temperature

(c) Reduction in volume when solid metal temperature is reduced from liquidus to room temperature
(d) Reduction in volume when the metal changes from liquid to solid state at the solidus temperature

3.7 Shrinkage allowance is added to pattern dimensions to take care of
(a) Liquid shrinkage
(b) Liquid shrinkage or solid shrinkage
(c) Solid shrinkage
(d) Liquid shrinkage and solid shrinkage

3.8 In sand casting, shrinkage allowance is added to
(a) All linear dimensions
(b) Only the interior dimensions
(c) Only the exterior dimensions
(d) Only to the dimensions that are perpendicular to the parting plane

3.9 In sand casting, draft allowance is added to
(a) All linear dimensions
(b) Only the interior dimensions
(c) Only the exterior dimensions
(d) Only to the dimensions that are perpendicular to the parting plane

3.10 In sand casting, machining allowance is added to
(a) All linear dimensions
(b) Only the interior dimensions
(c) All the dimensions which have tolerances assigned
(d) Only to the dimensions that are perpendicular to the parting plane

3.11 The following pattern allowance can be specified quantitatively
(a) Machining allowance
(b) Shake allowance
(c) Draft allowance
(d) Shrinkage allowance

3.12 Aluminium is used as pattern material for making the master pattern because of
(a) High strength
(b) Low cost
(c) Low density
(d) None of the above

3.13 Main disadvantage of aluminium as a pattern material for making the master pattern is
(a) High strength
(b) Low cost
(c) Low density
(d) High cost

3.14 Main disadvantage of gray cast iron as a pattern material for making the master pattern is
(a) High density
(b) Good machinability
(c) Low density
(d) High cost

3.15 Gray cast iron is used as pattern material for making the master pattern because of
(a) High corrosion resistance
(b) Low cost
(c) Low density
(d) None of the above

3.16 Wood is used as pattern material for sand casting because
(a) Long life
(b) Easily shaped
(c) Less distortion due to moisture absorption
(d) Higher strength

3.17 The main disadvantage of wood is used as pattern material for sand casting is
(a) Long life
(b) Difficult to shape
(c) Distortion due to moisture absorption
(d) High density

3.18 Main disadvantage of brass as a pattern material for making the master pattern is
(a) High density
(b) Low surface finish
(c) Low strength
(d) Low corrosion resistance

3.19 Brass is used as pattern material for making the master pattern because of
(a) Low cost
(b) Low density
(c) High corrosion resistance
(d) None of the above

3.20 Permeability of green moulding sand indicates its
(a) Green strength
(b) Dry strength
(c) Gas evolution capability
(d) Moisture content

3.21 Permeability of a moulding sand is increased with
(a) Increasing the moisture content
(b) Coarse grain size
(c) Increasing the compacting pressure
(d) Fine grain size

3.22 Surface finish of a part that is sand cast increases by using
(a) Higher moisture content
(b) Coarse grain sand
(c) Fine grain sand
(d) Lower moisture content

3.23 Refractoriness of moulding sand is increased with
(a) Increasing the moisture content
(b) Coarse grain size
(c) Increasing the compacting pressure
(d) Fine grain size

3.24 Strength of moulding sand is increased with
(a) Increasing the moisture content
(b) Coarse grain size
(c) Decreasing the moisture content
(d) Fine grain size

3.25 Permeability of a moulding sand is increased with
(a) Increasing the clay content
(b) Decreasing the clay content
(c) Increasing the compacting pressure
(d) Fine grain size

3.26 The following is **not** an additive that is mixed with moulding sand
(a) Coal dust (b) Saw dust
(c) Clay (d) Iron oxide

3.27 To improve the surface finish of a sand casting, the following is applied to the mould cavity
(a) Coal dust (b) Facing sand
(c) Saw dust (d) Iron oxide

3.28 Fluidity of molten metal is increased by
(a) Increasing the moisture content in moulding sand
(b) Increasing metal temperature
(c) Decreasing the clay content in moulding sand
(d) None of the above

3.29 Problem with green sand moulds is
(a) The process is more expensive
(b) The process produces sound castings without any defects
(c) Erosion of the mould takes place
(d) None of the above

3.30 Problem with dry sand moulds is
(a) The process is more expensive
(b) The process produces castings with defects
(c) Erosion of the mould takes place
(d) None of the above

3.31 Problem with skin dried sand moulds is
(a) Mould distortion takes place
(b) The process produces sound castings without any defects
(c) Erosion of the mould takes place
(d) None of the above

3.32 In a jolt moulding machine, the sand in the moulding flask is compacted by
(a) A plate that rams the sand thoroughly
(b) Impact force by dropping the flask on the machine table
(c) Combining the above two (a) and (b)
(d) Throwing sand into the flask with high force

3.33 In a squeeze moulding machine, the sand in the moulding flask is compacted by
(a) A plate that rams the sand thoroughly
(b) Impact force by dropping the flask on the machine table
(c) Combining the above two (a) and (b)
(d) Throwing sand into the flask with high force

3.34 In a jolt squeeze moulding machine, the sand in the moulding flask is compacted by
(a) A plate that rams the sand thoroughly
(b) Impact force by dropping the flask on the machine table
(c) Combining the above two (a) and (b)
(d) Throwing sand into the flask with high force

3.35 In a sand slinging moulding machine, the sand in the moulding flask is compacted by
(a) A plate that rams the sand thoroughly
(b) Impact force by dropping the flask on the machine table
(c) Combining the above two (a) and (b)
(d) Throwing sand into the flask with high force

3.36 Desirable property for a core in a sand mould is
(a) Good green strength
(b) Good collapsibility
(c) Good friability
(d) All of the above

3.37 The binder that is more commonly used in core sands is
(a) Clay (b) Bentonite
(c) Linseed oil (d) Starch

3.38 The binder used in the carbon dioxide core making is
(a) Clay (b) Sodium silicate
(c) Linseed oil (d) Sodium nitrate

3.39 A chaplet is used in a sand mould to
(a) Support the core during the molten metal pouring
(b) Support the mould from breaking
(c) Support the mould from moving mould walls
(d) Prevent the formation of cold shuts

3.40 Metallostatic force in the sand mould is caused by
(a) Weight of the moulding sand in the cope
(b) Buoyant force of the molten metal in the mould cavity
(c) Weight of the moulding sand in the drag
(d) Combined weight of the moulding sand in the cope and drag

Answers to MCQs

3.1 (b)	3.2 (c)	3.3 (a)	3.4 (b)	3.5 (b)
3.6 (d)	3.7 (c)	3.8 (a)	3.9 (d)	3.10 (c)
3.11 (b)	3.12 (c)	3.13 (d)	3.14 (a)	3.15 (b)
3.16 (b)	3.17 (c)	3.18 (a)	3.19 (c)	3.20 (c)
3.21 (b)	3.22 (c)	3.23 (b)	3.24 (d)	3.25 (b)
3.26 (c)	3.27 (b)	3.28 (b)	3.29 (c)	3.30 (a)
3.31 (d)	3.32 (b)	3.33 (a)	3.34 (c)	3.35 (d)
3.36 (d)	3.37 (c)	3.38 (b)	3.39 (a)	3.40 (b)

Gating Systems for Casting

CHAPTER 4

Objectives

After completing this chapter, the reader will be able to

- Understand the importance of various elements present in a gating system
- Design gating system for different castings
- Appreciate the risering requirements and design them for different castings

4.1 INTRODUCTION

As defined earlier, gating systems refer to all those elements, which are connected with the flow of molten metal from the ladle to the mould cavity. The various elements that are connected with a gating system (Fig. 4.1) are:

- Pouring basin
- Sprue
- Sprue-base well
- Runner
- Runner extension
- In-gate
- Riser

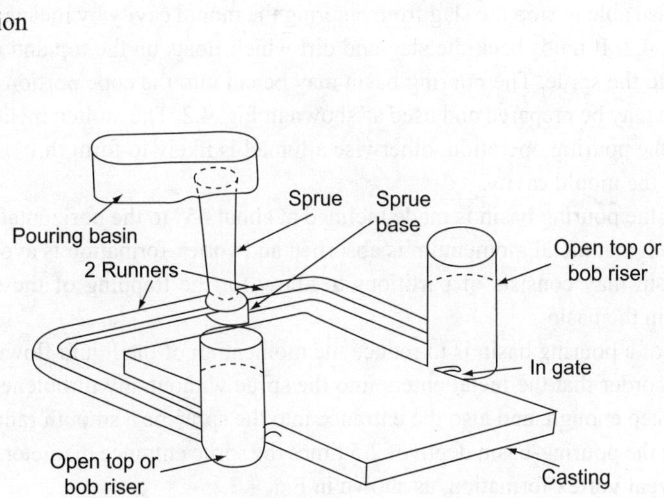

Fig. 4.1 *Typical gating system*

In this chapter, the functions and the design of the various elements of a gating system will be discussed. Any gating system designed should aim at providing a defect-free casting. This can be achieved by making provision for certain requirements while designing the gating system. These are as follows:

1. The mould should be completely filled in the smallest time possible without having to raise the metal temperatures or use higher metal heads.
2. The metal should flow smoothly into the mould without any turbulence. A turbulent metal flow tends to form dross in the mould.
3. Unwanted material such as slag, dross and other mould material should not be allowed to enter the mould cavity.
4. The metal entry into the mould cavity should be properly controlled in such a way that aspiration of the atmospheric air is prevented.
5. A proper thermal gradient be maintained so that the casting is cooled without any shrinkage cavities or distortions.
6. Metal flow should be maintained in such a way that no gating or mould erosion takes place.
7. The gating system should ensure that enough molten metal reaches the mould cavity.
8. The gating system design should be economical and easy to implement and remove after casting solidification.
9. Ultimately, the casting yield should be maximised.

To have all these requirements together is a tall order, still mould designer should strive to achieve as many of the above objectives as possible. Before going into the mechanics of gating design, let us describe some of the functions and types of the various gating-system elements.

4.1.1 Elements of a Gating System

Pouring Basin

The molten metal is not directly poured into the mould cavity because it may cause mould erosion. Molten metal is poured into a pouring basin, which acts as a reservoir from which it moves smoothly into the sprue. The pouring basin is also able to stop the slag from entering the mould cavity by means of a skimmer or skim core, as shown in Fig. 4.2. It holds back the slag and dirt which floats on the top and only allows the clean metal underneath it into the sprue. The pouring basin may be cut into the cope portion directly or a separate dry sand pouring basin may be prepared and used as shown in Fig. 4.2. The molten metal in the pouring basin should be full during the pouring operation, otherwise a funnel is likely to form through which atmospheric air and slag may enter the mould cavity.

One of the walls of the pouring basin is made inclined at about 45° to the horizontal. The molten metal is poured on this face such that metal momentum is absorbed and vortex formation is avoided. In some special cases, the pouring basin may consists of partitions to allow for the trapping of the slag and maintaining constant metal height in the basin.

The main function of a pouring basin is to reduce the momentum of the liquid flowing into the mould by settling first into it. In order that the metal enters into the sprue without any turbulence, it is necessary that the pouring basin be deep enough, and also the entrance into the sprue be a smooth radius of at least 25 mm. Experience shows that the pouring-basin depth of 2.5 times the sprue entrance diameter is enough for smooth metal flow and to prevent vortex formation, as shown in Fig. 4.3.

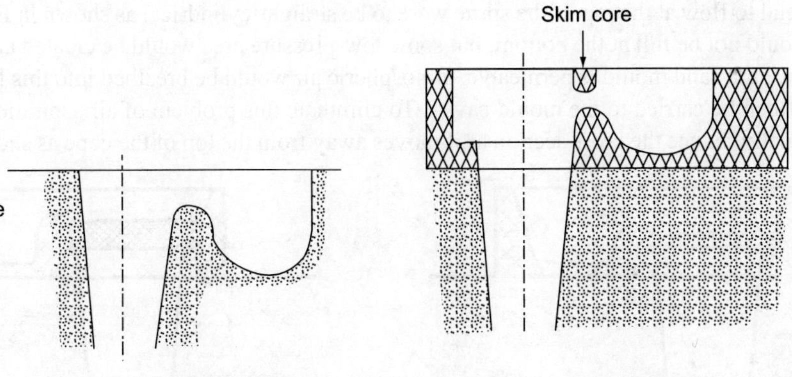

FIG. 4.2 *Pouring basin*

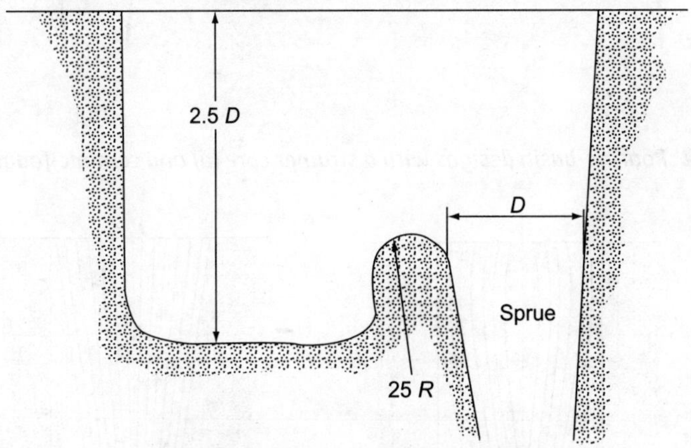

FIG. 4.3 *Pouring-basin proportions*

In order that a vortex is not formed during pouring, it is necessary that the pouring basin be kept full. Further provision should be made in the pouring basin so that constant conditions of flow are established. This can be achieved by using a strainer core, as in Fig. 4.4(a). A strainer core is a ceramic-coated screen with many small holes. The strainer restricts the flow of metal into the sprue and thus helps in quick filling of the pouring basin. It helps in providing a smooth flow while restricting the flow of slag into the mould. A ceramic filter as shown in Fig. 4.4(b) completely stops slag allowing only clean metal to go into the sprue, and also ensures a constant flow of metal.

The metal should be poured steadily into the pouring basin keeping the lip of the ladle as close as possible. Pouring basins are most desirable with castings in alloys which form troublesome oxide skins (aluminium, aluminium bronze, etc.).

Sprue

Sprue is the channel through which the molten metal is brought into the parting plane, where it enters the runners and gates to ultimately reach the mould cavity. The molten metal when moving from top of the cope to the parting plane gains in velocity, and as a consequence, requires a smaller area of cross section for the same amount of metal to flow at the top. If the sprue were to be straight-cylindrical as shown in Fig. 4.5(a) then the metal flow would not be full at the bottom, but some low-pressure area would be created around the metal in the sprue. Since the sand mould is permeable, atmospheric air would be breathed into this low-pressure area, which would then be carried to the mould cavity. To eliminate this problem of air aspiration, the sprue is tapered to gradually reduce the cross section as it moves away from the top of the cope as shown in Fig. 4.5(b).

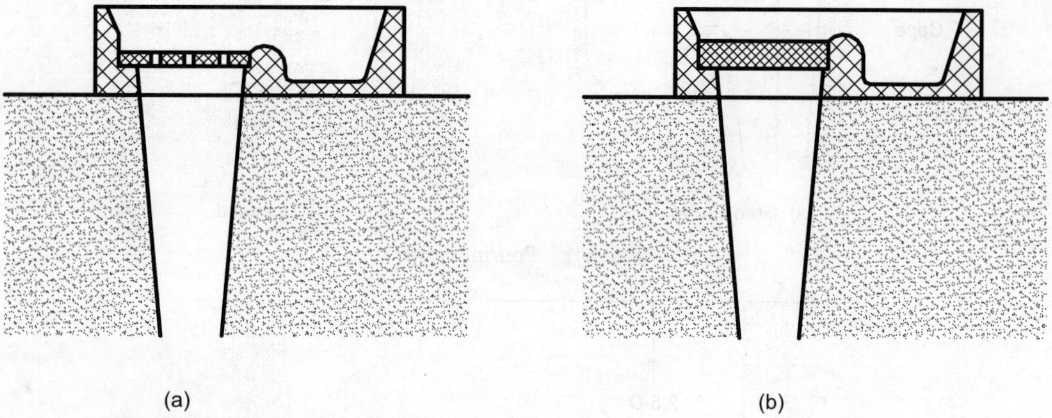

FIG. 4.4 *Pouring-basin designs with a strainer core (a) and ceramic-foam filter (b)*

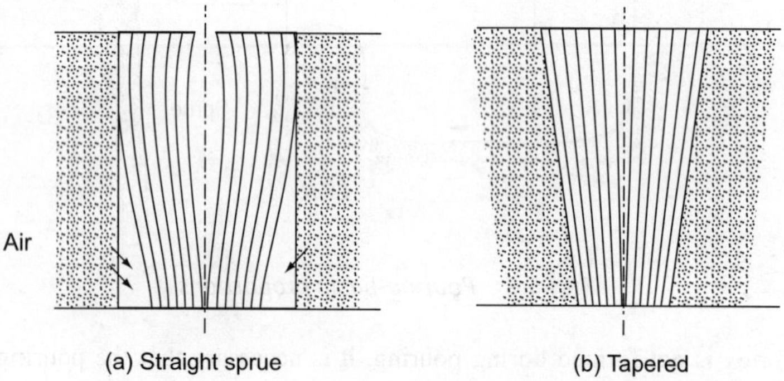

FIG. 4.5 *Sprue designs: (a) Straight (b) Tapered*

The exact tapering can be obtained by the equation of continuity. Denoting the top and choke sections of the sprue by the subscripts *t* and *c* respectively, we get

$$A_t V_t = A_c V_c$$

or
$$A_t = A_c \frac{V_c}{V_t}$$

Since the velocities are proportional to the square of the potential heads, it can be derived from Bernoulli's equation,

$$A_t = A_c \sqrt{\frac{h_c}{h_t}}$$

The square root suggests that the profile of the sprue should be parabolic if exactly done as per the above equation. But making a parabolic sprue is too inconvenient in practice and, therefore, a straight taper is preferable. It has been found in practice that a straight tapered sprue is able to effectively reduce the air aspiration, as well as increase the flow rate, compared to a cylindrical sprue.

In order to arrive at the dimensions of the sprue at the top and the subsequent taper, one has to consider the head of the metal in the pouring basin as shown in Fig. 4.6. Metal at the entry of the sprue would be moving with a velocity of $\sqrt{2gh}$ and, hence,

$$A_t = \sqrt{\frac{h_t}{h}}$$

where H = actual sprue height

and $h_t = h + H$

Table 4.1 shows the theoretical values of area ratios of top and choke portions of the sprue based on sprue height and metal head in the pouring basin.

Though these ratios are theoretically correct, often it is not possible to control exactly the amount of the head in the pouring basin during the pouring. Hence, it is a general practice to neglect the effect of the pouring-basin head, and proportion the sprue top solely based on the sprue height alone.

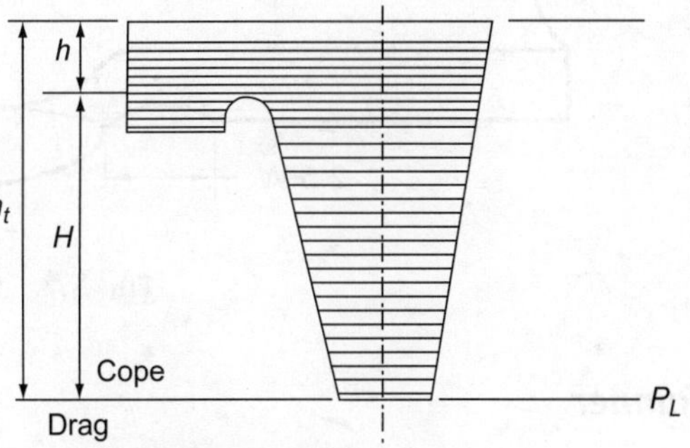

Fig. 4.6 *Sprue proportions*

Table 4.1 **Theoretical ratios of sprue top and choke areas based on pouring basin depth**

Sprue Height, (mm)	Depth in Pouring Basin, (mm)				
	50	100	150	200	250
50	1.414	1.225	1.155	1.118	1.095
100	1.732	1.414	1.291	1.225	1.183
150	2.000	1.581	1.414	1.323	1.265
200	2.236	1.732	1.528	1.414	1.342
250	2.450	1.871	1.633	1.500	1.414
375	2.915	2.179	1.871	1.696	1.581
500	3.317	2.450	2.082	1.871	1.732
600	3.742	2.739	2.309	2.062	1.897

Spruce Base Well

This is a reservoir for metal at the bottom of the sprue, to reduce the momentum of the molten metal. The molten metal as it moves down the sprue gains in velocity, some of which is lost in the sprue-base well by which the mould erosion is reduced. This molten metal then changes direction and flows into the runners in a more uniform way.

Reasonable proportions for a sprue-base well are presented in Fig. 4.7. A general guideline could be that the sprue-base-well area should be five times that of the sprue choke area, and the well depth should be approximately equal to that of the runner. For a narrow and deep runner, the well diameter should be 2.5 times the width of the runner in a two-runner system, and twice its width in a one-runner system.

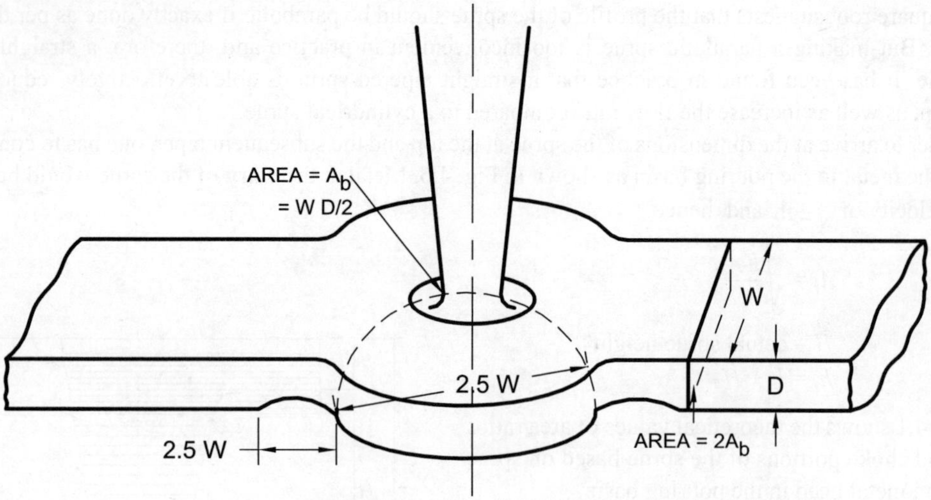

Fig. 4.7 *Sprue-base-well design*

Runner

It is generally located in the horizontal plane (parting plane), which connects the sprue to its in-gates, thus allowing the metal to enter the mould cavity. The runners are normally made trapezoidal in cross section. It is a general practice for ferrous metals to cut the runners in the cope and the in-gates in the drag. The main reason for this is to trap the slag and dross, which are lighter and thus trapped in the upper portion of the runners. For effective trapping of the slag, runners should flow full as shown in Fig. 4.8(a). When the amount of molten metal coming from the down sprue is more than the amount flowing through the in-gates, the runner would always be full and thus slag-trapping would take place. But when the metal flowing through the in-gates is more than that flowing through the runners then the runner would be filled only partially as shown in Fig. 4.8(b), and the slag would then enter the mould cavity.

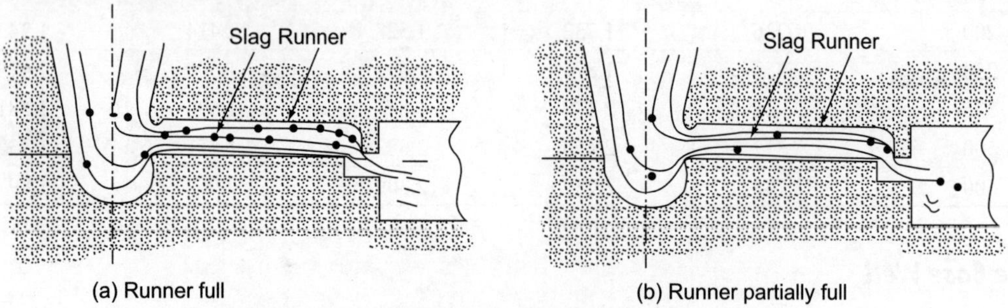

Fig. 4.8 *Runner: (a) Full runner (b) Runner partially full*

Runner Extension

The runner is extended a little further after it encounters the in-gate. This extension is provided to trap the slag in the molten metal. The metal initially comes along with the slag, floating at the top of the ladle, and flows straight going beyond the in-gate, and then trapped in the runner extension.

4.1.2 Gates or In-gates

These are the openings through which the molten metal enters the mould cavity. The shape and the cross section of the in-gate should be such that it can readily be broken off after casting solidification, and also that it allows the metal to enter quietly into the mould cavity.

Depending on the application, various types of gates are used in the casting design. They are the following:

Top Gate

This is the type of gating through which the molten metal enters the mould cavity from the top as shown in Fig. 4.9. Since the first metal entering the gate reaches the bottom and hotter metal is at the top, a favourable temperature gradient towards the gate is achieved. Also, the mould is filled very quickly. But as the metal falls directly into the mould cavity through a height, it is likely to cause mould erosion. Also, because it causes turbulence in the mould cavity, it is prone to form dross and as such the top gate is not advisable for those materials, which are likely to form excessive dross. It is not suggested for nonferrous materials and is suggested only for ferrous alloys. It is suitable only for simple casting shapes, which are essentially shallow in nature. To reduce the mould erosion, pencil gates Fig. 4.9(a) are provided in the pouring cup. This type of gate requires minimum of additional runners to lead the liquid metal into the cavity, and as such provides higher casting yield. Capabilities of top-gating system are significantly extended by using direct pour-top gating system, incorporating insulating sleeve with a ceramic foam filter as shown in Fig. 4.9(b).

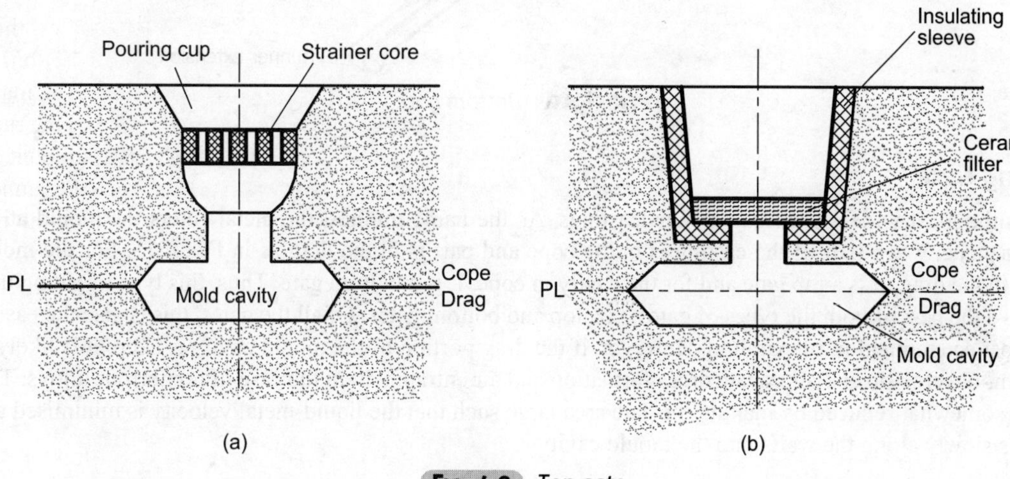

Fig. 4.9 *Top gate*

Bottom Gate

When molten metal enters the mould cavity slowly as shown in Fig. 4.10, it would not cause any mould erosion. The bottom gate is generally used for very deep moulds. It takes somewhat higher time for the filling of the mould. These gates may cause unfavourable temperature gradients compared to the top gating. Thus, the system may have to use additional padding of sections towards risers, and large riser sizes to compensate for the unfavourable temperature distribution. Bottom gating may sometimes be preferable in conjunction with the use of side risers, since the metal enters the riser directly without going through the mould cavity.

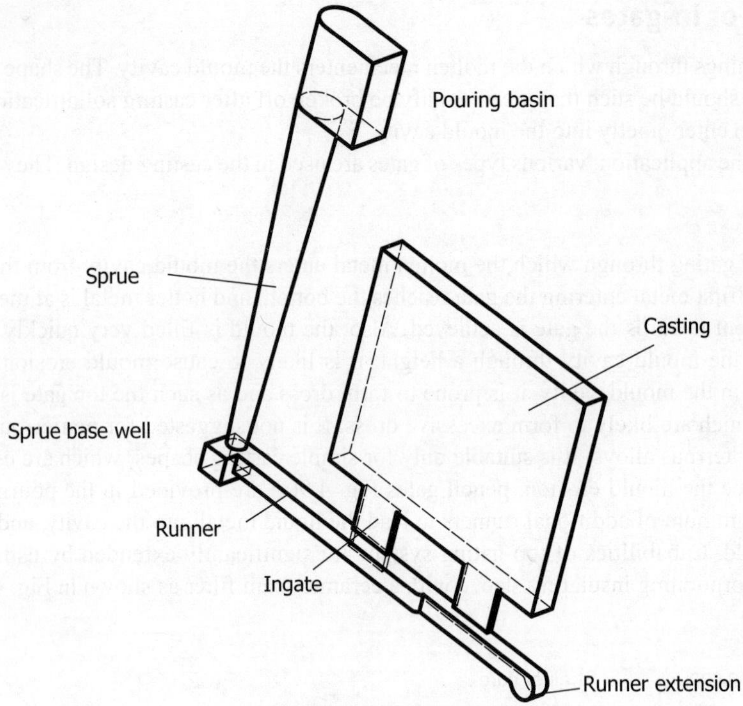

FIG. 4.10 *Bottom gate*

Parting Gate

This is the most widely used gate in sandcastings. As the name implies, the metal enters the mould at the parting plane, when part of the casting is in the cope and part is in the drag as in Fig. 4.11. For the mould cavity in the drag, it is a top gate and for the cavity in cope, it is a bottom gate. Thus, this type of gating tries to derive the best of both the types of gates, viz. top and bottom gates. Of all the gates, this is also the easiest and most economical in preparation. However, if the drag portion of the mould cavity is deep, it is likely to cause mould erosion and aggravate dross formation and air entrapment in the case of nonferrous alloys. This can be somewhat reduced by making the gate area large such that the liquid-metal velocity is minimised and it flows slowly along the walls into the mould cavity.

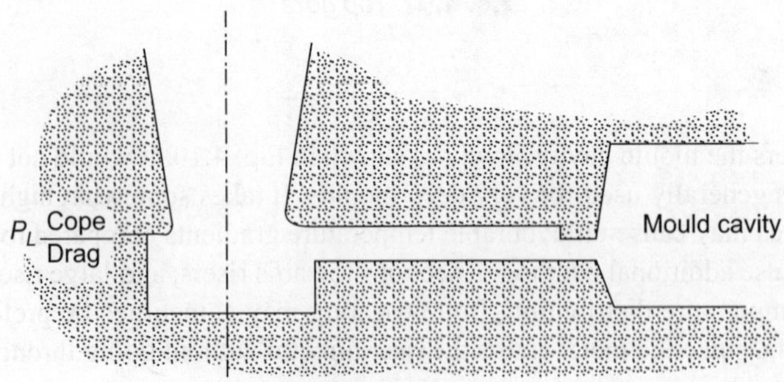

FIG. 4.11 *Parting gate*

Step Gate

Such gates are used for heavy and large castings. The molten metal enters (Fig. 4.12) mould cavity through a number of in-gates, which are arranged in vertical steps. The size of in-gates are normally increased from top to bottom such that metal enters the mould cavity from the bottom-most gate and then progressively moves to the higher gates. This ensures a gradual filling of the mould without any mould erosion and produces a sound casting.

In designing a casting, it is essential to choose a suitable gate, considering the casting material, casting shape and size so as to produce a sound casting.

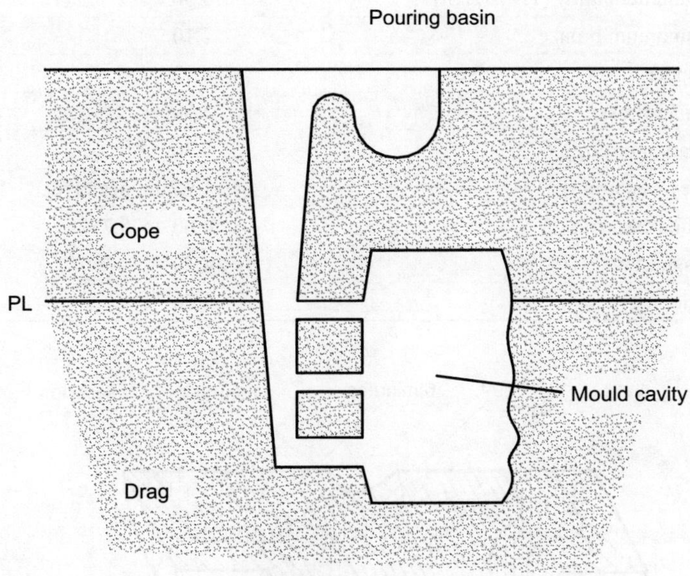

Fig. 4.12 *Step gate*

Riser

Most foundry alloys shrink during solidification. Table 4.2 shows the various volumetric shrinkages for typical materials. As a result of this volumetric shrinkage during solidification, voids are likely to form in the castings, as shown in Fig. 4.13, unless additional molten metal is fed into these places which are termed **hot spots**, since they remain hot till the end. Hence, a reservoir of molten metal is to be maintained from which the metal can flow readily into the casting when the need arises. These reservoirs are called risers.

As shown in Table 4.2, different materials have different shrinkages and hence the risering requirements vary for the materials. In grey cast iron, because of graphitisation during solidification, there may be an increase in volume sometimes. This of course, depends on the degree of graphitisation in grey cast iron, which is controlled by the silicon content.

In order to make them effective, the risers should be designed keeping the following in mind.

- The metal in the riser should solidify in the end.
- The riser volume should be sufficient for compensating the shrinkage in the casting.

TABLE 4.2 Volumetric liquid shrinkages

Material	Shrinkage, (%)
Medium-carbon steel	2.50 to 3.50
High-carbon steel	4.00
Nickel	6.10
Monel	6.30
Aluminium	6.60
Aluminium alloy (11–13% Si)	3.50
Aluminium bronze	4.10
Copper	4.92
70–30 Brass	4.50
Bearing bronze	7.30
Grey cast iron	1.90 to negative
White cast iron	4.00 to 5.75
Magnesium	4.20
Zinc	6.50

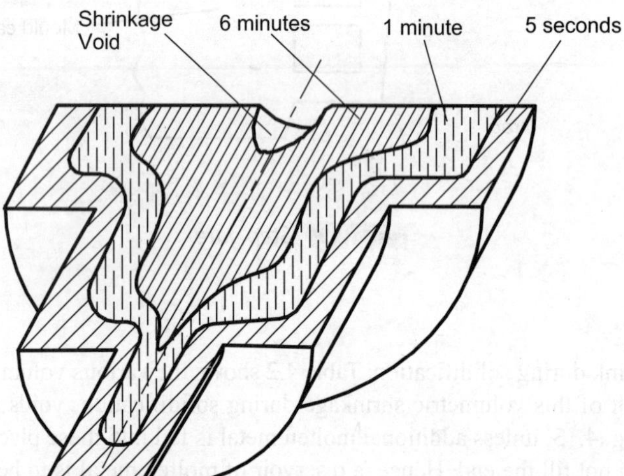

FIG. 4.13 *Shrinkage cavity formation*

In order to satisfy the above requirements, risers of large diameters are generally used. But it proves to be a very expensive solution since the solidified metal in the riser is to be cut off from the main casting and is to be melted for reuse. Higher the riser volume, lower is the casting yield and as such it is very uneconomical.

The risers are normally of the following types: top risers which are open to the atmosphere; blind risers which are completely concealed inside the mould cavity itself and internal risers which are enclosed on all sides by the casting.

The top riser is the most conventional and convenient to make. But the position where it can be placed is limited. The top being open loses heat to the atmosphere by radiation and convection. To reduce this, often insulation is provided on the top such as plaster of paris, asbestos sheet, etc. The blind riser, since it is surrounded by the moulding sand, would lose heat slowly and thus would be more effective. Also, it can be located more conveniently than an open riser. The best is the internal riser, which is surrounded on all sides by the casting such that heat from the casting keeps the metal in the riser hot for a longer time. These are normally used for castings, which are cylindrically shaped or have a hollow cylindrical portion.

Chill

In a casting, metallic chills are used in order to provide progressive solidification or to avoid the shrinkage cavities. Chills are essentially large heat sinks. Whenever it is not possible to provide a riser for a part of the casting which is heavy, a chill is placed close to it as shown in Fig. 4.14, so that more heat is quickly absorbed by the chill from the larger mass, making the cooling rate equal to that of the thin sections. Thus, this does not permit the formation of a shrinkage cavity. But use of a chill essentially means providing higher cooling rate, which is also likely to form a hard spot at the contact area with the chill, and may, therefore, cause a problem if that areas need further processing by way of machining.

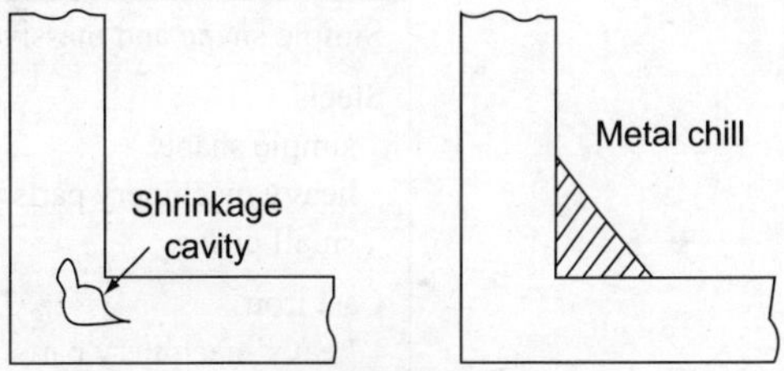

Fig. 4.14 *Chill*

4.1.3 Casting Yield

All the metal that is used while pouring is not finally ending up as a casting. Typical routes the metal would take in a foundry are shown in Fig. 4.15. There will be some losses in the melting. Also, there is a possibility that some castings may be rejected because of the presence of various defects. On completion of the casting process, the gating system used is removed from the solidified casting and re-melted to be used again as raw material. Hence, the casting yield is the proportion of the actual casting mass, W, to the mass of metal poured into the mould, w, expressed as a percentage as follows.

$$\text{Casting yield} = \frac{W}{w} \times 100$$

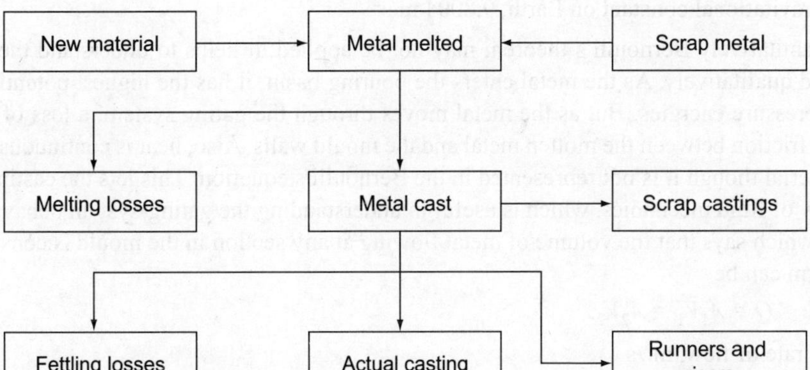

Fig. 4.15 *Utilization of the metal in the foundry*

Higher the casting yield, higher is the economics of the foundry practice. It is therefore, desirable to give consideration to maximising the casting yield, at the design stage itself.

Casting yield depends to a great extent on the casting materials and the complexity of the shape. Generally, those materials which shrink heavily have lower casting yields. Also massive and simple shapes have higher casting yield compared to small and complex parts. Typical casting yields are presented in Table 4.3 as a guide.

TABLE 4.3 Casting yields

Casting Description	Yield Range, %
Simple shape and massive	85 to 95
Steels	
simple shape	75 to 85
heavy machinery parts	55 to 65
small pieces	35 to 45
Cast iron	
heavy machinery parts	65 to 75
small pieces	45 to 55
Aluminium	25 to 45

4.2 GATING-SYSTEM DESIGN

The liquid metal that runs through the various channels in the mould obeys Bernoulli's theorem, which states that the total energy head remains constant at any section. The same stated in the equation form ignoring frictional losses is,

$$h + \frac{P}{w} + \frac{V^2}{2g} = \text{constant}$$

where h = potential head, m
P = pressure, Pa
V = liquid velocity, m/s
w = specific weight of liquid, N/m^3
g = gravitational constant on Earth, 9.8004 m/s^2

Though quantitatively Bernoulli's theorem may not be applied, it helps to understand the metal flow in the sand mould qualitatively. As the metal enters the pouring basin, it has the highest potential energy with no kinetic or pressure energies. But as the metal moves through the gating system, a loss of energy occurs because of the friction between the molten metal and the mould walls. Also, heat is continuously lost through the mould material though it is not represented in the Bernoulli's equation. This lets the casting solidify.

Another law of fluid mechanics, which is useful in understanding the gating-system behaviour, is the law of continuity, which says that the volume of metal flowing at any section in the mould is constant. The same in equation form can be

$$Q = A_1 V_1 = A_2 V_2$$

where Q = rate of flow, m^3/s
A = area of cross section, m^2
V = velocity of metal flow, m/s

It was suggested earlier that sprues are tapered to reduce the aspiration of air due to the increased velocity as the metal flows down the sprue. This conclusion was drawn by applying the above equation of continuity along with Bernoulli's equation.

4.2.1 Pouring Time

One of the objectives as mentioned earlier, for the gating-system design is to fill the mould in the smallest time. The time for complete filling of a mould termed pouring time, is a very important criterion for design. Too long a pouring time requires a higher pouring temperature and too less a pouring time means turbulent flow in the mould, which makes the casting defect-prone. There is thus an optimum pouring time for any given casting.

The pouring time depends on the casting materials, complexity of the casting, section thickness and casting size. The various relations used are not theoretically obtained but established generally by the practice at various foundries and by experimenters. The general considerations for choosing pouring time for grey cast iron may not be much relevant for steels since they lose heat very fast and, therefore, the pouring time should be very less. For nonferrous materials, a longer pouring time would be beneficial since they lose heat slowly and also tend to form dross if metal is poured too quickly.

Since the thickness of casting is effected to a great extent, by the ratio of surface area to volume of the casting, it is an important variable in calculating the optimum pouring time in addition to the mass of the casting itself. Normally, while considering the mass of the casting, it may not be necessary to consider the mass of the gating system because the gating system is completely filled before metal starts entering the mould cavity. However, if the gating systems are in comparable size to the actual casting, it may be desirable to include them for the calculation.

The following are some standard methods to calculate the pouring time for different casting materials.

Grey Cast Iron, Mass less than 450 kg

Pouring time, $$t = K\left(1.41 + \frac{T}{14.59}\right)\sqrt{W}\ \text{s}$$

where $$K = \frac{\text{Fluidity of iron in inches}}{40},$$

T = average section thickness, mm

W = mass of the casting, kg

Grey Cast Iron, Mass greater than 450 kg

Pouring time, $$t = K\left(1.236 + \frac{T}{16.65}\right)\sqrt[3]{W}\ \text{s}$$

Typical pouring times for cast iron are

Casting mass	Pouring time in seconds
20 kg	6 to 10
100 kg	15 to 30
100 000 kg	60 to 180

Steel Castings

Pouring time, $t = (2.4335 - 0.3953 \log W)\sqrt{W}$ s

Shell-moulded Ductile Iron (vertical pouring)

Pouring time, $t = K_1\sqrt{W}$ s

where $K_1 = 2.080$ for thinner sections
$= 2.670$ for sections 10 to 25 mm thick
$= 2.970$ for heavier sections

Copper Alloy Castings

Pouring time, $t = K_2\sqrt[3]{W}$ s

K_2 is a constant given by

Top gating	1.30
Bottom gating	1.80
Brass	1.90
Tin bronze	2.80
Russian practice	

Intricately Shaped Thin-walled Castings of Mass up to 450 kg

Pouring time, $t = K_3\sqrt[3]{W'}$ s

where W' = mass of the casting with gates and risers, kg and

K_3 = a constant as given below

T (mm)	K_3
1.5 to 2.5	1.62
2.5 to 3.5	1.68
3.5 to 8.0	1.85
8.0 to 15.0	2.20

For Castings above 450 kg and up to 1000 kg

Pouring time, $t = K_4\sqrt[3]{W'\,T}$ s

where K_4 is a constant given by

T (mm)	K_4
up to 10	1.00
10 to 20	1.35
20 to 40	1.50
above 40	1.70

Typical pouring times for castings whose mass is less than 200 kg and average section thickness of 25 mm are

Grey cast iron	40 s
Steel	20 s
Brass	15 to 45 s

Typical pouring rates used for different casting alloys are given in Table 4.4. The calculated values from any of the given formulae may be checked from Table 4.4.

TABLE 4.4 Typical pouring rates for sandcastings

Metal	Pouring Rate in kg/s for Castings of Mass, kg			
	upto 10	10–50	50–100	100–500
Pig iron	1.1	1.5–2.0	3.0–4.0	3.5–6.0
Steel	1.2–1.4	1.9–2.5	4.0–5.0	4.5–7.0
Aluminium alloys	0.25–0.3	0.5–0.7	1.0–1.3	1.2–2.0

Example 4.1

Calculate the optimum pouring time for a casting whose mass is 20 kg and having an average section thickness of 15 mm. The materials of the casting are grey cast iron and steel. Take the fluidity of iron as 28 inches.

Grey cast iron

Pouring time, $t = K\left(1.41 + \frac{T}{14.59}\right)\sqrt{W}$ s

$$= \frac{28}{40}\left(1.41 + \frac{15}{14.59}\right)\sqrt{20} = 7.632 \text{ s}$$

Steel

Pouring time, $t = (2.4335 - 0.3953 \log W)\sqrt{W}$ s

$$= (2.4335 - 0.3953 \log 20)\sqrt{20} = 8.5825 \text{ s}$$

Example 4.2

Calculate the optimum pouring time for a casting whose mass is 100 kg and a thickness of 25 mm. Fluidity of iron is 32 inches. Calculate both for cast iron and steel.

Grey cast iron

Pouring time, $t = \frac{32}{40}\left(1.41 + \frac{25}{14.59}\right)\sqrt{100} = 24.988$ s

Steel

Pouring time, $t = (2.4335 - 0.3953 \log 100)\sqrt{100} = 16.429$ s

4.2.2 Choke Area

Having calculated the optimum pouring time, it is now required to establish the main control area which meters the metal flow into the mould cavity, so that the mould is completely filled within the calculated pouring time. This controlling area is called choke area. Normally, the choke area happens to be at the bottom of the sprue and hence, the first element to be designed in the gating system is the sprue size and its proportions. The main advantage in having sprue bottom as the choke area is that proper flow characteristics are established early in the mould.

The choke area can be calculated using Bernoulli's equation as

$$A = \frac{W}{dtC\sqrt{2gH}}$$

where A = choke area, mm^2
W = casting mass, kg
t = pouring time, s
d = mass density of the molten metal, kg/mm^3
g = acceleration due to gravity, mm/s^2
H = effective metal head (sprue height), mm
C = efficiency factor which is a function of the gating system used

The effective sprue height ***H***, of a mould depends on the casting dimensions and the type of gating used. The effective sprue heads can be calculated using the following relations.

Top gate, $H = h$

Bottom gate, $H = h - \frac{c}{2}$

Parting gate, $H = h - \frac{p^2}{2c}$

where h = height of sprue,
p = height of mould cavity in cope, and
c = total height of mould cavity.

The values of h, p and c are shown in Fig. 4.16 for the various types of gating.

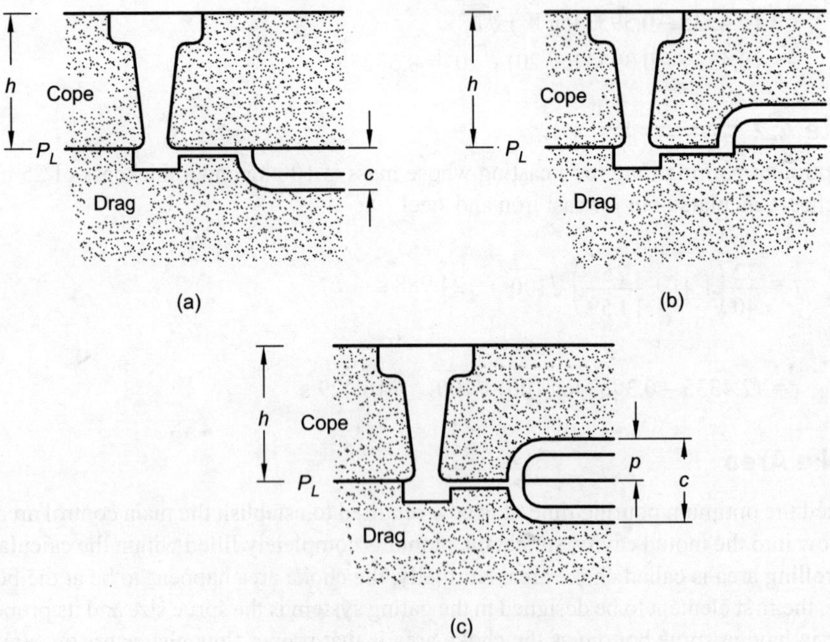

Fig. 4.16 *Different gating systems*

The efficiency coefficient of the gating system depends on the various sections that are normally used in a gating system. The elements of a gating system should normally be circular in cross section since they have lower surface area to volume ratio, which would reduce heat loss and have less friction. Moreover, streamlining the various gating elements would greatly increase the volumetric efficiency of the gating system and allow for smaller size gates and runners which would increase the casting yield.

Whenever a runner changes direction or joins with another runner or gate, there is some loss in the metal head, all of which when taken properly into consideration would give the overall efficiency of the gating system. Hence,

$$C = \frac{1}{\sqrt{1 + K_1 \dfrac{A^2}{A_1^2} + K_2 \dfrac{A^2}{A_2^2} + \ldots}}$$

where $K_1, K_2, \ldots$ are loss coefficients occurring at changes in direction or area as given in Table 11.2
$A_1, A_2,$ are areas down stream from changes
A is the choke area

Though this is the most rigorous way of calculating the efficiency factor, it may not be necessary to go to this length all the time. Average values of the efficiency factor are provided for typical gating systems in Table 4.5, which may be used for calculating the gating.

TABLE 4.5 Values of loss coefficients for various gate elements

Gate Element	Sharp	Round
Sprue entry from pouring cup	0.75	0.20
Bend of sprue into runner	2.00	1.00
Right angle bend in runner:		
square cross section	2.00	1.50
round cross section	1.50	1.00
Junction at right angles to runners	4.0 to 6.0	—
Junction with 25 % or more area reduction from runner into in-gates		0.50
Runner choke when choke area approximately one-third runner area, plus bend of sprue into runner	2.00 13.00	—
Losses from wall friction Round channel loss = $0.02\ L/D$ Square channel loss = $0.06\ L/D$ Rectangular channel loss = $\dfrac{0.07\ L\ (A+B)}{2AB}$ where L = length, D = diameter of round or side of square, A = one side of rectangle, and B = other side of rectangle.		

Though it is preferable to have the choke in the sprue, it may sometimes be convenient to mould a straight sprue in which case the choke is provided in the runners. The efficiency factors for such systems are also provided in Table 4.6.

For aluminium castings, a metal flow rate of 4.044 g/min for 1 sq mm of sprue area in an unpressurised gating is found to be sufficient for achieving sound castings.

TABLE 4.6 Efficiency coefficients, C for various types of gating systems

Type of System	Tapered Choked Sprue	Straight Sprue Runner Choke
Single runner entering runner	0.90	0.73
Two runners with multiple in-gates, no bends in runners	0.90	0.73
Two runners with multiple in-gates, 90o bends in runners	0.85	0.70

Example 4.3

For the casting shown in Fig. 4.17, which is to be made in cast iron, calculate the choke area.

Volume of the casting = $500 \times 250 \times 50 = 6.25 \times 10^6$ mm^3

Weight of the casting = $7.86 \times 10^{-6} \times 6.25 \times 10^6 = 49.125$ kg

Assuming a composition factor of 4.0 and a pouring temperature of 1300°C, the fluidity, obtained from Fig. 3.32, is 22 inches.

$$\text{Pouring time, } t = \frac{22}{40}\left(1.41 + \frac{50}{14.59}\right)\sqrt{49.125} = 18.646 \text{ s} \approx 19 \text{ s}$$

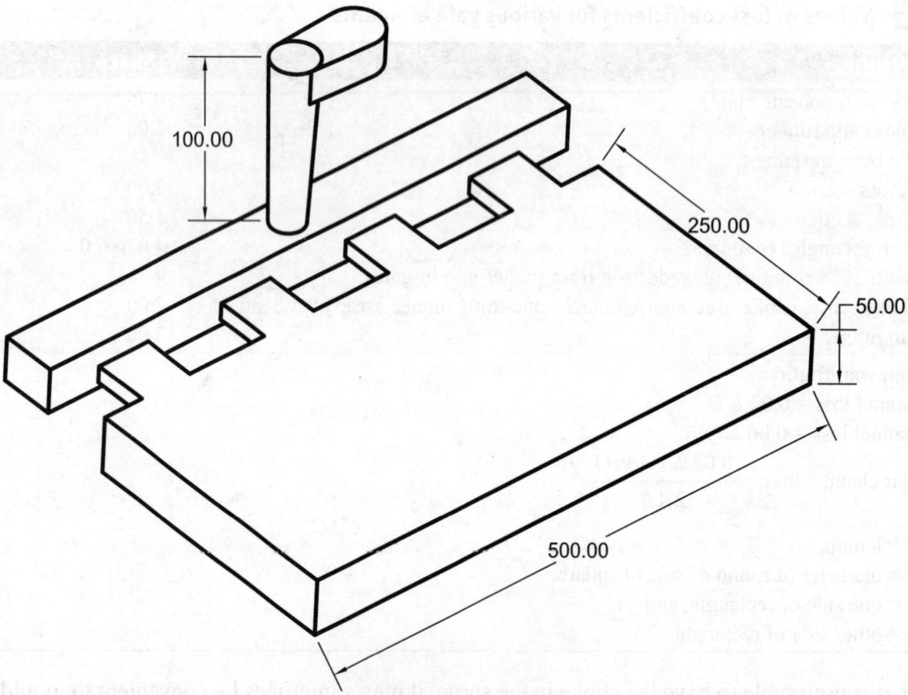

FIG. 4.17 *Example*

Calculate effective sprue height. Assuming a top gating system with 100 mm cope height, effective sprue height = 100 mm (Figure 5.21)

Select efficiency factor C from Table 4.6, assuming that gating system is pressurized and consists of two runners with four in-gates; $C = 0.73$

Assuming the density of the liquid metal = 6.90×10^{-6} kg/mm^3

$$\text{Choke area, } A = \frac{49.125}{6.90 \times 10^{-6} \times 19 \times 0.73\sqrt{2 \times 9800 \times 100}} = 366.648 \text{ mm}^2$$

In a pressurised gating system, the choke is located in in-gates, with four in-gates, each of which has an area of 90 mm^2. Assuming rectangular cross sections for in-gates, the in-gate dimensions would be 15 × 6 mm.

4.2.3 Gating Ratios

The gating ratio refer to the proportion of the cross-sectional areas between the sprue, runner and in-gates, and is generally denoted as sprue area, runner area, and in-gate area. Depending on the choke area there can be two types of gating systems:

- Nonpressurised
- Pressurised

A **nonpressurised gating system** has choke at the bottom of the sprue base, and have total runner area and in-gate areas higher than the sprue area. In this system, there is no pressure existing in the metal-flow system and thus it helps to reduce turbulence. This is particularly useful for casting drossy alloys such as aluminium and magnesium alloys. These have tapered sprues, sprue-base wells and pouring basins. When the metal enters the mould cavity through multiple in-gates, the cross section of the runner should accordingly be reduced at each runner break up, to allow equal distribution of metal through all the in-gates. The gating ratio of a typical example is

Sprue : runner : in-gate :: 1 : 4 : 4

The disadvantages of nonpressurised gating are, the gating system needs to be carefully designed to see that all parts flow full. Otherwise some elements of the gating system may flow partially allowing the air aspiration. Tapered sprues are invariably used with nonpressurised system. Also, the runners are maintained in drag while the gates are kept in cope to ensure that the runners are full.

Casting yield gets reduced because of the large metal involved in the runners and gates.

In case of a **pressurised gating system**, normally the in-gate area is the smallest, thus maintaining a back pressure throughout the gating system. Because of this back pressure in the gating system, the metal is more turbulent and generally flows full and thereby, can minimise the air aspiration even when a straight sprue is used (after the initial stages of pouring). When multiple gates are used, this system allows all the gates to flow full. These systems generally provide a higher casting yield, since the volume of metal used up in the runners and gates is reduced. Because of the turbulence and the associated dross formation, this type of gating system is not used for light alloys but can be advantageously used for ferrous castings. Gating ratio of a typical pressurised gating system is

sprue: runner : in-gate :: 1 : 2 : 1

These are the general considerations on the choice of a gating system. But a lot depends on the specific foundry practice, as evidenced from Table 4.7, of the various gating ratios recommended or used in practice.

While designing the runner system, care should be taken to reduce sharp corners or sudden change of sections since they tend to cause turbulence and gas entrapment. Though from the heat-loss factor circular cross-section runners are preferable, traditionally trapezoidal runner sections are employed to reduce the turbulence. The approximate proportions are from a square to rectangle with width twice as that of the depth of the runner. When multiple in-gates are used, the runner cross section should be suitably restricted at the separation of each runner in the interest of uniform flow through all the sections.

TABLE 4.7 Some gating ratios used in practice

Aluminium	1:2:1 1:3:3 1:4:4
Aluminium bronze	1:2.88:4.8
Brass	1:1:1 1:1:3 1.6:1.3:1
Copper	2:8:1 3:9:1
Ductile iron	1.15:1.1:1 1.25:1.13:1
Grey cast iron	2:1.5:1 2:3:1
Magnesium	1:2:2 1:4:4
Malleable iron	1:2:9.5 1.5:1:2.5 2:1:4.9
Steels	1:1:7 1:2:1 1:2:1.5

Also, it is a general practice to cut the runner in the cope and the in-gate in the drag to help in the trapping of slag. In some cases it was also found to be good to have half of the runner in the cope and the rest with in-gate in the drag, which effectively reduces slag inclusions. But for aluminium alloy castings, it is recommended that the runners be placed in the drag and the in-gates in the cope so that the dross which is heavier (3.99 g/cm^3) compared to aluminium (2.70 g/cm^3) is restricted. Also, the entry into runners from sprue-base well should be made as smooth as possible in such castings; otherwise the direction the flow would tend to be turbulent and leads to drossing when any change abruptly occurs in the cross-sectional area.

For cylindrical castings, the sprues may be located on the axis of rotation with sufficient number of radial runners feeding the casting. An alternative arrangement is that the sprue is located to one side of the casting and a runner around the periphery with the properly positioned in-gates. In case of thin castings, misruns are a problem and, therefore, they should be fed as quickly as possible with a number of in-gates all around the casting. A preferred gating system for grey cast iron plate like casting is presented in Fig. 4.18.

In-gate Design

The in-gates are generally made wider compared to depth, up to a ratio of 4. This facilitates in the severing of gating from the casting after solidification. It may sometimes be preferable to reduce the actual connection between the in-gate and the casting by means of a neck-down, wash-burn or dry-sand core, so that the removal of the gating is simplified.

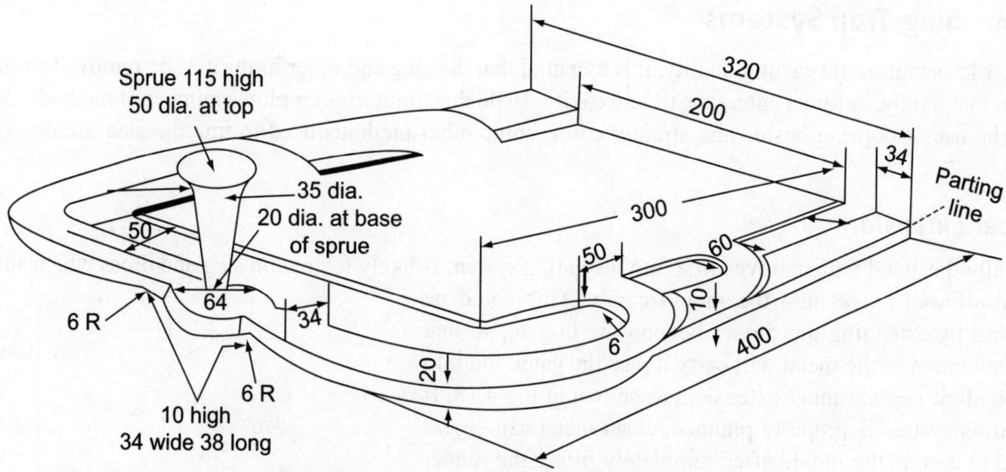

FIG. 4.18 *Gating system for platelike castings*

The following points should be kept in mind while choosing the positioning of the in-gates.

- In-gate should not be located near a protruding part of the mould, to avoid the striking of vertical mould walls by the molten metal stream.
- In-gates should preferably be placed along the longitudinal axis of the mould wall.
- In-gates should not be placed near a core print or a chill.
- In-gate cross-sectional area should preferably be smaller than the smallest thickness of the casting, so that the in-gates solidify first and isolate the castings from the gating system. This would reduce the possibility of air aspiration through the gating system in cases of metal shrinkage.

Small castings may be designed with a single in-gate. However, large or complex castings require multiple in-gates to completely fill all the sections of the castings effectively. In the case of multiple in-gates, care has to be taken to see that all the gates would be distributing the molten metal uniformly. To make for more uniform flow through all the gates, progressively the runner area should be reduced after each in-gate, such that restriction on the metal flow would be provided. A typical method followed in the case of plate like castings with a tapered runner is shown in Fig. 4.19(a).

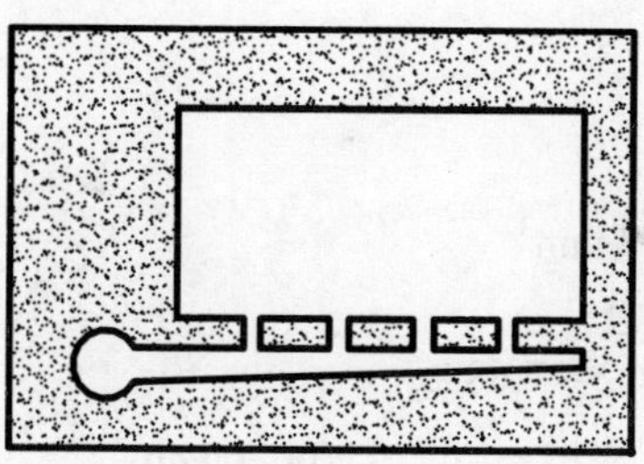

(a)

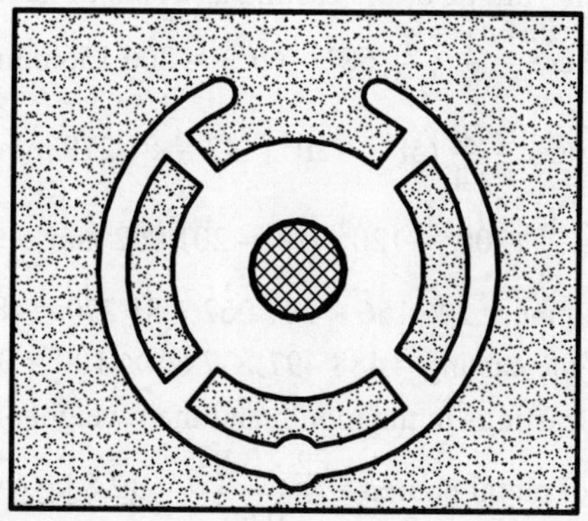

(b)

FIG. 4.19 *Multiple in-gates designed to induce uniform flow through all the gates for various type of castings: (a) Flat, rectangular; (b) Hollow cylinder*

4.2.4 Slag-Trap Systems

In order to obtain sound casting quality, it is essential that the slag and other impurities be removed from the molten metal fully, before it enters the mould cavity. To do this, foundries employ a number of methods. Apart from the use of pouring basins and strainer cores, some other methods used to trap the slag are described below.

Runner Extension

Normally the metal which moves first into the gating system is likely to contain slag and dross which should not be allowed to get into the mould cavity. This could be achieved by extending the runner beyond the in-gate, so that the momentum of the metal will carry it past the gates and into a blind alley, i.e the runner extension, as shown in Fig 4.18. If the gating system is properly planned, clean metal can be expected to go into the mould after completely filling the runner extension. A runner extension having a minimum of twice the runner width is desirable.

Whirl Gate

Another method employed successfully to trap the slag from entering steel castings is a whirl gate, the design of which is presented in Fig. 4.20. This utilises the principle of centrifugal action to throw the dense metal to the periphery and retain the lighter slag to be retained at the centre. In order to achieve this action, it is necessary that entry area should be at least 1.5 times the exit area so that the metal is built up at the centre quickly. Also, the metal should revolve 270° before reaching the exit gate, so as to gain enough time for separating the impurities.

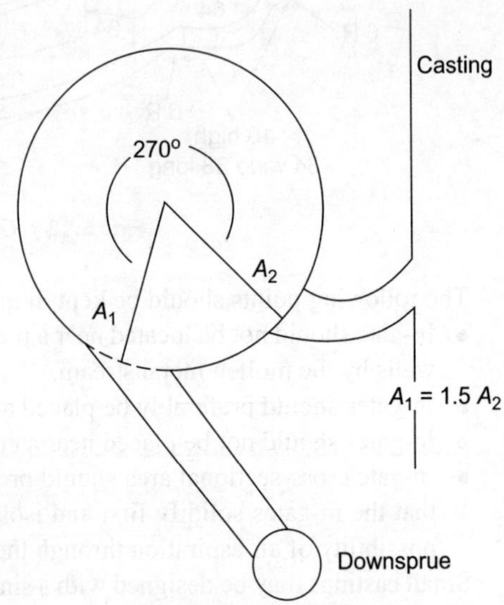

Fig. 4.20 *Whirl gate*

Example 4.4

Calculate the gating requirements for the casting shown in Fig. 4.21 to be cast in C30 steel.

Calculate the volume of the casting as a sum of hemisphere, cylinder and flange volumes.

$$\text{Hemisphere} = \frac{4}{3} \times \frac{\pi}{2}(68^3 - 60^3) = 206\,156 \text{ mm}^3$$

$$\text{Top cylinder} = \frac{\pi}{4}(58^2 - 30^2) \cdot 42 = 81\,279 \text{ mm}^3$$

$$\text{Flange} = \frac{\pi}{4}(200^2 - 120^2)10 = 201\,062 \text{ mm}^3$$

$$\text{Total volume} = 206\,156 + 201\,062 + 81\,279 = 488\,497 \text{ mm}^3$$

$$\text{Mass of the casting} = 488\,497 \times 7.86 \times 10^{-6} = 3.84 \text{ kg}$$

There are two castings in the mould and their weight $3.84 \times 2 = 7.68$ kg

$$\text{Weight of metal poured} = \frac{2 \times 3.84}{0.60} = 12.8 \text{ kg (Assuming a casting yield of 0.60)}$$

$$\text{Pouring time} = (2.4335 - 0.3953 \cdot \log 12.80)\sqrt{12.80} = 7.14 \text{ s}$$

$$\text{Pouring rate} = \frac{12.8}{7.14} = 1.793 \text{ kg/s}$$

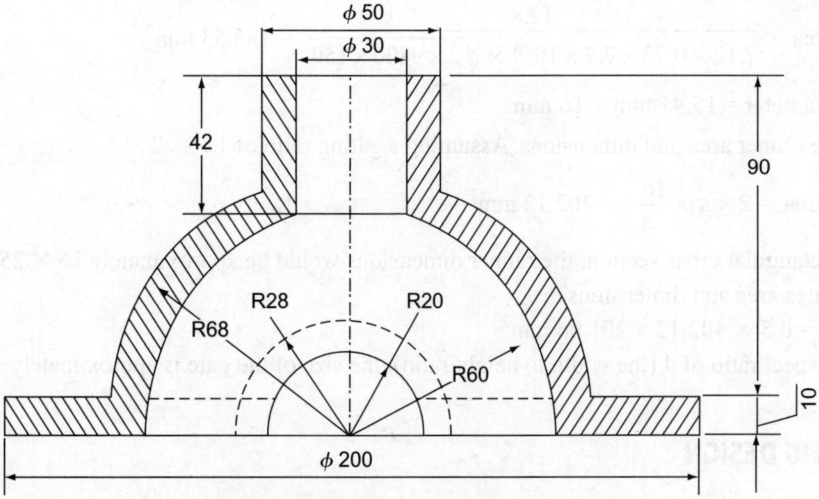

Fig. 4.21 *Example for gating calculations*

Choosing parting ingates and two castings in the mould, the casting arrangement would be as shown in Fig. 4.22 and effective sprue height equal to 150 mm.

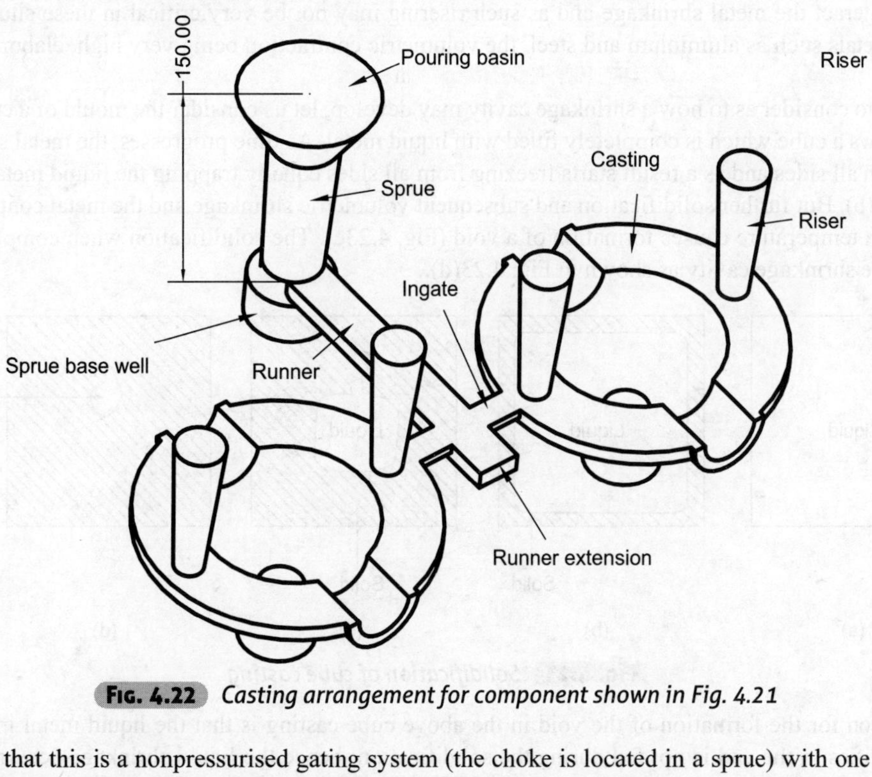

Fig. 4.22 *Casting arrangement for component shown in Fig. 4.21*

Assume that this is a nonpressurised gating system (the choke is located in a sprue) with one runner and two in-gates. For this combination, the efficiency coefficient of the gating system from Table 4.6 is 0.73.

$$\text{Choke area} = \frac{12.8}{7.12 \times 0.73 \times 7.7 \times 10^{-6} \times \sqrt{2 \times 9800 \times 150}} = 186.53 \text{ mm}^2$$

Choke diameter = 15.41 mm ≈ 16 mm

Calculate runner area and dimensions. Assuming a gating ratio of 1 : 2 : 2

$$\text{Runner area} = 2 \times \pi \times \frac{16^2}{4} = 402.12 \text{ mm}^2$$

Assuming a rectangular cross section, the runner dimensions would be approximately 16 × 25 mm.
Calculate in-gates area and dimensions.
Each gate area = 0.5 × 402.12 = 201.06 mm^2
Assuming an aspect ratio of 4 (the width to height ratio), the size of the gate is approximately 7 × 28 mm.

4.3 RISERING DESIGN

The function of a riser is to feed the casting during solidification so that no shrinkage cavities are formed. The requirement of risers depends to a great extent upon the type of metal poured and the complexity of the casting. As shown in Table 4.2, various materials have different volumetric shrinkages. Of particular interest from among them is grey cast iron, which sometimes may have a negative shrinkage. This happens because with higher carbon and silicon contents, graphitisation occurs which increases the volume and, therefore, would counteract the metal shrinkage and as such risering may not be very critical in these situations. But for some metals such as aluminium and steel, the volumetric contraction being very high, elaborate risering is required.

In order to consider as to how a shrinkage cavity may develop, let us consider the mould of a cube. Figure 4.23(a) shows a cube which is completely filled with liquid metal. As time progresses, the metal starts losing heat through all sides and as a result starts freezing from all sides equally trapping the liquid metal inside, as in Fig. 4.23(b). But further solidification and subsequent volumetric shrinkage and the metal contraction due to change in temperature causes formation of a void (Fig. 4.23c). The solidification when complete, finally results in the shrinkage cavity as shown in Fig. 4.23(d).

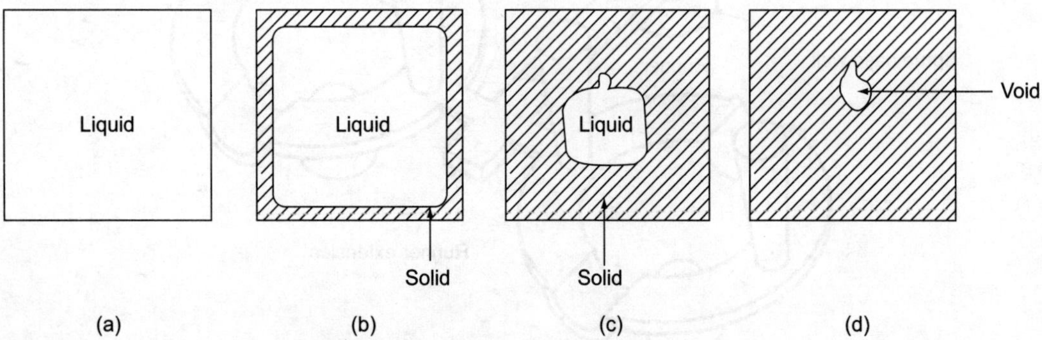

Fig. 4.23 *Solidification of cube casting*

The reason for the formation of the void in the above cube casting is that the liquid metal in the centre which solidifies in the end is not fed during the solidification; hence the liquid shrinkage occurred ends up as a void. Such isolated spots, which remain hot till the end, are called 'hot spots'. It is the endeavour of the casting designer to reduce all these hot spots so that no shrinkage cavities occur.

In this connection, the term **directional solidification** is normally used in the casting terminology. It means that the solidification of the metal should start at the remotest point of the casting from the feeder. Since the cooling is achieved by the removal of heat from all surfaces which are exposed to the atmosphere or sand, cooling normally starts from point which is the thinnest or is exposed over a larger surface area.

4.3.1 Caine's Method

Since solidification of the casting occurs by losing heat from the surfaces and the amount of the heat is given by the volume of the casting, the cooling characteristics of a casting can be represented by the surface area to volume ratio. Since the riser is also similar to the casting in its solidification behaviour, the riser characteristic can also be specified by the ratio of its surface area to volume.

If this ratio of the casting is higher then it is expected to cool faster. Chvorinov has shown that the solidification time of a casting is proportional to the square of the ratio of volume-to-surface area of the casting. The constant of proportionality called mould constant depends on the pouring temperature, casting and mould thermal characteristics.

$$t_s = k\left(\frac{V}{SA}\right)^2$$

where t_s = solidification time, s
V = volume of the casting
SA = surface area
k = mould casting

The 'freezing ratio', X of a mould is defined as the ratio of cooling characteristics of casting to the riser.

$$X = \frac{SA_{\text{casting}}\,/\,V_{\text{casting}}}{SA_{\text{riser}}\,/V_{\text{riser}}}$$

In order to be able to feed the casting, the riser should solidify last and hence its freezing ratio should be greater than unity. It may be argued that the sphere has the lowest surface-area-to-volume ratio and hence, it should be used as a riser. But in a sphere, the hottest metal being at the centre, it is difficult to use for feeding the casting. The next best is the cylindrical type which is most commonly used for their ease in moulding.

Based on the Chvorinov's rule, Caine developed a relationship empirically for the freezing ratio as follows:

$$X = \frac{a}{Y - b} + c$$

where Y = riser volume/casting volume, and a, b and c are constants whose values for different materials are given in Table 4.8.

TABLE 4.8 **Constants for Caine's equation**

	a	b	c
Steel	0.10	0.03	1.00
Aluminium	0.10	0.06	1.08
Cast iron, brass	0.04	0.017	1.00
Grey cast iron	0.33	0.030	1.00
Aluminium bronze	0.24	0.017	1.00
Silicon bronze	0.24	0.017	1.00

The following equations are used for calculating the risering requirements for aluminium alloys LM4 (Cu_2–4%; Si_4–6%) and LM11 (Cu 4–5%).

When height to diameter ratio of the riser is equal to 1,

$$\text{LM4: } X = \frac{33.45}{23.98 - Y}$$

$$\text{LM11: } X = \frac{17.11}{12.58 - Y}$$

When height-to-diameter ratio of the riser is equal to 1.5,

$$\text{LM4: } X = \frac{32.09}{23.98 - Y}$$

$$\text{LM11: } X = \frac{16.75}{12.58 - Y}$$

For nodular iron in CO_2 moulds, the following equation may be used.

$$X = \frac{39.44}{27.78 - Y}$$

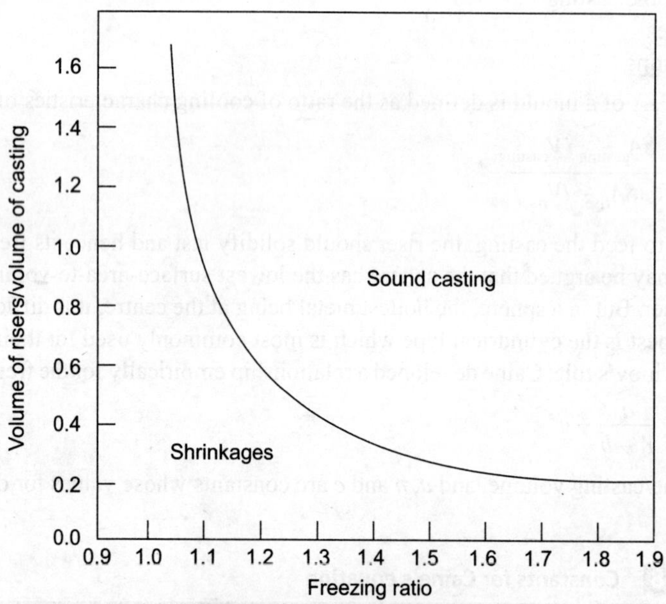

Fig. 4.24 *Caine's equation*

The above equation when plotted will be as shown in Fig. 4.24. The line shows the locus of the points that separate the sound castings and castings with shrinkages in steel castings.

Example 4.5

Calculate the size of a cylindrical riser (height and diameter equal) necessary to feed a steel slab casting 25 × 25 × 5 cm, with a top riser casting poured horizontally into the mould.

$$\text{Volume of the casting} = 25 \times 25 \times 5 = 3125 \text{ cm}^3$$

$$\text{Surface area of the casting} = 2 \times 25 \times 25 + 4 \times 25 \times 5 = 1750 \text{ cm}^2$$

$$\text{Volume of the riser} = \frac{\pi \cdot D^3}{4}$$

where D is the riser diameter.

$$\text{Surface area of the riser} = \pi \cdot D^2 + \frac{\pi \cdot D^2}{4} = 1.25\,\pi D^2$$

$$\text{Freezing ratio, } X = \frac{1750/3125}{1.25\,\pi D^2 / 0.25\,\pi D^3} = 0.112\,D$$

$$Y = \frac{\text{Volume of riser}}{\text{Volume of casting}} = \frac{0.25\,D^3}{3125} = 0.000\,251\,D^3$$

Substituting this in the Caines' equation for steels

$$0.112\,D = \frac{0.10}{0.000251\,D^3 - 0.03} + 1.0$$

On simplification, we get

$$D^4 - 8.9286\,D^3 - 119.52\,\mathrm{D} = 2490$$

By trial and error, we get

$$D = 11.44 \text{ cm} \approx 12 \text{ cm}$$

4.3.2 Modulus Method

Another method for finding the optimum riser size is the 'modulus method', extensively documented by Wlodawer. It has been empirically established that if the modulus of the riser exceeds the modulus of the casting by a factor of 1.2, the feeding during solidification would be satisfactory.

The modulus is the inverse of the cooling characteristic (surface area/volume) as defined earlier. In steel castings, it is generally preferable to choose a riser with a height to diameter ratio of 1.

$$\text{Volume} = \frac{\pi D^3}{4}$$

where D = diameter of the riser.

The bottom end of the riser is in contact with the casting and thus does not contribute to the calculation of surface area.

$$\text{Surface area} = \frac{\pi D^2}{4} + \pi D^2$$

The modulus of such a cylindrical riser, M_r would be

$$M_r = 0.2\,D$$

Since $M_r = 1.2\,M_c$

$$D = 6\,M_c$$

where M_c = modulus of the casting.

Thus, in this method, the calculation of the riser size is simplified to the calculation of the modulus of the casting itself and no trial and error solution as is given in the previous case. Though this takes into account the cooling effect of the riser, it does not consider exactly the amount of feeding metal required to compensate for the shrinkage of the casting. If allowance is made for the volume of metal to be fed to counteract the contraction of the casting, the equation would change to

$$D^3 - 5.46\, M_c D^2 - 0.050\,93\, V_c = 0$$

where V_c = volume of the casting.

The above is valid when the height to diameter ratio of the riser is one. When the third term in the equation relating to feed volume is neglected, the previous simplified equation would be arrived at.

With 'chunky' castings, e.g. cubes, the volume component may be negligible, but for those 'rangy' castings, similar to plate like, the influence of volume component becomes increasingly significant. It is sometimes useful to have a parameter called **"ranginess factor" R** to define the casting type. It may be defined as

$$R = \frac{\text{Modulus of a cube of same volume as casting}}{\text{Modulus of casting}}$$

Table 4.9 gives the moduli of simple geometrical shapes.

For calculating the modulus of a complex shape, it is useful to consider it as a combination of the above simple shapes shown in Table 4.9 or by a suitable approximation. For example, a ring on a plate of thickness t can be considered as a ring with noncooling surface of thickness t, as shown in Fig. 4.25.

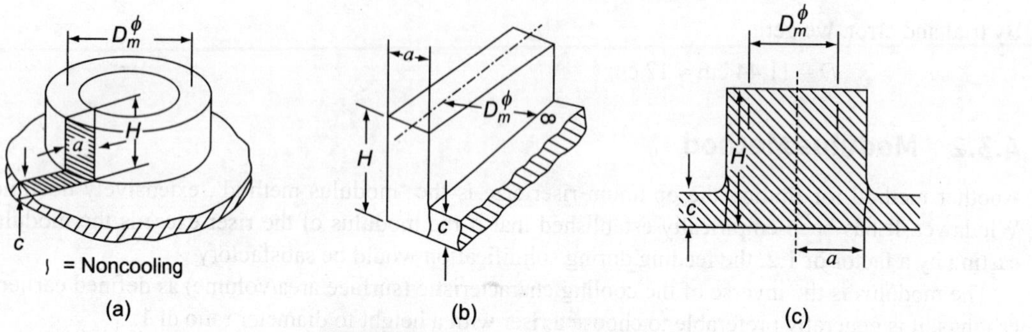

FIG. 4.25 *Modulus method*

Let the mean ring diameter, $Dm = n\,a$

where n is a suitable constant based on the geometry.

$$M_C = \frac{\text{volume}}{\text{area}} = \frac{\pi D_m a H}{2\pi a^2 n + a\pi\,(1+n)(H-c) + a\pi\,(n-1)H}$$

$$= \frac{aH}{2\,(a+H) - c\,(n+1)/n}$$

In the limiting case when the ring becomes a bar as shown in Fig. 4.25(b) then n tends to infinity and thus

$$M_C = \frac{aH}{2\,(a+H) - c}$$

TABLE 4.9 Moduli of simple geometric shapes

	Casting Shape	Modulus, Mc
Plate	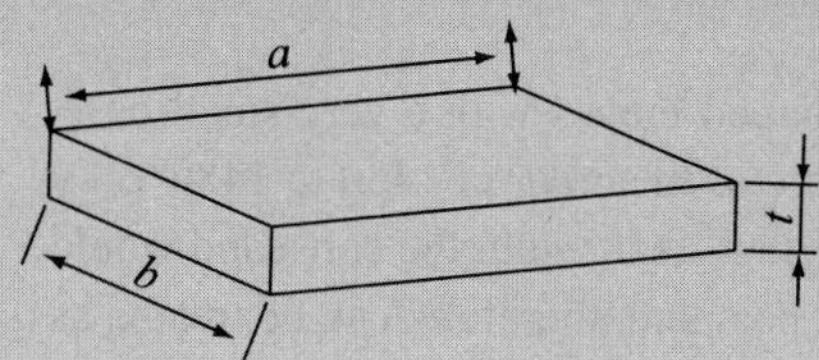	$0.5\,t$ $(a < 5\,t)$
Long bar	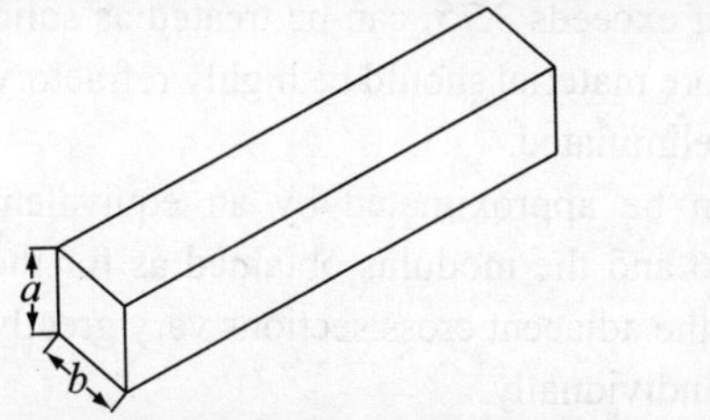	$\frac{a\,b}{2(a+b)}$
Cube	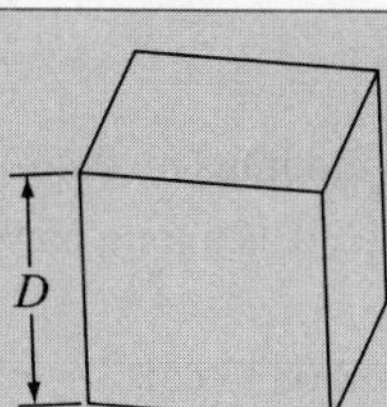	$\frac{D}{6}$
Cylinder	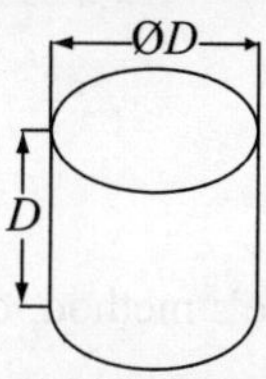	$\frac{D}{6}$
Sphere	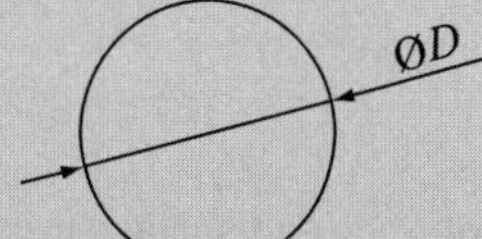	$\frac{D}{6}$
Hollow cylinder	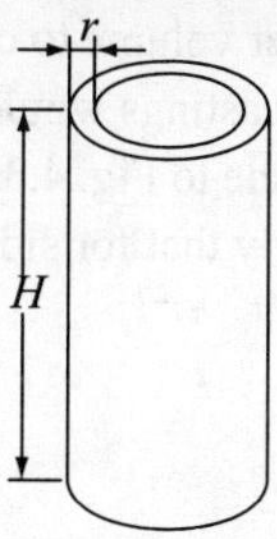	$\frac{r\,H}{2\,(r+H)}$

Alternatively, when the boss becomes solid as in Fig. 4.25(c) then $n = 1$, and

$$M_C = \frac{aH}{2\,(a + H - c)}$$

In the case of massive ring-shaped bodies with a very small bore, the core would reach a temperature of the order of 1450 to 1480°C by the time the casting starts solidifying. As a result, the core sand would not be able to extract any further heat and, therefore, can be treated as a solid body itself for risering purposes. The ratio of outer diameter to core diameter of such bodies, if exceeds 3.75, can be treated as solid bodies. In such situations, the core material should be highly refractory in nature or the core should be eliminated.

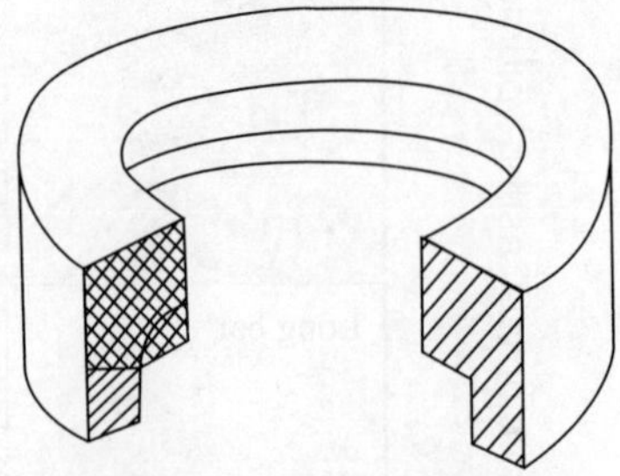

FIG. 4.26 *Modulus of joined sections*

Irregular cross sections can be approximated by an equivalent rectangle as shown in Fig. 4.26 and the modulus obtained as for the long bar from Table 4.9. But if the adjacent cross sections vary greatly then they should be calculated individually.

Example 4.6

Calculate the riser size for the Example 4.5 using the modulus method.

Since it is a slab of 25 × 25 × 5 cm, it can be considered as a long bar with cross section 25 × 5.

$$\text{Modulus, } M_c = \frac{25 \cdot 5}{2\,(25 + 5)} = \frac{125}{60} = 2.0833 \text{ cm}$$

The riser diameter, $D = 6\,Mc = 6 \times 2.0833 = 12.5$ cm.

4.3.3 Naval Research Laboratory Method

This method, which is essentially a simplification of Caine's method, defines a shape factor to replace the freezing ratio. The shape factor is defined as,

$$\frac{\text{Length + Width}}{\text{Thickness}}$$

The underlying argument is that, calculating volumes and surface areas is too complicated and, therefore, simplification would be desirable. The length, width and thicknesses are computed from the maximum dimensions of the casting section. Then, the ratio of the riser volume to casting volume can be obtained from the graph shown in Fig. 4.27. This shows when the sound castings would be obtained for C20 to C50 steels. Having obtained the riser volume, the reference may be made to Fig. 4.30 to obtain riser diameter and height for the obtained riser volume. It has been proven empirically that for side risers, the height-to-diameter ratio be 1 and for top risers it be 0.5.

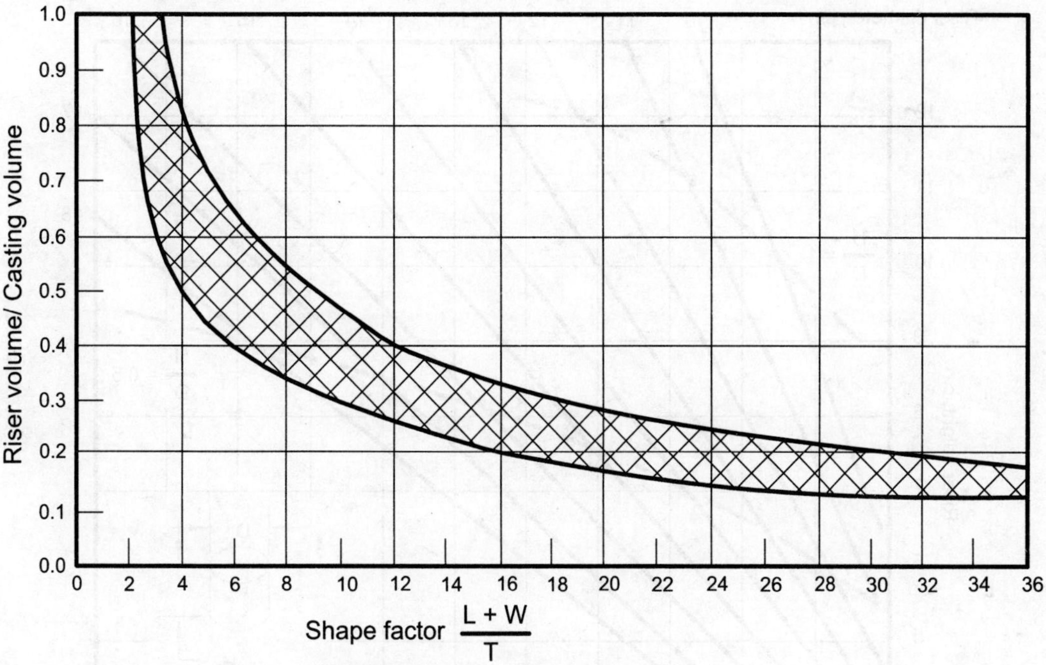

FIG. 4.27 *Riser volume selection chart*

Example 4.7

Recalculate the riser dimensions for Example 4.5.

$$\text{Shape factor} = \frac{25 + 25}{5} = 10$$

From the chart in Fig. 4.29,

$$V_r\text{, Riser volume} = 0.47 \times \text{Casting volume}$$
$$= 0.47 \times 25 \times 25 \times 5 = 1468.75 \text{ cm}^3$$

For a cylindrical riser of equal diameter and height

$$V_r = 0.25 \times \pi \times D^3$$

$$D = \sqrt[3]{\frac{4 \times 1468 \times 75}{\pi}} = \sqrt[3]{1870} = 12.32 \text{ cm}$$

The same can also be directly read off from Fig. 4.28.

For circular plates, the length and width are same as that of the diameter. But for cylinders, the width and thickness are same as the diameter for calculating the shape factor. But for calculating the riser volume, the actual casting volume is to be used. The values obtained by NRL method are generally conservative in nature and correspond to Caine's method for most of the simple geometries.

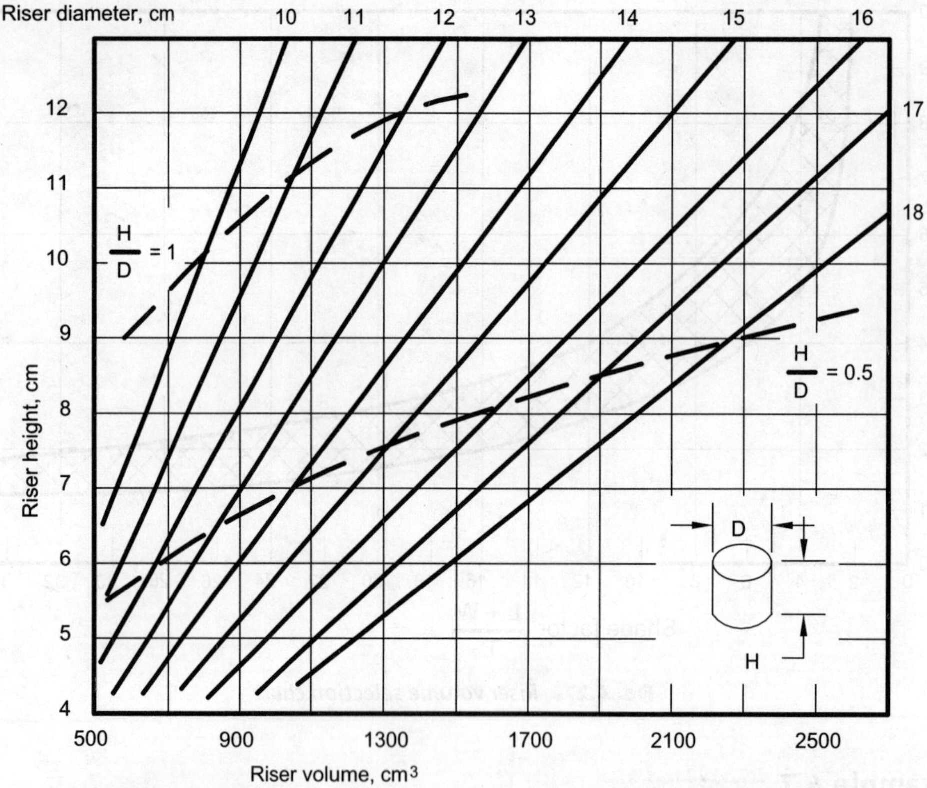

Fig. 4.28 *Selection chart for riser dimensions based on NRL method*

Normally, the risers should be located at the heaviest sections as they themselves act as feeders for the thin sections. But when smaller sections are connected to the thicker sections, the riser should have larger volume to cater to this appendage. But if this branch is thin, no change in riser volume is required.

Example 4.8

Calculate the risering requirement for the casting shown in Fig. 4.29.

First neglect the branch shown and calculate the shape factor for the main plate.

$$\text{Shape factor} = \frac{25 + 12.5}{5} = 7.5$$

Volume of the casting = 25 × 12.5 × 5 = 1562.5 cm^3

Volume of the riser = 0.575 × 1562.5 = 898.4375 cm^3

The branch volume = 2.5 × 2.5 × 10 = 62.5 cm^3

This is a plate feeding the bar with a thickness ratio of 0.5. Hence, from Fig. 4.30, we get parasitic volume as 30%.

Hence, riser volume = 0.30 × 62.5 + 898.4375 = 917.1875 cm^3

Riser diameter, D = 10.531 cm

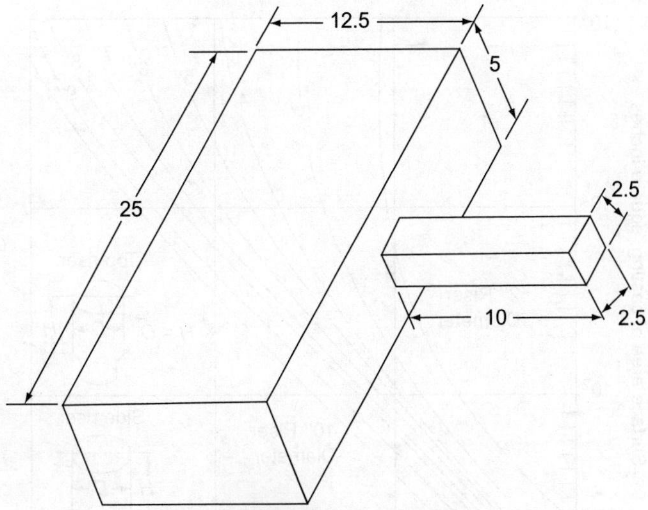

Fig. 4.29 *Example*

The other shapes of interest are the hollow cylindrical shapes. In these, since the heat removal is restricted, a correction factor, k needs to be applied to get the effective plate thickness. If T is the true wall thickness then the correction factor is given by

core diameter	0.5 T	T	2T	4T
correction factor,	k1.17	1.14	1.02	1.00

$$\text{Shape factor} = \frac{L+W}{kT}$$

Example 4.9

Calculate the riser diameter for an annular cylinder of 30 cm outside diameter, 10 cm inside diameter and 30 cm height.

Core diameter = 10 cm

Plate thickness = 10 cm

Correction factor = 1.14

$$\text{Shape factor} = \frac{20\pi + 30}{1.14 \cdot 10} = 8.143$$

Casting volume = $0.25 \times \pi \times (30^2 - 10^2) \times 30 = 18\,849\ \text{cm}^3$

Riser volume = $0.48 \times 18\,849 = 9047.52\ \text{cm}^3$

Riser diameter = 22.585 cm

As explained earlier, during solidification, grey cast iron has different rates of volumetric changes depending on the carbon and silicon compositions in the final melt. Hence, not enough data is available on the risering practice. Wallace and Evans have calculated the riser sizes required, based on the surface area and volume of the casting for high-strength grey cast iron (3% C; 1.5% Si; 0.05% P; 0.10% S; 0.8% Mn) being a representative composition. The charts for green sand mould and dry sand mould are presented in Fig. 4.30 and 4.31. The soft grey cast iron (3.5% C; 2.2% Si) may not require any risering. If any doubt persists then a small shrink bob of 2 to 4 % of casting volume may be provided at the necessary point.

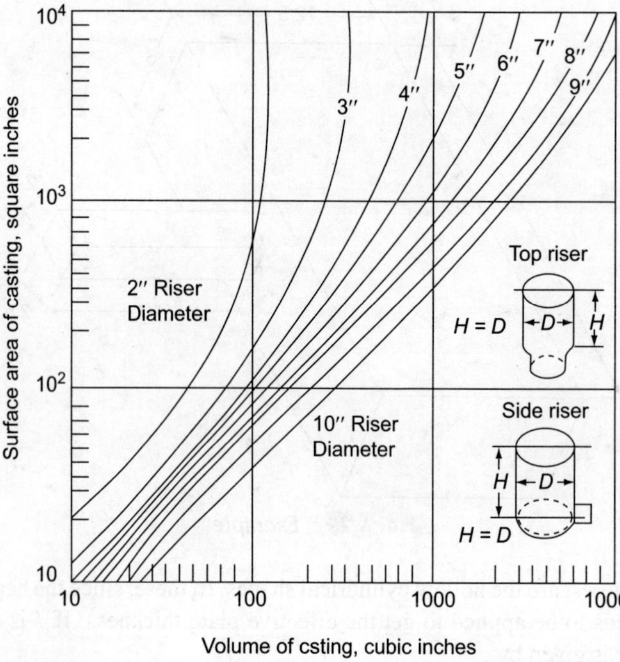

FIG. 4.30 *Risering curves for grey cast iron in green-sand moulds*

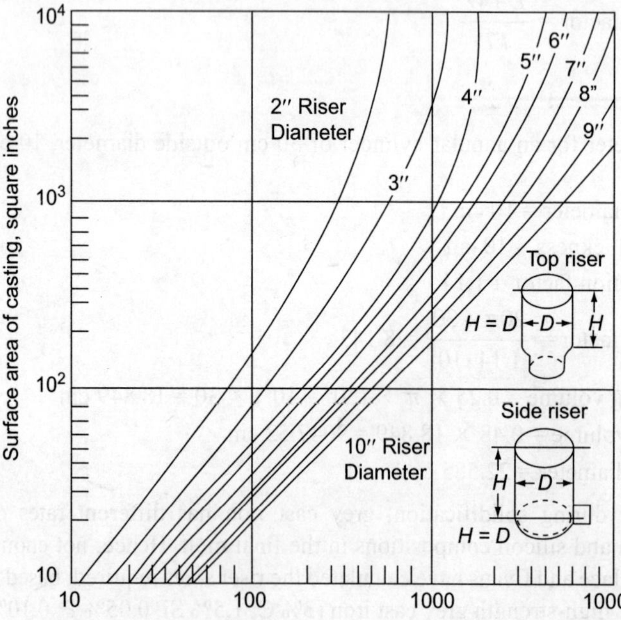

FIG. 4.31 *Risering curves for grey cast iron in dry-sand moulds*

Whenever more than one riser is to be used for any casting, the casting is divided into a number of parts and the risering requirements are to be calculated separately for each part.

The way the riser is connected to the casting is very important. The connection called the **neck**, should solidify first so that the shrinkage cavity is located in the riser itself. Also the removal of the riser should be made simpler by making the connection smaller. But if the neck size is too small then it may solidify before the casting itself and thus defeats the function of the riser. It is a normal practice to provide a thin ceramic or core material for the neck to reduce the cross-sectional area.

These are called **washburn cores.** The neck dimensions for various types of risers are presented in Fig. 4.32.

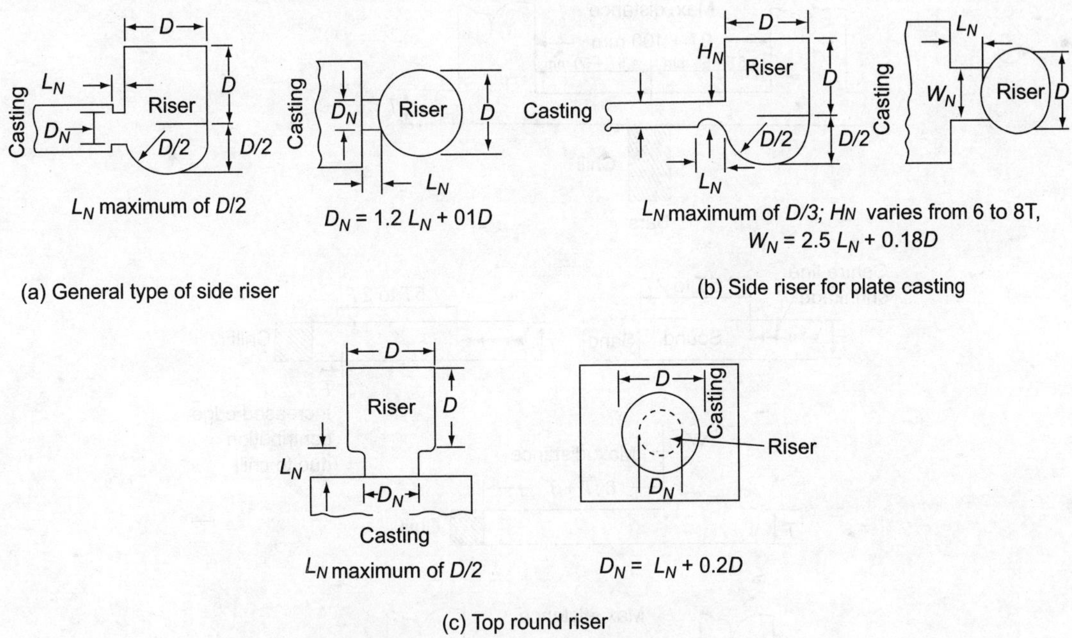

Fig. 4.32 *Schematic riser-neck dimensions*

4.3.4 Chills

Chills are provided in the mould so as to increase the heat-extraction capability of the sand mould. A chill normally provides a steeper temperature gradient so that directional solidification as required in a casting can be obtained. The chills are metallic objects having a higher heat-absorbing capability than the sand mould. The chills can be of two types:

- external, and
- internal

The external chills are placed in the mould cavity adjoining the mould cavity at any required position. Providing a chill at the edge may not normally have the desired effect as the temperature gradient is steeper at the end of the casting, since heat is removed from all sides as shown in Fig. 4.33. However, if it is placed between two risers, it would have maximum effect.

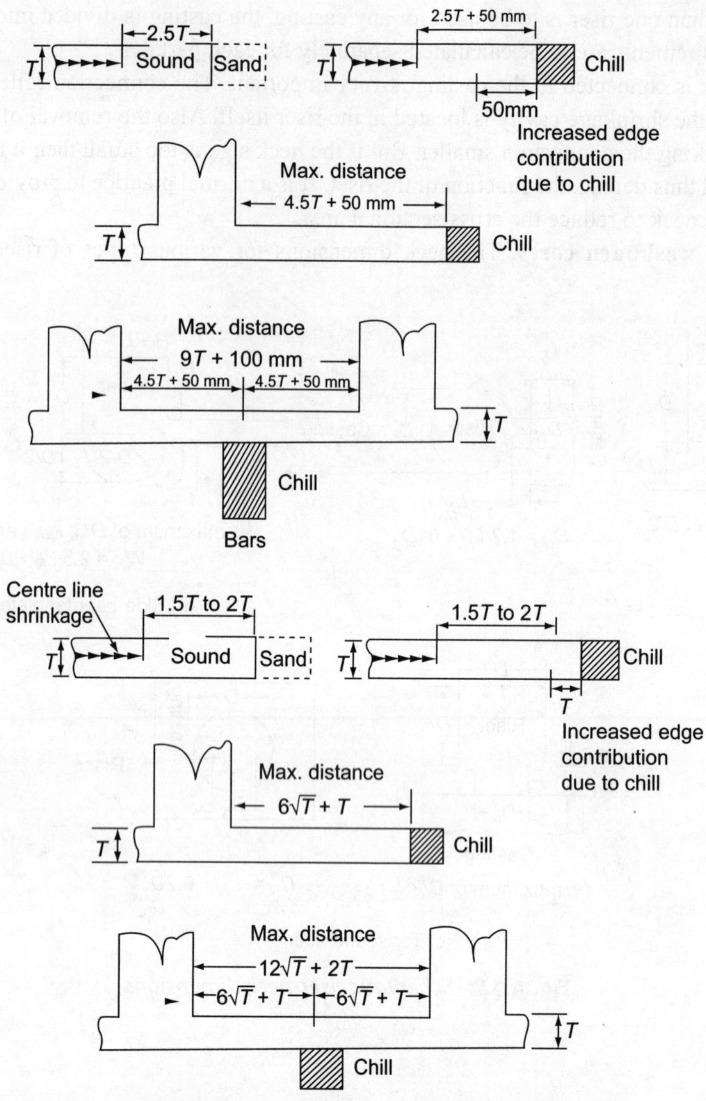

Fig. 4.33 *Effect of chill on feeding distance*

The chills when placed in the mould should be clean and dry, otherwise gas inclusions will be left in the castings. Also, after placing the chills in the mould, they should not be kept for long, since moisture may condense on the chills causing blow holes in the casting.

The internal chills are placed inside the mould cavity where an external chill cannot be provided. The material of chill should approximately resemble the composition of the pouring metal for proper fusing. Cleanliness of internal chills is far more important because they are surrounded on all sides by the molten metal. Because of the larger problems associated with the internal chills they should be sparingly used.

In those castings which have ribs or bosses, the joint is a source of hot spot. This hot spot could be eliminated by using a chill of appropriate size at the end of the rib or the appendage. To calculate the proportions of the chill, the results obtained by Myskowski are used, as shown in Fig. 4.34, where '*T*' is the thickness of the parent casting.

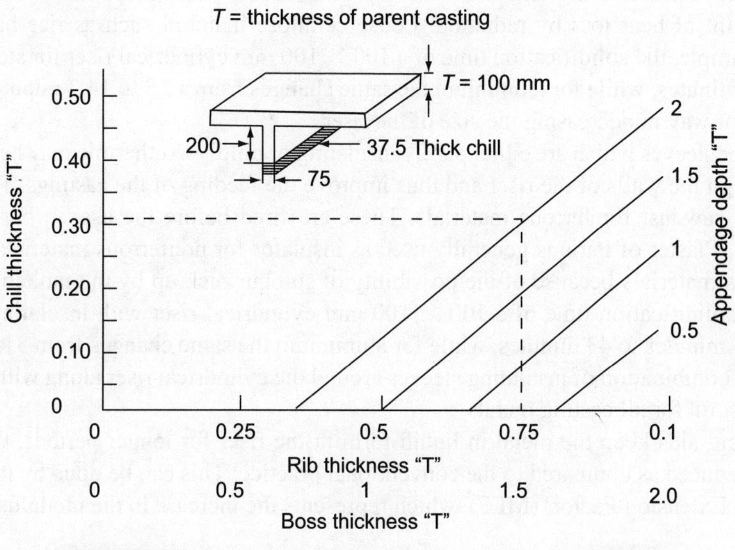

FIG. 4.34 *Chill proportions required in plate steel castings*

Grouping Castings

Grouping of several castings around a single riser helps in increasing the casting yield, since the same riser will be able to feed to more than one casting, as shown in Fig. 4.37(b). Also, by a small variation in the moulding practice, it is possible to reduce the risering requirement as shown in Fig. 4.35(c) as the heat from the castings would keep the metal hot in the riser for a longer period.

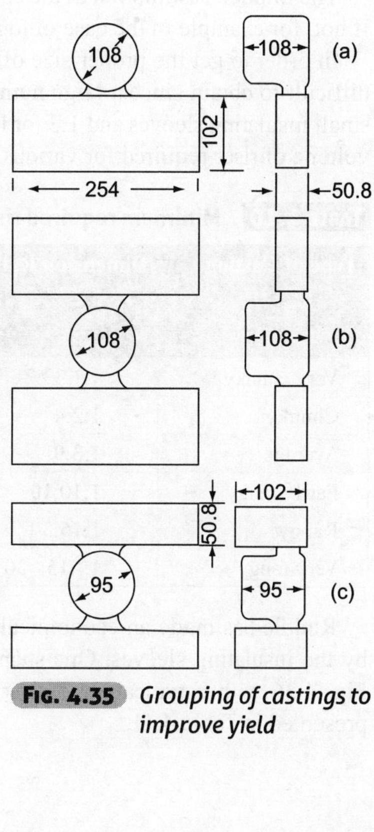

FIG. 4.35 *Grouping of castings to improve yield*

4.3.5 Feeding Aids

To increase the efficiency of a riser, it is necessary to keep the metal in the riser in liquid form for as long a period as required, so that it would feed the casting till it solidifies. When this is done, the riser volume decreases, resulting in a higher yield. The aids used for this purpose are called feeding aids. They can be either exothermic materials or insulators.

Exothermic materials when added at the top of the riser would give out heat and also act as heat insulators (through the reaction products) to the atmosphere, thus reducing the heat loss through radiation. The exothermic materials that can be used are graphite or charcoal powder, rice hulls and thermit mixtures. The carbonaceous materials are called **anti-piping compounds,** besides being cheap, provide the insulation and also get oxidised slowly. But thermit mixtures get oxidised very quickly and release heat over a very small period and subsequently, the effect would be that of only the residue acting as an insulator. As such, these are not generally used.

For steel casting, an insulating shield on the top of an open riser is very effective since it reduces considerable amount of heat loss by radiation. Loose cellulose material such as rice husk is used as top insulation. For example, the solidification time of a 100 × 100 mm cylindrical riser for steels increases from 5 minutes to 13.4 minutes, while for aluminium the same changes from 12.3 to 14.3 minutes. Thus, for steels this is the important way in decreasing the size of the riser.

Alternately, riser sleeves which are either purely insulating or mildly exothermic may be used to reduce the heat transfer through the walls of the riser and thus improve the feeding of the castings. These are generally made of fire clay–sawdust for ferrous materials. These are fired before the use, to burn out the sawdust and leave air cells. Plaster of Paris is generally used as insulator for nonferrous materials, though it cannot be used for ferrous materials because of the possibility of sulphur pick up by the molten metal in the riser. For example, the solidification time of a 100 × 100 mm cylindrical riser with insulating sleeve for steels increases from 7.5 minutes to 43 minutes, while for aluminium the same changes from 31.1 to 45.6 minutes. Thus, an optimum combination of insulating sleeves around the cylindrical riser along with the top insulating shield would be useful for all casting metals.

Since the feeding aids keep the metal in liquid form in the riser for longer periods, the modulus of the riser needs to be reduced as compared to the conventional practice. This can be done by introducing a factor *f*, called 'Modulus Extension Factor' (MEF), which represents the increase in the modulus of the riser. Thus,

$$M_{r'} = \frac{1.2}{f} M_c$$

The implicit assumption in the above equation is that the insulation effect at the sides and the top are same. If not, for example in the case of top risers, proper care needs to be taken to get an average value of *f*.

In order to get the proper size of the riser, it is necessary to get the correct value of the factor *f*, which is difficult to obtain since a large number of proprietary sleeves are available. Typical values could be 1.8 for small insulating sleeves and 1.3 for large sleeves. The effect of the insulating sleeve in reducing the minimum volume of riser required for various types of castings is presented in Table 4.10.

TABLE 4.10 Minimum required riser volume

Type of Casting	Dimension Ratio	Volume of Riser/Volume of Casting			
		Sand Riser		Insulated Riser	
		$H = D$	$H = 2D$	$H = D$	$H = 2D$
Very chunky	1:1.33:2	1.40	1.98	0.32	0.40
Chunky	1:2:4	1.06	1.40	0.26	0.32
Average	1:3:9	0.58	0.75	0.19	0.22
Fairly rangy	1:10:10	0.30	0.38	0.14	0.16
Rangy	1:15:30	0.13	0.15	0.09	0.10
Very rangy	1:>15:>30	0.11	0.13	0.08	0.08

Ruddle has made an economical analysis for various steel castings to demonstrate the savings achieved by the insulating sleeves. One such example a carrier bracket whose shipped mass is 143 kg, is shown in Fig. 4.36 with a typical sand riser and the insulated sleeve riser. The various savings thus obtained are presented in Table 4.11.

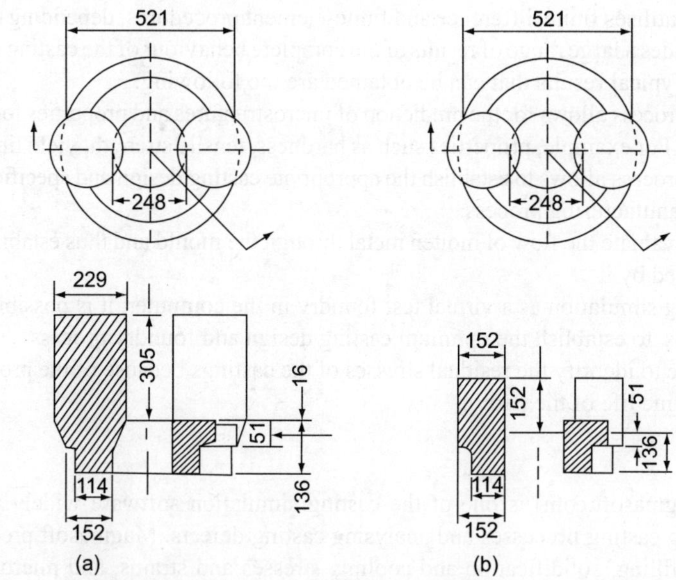

Fig. 4.36 *Carrier bracket with old and new practice*

Table 4.11 Estimated savings for carrier bracket casting shipped mass, W = 143 kg

	Original Practice	New Practice	Savings
(a) Mass of liquid metal in riser, kg	176	32.4	143.6
(b) Mass of gates, kg	10	10	—
(c) Estimated feed out (7% of W), kg	10	10	—
(d) Total mass in the mould ($W + a + b - c$), kg	319	175	144
(e) Metal melted (assuming 6% loss), kg	339	186	153
(f) Riser cut-off loss ($0.112 \times a$), kg	14	6	8
(g) Returns ($a + b - c - f$), kg	162	26.5	135

4.4 SOLIDIFICATION SIMULATION

(With contributions from Scott Giese and Sairam Ravi, Metal Casting Center, University of Northern Iowa)

With rapid developments in computer hardware and software technologies, the use of simulation tools in foundries has increased. Casting-process simulation is now a well-established tool for the optimisation of casting design as well as process design. They have reached a point where the simulation tools have now become an integrated part of the entire casting-production process, from casting design to adjusting production parameters. There are many aspects of the casting process that can be evaluated by using the casting-simulation process. As seen earlier, the casting process is extremely complex, and analytical (thermo-physical) procedures are very difficult to apply without using a large number of assumptions. Many of the assumptions, though simplify the calculation procedures, do not guarantee accurate results. This calls for some trial-and-error methods to achieve sound casting which is both time-consuming and expensive.

Casting simulation utilises finite difference and finite-element procedures, depending upon the complexity of the problem. It provides a large range of results of the complete behaviour of the casting during solidification as well as in service. Typical results that can be obtained are the following:

- The simulation process allows for the prediction of microstructures and properties for different alloys and their metallurgy. For example, properties such as hardness, tensile strength, yield limit can be estimated.
- The simulation process allows to establish the appropriate casting design and specific process parameters to control the manufacturing process.
- It is possible to evaluate the flow of molten metal through the mould and thus establish the likely defects that will be caused by it.
- Using the casting simulation as a virtual test foundry in the computer, it is possible to do a number of what-if-scenarios, to establish the optimum casting design and foundry process.
- It is also possible to identify the residual stresses of the castings because of the process parameters and estimate the fatigue life of the part.

MagmaSoft

Magmasoft (www.magmasoft.com) is one of the casting simulation software widely used in the foundry industry for simulating casting processes and analysing casting defects. Magmasoft predicts casting quality by simulating mould filling, solidification and cooling, stresses and strains, and microstructure formation, utilising finite element formulation. A few examples of Magmasoft usage are presented here.

Magmasoft is extremely effective in predicting casting defects that arise from misruns, turbulence, gas porosity, slag entrainment, metal penetration and shrinkage. For example, shrinkage defects can be easily determined by viewing the solidification time, hot spot, solid fraction, liquid fraction, temperature, and porosity results. Shrinkage porosity occurs in locations with hot spots where liquid metal still remains after feeding becomes difficult.

Figure 4.37 illustrates an example of the way Magmasoft 5.2 is utilised to perfect the gating design to avoid shrinkage cavities. The casting was simulated using steel as the pouring metal and the design of the casting is shown in Fig. 4.37(a). The green-sand mould and the cores were made using zircon phenolic urethane cold-box process. As shown in Fig. 4.37(b), when the casting is simulated without any riser, prominent shrinkage cavities were noticed in the bigger end of the casting. Then a riser with feeding aids was added to remove the shrinkage cavities and its simulation is shown in Fig. 4.37(c). It can be noticed that the shrinkage cavities were not completely removed by riser alone. Then copper chills were used at the location to achieve the directional solidification and to move the shrinkage cavities into the riser as shown in Fig. 4.37(d).

Though predicting solidification defects is an important aspect of casting quality, defects that arise during mould filling also need to be taken into consideration. One important concern in the foundry industry is the turbulent flow of liquid metal while it travels through the gating system to the casting cavity. Turbulent flow of the molten metal can cause entrapped gases in the casting and mould erosion. Turbulence can be reduced by designing a good gating system which ensures a laminar flow of the molten metal. For example, sharp corners and abrupt changes in sections will cause a turbulent flow of metal. The filling results in Magmasoft can be used to predict and analyse defects caused due to turbulence. These results include air entrapment, air pressure, metal velocity et cetera. Figure 4.38 shows an example of a turbulent flow.

The two case studies discussed above are only a small part of the Magmasoft casting-simulation software capabilities. Magmasoft is capable of helping the foundry industry from designing the pattern to machining the final casting. Also, Magmasoft is very useful for simulating different alloys of a metal and predicting many of the properties of the alloy such as liquidus temperature and solidus temperature. This can be done by placing thermocouples in the metal before simulation. Thermocouples can also be placed in the mould and the mould-metal interface to effectively calculate thermal diffusivity of the mould aggregate used.

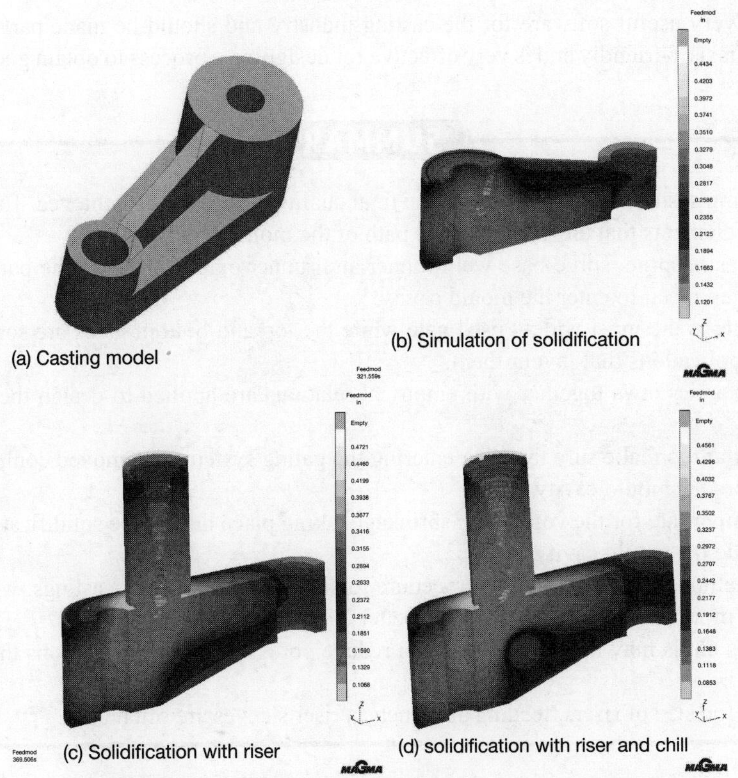

FIG. 4.37 *Use of Magmasoft for casting-process design*

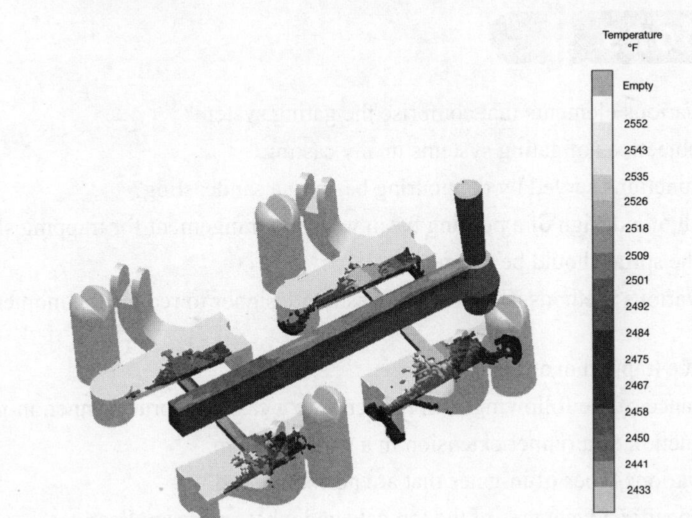

FIG. 4.38 *Magmasoft representation of the casting showing the molten metal flow (laminar or turbulent) during the casting filling process*

Magmasoft is very useful software for the casting industry and should be made part of the casting and process design. It is user-friendly and is very effective for designing a process to obtain good quality castings.

SUMMARY

Gating and risering systems are important for the final quality of the casting achieved. To this extent, there are a number of elements that are present in the path of the molten metal.

- Pouring basin, sprue, sprue-base well, runner and runner extension serve the purpose of allowing clean molten metal to enter the mould cavity.
- Parting gate is the most widely used gate while the top and bottom gates are sometimes used for specific applications that favour them.
- Fluid mechanics laws together with empirical relations are applied to design the optimum gating system.
- It is important to make sure that slag entering the gating system be removed completely before the metal enters the mould cavity.
- Risers compensate for the volumetric shrinkage taking place during the solidification of the molten metal inside the mould cavity.
- Caine's method provides a simple procedure to size the risers in sandcastings, while the modulus method is more elaborate and provides a better design.
- Sometimes chills may need to be added to reduce porosity at isolated sections that are not fed by risers.
- To reduce the size of risers, feeding aids such as riser sleeves are often used.

Questions

4.1 What are the various elements that comprise the gating system?

4.2 Describe the objectives of gating systems in any casting.

4.3 What are the functions served by the pouring basin in a sandcasting?

4.4 Give the sketch of a design of a pouring basin with an arrangement for trapping slag.

4.5 Explain why the sprue should be tapered?

4.6 What are the various methods available to a casting designer to reduce the momentum of the molten metal?

4.7 How can slag be trapped in a runner?

4.8 Give the relevance of the following with reference to a casting: sprue, runner, in-gate.

4.9 Explain the function of a runner extension in a gating system.

4.10 What are the various types of in-gates that are normally used?

4.11 What are the specific advantages of the top gate and what are its applications?

4.12 Compare the bottom gate with the top gate vis a vis its merits and demerits.

4.13 What is the most commonly used type of gate? Explain the reasons for its choice.

4.14 Explain the reasons which make a riser a necessity in some castings.

4.15 Describe the types of risers and their uses with suitable sketches.
4.16 What are the essential conditions that are to be kept in mind while designing risers?
4.17 Explain the use of the chill with an example.
4.18 Define casting yield and explain its importance.
4.19 State two functions served by chilling of a casting.
4.20 'The size of gates should be neither too small nor too large'. Discuss the statement.
4.21 Show by means of sketches the bottom and top gating systems.
4.22 Explain the parameters which cause the pouring time of a given casting to vary.
4.23 Define choke area in castings. Explain reasons why normally sprue base is used as the choke area for ferrous castings.
4.24 Why is it necessary to streamline the gating-system elements? Explain with examples.
4.25 What is the ideal profile of a sprue? Explain how it is actually practised.
4.26 What are the criteria to be used for designing the pouring basin?
4.27 Explain about the various methods available for trapping slag in pouring basin.
4.28 Differentiate between pressurised and unpressurised gating systems with reference to the applications.
4.29 Why is it necessary to use unpressurised gating system for the nonferrous metals?
4.30 What considerations should be observed while positioning sprues based on the casting geometry and size?
4.31 How are the in-gate position and size determined?
4.32 Describe how runner extension is helpful for good casting quality.
4.33 Write a short note on whirl gate.
4.34 Explain the formation of shrinkage cavities in steel castings.
4.35 Briefly explain the rationale of Caine's empirical rules for risering.
4.36 Compare the modulus method with that of Caine's method of fixing riser dimensions.
4.37 "NRL Method is a simplification of Caine's". Comment on this statement with a simple example.
4.38 How is the shape factor obtained in the case of NRL method for cylindrical objects?
4.39 What is the methodology for choosing the riser connection to the casting?
4.40 What are the methods available to a casting designer to increase the casting yield?
4.41 What precautions should one take while choosing chills?
4.42 Describe briefly the feeding aids used in steel castings.

Problems

4.1 Examine the following components from the sandcasting point view, and comment as to the features good and bad from processing. Design the pattern, core and core boxes by considering the various allowances and other casting requirements. Work out the details of gating design, explaining at each stage the choice of parameters and values chosen. Give the details of sand mould with dimensions and the arrangement of gating elements.

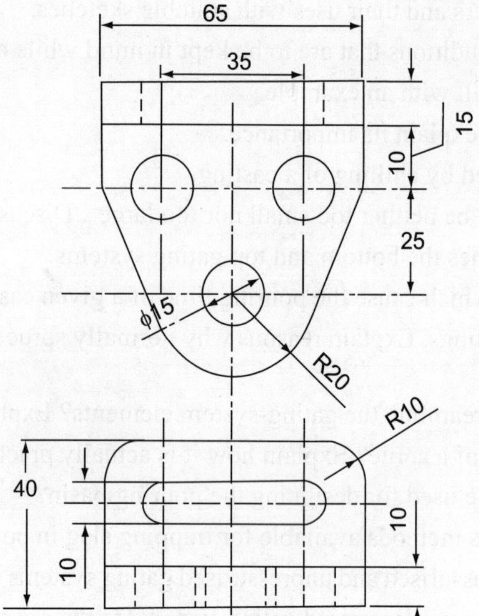

Fig. 4.39 *Pattern example*

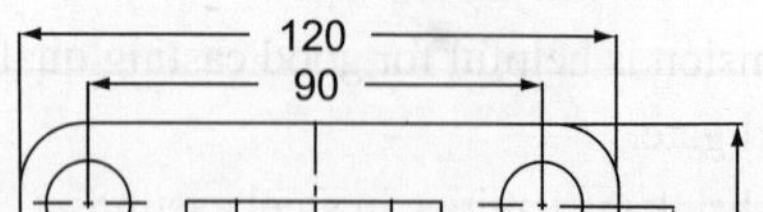

Fig. 4.40 *Pattern example 2*

4.2 Two castings are moulded in green sand. They differ in weight by a factor of 3.8 but they are both cubes. An experiment has shown that the lighter casting solidifies in 8.7 minutes. How much time would you estimate that it would take for the larger casting to solidify. [21.18 min]

4.3 Below are given the solidification times for top cylindrical risers which are 100 mm diameter × 100 mm high for steel, copper and aluminium.

Solidification times, min.

Metal	No Treatment	Insulating Leeve	Radiation Shield	Insulation Shield
Steel	5	7.5	13.4	43.0
Copper	8.2	15.1	14.0	45.0
Aluminium	12.3	31.1	14.3	45.6

(a) Determine the effective constant for Chvorinov's rule from the data for each of the metals listed above.
(b) Discuss the relative effect of radiation shielding for each metal listed, in terms of specific heat, latent heat of fusion and pouring temperature.
(c) Repeat item (b) in terms of using an insulating sleeve only.
(d) Discuss why the solidification times for steel and aluminium risers are about the same when they have both insulating sleeves and radiating shielding.

Multiple Choice Questions

4.1 Pouring basin is used in sandcastings to
(a) reduce the momentum of the molten metal as it enters the sprue
(b) remove the slag and dirt present in the molten metal
(c) reduce the mould erosion
(d) all of the above

4.2 The best shape for sprue in sandcasting is
(a) straight cylindrical
(b) tapered
(c) either straight cylindrical or tapered
(d) shape will not matter

4.3 To reduce the mould erosion in a sandcasting,
(a) sprue is tapered
(b) cylindrical sprue is used
(c) sprue base well is provided
(d) runner is made trapezoidal

4.4 The preferred shape of a runner in sandcasting is
(a) spherical (b) cylindrical
(c) trapezoidal (d) rectangular

4.5 Advantage of providing a top gate in a sandcasting is
(a) favourable temperature gradient is achieved
(b) slow filling of the mould cavity
(c) smooth flow of metal into the mould cavity
(d) none of the above

4.6 The advantage of providing a bottom gate in a sandcasting is
(a) favourable temperature gradient is achieved
(b) fast filling of the mould cavity
(c) smooth flow of metal into the mould cavity
(d) none of the above

4.7 The advantage of providing a parting gate in a sandcasting is
(a) favourable temperature gradient is achieved
(b) best gating with minimum defects

(c) smooth flow of metal into the mould cavity
(d) none of the above

4.8 Riser is used in a sandcasting to
(a) reduce the slag inclusion
(b) reduce defects due to air aspiration
(c) reduce mould erosion
(d) reduce shrinkage cavities

4.9 To improve the directional solidification for difficult casting geometries, use a
(a) chill
(b) parting gate
(c) step gate
(d) runner extension

4.10 Pouring time for a given casting is increased by
(a) using green sand mould
(b) using sprue base well
(c) increased section thickness of the casting
(d) using wide runners

4.11 The following is a point to be considered during the design of in-gate in a sand mould:
(a) In-gate should not be located near a protruding part of the mould to avoid the striking of vertical mould walls by the molten metal stream.
(b) In-gates should preferably be placed along the longitudinal axis of the mould wall.
(c) In-gates should not be placed near a core print or a chill.
(d) All of the above

4.12 The following type of gate is used for trapping the slag in a sand mould:
(a) Bottom gate
(b) Whirl gate
(c) Parting gate
(d) Step gate

4.13 Shrinkage cavity in a sandcasting can be reduced by
(a) using a riser
(b) pouring more molten metal into the mould cavity
(c) pouring just sufficient molten metal into the mould cavity
(d) using a runner extension

4.14 Riser for a sandcasting can be designed using
(a) Caine's method
(b) naval research laboratory method
(c) modulus method
(d) all of the above

4.15 To reduce the risering requirement in a sandcasting,
(a) use a runner extension
(b) use a strainer core
(c) use a riser sleeve
(d) none of the above

Answers to MCQs

4.1 (d)	4.2 (b)	4.3 (c)	4.4 (c)	4.5 (a)
4.6 (c)	4.7 (b)	4.8 (d)	4.9 (a)	4.10 (c)
4.11 (d)	4.12 (b)	4.13 (a)	4.14 (d)	4.15 (c)

CASE STUDY

DUCTILE IRON CASTING

Ductile iron is used in several engineering applications like gearboxes, cable drums, etc., due to its high strength and ductility. Unlike the gray cast iron, ductile iron has good impact and fatigue resistance owing to its nodular graphite content. However, due to the presence of magnesium in ductile iron, its casting is difficult. Gray iron castings have a problem of low elongation value and any poor casting procedures can lead to large stresses in the castings. In this case study the results of a research that was carried out to obtain thin walled casting of ductile iron that is lighter than the forged aluminum part but possesses higher strength, damping and is cheap is presented.

In this study moulds were prepared using silica sand and ductile iron was produced using an electric induction furnace with pig iron, scrap and silicon as raw materials. The mould had a mixing chamber to ensure proper mixing of the liquid iron after adding magnesium. The volume of the mixing chamber was kept equal to or more than the volume of the casting. The produced castings were tested for their mechanical properties (Table 1). In addition, it was found that thinner sections had higher nodule count. Thus, it was possible to control the formation of nodules without actually placing chills and controlling the cooling/solidification.

TABLE 1 Mechanical properties of castings

Material	Ultimate Tensile Strength, MPa	Hardness, BHN	Yield Strength, MPa
Aluminium Forging	290	118	149
Ductile iron casting	530	188	325

It may be clearly noted that all the properties are better for the ductile iron castings than aluminum castings. The control arm whose weight of the thin wall section made of ductile iron was less (480 g) compared to the forged aluminum part (585 g).

Computer simulations (FEA) were carried out using Solid Works® software. At the same loading conditions, the maximum von Mises stress in aluminium forging was 82 MPa, while for ductile iron casting it was 177 MPA. It may be noted that the highest permissible stress for ductile iron is 325 MPa, so the stresses are within limits.

Thus, the use of ductile iron in thin section castings resulted in improved strength and hardness of the castings and reduced weight of the member as well as the cost.

Górny, M. (2009). Thin wall ductile iron casting as a substitute for aluminum alloy casting in automotive industry. *Archives of Foundry Engineering*, *9*(1), 143–146.

CHAPTER 5

Melting and Casting Quality

Objectives

After completing this chapter, the reader will be able to

- Understand the functioning of cupola furnace for melting cast iron
- Calculate the charge quantities to be used in cupola for the required final metal composition
- Identify different types of furnaces used in foundries
- Understand the various requirements for casting cleaning
- Learn different types of casting defects and their remedial measures
- Understand the application of different non-destructive testing methods
- Design castings to ensure simple moulding with less number of defects

5.1 MELTING PRACTICES

After moulding, melting is the major factor which controls the quality of the casting. There are a number of methods available for melting foundry alloys such as pit furnace, open hearth furnace, rotary furnace, cupola furnace, etc. The choice of the furnace depends on the amount and the type of alloy being melted. For melting cast iron, cupola in its various forms is extensively used basically because of its lower initial cost and lower melting cost. We will see more details of the cupola in the following article.

5.1.1 Cupola

Cupola consists of a cylindrical steel shell with its interior lined with heat resisting fire bricks. It consists of drop doors at the bottom after closing which a proper sand bed could be prepared. This sand bed provides the necessary refractory bottom for the molten metal and the coke. Immediately above the sand bed is the metal tapping hole which is initially closed with clay called 'bot' till the molten metal is ready for tapping. Above the metal tap hole normally in a position opposite to it is the slag hole through which the slag generated during the melting process is tapped.

Above the slag hole is the wind box which is connected to the air blowers supplying the requisite air at a given pressure and quantity. The air enters the cupola through the tuyeres. A little above the charging platform is the charging hole in the shell from where the charge consisting of a combination of pig iron, scrap iron, coke and fluxes, is put into the cupola. The refractory lining above the charge door need not necessarily be as thick as that below, since it is not exposed to much heat, as represented in Fig. 5.1.

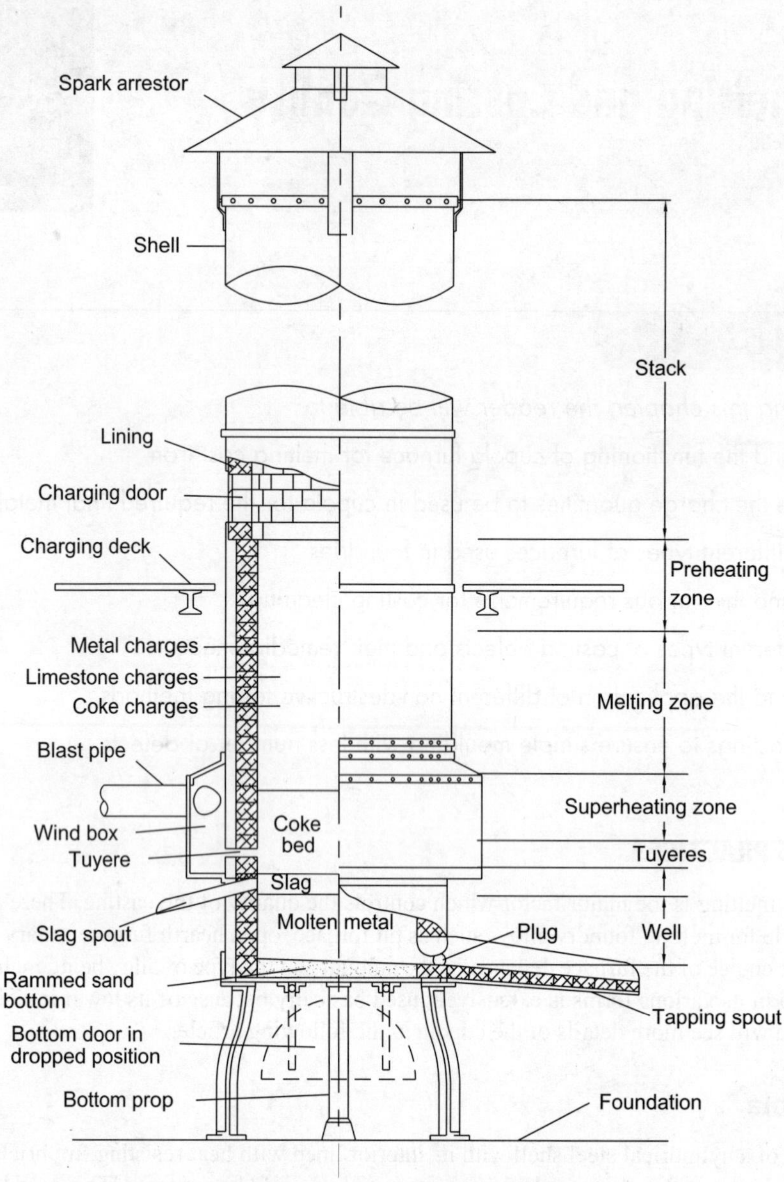

Fig. 5.1 *Cross-section of a cupola*

To operate the cupola, first, the drop doors at the bottom are closed and a sand bed with a gentle slope towards the tap hole is rammed. Then a coke bed of suitable height is prepared above the sand bottom and ignited through the tap hole or any other hole. When the coke bed is properly ignited, alternate layers of charge, flux and coke are alternately fed into the cupola through the charge door maintaining the necessary proportions and rate of charging. The charge is then allowed to soak in the heat for a while, and then the air blast is turned on. Within about 5 to 10 minutes, the molten metal is collected near the tap hole. When enough molten metal is collected in the well of the cupola, the slag is drained off through the slag hole before opening the tap hole. The molten metal is collected in the ladles and then transported to the moulds into which it is poured with a minimum time loss.

The charge needed to produce cast iron, essentially consists of pig iron, cast iron scrap and steel scrap when alloy cast iron is needed. The proportions of these depend on their chemical compositions and on the final chemical composition of cast iron desired.

The fluxes are added in the charge to remove the oxides and other impurities present in the metal. The flux most commonly used is lime stone ($CaCO_3$) in a proportion of about 2 to 4% of the metal charge. Some of the other fluxes that may also be used are dolomite, sodium carbonate and calcium carbide. The flux is expected to react with the oxides and form compounds which have low melting point and also are lighter. As a result, the molten slag tends to float on the metal pool and thus, can very easily be separated. Typical operation characteristics of cupola are presented in Table 5.1.

TABLE 5.1 **Cupola operation Data**

Diameter of Melting Zone, m	Melting Rate Tonnes/h, Metal: Coke Ratio		Blast Rate, m^3/h	Blast Pressure, kPa	Typical Charge, kg			Bed Height above Tuyeres, m	Shaft Height from Tuyeres to Charge Door Sill, m
	10:01	08:01			Coke	Iron	Limestone		
0.50	1.97	1.57	1,340	10.20	20	200	7	1.00	2.50
0.60	2.84	2.46	1,940	10.50	28	284	9	1.00	3.00
0.80	5.11	4.36	3,450	11.20	51	510	17	1.05	3.00
1.00	7.99	6.83	5,380	11.70	80	800	26	1.05	3.50
1.20	11.50	9.79	7,750	12.70	115	1150	38	1.10	4.00
1.40	15.60	13.33	10,600	13.40	157	1570	52	1.10	4.00
1.60	20.44	17.41	13,800	14.40	200	2040	67	1.10	4.50
1.80	25.88	22.05	17,450	15.40	260	2590	85	1.15	5.00
2.00	31.95	27.22	21,550	17.20	320	3200	106	1.15	5.00

A variation of the cupola is called "hot blast cupola". In this, the air supply is pre heated to a temperature of 200 to 400°C with the help of the hot gases coming out of the stack or by a separate heat input. In either case, the equipment gets complicated by the addition of the extra pre heaters and the air circulation equipment. The main advantage gained is that the amount of heat required by the cupola gets reduced. This in turn, reduces the contact of the metal with the coke and air thus reducing the carbon and sulphur pickups as well as the oxidation losses. Because of the additional equipment and the extra care needed for operation, the hot blast cupolas are used only in shops that require large amounts of metal to be melt in a continuous basis.

5.1.2 Charge Calculations

It is very important in the foundry to know the final composition of the metal being obtained, so as to control it properly. The elements in the final analysis are essentially the sum total of what is contained in each of the charge ingredients, with some losses or pick up in the cupola. Out of the various elements, the ones that are relevant are carbon, silicon, manganese and sulphur.

As the charge comes through the coke bed, some amount of carbon is picked up by the metal depending on the temperature and the time when the metal is in contact with the coke. However, it may be reasonable to assume a pickup of the order of 0.15% carbon.

Silicon is likely to get oxidised in the cupola and therefore, a loss of 10% of total silicon contained in the charge is normal. Under the worst conditions, it may go as high as 30%. If the silicon content in the charge is not high, extra silicon can be added by inoculating the metal in the ladle with ferro silicon.

Manganese is also likely to be lost in the melting process. The loss could be of the order of 15 to 20%. Loss of manganese in the final analysis can be made up by the addition of ferro manganese.

Similar to carbon, sulphur is also likely to be picked up from coke during melting. The pickup depends on the sulphur content of the coke but a reasonable estimate could be 0.03 to 0.05%.

Examples are presented below for estimating the final analysis of the melt.

Example 5.1

Estimate the final composition of the cast iron produced with the following charge compositions and proportions.

Solution

	Carbon	Silicon	Manganese	Sulphur	Phosphorous	% In the Charge
Pig iron 1	3.50	2.50	0.40	0.01	0.40	40
Pig iron 2	3.20	1.50	1.00	0.02	0.60	35
Gray iron scrap	3.20	2.50	0.50	0.10	0.40	25

Let us now analyse the total amount of elements present in 1 ton (1000 kg) of charge, assuming carbon pick up as 0.15%, sulphur pick up as 0.05%, silicon loss as 10% and manganese loss as 20%.

Charge Materials	kg	%	Carbon, %		Silicon, %		Manganese, %		Sulphur, %	
			in Material	in Charge	in Material	in Charge	in Material	in Charge	in Material	in Charge
Pig iron 1	400	40	3.50	3.5 × 0.40 = 1.40	2.50	2.5 × 0.40 = 1.00	0.40	0.4 × 0.40 = 0.160	0.01	0.01 × 0.40 = 0.004
Pig iron 2	350	35	3.20	3.2 × 0.35 = 1.12	1.50	1.5 × 0.35 = 0.525	1.00	1.00 × 0.35 = 0.35	0.02	0.02 × 0.35 = 0.007
Grey iron scrap	250	25	3.20	3.2 × 0.25 = 0.80	2.5	0.25 × 0.25 = 0.625	0.50	0.50 × 0.25 = 0.125	0.10	0.10 × 0.25 = 0.025
Total	1000	100								
Total in charge, %				3.32		2.15		0.635		0.036
Change in cupola				+ 0.15		– 0.215		– 0.127		+ 0.050
Estimated composition, %				3.47		1.935		0.508		0.086

The foundry engineer can actually control the final analysis by actually trying with various mixes of charge materials available in the foundry to obtain an economical melt. The following example presents one such case.

Example 5.2

In a foundry, it was required to obtain a cast metal with the following composition: carbon 3.20 to 3.60%, silicon 2.30 to 2.60%, manganese 0.60 to 0.80%, sulphur 0.08% maximum, and phosphorous 0.40 to 0.60%. If the following raw materials are available, estimate the best charge proportions.

Solution

	Carbon	Silicon	Manganese	Sulphur	Phosphorous
Pig iron 1	3.50	3.00	1.00	0.02	0.40
Pig iron 2	3.20	1.50	0.50	0.01	0. 80
Pig iron 3	3.50	2.50	0.80	0.02	0.50
Scrap 1	3.50	1.80	0.60	0.08	0.50
Scrap 2	3.20	1.20	0.60	0.10	0.40
Ferro silicon		50.00			

Looking at the raw materials available, pig iron 2 has got very low silicon content and therefore it should not be used. Similarly the scrap 2 which has low silicon and high sulphur may also be eliminated. So the final choices are the pig iron 1 and 3 and scrap 1. Let the charge be 40% of scrap with the pig irons in equal proportions of 30% each.

Let us do the analysis for 1000 kg with the carbon pick up assumed as 0.15%, sulphur as 0.05%, the oxidation losses of silicon as 10% and that of manganese as 20%.

Charge Materials	kg	%	Carbon, %		Silicon, %		Manganese, %		Sulphur, %	
			in Material	in Charge	in Material	in Charge	in Material	in Charge	in Material	in Charge
Pig iron 1	300	30	3.50	$3.5 \times 0.30 = 1.05$	3.00	$3.0 \times 0.30 = 0.90$	1.00	$1.0 \times 0.30 = 0.30$	0.02	$0.02 \times 0.30 = 0.006$
Pig iron 3	300	30	3.50	$3.5 \times 0.30 = 1.05$	2.50	$2.5 \times 0.30 = 0.75$	0.80	$0.80 \times 0.30 = 0.24$	0.02	$0.02 \times 0.30 = 0.006$
Scrap 1	400	40	3.50	$3.5 \times 0.40 = 1.40$	1.8	$1.8 \times 0.40 = 0.72$	0.60	$0.60 \times 0.40 = 0.24$	0.08	$0.08 \times 0.40 = 0.032$
Total	1000	100								
Total in charge, %				3.50		2.37		0.78		0.044
Change in cupola				+ 0.15		– 0.237		– 0.156		+ 0.050
Estimated composition, %				3.65		2.133		0.624		0.094
Required composition, %				3.2 to 3.6		2.3 to 2.6		0.6 to 0.8		0.08 max

With the chosen charge composition, the final analysis differs from the required one with carbon, silicon and sulphur percentages. To reduce the carbon and sulphur percentages, we may have to use pig iron 2 which may reduce the silicon percentage but that can be made up by adding suitable amount of ferro silicon.

The reduction in sulphur that could be achieved by the addition of 1 kg of pig iron 2 for scrap 1 is

$$\frac{0.08-0.01}{100} = 0.0007 \text{ kg}$$

The reduction in carbon that could be achieved by the addition of 1 kg of pig iron 2 for scrap 1 is

$$\frac{3.50-3.20}{100} = 0.003 \text{ kg}$$

The excess sulphur present in 1000 kg is

$$\frac{1000\,(0.094-0.080)}{100} = 0.14 \text{ kg}$$

The excess carbon present in 1000 kg is

$$\frac{1000\,(3.65-3.60)}{100} = 0.50 \text{ kg}$$

To reduce the sulphur to the desirable limit, the amount of pig iron 2 to be substituted is

$$\frac{0.14}{0.0007} = 200 \text{ kg}$$

To reduce the carbon to the desired limit, the amount of pig iron 2 to be substituted is

$$\frac{0.50}{0.003} = 167 \text{ kg}$$

Hence, substitution of 200 kg of pig iron would be sufficient to give the desired analysis. To confirm this let us do a fresh analysis.

Charge Materials	kg	%	Carbon, %		Silicon, %		Manganese, %		Sulphur, %	
			in Material	in Charge	in Material	in Charge	in Material	in Charge	in Material	in Charge
Pig iron 1	300	30	3.50	3.5 × 0.30 = 1.05	3.00	3.0 × 0.30 = 0.90	1.00	1.0 × 0.30 = 0.30	0.02	0.02 × 0.30 = 0.006
Pig iron 2	200	20	3.20	3.2 × 0.20 = 0.64	1.50	1.5 × 0.20 = 0.30	0.50	0.5 × 0.20 = 0.10	0.01	0.01 × 0.20 = 0.002
Pig iron 3	300	30	3.50	3.5 × 0.30 = 1.05	2.50	2.5 × 0.30 = 0.75	0.80	0.80 × 0.30 = 0.24	0.02	0.02 × 0.30 = 0.006
Scrap 1	200	20	3.50	3.5 × 0.20 = 0.70	1.8	1.8 × 0.20 = 0.36	0.60	0.60 × 0.20 =0.12	0.08	0.08 × 0.20 = 0.016
Total	1000	100								
Total in charge, %				3.44		2.31		0.76		0.03
Change in cupola				+ 0.15		–0.231		– 0.152		+ 0.050
Estimated composition, %				3.59		2.079		0.608		0.08
Required composition, %				3.2 to 3.6		2.3 to 2.6		0.6 to 0.8		0.08 max

It is not possible to increase silicon any further, and therefore, the only way out is to add ferro silicon of requisite amount to make up the short fall.

Amount of silicon to be added for 1000 kg is

$$\frac{1000(2.300 - 2.079)}{100} = 2.21 \text{ kg}$$

This can be made up by adding 5 kg of ferro silicon in place of scrap 1.

$$\text{Extra silicon added} = \frac{5(50.0 - 1.80)}{100} = 2.41 \text{ kg}$$

So, the final mix of the charge is

Pig iron 1 = 300 kg = 30 %
Pig iron 2 = 200 kg = 20 %
Pig iron 3 = 300 kg = 30 %
Scrap 1 = 195 kg = 19.5 %
Ferro silicon = 5 kg = 0.5 %

In the previous example, we have proceeded with a trial and error method and have not considered the cost of various constituents of the charge. A more appropriate analysis would be to take into account the cost of each of the constituents of the charge and the chemical compositions. It is possible to develop this methodology in a spread sheet thereby making possible to explore various scenarios before finalizing the optimum mix of the charge. This can be done by formulating the linear programming problem and solving it by the simplex method.

5.1.3 Other Furnaces

Besides cast iron, foundries melt a large number of different metals and alloys. The melting requirements and characteristics required of furnaces greatly differ for different metals. In Table 5.2, the heat required for melting various foundry materials is given.

TABLE 5.2 Properties and Heat required for melting

Metal	Melting Point °C	Mean Specific Heat cal/g °C	Latent Heat of Fusion cal/g	Heat Required for Melting 1 kg kJ
Iron	1537	0.141	65.0	1168
Nickel	1453	0.131	72.1	1084
Copper	1083	0.105	48.9	674
Aluminium	660	0.237	92.7	1022
Magnesium	650	0.289	85.6	1122
Zinc	420	0.101	26.3	285
Lead	327	0.034	5.7	67
Tin	232	0.058	14.2	113

Most of the foundries operate on a batch basis. That means a number of sand moulds are prepared and kept ready for pouring before the molten metal is prepared. This process may take a few days to weeks depending upon the size and nature of the foundry plant. Thus it becomes necessary only to start melting may be once a week or so.

The main types of furnaces that are generally used in foundries are:

- Open hearth
- Rotary
- Crucible, and
- Immersion heated

Based on the source of heating, they can be classified as

- Electrical heating (Arc, Resistance or Induction)
- Fossil fuel fired (Solid, Oil or Gaseous fuel)

Cupola has been the most widely used furnace for melting cast iron. This is because of the low cost involved in its operation. However, as seen earlier, less control of the final quality and the losses involved would call for some change in the choice. Hence for other material melting in place of solid fuel, liquid and/or gas fuel and electric furnaces are being increasingly used in view of the better control of molten metal provided by them and lower melting losses. However, these are more expensive compared to the solid fuel fired furnaces and therefore the higher cost is to be justified based on the better control of quality achieved in terms of the composition and temperature.

In the electric furnaces the resistance type heating is generally used for holding furnaces to maintain the liquid metal at a certain temperature for nonferrous alloys such as for die casting. However, the electric furnaces in view of their high degree of temperature control and flexibility of operation have been widely used for melting of small to medium sized castings in ferrous as well as nonferrous alloys.

Electric Arc Furnace

For heavy steel castings, the open hearth type of furnaces with electric arc or oil fired would be generally suitable in view of the large heat required for melting. Electric arc furnaces are more suitable for ferrous materials and are larger in capacity. This type of furnace draws an electric arc that rapidly heats and melts the charge material, as shown in Fig. 5.2. The bowl-shaped bottom of the furnace, called the hearth, is lined with refractory bricks and granular refractory material. Heat is directly transferred to the charge metal from the electrode arc. The furnace has a tilting mechanism allowing it to be tilted forward for metal tapping or backward for deslagging. Once the melt is ready to pour, the electrodes are raised through the roof and the furnace is tilted to pour the molten metal into a receiving ladle.

Crucible Furnace

Smaller foundries generally prefer the crucible furnace. The crucible is generally heated by electric resistance or gas flame. In these the metal is placed in a crucible of refractory metal and the heating is done to the crucible thus there is no direct contact between the flame and the metal charge. This type of melting is very flexible since it suits a variety of casting alloys. Degassing and any metal treatment can be completed in the crucible before it is removed for pouring. Melt quality and temperature can also be controlled reasonably well.

Induction Furnace

The induction furnaces are used for all types of materials, the chief advantage being the heat source is isolated from charge, enabling slag and flux getting necessary heat directly from the charge instead of the heat source. The stirring effect of the electric current would cause fluxes to be entrained in the melt if they are mixed along with the charge. Therefore, flux is generally added after switching off the current to the furnace. Then sufficient time must be allowed for the oxides to be removed by the flux as slag before transferring the metal for pouring. High frequencies help in stirring the molten metal and thus help in using the metal swarf (chips). Thus low cost raw materials could be used and at the same time better control of temperature and composition can be achieved.

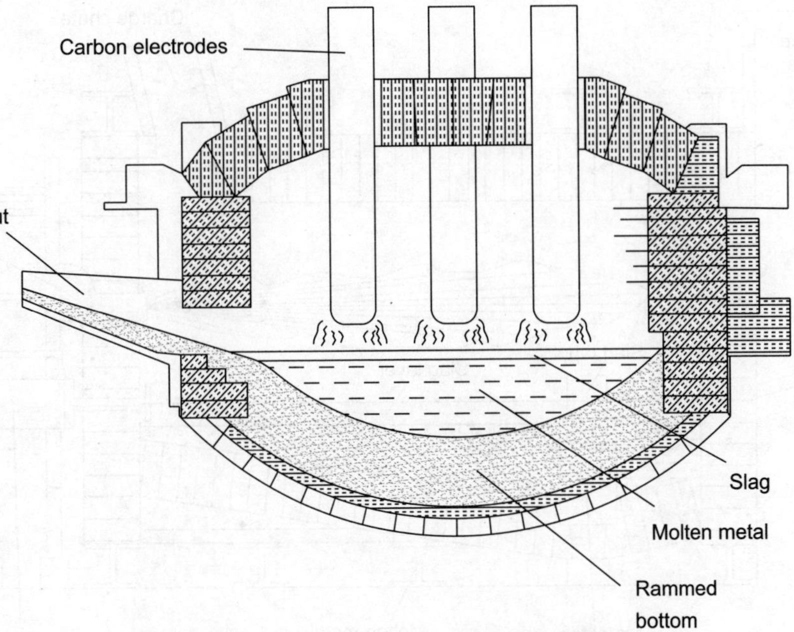

FIG. 5.2 *Cross-section of an Electric arc furnace*

Coreless induction furnaces use a refractory envelope that contains the metal and surround that by the electric coil. The charge acts as a single secondary turn, producing heat through eddy current flow when power is applied to the primary coil surrounding the charge. When the metal melts, these electromagnetic forces also produce a stirring action. Mixing and melting rates can be controlled by carefully selecting frequency and power.

Advantages offered by induction furnaces are higher yield, faster start up, cleaner melting, natural stirring and compact installation.

Reverberatory Furnace

In these furnaces the fuel burners fire within a refractory hood above the metal bath. These are generally used to melt large amounts of metal for example, aluminium to supply to holding furnaces such as those used with pressure die casting machines. These use gas fired burners located generally high in the furnace transferring the heat by radiation to the walls and roof (Fig. 5.3). As the walls and roof become incandescent they radiate the heat to the metal bath. These furnaces are simple and have relatively low capital cost. Thus these are generally used for melting large volumes of metal.

5.1.4 Inoculation Practice for Gray and Ductile Iron

The inoculation of cast iron is the addition of certain silicon alloys to molten iron to produce changes in graphite distribution, improvements in mechanical properties and a reduction of the chilling speed. Inoculation can be broadly defined as that being carried out while the molten metal is in the furnace or while it is in transit to the mould, known as post-inoculation. Post-inoculation is more commonly used. The post-inoculation is further divided into three types; the ladle, in-stream or in the mould.

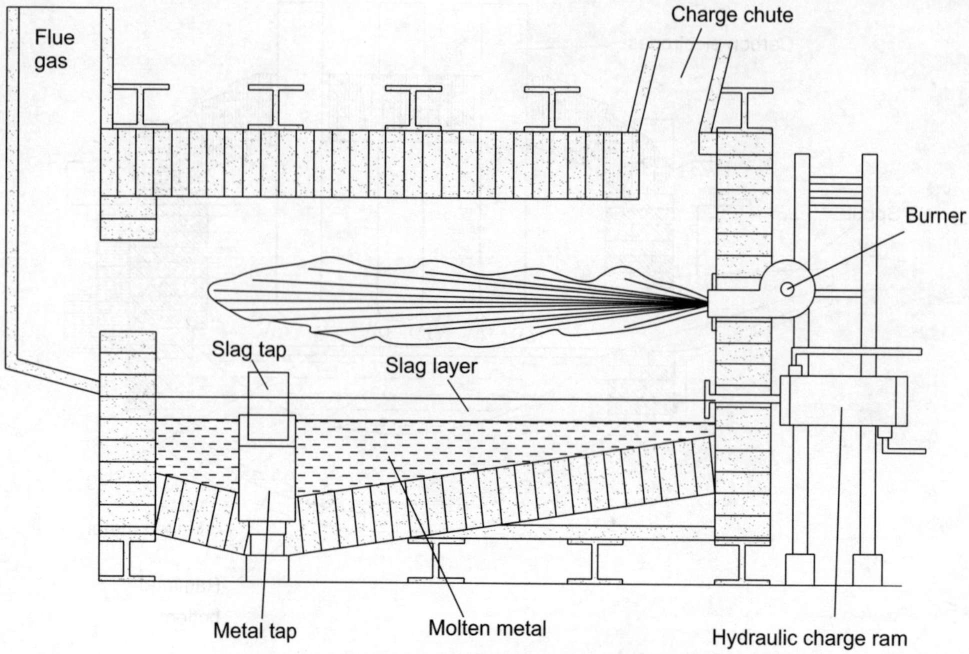

Fig. 5.3 *Cross-section of a reverberatory furnace*

The inoculation requirement becomes more important for the production of ductile iron than grey iron. The requirement of inoculants for gray iron depends upon the section size. While the ductile iron needs considerably more inoculation due to the removal of most of the sulphur and oxygen from the melt during the treatment step and the carbide - promoting tendency of the magnesium alloy. Most inoculants have some Al and Ca with small addition of stronger inoculating elements such as Sr, Ba, etc. The stronger inoculants for gray and ductile are different and should not be interchanged. Sr base inoculants are not used in ductile iron as the Ca in the metal retards the effectiveness of the Sr. Table 5.3 gives some indication of the type of composition of a few of the commercially available inoculants.

Table 5.3 Some typical commercial inoculant compositions

Inoculant	%Si	%Ca	%Ba	%Sr	%Zr	%RE	%Al
Superseed®75	73–78	0.1 max	—	0.6–1.0	—	—	0.5 max
Superseed®50	46–50	0.1 max	—	0.6–1.0	—	—	0.5
Superseed®Extra	73–78	0.1 max	—	0.6–1.0	1.0–1.5	—	0.5
Superseed®Quantum	72–78	0.1 max	—	0.8–1.2	—	1.8–2.2	0.5
Ultraseed®	70–76	0.75–1.25	—	—	—	1.5–2.0*	0.75–1.25
Foundrisil®	73–78	0.75–1.25	0.75–1.25	—	—	—	0.75–1.25
Barinoc®	72–78	1.0–2.0	2.0–3.0	—	—	—	1.5
Zircinoc®	73–78	2.0–2.5	—	—	1.3–1.8	—	1.0–1.5
Reseed®	72–78	0.5–1.0	—	—	—	1.8–2.1	0.5–1.25
Alinoc®	70–75	0.5–1.5	—	—	—	—	3.5–4.5

*Rare Earth present as Cerium, total added sulphur and oxygen content of this inoculant is less than 1%.

® Elkem Foundry products

Ladle Inoculation

Ladle inoculation is the simplest and most flexible method to inoculate iron and being used while the iron is tapped from the furnace into the pouring ladle. The turbulence created during the pouring of the metal into the ladle will help in fast and homogeneous distribution of the alloy. Fading is the decrease in the effect of inoculant with time. Typically the inoculant effect will be good for the first 10 minutes and beyond that its effect will be reduced. If the metal in the ladle is not immediately poured after inoculation, then mould inoculation should be considered.

In-stream Inoculation

In-stream inoculation is normally used when automatic pouring lines are used in a foundry. In this system the inoculant is dispensed directly into the metal stream as it is being poured into the mould through a carrier such as compressed air. A hopper for the inoculant is mounted on the ladle with a limit switch to start when the ladle tilts. The inoculant is gravity fed into the stream as it enters the mould. In-stream units meter inoculant either by flow rate (grams/sec) or fixed mass (grams/mould).

In-mould Inoculation

The in-mould inoculation involves the placement of the inoculation alloy directly into the gating system, such as in a pouring basin, in the sprue well area or in suitable chambers in the running system as shown in Fig. 5.4. Inoculants used for this method may be in the form of crushed material, fine granular material bonded into pellets, as pre-cast slugs or blocks. Typically, crushed inoculant alloy of 20–70 mesh in size is used for this application and the recommended addition rate varies from 0.05 to 0.1%.

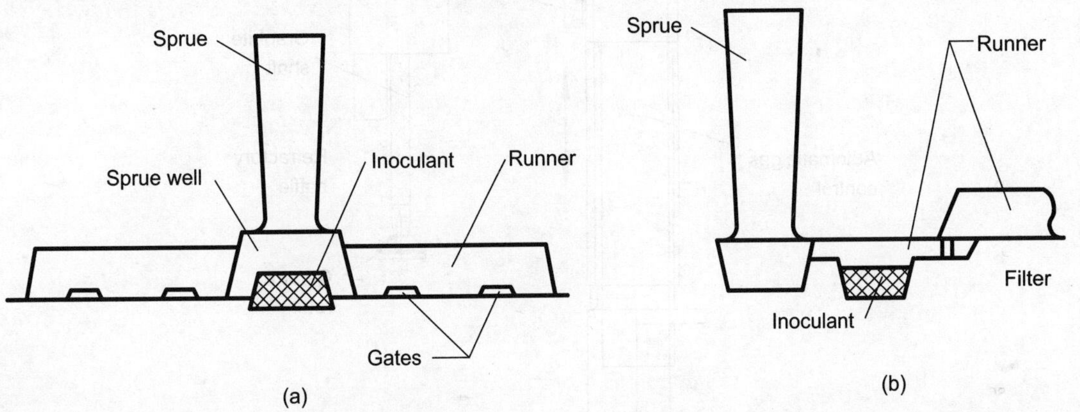

Fig. 5.4 *In-mould inoculation; (a) inoculant block placed in sprue well area; (b) granular inoculant placed in the reaction chamber in the runner.*

5.1.5 Degassing

Molten aluminium contains a large amount of dissolved hydrogen which should be expelled before it is poured in the moulds. This process is called degassing. During the solidification of aluminium alloys, dissolved hydrogen creates porosity that, left unchecked, is detrimental to the mechanical properties of aluminium alloy castings. The inert gas when purged through the melt collects the soluble hydrogen atoms, allowing a hydrogen molecule to form inside the lower pressure of the collector gas bubble. Pure argon is frequently injected into the liquid alloy through a submerged lance or bubbler, so that dissolved hydrogen enters the argon bubble prior to discharge into the environment. As these bubbles break the surface, aluminium is lost to oxidation by the furnace gases and entrapment in dross. Additionally, the use of chlorine creates environmental issues.

Another technique, rotor degassing system as shown in Fig. 5.5, proved to be more effective by producing smaller bubbles and hence in degassing. One of the advantages of this technique is that flux may be injected simultaneously with the degassing when gas is used as a carrier for different types of fluxes. The Foseco Metal Degassing Unit (MDU) has a graphite rotor, which will introduce inert gases near the bottom of the vessel generating very small bubbles and blending them with molten aluminium drawn into the rotor. As the bubbles float to the surface, they enlarge with a decrease in pressure that attracts the dissolved hydrogen to the bubble surface and into the bubbles that escape from the metal.

The Foseco MDU rotor operates as a pump, pulling the metal into the rotor chamber as shown in Fig. 5.6. The inert gas is transported to the rotor through the centre of the shaft which is then sheared into tiny gas bubbles that interact with the metal. The mixture is then ejected through quadrants around the perimeter of the rotor. The rotor, since it rotates continuously, assures fast and complete treatment. A refractory baffle board at the centre eliminates vortexing to minimize oxidation.

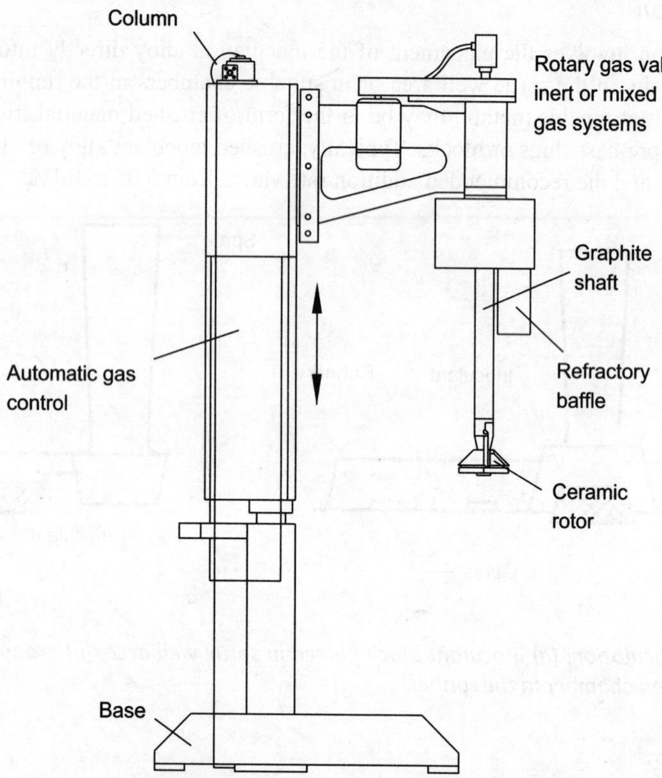

Fig. 5.5 *Rotary degassing unit (Foseco)*

5.1.6 Ladles

The molten metal from the furnace is tapped into the ladles at requisite intervals and then poured into the moulds. Depending on the amount of metal to be handled, there are different sizes of ladles. They may range

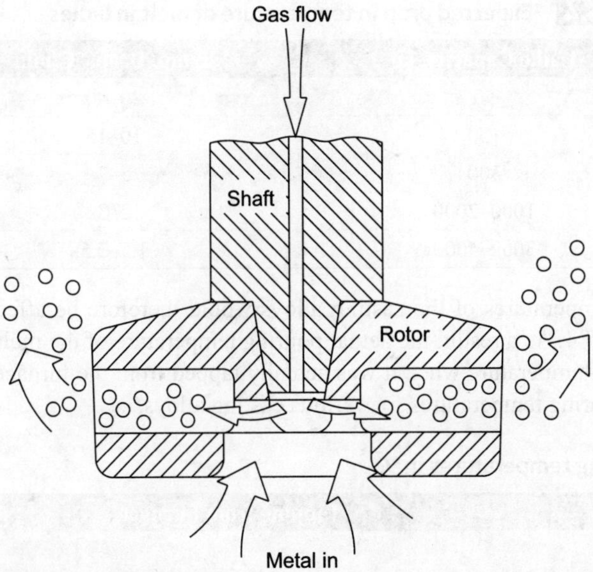

FIG. 5.6 *Close up view of the rotor action of the Foseco Rotary degassing unit*

between 50 kg to 30 tons depending upon the casting size. For gray cast iron, since the slag can be easily separated, top pouring ladles such as shown in Fig. 5.7(a) would be enough. But for steels, to separate the slag effectively, the metal is to be poured from the bottom with the help of the bottom pour ladle as shown in Fig. 5.7(a). The bottom pour ladle has an opening in the bottom that is fitted with a refractory nozzle. A stopper rod, suspended inside the ladle, pulls the stopper head up from its position thus allowing the molten alloy to flow from the ladle.

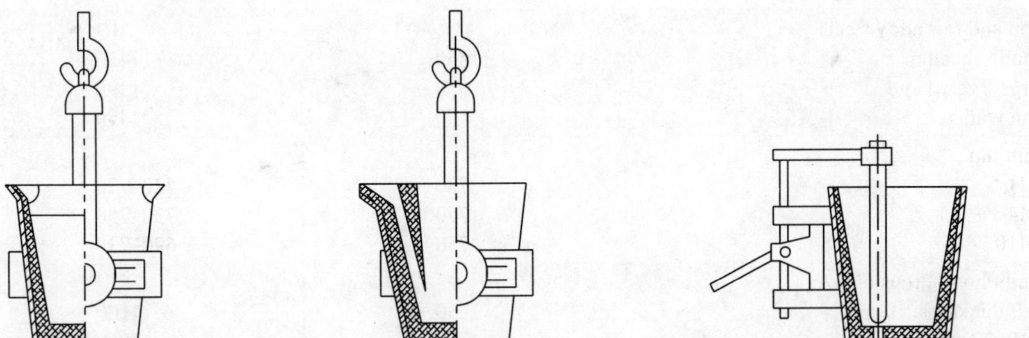

FIG. 5.7 *Different types of ladles used in the foundry*

As the metal in the ladle loses a large amount of heat to the surrounding atmosphere by radiation it is necessary to account for this drop in the temperature of the casting metal. Expected drop in temperatures as a basis of ladle capacity is given in Table 5.4. In the large ladles in view of the larger heat content, there is relatively small drop in temperature while in the small ladles the drop is appreciable. Hence, more speed in operation of the small ladles is desirable particularly in manual operations.

TABLE 5.4 Expected drop in temperature of melt in ladles

Ladle Capacity, kg	Temp. Drop, °C/min
50	20–40
150	10–15
300	5–7
1000–2000	2–3
3000–4000	1.5–2.5

The actual pouring temperatures of the casting alloys would therefore be affected by their heat loss in ladles as shown in Table 5.4. As a result, the actual pouring temperature of the melt as it enters the mould is different from that of the temperature when it was actually tapped from the furnace. The Table 5.5 gives an approximate range of pouring temperatures for various casting alloys.

TABLE 5.5 Metal pouring temperatures in °C

	Tapping from Furnace	Pouring into Moulds
Grey cast iron		
Small castings	1380	1300
Medium size castings	1360	1300
Large & very large castings	1360	1290
Thin walled castings	1380	1320
Malleable iron		
Based on wall thickness	1480	1380
Up to 4 mm	1450	1350
4–10 mm	1430	1350
10–20 mm	1410	1320
> 20 mm		
Carbon and low alloy steels		
Small - medium	1550	1420
Large - very large	1520	1390
Thin walled	1550	1450
Aluminium		
LM1	780	720–770
LM4	770	730–750
LM10	750	690–730
Tin and Phosphorous bronze		
Wall thickness: 10 mm	1150	1100
10–20 mm	1100	1050
20 mm	1050	1000

5.2 CASTING CLEANING

Having poured the metal into the mould, the casting is allowed to solidify and cooled in the mould itself. The sand mould is to be broken to extract the casting. But the breaking of the sand mould is to be done only when the casting is sufficiently cooled, since the metal at high temperatures has very little strength. The cooling time depends upon the casting section thickness, the total mass as well as the type of mould. Approximate values for cooling times are given in Table 5.6.

TABLE 5.6 Cooling times for sand castings

Mould Description	Mass, kg	Cooling Time, Hours
Green sand mould: simple to moderate complexity	< 20	0.4–0.75
Green sand mould: moderate to high complexity	21–100	0.75–1.5
Dry sand mould: moderate to high complexity	< 20	0.6–1.0
Dry sand mould: Moderate size	101–500 501–1000	2–6 6–9
Dry sand mould: complex shapes	1001–3000 3001–5000	8–18 18–30

Also, if the hot casting is exposed to air, there is likely to be faster and uneven cooling because of which the casting may warp, crack or induce thermal stresses beneath the skin. The moulding sand provides a uniform cooling medium for the casting while producing least amount of internal stresses.

Ideally the moulds should be broken at a temperature when no transformation occurs. For example, the breaking should be done at a temperature below 700°C for ferrous alloys. If the castings are thin and fragile, they should be removed at a temperature as low as 400°C, whereas for the heavier castings, a little higher temperature of 500°C may be suitable.

The casting as comes from the sand mould is attached with the gates, runners and risers. It is also possible that sand will adhere to the casting at a number of places that needs to be completely cleaned. The complete process of the cleaning of castings, called 'fettling', involves the removal of the cores, gates and risers, chipping of any of the unnecessary projections on surfaces and cleaning of the casting surface.

The dry sand cores can be removed simply by knocking off with an iron bar, by means of a core vibrator or by means of hydro blasting. The method depends on the size, complexity and the core material used.

The gates and risers can be removed by hammering, chipping, hack sawing, abrasive cut off or by flame or arc cutting. Removal of gates and risers can be simplified by providing a reduced metal section at the casting joint as shown in Fig. 4.36. For brittle materials such as gray cast iron, the gates can be easily broken by hitting with a hammer. For steel and other similar materials, sawing with any metal cutting saw like hack saw or band saw would be more convenient. For large size gates and risers, it may be necessary to use flame or arc cutting to remove them. Similarly, abrasive cut off may also be used for removal of gates. Most of the abrasive cut off can be carried out by portable grinding machines with an angled grinding head. Typical wheel speeds used are in the range of 45 to 80 m/s.

The casting surface after removal of the gates may still contain some rough surfaces left at the time of removal of gates, or sand that is fused with the surface, or some fins and other projections on the surface near the parting line. These need to be cleaned thoroughly before the casting is put to use. The fins and other small projections may easily be chipped off with the help of either hand tools or pneumatic tools. For smoothening the rough cut gate edges either the pedestal or swing frame grinder is used depending on the size of the casting.

For cleaning the sand particles sticking to the casting surface, sand blasting is normally used. The casting is kept in a closed box and a jet of compressed air with a blast of sand grains or steel grit is directed against the casting surface, which thoroughly cleans the casting surface. The typical shot speed reached of the order of 80 m/s. The shots used are either chilled cast iron grit or steel grit. Chilled iron is less expensive but is likely to be lost quickly by fragmentation. In this operation, the operator should be properly protected.

Another useful method for cleaning the casting surface is the tumbling. Here the castings are kept in a completely closed barrel and then is slowly rotated on a horizontal axis at 30 to 40 rpm. The barrel is reasonably packed with enough room for castings to move so that they will be able to remove the sand and unwanted fins and projections. However one precaution to be taken for tumbling is that the castings should all be rigid with no frail or overhung segments, which may get knocked off during the tumbling operation.

After the cleaning of the casting, it needs to be inspected to make sure that it has all the necessary characteristics as specified in the production documentation. This involves the dimensional accuracy, mechanical properties, chemical composition and casting soundness. Dimensional accuracy, mechanical properties, and chemical composition are measured by the various standard methods that are used in casting shops. Casting soundness, however, needs to be tested both by visual inspection and non-destructive testing. Visual inspection involves the use of human eye to identify many common casting defects such as gas defects, pouring metal defects, some metallurgical defects and moulding material defects that are explained below.

On the other hand, Non-Destructive testing (NDT) helps to identify defects that are hidden in the casting and can be identified without destroying the casting. Some of the methods generally employed are described in 5.4.

5.3 CASTING DEFECTS

Any irregularity in the moulding process causes defects in castings which may sometimes be tolerated, sometimes eliminated with proper moulding practice or repaired using methods such as welding and metallisation. The following are the major defects which are likely to occur in sand castings:

(i) Gas defects
(ii) Shrinkage cavities
(iii) Moulding material defects
(iv) Pouring metal defects
(v) Metallurgical defects

5.3.1 Gas Defects

The defects in this category can be classified into blow holes and open blows, air inclusion and pin hole porosity. All these defects are caused to a great extent by the lower gas passing tendency of the mould which may be due to lower venting, lower permeability of the mould and/or improper design of the casting. The lower permeability of the mould is, in turn caused by finer grain size of the sand, higher clay, higher moisture, or by excessive ramming of the moulds.

Blow holes and open blows These are the spherical, flattened or elongated cavities present inside the casting or on the surface as shown in Fig. 5.8. On the surface they are called open blows and inside, they are called blow holes. These are caused by the moisture left in the mould and the core. Because of the heat in the molten metal, the moisture is converted into steam, part of which when entrapped in the casting ends up as blow hole or ends up as open blow when it reaches the surface. The main reason for this is the low permeability of the sand mould. Low permeability is caused by the use of too fine sand grains, higher amount of binder or over ramming of the mould. This can also be caused by insufficient venting practice. In green sand moulds it is very difficult to get rid of the blow holes, unless proper venting and high permeability is provided.

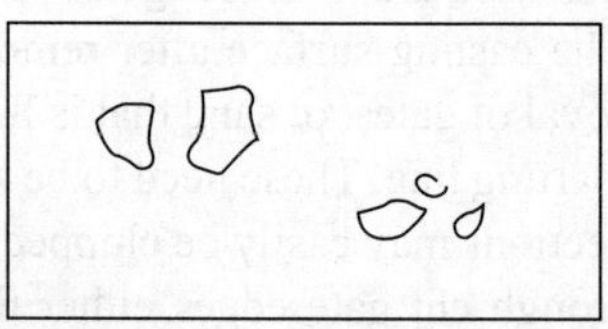

Fig. 5.8 *Open blows on the surface of the casting*

Air inclusions The atmospheric and other gases absorbed by the molten metal in the furnace, in the ladle, and during the flow in the mould, when not allowed to escape, would be trapped inside the casting and weaken it. The main reasons for this defect are:

- The higher pouring temperatures, which increase the amount of gas absorbed,
- Poor gating design such as straight sprues in unpressurised gating, abrupt bends and other turbulence causing practices in the gating and increasing the air aspiration and finally the low permeability of the mould itself.

The remedies would be to choose the appropriate pouring temperature and improve gating practices by reducing the turbulence.

Pin hole porosity This is caused by hydrogen in the molten metal. This could have been picked up in the furnace or by the dissociation of water inside the mould cavity. As the molten metal gets solidified, it loses the temperature which decreases the solubility of gases and there by expelling the dissolved gases. The hydrogen while leaving the solidifying metal would cause very small diameter and long pin holes showing the path of escape. These series of pin holes cause leakage of fluids under high operating pressures. The main reason for this is the high pouring temperature which increases the gas pick up. This is particularly severe in aluminium alloys or steels and irons having aluminium.

Shrinkage cavities These are caused by the liquid shrinkage occurring during the solidification of the casting. To compensate this, proper feeding of liquid metal is required as also proper casting design.

5.3.2 Moulding Material Defects

Under this category are those defects which are caused because of the characteristics of the moulding materials. The defects that can be put in this category are:

- Cuts and washes
- Metal penetration
- Fusion
- Run out
- Rat tails and buckles
- Swell
- Drop

These defects occur essentially because the moulding materials are not of requisite properties or due to improper ramming.

Cuts and washes These appear as rough spots and areas of excess metal, and are caused by the erosion of moulding sand by the flowing molten metal. This may be caused by the moulding sand not having enough strength or the molten metal flowing at high velocity. The former can be remedied by proper choice of moulding sand and using appropriate moulding method. The latter can be taken care of by altering the gating design to reduce the turbulence in the metal, by increasing the size of gates or by using multiple in-gates.

Metal penetration When the molten metal enters the gaps between the sand grains, the result would be a rough casting surface. The main reason for this is that, either the grain size of the sand is too coarse or no mould wash has been applied to the mould cavity. This can also be caused by higher pouring temperatures. Choosing appropriate grain size, together with a proper mould wash should be able to eliminate this defect.

Fusion This is caused by the fusion of sand grains with the molten metal, giving a brittle, glassy appearance on the casting surface. The main reason for this defect is that the clay in the moulding sand is of lower refractoriness or that the pouring temperature is too high. The choice of an appropriate type and amount of bentonite would cure this defect.

Run out A run out is caused when the molten metal leaks out of the mould. This may be caused either due to faulty mould making or because of the faulty moulding flask.

Rat tails and buckles Rat tail is caused by the compression failure of the skin of the mould cavity because of the excessive heat in the molten metal. Under the influence of the heat, the sand expands, thereby moving the mould wall backwards and in the process when the wall gives away, the casting surface may have this marked as a small line, as shown in Fig. 5.9. With a number of such failures, the casting surface may have a number of criss crossing small lines. Buckles are the rat tails which are severe.

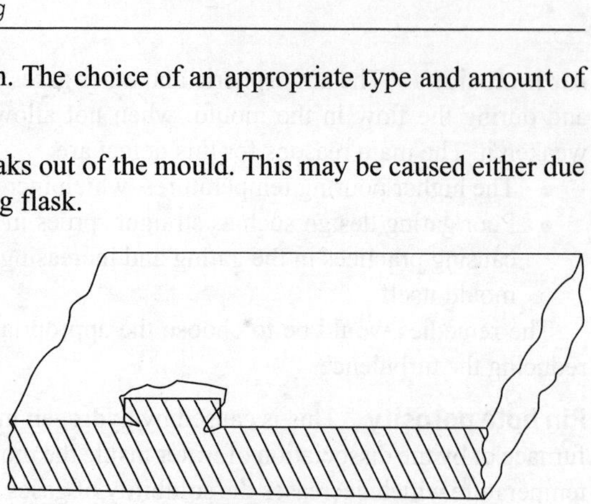

Fig. 5.9 *Rat tail*

The main cause for these defects are: the moulding sand has got poor expansion properties and hot strength or the heat in the pouring metal is too high. Also the facing sand applied does not have enough carbonaceous material to provide the necessary cushioning effect. Proper choice of facing sand ingredients and the pouring temperature are the measures to reduce the incidence of these defects.

Swell Under the influence of the metallostatic forces, the mould wall may move back causing a swell in the dimensions of the casting. As a result of the swell, the feeding requirements of castings increase which should be taken care of by the proper choice of risering. The main cause of this is the faulty mould making procedure adopted. A proper ramming of the mould should correct this defect.

Drop The dropping of loose moulding sand or lumps normally from the cope surface into the mould cavity is responsible for this defect. This is essentially due to improper ramming of the cope flask.

5.3.3 Pouring Metal Defects

The likely defects in this category are mis runs, cold shuts and slag inclusions.

Mis runs and cold shuts Mis run is caused when the metal is unable to fill the mould cavity completely and thus leaving unfilled cavities as shown in Fig. 5.10. A cold shut is caused when two metal streams, while meeting in the mould cavity, do not fuse together properly thus causing a discontinuity or weak spot in the casting as shown in Fig. 5.11. Sometimes a condition leading to cold shuts can be observed when no sharp corners are present in a casting. These defects are caused essentially by the lower fluidity of the molten metal or that the section thickness of the casting is too small. The latter can be rectified by proper casting design. The remedy available is to increase the fluidity of the metal by changing the composition or raising the pouring temperature. This defect can also be caused when the heat removal capacity is increased such as in case of green sand moulds. The castings with large surface area to volume ratio are more likely to be prone to these defects. This defect is also caused in moulds which are not properly vented because of the back pressure of the gases. The remedies are basically improving the mould design.

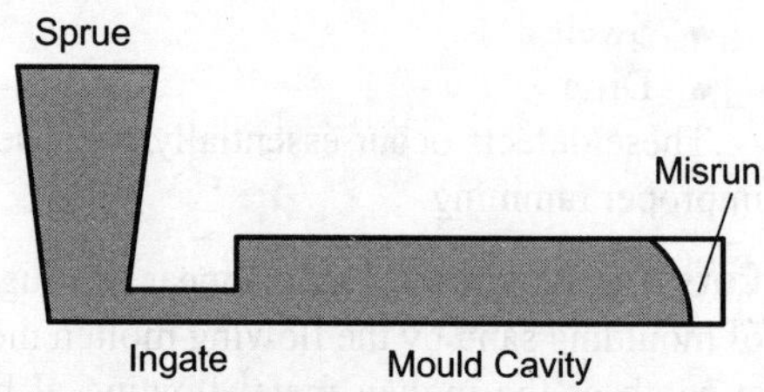

Fig. 5.10 *Mis run*

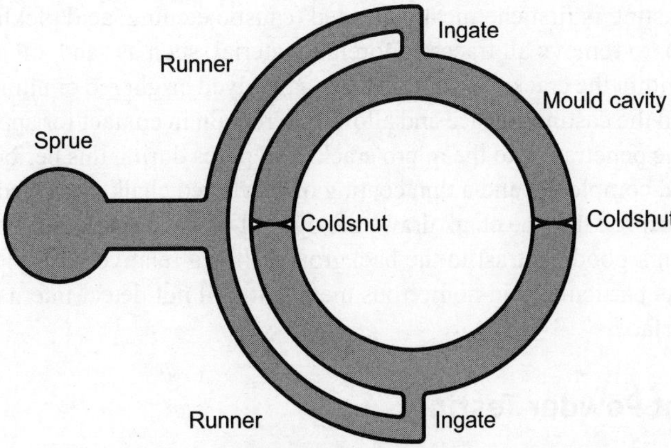

Fig. 5.11 *Cold shut*

Slag inclusions During the melting process, flux is added to remove the undesirable oxides and impurities present in the metal. At the time of tapping, the slag should be properly removed from the ladle before pouring the metal into the mould. Otherwise any slag entering the mould cavity will weaken the casting and also spoil the surface of the casting. This can be eliminated by some of the slag trapping methods discussed in Chapter 4 such as pouring basin screens or runner extensions.

5.3.4 Metallurgical Defects

The defects that can be grouped under this category are hot tears and hot spots.

Hot tears Since metal has low strength at higher temperatures, any unwanted cooling stress may cause the rupture of the casting. The main cause for this is the poor casting design which is discussed later.

Hot spots These are caused by the chilling of the casting. For example, with gray cast iron having small amounts of silicon, very hard white cast iron may result at the chilled surface. This hot spot will interfere with the subsequent machining of this region. Proper metallurgical control and chilling practices are essential for eliminating the hot spots.

As seen from earlier paragraphs, the remedies of some defects are also the causes of others. Therefore the foundry engineer has to analyse the casting from the view point of its final application and thus arrive at a proper moulding procedure to eliminate or minimise the most undesirable the casting defects.

5.4 NON DESTRUCTIVE TESTING (NDT)

A brief description of various Non-destructive Testing (NDT) methods used for inspection of castings is provided here. The effectiveness of any particular method of NDT depends upon the skill, experience, and training of the person performing the inspection process. Each process is limited in its usefulness by its adaptability to the particular component to be inspected.

5.4.1 Dye Penetrant Testing

This method is used to find minute cracks and pores on surface of the casting that are hard to detect by visual inspection. It can be used for all materials. For the use of dye penetrant testing, proper preparation of the test surface is critical. The casting is first chemically cleaned (caustic etching, acid pickling or a combination of these surface treatments), to remove all traces of foreign material, such as sand, oil or grease, etc., from the surface as well as from within the cracks. Next, thin oil usually dyed bright red or ultra-violet fluorescent used as penetrant, is applied on the casting surface and allowed to remain in contact for approximately 15 minutes. Capillary action draws the penetrant into the micro cracks and pores during this period. Any excess penetrant on the surface is removed completely and a thin coating of powdered chalk is applied as a developer. After a few minutes it can be observed that the chalk draws the dye out of the crack to provide the crack location by magnifying in width with a good contrast to the background. It is a relatively simple manual operation and is good for surface cracks particularly in nonferrous metals. It will not detect internal porosity or shrinkage that is not open to the surface.

5.4.2 Fluorescent Powder Testing

This process uses a fluorescent powder that glows bright yellow when reacted with ultraviolet radiation. Similar to the liquid penetrant testing, the fluorescent powder testing involves the application of fluorescent powder suspended in penetrating oil. Similar to the above, the solution penetrates the defects given sufficient dwell time. After the dwell time, carefully clean the surface of the excess powder. Then a contrasting developer is applied on the surface which draws the solution from the defect. Then under an ultraviolet light of suitable strength is used to examine where defects have occurred. Fluorescent powder testing is more effective and economical than radiographic testing discussed later.

5.4.3 Magnetic Particle Inspection

This test is used for detecting cracks in ferrous castings such as cast iron and steel that can be magnetized. Cracks have magnetic properties that are different from the surrounding material, which will interrupt the magnetic field, causing distortion as shown in Fig. 5.12 and a leakage field will occur. The field strength is increased in the area of the defect and opposite magnetic poles North and South form on either side of the defect as shown in Fig. 5.12. Small magnetic particles, when sprinkled near this gap, get attracted to these poles and show the path of the flux line that spreads out in order to detour around the crack, thereby indicating the shape and position of the crack or void. Magnetic particles are applied over a surface either dry, as a powder, or wet, as particles in a liquid carrier such as oil or water. The casting surface should be clean and dry before inspection. To locate a defect, it is necessary to control the direction of magnetization and flux lines must be perpendicular to the longitudinal axes of expected defects. A ferromagnetic casting is magnetized with a strong magnetic field created by a permanent magnet or special equipment. Typically a high-amperage (up to 6000 A), low-voltage current is passed through the casting, which establishes a magnetic field. Finely milled iron particles coated with a suitable dye pigment are applied to the casting. These particles are attracted to leakage fields and will cluster to form an indication directly over the crack. The particles can be red or black oxide, or they can be coated with a substance, which fluoresces brilliantly under ultra-violet illumination (black light). The objective is to present as great a contrast as possible between the crack indication and the material background.

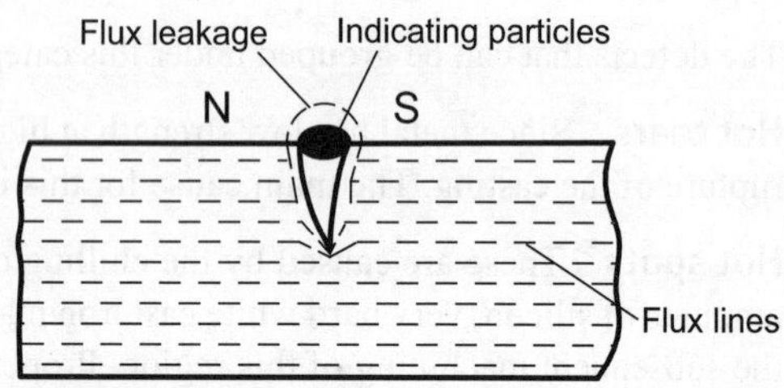

Fig. 5.12 *Surface crack identified by the magnetic particles*

Very small cracks or voids at or near the surface which may not even be detected by radiography are easily revealed by this method. Magnetic particle inspection is quick, inexpensive and sensitive to defects, particularly shallow (0.08 mm) surface cracks.

5.4.4 Radiographic Inspection

Radiography is used to identify and measure the internal defects in the casting such as shrinkage and inclusions. Electrically generated X-rays are used for radiographic inspection. The radiation is differentially absorbed by the material through which it passes; the greater the thickness, the greater the absorption. Furthermore, the denser the material the greater is the absorption. The casting absorbs part of the radiation and the remaining portion of the radiation exposes the radiographic film. In X-radiography the penetrating power is determined by the number of volts applied to the X-Ray tube. In steel, approximately 1000 volts per 25 mm thickness is necessary. When the radiation passes through a sound casting most of it is absorbed by the casting and the resulting image on the photographic film is light. However if a section of the casting is with a shrinkage cavity, less X-rays are absorbed and therefore a dark image will result on the film. Thus any hole, crack or inclusion that is less dense than the casting alloy, is revealed as a dark area. Radiographic inspection is the best non-destructive method for detecting internal defects for any material and provides a permanent record in the form of a film. It is generally good for thin sections. Care has to be taken because of health hazards of radiation. It is also possible to have 3-D imaging of the casting using the computerized axial tomography (CAT scanning).

5.4.5 Ultrasonic Testing

Ultrasonic testing is used for the detection of internal and surface defects in sound conducting castings. Piezo electric crystal vibrates for a very short period at a frequency related to the thickness of the crystal. The crystal does not oscillate continuously but in short pulses, between each pulses it is quiet. Piezo electric materials can convert electrical pulses to mechanical oscillations as well as mechanical oscillations into electrical pulses. In Ultrasonic testing, this frequency is usually in the range of one million to six million times per second (1 MHz to 6 MHz). Ultrasonic waves have the ability to travel a considerable distance in homogeneous elastic material, such as many metals with little attenuation. The velocity at which these waves propagate is related to the Young's Modulus of the material. For example, the velocity in steel is 5900 metres per second and in water 1400 metres per second. A probe that generates the pulse taking a finite time to travel through the casting material to the interface will be reflected back to the probe. The interface or defect reflects portions of the energy which are collected in a receiving unit and is displayed for analysis. The pattern of the energy deflection can indicate the location and size of an internal defect, as well as wall thickness and the nodule count of ductile iron. Ultrasonic testing is versatile and can help to determine the position, size and type of defect without using any consumables. Considerable degree of skill is necessary to obtain the fullest information from the test.

5.4.6 Eddy Current Inspection

The eddy current inspection method is applied to the detection of cracks at or near the surface. Eddy currents can be produced in any electrically conducting material that is subjected to an alternating magnetic field (typically 10 Hz to 10 MHz). The alternating magnetic field is normally generated by passing an alternating current through a coil. The coil can have many shapes and can between 10 and 500 turns of wire. The magnitude of the eddy currents generated in the product is dependent on conductivity, permeability and the set up geometry. Any change in the material or geometry can be detected by the excitation coil as a change in the coil impedance. The depth to which the eddy currents penetrate a material depends on the frequency. Higher frequency causes lower penetration. Lower frequency has lower sensitivity to small defects. Larger coils are less sensitive to surface roughness and vice versa. Eddy current inspection is accurate for detection of small flaws or material changes that may not be detected with other inspection methods, and the discontinuities in the casting will give an immediate response on the monitoring equipment. The operator should have knowledge and experience to properly interpret the results.

5.5 PRODUCT DESIGN FOR SAND CASTINGS

In designing a casting, the responsibility of the designer includes much more than just ensuring proper functioning. Economy of production is a very important factor. Consideration for economy in moulding procedures should be constantly incorporated in the development of the design. Variations in the foundry procedures, in the form and construction of patterns and core boxes, should be thoroughly investigated to ensure the most economical plan for producing castings of specific quality. Furthermore, operations other than those of the pattern shop and foundry must also be considered. The design of the casting and the pattern equipment should be such that the cost of all other operations is reduced to the minimum. Such operations may include finishing, or the elimination of finishing where possible, in whole or in part; assembly of one part with another part; and economical servicing during the life of the casting.

Accordingly, the product design process should be studied under the following categories:

1. Designing for economical moulding
   (a) Parting line
   (b) Bosses and undercuts
   (c) Coring
   (d) Simplified moulding
2. Designing for eliminating defects
   (a) Shrinkage defects
   (b) Distortions
   (c) Hot tears
   (d) Escape of gases
3. Designing for features to aid handling of castings

5.5.1 Designing for Economical Moulding

Parting Line

Parting line is determined by the shape of the casting. The simplest parting line is that running through the centre line of the casting as shown in Fig. 5.13. Unnecessary complexities in the parting line increase the cost. Hence the designer should strive to design castings with parting planes that are straight or as nearly straight as possible.

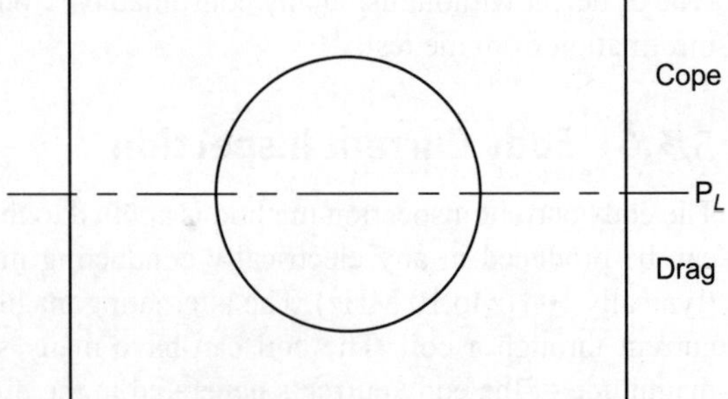

Fig. 5.13 *Simple parting line*

As an example, Fig. 5.14(a) shows the possible redesigning of a component. The portions of the hubs projecting above the parting plane at the centre of the casting and the end of one arm are loose pattern parts and are assembled to the main pattern with the help of the dowel pins. But modifying the component as shown in Fig. 5.14(b) would make the parting line straight and thus simplifying the moulding procedure.

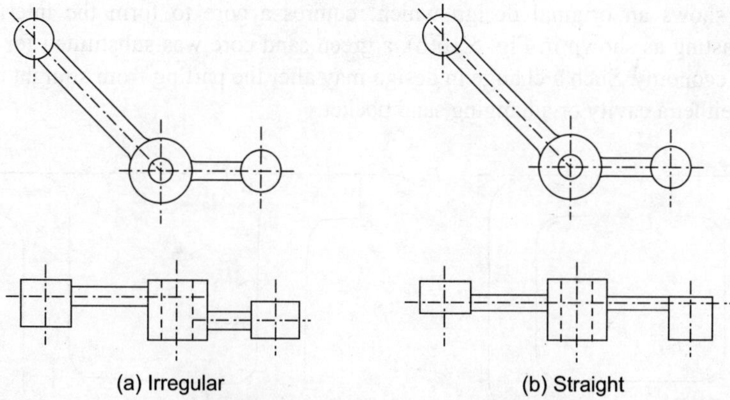

Fig. 5.14 *Parting line modification*

When an irregular parting results in a deep mould pocket, it may be more economical to redesign the pattern equipment and change over to a mould method with a straight parting involving the use of cores.

Bosses

Bosses are frequently used to increase the sectional thickness of the housing and the like in order to provide longer bolt, or tap holes, or to improve the strength of certain parts of the casting. This may be cast satisfactorily if the axis of the cylindrical boss is parallel to the direction in which the pattern is drawn out of the mould or if the centre line of the boss is in the parting plane. When this is not the case, the boss on the pattern must be loose, and the skilled technique of moulding loose pattern must be employed. The section shown in Fig. 5.15(a) illustrates the positioning of a boss well below a flange whose upper surface is chosen as a parting line. To mould this design a core is required to permit removal of the pattern from the mould. In producing such a casting as shown, accurate positioning of the core is difficult and any shifting of the core results in surface irregularities. A somewhat less complicated design in Fig. 5.15(b) extends the boss to the flange, eliminating the undercut and the need for a core.

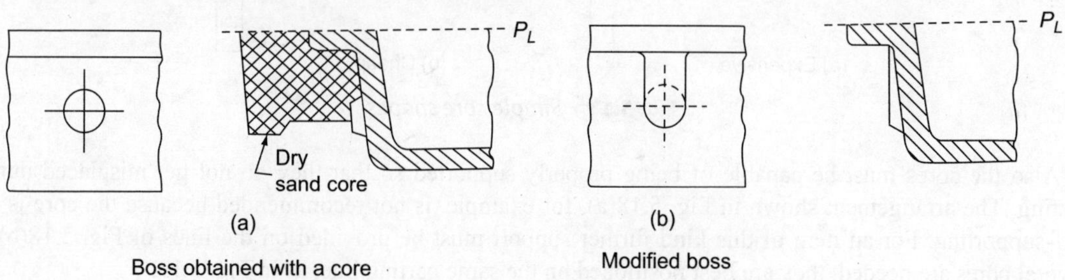

Fig. 5.15 *Modification of bosses to reduce dry sand core*

Coring

Cores are placed in the mould to provide castings with contours, cavities and passages not possible otherwise to obtain by normal moulding. Frequently, a minor change in the casting design may economise on production cost by eliminating cores. For example, if there are any back drafts, cores must be used so that the pattern can be withdrawn from the mould.

Figure 5.16(a) shows an original design which requires a core to form the interior of the casting. Redesigning the casting as shown in Fig. 5.16(b), a green sand core was substituted for the dry sand core, thus achieving the economy. Such a change in design may alter the parting from straight to irregular and the cope may contain either a cavity or a hanging sand pocket.

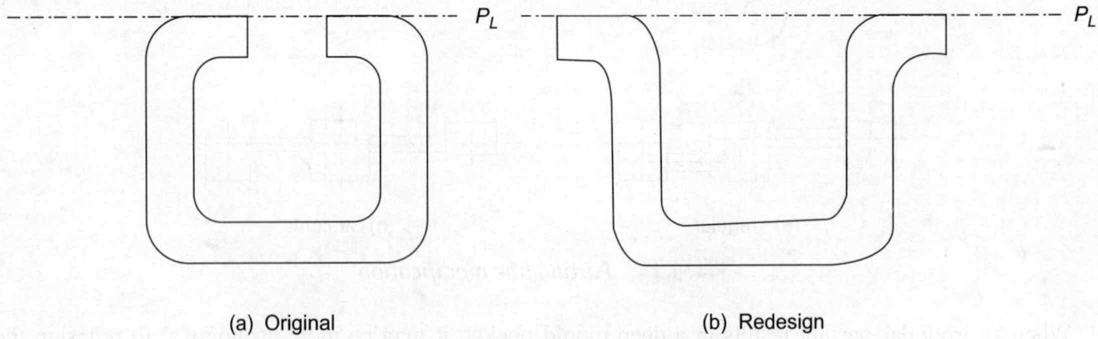

Fig. 5.16 *Eliminating dry sand core by modifying the draft angle*

When the cores cannot be avoided, the designer should strive to make them as simple as possible in the interest of economy, by using simple surfaces which are easy to produce. Examples of such surfaces for cores are presented in Fig. 5.17(a) and (b).

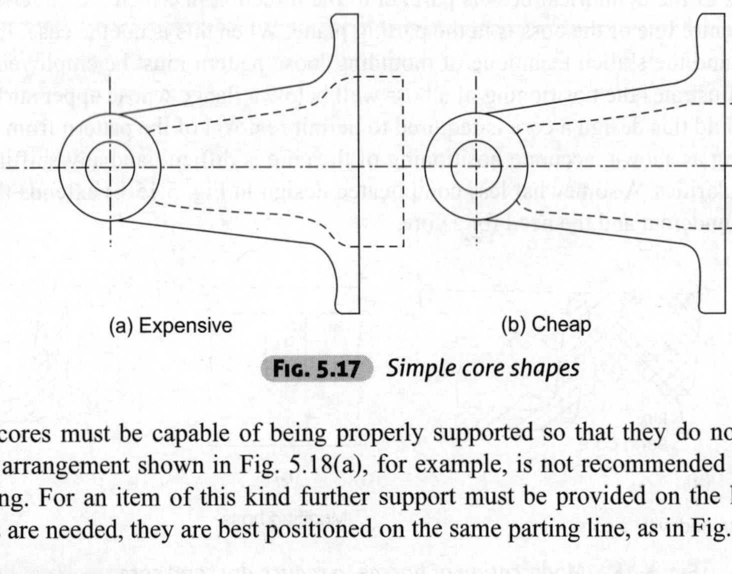

Fig. 5.17 *Simple core shapes*

Also the cores must be capable of being properly supported so that they do not get misplaced during casting. The arrangement shown in Fig. 5.18(a), for example, is not recommended because the core is left self-supporting. For an item of this kind further support must be provided on the lines of Fig. 5.18(b). If several cores are needed, they are best positioned on the same parting line, as in Fig. 5.19.

Simplified Moulding

Generally a two-part moulding is simpler and economical compared to a three part moulding. But when the part configuration, as shown in Fig. 5.20(a), necessitates, it may be desirable to modify the moulding procedure by providing an external ring core, as shown in Fig. 5.20(b) to avoid the intermediate flask. But this also is an expensive alternative. The best choice would be to redesign by eliminating the bottom flange, as shown in Fig. 5.20(c), which completely avoids the external core and thus is economical.

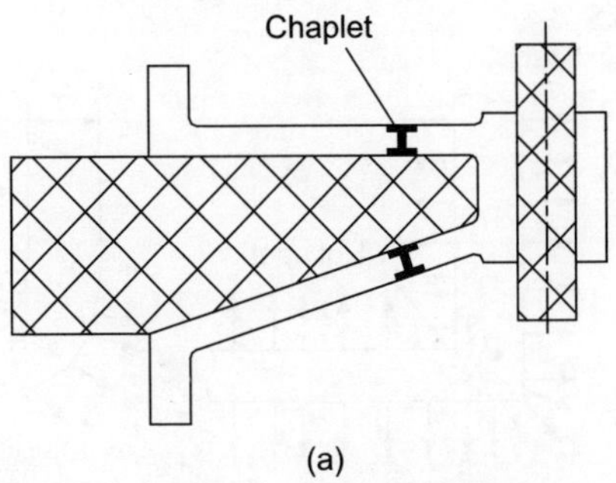

Fig. 5.18 *Providing proper core support*

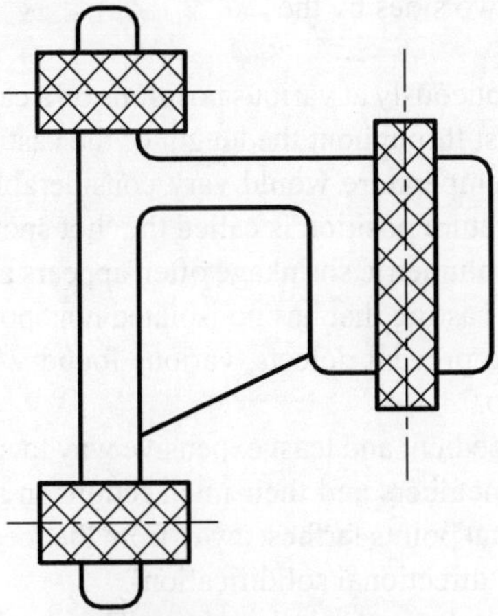

Fig. 5.19 *Several cores*

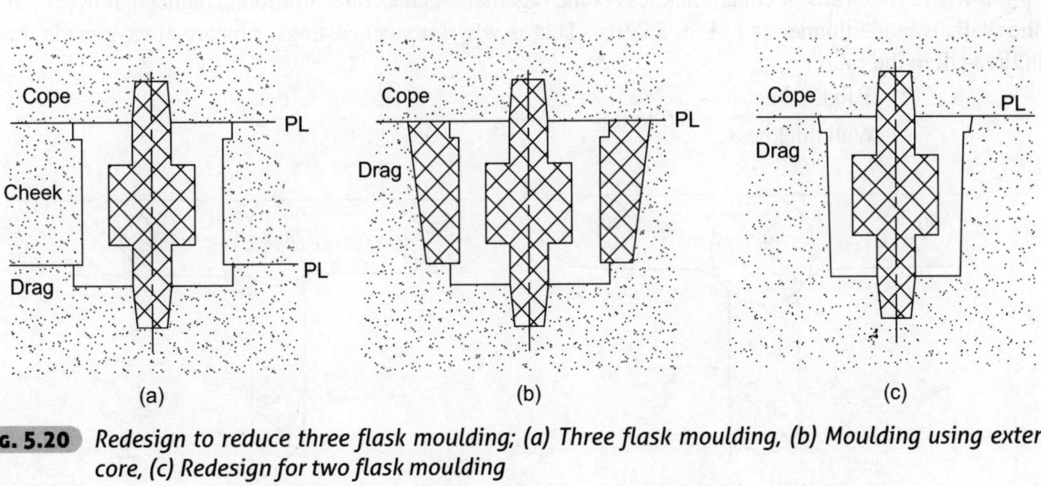

Fig. 5.20 *Redesign to reduce three flask moulding; (a) Three flask moulding, (b) Moulding using external core, (c) Redesign for two flask moulding*

5.5.2 Designing for Eliminating Defects

Shrinkage Defects

The main reasons for these defects are:

(i) Volumetric contraction both in liquid and solid state;

(ii) Low strength at high temperature.

Cracks often begin at the shrinkage cavities and work their way outward as the casting is stressed in service. It should be noted that solidification progresses from thin to thick sections and that external angles have a greater cooling rate than re-entrant angles,,the reason being that sand around a re-entrant angle is surrounded on two sides by the heat sources, as shown in Fig. 5.21.

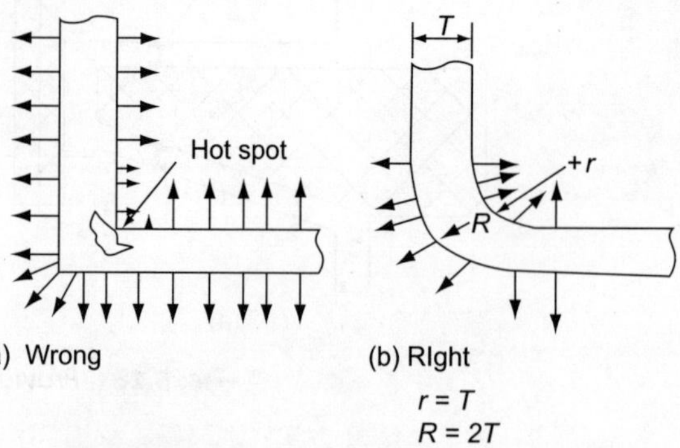

Fig. 5.21 *Re-entrant angles at junctions*

If temperatures were taken simultaneously at various positions on a casting of uniform width and thickness, no temperature variations would exist throughout the length of the casting. However, if the casting had non-uniform cross sectional areas, the temperature would vary considerably depending on the variation in the sectional thicknesses. A high temperature position is called the 'hot spot'.

Unless a casting is properly fed, volumetric shrinkage often appears at the hot spots. The best way to avoid volumetric shrinkages is to design a casting that has no isolated hot spots and which cannot be properly fed. If good designing is not sufficient to prevent defects, various foundry techniques such as the use of chills, feeders and cores must be resorted to.

Redesign is the most effective expedient and least expensive way to avoid shrinkage defects. The designer should try to place and proportion members and their intersections in such a way as to establish a positive temperature gradient which is lowest at points farthest away from the feed head and which gradually increases toward the feed head. This is called 'directional solidification'.

The shrinkage problem is particularly severe in junctions. The check circle method is normally used for detecting the concentration of the metal. As shown in Fig. 5.22(a), a concentration of metal always occurs at the point where two walls of equal thickness come together. Such a concentration is reduced, however, if one of the walls is made thinner as in Fig. 5.22(b). That is why for steel castings, ribs are always made thinner than the wall in the ratio

$$\frac{\text{Rib thickness}}{\text{Wall thickness}} = 0.6 \text{ to } 0.8$$

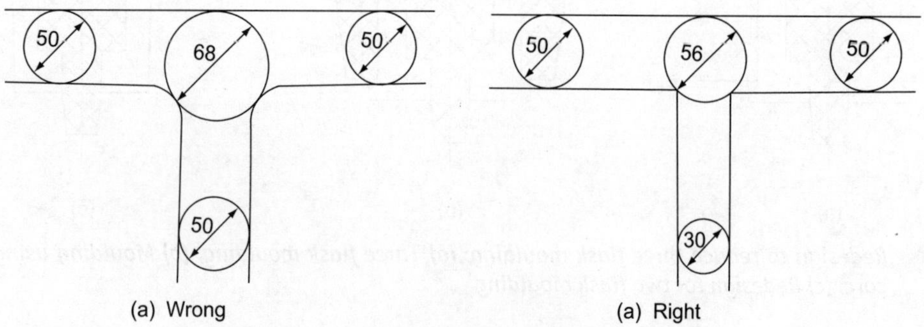

Fig. 5.22 *Avoiding concentration of metal at joints*

A rib which enters a wall at an angle gives a bigger concentration of metal than one which enters perpendicularly (Fig. 5.23). The former should, therefore, be avoided where possible.

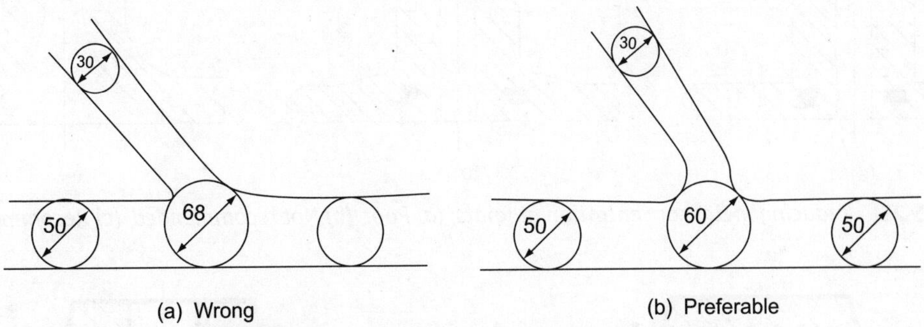

Fig. 5.23 *Concentration of metal at angular joint*

Extensive investigations have been carried out to reach a common understanding of the problems of junction and fillet design and of the specific effects of design variables on the soundness of the metal in different types of intersections. The castings for these studies were deliberately produced in such a way that defects were caused to occur at the junctions. Thus the influence of definite junction and corner configuration on the size and number of defects could be observed systematically.

All casting sections were 3 × 3 inches (76.2 mm). The arms were about 24 inches (609.6 mm) long, with the risers placed at the extreme ends. The risers were utilised as sprues for pouring. The five types of junctions that were considered are represented by the letters L, T, V, X and Y, as by their shapes.

An example of results obtained for L sections are shown in Fig. 5.24. It shows that defects are eliminated by using a 0.5 inch (12.7 mm) fillet and reducing the wall thickness at the corner. No detectable defects were present in the L junction where the radius of the fillet was equal to the wall thickness and a uniform wall was maintained at the corner. In such a design, however, a centre line weakness similar to that in a long uniform wall may be present at the interface, where the two freezing fronts meet. Similar results were obtained for other junctions also.

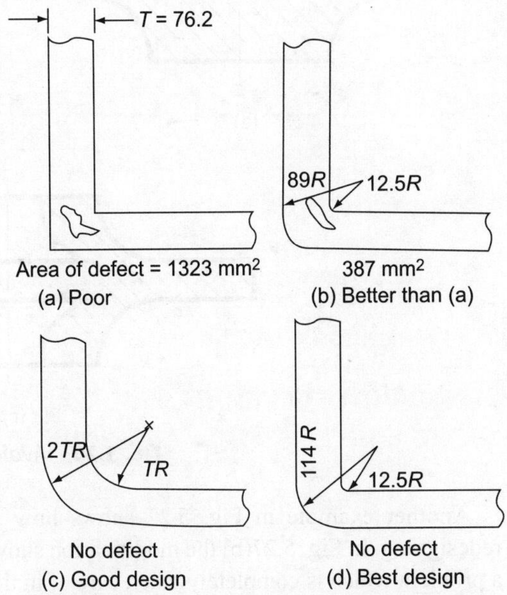

Fig. 5.24 *Results of experimental study on development of shrinkage*

The application of these results as applied to a cylindrical opening into the flange is shown with the original design in Fig. 5.25(a) and recommended practice with no cavities in L sections shrinkage cavities is presented in Fig. 5.25(c).

The hubs of wheels are other places where the concentration of metal is likely to occur as shown in Fig. 5.26(a). An acceptable practice is to core the hub as in Fig. 5.26(b) which is better but still is likely to cause shrinkage cavities. The best design would be as shown in Fig. 5.26(c) where the excess metal is completely removed.

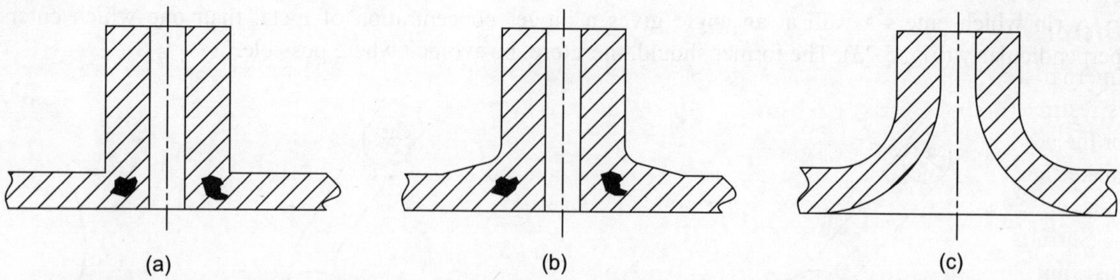

Fig. 5.25 *Reducing metal concentration at joints; (a) Poor, (b) Not recommended, (c) Recommended*

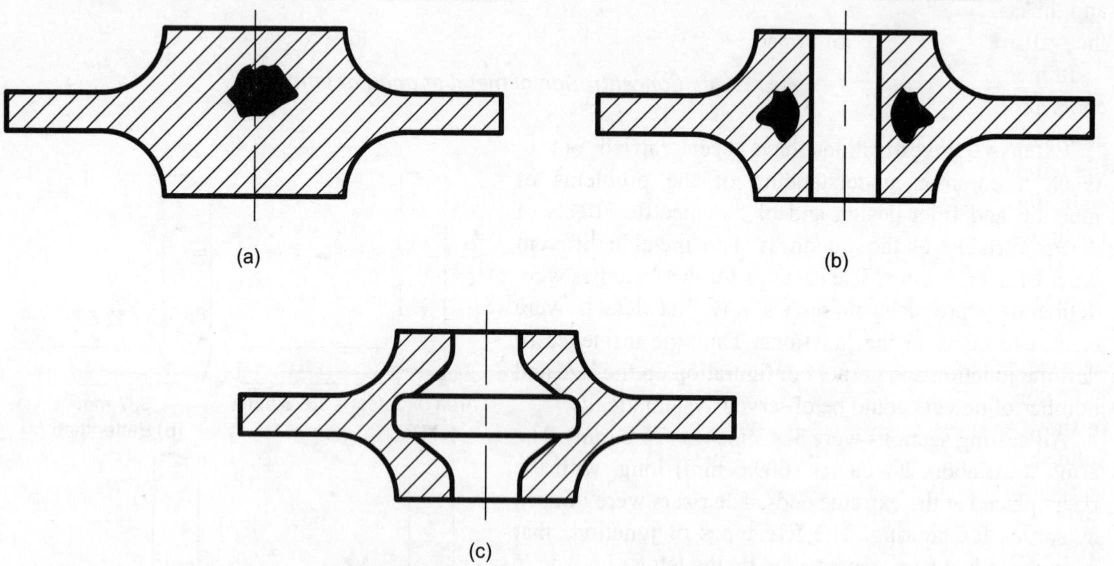

Fig. 5.26 *Avoiding concentration of metal*

Another example in Fig. 5.27 shows how massive elements in a cast frame can be eliminated by redesigning. In Fig. 5.27(b) the modification shown is comparatively better, but venting of core gases may be a problem which is completely taken care of in the design shown in Fig. 5.27(c). Rapid cooling by the use of chills should be provided in the sections where massive elements are inevitable.

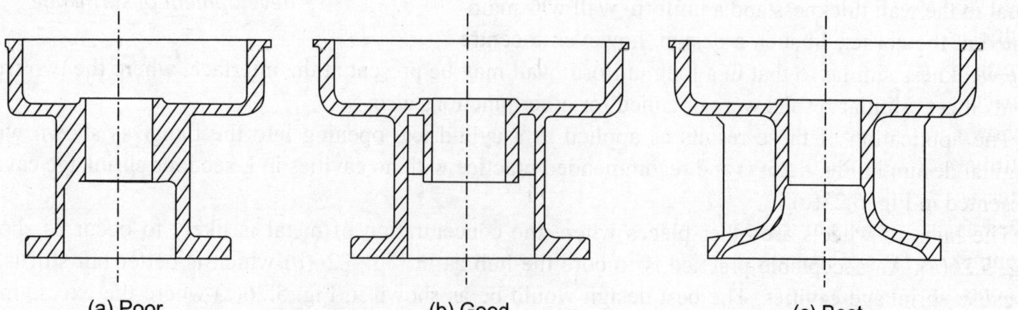

Fig. 5.27 *Concentration of metal in housings*

Distortions

Internal stresses appear in the casting walls when shrinkage is restricted because of the resistance of the mould elements or the action of the adjacent walls. Increased internal stresses make the casting warp and may lead to the development of cracks.

Shrinkage stresses develops during cooling when the metal loses its plasticity (within 500–600°C for cast iron and 600–700°C for steel). At higher temperatures, the change in dimensions is readily compensated by the plastic flow of the metal and thus the shrinkage manifests itself only in the thinning of the walls.

In the box shaped casting (Fig. 5.28), the internal partition cools at a slower rate than the horizontal walls because of the core sand being heated from all sides. While cooling below the temperature at which metal passes from plastic to elastic state, the partition material hardens and contracts and as a result, it undergoes tension. If the tension exceeds the strength of vertical walls, then they are likely to warp and introduce distortions in the casting.

For example, as shown in Fig. 5.29, if the casting walls have non-uniform thicknesses, then the thinner walls would have cooled down very quickly leaving the thicker wall still in the plastic state. When the thinner walls contract due to solid shrinkage, because of the plasticity, the thicker rib would warp.

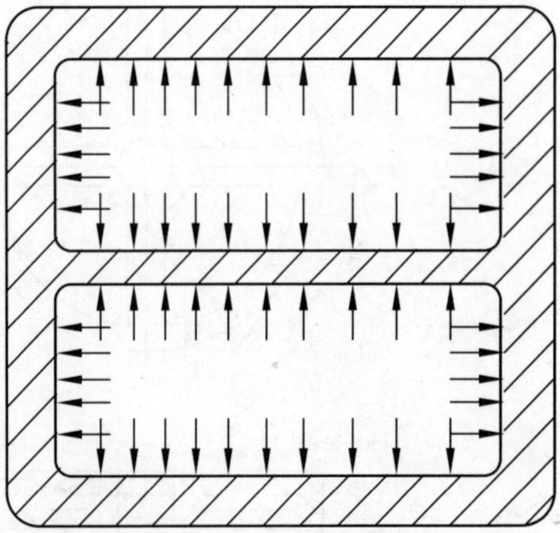

Fig. 5.28 *Distortion in sand castings*

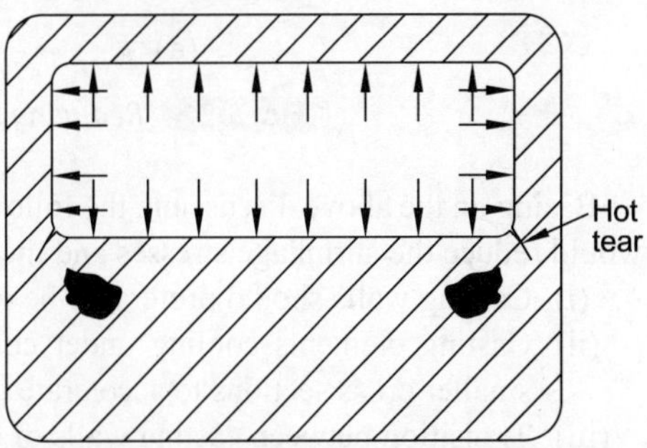

Fig. 5.29 *Warping due to uneven section thickness*

Hot Tears

Hot tears are formed in the castings because of the differential cooling rates and the low strength of metal at higher temperatures. This in fact, is the extension of warping. In the example shown in Fig. 5.30, the horizontal members being thinner cool fast and try to bring the thick vertical members closer which would is resisted by the core. This resistance would cause the tearing of metal to take place at the joint with the thicker rib.

In the hand wheels or big gear wheels where the rims are connected to the hub through spokes, hot tears are likely, as shown in the design in Fig. 5.31. But by increasing the ductility of the spokes, it is possible to reduce the formation of hot tears. The ductility can be increased by making the spokes curved and having them in odd numbers as in Fig. 5.31. The odd number of spokes ensures that the restraining force acts in only one direction making the wheel more ductile.

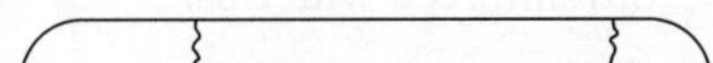

Fig. 5.30 *Hot tear formation*

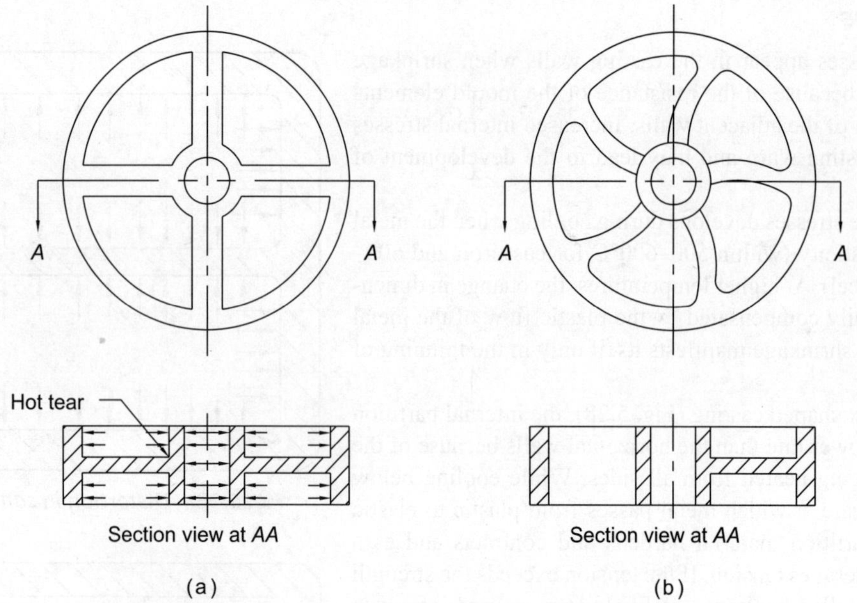

FIG. 5.31 *Reducing hot tear by making curved spokes in wheels*

Basing on the above discussion, the following rules which would aid the directional solidification and thus would reduce the shrinkage stresses and the resulting distortions may be laid down.

(i) Casting walls should preferably be of uniform thickness

(ii) Casting elements cooling under conditions of reduced heat removal (internal walls) should have smaller cross sections to accelerate their solidification.

(iii) Transition between casting walls of different thicknesses should be smooth.

(iv) Casting walls should have no abrupt changes, but be connected by smooth transitions.

(v) Local metal accumulations and massive elements should be avoided, if possible.

(vi) Sections where casting walls join massive elements should be gradually thickened towards the latter or reinforced with ribs.

Escape of Gases

The internal cavities should be so designed as to permit the escape of gases evolving from the cores when the molten metal is poured in. Internal cores, which are small and long, are likely to pose difficulties in cleaning also. Sand burned into core holes, and fins or veins, are very difficult to remove when they are hard to reach. Providing access holes or clean out holes, makes available additional core prints for support, to vent core gases as well as to permit the core sand to be removed.

For example, an unsatisfactory design is illustrated in Fig. 5.32(a). The gases accumulating in the upper part of the core form blow holes. This problem may be reduced by making small vent holes (plugged afterwards if necessary) for the escape of gases, as shown in Fig. 5.32(b). The redesigned vaulted shape of the upper portion of the casting as in Fig. 5.32(c) would be the best way to ensure the escape of gases through the top core print.

Another way of solving the problem is to change the position of the casting with reference to the parting plane. For example, in Fig. 5.33(a), the casting being in the cope, the core gases are not vented at all. But by bringing the casting into the drag as in Fig. 5.33(b), the core gases are properly vented.

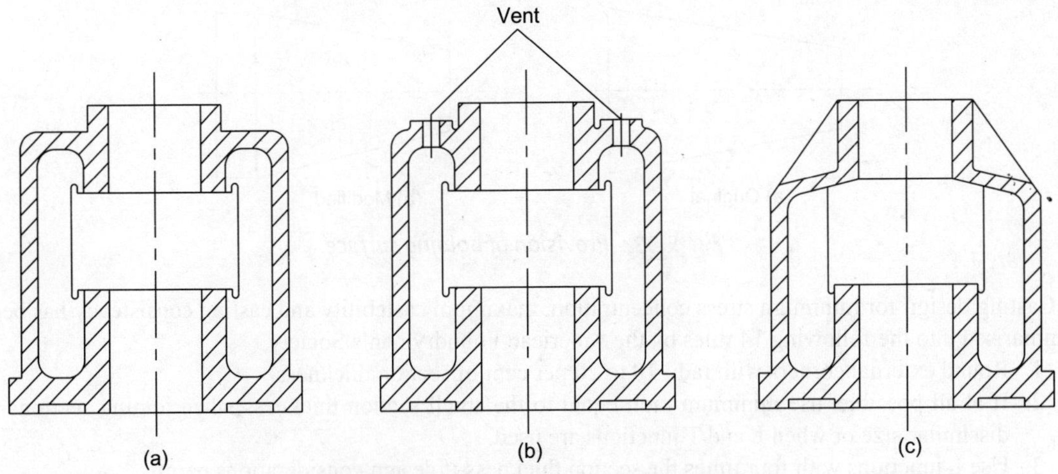

Fig. 5.32 *Escape provision for core gases*

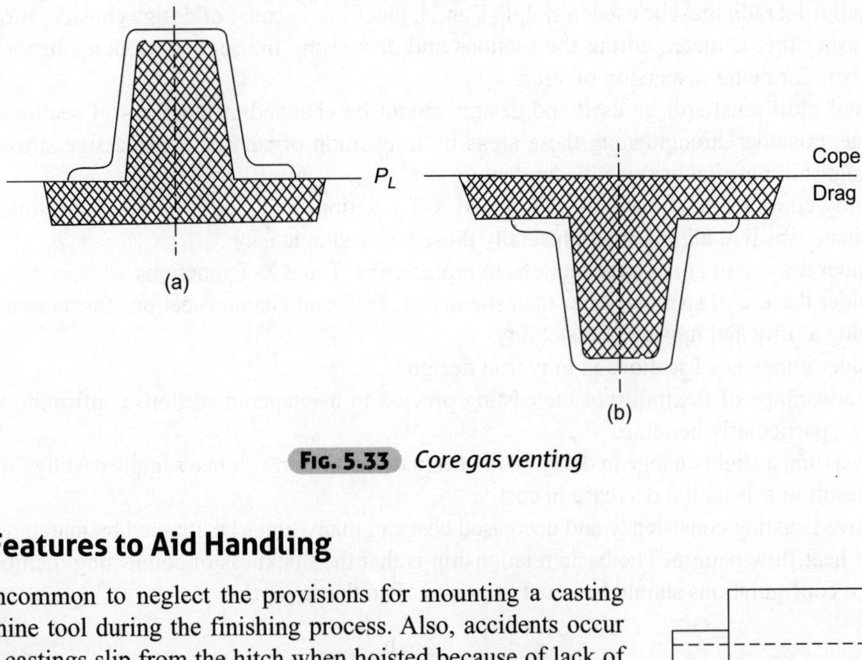

Fig. 5.33 *Core gas venting*

5.5.3 Features to Aid Handling

It is not uncommon to neglect the provisions for mounting a casting in the machine tool during the finishing process. Also, accidents occur when large castings slip from the hitch when hoisted because of lack of hoisting facilities.

When a casting is to be machined to a close tolerance, it is often desirable to do as much finishing as possible without removing the piece from the machine tool. Figure 5.34 shows a chucking extension on a casting which will permit the entire casting to be machined on a lathe in one setting, the last operation being the cut off.

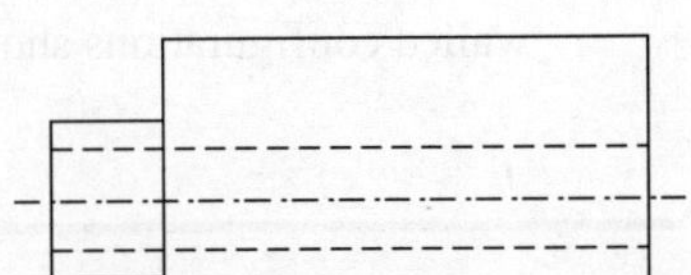

Fig. 5.34 *Chucking extension*

Similarly, castings with tapered sides are difficult to chuck in a lathe and if possible, should be provided with pads or flats as shown in Fig. 5.35.

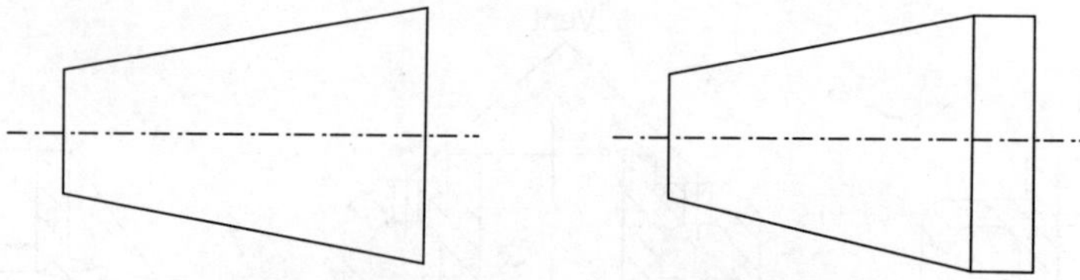

(a) Original (b) Modified

FIG. 5.35 *Provision of holding surface*

Casting design for minimum stress concentration, maximum castability and casting consistency has been summarised into the following 14 rules by the American Foundrymen's Society.

1. Round external corners with radii 10 to 20 per cent of section thickness.
2. If at all possible, use minimum radii equal to the small section thickness when joining sections of dissimilar size or when L and T junctions are used.
3. Use L junctions with four times the section thickness if design considerations permit.
4. Use even larger radii when joining connecting members to sections with much larger section moduli. Radii of 10 or more times the section thickness are beneficial.
5. If small fillet radii must be used in simple T and L junctions because of design considerations, consider increasing the diameter, coring the sections and decreasing the section modulus of the connecting members for better dispersion of stress.
6. If small radii must still be used and design cannot be changed at junctions of sections stressed in fatigue, consider strengthening these areas by imposition of surface compressive stress by rolling, coining, shot peening or selective hardening.
7. Simplify complex junctions like X, V, Y and X-T junctions to T, and if possible to L junctions.
8. Eliminate ribs if at all possible, especially those stressed in tension.
9. Consider the use of corrugated sections to replace ribs, T and X-T junctions.
10. Consider the use of sections other than standard I, H, Z and channel sections for more efficient load carrying ability and improved castability.
11. Consider unbalanced sections in gray iron design.
12. Take advantage of flexibility of the casting process to use tapered sections confirming to the stress pattern, particularly bending.
13. Many a time a slight change in design will eliminate cores or result in a sample casting. Such changes will result in substantial decrease in cost.
14. Improved casting consistency and decreased cost can many times be attained by matching the design to the heat flow pattern. The basic relationship is that the thickness of connecting members of multi walled configurations should be equal or greater than the strength.

SUMMARY

Cupola is used for melting cast iron. Cupola is simple in operation and is also less expensive compared to the other furnaces. Charge in the cupola consists of iron, limestone (flux) and coke (fuel). The charge for cupola can be virgin metal along with some scrap to reduce the overall cost of the metal.

- Hot blast cupola reduces the heat required for melting by extracting the heat from the hot stack gases.

- In addition to the cupola, other furnaces such as arc furnace, induction furnace and reverberatory furnaces are also used for melting metals in the foundry depending upon the requirements.
- Different types and sizes of ladles are used for transferring the molten metal from the furnace to the mould.
- After the casting is solidified in the mould, it needs to be extracted by breaking the mould. Then the casting is cleaned and inspected for any defects.
- There are a number of defects such as blow holes, rat tails, run out, swell, mis run, etc. which need to be avoided for the sake of casting quality. For this purpose a number of precautions need to be taken during the preparation of the sand mould.
- Non-destructive testing is used in castings to identify the defects that cannot be directly identified by visual inspection alone. For this purpose, a number of testing methods such as dye penetrant testing, magnetic particle testing, radiographic inspection, etc. are used.
- Casting geometry can be modified to suit the mould preparation for economical and simplified moulding or reduce defects.

Questions

5.1 Describe the operation of a cupola furnace for melting cast iron.

5.2 What are the functions of flux in melting metals and alloys?

5.3 What does normally constitute the charge in a cupola furnace?

5.4 Describe the constructional features of a cupola furnace.

5.5 What are the causes for the change in the composition of the melt obtained from cupola?

5.6 How are the types of ladles chosen for any particular operation?

5.7 What are the ideal conditions for breaking the mould to extract the casting?

5.8 Explain the term fettling.

5.9 What are the methods available for the removal of gates and risers from the casting?

5.10 How is the casting surface cleaned?

5.11 What are the possible casting defects that may be caused by the improper gating system design? State at least four defects.

5.12 Mention the causes and remedies of the following sand casting defects: Blow holes, hot tears and mis runs.

5.13 State any three common green sand casting defects. Give their causes and remedies.

5.14 Explain how shrinkage cavities are sometimes formed in a casting.

5.15 Give reasons for the occurrence of the following casting defects: Pin hole porosity, run out and drop.

5.16 Differentiate between the following casting defects with reference to the causes and methods of prevention:
(a) Cold shut and mis run,
(b) Blow hole and pin hole porosity.

5.17 Describe the various defects which are likely to be caused in sand castings because of higher pouring temperatures.

5.18 Give the differences between the following sand casting defects from the stand point of their causes and remedies.
 (a) Metal penetration and fusion
 (b) Rat tails and cuts and washes

5.19 Explain the purpose of dye penetrant testing used for non-destructive inspection of castings.

5.20 Describe the magnetic particle inspection procedure for inspecting the castings. Give the precautions to be taken while applying this method.

5.21 Write a short note on the application of various non-destructive testing methods for castings.

5.22 Compare magnetic particle inspection and radiographic inspection in terms of the process and application.

5.23 State any three considerations for choosing a proper parting line in the case of castings.

5.24 In sand castings, apart from the stress concentration is there any reason for rounding of the corners. Explain.

5.25 Internal corners are more prone to solidification shrinkages than the external corners. Explain.

5.26 What are the problems caused by bosses in sand castings? Explain the methods available to solve them.

5.27 What precautions should a casting designer take when using cores?

5.28 Why the shrinkage cavities are caused in castings? Discuss the methods used for eliminating them.

5.29 As far as possible most of the casting should be in the drag instead of the cope. Is this statement correct? Explain your answer.

5.30 What are the considerations for choosing or avoiding cored holes in sand casting?

5.31 Examine the components shown in Fig. 5.36. Decide the method of moulding for these components. Discuss how these castings can be redesigned to improve the casting quality.

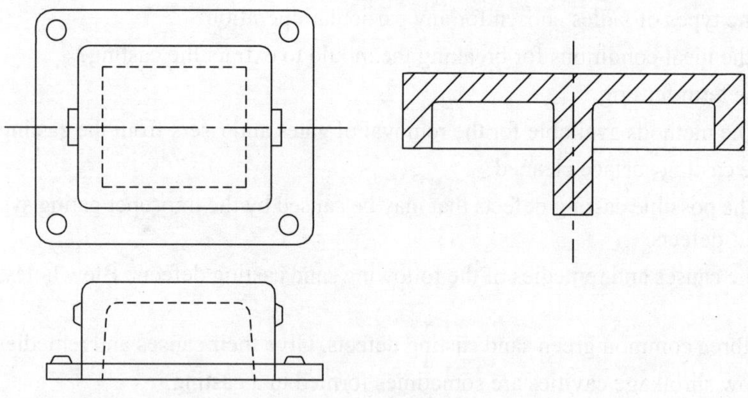

(a) (b)

Fig. 5.36 *Examples*

5.32 From the castings shown in Fig. 5.37, how many of them are likely to be prone to shrinkage cavity formation? Are any other defects likely in these and other castings? Suggest the possible redesigning for these castings.

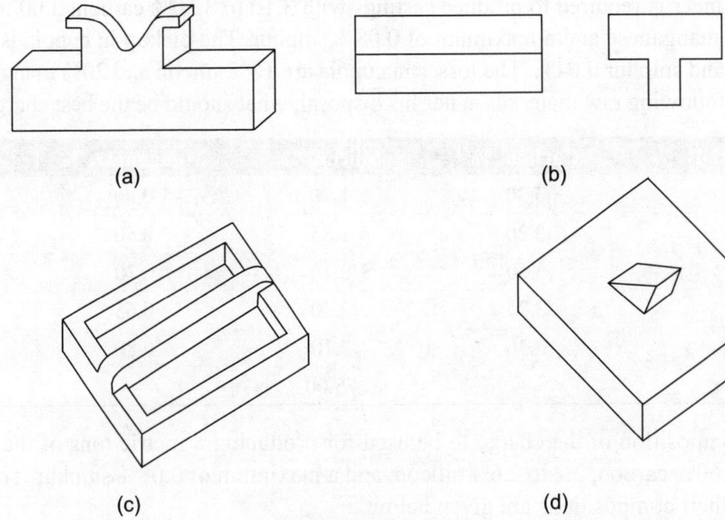

FIG. 5.37 *Examples*

5.33 Figure 5.38 shows some cases of hot tear formation in sand castings. Discuss each case as to how the hot tear formed and give remedial measures to be taken to reduce its incidence.

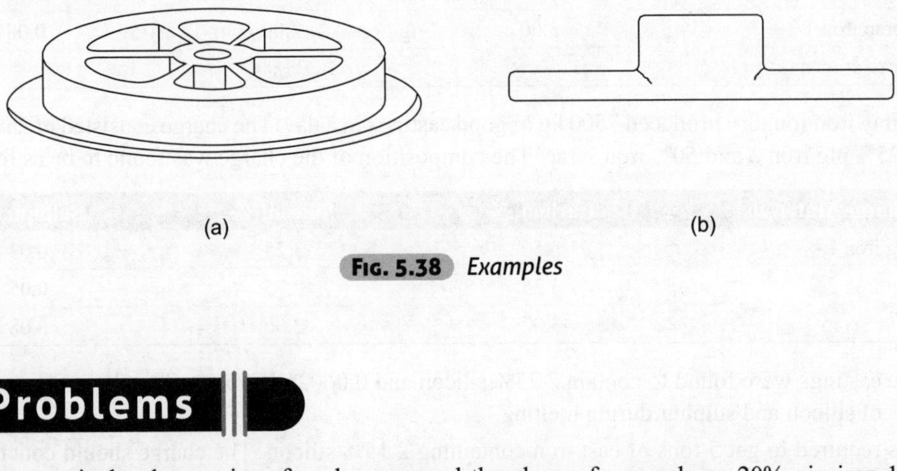

FIG. 5.38 *Examples*

Problems

5.1 On a particular day, an iron foundry prepared the charge for cupola as 20% pig iron 1, 25% pig iron 2, and 55% scrap iron. Find the final composition of the produced melt, given the following compositions for the charge elements. Assume suitable losses or gains in the cupola.

Charge Materials	Carbon, %	Silicon, %	Manganese, %	Sulphur, %
Pig iron 1	3.20	1.70	0.80	0.03
Pig iron 2	3.50	2.40	0.60	0.01
Scrap iron	3.25	2.30	0.65	0.08

5.2 A foundry engineer is required to produce castings with 3.10 to 3.50% carbon, 2.00 to 2.40% silicon, 0.60 to 1.00% manganese and a maximum of 0.08% sulphur. The pickup in cupola is approximately: carbon 0.15% and sulphur 0.04%. The losses in cupola are 10% silicon and 20% manganese. If the engineer has the following raw materials at her/his disposal, what should be the best charge composition?

Charge Materials	Carbon, %	Silicon, %	Manganese, %	Sulphur, %
Pig iron 1	3.20	1.70	0.80	0.03
Pig iron 2	3.20	1.85	0.60	0.01
Pig iron 3	3.50	2.10	0.70	0.02
Scrap iron 1	3.25	2.30	0.65	0.08
Scrap iron 2	3.30	2.10	0.80	0.10
Ferro silicon	—	75.00	—	—

5.3 Specify the composition of the charge to be used for producing 8 metric tons of the following gray iron: 3.20 to 3.60% carbon, 2.2 to 2.6% silicon, and a maximum of 0.08% sulphur. The available raw materials and their compositions are given below:

Charge Materials	Carbon, %	Silicon, %	Sulphur, %
Pig iron 1	3.25	2.75	0.05
Pig iron 2	3.50	2.40	0.01
Pig iron 3	3.45	2.50	0.01
Scrap iron 1	3.00	2.00	0.08
Ferro silicon	—	75.00	—

5.4 A gray iron foundry produced 7500 kg of good castings in a day. The charge consisted of 25% pig iron 1, 25% pig iron 2 and 50% iron scrap. The composition of the charge was found to be as follows:

Charge Materials	Carbon, %	Silicon, %	Sulphur, %
Pig iron 1	3.50	3.25	0.03
Pig iron 2	3.25	2.75	0.05
Iron Scrap	3.00	2.00	0.08

The castings were found to contain 2.25% silicon and 0.0602% sulphur. Find the percentage gain or loss of silicon and sulphur during melting.

5.5 It is required to get 5 tons of cast iron containing 2.15% silicon. The charge should contain at least 2 tons of scrap with 2.2% silicon and 0.8% phosphorous. Two pig irons A and B are available with A having 3.0% silicon and 1.5% phosphorous while B has 2.0% silicon and 0.4% phosphorous. Allowing 10% loss of silicon and no loss in phosphorous in melting, calculate the percentage of phosphorous in the final melt.

5.6 In a foundry, the following raw materials are available:

Charge Materials	Carbon, %	Silicon, %	Manganese, %
Pig iron A	3.60	2.50	0.40
Pig iron B	3.20	1.90	0.70
Iron Scrap	3.45	1.90	0.60

The final charge should use at least 50% of scrap and the silicon content should be at least 2.0%. Calculate the final composition assuming no change in carbon, 10% loss in silicon and 25% in manganese.

5.7 It is required to make castings with at least 1.80% silicon using 50% of scrap containing 2.20% silicon. Two pig irons are available, one with 2.50% silicon and other with 1.20% silicon. Allowing 20% loss of silicon in melting, estimate the final composition of the charge.

5.8 In a foundry, the following raw materials are included in the charge:

Charge Materials	Charge, kg	Carbon, %	Silicon, %	Manganese, %	Sulphur, %	Phosphorous, %
Pig iron A	250	3.82	3.17	0.90	0.01	1.40
Pig iron B	250	3.50	2.50	0.60	0.04	0.80
Scrap 1	500	3.30	2.50	0.60	0.08	1.00

Assuming a carbon pickup of 0.30%, sulphur pickup of 0.03%, 10% loss of silicon, 25% loss of manganese and no change in phosphorous, calculate the final melt composition.

Multiple Choice Questions

5.1 The charge of a cupola contains
(a) Pig iron, coal and lime stone
(b) Pig iron and coke
(c) Pig iron, coke and lime stone
(d) None of the above

5.2 During the cupola melting of cast iron, the following element is lost due to oxidation
(a) Sulphur (b) Carbon
(c) Zinc (d) Nickel

5.3 During the cupola melting of cast iron, the following element is lost due to oxidation
(a) Carbon (b) Phosphorous
(c) Silicon (d) All of the above

5.4 During the cupola melting of cast iron, the following element is picked up from the furnace
(a) Carbon (b) Phosphorous
(c) Sulphur (d) Silicon

5.5 The most economical furnace to melt cast iron in large volume is
(a) Reverberatory furnace
(b) Cupola furnace
(c) Induction furnace
(d) Electric furnace

5.6 To get a better control of molten metal quality the following furnace is preferred
(a) Cupola furnace
(b) Electric Induction furnace
(c) Crucible furnace
(d) None of the above

5.7 Inoculation is done in foundry to help in
(a) Improving the mechanical strength of steel
(b) Producing changes in graphite distribution in the case of ductile iron
(c) Reducing the slag from molten metal
(d) Improving the flow characteristics of copper alloys

5.8 Inoculation is more important for this foundry material
(a) Gray cast iron
(b) Ductile iron
(c) White cast iron
(d) Mild steel

5.9 The type of inoculation method practiced in automatic pouring lines
(a) Ladle inoculation
(b) In-mould inoculation

(c) In-stream inoculation
(d) In-furnace inoculation

5.10 Surface of large castings taken out from sand mould can be effectively cleaned by
(a) Tumbling
(b) Sand blasting
(c) Grinding
(d) None of the above

5.11 Blow holes in sand casting are caused by
(a) Poor casting design
(b) Lower strength of the solidified metal
(c) Higher moisture in the moulding sand
(d) Faulty moulding flask

5.12 Open blows in sand casting are caused by
(a) Poor casting design
(b) Higher moisture in the moulding sand
(c) Lower strength of the solidified metal
(d) Faulty moulding flask

5.13 Air inclusions in sand casting are caused by
(a) Poor casting design
(b) Faulty moulding flask
(c) Lower strength of the solidified metal
(d) Higher pouring temperature

5.14 Remedy for reducing air inclusions in sand castings is
(a) Improve the casting design
(b) Use high strength materials
(c) Increase the permeability of the sand mould
(d) Use better moulding flasks

5.15 Pin hole porosity in sand casting is caused by
(a) Poor casting design
(b) Faulty moulding flask
(c) Lower strength of the solidified metal
(d) Higher pouring temperature

5.16 Shrinkage cavities in sand casting are caused by
(a) Poor casting design
(b) Faulty moulding flask
(c) Lower strength of the solidified metal
(d) Higher pouring temperature

5.17 Cuts and washes in sand casting are caused by
(a) Poor casting design
(b) Lower strength of the solidified metal
(c) Lower strength of the moulding sand
(d) Higher pouring temperature

5.18 Metal penetration in sand casting is caused by
(a) Poor casting design
(b) Lower strength of the solidified metal
(c) Lower strength of the moulding sand
(d) Higher pouring temperature

5.19 Fusion in sand casting is caused by
(a) Lower refractoriness of the moulding sand
(b) Lower strength of the solidified metal
(c) Lower strength of the moulding sand
(d) Lower hardness of the moulding sand

5.20 Runout in sand casting is caused by
(a) Poor casting design
(b) Faulty moulding flask
(c) Lower strength of the solidified metal
(d) Higher pouring temperature

5.21 Rat tails and buckles in sand casting are caused by
(a) Lower refractoriness of the moulding sand
(b) Lower strength of the solidified metal
(c) Lower strength of the moulding sand
(d) Lower hardness of the moulding sand

5.22 Swell in sand casting is caused by
(a) Lower refractoriness of the moulding sand
(b) Lower strength of the solidified metal
(c) Lower strength of the moulding sand
(d) Faulty mould making procedure (lower ramming)

5.23 Mis runs and cold shuts in sand casting are caused by
(a) Lower pouring temperature
(b) Faulty moulding flask
(c) Lower strength of the solidified metal
(d) Higher pouring temperature

5.24 Slag inclusions in sand casting can be reduced by
(a) Increasing pouring temperature
(b) Using better moulding flasks
(c) Using trapezoidal runners
(d) Adding a runner extension

5.25 Parting line in sand castings should be
(a) Any type will not make a difference
(b) Simple and straight
(c) Irregular parting line decreases casting cost
(d) Multiple parting lines decrease casting cost

5.26 If cores are to be used in sand casting
(a) Either a green sand core or a dry sand core cost the same
(b) A dry sand core is better than a green sand core
(c) A green sand core is better than a dry sand core
(d) A dry sand core is preferred than a green sand core

5.27 A hot spot in sand casting is caused by
(a) Pouring hot metal into the mould
(b) The higher concentration of metal in the joints
(c) Poor refractory property of the moulding sand
(d) Low thermal conductivity of the moulding sand

5.28 Distortion in sand casting is caused by
(a) Pouring hot metal into the mould
(b) Low thermal conductivity of the moulding sand
(c) Poor refractory property of the moulding sand
(d) Non uniform thickness of walls joining in a casting

5.29 Hot tear in sand casting is caused by
(a) Pouring hot metal into the mould
(b) Non uniform thickness of walls joining in a casting
(c) Poor refractory property of the moulding sand
(d) Low thermal conductivity of the moulding sand

5.30 To achieve directional solidification in a casting
(a) Casting walls should preferably be of uniform thickness
(b) Local metal accumulations and massive elements should be avoided, if possible.
(c) Transition between casting walls of different thicknesses should be smooth
(d) All of the above

Answers to MCQs

5.1 (c)	5.2 (b)	5.3 (d)	5.4 (c)	5.5 (b)
5.6 (b)	5.7 (b)	5.8 (b)	5.9 (c)	5.10 (b)
5.11 (c)	5.12 (b)	5.13 (d)	5.14 (c)	5.15 (d)
5.16 (a)	5.17 (c)	5.18 (d)	5.19 (a)	5.20 (b)
5.21 (c)	5.22 (d)	5.23 (a)	5.24 (d)	5.25 (b)
5.26 (c)	5.27 (b)	5.28 (d)	5.29 (b)	5.30 (d)

CHAPTER 6

Special Casting Processes

Objectives

Sand casting processes described so far are not suitable and economical in many applications. In such situations special casting processes would be more appropriate. After completing this chapter, the reader will be able to

- Understand the shell moulding as a process for higher quality castings with thin surface details
- Know precision investment casting used for complex shapes
- Use permanent mould casting for mass production of relatively simple shapes
- Apply die casting process for complex parts
- Understand the advantages of using centrifugal casting process
- Learn the advantages and applications of various other special casting process

6.1 SHELL MOULDING

It is a process in which the sand mixed with a thermosetting resin is allowed to come into contact with a heated metallic pattern plate, so that a thin and strong shell of mould is formed around the pattern. Then the shell is removed from the pattern, the cope and drag are removed together and kept in a flask with the necessary back up material and the molten metal is poured into the mould.

Generally, dry and fine sand (90 to 140 GFN) that is completely free of the clay is used for preparing the shell moulding sand. The grain size to be chosen depends on the surface finish desired on the casting. Too fine a grain size requires large amount of resin which makes the mould expensive.

The synthetic resins used in shell moulding are essentially thermosetting resins which get hardened irreversibly by heat. The resins most widely used are the phenol formaldehyde resins. Combined with sand, they have very high strength and resistance to heat. The phenolic resins used in shell moulding usually are of the two stage type, that is, the resin has excess phenol and acts like a thermoplastic material. During coating with the sand the resin is combined with a catalyst such as hexa-methylene-tetramine (hexa) in a proportion of about 14 to 16% so as to develop the thermosetting characteristics. The curing temperature for these would be around 150°C and the time required would be 50 to 60 s.

Typical sand compositions for shell moulding are given in Table 6.1.

Additives may sometimes, be added into the sand mixture to improve the surface finish and avoid thermal cracking during pouring. Some of the additives used are coal dust, pulverised slag, manganese dioxide, calcium carbonate, ammonium boroflouride and magnesium silicofluoride. Some lubricants such as calcium stearate and zinc stearate may also be added to the resin sand mixture to improve the flowability of the sand and permit easy release of the shell from the pattern.

TABLE 6.1 Shell sand compositions in percentages

Casting Material	Silica Sand	Zircon Sand	Resin	Additives
Low carbon and alloy steels	63	30	5	2
Medium and high carbon steel	—	96	3	1
Gray cast iron	60	35	4	1
Brass and bronze	90	—	6	4
Aluminium alloys	95	—	4	1
Magnesium alloys	—	95	4	1

The first step in preparing the shell mould is the preparation of the sand mixture in such a way that each of the sand grain is thoroughly coated with resin. To achieve this, first the sand, hexa and additives, which are all dry, are mixed inside a mueller for a period of 1 min. Then the liquid resin is added and mixing is continued for another 3 min. To this cold or warm air is introduced into the Mueller and the mixing is continued till all the liquid is removed from the mixture and coating of the grains is achieved to the desired degree.

Since the sand resin mixture is to be cured at about 150°C temperature, only metal patterns with the associated gating are used. The metal used for preparing patterns is gray cast iron, mainly because of its easy availability and excellent stability at the temperatures involved in the process. Sometimes additional risering provision is required as the cooling in shell moulds is slow.

The metallic pattern plate is heated to a temperature of 200 to 350°C depending on the type of the pattern. It is essential that the pattern plate is uniformly heated so that the temperature variation across the whole pattern is within 25 to 40°C depending on the size of the pattern. A silicone release agent is sprayed on the pattern and the metal plate. The heated pattern is securely fixed to a dump box, as shown in Fig. 6.1(a), wherein the coated sand in an amount larger than required to form the shell of necessary thickness is already filled in.

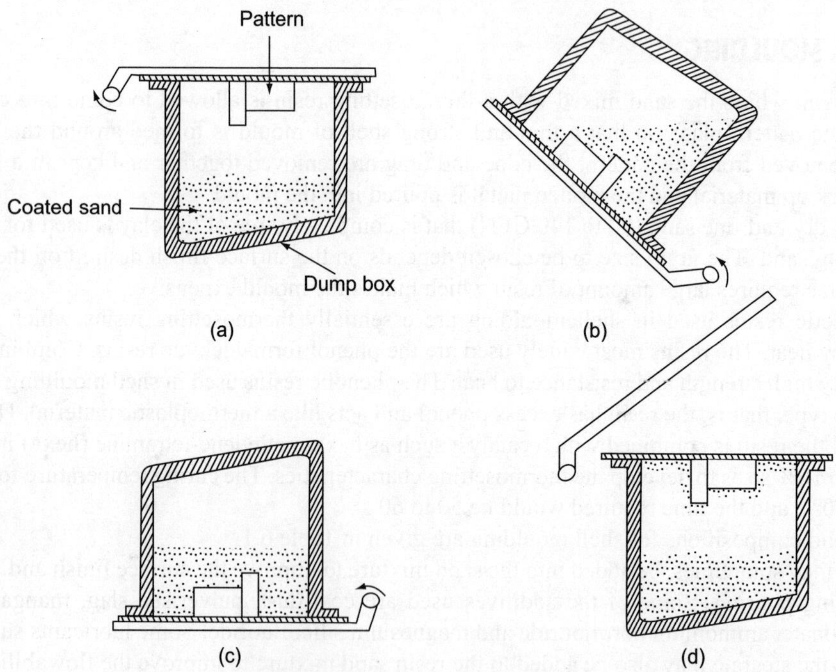

FIG. 6.1 *Shell moulding procedure*

Then the dump box is rotated as shown in Fig. 6.1(b) so that the coated sand falls on the heated pattern. The heat from the pattern melts the resin adjacent to it thus causing the sand mixture to adhere to the pattern.

When a desired thickness of shell is achieved, the dump box is rotated backwards by 180 degrees so that the excess sand falls back into the box, leaving the formed shell intact with the pattern as in Fig. 6.1(d). The average shell thickness achieved depends on the temperature of the pattern and the time for which the coated sand remains in contact with the heated pattern. Figure 6.2 shows typical shell thicknesses that can be obtained with various pattern temperatures and contact times. The actual shell thicknesses required depends on the pouring metal temperature and the casting complexity. This may normally be achieved by trial and error method.

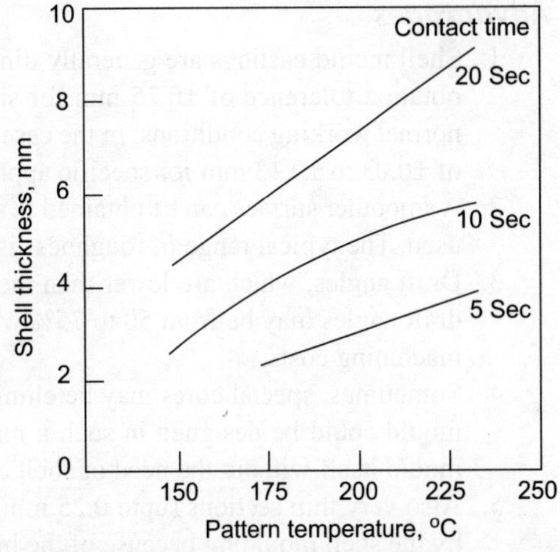

Fig. 6.2 *Variation of shell thickness with contact pattern temperature*

The shell along with the pattern plate is kept in an electric or gas fired oven for curing the shell. The curing of the shell should be done as per requirements only because over curing may cause the mould to breakdown as the resin would burn out. The under curing may result in blow holes in the casting or the shell may break during handling because of the lack of strength.

The shells thus prepared are joined together by either mechanical clamping or by adhesive bonding. The resin used as an adhesive may be applied at the parting plane before mechanical clamping and then allowed for 20 to 40 s for achieving the necessary bonding. A finished shell mould ready for pouring is presented in Fig. 6.3.

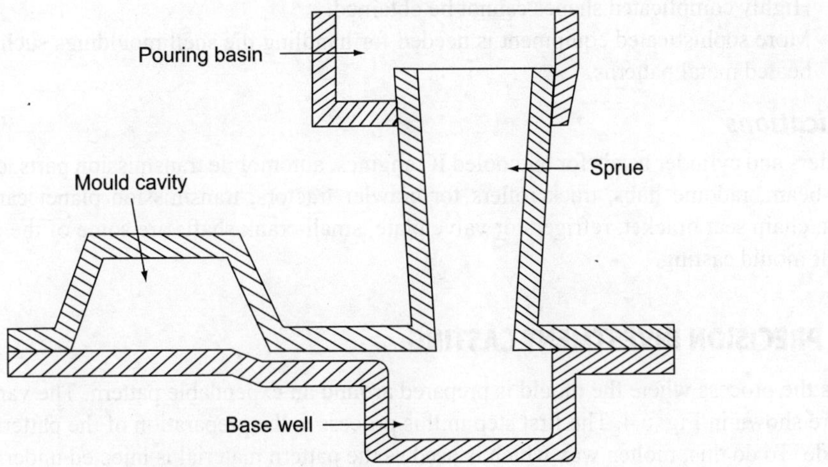

Fig. 6.3 *Shell mould ready for pouring*

Since the shells are thin, they may require some outside support so that they can withstand the pressure of the molten metal. A metallic enclosure to closely fit the exterior of the shell is ideal but is too expensive and therefore impractical. Alternatively, a cast iron shot is generally preferred as it occupies any contour without unduly applying any pressure on the shell. With such a backup material, it is possible to reduce the shell thickness to an economical level.

Advantages

1. Shell mould castings are generally dimensionally more accurate than sand castings. It is possible to obtain a tolerance of ±0.25 mm for steel castings and ±0.35 mm for grey cast iron castings under normal working conditions. In the case of close tolerance shell moulds, one may obtain it in the range of ±0.03 to ±0.13 mm for specific applications.
2. A smoother surface can be obtained in shell castings. This is primarily achieved by the finer size grain used. The typical range of roughness is of the order of 3 to 6 microns.
3. Draft angles, which are lower than the sand castings are required in shell moulds. The reduction in draft angles may be from 50 to 75% which considerably saves the material costs and the subsequent machining costs.
4. Sometimes, special cores may be eliminated in shell moulding. Since the sand has high strength the mould could be designed in such a manner that internal cavities can be formed directly with shell mould itself without the need of shell cores.
5. Also very thin sections (upto 0.25 mm) of the type of air cooled cylinder heads can be readily made by the shell moulding because of the higher strength of the sand used for moulding.
6. Permeability of the shell is high and therefore no gas inclusions occur.
7. Very small amount of sand needs to be used.
8. Mechanisation is readily possible because of the simple processing involved in shell moulding.

Limitations

1. The patterns are very expensive and therefore are economical only if used in large scale production. In a typical application, shell moulding becomes economical over sand moulding above 15000 pieces because of the higher pattern cost.
2. The size of the casting obtained by shell moulding is limited. Generally castings weighing up to 200 kg can be made, though in smaller quantity castings upto a weight of 450 kg were made.
3. Highly complicated shapes cannot be obtained.
4. More sophisticated equipment is needed for handling the shell mouldings such as those required for heated metal patterns.

Applications

Cylinders and cylinder heads for air cooled IC engines, automobile transmission parts, cast tooth bevel gears, brake beam, radome hubs, track rollers for crawler tractors, transmission planet carrier, steel eyes, gear blanks, chain seat bracket, refrigerator valve plate, small crank shafts are some of the common applications of shell mould casting.

6.2 PRECISION INVESTMENT CASTING

This is the process where the mould is prepared around an expendable pattern. The various steps in the process are shown in Fig. 6.4. The first step in this process is the preparation of the pattern for every casting to be made. To do this, molten wax which is used as the pattern material is injected under pressure of about 2.5 MPa into a metallic die and has the cavity of the casting to be made as shown in Step 1 of Fig. 6.4. The wax when allowed to solidify would produce the pattern. The pattern is ejected from the die as shown in step 2. Then the cluster of wax patterns is attached to the gating system by applying heat as shown in step 3.

To make the mould, the prepared pattern is dipped into a slurry made by suspending fine ceramic materials in a liquid such as ethyl silicate or sodium silicate (step 4). The excess liquid is allowed to drain off from the pattern. Dry refractory grains such as fused silica or zircon are "stuccoed" on this liquid ceramic coating (step 5). Thus a small shell is formed around the wax pattern. The shell is cured and then the process of

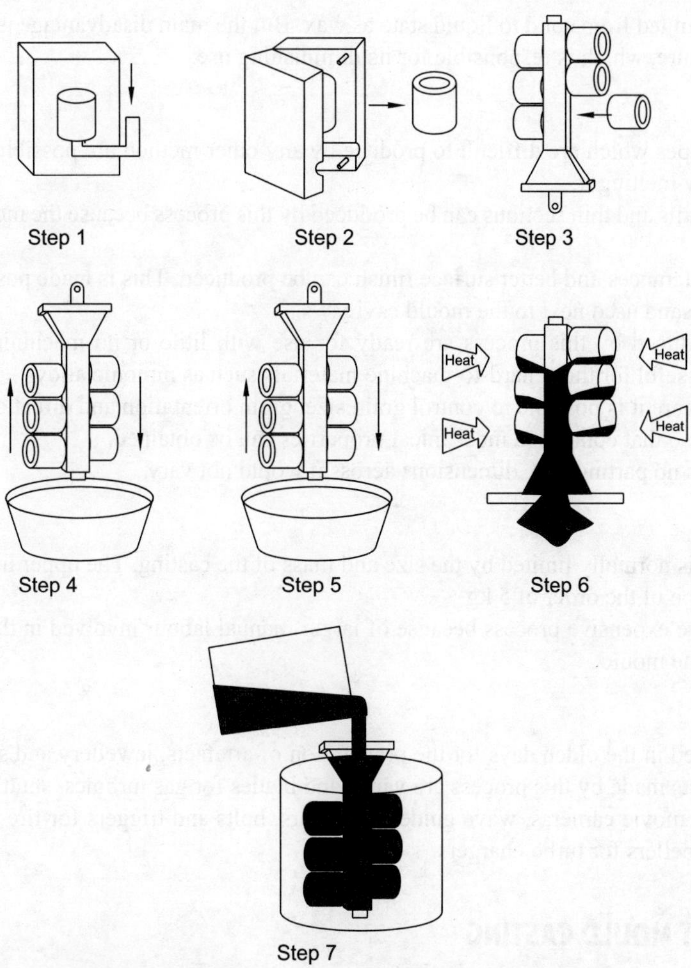

FIG. 6.4 ***Steps in the precision investment casting process***

dipping and stuccoing is continued with ceramic slurries of gradually increasing grain sizes. Finally when a shell thickness of 6 to 15 mm is reached, the mould is ready for further processing. The shell thickness required depends on the casting shape and mass, type of ceramic and the binder used.

The next step in the process is to remove the pattern from the mould which is done by heating the mould to melt the pattern (step 6). The melted wax is completely drained through the sprue by inverting the mould. Any wax remnants in the mould are dissolved with the help of the hot vapour of a solvent such as trichloroethylene.

The moulds are then pre heated to a temperature of 100 to 1000°C, depending on the size, complexity and the metal of the casting. This is done to reduce any last traces of wax left off and permit proper filling of all mould sections which are too thin to be filled in a cold mould.

The molten metal is poured into the mould under gravity and under slight pressure, by evacuating the mould first (step 7). The method chosen depends on the type of casting.

Other pattern materials used are plastics and mercury in place of wax. In the process called 'Mercast', the mercury is kept under –57°C where the mercury is frozen. The complete mould preparation is to be undertaken at a temperature below –38°C. The main advantage of mercury as a pattern material is that it does not expand when changed from solid to liquid state as wax. But the main disadvantage is keeping the pattern at such low temperature, which is responsible for its diminishing use.

Advantages

1. Complex shapes which are difficult to produce by any other method are possible since the pattern is withdrawn by melting it.
2. Very fine details and thin sections can be produced by this process because the mould is heated before pouring.
3. Very close tolerances and better surface finish can be produced. This is made possible because of the fine grain of sand used next to the mould cavity.
4. Castings produced by this process are ready for use with little or no machining required. This is particularly useful for those hard-to-machine materials such as nimonic alloys.
5. With proper care it is possible to control grain size, grain orientation and directional solidification in this process, so that controlled mechanical properties can be obtained.
6. Since there is no parting line, dimensions across it would not vary.

Limitations

1. The process is normally limited by the size and mass of the casting. The upper limit on the mass of a casting may be of the order of 5 kg.
2. This is a more expensive process because of larger manual labour involved in the preparation of the pattern and the mould.

Applications

This process was used in the olden days for the preparation of artefacts, jewellery and surgical instruments. Presently the products made by this process are vanes and blades for gas turbines, shuttle eyes for weaving, pawls and claws for movie cameras, wave guides for radars, bolts and triggers for fire arms, stainless steel valve bodies and impellers for turbo chargers.

6.3 PERMANENT MOULD CASTING

In all the processes that have been covered so far, a mould needs to be prepared for each of the casting produced. For large scale production, making a mould for every casting to be produced may be difficult and expensive. Therefore, a permanent mould, called 'die' may be made from which a large number of castings, any where from 100 to 250 000, can be produced depending on the alloy used and the complexity of the casting. This process is called permanent mould casting or gravity die casting, since the metal enters the mould under gravity.

The mould material is selected on the consideration of the pouring temperature, size of the casting and frequency of the casting cycle. They determine the total heat to be borne by the die. Fine grained gray cast iron is the most generally used die material. Alloy cast iron, C20 steel and alloy steels (H11 and H14) are also used for very large volumes and large parts. Graphite moulds may be used for small volume production from aluminium and magnesium. The die life is less for higher melting temperature alloys such as copper or grey cast iron.

For making any hollow portions, cores are also used in permanent mould casting. The cores can be made out of metal or sand. When sand cores are used, the process is called semi permanent moulding. The metallic core cannot be complex with under cuts and the like. Also the metallic core is to be withdrawn immediately after solidification; otherwise, its extraction becomes difficult because of shrinkage. For complicated shapes, collapsible metal cores (multiple-piece cores) are some times used in permanent moulds. Their use is not extensive because of the fact that it is difficult to securely position the core as a single piece as also due to the dimensional variations that are likely to occur. Hence, with collapsible cores, the designer has to provide coarse tolerance on these dimensions. A typical permanent mould is shown schematically showing the various components present in Fig. 6.5 which is very similar to the sand mould.

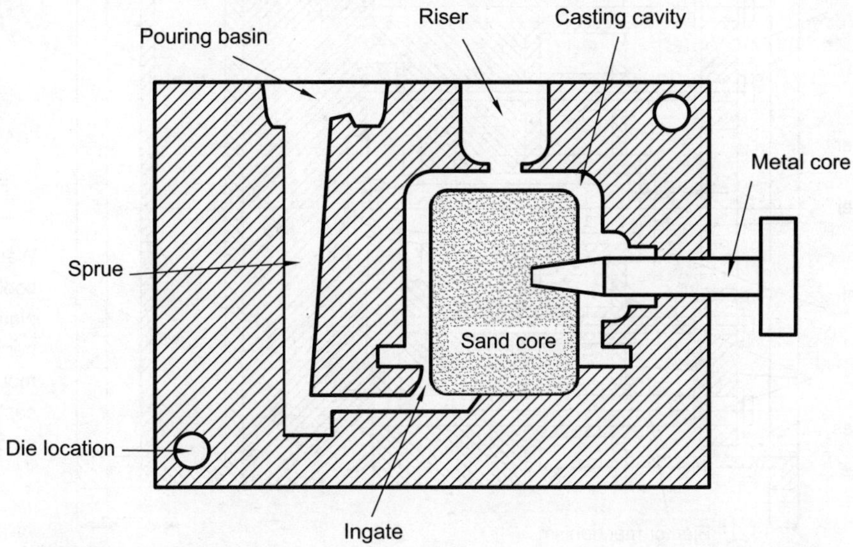

Fig. 6.5 *Schematic of a permanent mould die with various possible elements*

The mould cavity should normally be simple without any undesirable drafts or undercuts, which interfere with the ejection of the solidified castings. In designing the permanent moulds, care should be taken to see that progressive solidification towards the riser is achieved. The gating and risering systems used are very similar to that of the sand casting. In fact to get the proper gating arrangements, it may be desirable first to experiment with various gating systems in sand casting and then finally arrive at the correct gating system for the metallic mould.

Since the metallic mould would be used continuously, a large amount of heat will be absorbed by the mould material. It is, therefore, necessary to make provision in the construction of the permanent mould to extract this heat continuously to ensure higher mould life and better casting cycle times. Air cooling is not sufficient except for very small and thin walled castings. Air cooled dies will be provided with fins on the outer surfaces to increase the heat extraction. Cooling water may be circulated through the portions of the mould that are likely to have large heat concentration. Figure 6.6 shows the cross section of a typical metallic mould for casting a gear blank. The location of the cooling passages for the mould as well as the metallic core can be seen from this figure. The likely problems with the cooling water circulation are the formation of scales inside the cooling channels and their subsequent blocking after some use.

If the casting has heavy sections, which are likely to interfere with the progressive solidification, mould section around that area may be made heavier around that area to extract more heat. Chills supported by heavy air blast may also be used to remove the excess heat. Alternatively, cooling channels may be provided at the necessary points to get proper temperature distribution.

The moulds are coated with a refractory material to a thickness of around 0.8 mm. The coatings are used to increase the mould life

- By preventing the soldering of metal to the mould
- By minimising the thermal shock to the mould material
- By controlling the rate and direction of the casting solidification

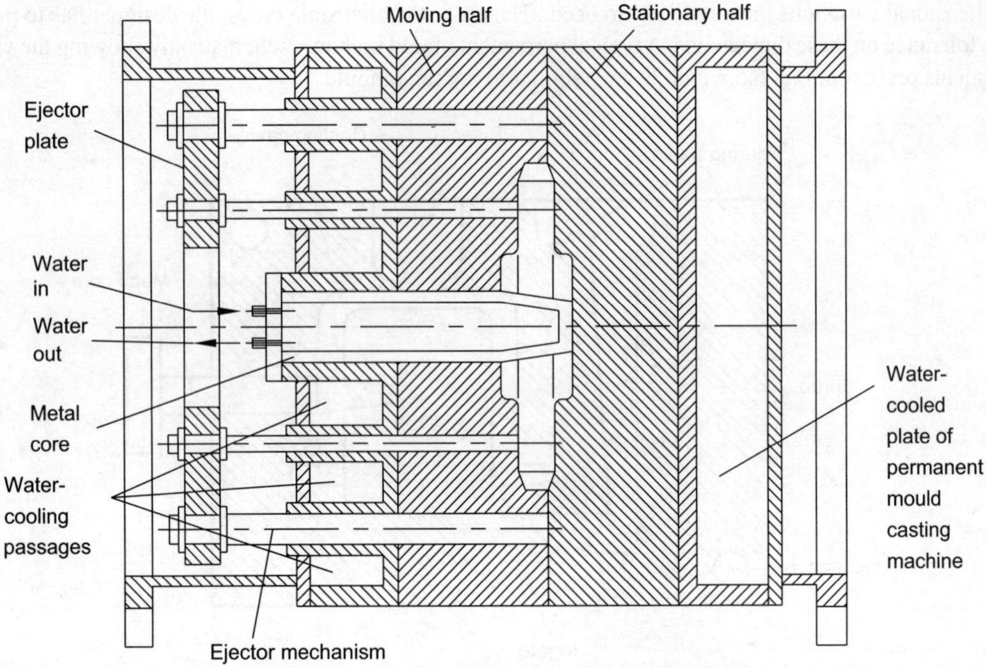

FIG. 6.6 *Cross section of a permanent mould of a gear blank showing the different elements and the cooling water passages*

The coatings normally are mixtures of sodium silicate, kaolin clay, soap stone and talc. The coatings are both insulating type and lubricating type. The main requirement of a coating is that it should be inert to the casting alloy. The coating may be applied by spraying or brushing. It must be thick enough to fill up any surface imperfections. The coatings can be applied thicker at surfaces, which need to be cooled slowly, for example, sprues, runners, risers and thin sections. The maximum thickness of a coating required is about 0.8 mm.

Under the regular casting cycle, the temperature at which the mould is used depends on the pouring temperature, casting cycle frequency, casting weight, casting shape, casting wall thickness, wall thickness of the mould and the thickness of the mould coating. If the casting is done with the cold die, the first few castings are likely to have mis runs till the die reaches its operating temperature. To avoid this, the mould should be pre heated to its operating temperature, preferably in an oven.

The materials which are normally cast in permanent moulds are aluminium alloys, magnesium alloys, copper alloys, zinc alloys and gray cast iron. The sizes of castings are limited to 15 kg in most of the materials. But, in case of aluminium, large castings with a mass of upto 350 kg have been produced. Permanent mould casting is particularly suited to high volume production of small, simple castings with uniform wall thickness and no intricate details.

Advantages

1. Because of the metallic moulds used, this process produces a fine grained casting with superior mechanical properties.
2. They produce very good surface finish of the order of 4 microns and better appearance.
3. Close dimensional tolerances can be obtained.
4. It is economical for large scale production as the labour involved in the mould preparation is reduced.
5. Small cored holes may be produced compared to sand casting.
6. Inserts can be readily cast in place.

Limitations

1. The maximum size of the casting that can be produced is limited because of the equipment.
2. Complicated shapes cannot be produced.
3. The cost of the die is very high and can only be justified for large scale production.
4. Not all materials are suited for permanent mould casting essentially because of the mould material.

Applications

Some of the components that are produced in permanent moulds are automobile pistons, stators, gear blanks, connecting rods, aircraft fittings, cylinder blocks, etc.

6.4 DIE CASTING

Die casting involves the preparation of components by injecting molten metal at high pressure into a metallic die. Die casting is closely related to permanent mould casting, in that both the processes use reusable metallic dies. In die casting, as the metal is forced in under pressure compared to permanent moulding, it is also called "pressure die casting". Because of the high pressure involved in die casting, any narrow sections, complex shapes and fine surface details can easily be produced.

In die casting, the die consists of two parts. One called the stationary half or cover die which is fixed to the diecasting machine. The second part called the moving half or ejector die is moved out for the extraction of the casting. The casting cycle starts when the two parts of the die are apart. The lubricant is sprayed on the die cavity manually or by the auto lubrication system so that the casting will not stick to the die. The two die halves are closed and clamped. The required amount of metal is injected into the die. After the casting is solidified under pressure the die is opened and the casting is ejected. The die casting die needs to have the provision of ejectors to push the casting after it gets solidified as shown in Fig. 6.7. It will also have cooling channels to extract the heat of the molten metal to maintain proper die temperature.

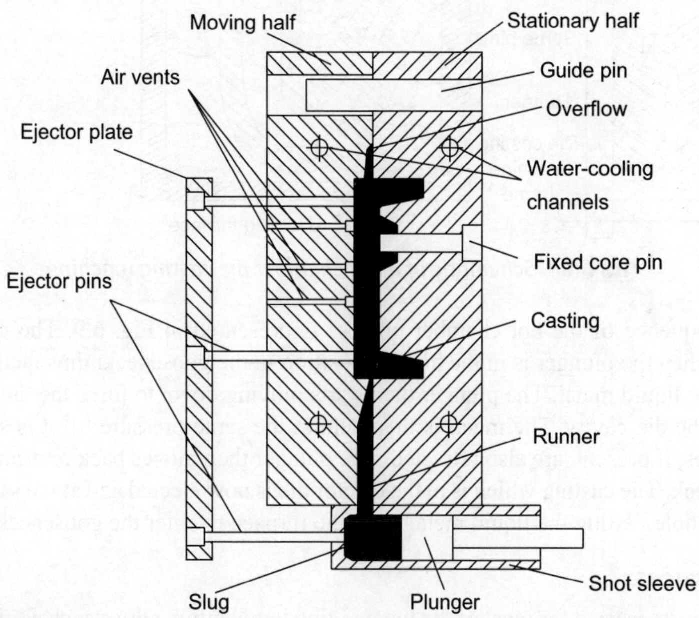

Fig. 6.7 *Cross section of a die casting die used for cold chamber die casting*

The die casting machines are of two types:

- Hot chamber die casting, and
- Cold chamber die casting

The main difference between these two types is that in hot chamber, the holding furnace for the liquid metal is integral with the die casting machine, whereas in the cold chamber machine, the metal is melted in a separate furnace and then poured into the die casting machine with a ladle for each casting cycle which is also called 'shot'.

Hot Chamber Process

A typical hot chamber diecasting machine is shown in Fig. 6.8. In this, a gooseneck is used for pumping the liquid metal into the die cavity. The gooseneck is submerged in the holding furnace containing the molten metal. The gooseneck is made of gray, alloy or ductile iron, or of cast steel. A plunger made of alloy cast iron and is hydraulically operated, moves up in the gooseneck to uncover the entry port for the entry of liquid metal into the goose neck. The plunger can then develop the necessary pressure for forcing the metal into the die cavity. A nozzle at the end of the gooseneck is kept in close contact with the sprue located in the cover die.

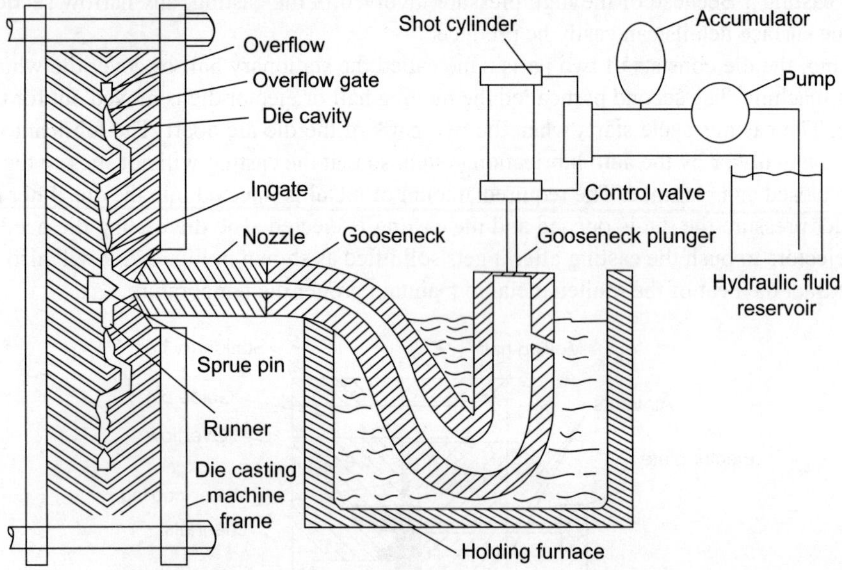

Fig. 6.8 *Schematic of a hot chamber die casting machine*

The operating sequence of the hot chamber process is presented in Fig. 6.9. The cycle starts with the closing of the die, when the plunger is in the highest position in the gooseneck, thus facilitating the filling of the gooseneck by the liquid metal. The plunger then starts moving down to force the metal in the gooseneck to be injected into the die cavity. The metal is then held at the same pressure till it is solidifies. The die is opened and any cores, if present, are also retracted. The plunger then moves back returning the unused liquid metal to the gooseneck. The casting which is in the ejector die is now ejected and at the same time the plunger uncovers the filling hole, letting the liquid metal from the furnace to enter the gooseneck.

Cold Chamber Process

The hot chamber process is used for most of the low melting temperature alloys such as zinc, lead and tin. For materials such as aluminium and brass, their high melting temperatures make it difficult to cast them by hot

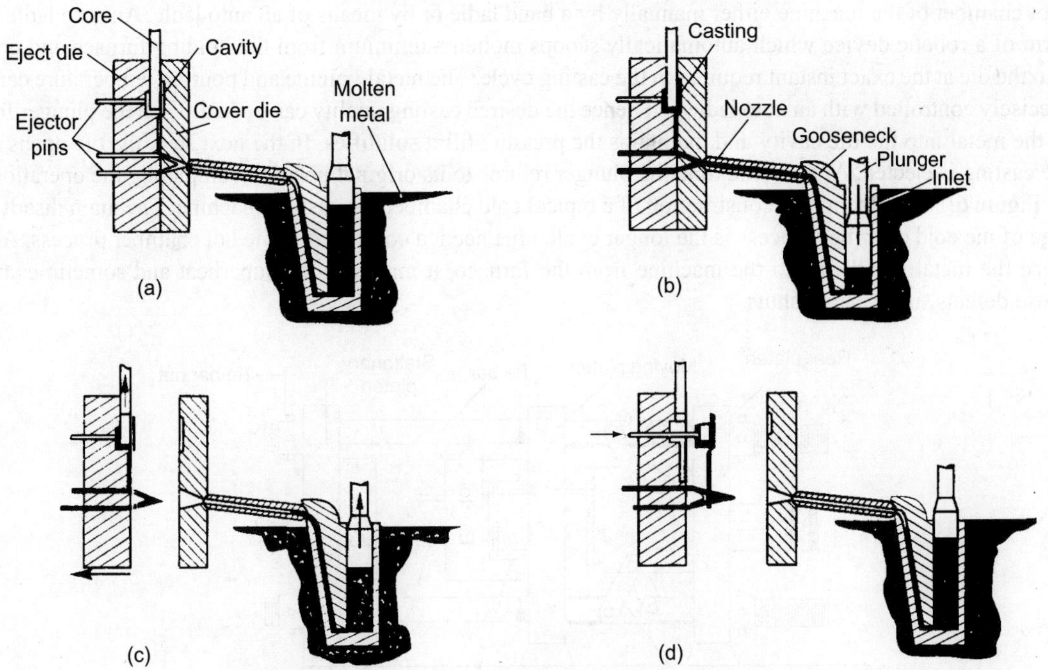

Fig. 6.9 *Operation sequence of hot chamber process*

chamber process, because gooseneck of the hot chamber machine is continuously in contact with the molten metal. Also liquid aluminium would attack the gooseneck material and thus hot chamber process is not used with aluminium alloys. In the cold chamber process, the molten metal is poured with a ladle into the shot chamber for every shot. This process reduces the contact time between the liquid metal and the shot chamber.

The operation sequence shown in Fig. 6.10 is similar to hot chamber process. The operation starts with the spraying of die lubricants throughout the die cavity and closing of the die when molten metal is ladled into the

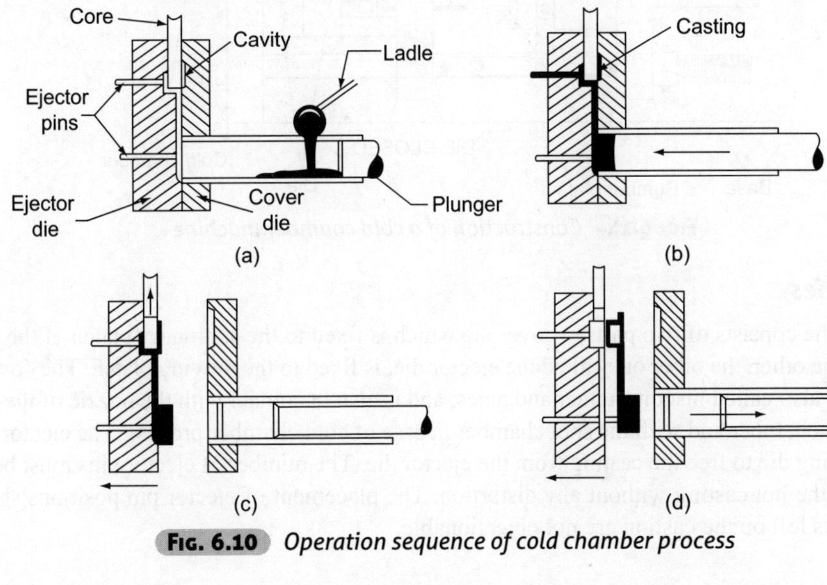

Fig. 6.10 *Operation sequence of cold chamber process*

shot chamber of the machine either manually by a hand ladle or by means of an auto ladle. An auto ladle is a form of a robotic device which automatically scoops molten aluminium from the holding furnace and pours into the die at the exact instant required in the casting cycle. The metal volume and pouring temperature can be precisely controlled with an auto ladle and hence the desired casting quality can be had. Then the plunger forces the metal into the die cavity and maintains the pressure till it solidifies. In the next step, the die opens and the casting is ejected. At the same time the plunger returns to its original position completing the operation.

Figure 6.11 represents the construction of a typical cold chamber die casting machine. The main disadvantage of the cold chamber process is the longer cycle time needed compared to the hot chamber process. Also, since the metal is ladled into the machine from the furnace, it may lose the superheat and sometimes may cause defects such as cold shuts.

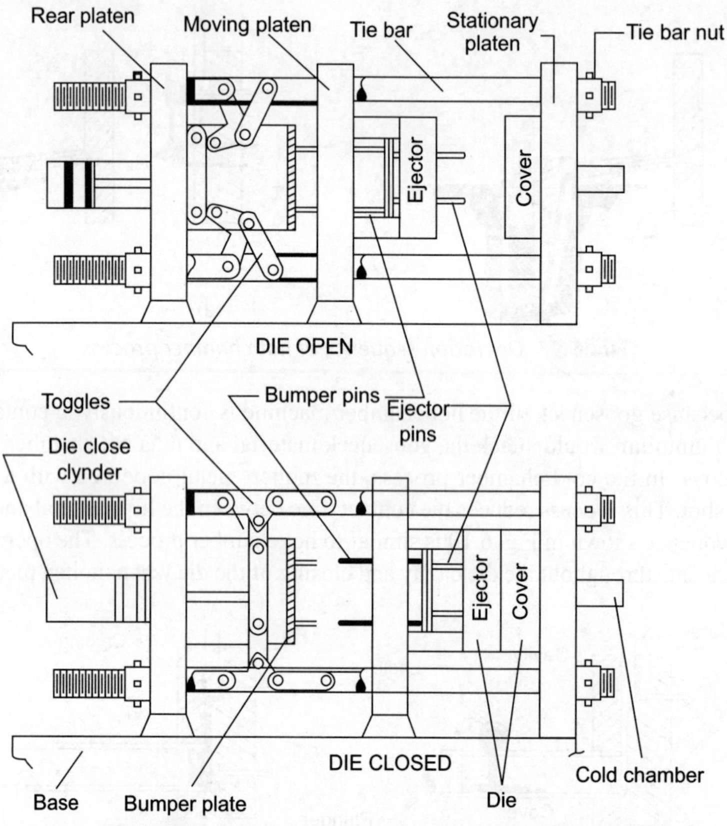

Fig. 6.11 *Construction of a cold chamber machine*

Die Casting Dies

The die casting die consists of two parts, a cover die which is fixed to the stationary platen of the die casting machine while the other, the other one called the ejector die, is fixed to the moving platen. The cover die consists of the sprue also called biscuit, runner and gates, and is also in contact with the nozzle of the gooseneck in the case of hot chamber and with the shot chamber in case of cold chamber process. The ejector pins move through the moving die to free the casting from the ejector die. The number of ejector pins must be sufficient so as to remove the hot casting without any distortion. The placement of ejector pin positions should be so that the pin marks left on the casting are not objectionable.

The cores used are all metallic and are of two types. The fixed cores are the ones which are fixed to the die halves. These are parallel to the die movement. The others called moving cores, are not parallel with the die movement and hence are to be removed before the casting is ejected from the die.

Sometimes, overflows are provided in the parting plane for the first metal which is normally cold, entering in the die cavity, to solidify. The overflow is a luxury and should be avoided in the design of a die. Overflows are primarily provided for small components to provide enough heat input to the die so that no cold shuts occur. Also the overflows can be utilised for positioning the ejector pins so that no objectionable ejector pin marks appear on the casting.

The cycle time in die casting being very small, the dies would readily get heated. Particularly in the sections such as sprues, or heavy sections of the casting, the temperature may be too high. To maintain uniform and requisite temperature for which the casting is designed, water is circulated through the identified hot regions of the die.

Hot working tool steels are normally used for the preparation of the dies, die inserts and cores. For zinc alloys, the normal die material is AISI P20 for low volume and H13 for high volume, whereas for aluminium and magnesium, H13 and H11 are used. For copper alloys H21, H20 and H22 are the usual die materials.

Some excess metal may be forced into the parting plane and is termed as flash. Before using the castings, the flash together with any overflow present in the casting is to be removed. The trimming of the flash is done either manually or more preferably in trimming dies. The trimming dies are similar to the sheet metal shearing dies which are discussed in chapter 8.

Advantages

1. Because of the use of the movable cores, it is possible to obtain fairly complex castings than that feasible by permanent mould casting.
2. Very small thicknesses can be easily filled because the liquid metal is injected at high pressure.
3. Very high production rates can be achieved. Typical rate could be 200 pieces per hour since the process is completely automated.
4. Because of the metallic dies, very good surface finish of the order of 1 micron can be obtained. The surfaces generated by die casting can be directly electroplated without any further processing.
5. Closer dimensional tolerances of the order of + 0.08 mm for small dimensions can be obtained compared to the sand castings.
6. The die has a long life which is of the order of 300 000 pieces for zinc alloys and 150 000 for aluminium alloys.
7. Die casting gives better mechanical properties compared to sand casting because of the fine grained skin formed during solidification.
8. Inserts can be readily cast in place.
9. It is very economical for large scale production.

Limitations

1. The maximum size of the casting is limited. The normal sizes are less than 4 kg with a maximum of the order of 15 kg because of the limitation on the machine capacity.
2. This is not suitable for all materials because of the limitations on the die materials. Normally zinc, aluminium, magnesium and copper alloys are diecast.
3. The air in the die cavity gets trapped inside the casting and is therefore a problem often with the die castings.
4. The dies and the machines are very expensive and therefore, economy in production is possible only when large quantities are produced.

Applications

The typical products made by die casting are carburettors, crank cases, magnetos, handle bar housings, and other parts of scooters, motor cycles and mopeds, zip fasteners, head lamp bezels and other decorative items on automobiles.

6.5 VACUUM DIE CASTING

The major problem with die casting is the air left in the cavity when the die is closed. Since air cannot escape, it will end up inside the casting. As a result, when the casting is heat treated blisters will appear on the surface. Also the molten metal when injected experiences this back pressure in the die cavity. This problem is solved by evacuating the air from the die after the die is closed and before the metal is injected. Thus the metal enters much faster into the die, thereby decreasing the fill time, and at the same time the parts do not experience any porosity due to the removal of all the air in the cavity. A typical arrangement for vacuum die casting process is shown in Fig. 6.12. The part is exposed to atmospheric air only after solidification and as such the oxidation

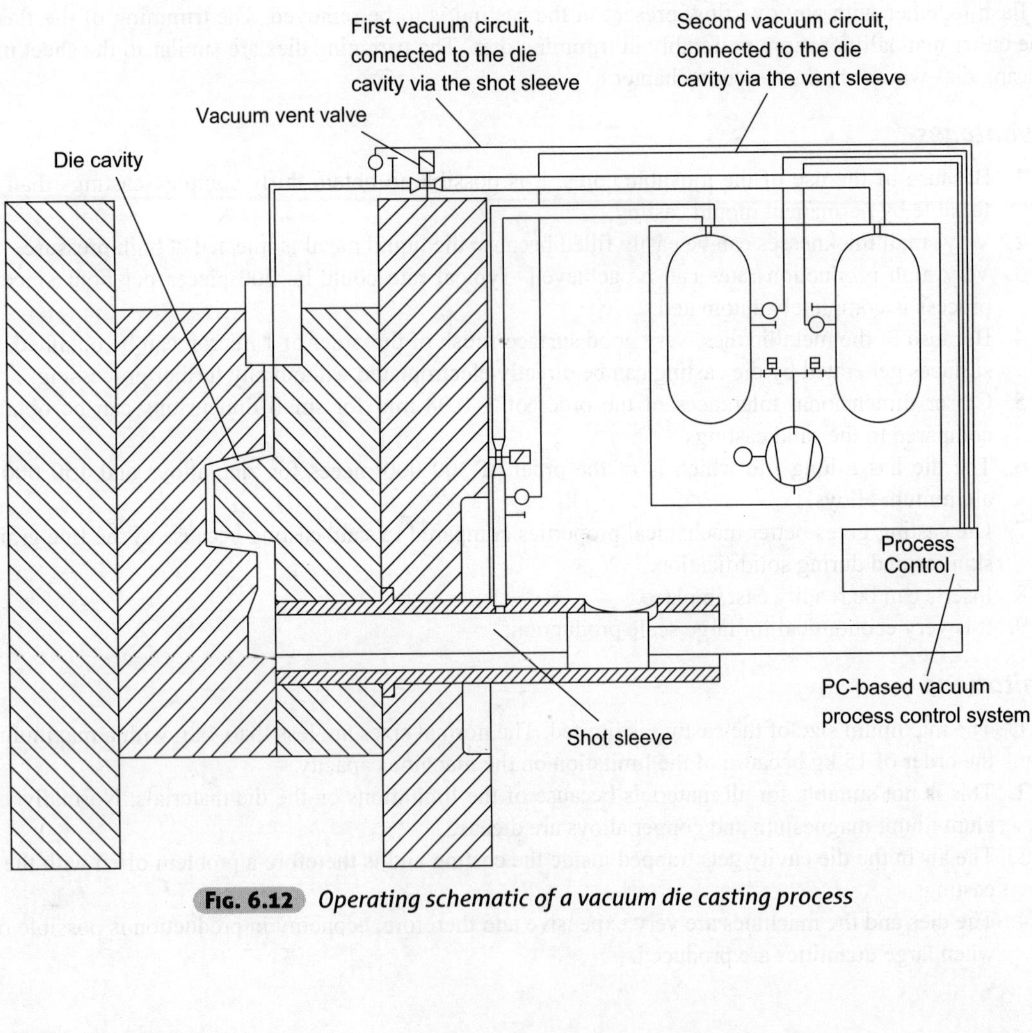

Fig. 6.12 *Operating schematic of a vacuum die casting process*

of the material is avoided. It would be possible with the vacuum die casting (VDC) process parts with very thin walls (1 to 12 mm), tight tolerances, fine microstructure due to the rapid solidification rates and therefore properties approaching that of wrought product, and with relatively short cycle times.

6.6 LOW PRESSURE DIE CASTING

Though this process is not new, it has been adopted generally for casting aluminium and magnesium-based alloys. In this process, the permanent mould and the filling system are placed over the furnace containing the molten alloy, as shown in Fig. 6.13. Then compressed gas is used, at a pressure typically ranging from 0.3 to 1.5 bars, to force the molten metal to rise slowly through the ceramic riser tube that is connected to the mould. Once the mould cavity is filled, the pressure in the crucible is removed and the residual molten metal in the tube flows back to the crucible. After the casting is solidified, the side die opens and the top die is raised vertically. The casting will move with the top die owing to the shrinkage and will be ejected onto a transfer tray.

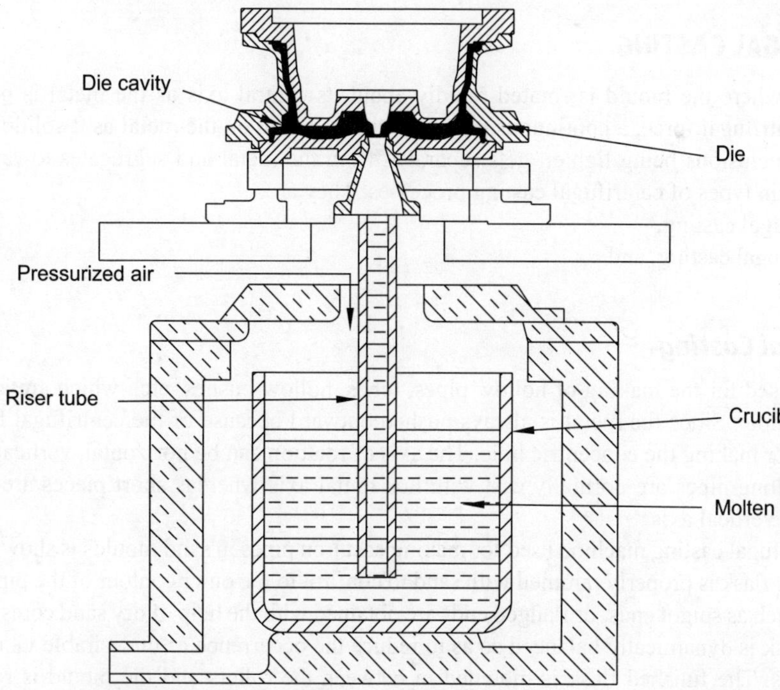

Fig. 6.13 *Operating schematic of a low pressure die casting process*

The quality of the casting is affected by the cooling rate, and therefore care has to be taken to see that the casting is properly cooled before ejecting from the die. The top and bottom dies are cooled by means of air jets. Care needs to be exercised during the design of the mould to provide proper cooling circuits, so that the heavier sections of the casting are located close to the feeding path acting as a riser. Since the metal enters the mould slowly compared to die casting with less turbulence, the casting quality is improved eliminating the defects. Close tolerance castings can be made using this process. Relative gas content in the castings due to the various processes is shown in Fig. 6.14.

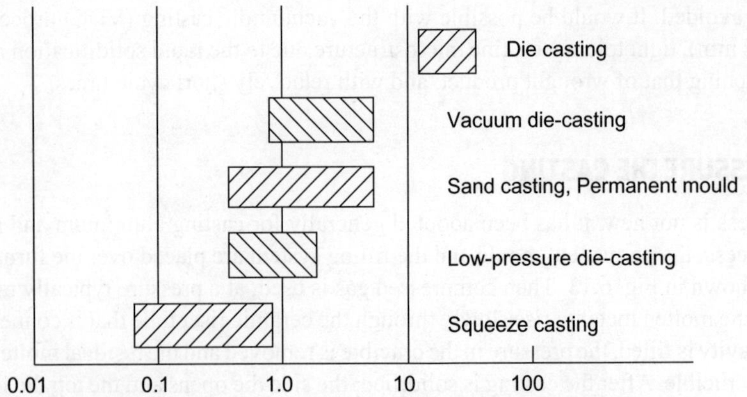

Fig. 6.14 *Relative gas content in various types of casting processes*

6.7 CENTRIFUGAL CASTING

This is a process where the mould is rotated rapidly about its central axis as the metal is poured into it. Because of the centrifugal force, a continuous pressure will be acting on the metal as it solidifies. The slag, oxides and other inclusions being lighter, gets separated from the metal and segregates toward the centre. There are three main types of centrifugal casting processes. They are:

- True centrifugal casting,
- Semi centrifugal casting, and
- Centrifuging

True Centrifugal Casting

This is normally used for the making of hollow pipes, tubes, hollow bushes, etc., which are axi-symmetric with a concentric hole. Since the metal is always pushed outward because of the centrifugal force, no core needs to be used for making the concentric hole. The axis of rotation can be horizontal, vertical or any angle in between. Very long pipes are normally cast with horizontal axis whereas short pieces are more conveniently cast with a vertical axis.

A normal centrifugal casting machine used for making cast iron pipes in sand moulds is shown in Fig. 6.15. First, the moulding flask is properly rammed with sand to confirm to the outer contour of the pipe to be made. Any end details, such as spigot ends, or flanged ends are obtained with the help of dry sand cores located in the ends. Then the flask is dynamically balanced so as to reduce the occurrence of undesirable vibrations during the casting process. The finished flask is mounted in between the rollers and the mould is rotated slowly. Now the molten metal in requisite quantity is poured into the mould through the movable pouring basin. The amount of metal poured determines the thickness of the pipe to be cast. After the pouring is completed,

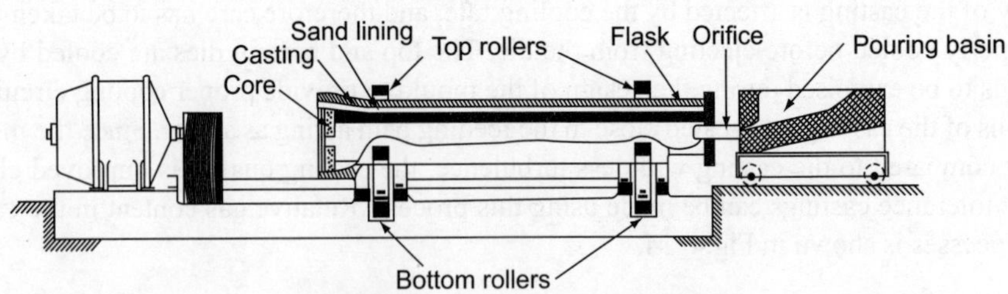

Fig. 6.15 *True centrifugal casting*

the mould is rotated at its operational speed till it solidifies, to form the requisite tubing. Then the mould is replaced by a new mould machine and the process continues.

Metal moulds can also be used in the true centrifugal casting process for large quantity production. A water jacket is provided around the mould for cooling it. The casting machine is mounted on wheels with the pouring ladle which has a long spout extending till the other end of the pipe to be made. To start, the mould is rotated with the metal being delivered at the extreme end of the pipe. The casting machine is slowly moved down the track allowing the metal to be deposited all along the length of the pipe. The machine is continuously rotated till the pipe is completely solidified. Afterwards, the pipe is extracted from the mould and the cycle repeated.

Advantages

1. The mechanical properties of centrifugally cast jobs are better compared to other processes because the inclusions such as slag and oxides get segregated towards the centre and can be easily removed by machining. Also the pressure acting on the metal throughout the solidification causes the porosity to be eliminated giving rise to dense metal.
2. Upto a certain thickness of objects, proper directional solidification can be obtained starting from the mould surface to the centre.
3. No cores are required for making concentric holes in the case of true centrifugal casting.
4. There is no need for gates and runners, which increases the casting yield, reaching almost 100%.

Limitations

1. Only certain shapes which are axi-symmetric and having concentric holes are suitable for true centrifugal casting.
2. The equipment is expensive and thus is suitable only for large quantity production.

Semi Centrifugal Casting

Semi centrifugal casting is used for jobs which are more complicated than those possible in true centrifugal casting, but are axi-symmetric in nature. It is not necessary that these should have a central hole which is to be obtained with the help of a core. The moulds made of sand or metal are rotated about a vertical axis and the metal enters the mould through the central pouring basin as in Fig. 6.16. For larger production rates, the moulds can be stacked one over the other, all feeding from the same central pouring basin. The rotating speeds used in this process are not as high as in the case of true centrifugal casting.

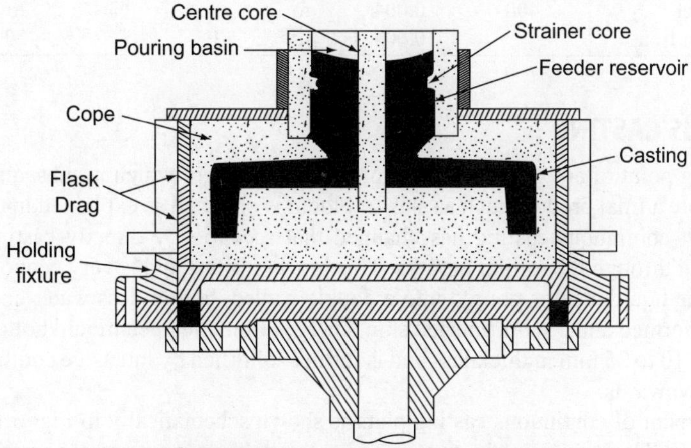

Fig. 6.16 *Semi centrifugal casting*

Centrifuging

In order to obtain higher metal pressures during solidification, when casting shapes are not axisymmetrical, the centrifuging process is used. This is suitable only for small jobs of any shape. A number of such small jobs are joined together by means of radial runners with a central sprue on a revolving table as in Fig. 6.17. The jobs are uniformly placed on the table around the periphery so that their masses are properly balanced. The process is similar to semi centrifugal casting.

A comparison of the various casting processes discussed above is presented in Table 6.2.

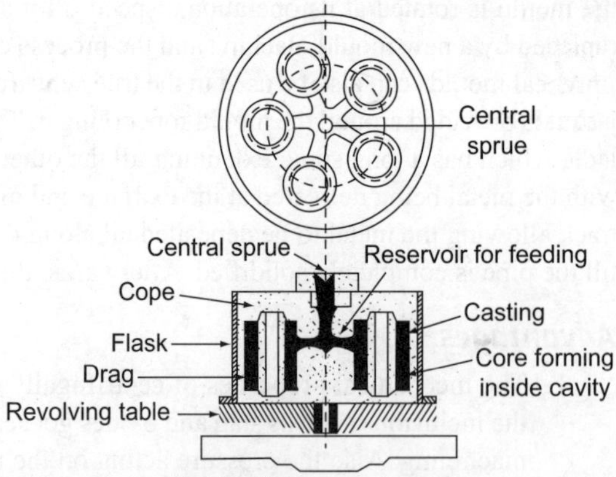

Fig. 6.17 *Centrifuging*

Table 6.2 Comparison of casting processes

	Metals that can be Cast	Min. and Max. Mass, kg	Tolerance on Dim., %	Surface Finish, m	Min. Section Size, mm	Draft Degrees	Production Rate, pcs/hr	Min. Cored Hole, mm	Porosity
Sand casting	Aluminium	0.03, 100	0.09	4	3	4	10		
	Steel	0.10, 200 000	to	8	6	to	to	6	5
	CI and others	0.03, 50 000	0.03	8	3.5	7	15		
Permanent mould	Cast Iron	0.01, 50	0.01	2	5			4.5 to	
casting	Al. & Mg	0.01, 10	min.		3	2	50	6	4
Die casting	Al	0.015, 35	0.0015	1	0.8	2	75 to	2.5	
	Mg	0.015, 35	0.0015	1	1.2	2	150	2.5	2
	Zn	0.05, 80		1	0.5	2	300 to 350	0.8	
Precision Invest-	Steel	0.005, 25	0.003 to	1	1	1 to	—	0.5 to	
ment casting	Aluminium	0.002, 10	0.005	1	0.8	3	—	1.25	1
Shell moulding	Steel	0.05, 120	0.01	6	3.5	2	30	3	
	Cast Iron	0.03, 50	to	6	3	to	to	to	1
	Aluminium	0.03, 15	0.003	2.5	1.5	3	80	6	
Centrifugal	Aluminium	up to	0.002	0.6	0.6		30		1
casting	Steel	400	0.004	to	to	3	to	—	or
	Cast Iron		0.004	3.5	1.2		50		2

6.8 CONTINUOUS CASTING

Generally the starting point of any structural steel product is the ingot which is subsequently rolled through number of mills before a final product such as slab or bloom is obtained as explained in Chapter 7. However, the wide adoption of continuous casting has changed that scenario by directly casting slabs, billets and blooms without going through the rolling process. This process is fast and is very economical.

In this process, the liquid steel is poured into a double walled, bottom less water cooled mould where a solid skin is quickly formed and a semi finished skin emerges from the open mould bottom. The skin formed in the mould is about 10 to 25 mm in thickness and is further solidified by intensive cooling with water sprays as casting moves downwards.

A typical arrangement of continuous casting plant is shown schematically in Fig. 6.18. The molten steel is collected in a ladle and kept over a refractory lined intermediate pouring vessel named 'tundish'. The steel is then poured into water cooled vertical copper moulds which are 450 to 750 mm long. Before starting the casting, a dummy starter bar is kept in the mould bottom as shown in Fig. 6.19. After starting the casting process, as the metal level rises in the mould to a desirable height, the starter bar is withdrawn at a rate equal to the steel pouring rate. The initial metal freezes onto the starter bar as well as the periphery of the

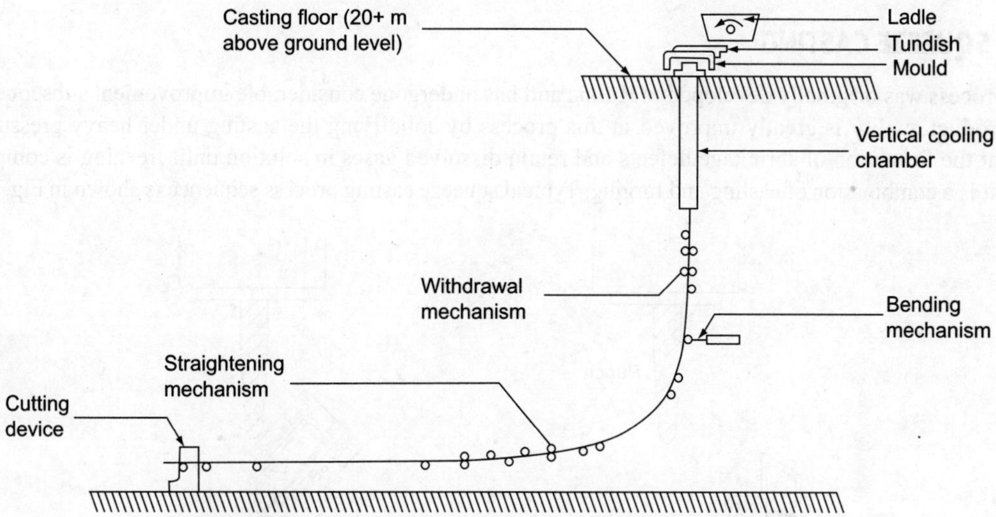

Fig. 6.18 *Typical arrangement of continuous casting plant*

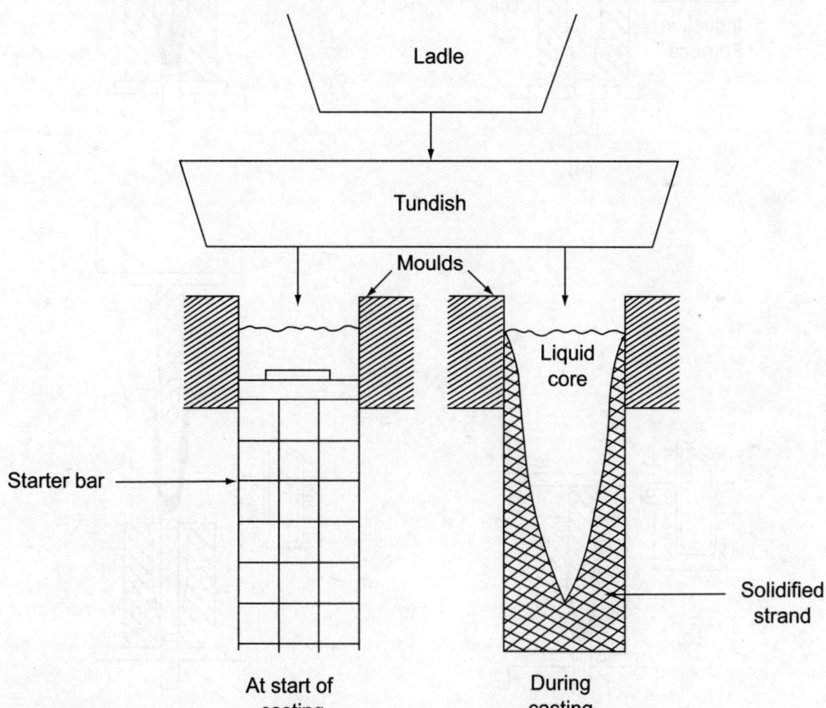

Fig. 6.19 *Use of dummy starter bar at the start of the continuous casting process*

mould. This solidified shell supports the liquid steel as it moves downwards. This steel shell is mechanically supported (rollers) as it moves down through the secondary cooling zone where water is sprayed onto the shell surface to complete the solidification process. After the casting is completely solidified, it is cut to the desired lengths by a suitable cut off apparatus.

6.9 SQUEEZE CASTING

This process was originally developed in Russia and has undergone considerable improvement subsequently. The product quality is greatly improved in this process by solidifying the casting under heavy pressure to prevent the formation of shrinkage defects and retain dissolved gases in solution until freezing is complete. Thus, it is a combination of casting and forging. Typical squeeze casting process sequence is shown in Fig. 6.20.

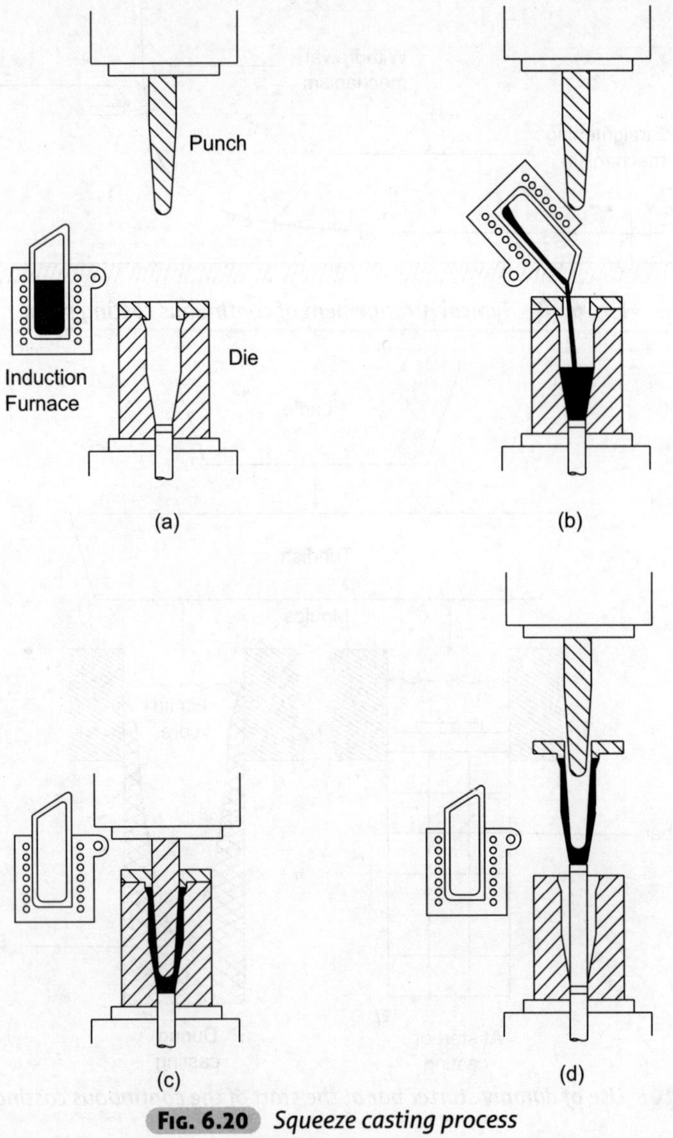

Fig. 6.20 *Squeeze casting process*

When the cycle is started the punch and die portion are separated as shown in Fig. 6.20(a). The holding furnace holds the molten metal at the requisite temperature. Then carefully metered charge of molten metal is poured into the die cavity as shown in Fig. 6.20(b). Then the punch is lowered into place forming a tight seal. The "punch" portion of the upper die is then forced into the cavity, displacing the molten metal under pressure until it fills the annular space between the die and punch as shown in Fig. 6.20(c). The metal is then under pressure (50 to 140 MPa) and at the same time loses heat rapidly because of its contact with the metallic die. The solidification under pressure is claimed to be responsible for the reduction in the shrinkage cavities in the resulting castings. Once the casting is completely solidified, the punch is retracted and the casting is ejected from the die as shown in Fig. 6.20(d).

The squeeze casting process has very low gas entrapment, lower shrinkage cavities, lower die costs and very high quality surface with fine details. Also it produces a fine grain size which improves mechanical properties. Aluminium, magnesium and copper alloy components can be readily manufactured using this process.

6.10 SLUSH CASTING

Slush casting also known as hollow casting is a special process first developed by William Britain in 1893 for the production of lead toy soldiers. This is a process used to produce hollow castings without the use of a core. The tools used are relatively cheap and produces a casting that is lighter and less expensive product. Typical process sequence for slush casting is shown in Fig. 6.21. Generally, it uses a permanent mould without any cores. It can be seen that in Fig. 6.21(b) the casting starts solidifying from the mould wall towards the inside of the casting. After a certain wall thickness is formed, the mould is inverted to let the excess liquid drains out from the mould. Then the mould is opened and the casting can be seen in Fig. 6.21(d).

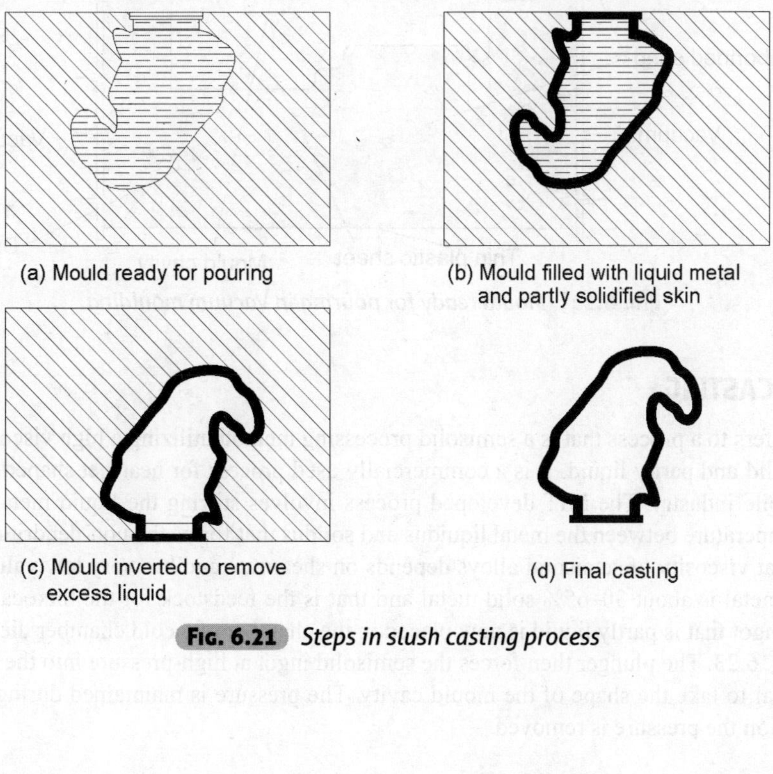

(a) Mould ready for pouring

(b) Mould filled with liquid metal and partly solidified skin

(c) Mould inverted to remove excess liquid

(d) Final casting

FIG. 6.21 *Steps in slush casting process*

6.11 VACUUM CASTING

Vacuum moulding or also called as V-process prepares moulds using dry unbonded sand, which is compacted through the application of vacuum. V- process utilizes either a match plate pattern or cope and drag pattern that have all the associated gating and risering system built in. This process was invented in Japan in 1970's.

The process starts with draping the drag pattern plate with a thin plastic film. The pattern utilized in V-process has very small holes for applying vacuum. When the vacuum is applied to the pattern, the plastic film adheres closely to the pattern. Then a special moulding flask that has the unbonded sand is kept over the pattern. A slight vibration then compresses the sand to achieve the required strength. Then the cope half of the mould is covered with a second sheet of plastic film and vacuum is applied to the unbonded sand, which becomes hard and rigid. Releasing the vacuum on the pattern plate allows the pattern to be easily removed from the mould. The two halves of the mould are assembled and if any cores are required are placed in the mould cavity before it is closed. Vacuum is maintained during the pouring as shown in Fig. 6.22. After the casting is solidified, the vacuum is released breaking the mould for easy removal of the casting.

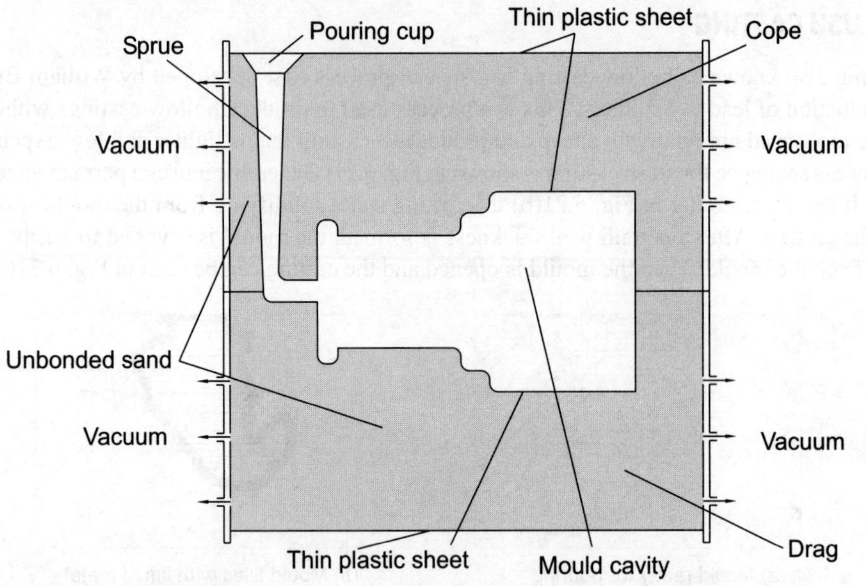

FIG. 6.22 *Mould ready for pouring in Vacuum moulding*

6.12 THIXOCASTING

Thixocasting refers to a process that is a semisolid processing method utilizing a high viscosity raw material that is partly solid and partly liquid. It is a commercially used process for near-net shaped aluminium parts by the automobile industry. The MIT developed process involves stirring the liquid metal with controlled cooling to a temperature between the metal liquidus and solidus that forms the non-dendritic structure. It has been noticed that viscosity of semisolid alloys depends on shear rate and has very low value upon vigorous agitation. The metal is about 30–65% solid metal and that is the feedstock for the thixocasting having low viscosity. The ingot that is partly liquid is now placed in the chamber of a cold chamber die casting machine as shown in Fig. 6.23. The plunger then forces the semisolid ingot at high-pressure into the preheated mould forcing the metal to take the shape of the mould cavity. The pressure is maintained during the cooling and after solidification the pressure is removed.

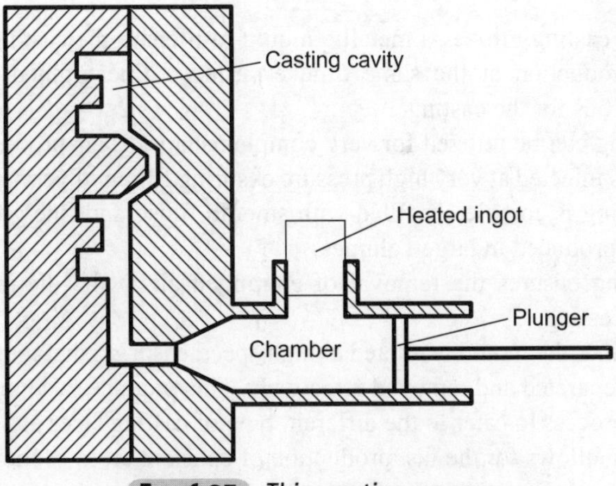

Fig. 6.23 *Thixocasting process*

Advantages

- The process allows for manufacturing components with complex shapes with thin walls.
- The thixocasting process provides good mechanical properties with a high dimensional accuracy and good surface finish.
- Good quality parts as the mould is filled smoothly with no gas entrapment and reduced solidification shrinkage because of the semisolid material.
- It is possible to obtain a near-net shaped product eliminating the expensive finishing operations.
- Because of the stirring used on the melt during the solidification of a continuous cast bar will ensure a globulitic (spherical) microstructure.
- It has a lower forming temperature that improves the die life significantly and improves production efficiency.

Limitations

- The process needs special feed stock with near spherical primary crystals which is expensive.
- It is necessary to maintain the operating conditions precisely to ensure the quality.

Applications

- Automotive parts such as engine mounts, tie rods, automobile wheels, hydraulic brake valve, motor brake cylinder, antilock brake valve, disk brake caliper, steering control arms, engine brackets, engine piston, etc.

SUMMARY

Sand casting is not suitable for applications where large volumes of parts are to be produced or complex geometries are involved. For this purpose a number of special casting methods are devised.

- Shell moulding is a special casting process used for specific applications that require higher quality castings with thin surface details. It uses resin coated sand for making a mould strong and porous.
- Precision investment casting utilizes an expendable pattern to create really complex objects without any parting line. This ensures very high precision and excellent properties.

- Permanent mould casting utilizes a metallic mould to prepare the casting. This helps in lowering costs for mass production at the same time ensuring a fine grained structure with improved mechanical properties for the casting.
- Pressure die casting can be utilized for very complex shapes. This process uses a metallic die and the molten metal is injected at very high pressure ensuring filling of very small cavities at a fast rate. The resulting casting is completely filled with smooth finish and fine grained structure. It is very economical when produced in large volumes.
- Vacuum die casting ensures the removal of entrapped air in the die inside the casting thereby ensuring a sound casting.
- In centrifugal casting, the mould is rotated at high speed ensuring that the slag and impurities in the molten metal are separated and removed effectively. This produces a sound casting. There are other variations in this process to cater to the different type of casting sizes and geometries produced.
- Continuous casting allows for the fast production of constant cross-section shapes in large volume.

Questions

6.1 Compare precision investment casting and shell moulding from the stand point of process, product and applications.

6.2 "Although sand casting is the most widely used process judging from the tonnage of castings produced, there are instances where one would choose die casting in preference to the sand casting." Discuss the statement.

6.3 Give examples of the typical products of the following processes: Die casting, centrifugal casting, permanent mould casting and shell moulding.

6.4 Specify the advantages of the precision investment casting process over other casting processes.

6.5 Describe the composition of the mould material in the shell moulding process.

6.6 What is the main difference in quality between a sand casting and a casting in a metal mould?

6.7 Why are most die castings not made out of high strength materials?

6.8 Is it possible to obtain a sound casting of a solid bar by centrifugal casting? Give reasons in support of your answer.

6.9. State the difference between centrifuging and true centrifugal casting.

6.10 What are the typical situations in which the following casting processes are used?
   (a) Precision investment casting
   (b) Shell moulding
   (c) True centrifugal casting
   (d) Pressure die casting

6.11 Give advantages and disadvantages of permanent mould casting compared to that of sand casting.

6.12 How is a semi permanent mould different from a permanent mould?

6.13 Why is aluminium preferred to be done by cold chamber die casting than hot chamber die casting?

6.14 "Large parts cannot be manufactured by the centrifuging process." Comment on the statement.

6.15 How will you compare a cold chamber die casting process with that of a hot chamber process?

6.16 Outline the precision investment casting process giving typical applications of the process.
6.17 Briefly enumerate the steps in sequence for producing castings from shell moulding.
6.18 What are the applications of thixocasting?

Multiple Choice Questions

6.1 The type of binder that is commonly used in shell moulding is
(a) Linseed oil
(b) Thermosetting resin
(c) Thermoplastic resin
(d) Core oil

6.2 Type of sand that is normally used in shell moulding is
(a) Synthetic sand
(b) River sand
(c) Dry and fine sand
(d) Moulding sand

6.3 The most widely used casting process for industrial components
(a) Squeeze casting
(b) Pressure die casting
(c) True centrifugal casting
(d) Vacuum pressure die casting

6.4 Thickness of the shell mould can be increased by
(a) Decreasing the grain size of the sand
(b) Decreasing the contact time between the shell sand and the metallic pattern
(c) Increasing the contact time between the shell sand and the metallic pattern
(d) Increasing the amount of thermosetting resin binder used with sand

6.5 Pattern used in shell moulding is normally made of
(a) Wood (b) Metal
(c) Plastic (d) Wax

6.6 Main advantage of using a shell moulding is
(a) Any complex shape can be obtained
(b) Cost of the pattern is low
(c) Very thin sections (up to 0.25 mm) can be readily made
(d) Any size of casting can be obtained

6.7 Pattern used in precision investment casting is normally made of
(a) Wood (b) Metal
(c) Plastic (d) Wax

6.8 Main advantage of using a precision investment casting is
(a) Any complex shape can be obtained
(b) Less expensive compared to other processes
(c) Very thin sections (up to 0.25 mm) can be readily made
(d) Any size of casting can be obtained

6.9 Pattern used in permanent mould casting is normally made of
(a) Wood (b) Metal
(c) Plastic (d) None of the above

6.10 Cores used in permanent mould casting are normally made of
(a) Sand (b) Metal
(c) Collapsible metal (d) All of the above

6.11 Coatings are normally used in permanent mould casting moulds to
(a) Improve the appearance of the mould cavity
(b) Improve the surface finish of the mould cavity
(c) Prevent the soldering of metal to the mould
(d) Improve the flow of the molten metal inside the mould cavity

6.12 Main advantage of using a permanent mould casting is
(a) Produces very good surface finish of the order of 4 microns and better appearance.
(b) Any complex shape can be obtained

(c) Any size of casting can be obtained
(d) Any material can be used

6.13 Pattern used in pressure die casting is normally made of
(a) Wood (b) Metal
(c) Plastic (d) None of the above

6.14 Main advantage of using a pressure die casting is
(a) Economical even in small quantities
(b) Any complex shape can be obtained
(c) Any size of casting can be obtained
(d) Any material can be used

6.15 Hot chamber die casting process is used for
(a) Aluminium (b) Zinc
(c) Cast iron (d) Steel

6.16 Cold chamber die casting process is used for
(a) Aluminium (b) Zinc
(c) Cast iron (d) Magnesium

6.17 Hot chamber die casting process when compared to cold chamber die casting process is
(a) Faster
(b) More expensive
(c) More versatile for the range of materials used
(d) Can use stronger materials

6.18 The main difference between hot chamber die casting process and cold chamber die casting process is
(a) Cold chamber process utilizes low melting temperature alloys compared to hot chamber process
(b) Hot chamber process utilizes a ladle to inject molten metal into the die
(c) Hot chamber process utilizes an integral furnace with the machine
(d) Cold chamber process utilizes an integral furnace with the machine

6.19 The casting process that do not require a core to produce a hollow casting is
(a) Shell moulding
(b) Permanent mould casting
(c) Hot chamber die casting
(d) True centrifugal casting

6.20 Main advantage of using a true centrifugal casting is
(a) Produce very good surface finish of the order of 4 microns and better appearance.
(b) Any axi-symmetrical shape can be obtained without using a core
(c) Economical even in small quantities
(d) None of the above

Answers to MCQs

6.1 (b)	6.2 (c)	6.3 (b)	6.4 (c)	6.5 (b)
6.6 (c)	6.7 (d)	6.8 (a)	6.9 (d)	6.10 (d)
6.11 (c)	6.12 (a)	6.13 (d)	6.14 (b)	6.15 (b)
6.16 (a)	6.17 (a)	6.18 (c)	6.19 (d)	6.20 (b)

CASE STUDY

GATING SYSTEM DESIGN FOR INVESTMENT CASTING*

Precision Investment Casting is used for manufacturing high precision complex parts. Hemodialysis machines are used for cleaning the blood of patients by passing through a filter. Due to the use of the machines in highly critical conditions, it is important that the rotors used in these machines be flawless and do not cause any problems during their operation. These rotors are made with 17-4 PH stainless steel (precipitation hardening martensitic stainless steel). In case of defects like shrinkage, incomplete filling and porosity, the rotor may have creep failure. Such critical parts are very carefully designed and optimized for precise functioning. This case study reports the results of a study that used five different gating systems for investment casting of the rotor to arrive at the best design. Since physical testing of different models is expensive, numerical simulations were carried out in the study using Moldflow® simulation software.

A reference gating system was used for comparing with five other gating systems models developed. The gating system consisted of a sprue with inlet diameter of 150 mm and the main runner connected to the two ends of the rotor. In the first improved model, the sprue diameter was reduced to 90 mm and gates were designed above the rotor arms. In the second improved model, risers (10×10 mm) were attached in the probable defective regions and ventilating system was added. In model 3, the risers enlarged to 60×60 mm and the ventilating systems were removed. In model 4, the cross-runners were replaced by a single runner. In model 5, the rib structure was modified at the ends. In all the models, pouring temperature was 1650°C.

Analysis showed that model 1 and 2 did not do much to reduce the shrinkage defects, however compared to other models, the location of the defect has moved up. This may be because of the location of the risers. In model 3, since the risers are larger, they ensured that the molten material was available for feeding in case of shrinkage. However, even in this model, the defects were not completely eliminated. This may be because the risers are not big enough and/or some sections of the casting are very narrow. Model 4 had similar results as it had the same riser dimensions.

Model 5, which had a modified rib design, showed the best results. Since the rib design was modified, better control of cooling rate was possible. Though the solidification sequence was similar to model 4, the solidification time reduced to 1059 s from 1176 s. Modified rib design has helped to reduce the resistance to the flow. This coupled with large riser helped eliminating the shrinkage defects. Compared to the reference model, the turbulence was less in model 5. Turbulence leads to the deposition of the mould particles and other impurities in the cavity and causes defects. In model 5, the metal flows smoothly and fills the cavity and the riser. This reduced the turbulence and solidification time helping in obtaining better quality of the cast part.

---

*Huang, P. H., & Lin, C. J. (2015). Computer-aided modeling and experimental verification of optimal gating system design for investment casting of precision rotor. *The International Journal of Advanced Manufacturing Technology*, *79*(5-8), 997-1006.

CHAPTER 7

Metal-Forming Processes

Objectives

Mechanical working processes are used to achieve optimum mechanical properties in the metal. Metal working reduces any internal voids or cavities present, and thus make the metal dense. The impurities present in the metal also get elongated with the grains and in the process they get broken and dispersed throughout the metal. This decreases the harmful effects of the impurities and improves the mechanical strength. After completing this chapter, the reader will be able to

- Understand the advantages of utilizing metal-working processes
- Learn about different rolling processes and applications
- Utilise different forging processes and applications
- Understand different extrusion processes and applications

7.1 NATURE OF PLASTIC DEFORMATION

Plastic deformation is the deformation which is permanent and beyond the elastic range of the material. Often, metals are worked by plastic deformation because of the beneficial effect that is imparted to the mechanical properties by it. The necessary deformation in a metal can be achieved by application of large amount of mechanical force only, or by heating the metal and then applying a small force.

The deformation of metals, which is caused by the displacement of the atoms, is achieved by one or both of the processes called **slip** and **twinning**. The details of the microscopic deformation methods can be found in the textbooks of metallurgy. On the macroscopic scale, when plastic deformation occurs, the metal appears to flow in the solid state along specific directions, which are dependent on the type of processing and the direction of applied force. The crystals or grains of the metal are elongated in the direction of metal flow. This flow of metal can be seen under microscope after polishing and suitable etching of the metal surface. These visible lines are called 'fibre flow lines', some representative specimens of which are presented in Fig. 7.1.

Since the grains are elongated in the direction of flow, they would be able to offer more resistance to stresses acting across them. As a result, the mechanically worked metals called wrought products would be able to achieve better mechanical strength in specific orientation, that of the flow direction. Since it is possible to control these flow lines in any specific direction by careful manipulation of the applied forces, as shown in Fig. 7.1, it is possible to achieve optimum mechanical properties. The metal, of course, would be weak along the flow lines.

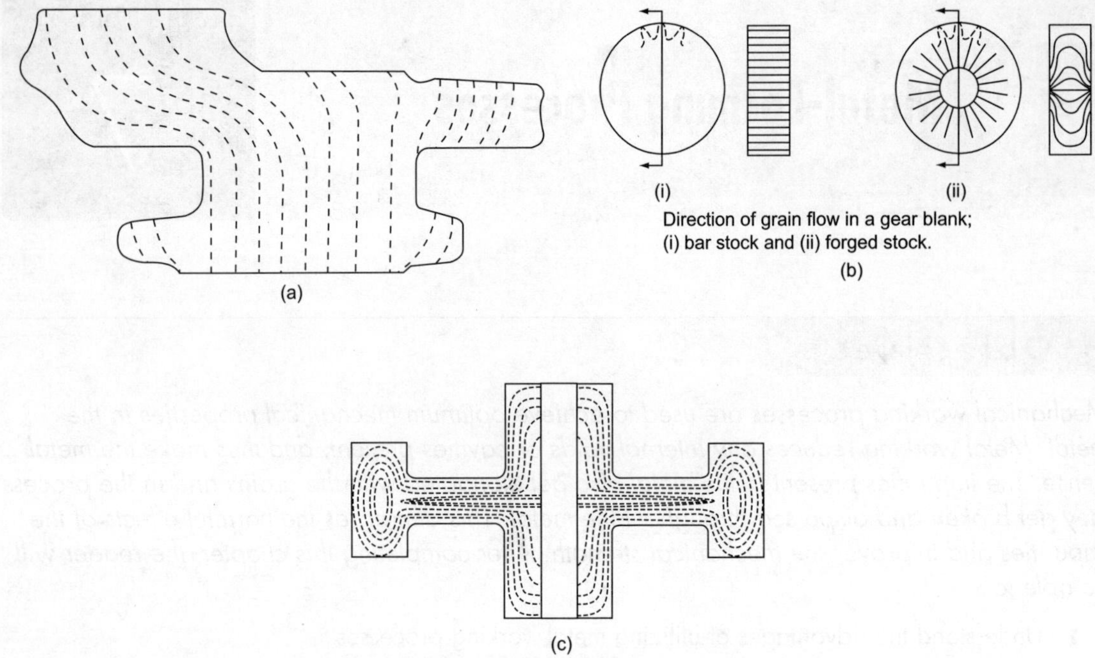

FIG. 7.1 *Fibre flow lines (courtesy B.W. Niebel and A.B. Draper: Product Design and Process Engineering, p417, McGraw-Hill, New York)*

The wastage of material in metal-working processes is either negligible or very small, and the production rate is in general very high. These two factors give rise to the economy in production.

7.1.1 Hot Working and Cold Working

The metal-working processes are traditionally divided into hot working and cold-working processes. The division is on the basis of the amount of heating applied to the metal before applying the mechanical force.

Those processes, working above the recrystallisation temperature, are termed as hot-working processes whereas those below are termed as cold-working processes.

Under the action of heat and the force, when the atoms reach a certain higher energy level, the new crystals start forming which is termed as recrystallisation. Recrystallisation destroys the old grain structure deformed by the mechanical working, and entirely new strain free crystals are formed. The grains in fact start nucleating at the points of severest deformation. Recrystallisation temperature as defined by American Society of Metals is "the approximate minimum temperature at which complete recrystallisation of a cold-worked metal occurs within a specified time".

The recrystallisation temperature generally varies between one third to one half the melting point of most of the metals. Typical values of recrystallisation temperatures are given in Table 7.1. The recrystallisation temperature also depends on the amount of cold work a material has already received. Higher the cold work, lower would be the recrystallisation temperature as shown in Fig. 7.2.

Though cold work affects the recrystallisation temperature to a great extent, there are other variables, which also affect as given in Table 7.2.

TABLE 7.1 Minimum recrystallisation temperatures

Material	Recrystallisation Temperature, °C	Material	Recrystallisation Temperature,°C
Lead	Below room temperature	Iron	450
Tin	Below room temperature	Nickel	600
Cadmium	Room temperature	Titanium	650
Zinc	Room temperature	Beryllium	700
Magnesium	150	Molybdenum	900
Aluminium	150	Tantalum	1000
Copper	200	Tungsten	1200

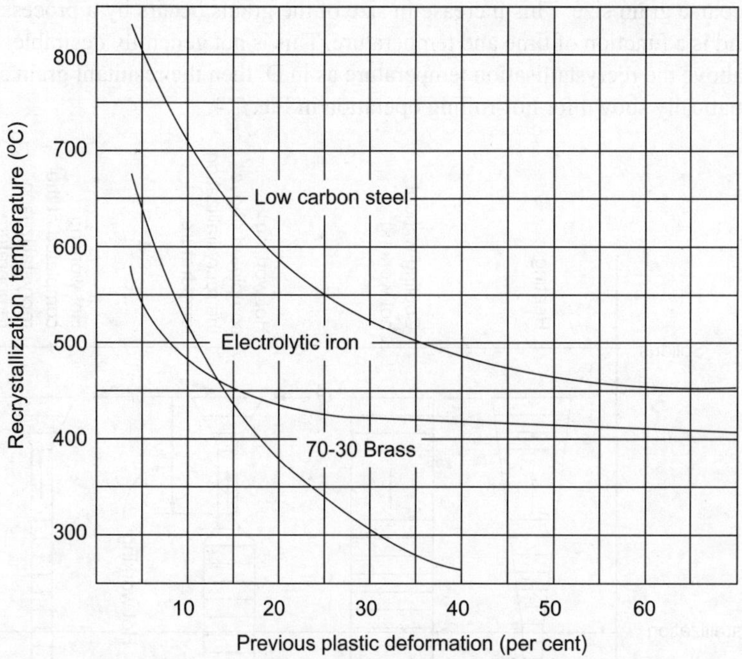

FIG. 7.2 *Effect of amount of plastic deformation on the recrystallisation temperature*

TABLE 7.2 Variables affecting recrystallisation temperatures

Variable	Effect on Recrystallisation Temperature
Solute concentration in solid solution	Generally increases temperature; depends upon effect on melting temperatures
Presence of second phases	Second phase particles decrease temperature
Original grain size before cold working	Decreasing grain size, decreases temperature
Rate of plastic deformation	Increasing rate, decreases temperature
Temperature of cold working	Decreasing temperature of cold working, decreases temperature

In hot working, the process may be carried above the recrystallisation temperature with or without actual heating. For example, for lead and tin the recrystallisation temperature is below the room temperature and hence working of these metals at room temperature is always hot working. Similarly for steels, the recrystallisation temperature is of the order 1000°C, and therefore working below that temperature is still cold working.

In hot working, the temperature at which the working is completed is important since any extra heat left after working will aid in the grain growth, thus giving poor mechanical properties. The effect of temperature of completion of hot working is shown schematically in Fig. 7.3. In A, it is shown a simple heating, where the grains start growing after the metal crosses the recrystallisation temperature. When it is cooled without any hot working as in B, the final grain size would be larger than when started in A. After heating, when the metal is worked, because of recrystallisation, the grain size is reduced. This is made possible because the working of metal gives rise to a large number of nucleation sites for the new crystals to form. But if the hot working is completed much above the recrystallisation temperature as in C, the grain size starts increasing and finally may end up with coarse grain size. This increase in size of the grains occurs by a process of coalescence of adjoining grains and is a function of time and temperature. This is not generally desirable. If the hot working is completed just above the recrystallisation temperature as in D, then the resultant grain size would be fine. The same is schematically shown for hot-rolling operation in Fig. 7.4.

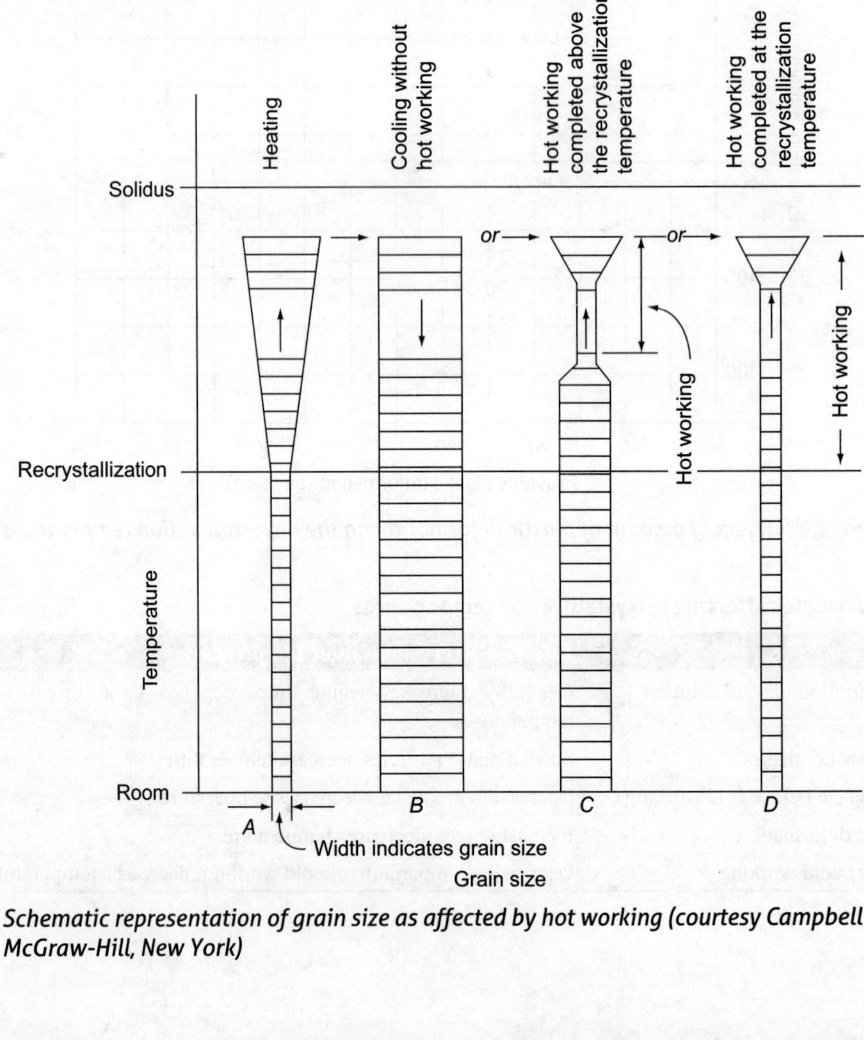

FIG. 7.3 *Schematic representation of grain size as affected by hot working (courtesy Campbell op cit, p 277, McGraw-Hill, New York)*

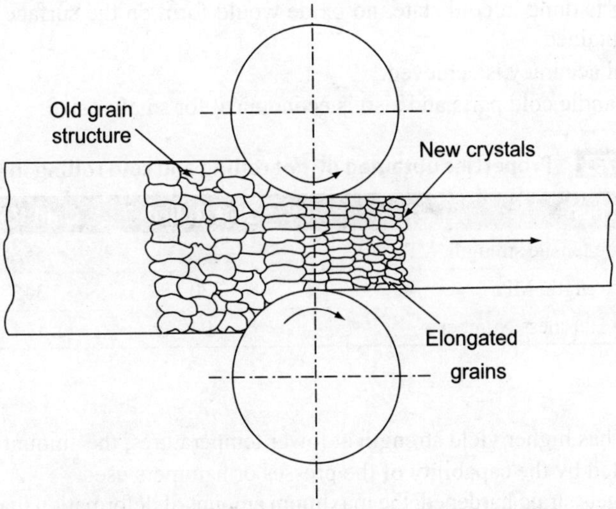

Fig. 7.4 *Grain reformation in hot rolling operation*

Hot Working

Advantages

- As the material is above the recrystallisation temperature, any amount of working can be imparted since there is no strain-hardening taking place.
- At a high temperature, the material would have higher amount of ductility and therefore, there is no limit on the amount of hot working that can be done on a material. Even brittle materials can be hot worked.
- Since the shear stress gets reduced at higher temperatures, the hot working requires much less force to achieve the necessary deformation.
- It is possible to continuously reform the grains in metal working and if the temperature and rate of working are properly controlled, a very favourable grain size could be achieved giving rise to better mechanical properties.

Disadvantages

- Some metals cannot be hot worked because of their brittleness at high temperatures.
- Higher temperatures of metal give rise to scaling of the surface and as a result, the surface finish obtained is poor. Also, there is a possibility of the decarburisation of skin in steels due to the high temperature.
- Because of the thermal expansion of metals, the dimensional accuracy in hot working is difficult to achieve since it is difficult to control the temperature of work pieces.
- Handling and maintaining of hot metal is difficult and troublesome.

Cold Working

Advantages

- Cold working increases the strength and hardness of the material due to strain hardening which would be beneficial in some situations. Table 7.3 gives some comparative values for the properties obtained after hot working and cold working. Further, there is no possibility of decarburisation of the surface.

- Since the working is done in cold state, no oxide would form on the surface and consequently, good surface finish is obtained.
- Better dimensional accuracy is achieved.
- It is far easier to handle cold parts and also is economical for smaller sizes.

TABLE 7.3 Properties obtained by hot rolling and cold rolling in C10 steel

Properties	Hot Rolled	Cold Rolled
Ultimate tensile strength, MPa	427	558
Yield strength, MPa	220.5	345
Brinell Hardness Number	94	174

Disadvantages

- Since the material has higher yield strength at lower temperatures, the amount of deformation that can be given to is limited by the capability of the presses or hammers used.
- Since the material gets strain hardened, the maximum amount of deformation that can be given is limited. Any further deformation can be given after annealing.
- Some materials, which are brittle, cannot be cold worked.

7.2 ROLLING

Rolling is a very economical process for producing large volume of material with constant cross section.

7.2.1 Principle

Rolling is a process where the metal is compressed between two rotating rolls for reducing its cross-sectional area (Fig. 7.5). It is one of the most widely used of all the metal-working processes, because of its higher productivity and low cost. Rolling would be able to produce components having constant cross section throughout its length. Many shapes such as I, T, L, and channel sections are possible as shown in Fig. 7.6, but not very complex shapes. It is also possible to produce special sections such as railway wagon wheels by rolling individual pieces.

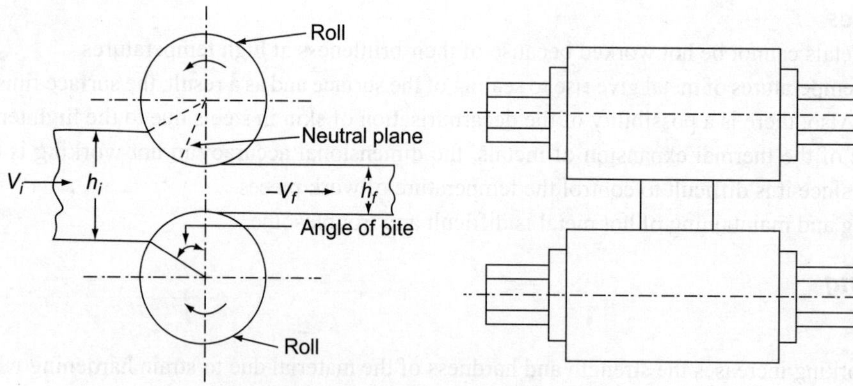

FIG. 7.5 *Schematic of the rolling process*

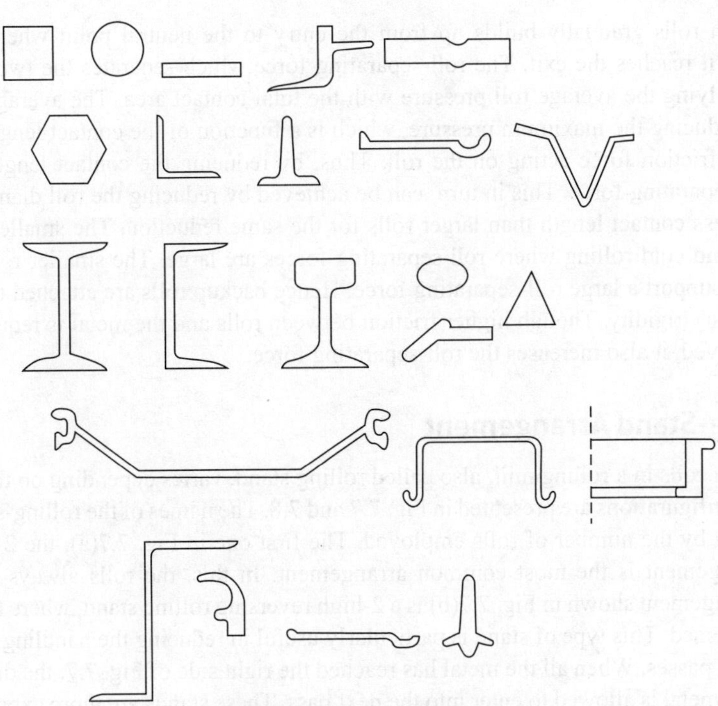

FIG. 7.6 *Typical rolling shapes*

Rolling is normally a hot working process unless specifically mentioned as cold rolling. The metal is taken into rolls by friction and subsequently compressed to obtain the final shape. The thickness of the metal that can be drawn into rolls depend on the roughness of the roll surface. Rougher rolls would be able to achieve greater reduction than smoother rolls. But, the roll surface gets embedded into the rolled metal thus producing rough surface.

The reduction that could be achieved with a given set of rolls is designated as the 'angle of bite' and is shown in Fig. 7.5. This depends on the type of rolling and the conditions of the rolls as shown in Table 7.4. The volume of the metal that enters the rolling stand should be same as that leaving it except in initial passes, when there might be some loss due to filling of voids and cavities in the ingots. Since the area of the cross section gets decreased, the metal leaving the rolls would be at a higher velocity than when it entered. Initially, when the metal enters the rolls, the surface speed of rolls is higher than that of the incoming metal, whereas, the metal velocity at the exit is higher than that of the surface speed of the rolls. Between the entrance and exit, the velocity of the metal is continuously changing, whereas the roll velocity remains constant. Somewhere in the contact length, the velocities of the metal and rolls are same, which is designated as 'neutral plane' in Fig. 7.5.

TABLE 7.4 **Angle of bites for various rolling mills**

Type of Rolling	Maximum Angle of Bite, (deg.)
Cold rolling with lubrication and smooth roll surfaces	3 to 4
Hot rolling of tubes	12 to 14
Hot rolling of sheets	15 to 22
Hot rolling with rough rolls (serrations on rolls)	27 to 34

The pressure on rolls gradually builds up from the entry to the neutral point where it is highest and then decreases till it reaches the exit. The roll-separating force which separates the two rolls apart can be obtained by multiplying the average roll pressure with the total contact area. The average roll pressure can be decreased by reducing the maximum pressure, which is a function of the contact length. Smaller contact length means less friction force acting on the roll. Thus, by reducing the contact length, it is possible to decrease the roll-separating force. This in turn, can be achieved by reducing the roll diameter, since smaller rolls would have less contact length than larger rolls for the same reduction. The smaller rolls are used for larger reductions and cold rolling where roll separating forces are large. The smaller rolls would not have enough rigidity to support a large roll-separating force. Hence backup rolls are attached to the small rolls to provide the necessary rigidity. Though higher friction between rolls and the metal is required for increasing the reduction achieved, it also increases the roll-separating force.

7.2.2 Rolling-Stand Arrangement

The arrangement of rolls in a rolling mill, also called rolling stand, varies depending on the application. The various possible configurations are presented in Fig. 7.7 and 7.8. The names of the rolling-stand arrangements are generally given by the number of rolls employed. The first one in Fig. 7.7(a), the 2-high nonreversing rolling-stand arrangement is the most common arrangement. In this, the rolls always move in only one direction. The arrangement shown in Fig. 7.7(b) is a 2-high reversing rolling stand, where the direction of roll rotation can be reversed. This type of stand is particularly useful in reducing the handling of the hot metal in between the rolling passes. When all the metal has reached the right side of Fig. 7.7, the direction of the rolls is reversed and the metal is allowed to enter into the next pass. These stands are more expensive compared to the nonreversible type because of the reversible drive needed.

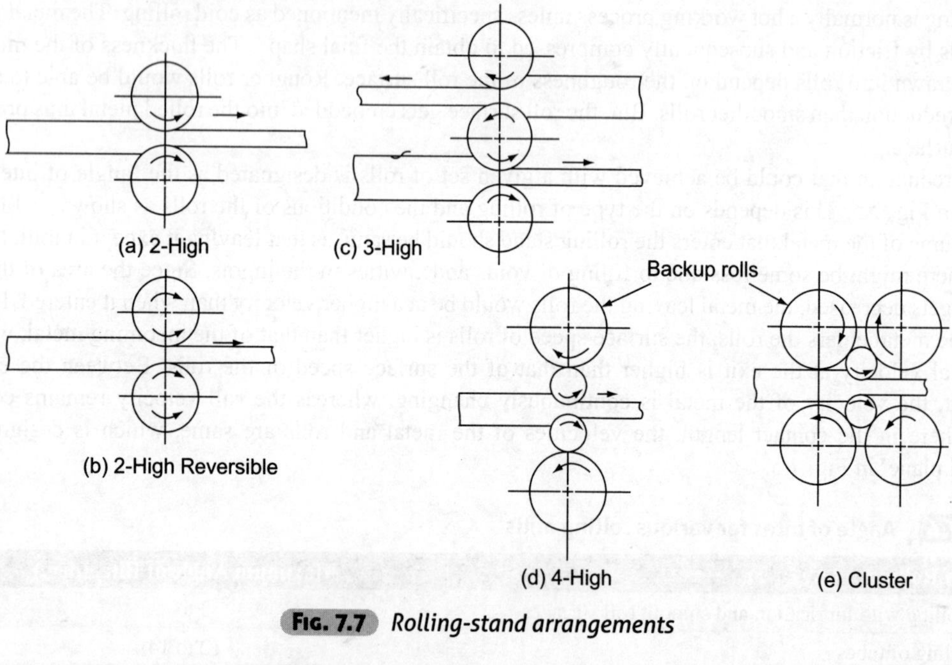

FIG. 7.7 *Rolling-stand arrangements*

The 3-high rolling-stand arrangement, shown in Fig. 7.7(c), is used for rolling of two continuous passes in a rolling sequence without reversing the drives. After all the metal has passed through the bottom roll set,

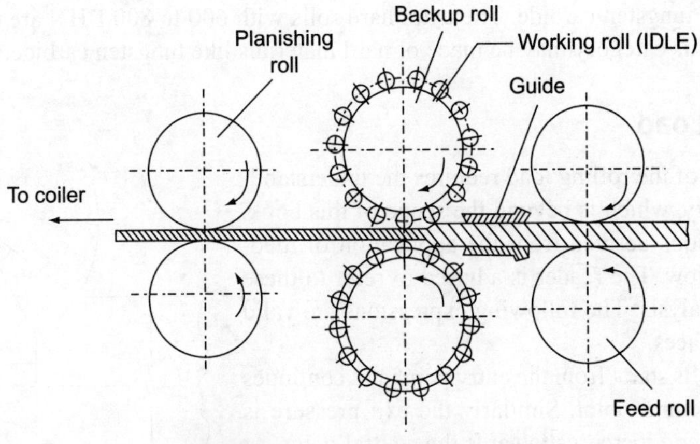

Fig. 7.8 *Planetary rolling mill*

the end of the metal is entered into the other set of the rolls for the next pass. For this purpose a table-tilting arrangement is required to bring the metal to the level with the rolls.

A 4-high rolling stand shown in Fig. 7.7(d) is essentially a 2-high rolling mill, but with the small-sized rolls. The other two rolls are the backup rolls for providing the necessary rigidity to the small rolls. A better backup can be provided to the small rolls with a cluster arrangement of rolls as shown in Fig. 7.7(e). For the rolling arrangements requiring large reduction, a number of free rotating wheels instead of a single small roll, are fixed to a large backup roll in the planetary rolling-mill arrangement shown in Fig. 7.8.

Since the final shape required cannot be obtained in a single pass, the rolling mills are generally arranged with rolls which have more than one pass machined into them so that, the material coming out of one pass is then to be fed back to the other pass by reversing the roll direction or by arranging the rolling mills in tandem. It is general practice to maintain tension in the strip between the stands in the interest of keeping it flat and also to control the thickness closely. This also helps in reducing the roll-separating force.

The steel coming out of a blast furnace or a remelting shop is in the form of an ingot which has a cross section of the order of 600 × 600 mm. These ingots are further processed in rolling mills to produce the intermediate shapes such as blooms, slabs and billets. Blooms are generally square cross section varying from 150 × 150 mm to 400 × 400 mm. Slabs are rectangular cross section with widths varying from 500 to 1800 mm and thickness varying from 50 to 300 mm. Billets are rectangular cross sections with sizes varying from 40 × 40 mm to 150 × 150 mm.

The ingot obtained by casting would normally have defects such as shrinkage cavities, blow holes and slag inclusions. When the ingot is rolled, the shrinkage cavity (called pipe) is towards the end, which is then cropped after a few initial passes to reduce the defects in the finished product.

The blooming and slabbing mills which produce blooms and slabs from the ingots are generally, the reversible 2-high rolling stands. The metal is turned 90 degrees in the initial passes to enable the material to be uniformly compressed from all sides, so that any of the internal voids such as blow holes are eliminated, and the slag inclusions are segregated.

The rolls are generally manufactured of chilled cast iron, carbon steel and alloy steels. The soft rolls with hardness range of 150 to 250 BHN are used in roughing stands of the heavy section mills and billet mills. The semi hard rolls with 250 to 400 BHN are used in roughing stands of rail structure mills, and finishing stands of heavy section mills and billet mills. Hard rolls with 400 to 600 BHN are used to make finishing stands of sheets, railway structures and backing rolls for small-diameter rolls. These may sometimes have a roll surface of hard alloy such as tungsten carbide. The extra hard rolls with 600 to 800 BHN are used for cold rolling. The smaller rolls in this category may be made of hard materials like tungsten carbide.

7.2.3 Rolling Load

Rigorous calculation of the rolling load requires the understanding of plasticity theory, which is beyond the scope of this book. The necessary equations derived from the application of theories are presented below. The reader is advised to refer to these books for detailed analysis. The following expressions are valid for rolling of flat surfaces.

The pressure on rolls starts from the entry point and continues to build up till the neutral point. Similarly the exit pressure is zero at the exit point and increases towards the neutral point. At any section i, between the entry point and exit point in the rolls as shown in Fig. 7.9, the pressure acting on the rolls is given by superscript – indicating from the entry side while + indicates from the exit side):

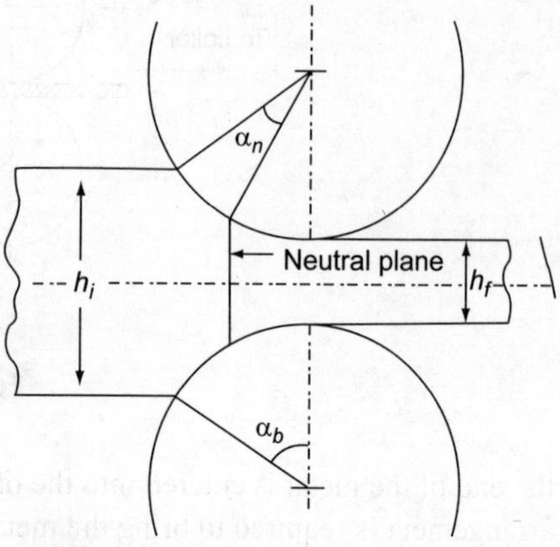

Fig. 7.9 *Schematic representation of rolling of slabs*

$$\frac{p^-}{S} = \frac{h}{h_i} e^{\mu(H_i - H)}$$

where, μ = Coefficient of friction between the work material and the rolls

S = Mean yield shear stress of the work material

$$H_i = 2\sqrt{\frac{R}{h_f}} \tan^{-1}\left(\sqrt{\frac{R}{h_f}}\right)\alpha_b$$

α_b = Angle of bite

$$\frac{p^+}{S} = \frac{h}{h_f} e^{\mu H}$$

The total rolling load acting on the rolls per unit width of the strip can be obtained by integrating the above two expressions for rolling loads with appropriate integration limits as follows:

$$\frac{P}{w} = R' \int_0^{\alpha_n} p^+ \, d\alpha + R' \int_{\alpha_n}^{\alpha_b} p^- \, d\alpha$$

where R' is the flattened radius of the rolls. The radius of rolls would be affected by the acting roll pressure. To account for this, the equation given by Hitchcock is generally used.

$$R' = R\left[1 + \frac{16(1 - \nu^2)P}{aE(h_i - h_f)}\right]$$

where ν = Poisson's ratio of the roll material

P = Average roll pressure per unit width

E = Young's modulus of the roll material

w = Width of the strip being rolled

To identify the neutral plane position (αn), the two expressions for the roll pressure $p-$ and $p+$ should be equated as the pressures from both ends peak at the neutral plane.

The solution of the above equations require a computer program to completely solve all the equations through a number of iterations before the values converge.

An approximate expression considering rolling similar to plane-strain-upsetting process, which gives a reasonably close results to those obtained by the earlier analysis, is given by Altan et al as follows:

$$\frac{P}{w} = \frac{2\sigma}{\sqrt{3}}\left(1 + \frac{ml}{4h}\right)l$$

where h = Average strip height = $\frac{h_i + h_f}{2}$,

$$\cos \alpha_b = 1 - \frac{h_i - h_f}{2R},$$

l = Average length of the deformed strip = $R\,\alpha_b$

σ = Flow stress of the work material

m = frictional shear factor, whose value is given in Table 7.5

TABLE 7.5 Values of frictional shear factor

m	Condition
0.05 to 0.015	Cold forming of steels, aluminium alloys and copper using lubricants or oils
0.2 to 0.4	Hot forming of steels, aluminium alloys and copper alloys using graphite-based lubricants
0.1 to 0.3	Hot forming of titanium and high temperature alloys with glass lubricants
0.7 to 1.0	Hot rolling of plates or slabs with no lubricants

7.2.4 Roll Passes

The final rolled products such as plates, flats, sheets, rounds and sections are obtained in a number of passes starting from billets or slabs. For rolling the flat products, plain cylindrical rolls are used, but for sections, grooved rolls are used. The type of grooving done is decided by the final section desired.

The roll-pass sequence can be broadly categorised into three types.

Breakdown Passes These are used for reducing the cross-sectional area nearer to what is desired. These would be the first to be present in the sequence.

Roughing Passes In these passes also, the cross section gets reduced, but along with it, the shape of the rolled material comes nearer to the final shape.

Finishing Passes These are the final passes, which give the required shape of the rolled section. Generally the finishing pass follows a leader pass.

Draught

In order to design the roll-pass schedule, the first to be considered is the draught or elongation in each of the passes. The total elongation can be calculated by dividing the initial area of the cross section by the final area of cross section of the rolled product.

$$\text{Total elongation} = \frac{\text{Initial cross-sectional area}}{\text{Final cross-sectional area}}$$

The elongation can be defined as total for all the passes or that for each pass separately. Draught of any given pass specifies the reduction achieved in the dimensions. Consider a rectangle, as shown in Fig. 7.10, as reduced from $A \times B$ to $a \times b$ mm. Then,

$$\text{Draught} = A + B - (a + b)$$

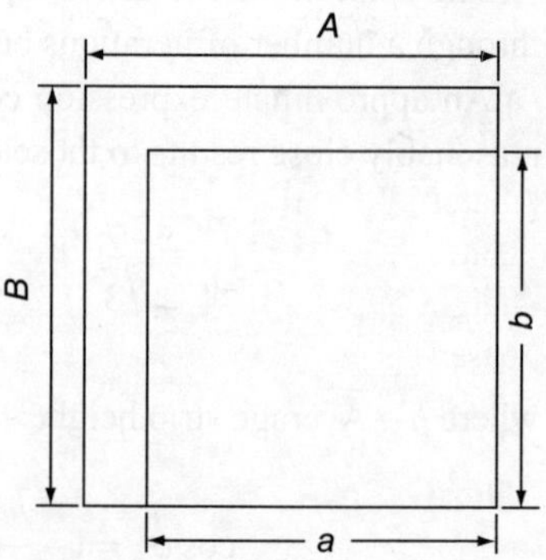

Fig. 7.10 *Draught definition*

If we know the mean draught to be provided in each pass then the number of passes can be estimated. But the draught provided in each pass also depends on the work material, angle of bite, roll strength, power of the rolling mill, and the condition of the rolls.

Besides, the draughts of all the passes in the sequence are not same. Normally, in the initial passes, it is necessary to reduce as much cross-sectional area as possible. This means that, the main criterion for choosing the draught is therefore to be the angle of bite or the maximum size the rolls can pull under them. In rolling plain shapes such as plates or sheets or intermediate passes, it is the strength of the rolls or the available power of the motors, which limit the draught. For finishing passes, roll wear may be the limiting factor in deciding the draught. Particularly for sections of complex shapes such as rails, it is necessary to reduce the roll wear to a minimum to obtain accurate dimensions and good surface finish.

The other reason for reducing draught in the later passes in hot rolling is the reduction in stock thickness which causes it to lose heat quickly and thus increases the rolling load. Similarly in cold rolling, the strain hardening of the stock material necessitates a reduction in draught in the succeeding passes.

7.2.5 Breakdown Passes

The principal breakdown pass sequences are:

(a) Box-pass series
(b) Diamond-square series
(c) Oval-square series

These breakdown pass sequences are schematically shown in Fig. 7.11.

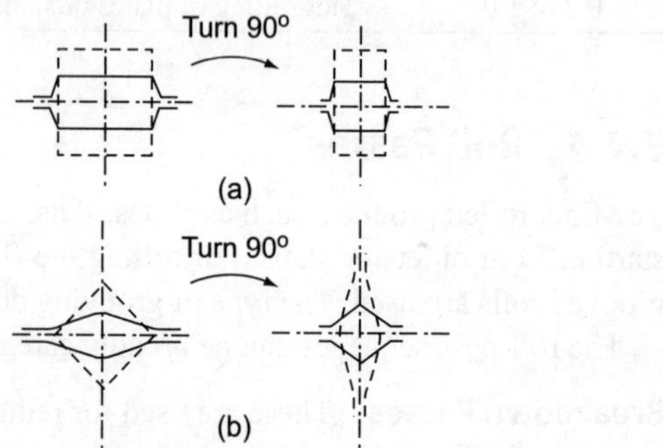

Fig. 7.11 *Breakdown pass sequence (a) Box pass (b) Diamond-square pass (c) Oval-square pass*

Box-Pass Series

Box passes are generally used for initial medium and large sections of blooming and billet mills. The rolls of box-pass series are stronger. They can be used for different sizes by screwing down the top roll. They provide for effective descaling and ensure uniform draught along the width of the bar. However, the main problems with these are the smaller elongation (1.05 to 1.15) and inaccurate square produced because the side spread is not controlled. They are generally used for larger products.

Diamond-square Series

In this, the square produced in a larger roll is turned over by 90 degrees and passed through the diamond rolls, the output of which is again turned over by 90 degrees and put through smaller square rolls, as shown in Fig. 7.11(b). The advantages of this series are the slow cooling of the entire material, accurate squares that are possible, and larger elongations that can be achieved (1.20 to 1.45). Because of these specific advantages, these are extensively used in the finishing passes of the comparatively smaller sections such as billets. The material being confined fully, these type of passes give a good stability and also uniform deformation of the material. The two principles that are generally used in designing the diamond-square series is to keep the height of one pass equal to the width of the next pass or the height of one pass less than the width of the next pass. The disadvantage with the series is that the roll gets weakened because of the deep impressions made into the rolls.

Oval-square Series

In this series, the diamond is replaced by oval shape. The stock after going through the oval pass, is turned over by 90 deg into the square pass, the output of which is then turned over by 45 deg to feed into the next oval pass. The main advantage of this series is the large elongation that can be obtained (1.35 to 1.80) compared to the other breakdown passes. The main problems of this are the difficulty to guide the stock in oval pass, the non-uniformity of the deformation along the width of the roll and higher roll wear in oval pass. When higher reductions are used, the material is likely to give rise to surface defect called 'fold' because of the wearing of the oval pass at points of maximum reduction, which cannot be eliminated.

The oval pass has some other variants as shown in Fig. 7.12.

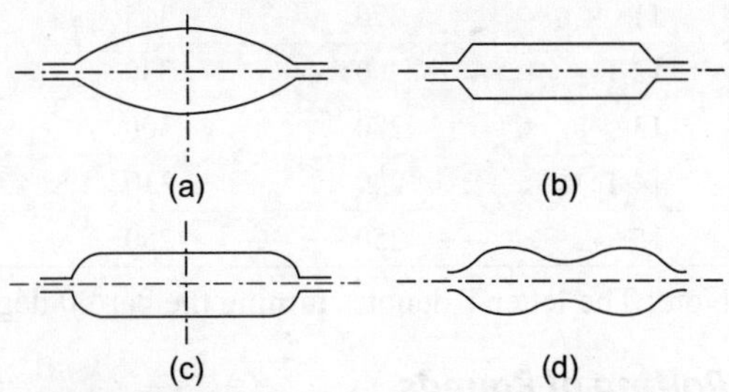

Fig. 7.12 *Different versions of oval passes*

7.2.6 Roll-Pass Sequences

Designing roll passes is a fairly complex and skilled art, the explanation of which is beyond the scope of the book. To this end specialist literature is available, the details of which are presented in references at the end of the book. Here, we will only refer to some salient points to be considered in the design of the roll passes.

Before deciding on the actual sequence, it is necessary to consider the requirements of the final product. Since hot rolling is finished at a high temperature of the order of 900 to 1100°C depending on the material, the dimension of the finishing pass should take into account the linear expansion of the metal to be used, at this temperature.

Pass Schedule for Blooms

The most normally used passes in blooming mills are the plain-sectioned rolls or bull-head passes. The main advantages of the bull-head passes are, strength of the roll not affected by any grooves, very good descaling of the stock in the initial pass itself, and higher draughts that can be achieved. If bull-head passes are chosen as initial passes, then large-size rolls can be applied.

While deciding the number of passes, the other factor to be considered is the number of times the bar is turned over. To increase the output of the mill, it is desirable that the bar be turned over as few times as possible. But to obtain high-quality blooms, it should be turned over as many times as possible, otherwise the stock is likely to form cracks and fissures, if reduced in only one direction.

The draughting schedule of a typical blooming mill is presented in Table 7.6

TABLE 7.6 Draughting schedule for rolling low-carbon steel ingots 760 × 680 mm into blooms 250 × 250 mm in cross section.

Pass	Dimension after Rolling, mm		Draught mm	Elongation	Spread mm	Angle of Contact Deg
	Thickness	Width				
1	680	685	80	1.109	5	23.58
2 T	600	690	80	1.125	5	23.58
3	615	605	75	1.113	5	22.92
4	540	610	75	1.130	5	22.92
5	470	615	70	1.140	5	22.08
6 T	400	620	70	1.166	5	22.08
7	540	410	80	1.120	10	23.83
8	460	420	80	1.146	10	23.83
9	380	435	80	1.169	15	23.83
10 T	300	450	80	1.224	15	23.83
11	370	315	80	1.158	15	23.33
12 T	290	330	80	1.218	15	23.33
13	280	300	50	1.139	10	17.67
14 T	240	310	40	1.129	10	16.50
15	250	250	60	1.190	10	20.50

Note: The letter T denotes turning the bar 90 deg after the corresponding pass.

Rolling of Rounds

The types of possible sequences for finishing passes for rolling round sections are presented in Fig. 7.13. Depending upon the starting stock size, the necessary break down passes, as explained earlier, are to be employed before this sequence. The most generally used pass sequence for smaller rounds from 5 to 20 mm is the one shown in Fig. 7.13(a). For higher size rounds of 20 to 100 mm, the pass sequence shown in (b) is used. Because of the flat box pass incorporated, descaling is done effectively and also the leader pass (the one just before the finish pass) is universal for a number of (5 to 6) round of sizes. The other advantage with this is that the edge pass (the one before the leader pass) causes a bulge in the material on the side, thereby decreasing the roll wear in the oval pass immediately after it.

Rounds requiring special steels and better quality are obtained by the rolling sequence shown in Fig. 7.13(c). The main feature here is that all the passes shown are without any sharp corners, thus maintaining uniform temperature throughout the material. For obtaining an exact round section in special steels, the oval-oval sequence as shown in Fig. 7.13(d) is used. The handling of material in the oval passes is difficult and hence the vertical oval is slightly modified to improve the stability. A typical sequence of roll passes to obtain a 12 mm rod from a billet of 100 × 100 mm is shown in Fig. 7.14.

Rolling of Sections

The rolling of structural sections is more difficult because of the presence of flanges. Shapes with flanges are generally rolled in closed passes. A closed pass is one where the complete shape of the pass is in one roll, as shown in Fig. 7.15(a). The open pass, as shown in Fig. 7.15(b), is the one where part of the pass is present in each of the two rolls.

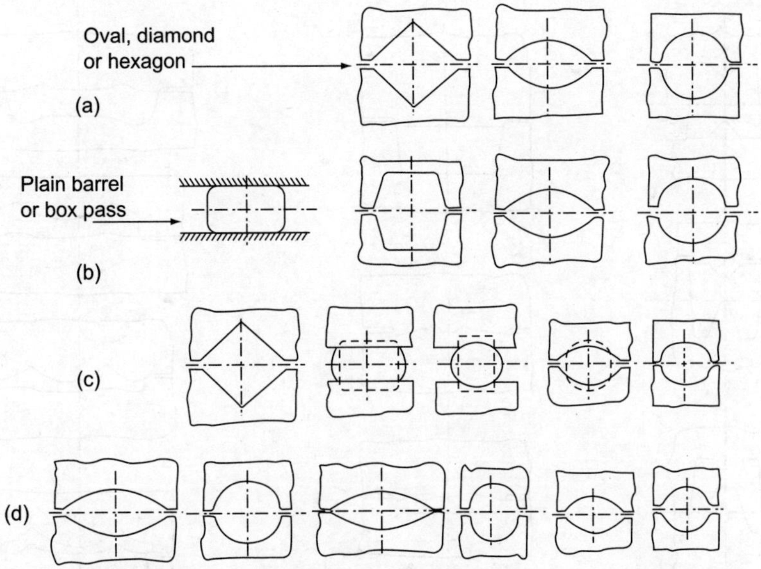

FIG. 7.13 *Various pass sequences for rolling round sections*

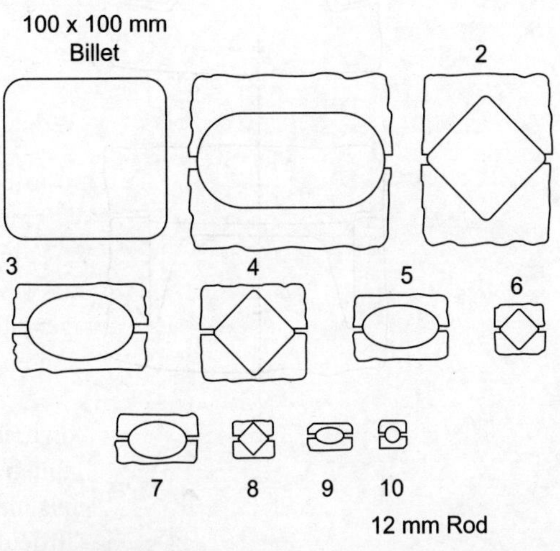

FIG. 7.14 *Roll pass sequence for 12 mm rod*

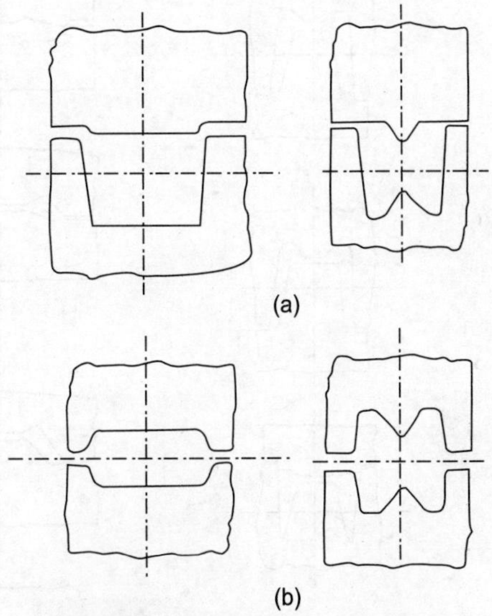

FIG. 7.15 *Close and open roll passes*

In these passes, the deformation is not uniform and also roll wear poses a serious problem. When any appreciable roll wear is noticed, the rolls are to be discarded because any restoration increases the pass size, thus changing the final dimensions.

The type of pass sequences that are used for channel sections are shown in Fig. 7.16. By the method called 'beam rolling', the flange sections of large thickness are formed to aid in the feeding of the metal into the rolls and also to reduce the cooling of the flanges. In the leader and finish passes, these are rolled down to the final shape as shown in Fig. 7.16(a).

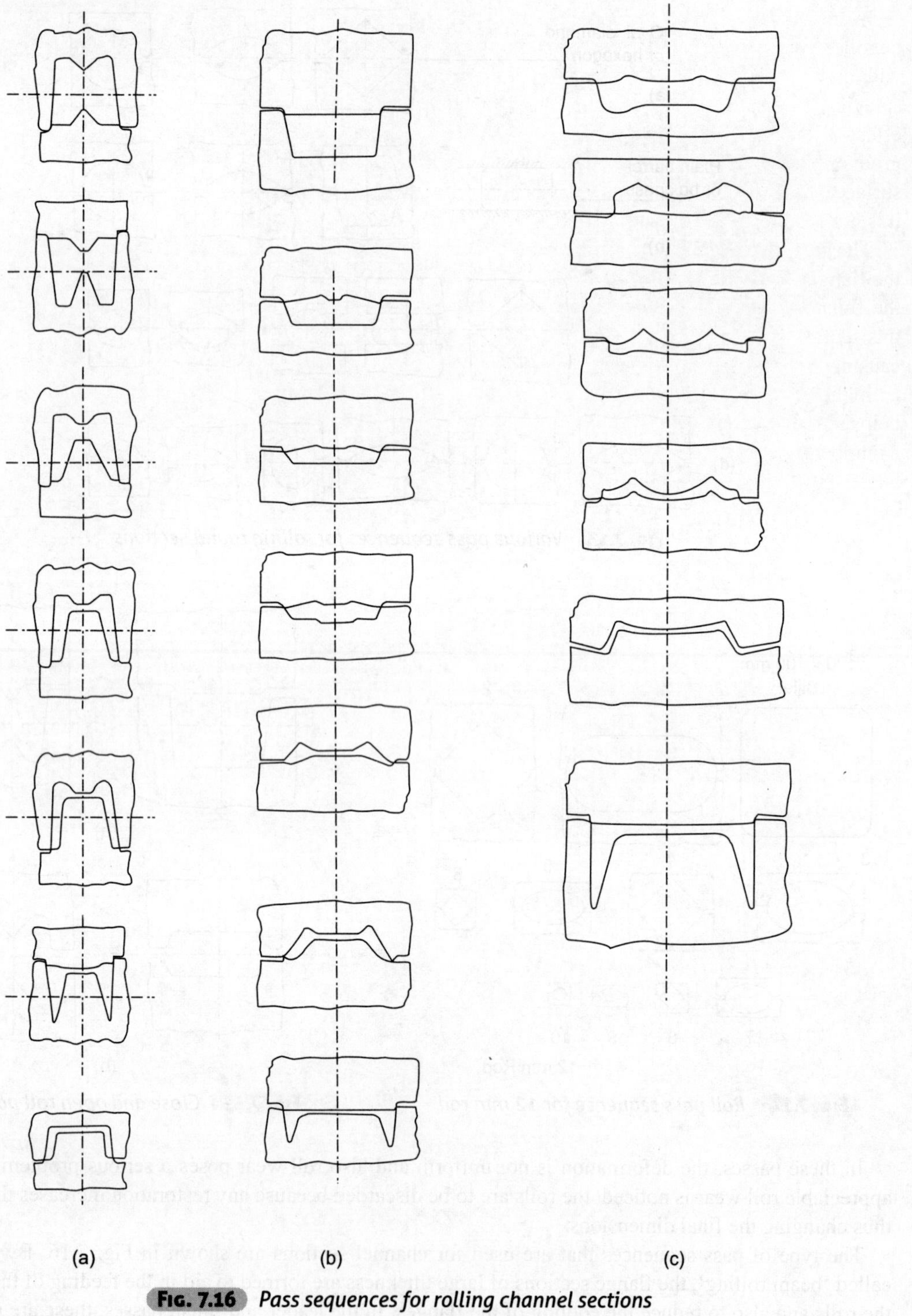

Fig. 7.16 *Pass sequences for rolling channel sections*

In an alternative method called 'butterfly method', the channel is considered as two angles. The stock can be rolled into splayed flanges in initial passes as shown in Fig. 7.16(b). This method helps in obtaining flanges with no taper in sides, and also a sharp corner between the web and the inner flange face. A different kind of splaying the flanges is shown in Fig. 7.16(c), which reduces the groove depth in rolls, reducing the weakening effect. This permits higher draught per pass and consequently less number of passes. Also, the wear is less in grooves because the surface speeds are more or less uniform at all points. This has a favourable effect on the surface quality of channels produced. But, due to splaying, the width of the groove increases and consequently, it is not used for heavier sections.

The pass sequences used for rolling angles are presented in Fig. 7.17. In the sequence shown at Fig. 7.17(a), the flanges are initially flat but are gradually bent in the roughing and leader passes such that, by the time the finishing pass comes, the flanges would have reached the 90 deg angle. The main disadvantage with this sequence is the deep grooves in the rolls which change the working speed at various points in the pass, causing higher roll wear, higher power requirement, and poor surface finish.

In the butterfly pass sequence shown in Fig. 7.17(b), which has the splayed sections as that of the channels, helps in minimising many of the above problems. The universal pass sequence shown in Fig. 7.17(c) utilises a number of passes, where the side spread is not controlled. Hence, the same rolls can be used for any size of angles required. These are economical and therefore generally used for high-production mills.

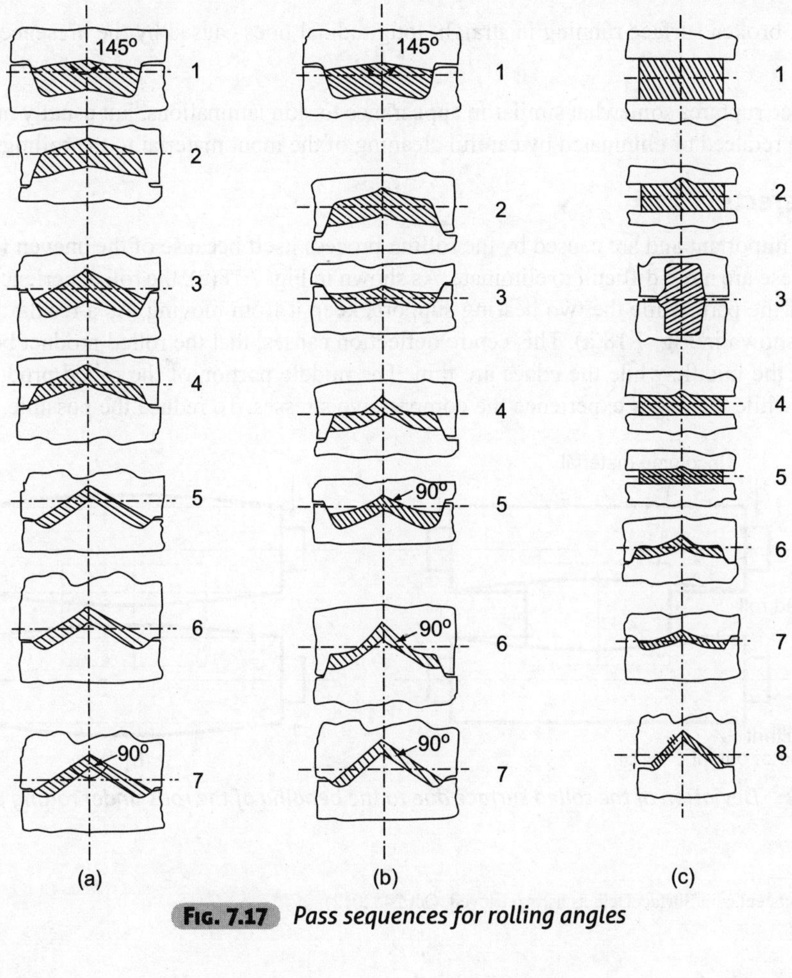

FIG. 7.17 *Pass sequences for rolling angles*

7.2.7 Rolling Defects

Being a very high production and often automated process, it is important to understand the likely defects in rolling and design methods, to reduce or eliminate them. In order to understand the causes and remedies of rolling defects, they may be divided into two classes:

Surface Defects

These are mainly caused by the improper preparation of the input material surface such as scales, rust and dirt. It may also be caused by the formation of scales during the hot-rolling operation. Some of the surface defects are*,

Lap This is a surface defect that occurs after multiple passes and would appear as a seam caused by folding over hot metal, fins, or sharp corners and then rolling or forging them into the surface, but not welding them.

Mill-shearing A defect which can be described as a feathering type light-surface lap.

Rolled-in Scale Surface oxides that are present on input surface that was improperly cleaned or formed on hot steel that would be pressed back into the surface without getting welded.

Scabs Elongated patches of loosened metal which have been rolled into the surface.

Seams Open, broken surface running in straight longitudinal lines caused by the presence of oxides near the surface.

Silvers Surface ruptures somewhat similar in appearance to skin laminations, but usually more prominent. These can be reduced or eliminated by careful cleaning of the input material to the rolling mill.

Structural Defects

These are more important and are caused by the rolling process itself because of the uneven forces acting on the material. These are more difficult to eliminate. As shown in Fig. 7.18(a), the roll experiences a heavy load at the middle of the part, while the two bearing supports keep it from moving. As a result, the roll deflects in a manner as shown in Fig. 7.18(a). This centre deflection causes, that the rolled product be barrel shaped with bulging in the middle while the edges are thin. The middle portion of the rolled product experiences tensile stresses while the edges experience the compressive stresses. To reduce the possible deformation of

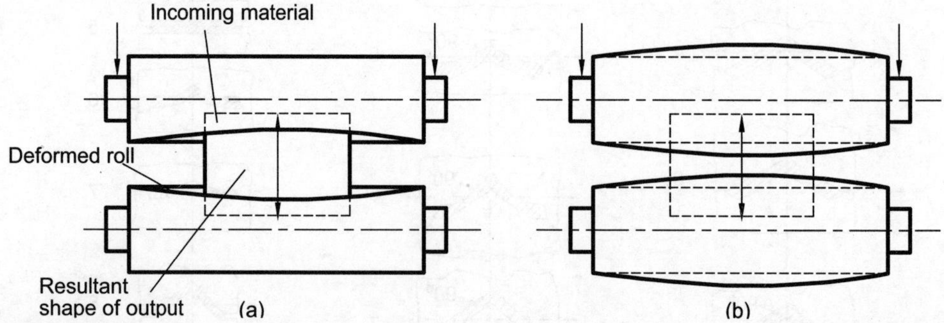

Fig. 7.18 *Deviation of the rolled surface due to the bending of the rolls under rolling pressure.*

* (http://www.summitsteel.com/SurfaceDefects.aspx retrieved, Oct 24, 2012)

the work piece due to the bending of the rolls, the rolls can be designed with camber as shown in Fig. 7.18(b), such that the rolled product will come out with uniform thickness throughout, if the designer knows exactly the amount of deflection for a given operation. This means that the rolling mill can only be used for a single product. Any change in the product will alter the force structure and thereby the accompanying stresses.

These defects include the following:

Wavy Edges

Wavy edges can be caused because of the elastic deflection of the rolls due to the rolling force acting on them. The edges of the rolled part gets thin and will experience compressive stresses. These compressive stresses cause the rolled sheet to be wavy at the edges.

Zipper Cracks

These are also caused due to the bending of rolls under the rolling pressure similar to wavy edges. However, zipper cracks will be seen at the centre due to the tensile stresses set-up in the rolled part due to the bending deformation of the rolls under the rolling pressure.

Edge Cracks

The main cause of edge cracking is the nonhomogeneous plastic deformation of metal across the width. Because of the surface deformation, while the centre of the slab remains fairly uniform the edges gets the plastic deformation in the form of lateral spread. This lateral spread is more prominent towards the edges which will cause the edge cracks.

Alligatoring

Since rolling involves the application of very high forces, there is considerable amount of friction between the rolled product and the rolled surface. Because of this friction, the top and bottom surface of the rolled part expands less compared to the middle portion of the part. If conditions become severe, then the rolled part may split, as shown in Fig. 7.19, with the upper part going up while the lower part follows in the opposite direction resembling the open mouth of an alligator. This defect is called alligatoring.

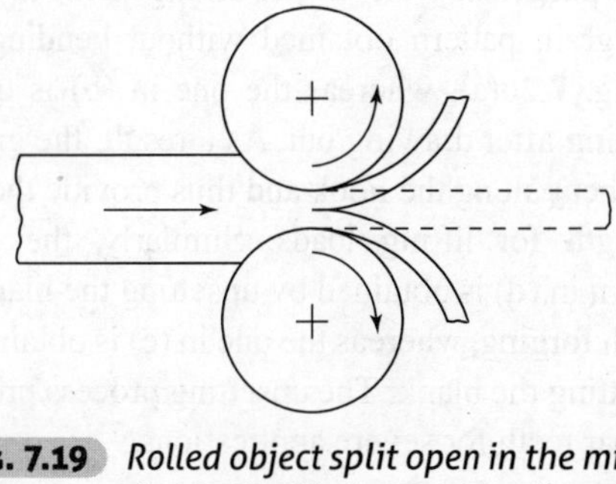

FIG. 7.19 *Rolled object split open in the middle due to the friction at the roll surface*

Laminations

This defect appears in the rolled object as sandwiched layers. This is caused when the ingot used for rolling is not sound and has a piping or blowholes. During rolling, these piping or blow holes do not get completely welded (e.g., if the piping has got oxidised material or nonmetallic inclusions it will not get welded) and cause laminations. The only way to eliminate this defect is by discarding the portion of the ingot where piping and other defects is present and selecting only good metal portion for rolling.

7.3 FORGING

Forging is the operation where the metal is heated and then a force is applied to manipulate the metal in such a way that the required final shape is obtained. This is the oldest of the metal-working processes known to mankind since the copper age. Forging is generally a hot-working operation though cold forging is used sometimes.

7.3.1 Forging Operations

Two types of operations are used in forging in order to arrive at the final object configuration. They are as follows:

Drawing Out

This is the operation in which the metal gets elongated with a reduction in the cross-sectional area. For this purpose, the force is to be applied in a direction, perpendicular to the length axis.

Upsetting

This is applied to increase the cross-sectional area of the stock at the expense of its length. To achieve the upsetting, force is applied in a direction parallel to the length axis.

Because of the manipulative ability of the forging process, it is possible to closely control the grain flow in the specific direction, such that the best mechanical properties can be obtained based on the specific application. Some examples of desirable grain flow directions are depicted in Fig. 7.20. For example, in the crane hook shown, it is possible to get the two types of grain patterns by the proper arrangement of operations. The grain pattern obtained without bending is shown in Fig. 7.20(a), whereas the one in (b) is obtained by bending after drawing out. As a result, the grain flow is also bent along the hook and thus provide the necessary strength for lifting loads. Similarly, the gear blank shown in (d) is obtained by upsetting the blank and then finish forging, whereas the one in (c) is obtained without upsetting the blank. The upsetting process provides the radial grain flow, which is essential for good strength in gear teeth for severe applications.

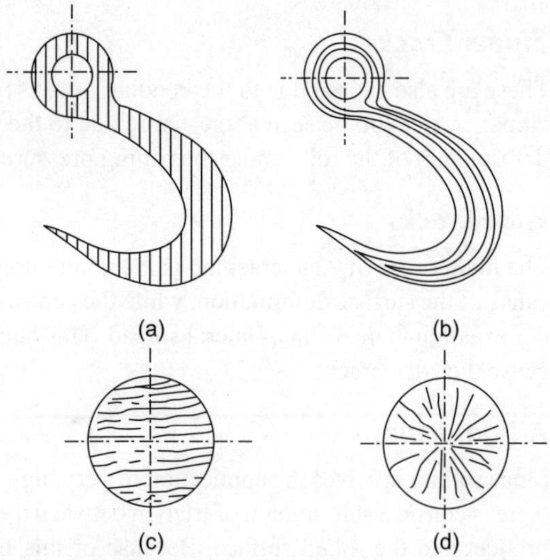

Fig. 7.20 *Grain-flow directions obtained in forging*

Forging Types

There are four types of forging methods, which are generally used.

Smith Forging This is the traditional forging operation done openly or in open dies by the village blacksmith or modern shop floor by manual hammering or by power hammers.

Drop Forging This is the operation done in closed impression dies by means of the drop hammers. Here the force for shaping the component is applied in a series of blows.

Press Forging Similar to drop forging, the press forging is also done in closed-impression dies with the exception that the force is a continuous squeezing type applied by the hydraulic presses.

Machine Forging Unlike the drop or press forging where the material is drawn out, in machine forging, the material is only upset to get the desired shape.

7.3.2 Smith Forging

The process involves heating the stock in the blacksmith's hearth and then beating it over the anvil. To get the desired shape, the operator has to manipulate the component in between the blows. The types of operations available are fullering, flattening, bending, upsetting and swaging.

In fullering, the material cross section is decreased and length increased. To do this, the bottom fuller is kept in the anvil hole with the heated stock over the fuller. The top fuller is then kept above the stock and then with the sledge hammer, the force is applied on the top fuller. The fullers concentrate the force over a very small area, thus decreasing the cross section at that point. Metal flows outward and away from the centre of the fullering die. Then the stock is advanced slightly over the fuller and the process repeated, as shown in Fig. 7.21.

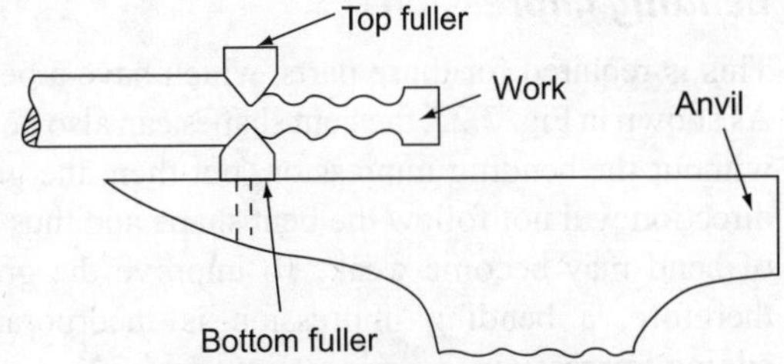

Fig. 7.21 *Fullering operation*

After fullering, the stock would have the fullering marks left which are then cleaned by means of flattening. To obtain specific shapes such as round, square, hexagon, etc., open general-purpose dies called swages are used. The force for shaping is applied by manual hammering or by means of the forging hammers, the latter being the industrial practice.

Smith forging involves a lot of skill on the part of the operator and also is more time consuming. But since no special dies are used, smith forging is more beneficial in the manufacture of small lots or in trial production, because the heavy cost of the closed-impression dies cannot be justified in these cases.

7.3.3 Drop Forging

Drop forging utilises a closed-impression die to obtain the desired shape of the component. The shaping is done by the repeated hammering given to the material in the die cavity. The equipment used for delivering the blows are called drop hammers.

The drop-forging die consists of two halves. The lower half of the die is fixed to the anvil of the machine, while the upper half is fixed to the ram. The heated stock is kept in the lower die while the ram delivers four to five blows on the metal, in quick succession so that the metal spreads and completely fills the die cavity. When the two die halves close, the complete cavity is formed.

The die impressions are machined in the die cavity, because of which more complex shapes can be obtained in drop forging, compared to smith forging. However, too complex shapes with internal cavities, deep pockets, re-entrant shapes, etc. cannot be obtained in drop forging due to the limitation of the withdrawal of the finished forging from the die. The typical products obtained in drop forging are crank, crank shaft, connecting rod, wrench, crane hook, etc.

The final shape desired in drop forging cannot be obtained directly from the stock in a single pass. Depending on the shape of the component, and the desired grain-flow direction, the material should be manipulated in a number of passes. The various passes used are the following:

Fullering Impression

Since drop forging involves only a reduction in cross section with no upsetting, the very first step is to reduce the stock to the desired size. The impression machined in the die to achieve this is called fullering impression. Sometimes it may be possible to obtain this without a special die impression, in open dies.

Edging Impression

Also called 'preform', this stage is required to gather the exact amount of material required at each cross section of the finished component. This is the most important stage in drop forging. Properly designed preform ensures a defect-free flow of metal, complete die fill and minimum flash loss.

Bending Impression

This is required for those parts, which have a bent shape. As shown in Fig. 7.22, the bent shapes can also be obtained without the bending impression, but then, the grain-flow direction will not follow the bent shape and thus the point of bend may become weak. To improve the grain flow, therefore, a bending impression is incorporated after edging impression.

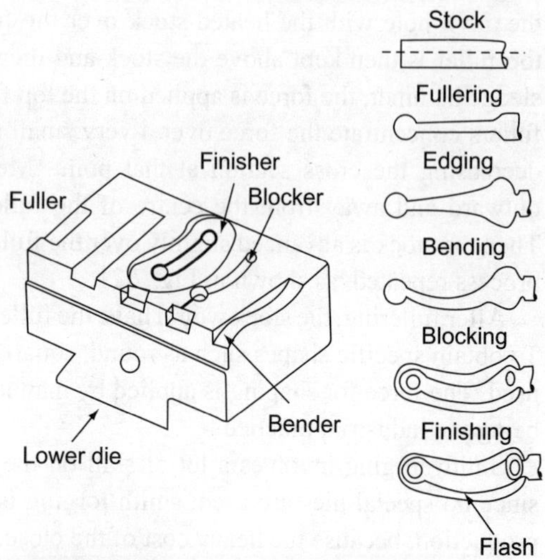

Fig. 7.22 *Stages in drop forging of a lever*

Blocking Impression

Also called 'semifinishing' impression, blocking is a step before finishing. In forging, it is very difficult for the material to flow to deep pockets, sharp corners, etc. Hence, before the actual shape is obtained, the material is allowed to have one or more blocking impressions where it acquires the shape very near to the final one. The blocking impression is characterised by large corner radii and fillets but no flash. For complex shapes, more than one blocking impression may be used.

Finishing Impression

This is the final impression where the actual shape required is obtained. In order to ensure that the metal completely fills the die cavity, a little extra metal is added to the stock. This extra metal will form the flash and surrounds the forging in the parting plane.

Trimming

In this stage, the extra flash present around the forging is trimmed to get the forging in the usable form.

The shape of a forging as it moves through the various stages is shown in Fig. 7.22.

7.3.4 Press Forging

Press-forging dies are similar to drop-forging dies as also the process. In press forging, the metal is shaped not by means of a series of blows as in drop forging, but by means of a single continuous squeezing action. This squeezing is obtained by means of hydraulic presses.

Because of the continuous action of the hydraulic presses, the material gets uniformly deformed throughout its entire depth. More hammer force is likely to be transmitted to the machine frame in drop forging, whereas in press forging it is absorbed fully by the stock. The impressions obtained in press forging are clean compared to that of the likely jarred impressions, which are likely in the drop-forged components. The draft angles used in press forging are less than in drop forging. But the press capacity required for deforming is higher and as a result somewhat smaller size components only are press forged in closed-impression dies. But there is no such limitation for press forging in open dies.

The presses may have capacities ranging from 5 MN to 50 MN for normal applications. For special heavy duty applications, higher capacity presses of the order of 150 MN are required. To get the equivalent weight of the falling parts in kg of drop hammer required, the capacity of the press expressed in tons is to be multiplied by a factor ranging between 2 to 2.5.

To provide the necessary alignment of the two die halves, die posts are attached to the bottom die so that the top die would slide only on the posts and thus register the correct alignment. This ensures better tolerances for the press-forged components.

The tong holds, which are normally required for manipulating dies in drop forging are not required in press forging, because metal is pressed only once in each of the impression and generally mechanical ejection is used for removing the piece from the die. But for this, the press forging die is similar to drop forging die with the various impressions such as fuller, bender, blocker and finisher impressions properly arranged.

7.3.5 Machine Forging

As it involves the upsetting operation, sometimes it is simply called **upset forging**. Though both drop and press forgings are also done by machines, historically, only upset forging is referred to as machine forging. Originally this was developed for making the bolt heads in a continuous fashion, but now there is fairly larger number of diverse uses for this process. Because of the beneficial grain flow obtained from upsetting, it is used for making gear blanks, shafts, axles and similar parts. Some typical examples of upset forgings are shown in Fig. 7.23.

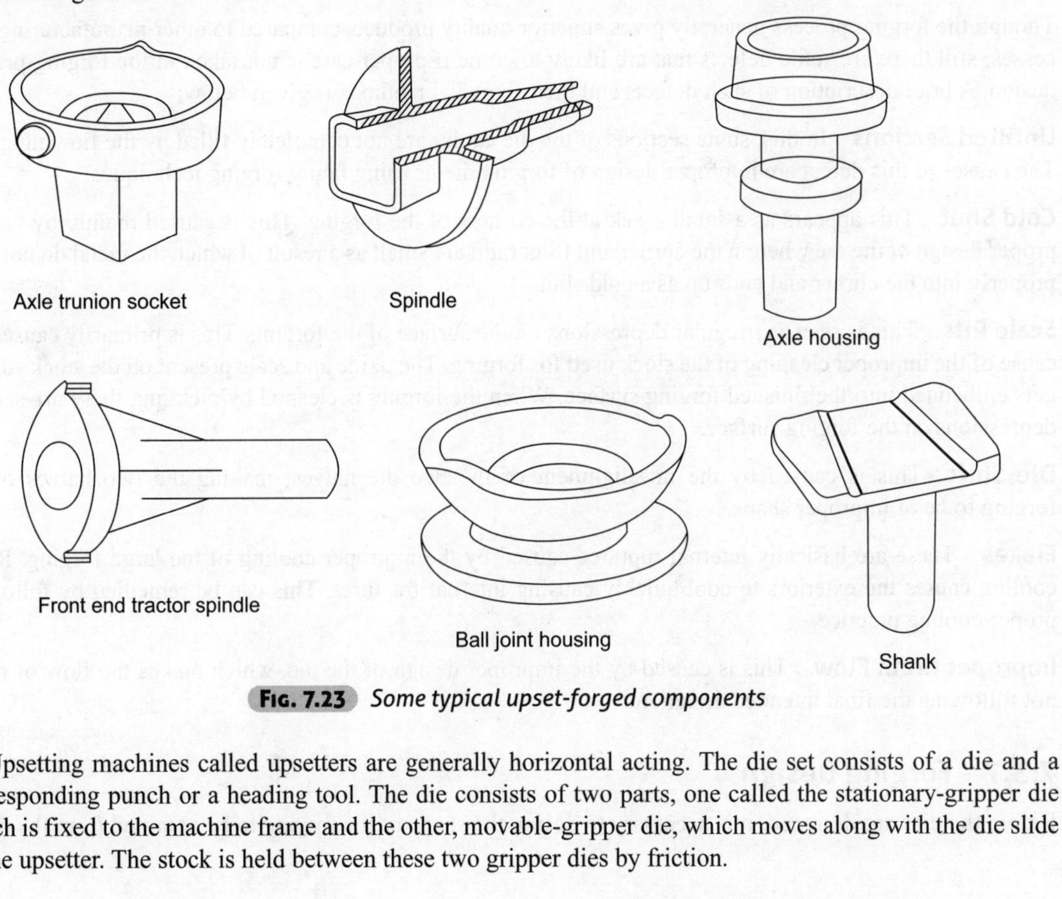

Fig. 7.23 *Some typical upset-forged components*

Upsetting machines called upsetters are generally horizontal acting. The die set consists of a die and a corresponding punch or a heading tool. The die consists of two parts, one called the stationary-gripper die which is fixed to the machine frame and the other, movable-gripper die, which moves along with the die slide of the upsetter. The stock is held between these two gripper dies by friction.

The upset-forging cycle starts with the movable die sliding against the stationary die, to grip the stock. The two dies when in closed position, form the necessary die cavity. Then the heading tool advances against the stock and upsets it to completely fill the die cavity. Having completed the upsetting, the heading tool moves back to its back position. Then the movable gripper die releases the stock by sliding backwards.

Similar to drop forging, it is not possible to get the final shape in a single pass in machine forging also. Therefore, the operation is carried out in a number of stages. The die cavities required for the various operations are all arranged vertically on the gripper dies. The stock is then moved from one stage to the other in a proper sequence till the final forging is ready. A heading tool each, for every upsetting stage is arranged on the heading slide of the upsetting machine. A typical upsetting die and heading tool are shown in Fig. 7.24.

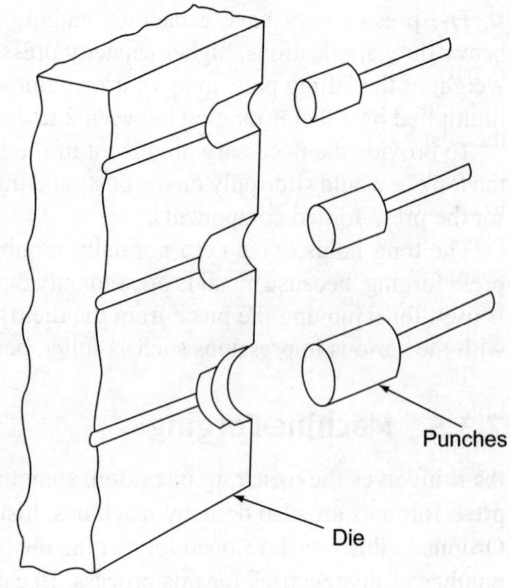

Fig. 7.24 *Upsetting die*

7.3.6 Forging Defects

Though the forging process generally gives superior quality products compared to other manufacturing processes, still there are some defects that are likely to come if proper care is not taken in the forging-process design. A brief description of such defects and their remedial methods is given below:

Unfilled Sections In this, some sections of the die cavity are not completely filled by the flowing metal. The causes of this defect are improper design of forging die or using faulty forging techniques.

Cold Shut This appears as a small crack at the corners of the forging. This is caused mainly by the improper design of the die wherein the corner and fillet radii are small as a result of which the metal do not flow properly into the corner and ends up as a cold shut.

Scale Pits This is seen as irregular depressions on the surface of the forging. This is primarily caused because of the improper cleaning of the stock used for forging. The oxide and scale present on the stock surface gets embedded into the finished forging surface. When the forging is cleaned by pickling, these are seen as depressions on the forging surface.

Die Shift This is caused by the misalignment of the two die halves, making the two halves of the forging to be of improper shape.

Flakes These are basically internal ruptures caused by the improper cooling of the large forging. Rapid cooling causes the exteriors to cool quickly causing internal fractures. This can be remedied by following proper cooling practice.

Improper Grain Flow This is caused by the improper design of the die, which makes the flow of metal not following the final intended directions.

7.3.7 Forging Design

Before the dies are designed, it is necessary to design the shape of the forging to be obtained from the die.

Parting Plane

A parting plane is the plane in which the two die halves of the forging meet. It could be a simple plane or irregularly bent, depending on the shape of the forging. The choice of a proper parting plane greatly influences the cost of the die as well as the grain flow in the forging.

In any forging, the parting plane should be the largest cross-sectional area of the forging, since it is easier to spread the metal than to force it into deep pockets. A flat parting plane is more economical. Also the parting plane should be chosen in such a way that equal amount of material is located in each of the two die halves, so that no deep die cavities are required. It may be required to put more metal into the top die half, since metal would more readily flow in the top half than in the bottom one. Some examples of proper parting plane choices are as below.

The example shown in Fig. 7.25(a) means that the required holes in the lever are to be obtained by drilling after forging. But the surface to be drilled is not flat but tapered because of the draft provided on the sides for withdrawal of the forging. This taper causes the drill to wander and therefore is not desirable. Therefore, a parting line, as shown in Fig. 7.25(b) would be desirable since it facilitates the generation of the holes by forging itself. But this brings in another problem of an inclined parting plane, which causes a sideward thrust (towards the left) as shown with an arrow. This sideward thrust may cause misalignment of the two die halves. To take care of this, the counter locking of the die is to be done as shown in Fig. 7.25(c), which is the final desirable parting plane.

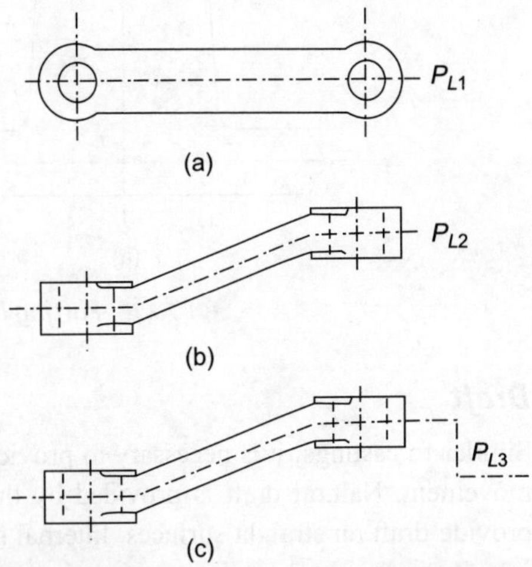

Fig. 7.25 *Parting planes in a bent lever*

Whenever the counter lock is provided, care should be taken to see that enough resisting area is provided, so that the sideward thrust generated is properly counter balanced. The thickness of counter lock should at least be 1.5 times the height for providing enough strength. The height of the counterlocking portion should at least be the same height as the forging detail which causes the sideward thrust. The counterlocked portions cannot be adequately lubricated and hence they wear out quickly, and call for reworking of the dies. Therefore, it is necessary that counter locking of dies should be eliminated as far as possible.

The elimination of counter locking can be done to a great extent by properly orienting the forging in the die. The web in the example shown in Fig. 7.26(a) causes the sideward thrust for the chosen parting plane. A slightly redesigned component as shown in Fig. 7.26(b) cannot be produced with a flat parting line but calls for a counter locking as shown. But orienting the component (as shown in Fig. 7.26(c)) so that the 90 degree angle is inclined to the parting line, makes for an easier forging of the angle as well as reduces the counter locking completely. For the design shown in Fig. 7.26(a) two components are produced in a die, the sideward thrust can be compensated. The components can be parted off after forging. This will facilitate making the die completely in one half only.

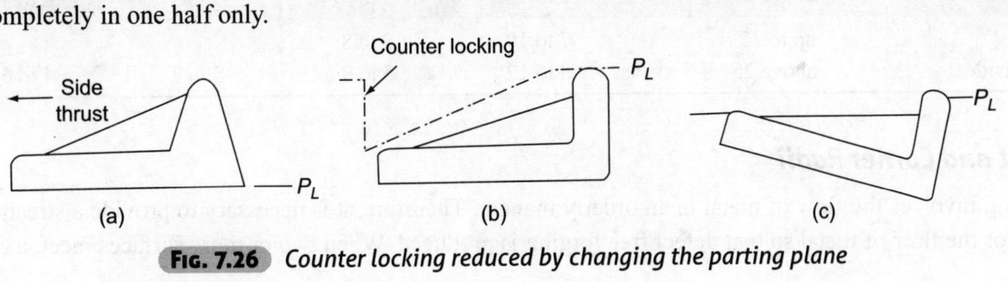

Fig. 7.26 *Counter locking reduced by changing the parting plane*

The other aspect to be considered in the choice of parting line is the punching of holes that are perpendicular to the parting plane. Depending on the depth of the hole compared to the diameter, it may be necessary to choose such a parting line that the hole is properly distributed, and the punch used for making the hole has enough strength. In Fig. 7.27(a) is shown one possibility of a parting plane, which simplifies the lower part of the die. But the punch in the upper die half becomes excessively long and hence higher draft. By changing the parting line as shown in Fig. 7.27(b), it is possible to punch from both sides, thus reducing the machining which was associated with the parting line as shown in Fig. 7.27(a). Also, this parting line gives a smaller height-to-diameter for the punch increasing its rigidity.

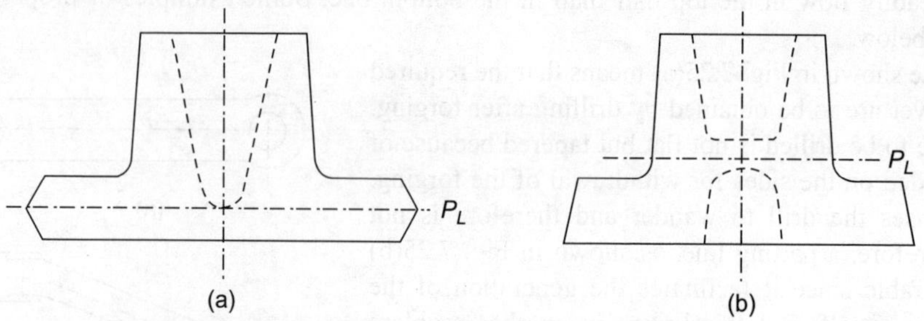

FIG. 7.27 *Parting line to reduce the depth of a punched hole*

Draft

Similar to castings, it is necessary to provide draft on forging surfaces, which are at right angles to the die movement. Natural draft is provided by the cylindrical or tapered surfaces. Otherwise it is necessary to provide draft on straight surfaces. Internal surfaces require more draft than external surfaces. During cooling, forging tend to shrink towards its centre and as a result, the external surfaces are likely to be separated, whereas the internal surfaces tend to cling to the die more strongly. The forged part is likely to be left in the die-half, which does the punching of the hole. In upset forgings, the draft problem is minimised because the part is held securely by the gripper die during the punch withdrawal and the gripper itself gets opened to release the component. Thus, in upset forgings a very small draft is normally used. The recommended draft angles are presented in Table 7.7. The tolerance applied on all draft angles is +2 deg and –1 deg.

TABLE 7.7 Recommended draft angle

Draft Position	Height or Depth (mm)	Drop Forgings		Upset Forgings	
		Normal (deg)	Close (deg)	Normal (deg)	Close (deg)
Outside	up to 25 above 25	5 to 7 5 to 10	3 to 7 3 to 7	3 to 5	2 to 4
Inside	up to 25 above 25	7 to 10 8 to 12	5 to 8 5 to 9	5 to 7	4 to 6

Fillet and Corner Radii

Forging involves the flow of metal in an orderly manner. Therefore, it is necessary to provide a streamlined path for the flow of metal so that defect free forging is produced. When two or more surfaces meet, a corner is formed that restricts the flow of metal. Therefore, these corners are to be rounded off to improve the flow of the metal. Fillets are for rounding off the internal angles, whereas corner is that of the external angle. For example, consider the flow of metal over a corner as shown in Fig. 7.28(a). Because of the large corner radius provided, metal is allowed to flow smoothly into the pocket. But when the corner radius is small, or not provided, as in Fig. 7.28(b), the metal flow is first hindered and when it enters finally the cavity, the metal would fold back against itself forming a defect called **lap** or **cold shut**.

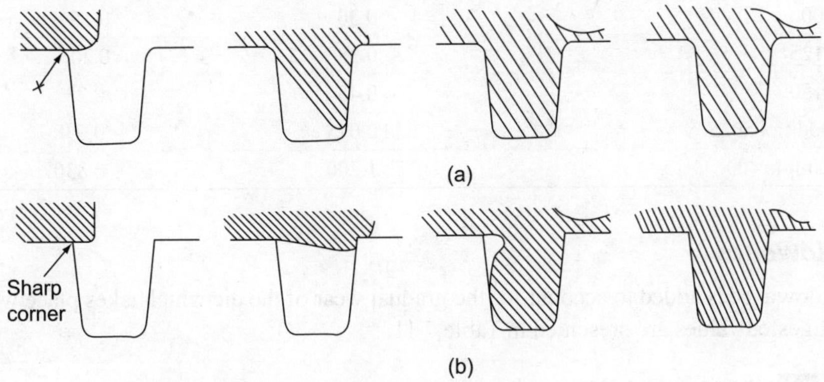

FIG. 7.28 *Effect of edge radius on the flow of metal*

Recommended fillet and corner radii are given in Tables 7.8 and 7.9.

TABLE 7.8 **Recommended fillet and corner radii for drop forgings**

Depth or Height, mm	Fillet Radius, mm	Corner Radius, mm
15	5	2.5
25	8	4.0
40	12	4.5
50	15	5.0
65	18	5.5
75	20	6.0

TABLE 7.9 **Recommended fillet and corner radii for upset forgings**

Upset Diameter / Stock Diameter	Fillet Radius, mm	Corner Radius, mm
up to 1.25	6.5	6.5
1.25 to 3.00	3.5	3.5
over 3.00	3.0	3.0

Shrinkage Allowance

The forgings are generally made at a temperature of 1150 to 1300°C. At this temperature, the material gets expanded and when it is cooled to the atmospheric temperature, its dimensions would be reduced. It is very difficult to control the temperature at which forging process would be complete, therefore we need to precisely control the dimensions. Hence a shrinkage allowance is added on all the linear dimensions as given in Table 7.10.

TABLE 7.10 Shrinkage allowance

Length or Width, mm	Commercial + or – mm	Close + or – mm
up to 25	0.08	0.05
26 to 50	0.15	0.08
51 to 75	0.23	0.13
76 to 100	0.30	0.15
101 to 125	0.38	0.20
126 to 150	0.45	0.23
Each additional 25	add 0.075	0.038
For example 400	1.200	0.830

Die-wear Allowance

The die-wear allowance is added to account for the gradual wear of the die which takes place with the use of the die. The suggested values are presented in Table 7.11.

TABLE 7.11 Die-wear tolerance

Net Mass of Forging (kg)	Commercial + or – (mm)	Close + or – (mm)
upto 0.45	0.80	0.40
0.46 to 1.35	0.88	0.45
1.36 to 2.25	0.95	0.48
2.26 to 3.20	1.03	0.53
3.21 to 4.10	1.11	0.55
4.11 to 5.00	1.18	0.60
Each additional 1 add	0.083	0.041
For example 15.00	2.010	1.010

Finish Allowance

Machining allowance is to be provided on the various forged surfaces, which need to be further machined. The amount of allowance to be provided should account for, besides the accuracy, the depth of the decarburized layer. Also, the scale pits that are likely to form on the component should also be removed during machining. A guideline for finish allowance is provided in Tables 7.12 and 7.13.

TABLE 7.12 Finish allowance for drop forgings

Greatest Dimension (mm)	Minimum Allowance per Surface (mm)
upto 200	1.5
201 to 400	2.5
401 to 600	3.0
601 to 900	4.0
above 900	5.0

TABLE 7.13 Finish allowance for upset forgings

Greatest Diameter (mm)	Minimum Allowance per Surface (mm)
up to 50	1.5
51 to 200	2.5
above 200	3.0

The component as affected by the various allowances is shown in Fig. 7.29.

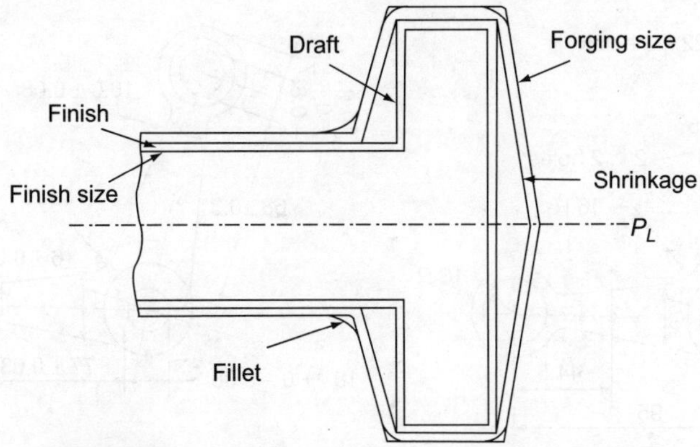

FIG. 7.29 *Allowances shown on forged component*

Cavities

The cavities and ribs including holes can be produced up to a certain depth only in drop forging. The main reason for the limitation is that the punch needs to have the necessary strength to withstand the forging load. Thin, long punches are likely to wear out quickly and need reconditioning of the die. The common limits for the depth of rib-to-web enclosures and for cavities are presented in Table 7.14 with reference to Fig. 7.30.

TABLE 7.14 Maximum limits of depth

Materials	Ratio of h : W	
	$L = W$	$L > 2W$
Aluminium, Magnesium	1.0	2.0
Steel, Titanium	1.0	1.5

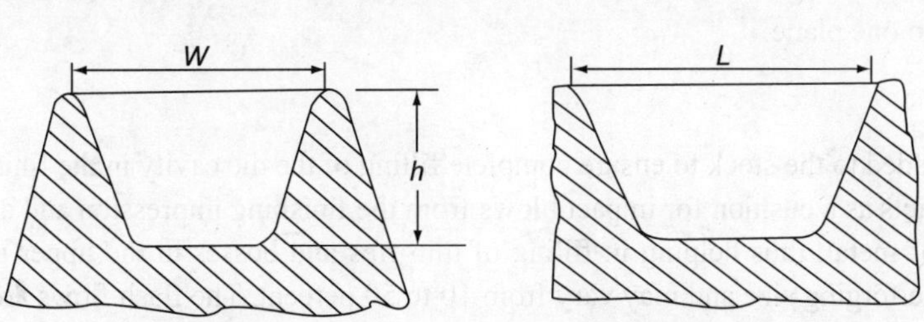

FIG. 7.30 *Cavity configuration in drop forging*

In addition to these allowances shown, the various tolerances that are applicable to forgings such as mismatch tolerance, weight tolerance, residual flash tolerance, thickness tolerance, burr tolerance, etc. are given in handbooks and standards which are listed at the end of this chapter in references. A sample component after providing the necessary tolerances and allowances is shown in Fig. 7.31.

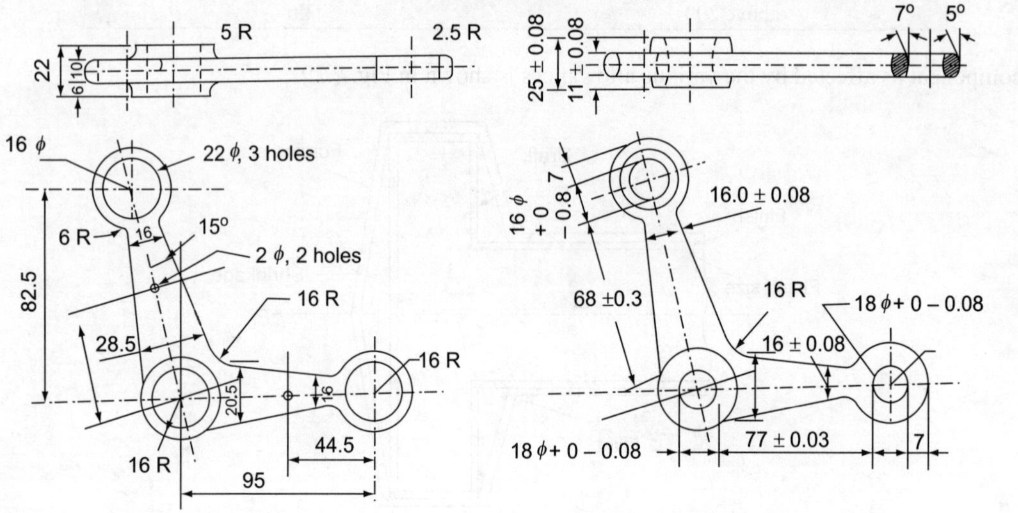

(a) Actual part (a) Final component showing draft and tolerances

Fig. 7.31 *Forging component as affected by allowances and tolerances*

7.3.8 Drop-Forging Die Design

The first step in the design of a drop-forging die is the decision regarding what impressions (or stages) are necessary to achieve the necessary fibre-flow direction so that the requisite strength is obtained. Normally fullering, edging and finishing impressions are necessary. The other types of impressions are only required in special situations.

A blocking impression becomes a necessity only when the component is to be accurately made or the component has deep pockets or thin ribs, which are difficult to be obtained in a single finishing impression. A bending impression is required when the part is of bent nature and the grain direction is to be along the bend line. In such a case, the bending impression is to be obtained before the blocking impression or finishing impression when no blocking is used. Similarly, a flattening impression is used when the component is thin and perpendicular to one plane.

Flash

The excess metal added to the stock to ensure complete filling of the die cavity in the finishing impression is called flash. Flash acts as a cushion for impact blows from the finishing impression and also helps to restrict the outward flow of metal, thus helping in filling of thin ribs and bosses in the upper die. The amount of flash depends on the forging size and may vary from 10 to 50 percent. The flash flows around the forging in the parting plane as shown in Fig. 7.32. The flash is provided uniformly around the periphery of the forging in the parting plane. The minimum flash allowances suggested are given in Table 7.15.

It has been found that the forging load is greatly influenced by the flash thickness and width. The forging load can thus be decreased by increasing the flash thickness. However, this increases the metal to be left in flash, increasing the scrap losses. Also, the forging load decreases with an increase in the average thickness of the component. Thus it is more difficult to forge thin components than the thicker ones.

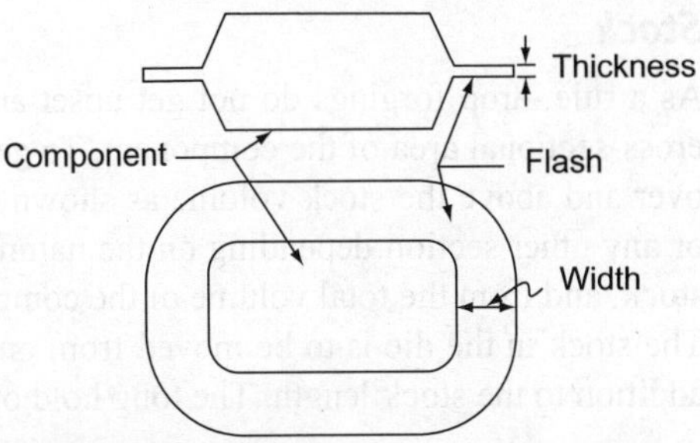

FIG. 7.32 *Proportions of flash in drop forging*

TABLE 7.15 **Flash and gutter sizes**

Stock Size (mm)	Flash		Gutter	
	Width (mm)	Thickness (mm)	Width (mm)	Thickness (mm)
upto 35	4.5	0.8	25	3.0
36 to 50	5.3	1.0	25 to 32	4.5
51 to 65	6.5	1.5	32 to 38	4.5
66 to 75	8.0	2.0	32 to 38	4.5
76 to 100	10.0	3.0	38 to 44	6.5

In addition to the flash, provision should be made in the die for additional space so that any excess metal can flow and help in the complete closing of the die. This is called **gutter**. Without gutter, flash may become excessively thick, not allowing the dies to close completely. The gutter as shown in Fig. 7.33 should be more than the flash provided. The preferred gutter sizes are presented in Table 7.15. The flash land provided in the die should be about 3% of the maximum forging thickness (0.5 to 8.0 mm). If the flash land is too small, then the energy required for the forging increases because of the excess metal trapped in the finishing impression and the flash land wears out quickly. Similarly, too high a flash land lets the work material to flow into the gutter and thus the die cavity gets unfilled.

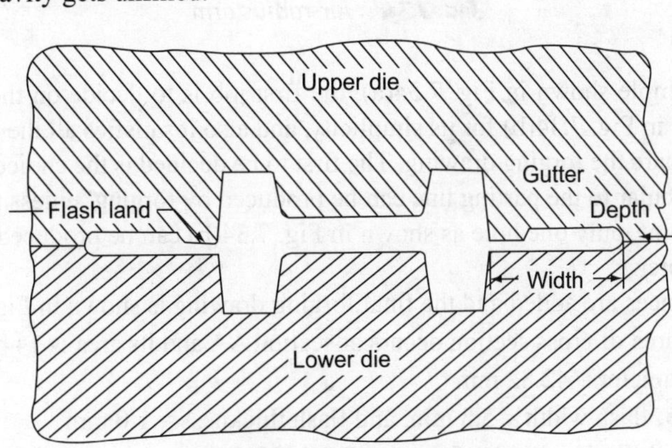

FIG. 7.33 *Gutter proportions*

Stock

As a rule, drop forgings do not get upset and therefore the stock size to be chosen depends on the largest cross-sectional area of the component. To get the stock size, the necessary flash allowance is to be provided over and above the stock volume as shown in Table 7.12. The stock to be used is either round, rectangular or any other section depending on the nature of the component. Having decided on the cross section of the stock, and from the total volume of the component and the flash, it is possible to find the length of the stock. The stock in the die is to be moved from one impression to the other, and hence a tong hold is provided in addition to the stock length. The tong hold of about 50 to 60 mm normally is required.

Example 7.1

Determine the stock size required for the air-radius arm shown in Fig. 7.34(a).

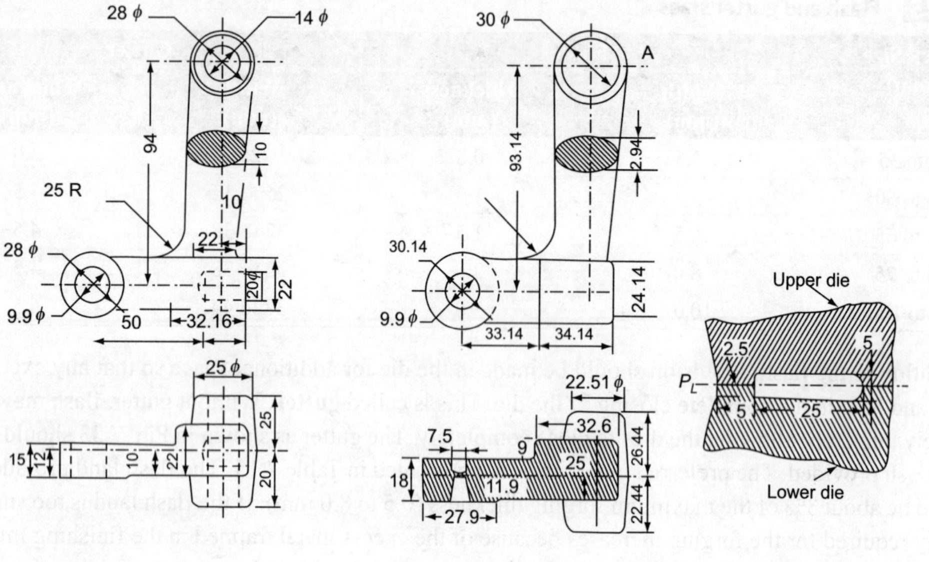

(a) Finished component (b) Forging (c) Gutter dimensions

FIG. 7.34 *Air-radius arm*

Solution For the example shown in Fig. 7.34(a), the first job is to decide on the parting line, which is chosen as the one shown in Fig. 7.34(b) for its simplicity, and also it satisfies all the criteria laid out earlier.

The next job is to prepare the forging drawing. The first to be decided is the choice of the holes. The holes whose axes are perpendicular to the parting line can be produced by forging, unless they are too deep as per Table 7.14. On this account, only one hole as shown in Fig. 7.34(b) can be produced whereas the rest are to be obtained by drilling later.

The necessary allowances are added and the final forging drawing is shown in Fig. 7.34(b).

The maximum area of cross section occurs at section AA and its area is 1440 mm^2.

Hence, stock diameter = 42.82 mm

From Table 7.15, flash width = 5.3 mm, and flash thickness = 1.0 mm

Area of stock with flash = $1440 + 5.3 \times 1 \times 2 = 1450.6$ mm^2

Stock diameter with flash = 42.976 mm ≈ 45 mm

$$\text{Volume of the forging} = 72\,870 \text{ mm}^3 \approx 73\,000 \text{ mm}^3$$

$$\text{Perimeter of forging} = \frac{\pi \times 30}{2} + 33 + 34 + 24 + 78 + \frac{\pi \times 30}{2} + 78 + 33 = 408.25 \text{ mm}$$

Flash volume = $5.3 \times 1 \times 408.25 = 2163.725 \text{ mm}^3$

This being too small, consider the flash to be 10% of forging volume.

Hence, flash volume = 7300 mm^3

Total stock volume = $73\,000 + 7300 = 80\,300 \text{ mm}^3$

Cross-sectional area of stock = 1590 mm^2

Length of stock = $\dfrac{80300}{1590} = 50.50 \approx 55$ mm

The shape of the gutter in the finishing impression would be as shown Fig. 7.34(c).

Fullering Impression

The first operation to be done in drop forging is to draw-out the stock at the necessary portions by striking the metal with sharp and rapid blows. The amount of drawing-out depends on the differences in the maximum and minimum cross sections of the component. If the drawing-out is large, an auxiliary helve hammer is used with standard round and square dies for drawing-out. After this drawing-out operation, the stock is then taken to the drop-forging die.

For smaller drawing-out applications, the fullering impression is made along with the other impressions in the same die. The stock is fullered to the required shape in approximately 4 to 5 blows in the fullering impression. The die shape required for fullering is shown in Fig. 7.35. The consideration in designing are to keep the two die-halves away from completely closing and to allow for the smooth flow of metal.

The gap between the crests of the fullering impressions in the two die halves is kept apart a distance of 1.5 to 5.0 mm less than the minimum area of the component. This reduction is made to ensure that the two die halves do not come into physical contact and thus provide cushioning effect.

The length of the impression could be either 0.25% of the fullered length or half of the stock length without flash and tong hold, whichever is smaller. The distance between the relief portions of the impressions should be twice the diameter of the stock.

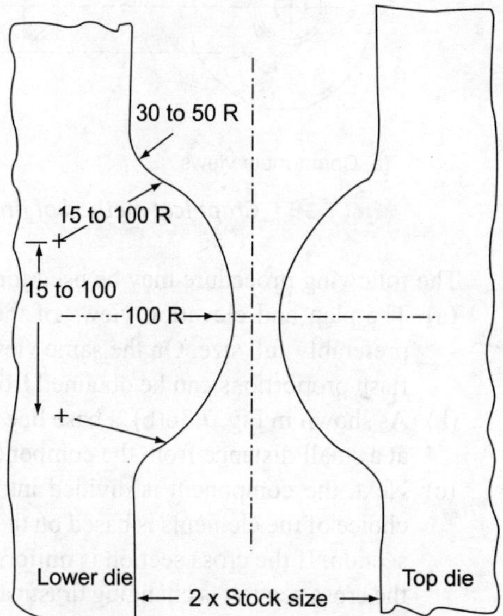

Fig. 7.35 *Fullering impression*

Edging Impression

The edging impression or preform, gathers the material as required in the final forging. As explained earlier this is the most important impression in a die. This gathering helps in the proper flow of metal and complete filling of the die cavities in the later impressions. The preform shape also helps in proper location of stock in the blocking impressions.

For irregular shapes with large variation in cross section, and which are extra-long compared to other dimensions, it is very difficult to gather the material. In such cases, it would be desirable to have two edging impressions.

In an edging impression, the area at any cross section should be same as that of the corresponding section in the component and the flash allowance. For simple shapes, it can be very easily calculated. For complex shapes or with continuous variation in cross section, it may be desirable to adopt a graphical approach as shown in Fig. 7.36.

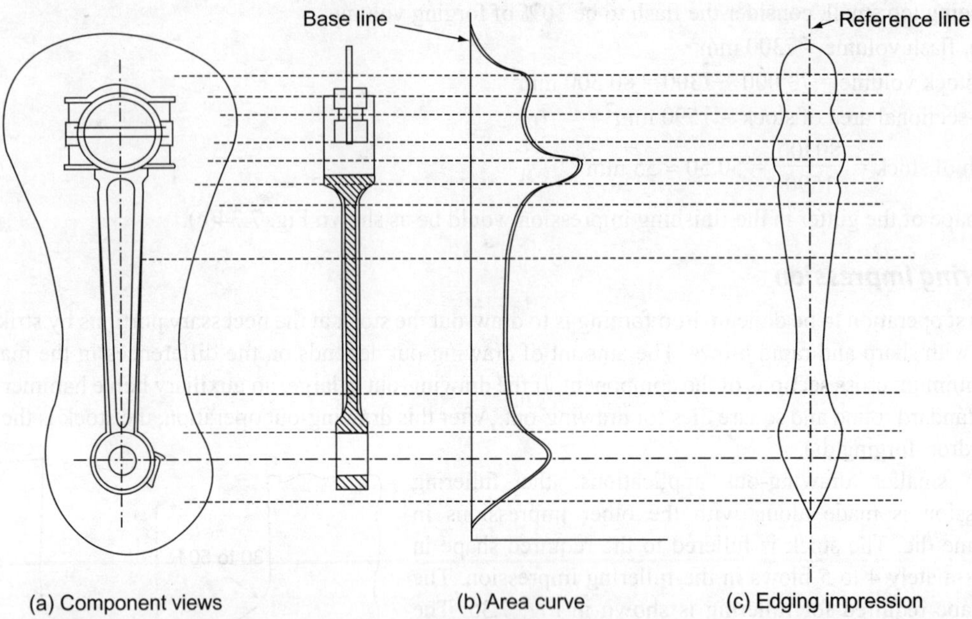

Fig. 7.36 *Graphical method of finding the edging impression of a drop-forged component*

The following procedure may be used for arriving at the preform shape.

(a) The plan and elevation views of the forging should be drawn side by side to a convenient scale, preferably full size. On the same views, the flash outline around the component is to be laid out. The flash proportions can be obtained from Table 7.14.

(b) As shown in Fig. 7.36(b), a base line is to be drawn parallel to the longitudinal axis of the component, at a small distance from the component views in Fig. 7.36(a).

(c) Next, the component is divided into a number of elements, as shown by the horizontal lines. The choice of the elements is based on the geometric shape of the component and the variation in the cross section. If the cross section is uniform over a length then only one section is enough. Whereas, when the cross section is changing drastically then it may be necessary to divide the component into a large number of sections over the change.

(d) The cross section of each of the elements chosen are calculated by simply multiplying the elements in the two views of the component. The areas of these various elements from the base line to any appropriate scale may be plotted as in Fig. 7.36(b). These plotted points are joined with a smooth curve. In this process, if there are any abrupt variations in the areas, then some more sections may be chosen to get a smooth curve.

(e) The flash area provided at each of these elements are calculated and added to the areas already plotted in Fig. 7.36(b).

(f) Having known the cross-sectional area of the component and flash at each of the elemental sections, the radius of preform at these elements given by the formula

$$\text{Radius, } R = \sqrt{\frac{\text{area}}{\pi}}$$

These radius values plotted on either side of a reference line drawn at Fig. 7.36(c) to the same scale as that used for drawing the component views at Fig. 7.36(a), provides an approximate contour that would promote smooth flow of metal into the final forged component.

Blocking Impression

Blocking or semi finishing impression resembles the final shape with liberal radii at corners. No gutter is provided in blocking. The area at each section is roughly 15 to 20% greater. The height of the blocked forging is large and breadth is smaller by an amount of the order of 0.8 to 1.5 mm. The length of the blocking impression remains the same and the centres correspond to that of the finishing impression. The edge and fillet radii are generously provided to aid the flow of metal in the blocking impression. For very complicated shapes with rapid changes in section, deep pockets or thin ribs, it may be necessary to include more than one blocking impression. A typical guideline for the blocking-impression dimensions in terms of finish dimensions is given in Table 7.16.

TABLE 7.16 Blocking impression proportions in terms of corresponding finishing dimensions

Dimension	Aluminium Alloys	Titanium Alloys
Web thickness	1 to 1.5 t	1.5 to 2.2 t
Fillet radius	1.2 to 2 R	2 to 3 R
Corner radius	1.2 to 2 R	2 to 3 R
Draft angle	2 to 5°	3 to 5°
Rib width	W – 0.8 mm	W – (1.6 to 3.2) mm

A few more rules derived from practical experience for blocking impression are summarised below:

- In plan view, the blocking impression is slightly narrower than the finishing impression (about 0.5 to 1 mm on each side), such that the blocked component will fit the finishing impression.
- For forging high ribs in the finishing impression, it is at times necessary to have lower ribs in the blocking impression. The web thickness in the blocking impression is larger than that in the finishing impression.
- In order to enhance metal flow towards the ribs, it is useful to provide an opening taper from the centre of the web towards the ribs.

Finishing Impression

The dimensions of the finishing impression are same as that of the final forging desired with the necessary allowances and tolerances. Gutter should be provided in the finishing impression as detailed earlier. The blocking and finishing impressions for a typical component are shown in Fig. 7.37.

Location of Impressions

The various forging impressions should be located in the die block in such a way that the forging force be as nearer to the centre as possible. This will minimise the likely mismatch of the two die halves, reduce the wear on the ram guides of the drop hammer and will help to maintain the thickness dimensions of the forging. To do this, the operation requiring the maximum forging force (usually blocking or finishing) should be placed at the centre with the other impressions distributed as nearly equal on either side of it in the die.

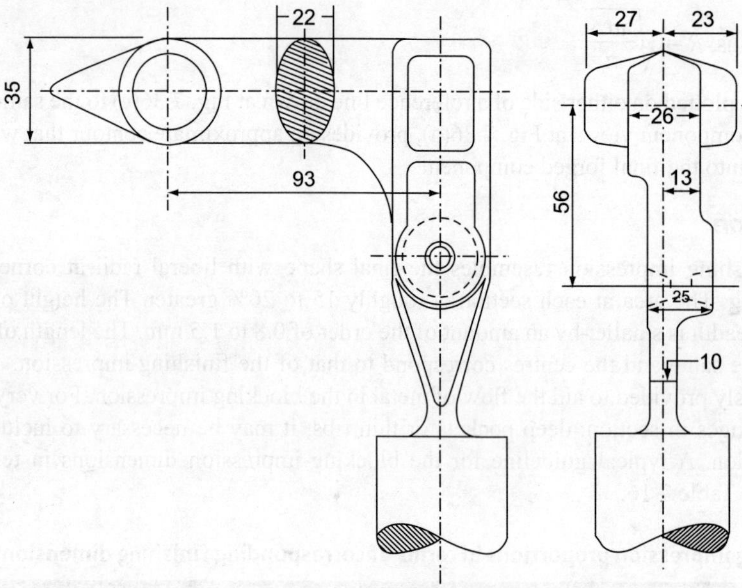

(a) Blocking impression

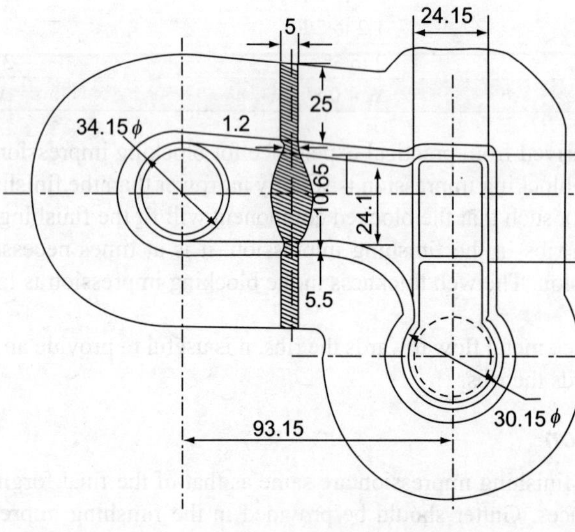

(b) Finishing impression

FIG. 7.37 ***Blocking and finishing impressions of air radius arm***

It is the normal practice to provide the fullering impression on the left-hand side and the edging impression on the right-hand side with the blocking and finishing impressions at the centre.

It is necessary to provide enough clearance of the order of 10 to 15 mm between the impressions in the die. If too little space is provided, then upsetting of the die block is likely to take place, which would decrease the thickness of the final forging.

Die Inserts

Die inserts are generally used for economy in the production of some forgings. Preparing inserts is cheaper than the whole die. Similarly, replacement of an insert is far more economical and less time consuming than the remaking of a complete forging die. Also, it is possible to use hard materials such as stellite or hard alloy steels for inserts, whereas they are not suitable or are very expensive to be of use for a complete die.

Die inserts are of two types. For certain forgings, specific portions of details such as deep pockets which are likely to be worn out quickly are used as inserts. These are called **plug-type inserts**, an example of which is shown in Fig. 7.38. A single plug type insert could be used or a number of inserts could be arranged with varying hardnesses related to the respective wearing tendencies.

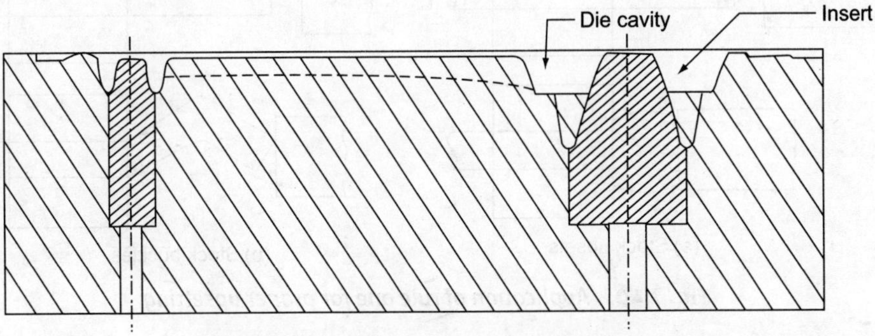

Fig. 7.38 *Plug-type insert in a drop-forging die*

A full insert is one where the complete forging impression is sunk into a harder insert which is then arranged in a softer die steel block (Fig. 7.39). These are generally used for shallow impressions. The main advantage of the full insert is that the same die block could be used for different forgings by changing the inserts. Also, any changes in forging design could be easily incorporated by changing the insert.

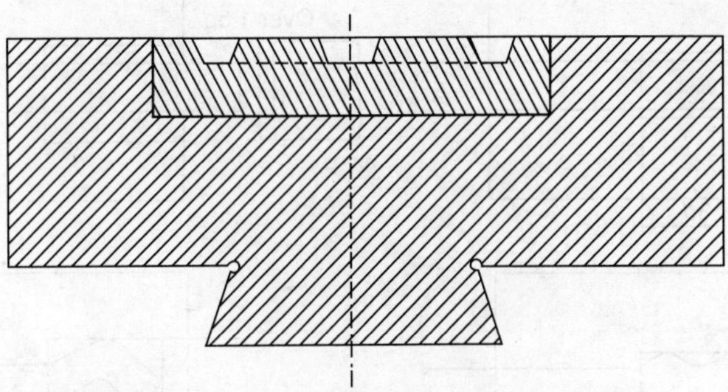

Fig. 7.39 *A full-die insert in a drop-forging die*

7.3.9 Upset-Forging Die Design

In upset forgings, as a rule, no reduction in cross section occurs and, therefore, the stock to be chosen is of the smallest area of cross section of the component. Also, very small or negligible flash is provided in upset forgings.

Depending on the shape of the upsetting to be done, the number of passes or blows in the die are to be designed. The amount of upsetting to be done in a single stage is limited. To arrive at the safe amount of upsetting in a given pass, the following three rules are to be satisfied, to achieve defect free upset forgings.

1. The maximum length of the unsupported stock that can be gathered or upset in a single pass is not more than three times the stock diameter. Beyond this length, the material is likely to buckle under the axial upsetting load rather than be upset, as shown in Fig. 7.40.

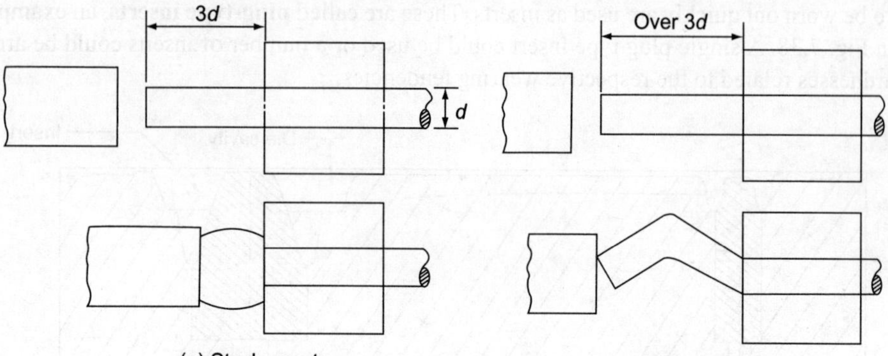

Fig. 7.40 *Application of rule one for proper upsetting*

2. If the stock longer than three times the diameter is to be upset in a single blow then the following conditions should be complied. The die cavity should not be wider than 1.5 times the stock diameter and the free length of the stock outside the die should be less than half the stock diameter. When these conditions are not complied, the stock would bend as shown in Fig. 7.41.

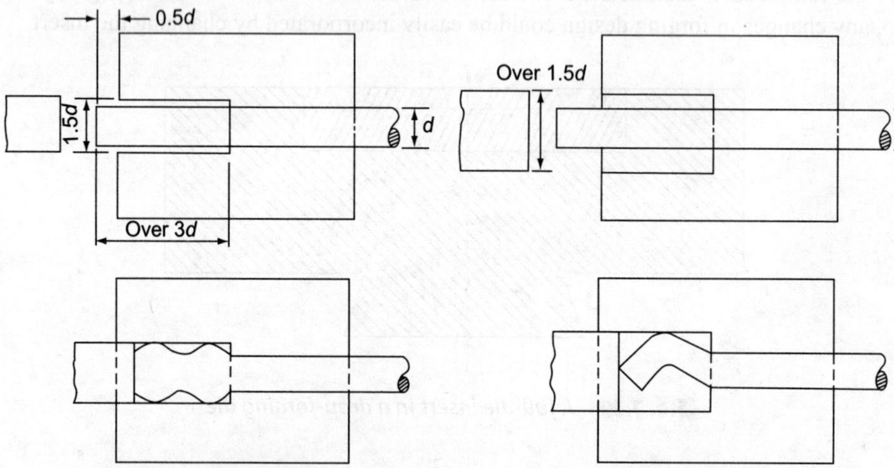

Fig. 7.41 *Application of Rule 2 for proper upsetting*

3. For upsetting the stock which is longer than three times the diameter and the free length of stock outside the die is up to 2.5 times the diameter, the following conditions should be satisfied. The material is to be confined into a conical cavity made in the punch with the mouth diameter not exceeding 1.5 times

the stock diameter and the bottom size being 1.25 times the stock diameter. Also, it is necessary that the heading tool recess be not less than two thirds the length of the working stock or not less than the working stock minus 2.5 times the stock diameter. The application of this rule is shown in Fig. 7.42.

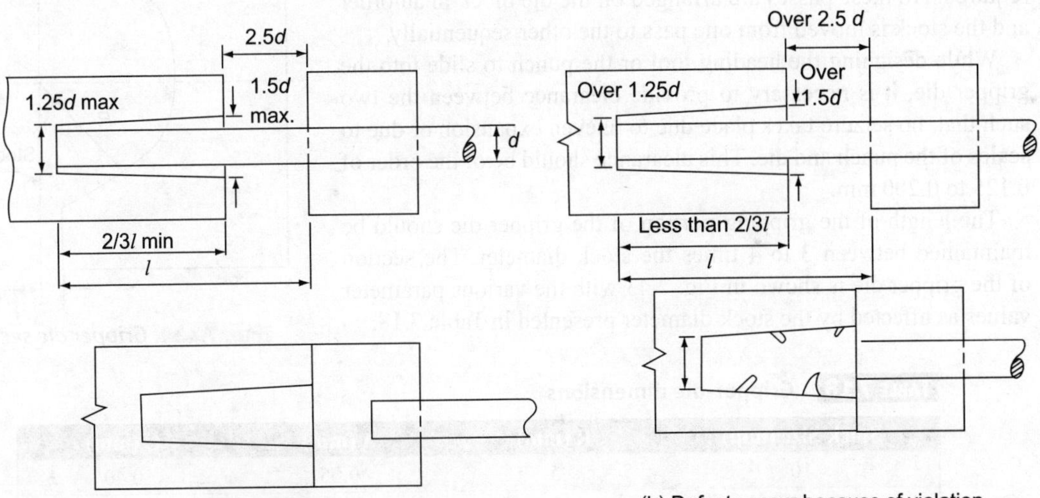

(a) Stock upsets

(b) Defects occur because of violation

FIG. 7.42 *Application of rule three for proper upsetting*

The above are the absolute limits for proper upsetting. But in practice, it may be possible to cross these absolute limits. The condition of the stock end as is cut, would also affect the maximum permissible upset lengths as presented in Table 7.17.

TABLE 7.17 Permissible stock lengths for free upsetting

Condition of Stock End	Angle of Cut (deg)	Shape of Heading Tool End Face	Permissible Length (Stock Diameter, d mm)		
			< 25	50	100
Even	< 1	Hollow	2.4 *d*	—	—
		Flat	2.2 *d*	2.5 *d*	3.0 *d*
		With preforming punch	1.8 *d*	2.0 *d*	2.0 *d*
	> 3	Flat	1.8 *d*	2.0 *d*	2.5 *d*
		With preforming punch	1.25 *d*	1.5 *d*	1.5 *d*
Uneven	< 1	Flat	1.8 *d*	—	—

Location of Cavities

The upset cavities may entirely be kept in the gripper dies, or in heading tool or part in gripper and part in heading tool. The choice of location mainly depends on the severity of the upsetting and the convenient location of flash for trimming. Simple forgings requiring smaller upsetting may contain the complete die cavity in the heading tool. Also complex shapes requiring severe upsetting are obtained with the die cavity completely in the gripper die.

For very simple shapes, one pass may be enough to obtain the final shape, but majority of forgings require three or more passes. If trimming of flash is also to be included, then an additional pass is required. All these passes are arranged on the die block in an order and the stock is moved from one pass to the other sequentially.

While designing the heading tool or the punch to slide into the gripper die, it is necessary to provide clearance between the two such that, no seizure takes place due to uneven expansion or due to scales of the punch and die. This clearance should be of the order of 0.125 to 0.200 mm.

The length of the gripping portion of the gripper die should be maintained between 3 to 4 times the stock diameter. The section of the gripper die is shown in Fig. 7.43 with the various parameter values as affected by the stock diameter presented in Table 7.18.

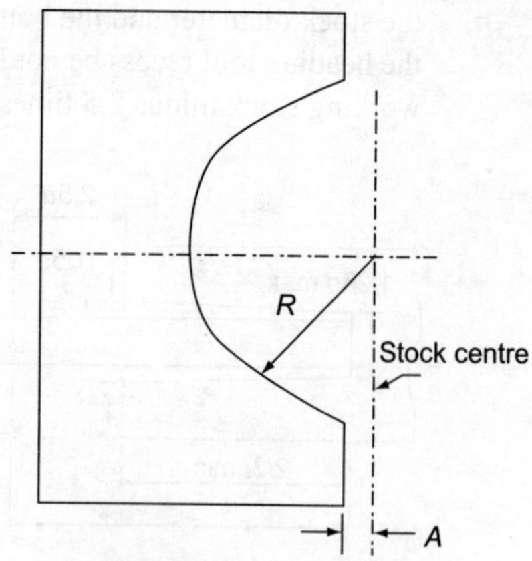

FIG. 7.43 Gripper die section

TABLE 7.18 Gripper-die dimensions

Stock Size (mm)	R (mm)	A (mm)	C (mm)
10	5	0.05	0.40
16	8	0.06	0.40
20	10	0.08	0.40
22	11	0.09	0.40
25	12.5	0.10	0.40
32	16	0.13	0.80
40	20	0.15	0.80
45	22.5	0.17	0.80
50	25	0.18	0.80
65	32.5	0.19	1.20
75	37.5	0.20	1.20

Example 7.2

Design the upsetting die required for the C24 shaft shown in Fig. 7.44.

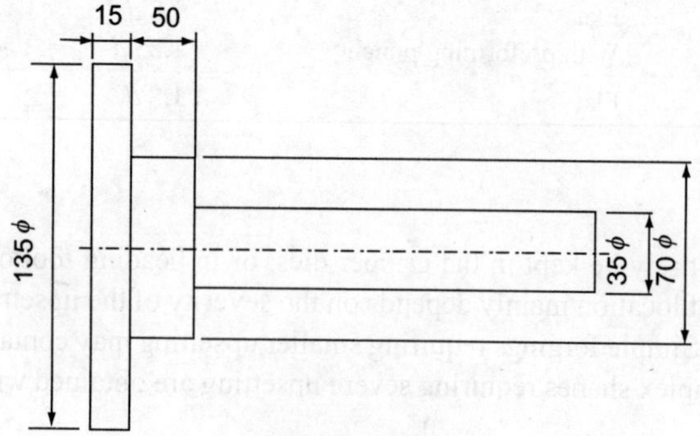

FIG. 7.44 Shaft

Solution The stock diameter is to be 35 mm based on the component shape.

$$\text{Area of stock cross section} = \frac{\pi \times 35^2}{4} = 962.1127 \text{ m}^2$$

$$\text{Volume of upset portion} = \frac{\pi}{4}(135^2 \times 15 + 70^2 \times 50) = 407130.773 \text{ mm}^3$$

$$\text{Stock length} = \frac{407130.773}{962.1127} = 423.16 \text{ mm} \approx 425 \text{ mm}$$

To arrive at the die design, assume that the final shape is to be done in a single stage. Applying Rule 1,

$$\frac{\text{Length of stock to be upset}}{\text{Diameter of stock}} = \frac{425}{35} = 12.1429$$

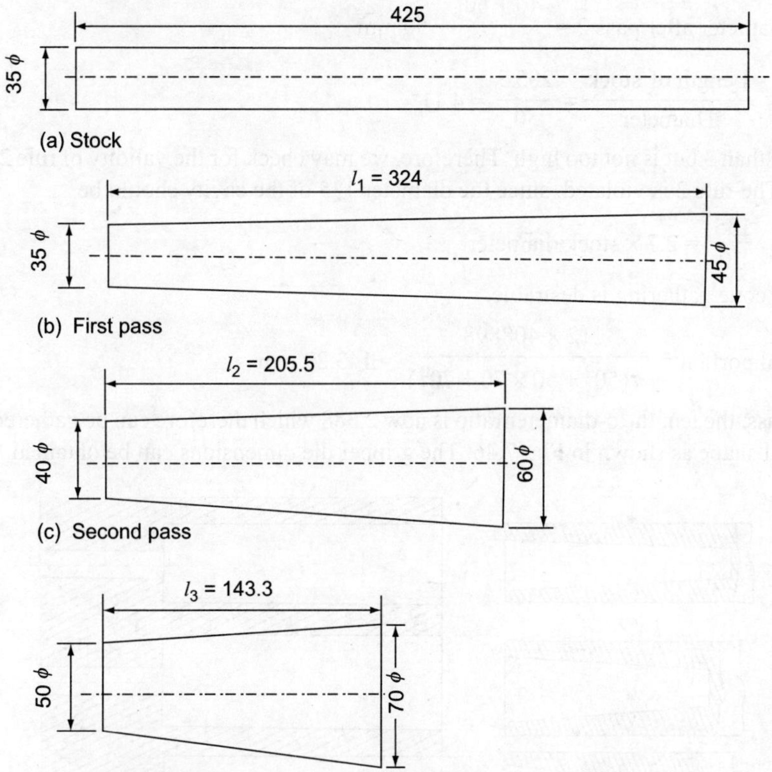

Fig. 7.45 *Various pass sequences for the component*

This is a severe upsetting operation and, therefore, cannot be done in a single pass. Therefore a conical gathering is tried as in Fig. 7.45(b).

$$\text{Stock volume} = \frac{\pi}{4}\, 35^2 \times 425 = 408\,898 \text{ mm}^3$$

$$\text{Length of conical portion} = \frac{12 \times 408898}{\pi(35^2 + 35 \times 45 + 45^2)} = 323.745 \text{ mm}$$

The length of the conical portion is within two thirds of the maximum working length. The unsupported stock beyond the die face is 101 mm which is 2.89 times the stock diameter and is acceptable since it is around 2.5.

The average stock diameter after pass 1 is $\frac{35+45}{2} = 40$ mm

$$\frac{\text{Length of stock}}{\text{Diameter}} = \frac{324}{40} = 8.1$$

This still is high and therefore one more conical gathering pass is desired as shown in Fig. 7.45(c).

$$\text{Length of conical portion} = \frac{12 \times 408898}{\pi(40^2 + 40 \times 60 + 60^2)} = 205.5 \text{ mm}$$

The unsupported length now is 118.5 mm which is 2.96 times the stock diameter and therefore can be acceptable.

$$\text{Average stock diameter after pass 2} = \frac{40+60}{2} = 50 \text{ mm}$$

$$\frac{\text{Length of stock}}{\text{Diameter}} = \frac{205.5}{50} = 4.11$$

This is still more than 3 but is not too high. Therefore, we may check for the validity of rule 2, since it already violated rule 1. The rule 2 is violated, since the diameter 135 of the cavity should be

$$\frac{135}{50} = 2.7 \times \text{stock diameter}$$

Hence one more cone gathering is desirable.

$$\text{Length of conical portion} = \frac{12 \times 408898}{\pi(50^2 + 50 \times 70 + 70^2)} = 143.29 \text{ mm}$$

After the third pass, the length-to-diameter ratio is now 2.388 which therefore can be gathered in a single pass which is the final shape as shown in Fig. 7.46. The gripper die dimensions can be obtained from Table 7.18.

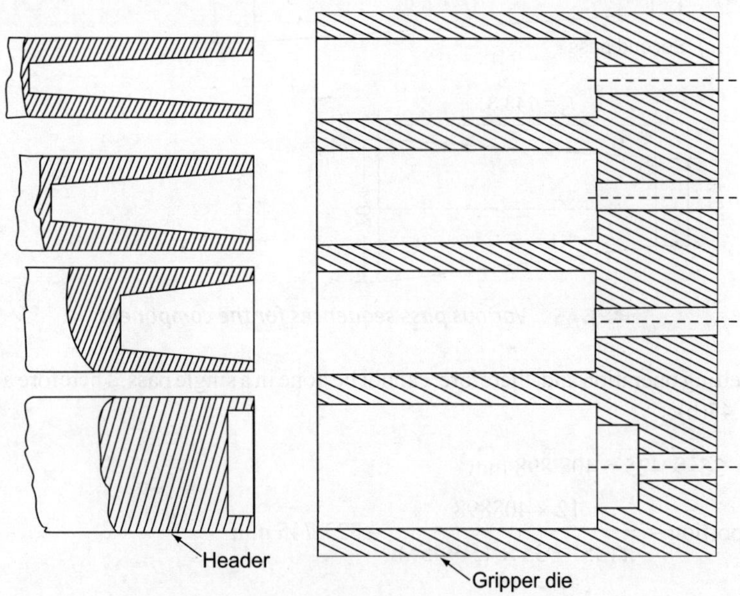

FIG. 7.46 *The die and the punch profiles for the upsetting die of the component shown in Fig. 7.42*

Example 7.3

Design the upsetting tools required for the finished component shown in Fig. 7.47.

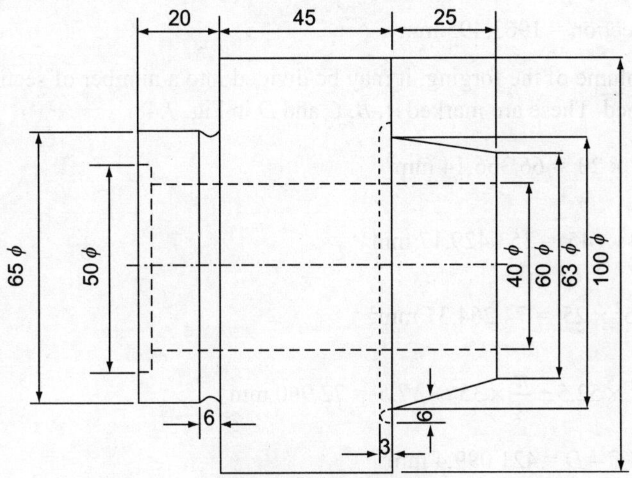

FIG. 7.47 *Example 2*

Solution The forging after providing for the necessary allowances and tolerances is shown in Fig. 7.48.

To find the stock size, we need to know the smallest size of the forging.

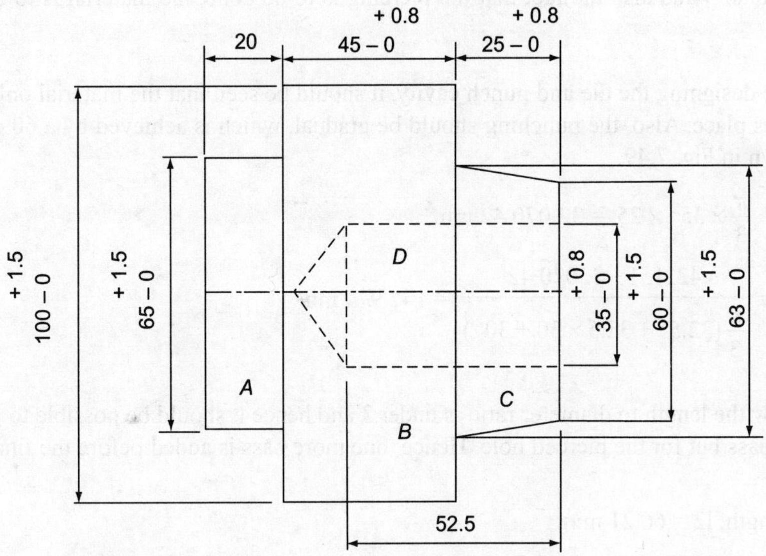

FIG. 7.48 *Forging drawing of component shown in Fig. 7.47*

The smallest cross-sectional area = $\frac{\pi}{4}(60^2 - 35^2) = 1865.32\ \text{mm}^2$

Hence, stock size = 48.73 mm ≈ 50 mm

Stock area of cross section = 1963.495 mm^2

To calculate the total volume of the forging, it may be divided into a number of sections whose volume can be analytically determined. These are marked *A*, *B*, *C* and *D* in Fig. 7.48.

$$\text{Volume of } A = \frac{\pi}{4} \times 65^2 \times 20 = 66\,366.14\ \text{mm}^3$$

$$\text{Volume of } B = \frac{\pi}{4} \times 100^2 \times 45 = 353\,429.17\ \text{mm}^3$$

$$\text{Volume of } C = \frac{\pi}{4} \times 61.5^2 \times 25 = 74\,264.31\ \text{mm}^3$$

$$\text{Volume of } D = \frac{\pi}{4} \times 35^2 \times 52.5 + \frac{\pi}{3} \times 35^2 \times 17.5 = 72\,960\ \text{mm}^3$$

$$\text{Total volume} = A + B + C - D = 421\,099.4\ \text{mm}^3$$

$$\text{Stock length} = \frac{421099.4}{1963.495} = 214.46\ \text{mm} \approx 215\ \text{mm}$$

Total stock volume = 1963.495 × 215 = 422 151 mm^3

Length-to-diameter ratio = 4.3

Since it is more than 3 and also the fact that the piercing is to be done, the material is to be upset in more than one pass.

First pass: While designing the die and punch cavity, it should be seen that the material only gets upset and no extrusion takes place. Also, the punching should be gradual, which is achieved by a 60 deg punch in the first pass as shown in Fig. 7.49.

$$\text{Volume of hole} = \frac{\pi}{3} \times 35^2 \times 25 = 32\,070.42\ \text{mm}^3$$

$$\text{Length of stock} = \frac{422151 + 32070.42}{\frac{\pi}{3}(32.5^2 + 32.5 \times 30 + 30^2)} = 147.974\ \text{mm}$$

Second pass: Now the length to diameter ratio is under 3 and hence it should be possible to get the upsetting done in a single pass but for the pierced hole. Hence, one more pass is added before the final pass as shown in Fig. 7.49.

As before, the length, l2 = 66.21 mm

Third pass: In the third pass, the requisite forging shape is acquired as in Fig. 7.48. The arrangement of the die and the punch cavities is presented in Fig. 7.49.

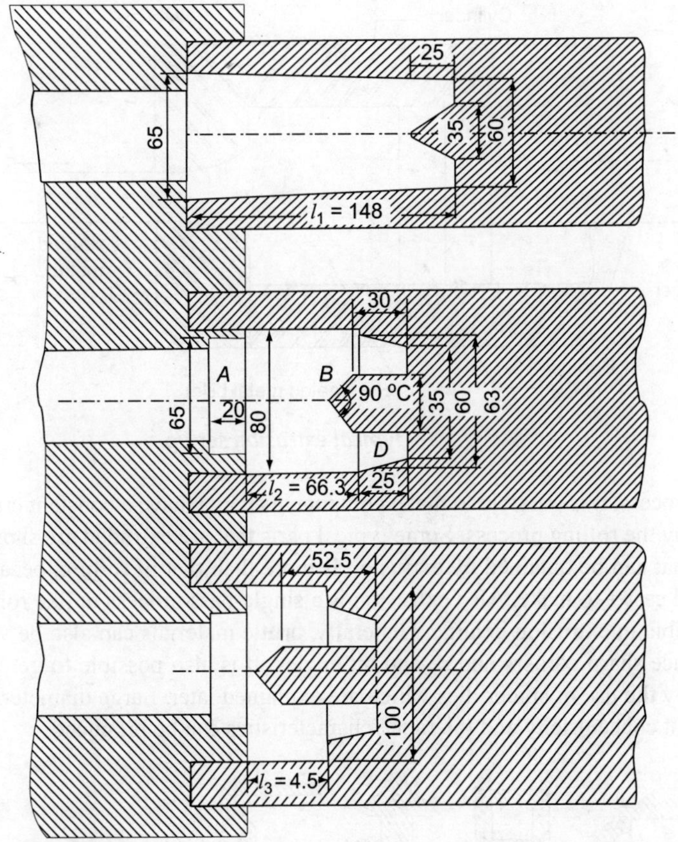

FIG. 7.49 *Arrangement of punch and die cavities for Example 7.3*

7.4 EXTRUSION

Extrusion is the process of confining the metal in a closed cavity and then allowing it to flow from only one opening so that the metal will take the shape of the opening. The operation is identical to the squeezing of toothpaste out of the tooth paste tube.

7.4.1 Extrusion Principle

A typical extrusion process is presented in Fig. 7.50. The equipment consists of a cylinder or container into which the heated metal billet is loaded. On one end of the container, the die plate with the necessary opening is fixed. From the other end, a plunger or ram compresses the metal billet against the container walls and the die plate, thus forcing it to flow through the die opening, acquiring the shape of the opening. The extruded metal is then carried by the metal handling system as it comes out of the die. A dummy block which is a steel disc of about 40 mm (0.50 to 0.75 of diameter) thick with a diameter slightly less than the container is kept between the hot billet and the ram to protect it from the heat and pressure.

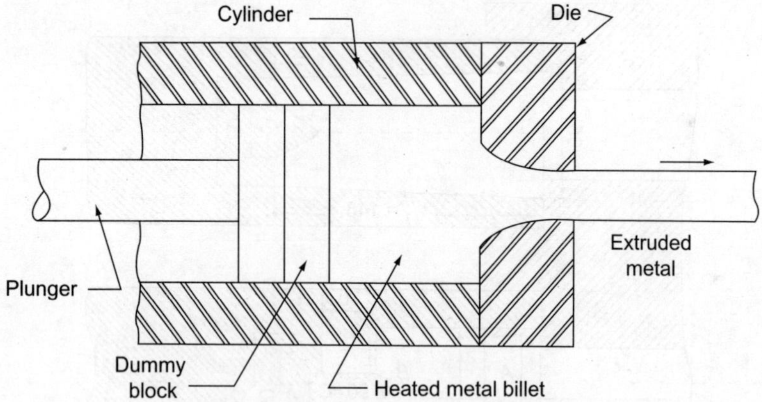

Fig. 7.50 *Typical extrusion set-up*

By the extrusion process, it is possible to make components which have a constant cross-section over any length as can be had by the rolling process. Some typical parts that are extruded are shown in Fig. 7.51. The complexity of parts that can be obtained by extrusion is more than that of rolling, because the die required being very simple and easier to make. Also extrusion is a single-pass process unlike rolling. The amount of reduction that is possible in extrusion is large. Generally, brittle materials can also be very easily extruded. It is possible to produce sharp corners and re-entrant angles. It is also possible to get shapes with internal cavities in extrusion by the use of spider dies which are explained later. Large diameter, thin walled tubular products with excellent concentricity and tolerance characteristics can be produced.

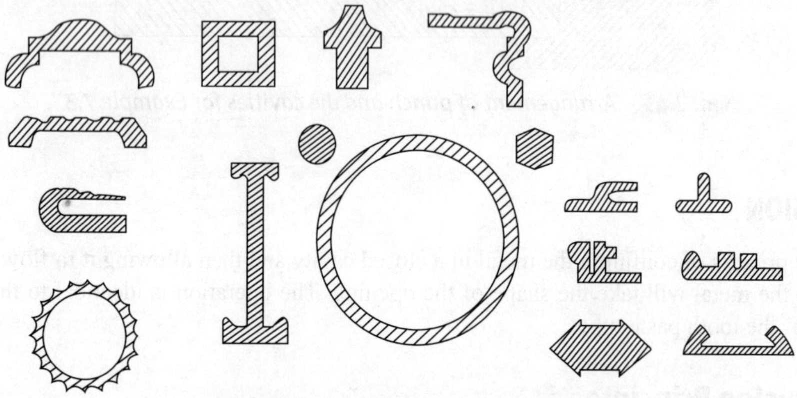

Fig. 7.51 *Typical extrusion shapes*

The flow of the metal in the extrusion process is shown schematically in Fig. 7.52. The extrusion ratio is defined as the ratio of cross-sectional area of the billet to that of the extruded section. The typical values of the extrusion ratio are 20 to 50. Low extrusion ratios are used for intermediate operations when the billets are extruded to a given diameter before the final extrusion. Since hot extrusion involves temperatures in the range of 500 to 1200°C depending on the work material extruded, the cylinder and ram are severely affected by the temperature as well as the stresses. The pressures applied may range from 35 to 1000 MPa. Typical extrusion pressures for various materials are presented in Table 7.19. The extrusion pressure for a given material depends on the extrusion temperature, the reduction in area and the extrusion speed.

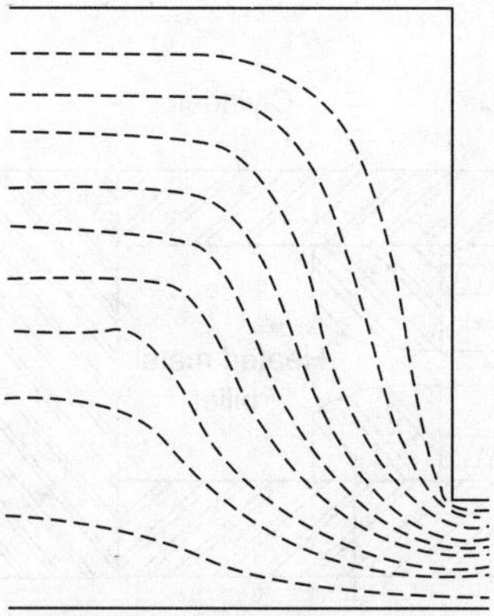

FIG. 7.52 *Flow of metal in extrusion*

TABLE 7.19 Extrusion pressures

Materials	Pressure, MPa
Soft lead	275 to 420
Copper bearing leads	300 to 500
Tin bearing alloys	420 to 620
Tin base alloys	275 to 700
Soft copper alloys	about 200
Hard copper alloys	up to 850
Silicon bronze	950
Aluminium base alloys	70 to 700
Magnesium base alloys	35 to 350
Zinc base alloys	700 to 850

The extrusion speed depends on the work material. Some of the light alloys may be extruded at a speed of 0.05 m/s, whereas for the copper alloys it may be as high as 4.50 m/s. Too high an extrusion speed would cause excessive heat generation in the extruded metal causing lateral cracks.

A classification of extrusion processes is presented in Fig. 7.53.

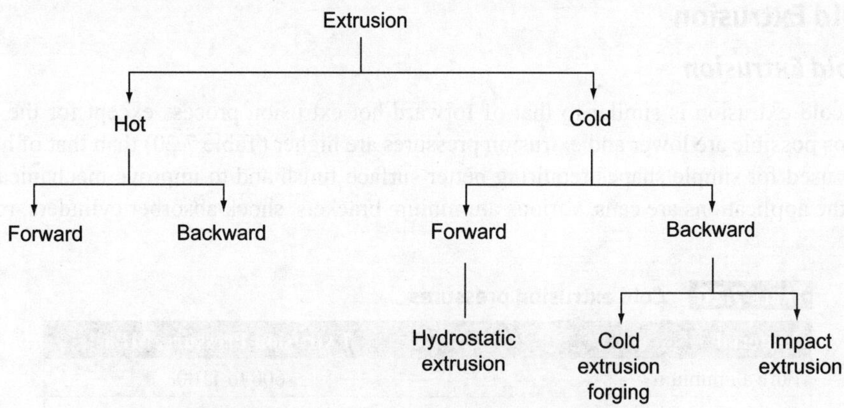

FIG. 7.53 *Classification of extrusion processes*

7.4.2 Hot Extrusion Processes

Forward Hot Extrusion

The process represented in Fig. 7.50 is called the forward hot extrusion, signifying the flow of metal in the forward direction, i.e. the same as that of the ram. In forward extrusion, the problem of friction is prevalent because of the relative motion between the heated metal billet and the cylinder walls. This is particularly severe in the case of steels because of their higher extrusion temperatures. To reduce this friction, lubricants are to be used. At lower temperatures, a mixture of oil and graphite is generally used. The problem of lubrication gets compounded at the higher operating temperatures. Molten glass is generally used for extruding steels. This stays in liquid form at the operating temperature and provides necessary heat insulation to the hot metal billet in addition to lubrication. To reduce the damage to equipment, extrusion is finished quickly and the cylinder is cooled before further extrusion.

Backward Hot Extrusion

In order to completely overcome the friction, the backward hot extrusion, as shown in Fig. 7.54 is used. In this, the metal is confined fully by the cylinder. The ram which houses the die, also compresses the metal against the container, forcing it to flow backwards through the die in the hollow plunger or ram. It is termed backward because of the opposite direction of the flow of metal to that of ram movement. Thus, the billet in the container remains stationary and hence no friction. Also, the extrusion pressure is not affected by the length of the billet in the extrusion press since friction is not involved. The surface quality achieved is generally good since there is no heat cracking due to the friction between the billet and the extrusion cylinder interface. The disadvantage of backward extrusion is that the surface defects of the billet would end up in the final product unlike direct or forward extrusion where these are discarded in the extrusion container. Though advantageous, this process is not extensively used because of the problem of handling extruding metal coming out through the moving ram.

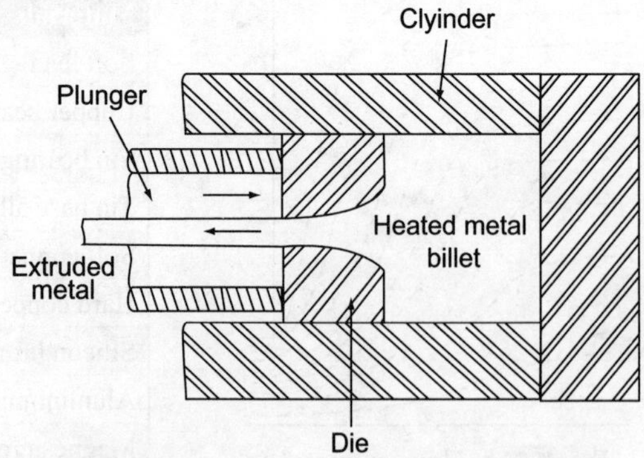

Fig. 7.54 *Backward hot extrusion process*

7.4.3 Cold Extrusion

Forward Cold Extrusion

The forward cold extrusion is similar to that of forward hot extrusion process except for the fact that the extrusion ratios possible are lower and extrusion pressures are higher (Table 7.20) than that of hot extrusion. It is normally used for simple shapes requiring better surface finish and to improve mechanical properties. Examples of the applications are cans, various aluminium brackets, shock absorber cylinders, rocket motors and heads, etc.

Table 7.20 **Cold extrusion pressures**

Material	Extrusion Pressure, MPa
Pure aluminium	600 to 1100
Soft brass	450 to 800
Soft copper	400 to 1100
C10 steel	800 to 2500
C20 steel	900 to 3100

Impact Extrusion

The backward cold extrusion is much more common particularly with softer materials such as aluminium and its alloys. In backward cold extrusion called the impact extrusion, the set up consists of a die and a punch as shown in Fig. 7.55. The slug for making the component is kept on the die and the punch strikes the slug against the die. The metal is then extruded through the gap between the punch and die opposite to the punch movement, as in Fig. 7.55. Because of the impact force, the side walls go straight along the punch though they are not confined. The height of the side walls is controlled by the amount of metal in the slug. This process is more commonly used for making the collapsible tubes for housing pastes, liquids and similar articles.

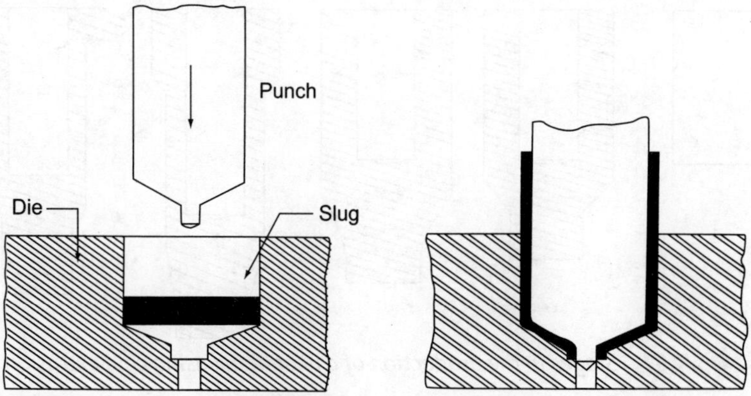

Fig. 7.55 *Impact extrusion*

Cold-Extrusion Forging

The cold extrusion forging is similar to impact extrusion but with the main difference that the side walls are much thicker and their height is smaller. This also contains a die and punch set as shown in Fig. 7.56. The punch slowly descends over the slug kept on the die, thus forging some metal between the punch and the die and the rest being extruded through the clearance between the punch and die side walls. The side walls thus generated are short and thick with any profile in the end unlike the impact extrusion. Afterwards, the component is ejected by means of the ejector pin provided in the die.

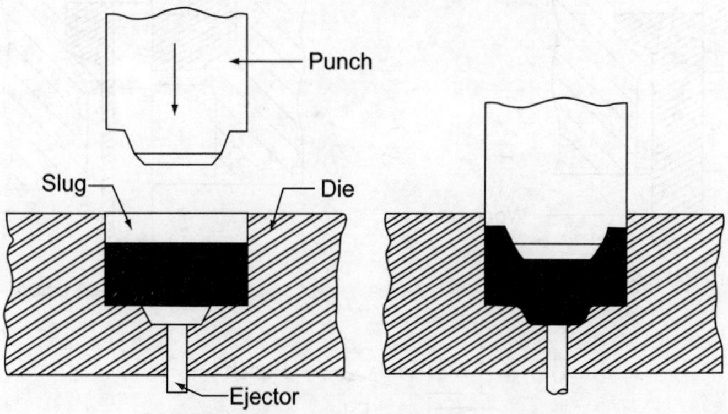

Fig. 7.56 *Cold-extrusion forging*

The backward cold-extrusion processes are different from other extrusion processes in that, each stroke of the punch prepares a directly usable single component which may not necessarily have a uniform cross section over its entire length. Also, these are limited to smaller sizes and for nonferrous alloys only.

7.4.4 Tooling for Cold Extrusion

Cold extrusion is one of the important processes in view of the fact that most of the material is used in forming the final shape and is convenient for automated mass production. The typical example is the body of spark plugs used in internal combustion engines. In the process both the forward as well as backward extrusions are used. A typical example of a gear blank as it is produced in cold forging with its various stages shown in Fig. 7.57.

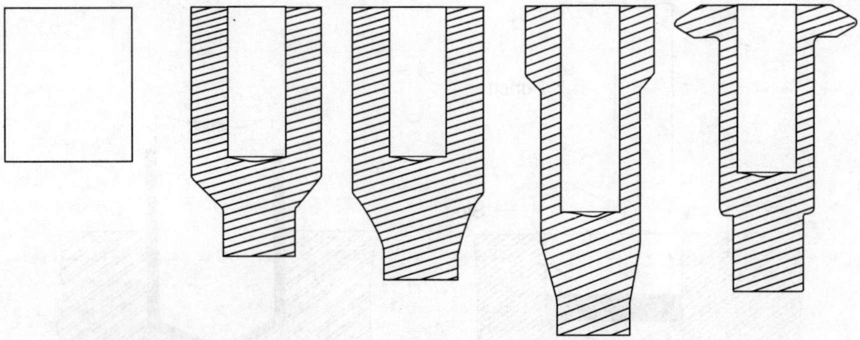

Fig. 7.57 *Stages in the production of a gear blank using cold extrusion*

The shapes that can be successfully cold extruded are the variants of the basic products such as rod, tube and can. The types of dies used for these variants are shown in Fig. 7.58.

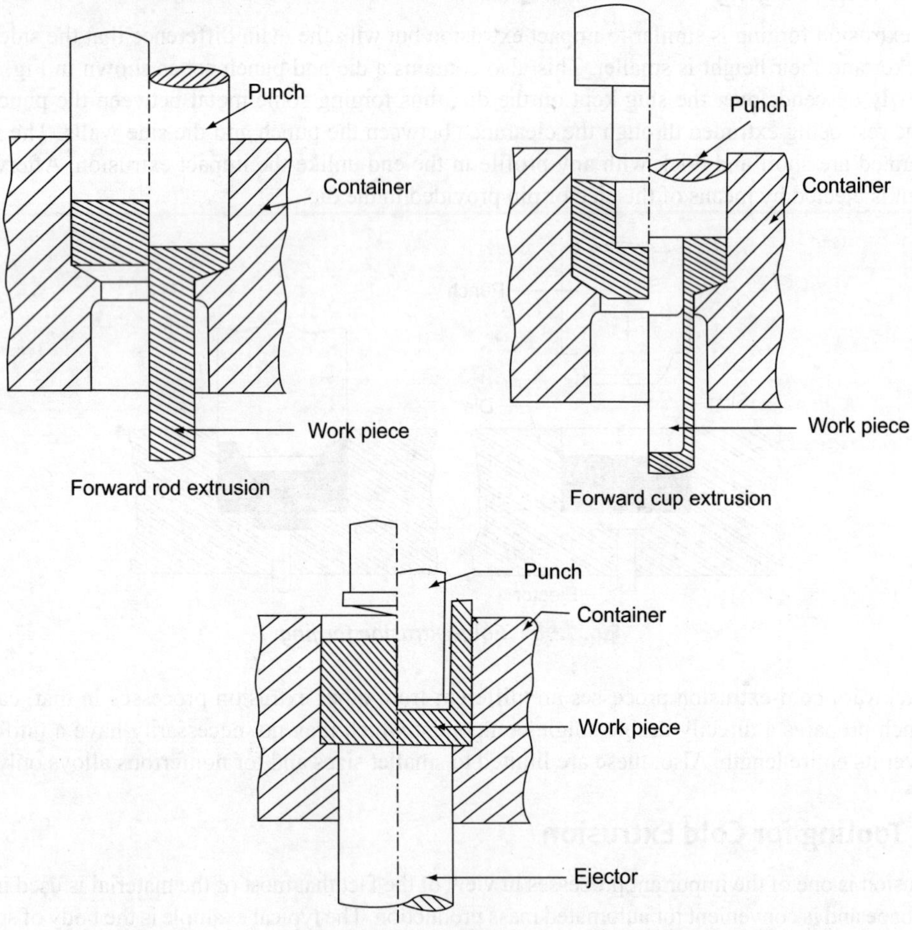

Fig. 7.58 *Typical die and punch shapes for different cold-extrusion processes*

The estimation of extrusion pressure in many cases is done with the help of empirical relationships. A number of relations which agree reasonably close with the experimental results are given in the literature. A typical relationship developed by PERA (Production Engineering Research Association, England) for calculating the maximum pressure for backward extrusion of carbon steels (0.1, 0.2 and 0.3% carbon) is given below:

$$p = \tau\left[3.45 \ln\frac{A_o}{A_b} + 1.15\right] \text{kN/mm}^2$$

where τ = the upper yield point, kN/mm^2,

	0.1% C steel	0.2% C steel	0.3% C steel,
τ, kN/mm^2	0.29	0.31	0.36

A_o = Cross-sectional area of the extruded component, and
A_b = Cross-sectional area of the billet

This expression is valid for extrusion ratios ranging from 1.65 to 4.25 using billets with 0.6 length-to-diameter ratio.

7.4.5 Extruding Tubes

Hollow objects such as tubes and other shapes can also be obtained by forward hot extrusion. One way of obtaining a tube is by means of a solid ram in a double-action press as is shown in Fig. 7.59. First, the solid ram moves through the heated metal billet creating a hole at the centre. Later, the hollow plunger moves the metal billet through the die. Because of the presence of the solid ram very close to the die, the necessary hole is made in the extruded metal.

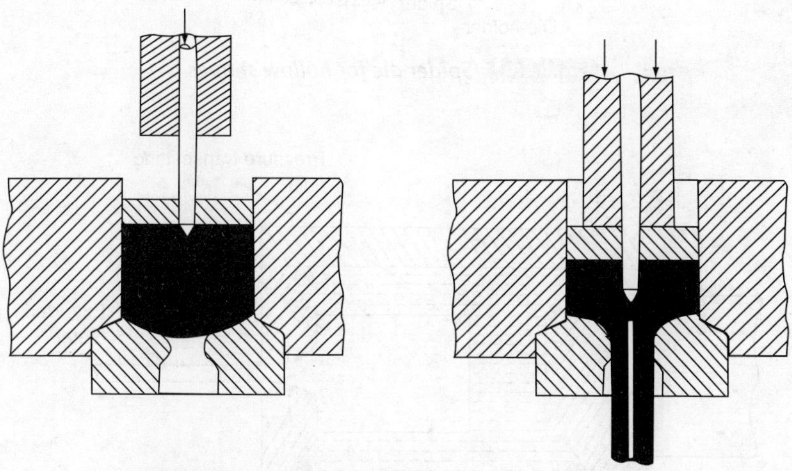

FIG. 7.59 *Extruding tubes in a double-action press*

Another way of obtaining hollow shapes is by the use of a spider extruding die. The spider die is essentially an extrusion die with a stub mandrel, for the hollow portion to be generated. It is held to the die by means of thin ribs simulating the spider legs. The material when extruded flows through the openings between the legs and form as the central opening because of the stub mandrel. The metal flowing out is actually separated but gets welded together since it is still in plastic state. A typical spider-die construction is shown in Fig. 7.60.

Hydrostatic Extrusion

Another extrusion process that is being used for special applications is the hydrostatic extrusion. In this, the metal billet is compressed from all sides by a liquid rather than the ram. The presence of liquid inside the container eliminates the need for any lubricant and also, the material is more uniformly compressed from all sides throughout the deformation zone. Because of this, highly brittle materials such as grey cast iron can also be extruded. A typical hydrostatic extrusion operation is shown in Fig. 7.61. Some of the pressure-transmitting fluids used are castor oil with 10% alcohol, SAE 30 mineral lubricating oil, glycerine, ethyl glycol and iso pentane. The hydrostatic pressure range is from 1110 to 3150 MPa. The commercial applications of the process are limited to the extrusion of reactor-fuel rods, cladding of metals, and making wires of less ductile materials.

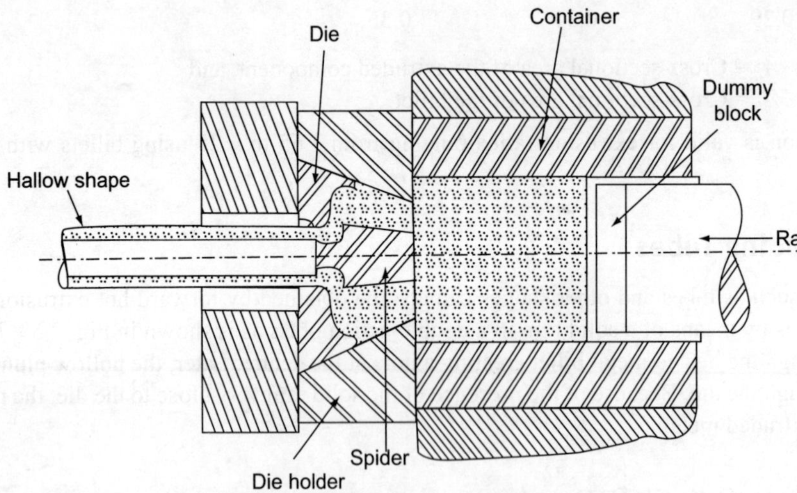

Fig. 7.60 *Spider die for hollow shapes*

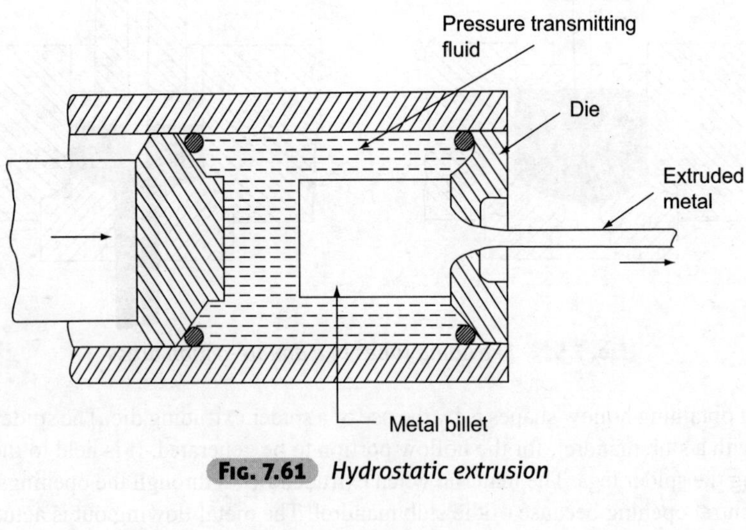

Fig. 7.61 *Hydrostatic extrusion*

7.5 WIRE DRAWING

A wire by definition, is circular with small diameters so that it is flexible. The process of wire drawing is to obtain wires from rods of bigger diameter through a die. Wire drawing is always a cold-working process.

A typical wire-drawing die is shown in Fig. 7.62 and the wire drawing machine in Fig. 7.61. The wire-drawing die is of conical shape. The end of the rod or wire, which is to be further reduced is made into a point shape and inserted through the die opening. This end is then gripped on the other side with a gripper, which would then pull the wire through the die. The wire thus drawn is then coiled round a power reel as shown in Fig. 7.63.

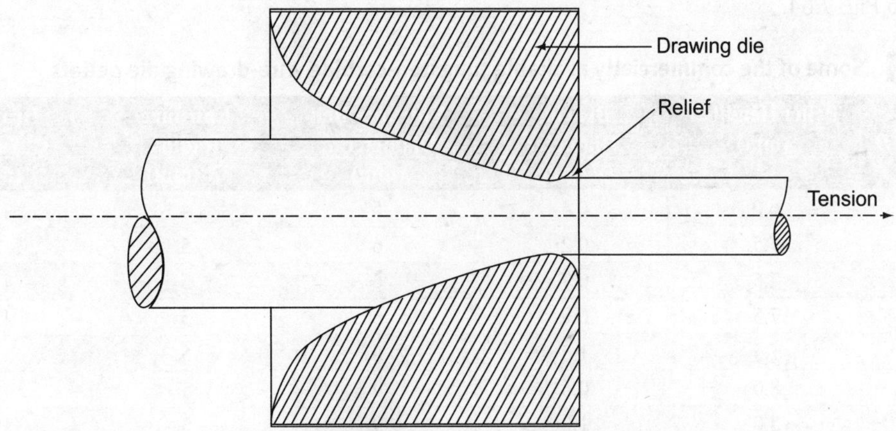

FIG. 7.62 *Wire-drawing die*

Before the wire is drawn, the stock needs to be prepared for wire drawing. The material should be sufficiently ductile since it is pulled by the tensile forces. Hence, the wire may have to be annealed properly to provide the necessary ductility. Further, the wire is to go through the conical portion and then pulled out through the exit by the gripper. In this process, there is no force applied for pushing the wire into the die from the entrance side. To make for an easier entrance of wire into the die, the end of the stock is made pointed to facilitate the entry. This pointing is done by means of rotary swaging or by simple hammering.

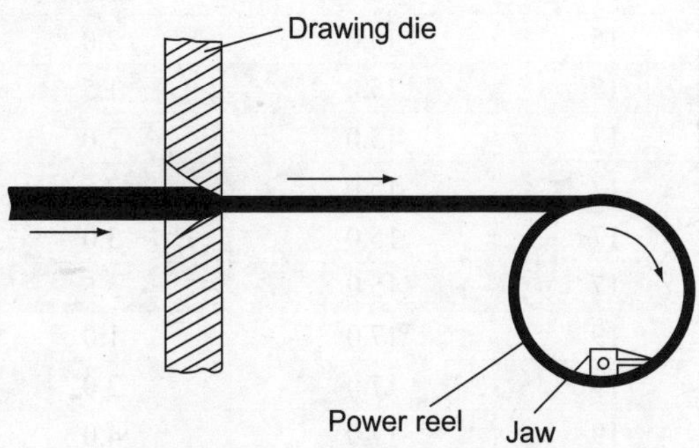

FIG. 7.63 *Wire drawing set-up*

The other aspect of preparation needed is the cleaning of the wire and lubricating it as it flows through the die. Cleaning is essentially done to remove any scale and rust present on the surface which may severely affect the die. It is normally done by **acid pickling**. The pressures acting at the interface of the die and the metal being very high, the lubrication of the die is a serious problem. Therefore, to carry the lubricant through the die, special methods such as sulling, coppering, phosphating and liming are used.

The wire is coated with a thin coat of ferrous hydroxide which when combined with lime acts as filler for the lubricant. This process is called **sulling**. In phosphating, a thin film of manganese, iron or zinc phosphate is applied on the wire which makes the lubricant to stick to the wire, thereby reducing the friction and consequently, the drawing load. Another lubricant vehicle that is used in wire drawing is a coating of lime.

After acid pickling, lime is applied and then allowed to dry. The lime neutralises any amount of acid left on the surface, and adsorbs the lubricant for carrying it to the die. The lubricant normally used is the soap solution. For very thin wires, electrolytic coating of copper is used to reduce friction.

The dies used for wire drawing are severely affected because of high stresses and abrasion. The various die materials that are used are chilled cast iron, tool steels, tungsten carbide and diamond. The cast-iron dies are used for small runs. For very large sizes, alloy steels are used in making the dies. The tungsten-carbide dies are used most commonly for medium size wires and large productions. The tungsten-carbide dies are preferred because of their long life which is 2 to 3 times that of alloy-steel dies. For very fine wires, diamond dies are used. Some of the commercially available tungsten-carbide die sizes are presented in Table 7.21 with reference to Fig. 7.64.

TABLE 7.21 Some of the commercially available tungsten-carbide wire-drawing die pellets

Pellet Diameter, D (mm)	Pellet Height H (mm)	Bearing Diameter d (mm)	Entrance Bell Diameter d_1 (mm)	Entrance Radius R (mm)	Drawing Angle (deg)
8	6.0	0.4	5	4	10
10	7.5	0.2	6	5	8
10	7.5	0.7	6	5	12
10	7.5	0.9	6	5	12
10	7.5	1.3	6	5	12
13	8.0	0.6	9	5	12
13	8.0	1.0	9	5	12
15	13.0	1.0	11	9	14
15	13.0	1.5	11	9	14
15	13.0	2.0	11	9	14
17	15.0	1.5	13	10	14
17	15.0	2.0	13	10	14
17	15.0	2.5	13	10	14
17	15.0	3.0	13	10	14
17	15.0	3.5	13	10	14
19	17.0	1.0	15	11	14
19	17.0	2.0	15	11	14
19	17.0	4.0	15	11	16
22	18.0	2.0	18	12	16
22	18.0	5.0	18	12	16
22	18.0	6.0	18	12	16
25	20.0	4.0	21	13	16
25	20.0	5.0	21	13	18
25	20.0	6.0	21	13	18
25	20.0	7.0	21	13	18
30	24.0	5.0	26	14	16
30	24.0	7.0	26	14	18
30	24.0	9.0	26	14	18

Wire drawing improves the mechanical properties because of the cold working. The material loses its ductility during the wire-drawing process and when it is to be repeatedly drawn to bring it to the final size, intermediate annealing is required to restore the ductility.

The drawing machines can be arranged in tandem so that, the wire coming from one die is coiled up to a sufficient length before it is re-entered into the subsequent die, and so on. This coiling of sufficient wire helps for any discrepancy in the speed of wire drawing in any die. Since there is no change in volume, successive drawings have to be done at higher speeds.

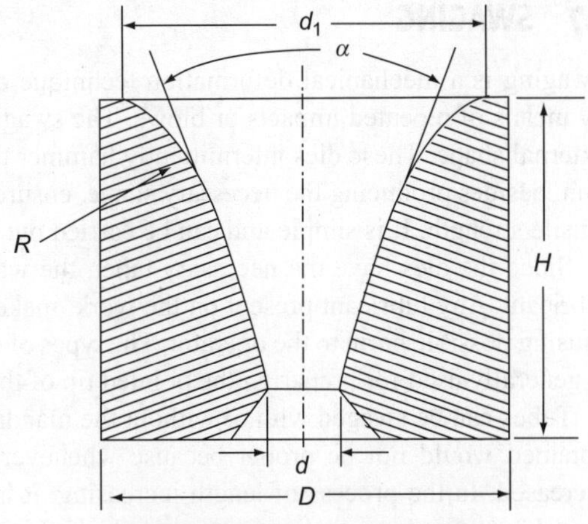

Fig. 7.64 *Wire-drawing pellet*

7.6 ROD AND TUBE DRAWING

Rod drawing is similar to wire drawing except for the fact that the dies are bigger because of the rod size being larger than the wire. But the rod drawn in coiled form is to be straightened and then cut into proper lengths.

For larger size stock, called **bars**, the heavy equipment which generally keeps the drawn product straight is used since bar cannot be coiled. The straight drawing equipment consists of a table which contains rollers on which the bar stock is fed into the die head after pointing. The point of the bar is then pulled out through the die and put on a carriage. The carriage will be somewhat like an endless chain with grips, which grasp the metal protruding from the die and pull along as the chain moves. The bars coming out of the die are generally of short lengths, so that when they are completely drawn, they can be transported to other place from the draw bench.

Tube drawing is also similar to the other drawing processes. The main difference is that it requires a mandrel of the requisite diameter to form the internal hole as shown in Fig. 7.65. The tubes are also first pointed and then entered through the die where the point is gripped in similar way as the bar drawing and pulled through in the form desired along a straight line. There may be more than one pass required to get the final size and when the final size is obtained, the tube may be annealed and straightened.

The practice of drawing tubes without the help of an internal mandrel is called **sinking**.

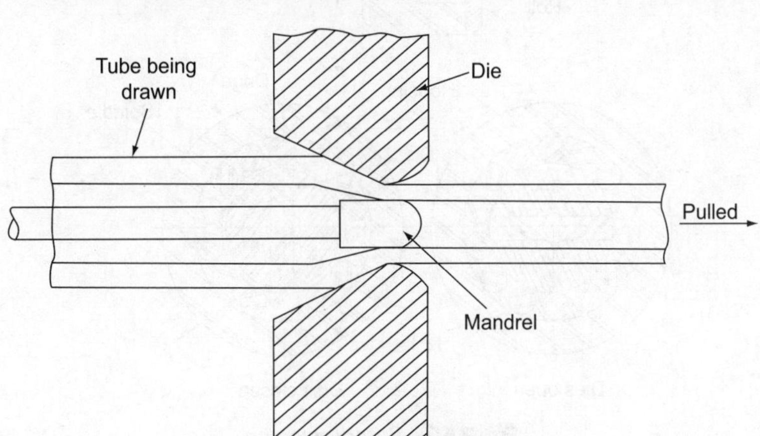

Fig. 7.65 *Tube drawing with a mandrel*

7.7 SWAGING

Swaging is a mechanical deformation technique of reducing or shaping the cross section of rods or tubes by means of repeated impacts or blows. The swaging process consists of dies which are given the requisite external shape. These dies intermittently hammer the stock to produce the deformation. This hammering action, besides producing the necessary shape, ensures good surface qualities, better grain structure and higher tensile strength. It is simple and can be carried out by any unskilled operator.

Since the dies have the necessary taper, the work should be completely clean and dry and without any lubricant. Any lubricant present on the work, makes it to slip under the tapered portion and feed backwards causing possible harm to the operator. The types of surfaces generated are external taperings or contours. This is generally used for preparing the pointed tip of the wire before it is put into the wire-drawing die.

Tubes can be swaged with or without the mandrels. In the case of swaging without mandrels, the surface obtained would not be proper because whenever a blow is applied, the diameter is reduced and length increased. In the process of length increasing, it is possible that the thickness may also increase, in which case the required inside diameter cannot be had and hence, a mandrel is desired.

Rotary swaging is the operation where the two dies which are free to move radially are held in a spindle which rotates continuously. On the internal side of the die is the shape to be obtained whereas the outside is simply cylindrical as shown in Fig. 7.66. The dies are free to move radially and held in position with the spindle by means of a retainer plate fixed to the housing. In between the housing and the spindle is a cage similar to the ball-bearing cage, which houses the rolls. As the spindle moves, the dies float radially in the spindle, the movement being controlled by the rolls in the cage. Whenever the rolls come to the top of the die a blow is given to the work, and when the die comes between the two rolls the die is opened. The work is fed between the dies continuously and for every rotation, the dies would hit the work as many times as the number of rolls. Rotary swaging is able to produce external surfaces which are axi-symmetric.

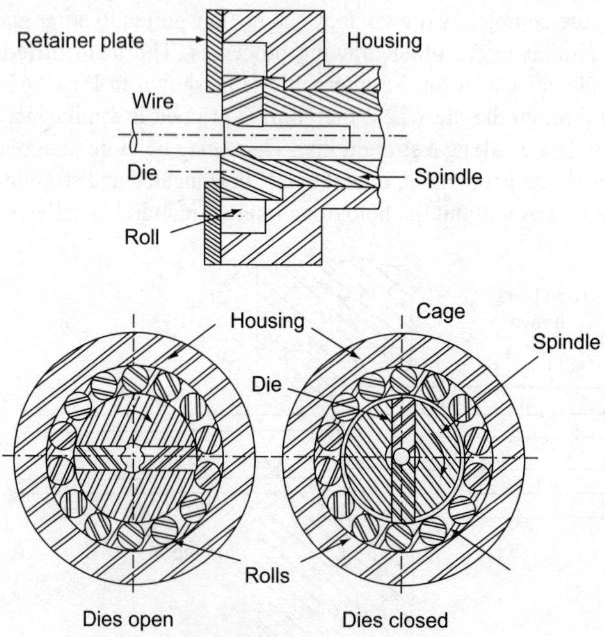

FIG. 7.66 *Rotary swaging*

7.8 TUBE MAKING

To obtain seamless tubes, extrusion can be used as described earlier. It is also possible to obtain seamless tubes by a variation of rolling called roll piercing. Here, the billet or round stock is rolled between two rolls, both of them rotating in the same direction with their axes at an angle of 4.5 to 6.5 deg as shown in Fig. 7.67. These rolls have a central cylindrical portion with the sides tapering slightly. There are two small side rolls, which help in guiding the metal.

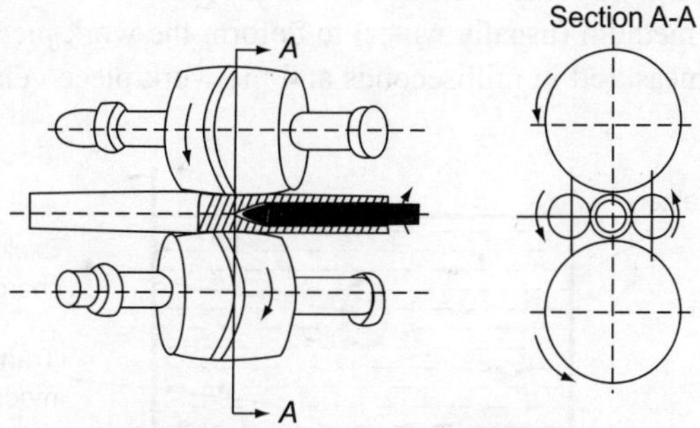

Fig. 7.67 *Roll-piercing mill arrangement*

Because of the angle at which the roll meets the metal, it gets in addition to a rotary motion, an additional axial advance, which brings the metal into the rolls. This cross-rolling action makes the metal friable at the centre, which is then easily pierced and given a cylindrical shape by the central piercing mandrel.

The tube obtained in the roll-piercing mill is further processed in a plug mill, as shown in Fig. 7.68, to obtain the desired size. Plug mill is usually a two-high reversing stand. It contains a central mandrel to form the tube inner diameter.

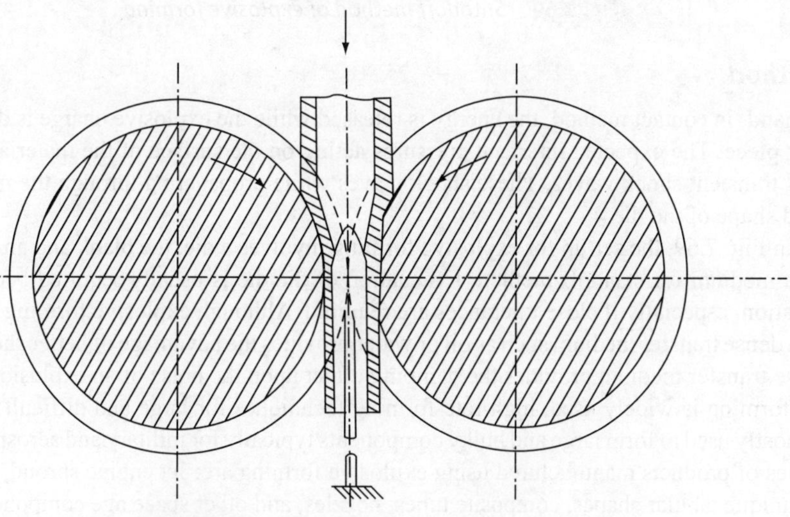

Fig. 7.68 *Principle of tube rolling in a plug mill*

Other method of making tubes is from sheet which is bent into round shape and then seam welded. The details of this process would be discussed in Chapter 9.

7.9 EXPLOSIVE FORMING

It is a very high-energy-rate manufacturing process also called **High Energy Rate Forming (HERF)**. In explosive forming, the explosive is employed to generate shock waves to directly deform the work piece at very high velocities (high strain rates). There are two types of possibilities:

Standoff Method

In this method, the explosive is at a specified distance from the work piece while the energy from the explosive is transmitted through a medium (usually water) to deform the work piece as shown in Fig. 7.69. The working times are normally measured in milliseconds and the work piece velocities are measured in terms of 100 m/s.

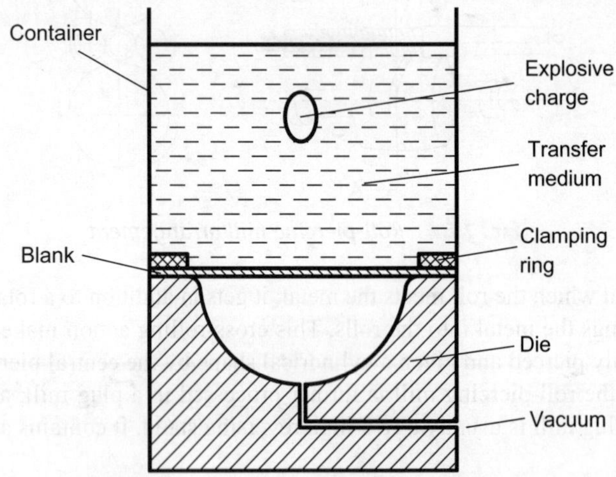

FIG. 7.69 *Satndoff method of explosive forming*

Contact Method

On the other hand, in contact method, the energy is released while the explosive charge is directly in contact with the work piece. The expected interface pressures acting on the surface of the material will experience high intensity, transient shock waves. These shock waves while propagating through the material deform it to the required shape of the die.

As shown in Fig. 7.69, the set-up for explosive forming involves a die, the blank and an explosive charge contained in a medium for standoff method. Evacuation of the die is usually required because of the high speed of operation, especially if close tolerances are required. Although explosive forming is possible in air, doing it with a dense transfer medium such as oil or water provides the advantage of better shock transmission. Also, the dense transfer medium provides the damping effect upon the noise of an explosion.

Explosive forming is widely used, high-rate-forming techniques for large and difficult geometries. This process was mostly used to form large and bulky components typically for military and aerospace applications. A few examples of products manufactured using explosive forming are: Jet engine shroud, Ariane V Rocket frame, many unique tabular shapes, composite tubes, nozzles, and other space age components.

Advantages

- Low tooling costs
- Simpler tooling as only one-sided tooling is required.

- Unlimited power (size)
- Extended formability and unique forming schemes
- Suitable for low-quantity production

Typical alloys that are used for explosive forming are aluminium alloys, Titanium and Ti alloys and nickel based super alloys

7.10 THREAD ROLLING

Thread rolling also called as thread milling is a cold-working process used to form the outer surface of round work pieces to form the threads. It utilises two rotating rolling dies with the external profile of the thread to exert dynamic force on the surface of the work piece. As the thread form on the rolling die is pressed into the surface of the work piece the thread form is transformed all the way to the root diameter of the thread. The thread rolling arrangement is shown in Fig. 7.70.

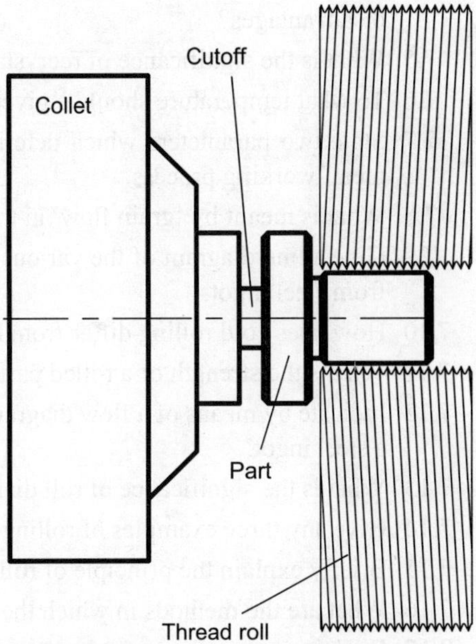

Fig. 7.70 *Thread-rolling operation*

Advantages

- Because of the cold-working process, the hardness of the thread is increased.
- Because of the cold-forming process, the fibre flow lines follows the thread profile in such a way that it has superior resistance to stripping compared to a cut thread.
- Excellent surface smoothness is achieved because of the smooth-forming process.
- Excellent dimensional accuracy is achieved.
- Economical when produced in relatively large-volume production batches.
- Any metal or alloy can be thread rolled except some very highly brittle materials such as cast iron or extremely soft materials such as lead.

SUMMARY

Metal-working processes create less scrap, and provide improved mechanical properties.

- Rolling is a method of producing constant cross section by passing the metal through cylindrical rolls. This is an economical process for producing structural shapes, and by far the largest tonnage of metal is produced using rolling processes
- Forging ensures directional properties be obtained by controlling the fibre-flow directions. As a result many engineering components that are required to withstand heavy forces are produced using forging process.
- Extrusion is similar to rolling in producing constant cross section shapes. However extrusion allows the production of complex shapes by using minimum number of passes.

Questions

7.1 List the various processes of metal working.
7.2 What are the advantages of hot working over cold working of metals?
7.3 What are specific merits of cold working over hot working?
7.4 For hot working it is often necessary to heat the work piece in a furnace and there are scale losses and other problems. Why is hot working sometimes preferred to cold working in spite of such disadvantages?
7.5 What is the significance of recrystallisation temperature in metal working?
7.6 To what temperature should a given plain-carbon steel be heated for hot working?
7.7 State two parameters which determine the extent of deformation possible in a given material in a metal-working process.
7.8 What is meant by 'grain flow' in the case of forged or rolled components?
7.9 Give a line diagram of the various processes involved in the manufacture of the steel sheet, starting from steel ingots.
7.10 How does cold rolling differ from hot rolling in terms of the process and product?
7.11 Why is the strength of a rolled part considered usually better than a cast piece?
7.12 Indicate by means of a flow diagram the different stages in manufacture of 50 mm diameter rod from a steel ingot.
7.13 What is the significance of roll diameter with reference to the roll-separating force in rolling?
7.14 Give any three examples of rolling-stand arrangements.
7.15 Briefly explain the principle of rolling with a neat sketch.
7.16 What are the methods in which the roll-separating force could be reduced in cold rolling.
7.17 Explain some rolling-stand arrangements involving small diameter rolls.
7.18 Explain what you understand by the terms ingot, slab, bloom and billet.
7.19 How does one choose the material for making rolls?
7.20 Briefly explain the meaning of draught and elongation as related to hot rolling.
7.21 What is meant by break down passes in rolling?
7.22 Briefly explain various methods available for break down passes in rolling. Explain their applications.
7.23 What are the differences in roll-pass sequences for billets and rounds?
7.24 Explain some defects that can be present in rolled products with their remedies.
7.25 Distinguish between open and closed die-forging processes.
7.26 Distinguish clearly between drop forging and press-forging processes with reference to the process and products obtained.
7.27 List the advantages of forging of metals. Why is press forging preferred over hammer-forging process?
7.28 List the different stages in the drop forging process in production of a component such as spanner.
7.29 What do you understand by the term flash in a forging? Explain with the help of a sketch.
7.30 How is upsetting different from fullering in forging?
7.31 Give examples of typical products manufactured by the following forging processes:
(a) Machine forging (b) Drop forging (c) Smith forging

7.32 How do you compare forged components with cast components?

7.33 Why are die inserts used in forging dies? Explain with an example.

7.34 What is counter locking of forging dies? Write the cause and effect of the same.

7.35 What is meant by balancing a die in drop forging? Explain with an example.

7.36 What are the technological sequences of operations for forging a crane hook for best mechanical properties? Sketch the various stages and name the operations.

7.37 What will happen if the fillet and corner radii are not provided in the case of a drop-forging die? Explain with sketches.

7.38 What types of drop forgings would necessitate a blocking impression?

7.39 What is the criterion for choosing the spacing of various impressions on the die block in drop forging?

7.40 The operation requiring the greatest forging force is generally kept at or near to the centre in drop-forging dies. Explain the validity of the statement.

7.41 How is the stock requirements determined in the case of upset forging?

7.42 How is the parting line chosen in the case of drop-forging operations? Explain with examples.

7.43 What function is served by die inserts?

7.44 What are the differences between the edging (preform) impression and blocking impression in case of drop forging.

7.45 What are the various types of forging methods available to a manufacturing engineer? Explain the application of each of them.

7.46 Explain the operations that are normally employed in forging.

7.47 Explain how forging improves the mechanical properties of components.

7.48 Why is draft provided in a drop-forging die?

7.49 Is it possible to get too complicated shapes by drop forging? Explain your answer with details.

7.50 What are the allowances that are normally provided in forging?

7.51 What is the basis for the design of upset-forging dies? State the various rules.

7.52 What methods are available to a die designer when the stock length is too long compared to its diameter while upsetting?

7.53 What is meant by trimming related to forging components?

7.54 Draw the shape of the parting line and the two die-halves for the following components (Fig. 7.71, 7.72) for drop forging. Explain your choice.

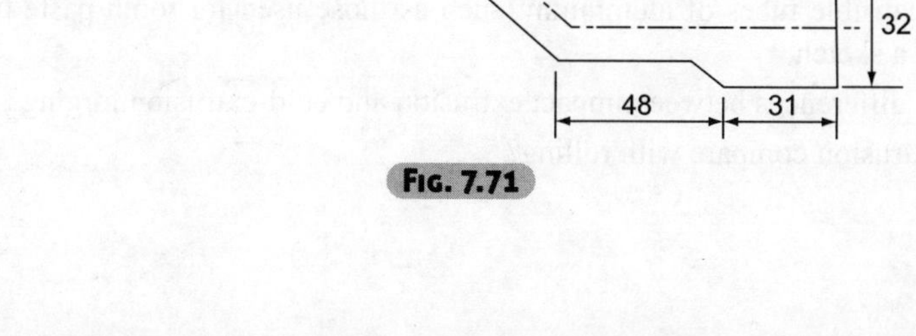

FIG. 7.71

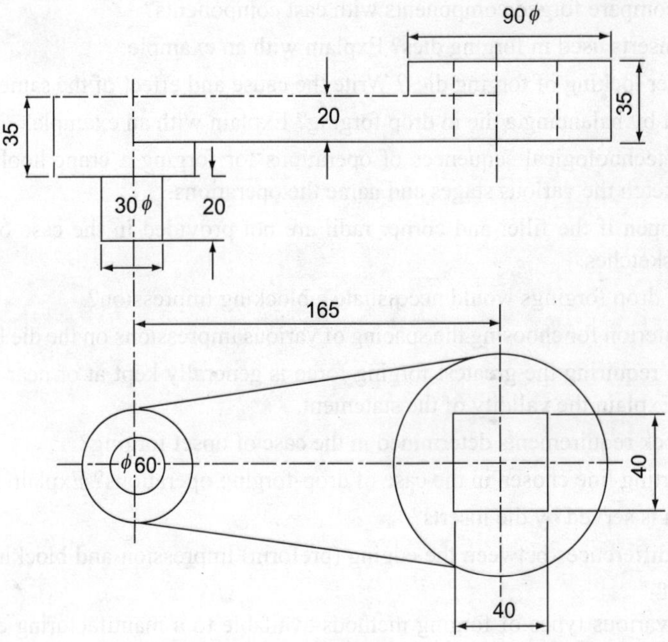

Fig. 7.72

7.55 The operation requiring the greatest is force is generally kept at or near to the centre in drop forging. Explain the validity of the statement.

7.56 What differences exists between drop forging and upset forging with reference to the allowances to be applied?

7.57 Draw the cross section of a drawing die, labelling various features of it.

7.58 What do you expect if during the wire-drawing operation, the reduction attempted is more than the recommended value?

7.59 Explain with sketches the difference between direct and indirect extrusion.

7.60 'Extrusion is a process involving three-dimensional compression'. Explain why brittle materials can be worked by extrusion more successfully than by some other metal-working methods.

7.61 Show by schematic sketches the process of forward and backward extrusion. Give two examples of components produced by extrusion.

7.62 Explain why the whole of the material (billet) put in the container for extrusion is not extruded.

7.63 What are the various materials used for making wire-drawing dies? Explain the application of each of them.

7.64 Why is lubrication difficult in wire drawing? Explain any one lubricating method that is generally employed in wire drawing.

7.65 How are collapsible tubes of aluminium (such as those used for tooth paste tubes) manufactured? Explain with a sketch.

7.66 What are the differences between impact extrusion and cold-extrusion forging?

7.67 How does extrusion compare with rolling?

7.68 How is lubrication done in hot extrusion?

7.69 What are the lubricating methods employed for the backward hot extrusion of steels?

7.70 Distinguish between wire drawing and tube drawing with sketches.

7.71 What are the specific applications of hydrostatic extrusion?

7.72 Explain how the variables that affect the extrusion pressure.

7.73 Briefly describe the method of swaging giving the main applications for the process.

7.74 What do you understand by the term 'roll piercing'? Explain with a simple sketch.

7.75 Is rolling useful for making tubes? Explain your answer with proper sketches.

7.76 Briefly explain the thread-rolling operation.

7.77 Describe the explosive forming process. Give its applications.

Multiple Choice Questions

7.1 Fibre-flow lines seen in the products produced by metal-working processes are
(a) the obstruction that was provided during the deformation process
(b) the direction in which the metal has actually moved during the deformation
(c) not really related to metal deformation process
(d) present in the metal before the deformation process

7.2 Hot working of metal in metal-forming processes is
(a) working below the melting temperature
(b) working below the recrystallisation temperature
(c) working above the recrystallisation temperature
(d) working above the ambient temperature

7.3 Recrystallisation temperature of an alloy is
(a) the approximate minimum temperature at which complete recrystallisation of a cold-worked metal occurs within a specified time
(b) the approximate maximum temperature at which complete recrystallisation of a cold-worked metal occurs within a specified time
(c) the approximate minimum temperature at which complete recrystallisation of a hot-worked metal occurs within a specified time
(d) the approximate maximum temperature at which complete recrystallisation of a hot-worked metal occurs within a specified time

7.4 Recrystallisation temperature of a metal is
(a) decreased by an increase in solute concentration in solid solution
(b) increases by an increase in original grain size before cold working
(c) increased by decreasing temperature of cold working
(d) decreases by a decrease in original grain size before cold working

7.5 A disadvantage of hot working of metals is
(a) any amount of working can be done
(b) surface finish obtained is poor
(c) it requires less force
(d) brittle metals can be worked

7.6 A disadvantage of cold working of metals is
(a) maximum deformation that can be given is limited
(b) dimensional accuracy obtained is less
(c) surface finish obtained is poor
(d) none of the above

7.7 Smaller diameter rolls in rolling are used to
(a) smaller reduction in cold rolling
(b) smaller reduction in hot rolling

(c) larger reduction in cold rolling
(d) for any reduction

7.8 The following steel section is generally a square cross section:
(a) Slab (b) Billet
(c) Bloom (d) Ingot

7.9 For effective descaling the following pass is suitable:
(a) Finishing pass
(b) Box pass
(c) Diamond-square pass
(d) Oval-square pass

7.10 Drawing out is a forging operation used to
(a) reduce the cross-section of a part
(b) increase the cross-section of a part
(c) finish the part
(d) flatten the part

7.11 Drop forging is used to make parts in
(a) open-impression dies that involves only upsetting operation
(b) closed-impression dies that involves only upsetting operation
(c) closed-impression dies that involves only drawing out operation
(d) open-impression dies that involves only drawing out operation

7.12 Machine forging is used to make parts in
(a) open-impression dies that involves only upsetting operation
(b) closed-impression dies that involves only upsetting operation
(c) closed-impression dies that involves only drawing out operation
(d) open-impression dies that involves only drawing out operation

7.13 Edging impression is an operation in drop forging used to
(a) increase the material cross section and decrease the length
(b) gather the exact amount of material required at each cross-section
(c) decrease the material cross section and increase the length
(d) acquires the shape very near to the final one

7.14 Press forging operation is characterised by
(a) larger draft angles than drop forging
(b) use of larger presses compared to drop forging
(c) no tong holds are required compared to drop forging
(d) poor surface finish compared to drop forging

7.15 Counterlocking of dies becomes a necessity
(a) in all press-forging dies
(b) in all drop-forging dies
(c) in all drop-forging dies that have irregular parting line
(d) in all drop-forging dies that have sectional-die construction

7.16 Extrusion is used to manufacture
(a) constant cross sections over its entire length
(b) any object
(c) only some complex cross sections
(d) none of the above

7.17 A lubricant used in forward hot extrusion of steels is
(a) soap solution (b) linseed oil
(c) molten glass (d) copper sulphate

7.18 The process used to make the aluminium collapsible tube used to keep the toothpaste is
(a) cold-extrusion forging
(b) impact extrusion
(c) forward hot extrusion
(d) forward cold extrusion

7.19 The following is always a cold-working operation:
(a) Rolling (b) Wire drawing
(c) Forging (d) Extrusion

7.20 The lubricant commonly used in wire-drawing operation is
(a) soap solution after lime coating
(b) linseed oil
(c) molten glass
(d) copper sulphate

Answers to MCQs

7.1 (b)	7.2 (c)	7.3 (a)	7.4 (a)	7.5 (b)
7.6 (a)	7.7 (c)	7.8 (c)	7.9 (b)	7.10 (a)
7.11 (c)	7.12 (b)	7.13 (b)	7.14 (b)	7.15 (c)
7.16 (a)	7.17 (c)	7.18 (b)	7.19 (b)	7.20 (a)

CASE STUDY

FORGING DEFECTS*

Defects in forged parts are seriously considered while selecting a supplier. Hence, continuous improvement is carried out in the design to eliminate the defects. In order to identify the defects and causes, it is important to analyze various scenarios. Since it is expensive to examine the forgings physically, numerical methods like finite element analysis (FEA) are commonly used. The results are usually validated with a limited number of physical tests and are then extended for untested scenarios.

In the present study, lever forging and yoke forging are examined at M/s Forge Jawor to identify the defects. The lever forging had several defects in the form of folds and under fills. In case of deep and narrow sections, the under fill was dominant. This occurred in locations like the forging head and lever pin. The structure of the plane normal to the crack was found to be hypoeutectoid steel. Non-metallic precipitates were observed on the surface of the forging. This led to further cracks on the surface. On examining the yoke forging, it was observed that the most common defects were folds and under fills. Further, small quantities of the unevaporated coolant collects on the backside of the yoke and increases the pressure. This leads to accelerated micro cracking. Usually, the dies are sprayed with graphite spray for lubrication. Even this can get collected and lead to defects.

Further, the number of cracks on the lower part of the die were more than the upper part. This is due to higher thermal loads on the lower part and remains of unevaporated coolants in the bottom die. Similar to lever, the yolk structure was also found to be hypoeutectoid.

In order to facilitate better and in-depth analysis, numerical simulations were done with FEA method and Forge 2011 software. The software uses 3D CAD models of the die. Forging temperature was assumed to be 1150°C. The work piece was taken as Carbon steel C45 while the die material was tool steel 1.2344 (AISI H13). It was found from the analysis that the location of the material had a significant influence on the quality of the parts. When the billet was moved by about 28 mm, a lap was formed in the lever. Further shifting of the material led to under fill. In the analysis, it was found that the tool shape may need to be modified in order to improve the material flow and reduce the effects of air pockets.

In the analysis of yoke, it was found that folds occur due the overlap of the front and side flash. This effect was also confirmed with the inspection of a forging. Also, the excess of the lubricant may get collected in the backside of the yoke causing increased pressures.

Improper positions of the material and improper flow of the material were found to the major causes for the defects.

---

*Hawryluk, M., & Jakubik, J. (2016). Analysis of forging defects for selected industrial die forging processes. *Engineering Failure Analysis*, *59*, 396-409.

CHAPTER 8

Sheet Metal Operations

Objectives

Sheet metal operations are cold working operations that manufacture low cost parts with very high volumes and at a fast rate. After completing this chapter, the reader will be able to

- Understand the different types of sheet metal operations
- Design sheet metal dies for different applications

8.1 PRESS TOOL OPERATIONS

Many of the consumer goods enjoyed today by the modern man owe their low cost to the press tools. But for the cheap way of making these sheet metal components, we possibly could not have even thought of having automobiles, type writers, mechanical toys, etc., at such a low cost. The press tool operation is by far one of the cheapest and fastest ways to completely manufacture a component. Sheet metal is generally considered to be a plate with thickness less than about 5 mm. In the following pages, important press tool operations are described.

The various operations carried out by press tools have been classified and shown in Table 8.1 on the basis of the type of stresses introduced into the component.

TABLE 8.1 Classification of press tool operations

Stresses Induced	Operations
Shearing	Shearing, blanking, piercing, trimming, shaving, notching, nibbling
Tension	Stretch forming
Compression	Coining, sizing, ironing, hobbing
Tension and compression	Drawing, spinning, bending, forming, embossing

One of the principal concerns in sheet metal operations is the spring back of the metal. When the metal is deformed, it is first elastically deformed and then plastically. When the applied load is removed, the plastic component of the deformation remains permanently while elastic part springs back to its original shape. This can be schematically observed from Fig. 8.1 where the stress strain diagram is shown. In the stress strain diagram, a stress *OA* is applied on the material so that it reaches the point *P* and has a strain of *OB* under the load. When the load is removed, the material springs back to the position *C*, finally with a permanent deformation of *OC* only. The amount *CB* is the amount of spring back. The amount of spring back is the property of the material which depends on the modulus of elasticity. Higher the modulus of elasticity, lower would be the spring back. However, it is very difficult to estimate theoretically, the actual amount of spring back and hence it is to be determined by trial and error method for any given case.

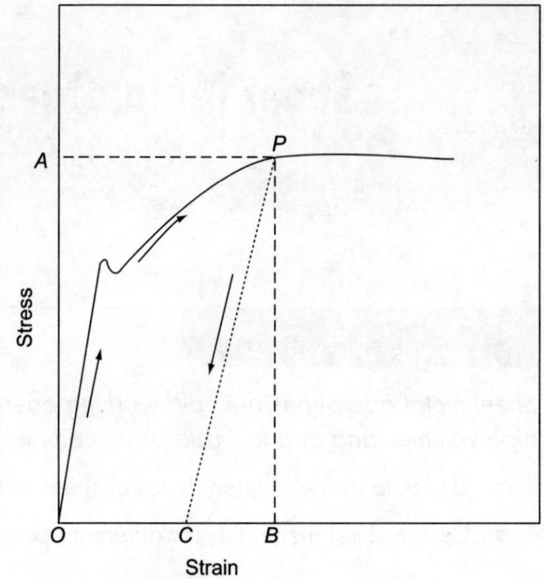

FIG. 8.1 *Spring back in metals*

8.2 SHEARING ACTION

The metal is brought to the plastic stage by pressing the sheet between two shearing blades so that fracture is initiated at the cutting points. The fractures on either side of the sheet further progressing downwards with the movement of the upper shear finally result in the separation of the slug from the parent strip.

The metal under the upper shear is subjected to both compressive and tensile stresses as shown in Fig. 8.2. In an ideal shearing operation, the upper shear pushes the metal to a depth equal to about one third of its thickness. Because of pushing of the material into the lower shear, the area of cross section of the metal between the cutting edge of the shears decreases and causes the initiation of the fracture. This portion of the metal which is forced into the lower shear is highly burnished and would appear as a bright band around the blank lower portion. The fractures which are initiated at both the cutting points would progress further with the movement of the upper shear and if the clearance is sufficient, would meet, thus completing the shearing action (Fig. 8.3).

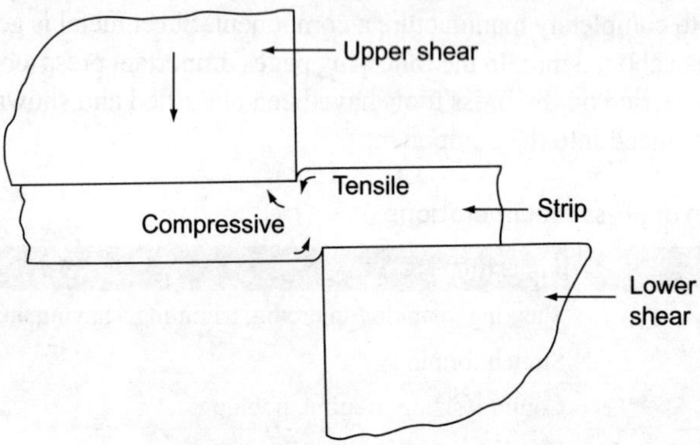

FIG. 8.2 *Stresses in shearing*

The appearance of the cut-edge of the blank is shown schematically in Fig. 8.4. When the plastic deformation starts, the material flows beneath the upper shear and would appear as an edge radius as shown at *A* in the stock. Similar flow of metal at the lower shear would result in the edge radius *B* in the slug. The metal pushed in by the upper shear before the separation would burnish the metal and result in the cut band as shown in *C* in the stock by the upper shear and *D* in the slug by the lower shear.

When correct clearances are used, a clean break would appear as a result of the extension of the upper and lower fractures towards each other. With an insufficient clearance additional cut bands would appear before the final separation. Ductile materials require smaller clearances and longer penetration of the punch compared to harder materials. The effects of clearances are shown schematically in Fig. 8.5.

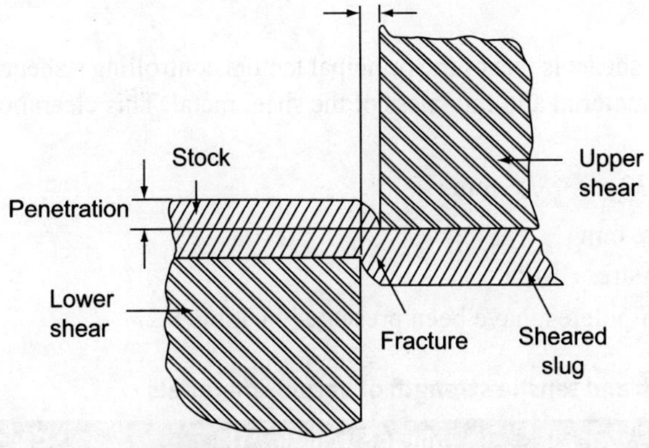

Fig. 8.3 *Material being sheared*

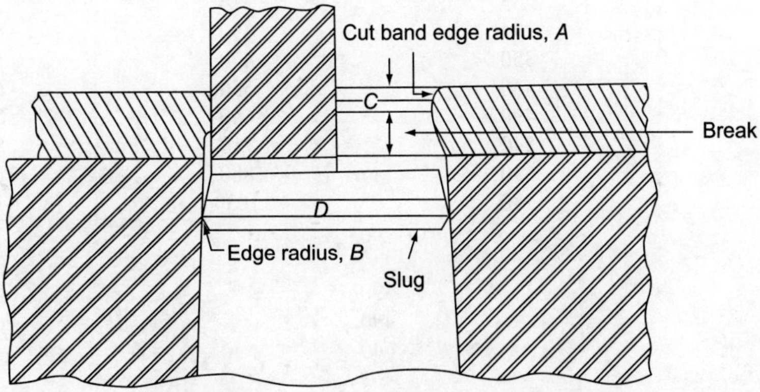

Fig. 8.4 *Characteristic of cut in shearing*

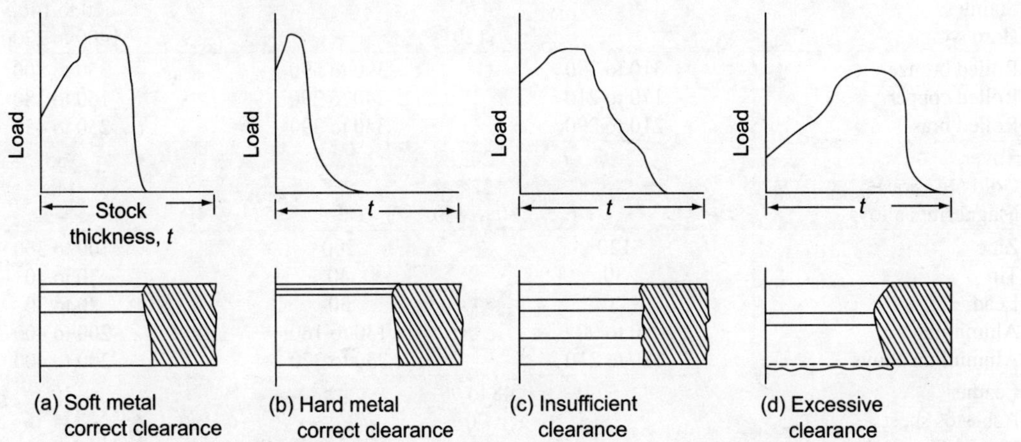

Fig. 8.5 *Effect of clearance on shearing load and edge characteristics*

Clearances

The clearance between two shears is one of the principal factors controlling a shearing process. This clearance depends essentially on the material and thickness of the sheet metal. This clearance can be approximated per side *C*, as

$$C = 0.0032 \times t < \sqrt{\tau} \text{ mm}$$

Where, t = sheet thickness, mm

And τ = material shear stress, MPa

The shear strength values of interest have been presented in Table 8.2.

Table 8.2 Shear strength and tensile strength of various materials

Material	Shear Strength, MPa		Tensile Strength, MPa
	Annealed	Not Annealed	
STEELS			
C10	235	314	390 to 490
C20	294	390	490 to 590
C30	350	470	590 to 690
C40	440	550	690 to 780
C50	490	640	780 to 880
C60	540	710	880 to 980
C80	690	880	1030 to 1180
C98	780	1030	1180 to 1370
St 32	290		310 to 390
St 37	340		360 to 440
St 50	390		490 to 590
St 63	490		620 to 740
St 78	640		770 to 880
Silicon	463		556
SAE	862		–
SAE	655		–
SAE	930		–
Stainless	470		600 to 1400
Hard spring	1120		1400 to 2200
Rolled bronze	310 to 390	390 to 590	350 to 800
Rolled copper	170 to 210	240 to 290	160 to 380
Rolled brass	210 to 290	340 to 390	250 to 450
Silver	208		285
Gold (14 carat)	324		448
Magnesium alloys	30 to 150		–
Zinc	120	200	100 to 360
Tin	30	40	30 to 40
Lead	20	30	20 to 30
Aluminium	70 to 90	130 to 160	200 to 400
Aluminium alloys	180 to 210	230 to 370	200 to 400
Leather	48 to 90		10
Asbestos sheet	70		–
Phenol fibre	180		–
Paper	44		–
Cellulose acetate	70		–
PVC	25 to 30		37 to 45

The clearances per side that are provided for normal working materials are presented in Table 8.3.

TABLE 8.3 Clearances as percentage of stock thickness

Material	Round	Other Contours
Soft aluminium < 1 mm	2	3
> 1 mm	3	5
Hard aluminium	4 to 6	5 to 8
Soft copper alloys	2	3
Hard copper alloys	4	5 to 6
Low carbon steel	2	3
Hard steel	3	5
Silicon steel	3	4 to 5
Stainless steel	4 to 6	5 to 8

8.3 SHEARING OPERATIONS

Straight line shears as described earlier are used for general purpose shearing work, for example, in cutting the small pieces from a large sheet. But the more useful are the die shearing operations where the shears take the form of the component to be made. The upper shear is called the punch and the lower shear is called the die. The two widely used processes are blanking and piercing.

Blanking

It is a process in which the punch removes a portion of material from the stock which is a sheet metal strip of the necessary thickness and width. The removed portion is called a blank and is usually further processed to be of some use, e.g., blanking of a pad lock key.

Piercing

Also sometimes called punching, the piercing is making holes in a sheet. It is identical to blanking except for the fact that the punched out portion coming out through the die in piercing is scrap. Normally a blanking operation will generally follow a piercing operation.

A typical set up used for blanking or piercing operation is shown in Fig. 8.6.

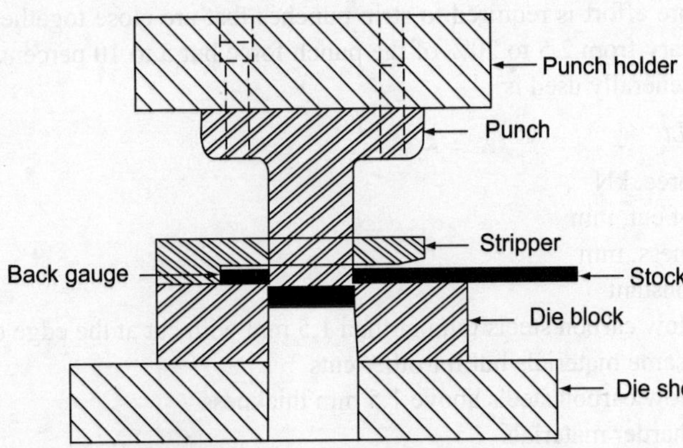

FIG. 8.6 *Simple blanking/piercing die*

Angular Clearance

In the shearing operation, first the material is elastically deformed, then plastically and finally removed from the stock strip. After the final breaking, the slug will spring back due to the release of stored elastic energy. This will make the blank to cling to the die face unless the die opening is enlarged. This enlargement is normally referred to as angular clearance or draft as shown in Fig. 8.7. The draft provided depends on the material, thickness and shape of the stock used. For thicker and softer materials, generally higher angular clearances are provided. The normal value is from 0.25 to 0.75 degree per side but occasionally a value as high as 2 degrees may be used.

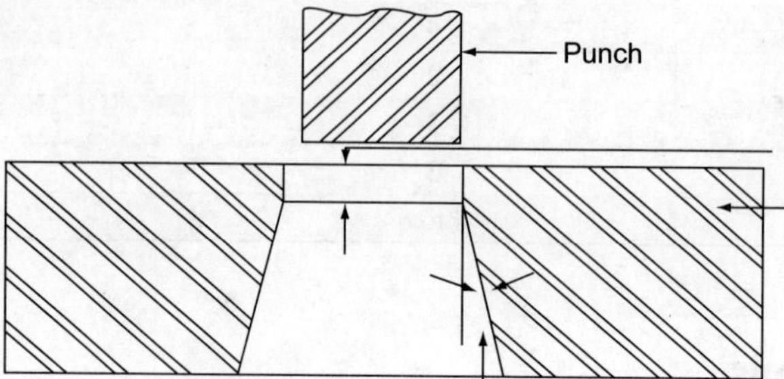

Fig. 8.7 *Angular clearance on die*

The die opening increases after every sharpening of the die because of the provision of the angular clearance. So to maintain the die size as per the design, the angular clearance is provided in the die opening along with a straight portion called as die land or cutting land. The length of the cutting land is about 3 mm for sheets which are less than 3 mm thick. For greater thicknesses, die land same as the material thickness has been found to be a good practice.

Stripper

Due to the release of the stored elastic energy in the stock left on the die, the stock tends to grip the punch as the punch moves upward. This necessitates the use of a stripper to separate the punch from the stock. The force required for stripping depends, besides the material, on other factors such as the position and size of the punched hole. For example, thicker materials or small hole in the middle of a strip require more stripping force than thin material or a hole towards one of the edges. A punch which has smooth side walls would strip very easily. Similarly more effort is required to strip punches that are close together. A general estimate of the stripping force may vary from 2.5 to 20% of the punch force but 5 to 10 percent is good for most of the applications. A formula generally used is

$$Ps = KLt$$

Where, Ps = stripping force, kN

L = perimeter of cut, mm

t = stock thickness, mm

K = stripping constant

= 0.0103 for low carbon steels thinner than 1.5 mm with cut at the edge or near a preceding cut

= 0.0145 for same materials but for other cuts

= 0.0207 for low carbon steels above 1.5 mm thickness

= 0.0241 for harder materials

Since the burr in the Fig. 8.3 represents the size of produced blank or hole, the clearances have to be accordingly apportioned to take into account the useful part obtained in the process. Thus, in blanking the die size is same as the component size whereas in piercing, the punch size is same as the actual hole size to be obtained.

Punching Force

The force required to be exerted by the punch in order to shear out the blank from the stock can be estimated from the actual shear area and the shear strength of the material. It is given by the following formula.

$$P = LtJ$$

Where, P = punching force, N

τ = shear strength, MPa

Sometimes the tensile strength may replace the shear strength in the above expression because shear is not the only active force the press has to overcome.

The punching force for holes which are smaller than the stock thickness may be estimated as follows:

$$P = \frac{dts}{\sqrt[3]{\frac{d}{t}}}$$

Where, d = diameter of the punch, mm

s = tensile strength of the stock, MPa

Shear

To reduce the required shearing force on the punch, for example to accommodate a large component on a smaller capacity punch press, shear is ground on the face of the die or punch. The effect of providing shear is to distribute the cutting action over a period of time depending on the amount of shear provided. Thus the shear is relieved of the punch or die face so that it contacts the stock over a period of time rather than instantaneously. It may be noted that providing the shear only reduces the maximum force to be applied but not the total work done in shearing the component. The effect of shear is shown schematically in Fig. 8.8.

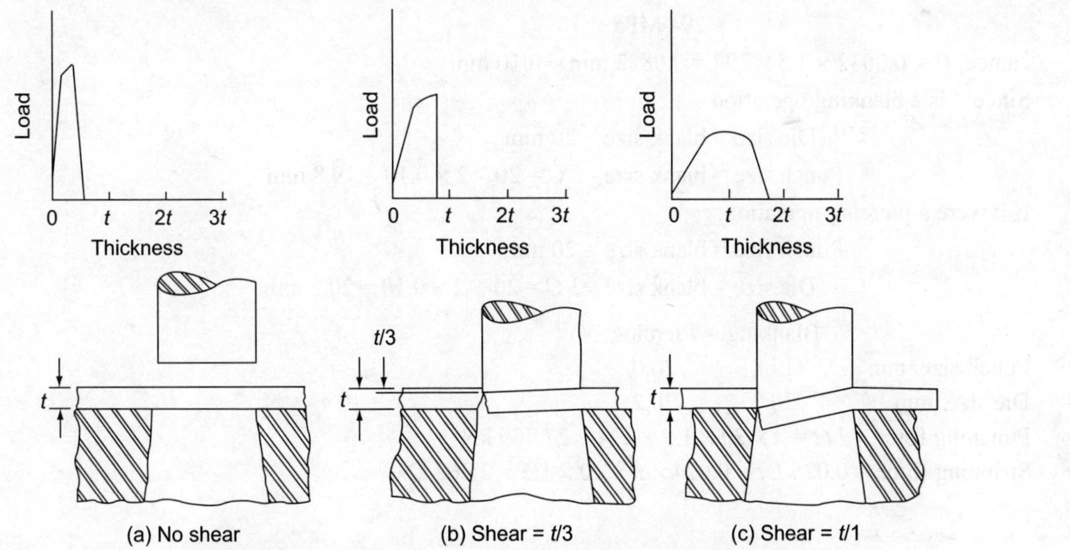

FIG. 8.8 *Effect of shear on the maximum load on punch*

For calculating the maximum shear force when shear is applied to the punch or the die, the following equation may be used.

$$P = \frac{Lt(tp)}{t_1}$$

Where, p = penetration of punch as a fraction
t_1 = shear on the punch or die, mm

The provision of the shear on the punch will change the slug whereas shear provided on the die would make the stock left on the die to bend, as shown in Fig. 8.9. Hence the shear is provided on the die for blanking and on the punch for piercing.

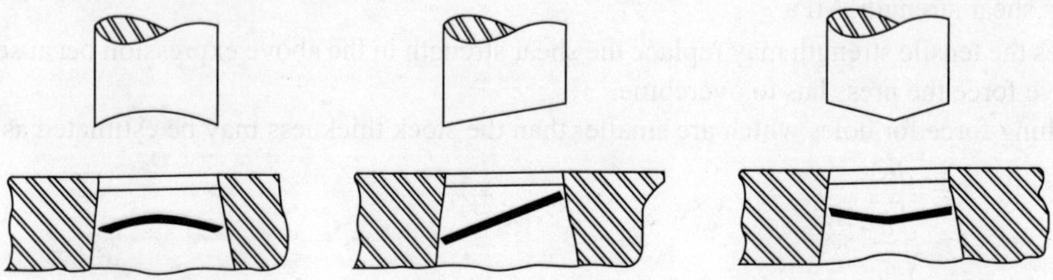

(a) Concave shear (b) Unbalanced shear (c) Balanced shear

Fig. 8.9 *Effect of shear on punch with the resultant distortion of the slug*

Example 8.1

Determine the die and punch sizes for blanking a circular disc of 20 mm diameter from a C20 steel sheet whose thickness is 1.5 mm.

The clearance to be provided is given by

$$C = 0.0032 \times t \times \tau$$

From Table 8.2, shear strength of annealed C20 steel is

$$= 294 \text{ MPa}$$

Hence, $C = 0.0032 \times 1.5 \times 294 = 0.0823 \text{ mm} \approx 0.10 \text{ mm}$

Since it is a blanking operation,

$$\text{Die size} = \text{blank size} = 20 \text{ mm}$$

$$\text{Punch size} = \text{blank size} - 2C = 20 - 2 \times 0.10 = 19.8 \text{ mm}$$

If it were a piercing operation,

$$\text{Punch size} = \text{blank size} = 20 \text{ mm}$$

$$\text{Die size} = \text{blank size} + 2C = 20 + 2 \times 0.10 = 20.2 \text{ mm}$$

	Blanking	Piercing
Punch size, mm	19.8	20.0
Die size, mm	20.0	20.2

Punching force = $Lt\tau = \pi \times 20 \times 1.5 \times 294 = 27.709$ kN

Stripping force = $0.024\, L\, t = 0.024 \times \pi \times 20 \times 1.5 = 2.262$ kN

Example 8.2

A hole with 100 mm diameter is to be punched in a steel plate of 6 mm thickness. The material is cold rolled C40 steel for which the maximum shear strength can be taken as 550 MPa. With normal clearance on the tools, cutting is complete at 40% penetration of the punch. Give suitable diameters for the punch and die, and shear angle on the punch in order to bring the work within the capacity of a 200 kN press available in the shop.

Clearance per side = $0.0032 \times 6 \times 550 = 0.45$ mm

Punch diameter = 100 mm

Die diameter = $100 + 2 \times 0.45 = 100.9$ mm

Without any shear, the maximum punching load is given by

$$= 550 \times \pi \times 100 \times 6 = 1036.725 \text{ kN}$$

Let p be the penetration of the punch in the form of shear reduces the punch load to the available 200 kN.

Total work done in punching = $1036.725 \times 6 \times 0.4 = 2488.14$ J

Work done with shearing and 200 kN maximum load

$$= 200\,(0.4 \times 6 + p)$$

Equating the two we get the value of p as

$$p = \frac{2488.14 - 480}{200} = 10.0407 \text{ mm}$$

Assuming a balanced shear on the punch, we get the shear angle as

$$= \tan^{-1}\left[\frac{10.0407}{50}\right] = 11.35°$$

While arriving at the shear angle, it was assumed that the maximum load is uniform over the entire punching stroke. This however is not true. Also, there is some additional load on the press due to the bending of the slug which was not included in the above calculations. Moreover, the shear on the punch being very large, of the order of twice the sheet thickness, it is desirable to go for a higher capacity press.

Besides blanking and piercing, there are a number of other shearing processes which are also used in industrial practice, but to a lesser extent. They are:

Trimming

In operations such as drop forging and die casting, a small amount of extra metal gets spread out near the parting plane, which is termed as flash. The flash is to be trimmed before the casting or forging is to be used. The dies used for this purpose are very similar to the blanking dies, the main difference being in the type of presses used for trimming which should normally have a larger table (bolster plate) area.

Shaving

In blanking or piercing operation, the edge of the blank or the hole is not perfectly clean because of the burr generated in the shearing process which is equal to the clearance on the die. For close tolerance work, the blanking or piercing process is followed by shaving (Fig. 8.10) which removes the burr left on the product. The clearances used in shaving dies are negligibly small to give the requisite tolerances.

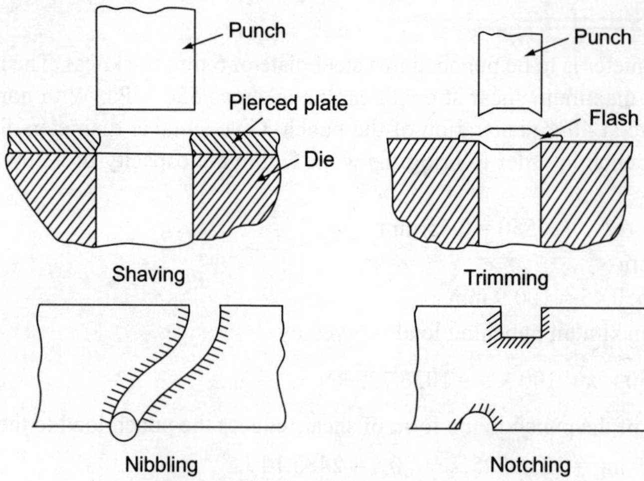

Fig. 8.10 *Typical shearing processes*

Nibbling

Nibbling is removing the metal in small increments. When a specific contour is to be cut in a sheet metal, a small punch is used to punch repeatedly along the necessary contour, generating the required profile. Nibbling is resorted to when the contour is long and a separate punch is impractical and uneconomical. The punches used can be round or square depending upon the applications.

Notching

Notching is a method to cut a specified small portion of metal towards the edge of the stock.

Lancing

In sheet metal, lancing is an operation in which a single line cut or slit is made part way across the strip stock so that part can be formed or bent into a required shape without removing any metal.

8.4 DRAWING

Drawing is the process of making cups, shells and similar articles from metal blanks. Typical tools used for drawing are shown in Fig. 8.11.

The setup is similar to that used in blanking except that the punch and die are provided with the necessary rounding at the corners to allow for the smooth flow of metal during drawing. The blank is first kept on the die plate. The punch slowly descends on the blank and forces it to take the cup shape formed by the end of the punch, by the time it reaches the bottom of the die. When the cup reaches the counter bored portion of the die, the top edge of the cup formed around the punch expands slightly due to the spring back. When the punch moves in the return stroke, the cup would be stripped by this counter bored portion.

This description is true in the case of shallow drawing operation only. Shallow drawing is defined as where the cup height is less than half the diameter. For drawing deeper cups it is necessary to make specific provisions to confine the metal in order to prevent excess wrinkling of the edges. For this purpose, a blank holder is normally provided on all deep drawing dies as shown in Fig. 8.12.

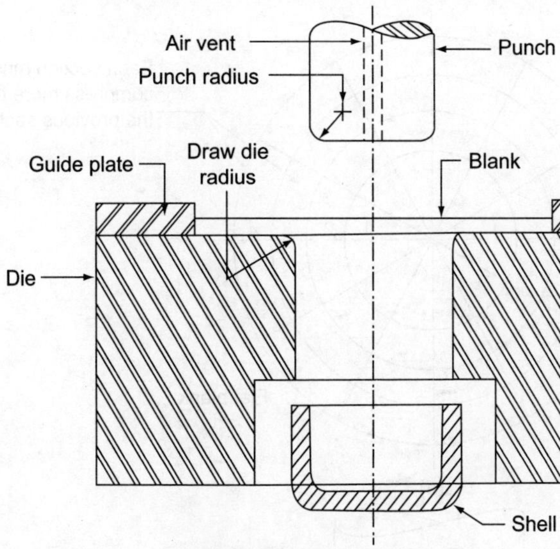

FIG. 8.11 *Drawing operation*

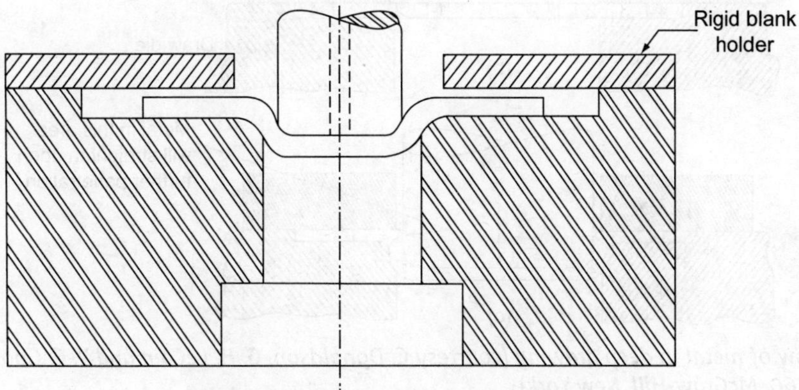

FIG. 8.12 *Blank holding in deep drawing*

The rigid blank holding as in Fig. 8.12(a), is normally used for thicker materials which are less likely to wrinkle. A more common usage is the spring loaded pressure pad shown in Fig. 8.12(b). The spring loaded pressure pad which moves with the punch, maintains a more uniform pressure on the blank throughout the drawing operation.

To understand the problem of wrinkling, consider the drawing of a cup as shown in Fig. 8.13 along with its blank. The blank has been divided into sections 1 to 4. The example is presented for a circular cup since these are more generally used. It is also possible to draw other sections, such as rectangle, but that would be much more difficult.

The first contact made by the punch with the blank is over the portion marked 1 which forms the bottom of the cup. In this portion, there is no deformation of the blank. As the punch further moves down, the metal shown in the ring 2 of the blank gets bent by the punch over the die radius. Upon further movement of the

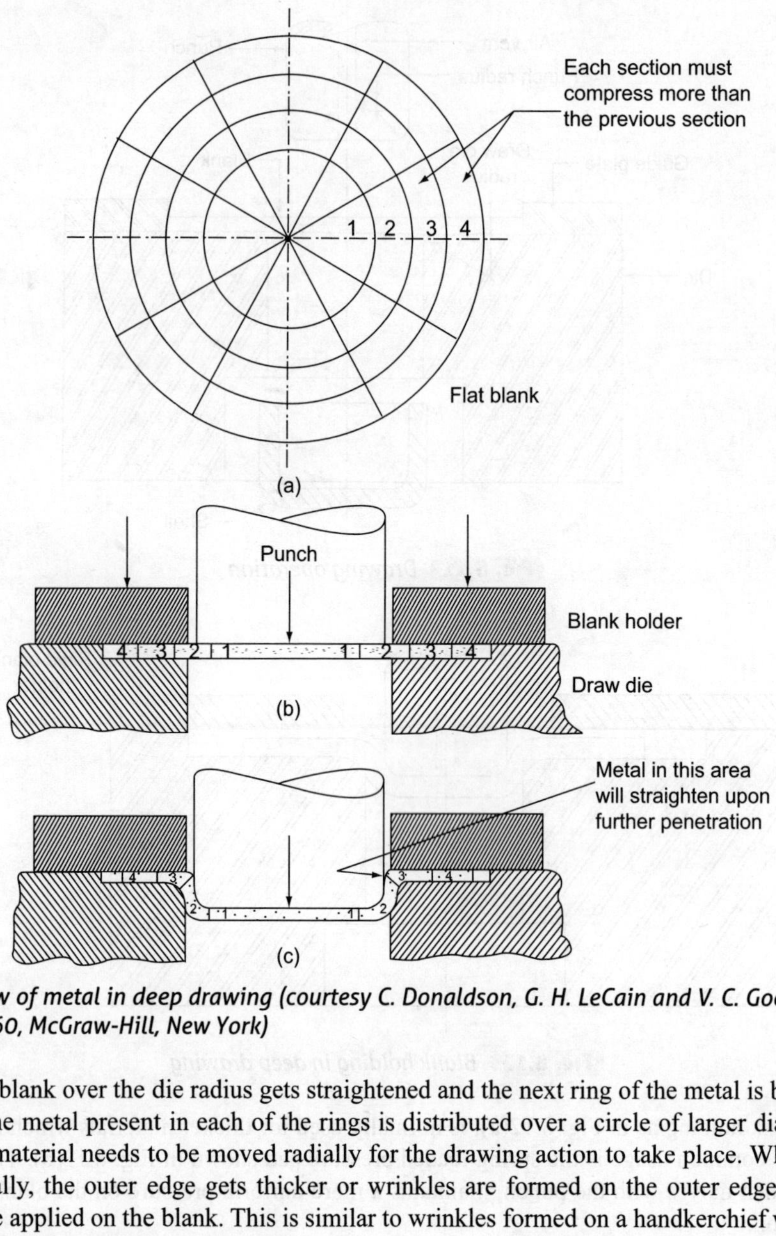

FIG. 8.13 *Flow of metal in deep drawing (courtesy C. Donaldson, G. H. LeCain and V. C. Goold: Tool Design, p760, McGraw-Hill, New York)*

punch, the bent blank over the die radius gets straightened and the next ring of the metal is bent over the die radius. Since, the metal present in each of the rings is distributed over a circle of larger diameter than that of the cup, the material needs to be moved radially for the drawing action to take place. While the material is flowing radially, the outer edge gets thicker or wrinkles are formed on the outer edges, if there is no restraining force applied on the blank. This is similar to wrinkles formed on a handkerchief when spread flat on a tumbler of small size and pushed into it.

The restraining force applied on the blank by way of the blank holder stops the blank from increasing in thickness beyond a limit but makes it to flow radially outward. The limiting thickness is controlled by the gap between the rigid blank holder and the die or by the spring pressure in the case of a spring loaded blank holder. Too high a blank holder pressure increases the friction and subsequently the drawing load. A lubricant is normally applied over the face of the blank to reduce this friction.

As explained earlier, shallow drawing is relatively simple. Drawing when cup height is more than half the diameter is termed as deep drawing. Ductile materials are easier to be drawn into deeper cups.

In deep drawing, because of the radial flow of material, the side walls increase in thickness as the height is increased, as shown in Fig. 8.14. There would be a slight thinning of metal at the bottom corners. For applications requiring uniform side walls, an operation called ironing is carried out on drawn cups.

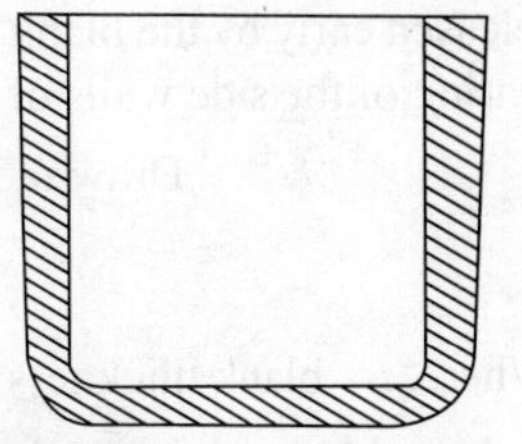

Fig. 8.14 *Side wall taper*

Ironing is the operation of thinning the side walls and increasing the height. The die and punch set used is similar to that of drawing operation except that the clearance between the die and punch is smaller than that used in the drawing operation. The material gets compressed between punch and die which reduces the thickness and increases the height, and thus is a severe operation. The wall thicknesses can be reduced to as much as 50 % in a single ironing operation.

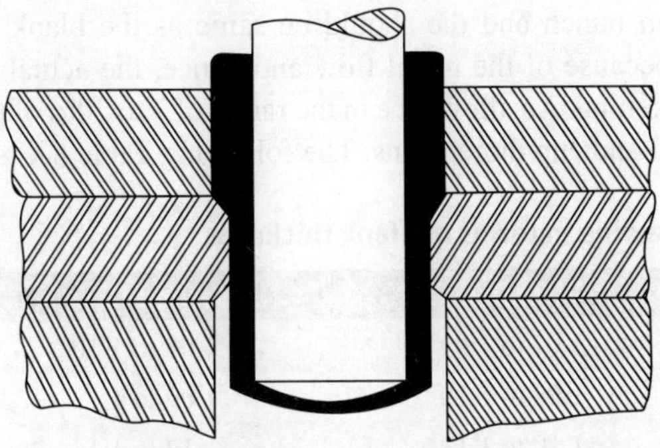

Fig. 8.15 *Ironing operation*

8.5 DRAW DIE DESIGN

In designing a die to work satisfactorily, one has to consider a number of variables which readily control the performance of the drawing die. In the following, we would consider a few such variables to help in the design of a draw die.

Corner Radius on Punch

Though there is no set rule for the provision of corner radius on the punch, it is customary to provide a radius of four to ten times the blank thickness. Too small a corner radius makes for the excessive thinning and tearing of the bottom of the cup. Ideally, the punch radius should be the same as the corner radius of the required cup because it takes its form. It is not possible in deep drawing to obtain the final cup size in a single drawing operation. Depending on the geometry of the cup, it may be necessary to go in for more than one drawing operations, in which case, the subsequent operations are called redrawing operations. For the initial draws, larger punch radii are used and for the final draw use the requisite radius. If the requisite radius is less than four times the blank thickness, it may be necessary to obtain it by restriking after the final draw with four times the thickness.

Draw Radius on Die

Since the draw radius on die does not contribute to the cup shape, it can be made as large as possible. Higher the radius, higher would be the freedom for the metal to flow. Too high a radius causes the metal to be released early by the blank holder and thus lead to edge wrinkling. Too small a radius causes the thinning and tearing of the side walls of the cups. A general guide is

$$\text{Draw radius} = 4t \text{ normal}$$
$$= 6 \text{ to } 8t \text{ when the blank holder is used}$$
$$= 0.8\,\pi\,(D - d)t$$

Where, t = blank thickness

In order to minimise the thinning of the stock, the centre point of the draw radius should be about 3 mm outside of the previous cup draw. Also the centre point of the corner radius on punch should be slightly inside the shell of the next draw.

Clearances

Ideally, the clearance between punch and die should be same as the blank thickness. But the blank gets thickened towards the edge because of the metal flow and hence, the actual clearance provided is slightly higher to account for this thickening. An allowance in the range of 7 to 20% of the blank thickness is provided depending on the cup material and cup dimensions. The following Table 8.4 gives the necessary details.

TABLE 8.4 Clearances in drawing in terms of blank thickness

Blank Thickness (mm)	First Draw	Second Draw	Sizing Draw
up to 0.40	1.07 to 1.09	1.08 to 1.10	1.04 to 1.05
0.41 to 1.25	1.08 to 1.10	1.09 to 1.12	1.05 to 1.06
1.30 to 3.00	1.10 to 1.12	1.12 to 1.14	1.07 to 1.09
above 3.01	1.12 to 1.14	1.15 to 1.20	1.08 to 1.11

The sizing draw refers to the situation when the cup is to be obtained with good surface finish and given diameter and wall thicknesses. Ironing may be required in sizing draws.

Blank Size

One of the first jobs of the draw die designer is to find the size of the blank to be used for making a given cup. It is often difficult to find a blank of the exact size required for making a given shell because of thinning and thickening of sheet during drawing. The calculation could be based on volume, surface area or by layout. The following gives the useful relations in calculating the blank diameter for cylindrical shells for relatively thin materials.

$$D = \sqrt{d^2 + 4dh} \quad \text{when } d \geq 20r$$

$$D = \sqrt{d^2 + 4dh - 0.5r} \quad \text{when } 15r \leq d \leq 20r$$

$$D = \sqrt{d^2 + 4dh - r} \quad \text{when } 10r \leq d \leq 15r$$

$$D = \sqrt{(d - 2r)^2 + 4d(h - r) + 2\pi r(d - 0.7r)} \quad \text{when } d < 10r$$

Where, r = corner radius on the punch, mm
h = height of the shell, mm
d = outer diameter of the shell, mm
D = blank diameter, mm

These are only the theoretical blank sizes, based on the surface area of the shell and blank. For practical purposes, it is necessary to add additional allowances to provide for trimming of the uneven and irregular rim of the deep drawn cup. This is termed as trim allowance. The trim allowance could be 3 mm for the first 25 mm cup diameter and additional 3 mm for each of the additional 25 mm of cup diameters.

For rectangular and other types of shells, reference may be made to literature that provides the necessary formulae for other generally used shapes.

Drawing Force

The drawing force depends on the cup material, its dimensions and the configuration. The drawing force can empirically be calculated using the following equation for cylindrical shells.

$$P = \pi dts\left[\frac{D}{d} - C\right]$$

Where, P = drawing force, N

t = thickness of the blank material, mm

s = yield strength of the metal, MPa

C = constant to cover friction and bending. Its value is between 0.6 and 0.7

For shapes other than cylindrical, the above formula would only give an approximation which can be used as a guide.

Blank Holding Force

The blank holding pressure required depends on the wrinkling tendency of the cup which is very difficult to determine and hence it is obtained more by trial and error. The maximum limit is generally to be one third of the drawing force.

Ironing Force

In ironing, the objective is only to reduce the wall thickness of the cup, and hence no blank holding is required because the punch is fitted closely inside the cup. Neglecting the friction and shape of the die, the ironing force can be estimated using the following equation.

$$F = \pi_{d_1 t_1 S_{av}} \log_e \frac{t_o}{t_1}$$

Where, F = ironing force, N

d_1 = mean diameter of the shell after ironing

t_1 = thickness of shell after ironing

t_o = thickness of the shell before ironing

S_{av} = average of tensile strength before and after ironing

Percent Reduction

The drawing operation relies on the ductility of the blank material. The ductility is affected by the amount of strain a material takes. But there is a limit to which it can be strained. The amount of straining or the drawability is represented by the percentage reduction which is expressed in terms of the diameter of the blank and the shell. The diameter of the shell to be used is either *ID* (internal diameter) or *OD* (outer diameter). But it is convenient to use *OD* since the cup is normally specified by *OD*. The percent reduction, *P* is given by

$$P = 100\left[1 - \frac{d}{D}\right]$$

Theoretically, it is possible to get a percentage reduction up to 50 but, is practically limited to 40. Also because of strain hardening, the percentage reduction possible gets reduced in the subsequent draws. Table 8.5 gives a general guide line for the percentage reduction to be used for various heights.

TABLE 8.5 Reduction in drawing with cup height

Height to Dia Ratio	No. of Draws	Percent Reduction			
		First Draw	Second Draw	Third Draw	Fourth Draw
up to 0.75	1	40	–	–	–
0.75 to 1.50	2	40	25	–	–
1.50 to 3.00	3	40	25	15	–
3.00 to 4.50	4	40	25	15	10

For cup heights more than that shown in the above table, annealing is to be done on the cup preferably after the third draw to restore the ductility, and then further drawings are to be carried on.

The recommended maximum reductions for various engineering materials are given in Table 8.6.

TABLE 8.6 Maximum reductions possible in a single draw

Materials	Percent Reduction
Aluminium alloys	45
Copper	45
Brass, Bronze	50
Low carbon steel	45
Stainless steel	50
Zinc	40

Air Vent

An air vent is normally provided in the punch to reduce the possibility of formation of vacuum in the cup, when it is stripped from the punch. For cylindrical shells, one vent located centrally would be enough, where as for other shapes it may be advantageous to provide two. The size of air vent depends on the punch diameter as shown in Table 8.7.

TABLE 8.7 Air vent sizes

Punch Diameter (mm)	Air Vent Diameter (mm)
up to 50	4.5
50 to 100	6.0
100 to 200	7.5
over 200	10.0

Drawing Speed

The speed with which the punch moves through the blank during drawing is termed as the drawing speed. This is a very important parameter in drawing because higher speeds are sometimes detrimental. Particularly harder and less ductile materials are likely to be excessively thinned out due to excessive drawing speeds. Suggested drawing speeds are given in Table 8.8 for various materials.

TABLE 8.8 Drawing speeds

Material	Drawing Speed (m/s)
Aluminium	0.90
Brass	1.00
Copper	0.75
Steel	0.28
Zinc	0.75

Example 8.3

A symmetrical cup of circular cross section with diameter 40 mm and height 60 mm, with a corner radius of 2 mm is to be obtained in C20 steel of 0.6 mm thickness. Make the necessary design calculations for preparing the die for the above cup.

$$\frac{\text{Corner radius}}{\text{Shell diameter}} = \frac{2}{40} = \frac{1}{20}$$

$$\text{Hence, blank diameter} = \sqrt{d^2 + 4dh}$$

$$= \sqrt{40^2 + 4 \times 40 \times 60} = 105.83 \text{ mm}$$

Trim allowance = 6 mm

Final blank diameter = 112 mm

Next step is to ascertain the amount of the percent reduction to estimate the number of draws.

$$\text{Percent reduction} = 100\left[1 - \frac{d}{D}\right] = 100\left[1 - \frac{40}{112}\right] = 64.2857$$

This is more than 45 and hence cannot be obtained in a single draw. From Table 8.5, it is known that two draws with the first one at 40% and the second one at 25% would be able to achieve the desired cup shape.

First draw:

$$40 = 100\left[1 - \frac{d}{112}\right]$$

Cup diameter after first draw = 44.8 mm

Corner radius on punch = $6t$ = 3.6 mm ≈ 4 mm

This also satisfies the criterion that the centre point of its radius should be within the cup of the next draw.

Draw radius on die = $0.8 \times \pi (112 - 44.8) \times 0.6 = 5.0798$ mm ≈ 5 mm

Die clearance from Table 8.4 = $1.09t = 1.09 \times 0.6 = 0.654$ mm ≈ 0.65 mm

Second draw:

Since the required drawing is only 2 mm.

Corner radius on punch = $4t = 2.4$ mm ≈ 2 mm

Draw radius on die = 6 mm

This is obtained by the criterion that the centre of the draw die radius should be outside the previous drawn cup by about 3 mm as shown in Fig. 8.16.

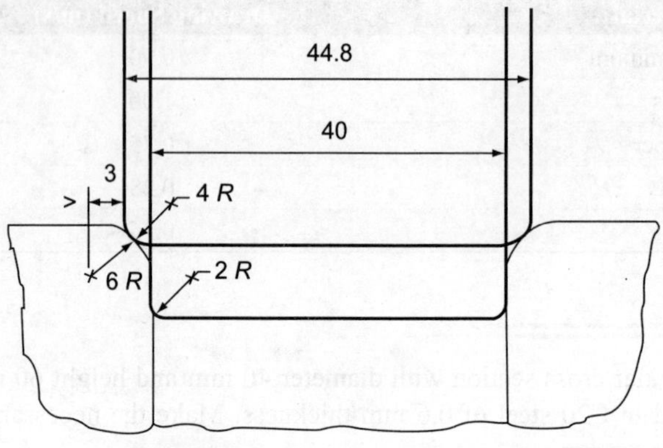

Fig. 8.16 *Draw die for the example*

Die clearance = $1.11\,t = 0.666$ mm ≈ 0.67 mm

$$\text{Drawing pressure} = \pi d t s \left[\frac{d}{D} - C\right]$$

$$\pi \times 40 \times 0.6 \times 550 \left[\frac{112}{40} - 0.6\right] = 91.23 \text{ kN}$$

8.6 SPINNING

Spinning is the process used for making cup shaped articles which are axisymmetric. The process of spinning consists of rotating the blank, fixed against the form block and then applying a gradually moving force on the blank so that the blank takes the shape of the form block. A typical setup is shown in Fig. 8.17.

The setup essentially consists of a machine similar to a centre lathe. In the head stock of the spinning machine, a hard wood form block which has the shape of the desired part is fixed. The blank is held against the form block by means of the freely rotating wooden block from the tail stock. After proper clamping, the blank is rotated to its operating speed. The spinning speed depends on the blank material, thickness and complexity of the desired cup. Then the hard wood or roller type metallic tool is pressed and moved gradually on the blank so that it conforms to the shape of the form block. The manipulation of spinning tools is a highly skilled art. The tool is to be moved back and forth on the blank so that no thinning takes place anywhere on the blank.

Spinning is comparable to drawing for making cylindrical shaped parts. Because of the simple tools used in spinning, it is economical for smaller lots. But the time required for making a cup is more in spinning

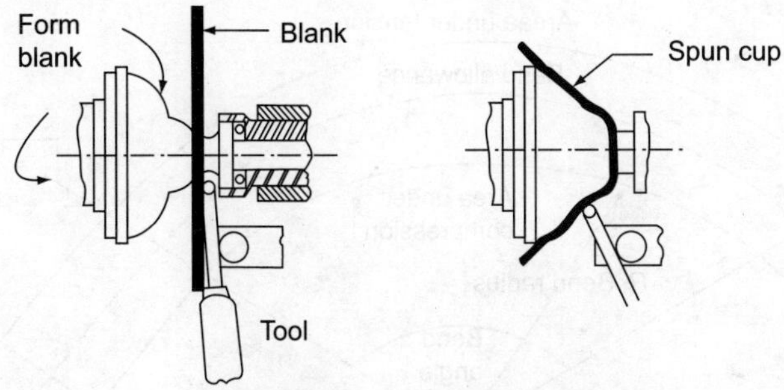

Fig. 8.17 *Metal spinning*

and also more skill is required in the process. Thus, it is not suitable for large scale production. Complicated shapes and re-entrant shapes are not feasible by drawing but can be made by spinning using the sectional and collapsible form blocks. Large parts are much more easily made in spinning than by drawing. When sheet thickness is more, for example, in making the dished ends of pressure vessels, cold spinning is not sufficient. Then the blank is heated to the forging temperature and so the process is called 'hot spinning'. Also in hot spinning, the tools are mechanically manipulated because of the higher magnitudes of forces required.

8.7 BENDING

Bending refers to the operation of deforming a flat sheet around a straight axis where the neutral plane lies. The disposition of the stresses in a bent specimen is shown in Fig. 8.18. Here, due to the applied forces, the top layers are in tension and bottom layers are in compression, as shown. The plane with no stresses is called the neutral axis. The neutral axis should be at the centre when the material is elastically deformed. But when the material reaches the plastic stage, the neutral axis moves downward, since the material oppose compression much better than tension.

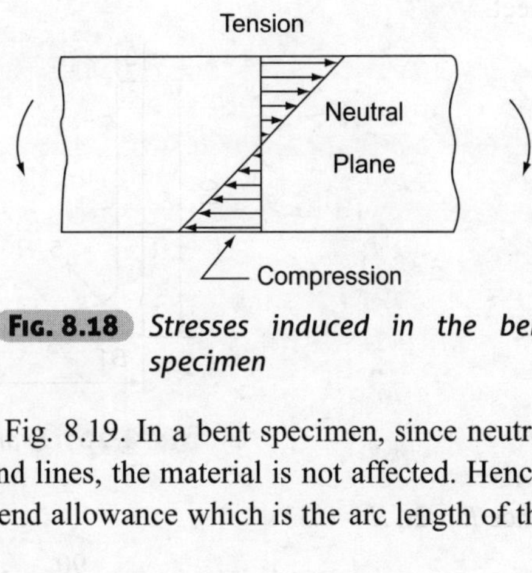

Fig. 8.18 *Stresses induced in the bent specimen*

The nomenclature normally used with bending is shown in Fig. 8.19. In a bent specimen, since neutral axis remains constant, it is the required length. Beyond the bend lines, the material is not affected. Hence, to calculate the length required it is necessary to find out the bend allowance which is the arc length of the neutral axis between the bend lines.

$$\text{Bend allowance, } B = \alpha(R + Kt)$$

Where, α = bend angle, radians

R = inside radius of the bend, mm

K = location of neutral axis from bottom surface

= 0.33 when $R < 2t$

= 0.50 when $R > 2t$

t = sheet thickness, mm

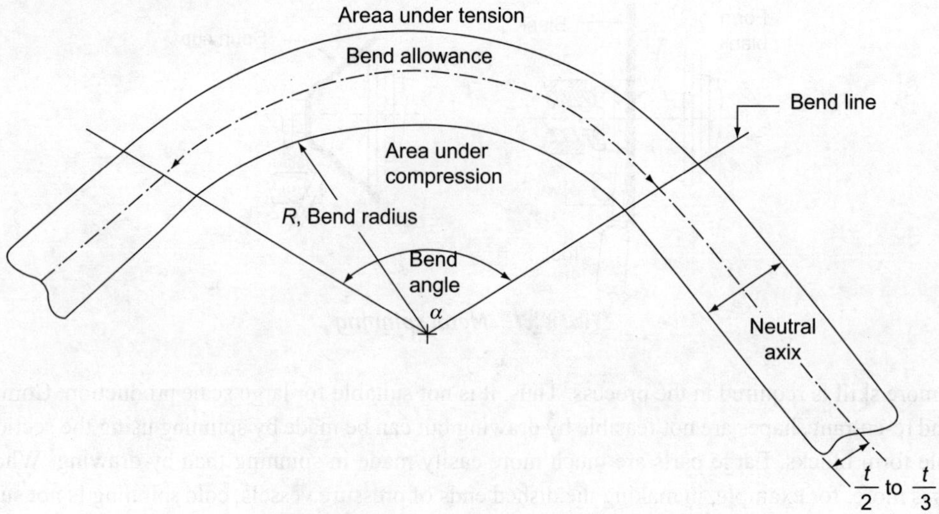

Fig. 8.19 *Nomenclature of bending*

Example 8.4

Calculate the length of sheet required for making the component shown in Fig. 8.20, out of 3 mm C20 steel sheet.

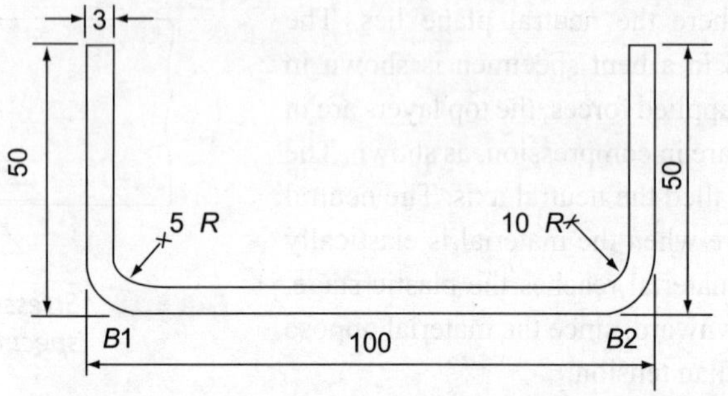

Fig. 8.20 *Example for bending length calculation*

Since $R < 2t$, $K = 0.33$

$$\text{Bend allowance} = \frac{90}{360} \times 2\pi\,(5 + 0.33 \times 3) = 9.42 \text{ mm}$$

Since $R > 2t$, $K = 0.50$

$$\text{Bend allowance} = \frac{90}{360} \times 2\pi\,(10 + 0.5 \times 3) = 18.06 \text{ mm}$$

$$\text{Total length} = 50 - (5 + 3) + 9.42 + 100 - (10 + 3) + 18.06 + 50 - (10 + 3) = 185.48 \text{ mm}$$

The outer layers which are under tension should not be stretched excessively; else there is the likelihood of rupture taking place. The amount of stretching depends on the sheet thickness and the bend radius. Lower the bend radius, higher is the strain in this zone. Hence there is a minimum bend radius to be specified, depending on the material characteristics.

Minimum bend radius = $0.5t$ soft materials
= t other materials
= $3.0t$ spring materials.

The other aspect to be considered in designing parts for bending is the grain orientation of the sheet which is bent. As far as possible, bending is to be done in a direction perpendicular to the grain direction in the metal. By virtue of the rolled sheets being used for bending, the grain direction usually is along the length axis, being the direction of rolling. There is a possibility of cracks appearing at the time of bending if the bending is done along the grains as shown in Fig. 8.21. But if two bendings are to be done on the same sheet at right angles, then it may be desirable to make them at 45 degrees to the grain direction so that the risk of cracking is minimised.

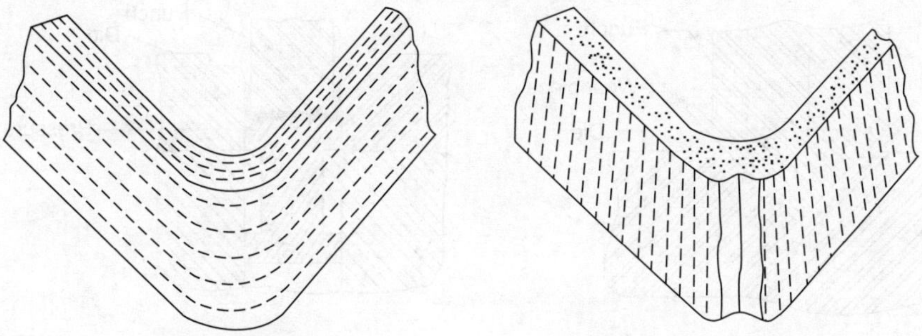

(a) Bent along grain (b) Bent across grain

Fig. 8.21 *Effect of bending along and across grain*

Spring back in bending is difficult to estimate theoretically. But it is essential to compensate it, because the bend geometry gets affected by the spring back directly. The data relating to the most generally used configurations are available in literature given at the end of the chapter.

The types of bending methods used are shown in Fig. 8.22. The first one is the wiping die which is used for simple 90 degree bends only. Here, the work is held firmly to the die, and the punch bends the extended portion of the blank as shown in Fig. 8.22(a). The V bending, as shown in Fig. 8.22(b), can be used for a wide range of angles as also the 90 deg. These are the ones that are most generally used.

The bending load may be calculated from the knowledge of material properties and the die characteristics as shown below (Fig. 8.23):

$$F_b = \frac{KLst^2}{W}$$

Where, F_b = bending force, N
K = 1.33 for die opening of $8t$
= 1.20 for die opening of $16t$
= 0.67 for U bending
= 0.33 for a wiping die
L = length of the bent part, mm
s = ultimate tensile strength, MPa
t = blank thickness, mm
W = width between the contact points, mm

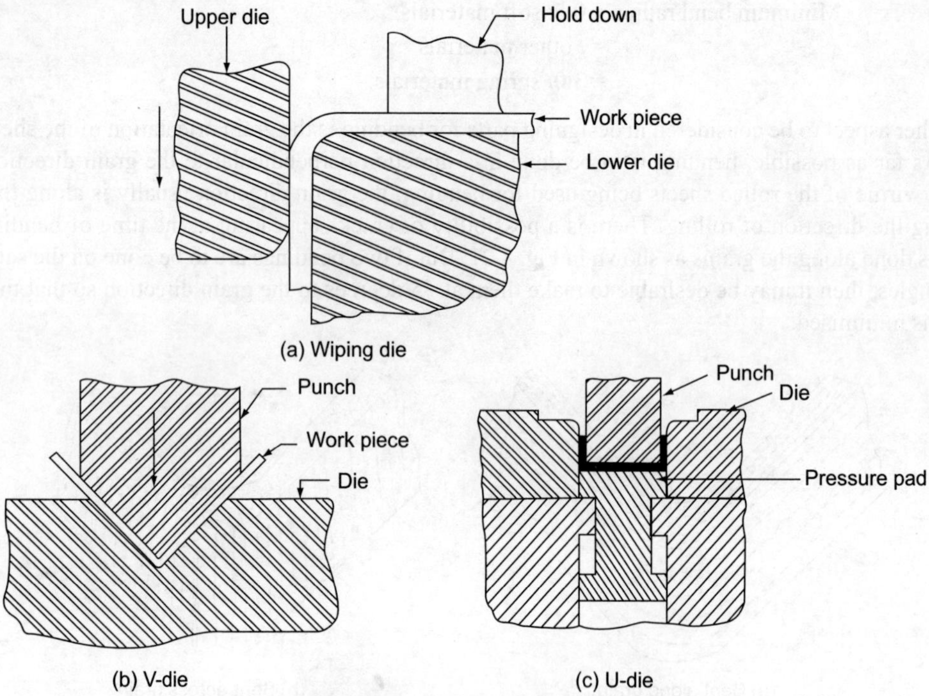

Fig. 8.22 *Types of bending dies*

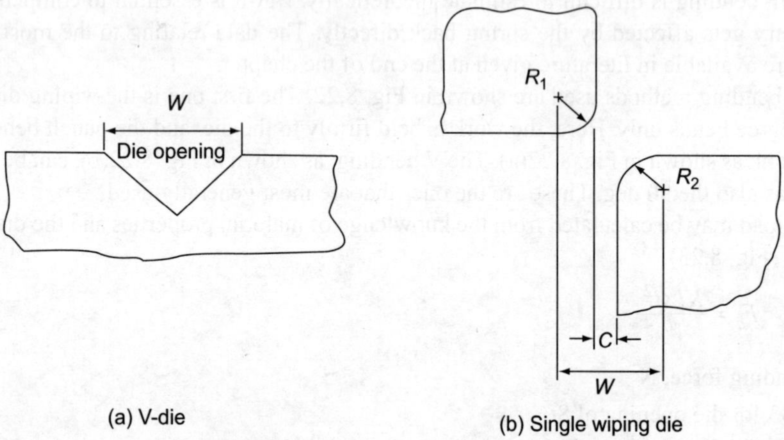

Fig. 8.23 *Bending die characteristics*

Example 8.5

Estimate the force required for a 90 degree bending of St 50 steel of thickness 2 mm in a V die. The die opening can be taken as eight times the thickness. The length of the bent part is 1 m.

From Table 8.1, for St 50 steel, the ultimate tensile strength can be taken as 500 MPa.

Die opening, $W = 8 \times 2 = 16$ mm.

$$\text{Bending force} = \frac{1.33 \times 1000 \times 500 \times 2^2}{16} = 166.25 \text{ kN}$$

Example 8.6

Calculate the bending force required for a C50 steel 1.5 mm sheet of width 1 m to be bent in a wiping die. The die radius used is 3 mm.

From Table 8.1, for C50 steel, ultimate tensile strength can be taken as 800 MPa.

Die opening, $W = 1.5 + 3 + 3 = 7.5$ mm

For a wiping die, $K = 0.33$

$$\text{Bending force} = \frac{0.33 \times 1000 \times 800 \times 1.5^2}{7.5} = 79.20 \text{ kN}$$

8.8 STRETCH FORMING

In all the previous processes considered, because the deformation of the metal is done in the elastic and plastic range, the spring back is always to be considered. However, in stretch forming, the complete deformation is carried out in plastic state only. The material is first brought into plastic state by stretching, hence the name stretch forming.

In the process, the sheet is held in the jaws of hydraulic cylinders and is stretched beyond elastic limit. Then the sheet is brought into contact with the die as shown in Fig. 8.24, so as to give it the shape of the die. Stretch forming is comparatively simple and inexpensive because it uses a single die. Many complicated shapes cannot be obtained but the component can have either singly curved or doubly curved surface. Also if the component is to have any holes, they may be punched after stretch forming otherwise the holes are likely to be enlarged. The physical properties are generally improved by the uniform stretching of the metal. The sheet used in stretch forming should have uniform thickness, otherwise the thinner portions are likely to be overstretched.

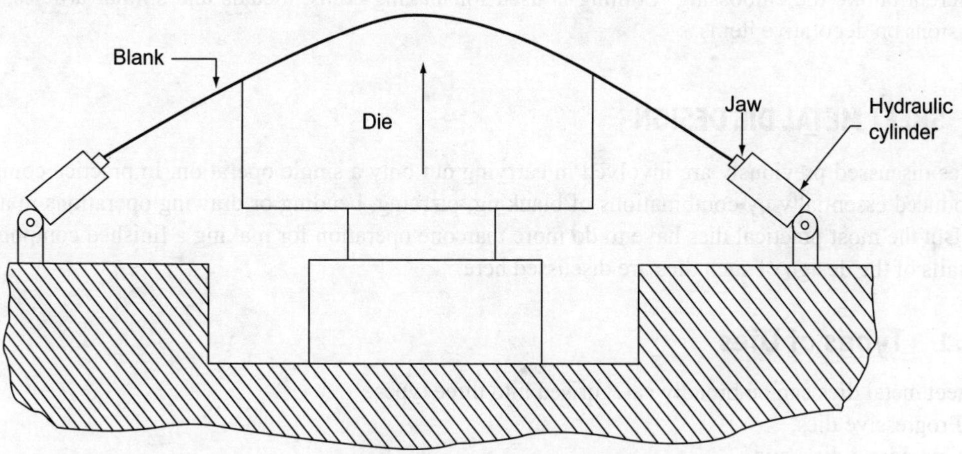

FIG. 8.24 *Stretch forming*

8.9 EMBOSSING AND COINING

Embossing is the operation used in making raised figures on sheets with its corresponding relief on the other side. The process essentially involves drawing and bending of the metal. An example of the embossing operation is shown in Fig. 8.25. The die set consists essentially of a die and punch with the desired contours, so that when the punch and die meet the clearance between them is same as that of the sheet thickness. Embossing operation is generally used for providing dimples on sheets to increase their rigidity and for decorative sheet work used for panels in houses and religious places. Stretch forming operation is relatively less severe operation.

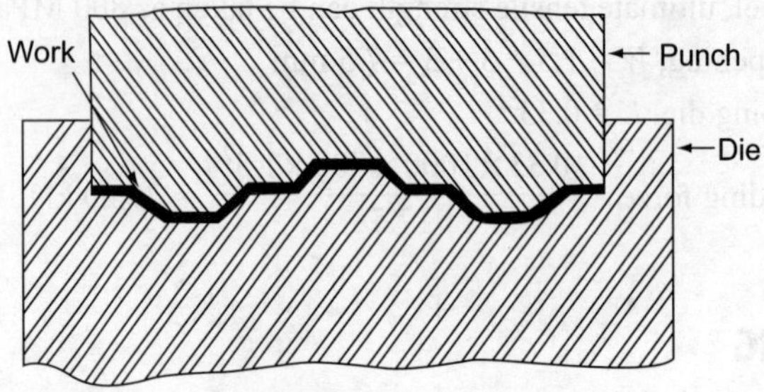

Fig. 8.25 *Embossing*

Coining is essentially a cold forging operation except for the fact that the flow of the metal occurs only at the top layers and not the entire volume. The coining die consists of punch and die which are engraved with the necessary details required on both sides of the final object. A blank, kept on the die, is compressed by it, as shown in Fig. 8.26. The pressures involved in coining are very high, of the order of 1600 MPa because of the very fine details that are normally desired in coining. The type of impression obtained on both sides would be different unlike the embossing. Coining is used for making coins, medals and similar articles, and for impressions on decorative items.

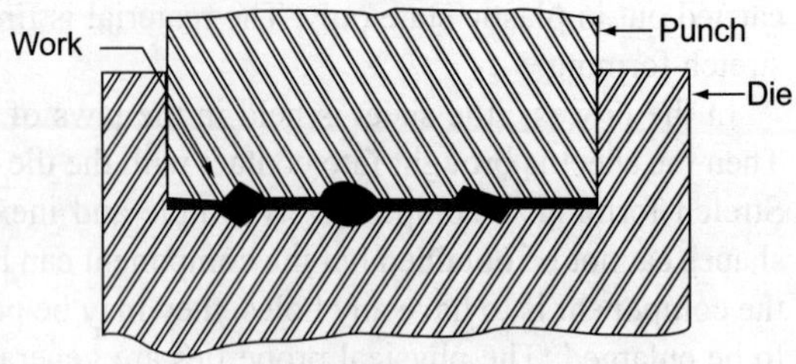

Fig. 8.26 *Coining*

8.10 SHEET METAL DIE DESIGN

The dies discussed previously are involved in carrying out only a single operation. In practice components are produced essentially by combinations of blanking, piercing, bending or drawing operations in a certain order. But the most practical dies have to do more than one operation for making a finished component and the details of the design of such dies are discussed here.

8.10.1 Types of Dies

The sheet metal dies can be broadly categorised into three types:

- Progressive dies,
- Compound dies, and
- Combination dies

Progressive Dies

The progressive dies perform two or more operations simultaneously in a single stroke of a punch press, so that a complete component is obtained for each stroke. The place where each of the operations is carried out is called stations. The stock strip moves from station to station undergoing the particular operation. When the strip finally leaves the last station, a finished component is ready. An example is presented in Fig. 8.27.

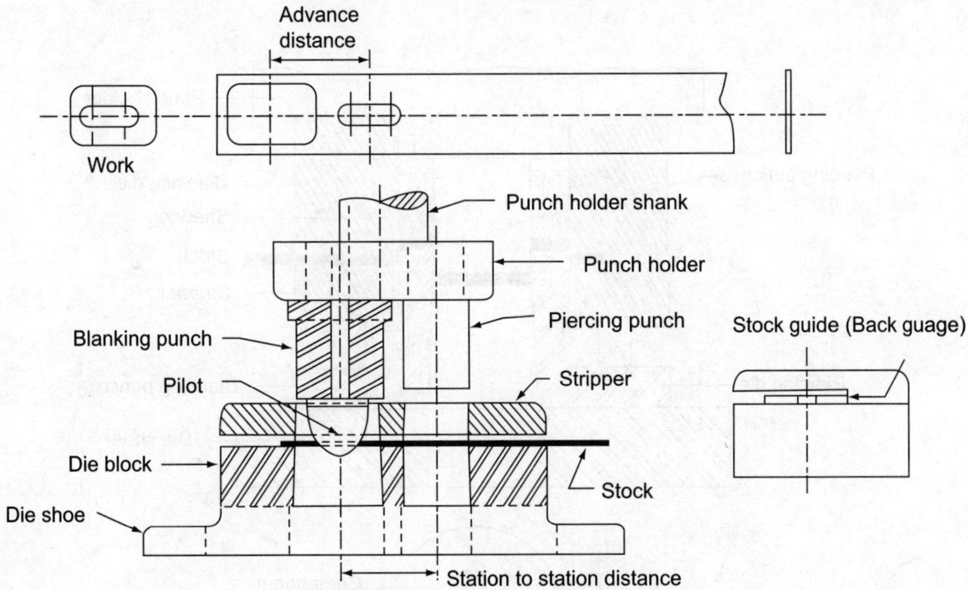

Fig. 8.27 *A simple progressive die*

At the start of the operation, the sheet is fed into the first station. After undergoing the operations at this station the ram of the press moves to the top and the stock is advanced from the first station to the second station, while a fresh portion of the stock comes under the first station. The distance moved by the strip from station one to two so that it is properly registered under the stations is called advance distance. This advance distance should be the same between any two stations in sequence. Another variable called the feed distance is the amount of stock fed under the punch when ram comes for the next stroke. The feed distance may or may not be the same as the advance distance. The reason for the variance between the feed distance and advance distance, that is sometimes observed, is because often the sheet is overfed against a stop. The strip is, therefore, positioned correctly under the punch by pulling it backwards with the use of pilots. This helps in preventing the stop from interfering with the part geometry.

Progressive dies contain large number of stations. It is generally preferred to have piercing operation first in the sequence and a blanking or cut-off operation in the end to get the final component. Any of the pierced holes may be advantageously used as a pilot hole. If none of the pierced holes are satisfactory as pilot holes, then special pilot holes may have to be made in the scrap part of the stock. It is also feasible to have blanking done in the first station, then the blank returned on to the die by means of spring plates and finally carried for subsequent operations.

The choice of progressive dies is made only when the production is of large numbers so that the handling costs are saved; stock material is not very thin, so that movement of the strip by pilots is convenient; stock material is not too thick so as to avoid the problems of stock straightening; and the overall size of the die or the press capacity are large and no available press in the shop can be utilised.

Compound Die

In a compound die, as distinct from the progressive die, all the necessary operations are carried out at a single station, in a single stroke of the ram. To do more than one set of operations, a compound die consists of the necessary sets of punches and dies. During the part of the stroke, piercing of holes is done in the stock and upon further travel, the blanking operation is done. For the blanking operation, the punch used for piercing becomes a die as shown in Fig. 8.28. In other words, blanking is done in a direction opposite to that of piercing.

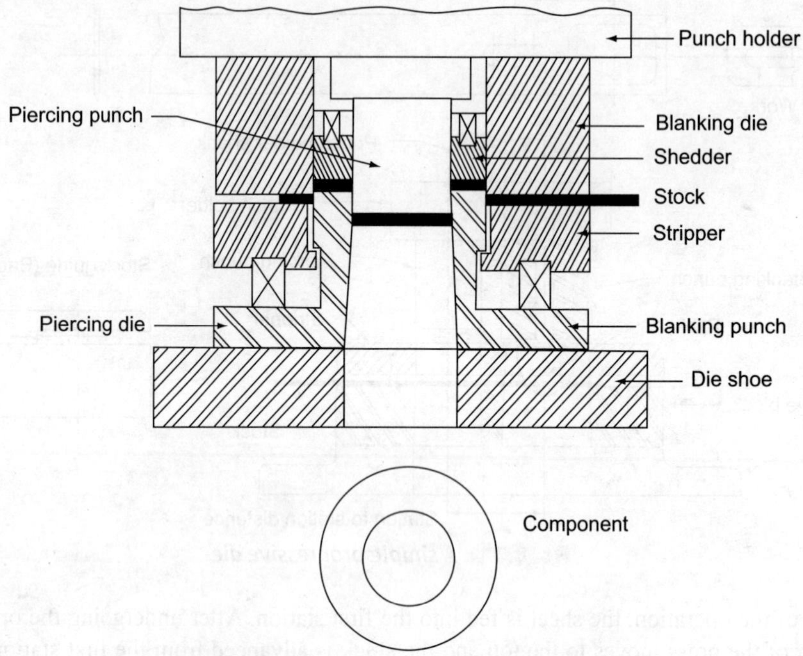

Fig. 8.28 *Compound die for making a washer*

Compound dies are somewhat slower than the corresponding progressive dies in operation. But higher tolerances can be achieved in them than in the progressive dies. This is mainly because the part located in one position undergoes all the operations. Also in compound dies, small strips can be advantageously used, whereas in progressive dies very long strips are required to cover all the stations.

Combination Die

A combination die is same as that of a compound die with the main difference that here non-cutting operations such as bending and forming are also included as part of the operation. Often the nomenclature compound and combination are interchangeably used for the same type of die.

8.10.2 Die Construction

Screws and Dowels

Socket head cap screws and dowels are universally used for joining the die components together. Dowels are used for alignment and screws for keeping them intact. The head of the cap screw should be kept in a recessed (counter bored) hole to eliminate any projection. The recess should be deeper than the head depth by about 3 mm to account for the resharpening of the die as shown in Fig. 8.29.

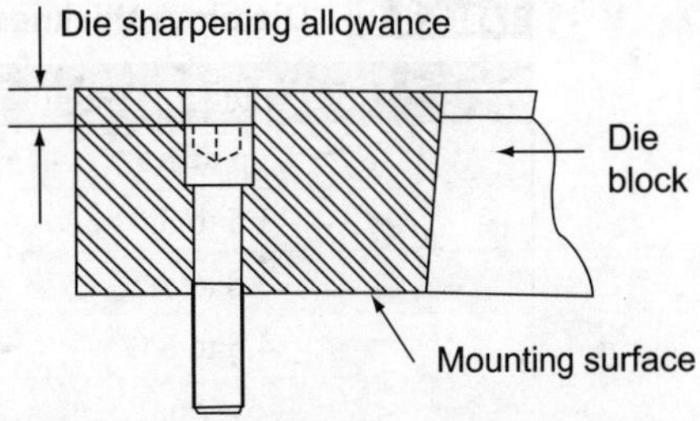

Fig. 8.29 *The use of screws and dowels in die construction*

For proper alignment of components, only two dowels are required. If the component is too small, the dowels would only weaken it and the construction would not offer much resistance to any lateral deflection which may be caused by the unbalanced side thrust. When the height of the component denoted as H in Fig. 8.30 is greater than four times the dowel diameter, d, it is good practice to relieve the dowel holes as shown. The relieving would help in the finishing of the dowel hole. The relieving is to be done before the reaming of the dowel hole and on the surface opposite to the mounting surface.

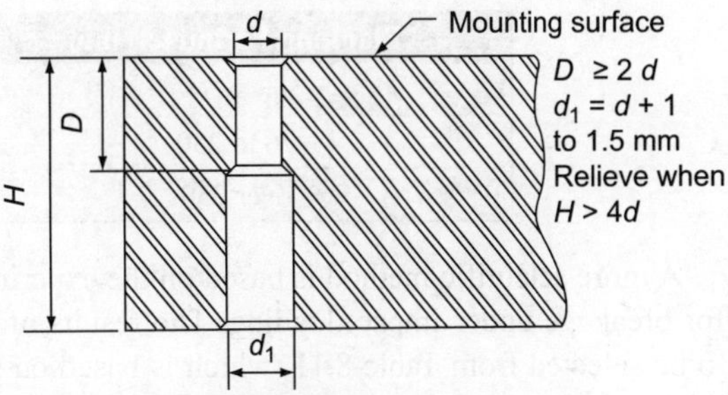

Fig. 8.30 *Dowel hole*

The dowel holes should always be extended through the component and never a blind one. Since dowels are press fitted into the component, it is practically impossible to remove a dowel from a blind hole by conventional methods. The dowels should be located as far apart as possible to increase the locational accuracy. As far as possible, the dowels and cap screws should be located towards the outer edge of the die block and away from the blank contour. The dowels and screws should be located from the outer edge at a distance of 1.5 to 2 times their diameter.

To keep the components in place, one cap screw may be sufficient, but it is general practice to keep two cap screws along with two dowels. But for small components, space may not be available for the second screw. The diameters of dowels and screws depend on the component size. The most commonly used size is 10 mm up to a component area of 4500 mm^2. For larger component areas up to 6500 mm^2, three cap screws are used. For heavier stock thicknesses, larger sizes up to 15 mm can be used. It is also a normal practice, to make the dowel size and screw size the same.

To avoid incorrect assembly of the symmetrical solid die blocks after repair, the dowels are positioned asymmetrically with respect to the screws, so that assembling can be done in only one position. This is termed as fool-proofing of the die.

Die Block

The die block size essentially depends on the work piece size and stock thickness. Though sometimes, the type of blank contour and the type of die may also influence the choice of die block size. A number of thumb rules are available which are essentially based on practical experience. These are summarised in Tables 8.9 and 8.10.

For softer materials such as non-ferrous alloys, die blocks can be made thinner and similarly for higher shear strength materials they can be made thicker. The minimum distance between the die opening and the outside edge of the die block should be from 30 mm to 1.25 times the die block thickness. This should be increased to 1.5 to 2 times for larger dies or when the component contour has sharp corners.

TABLE 8.9 Die block thicknesses for mild steel stock

Stock Thickness, mm	Die Block Thickness, mm
up to 1.5	20 to 25
1.5 to 3.0	25 to 30
3.0 to 4.5	30 to 35
4.5 to 6.0	35 to 40
over 6.0	40 to 50

TABLE 8.10 Die block thicknesses based on blanking perimeter

Blanking Perimeter, mm	Die Block Thickness, mm
up to 75	20
76 to 250	25
over 250	30

A more scientific method is based on the experimental results obtained from a series of tests on die plates for breakage under impact loading. The results of which are given in Table 8.11. First the die thickness is to be selected from Table 8.11, which is based on stock thickness and its shear strength. The minimum die thickness is to be 10 mm.

TABLE 8.11 Die block sizes

Stock Thickness, mm	Die Thickness for 1 MPa of Shear Strength, mm
0.25	0.05
0.50	0.10
0.75	0.14
1.00	0.18
1.25	0.21
1.50	0.24
1.75	0.27
2.00	0.29
2.25	0.31
2.50	0.32

The values given in Table 8.11 are for smaller blanks with cutting perimeters of less than 50 mm. For those with larger perimeters, a correction factor given in Table 8.12 is to be applied to the die thickness obtained from Table 8.11.

If the die plate is properly supported in a die shoe, the thickness can be reduced to a proportion of as much as 50 percent. A grinding allowance of 3 to 5 mm is to be added to the die thickness to account for the necessary die sharpenings during the life of the die.

The die block should be able to withstand the impact of punch striking the material. To this end, one should consider the smallest area of cross section of the die, A, as shown in Fig. 8.31. The value of *A* should be 1.25 times the die thickness for smaller dies and 1.5 to 2.0 times for larger dies. To check for the impact

TABLE 8.12 Expansion factors for die thickness

Cutting Perimeter, mm	Expansion Factor
up to 50	1.00
51 to 75	1.25
76 to 150	1.50
151 to 300	1.75
301 to 500	2.00

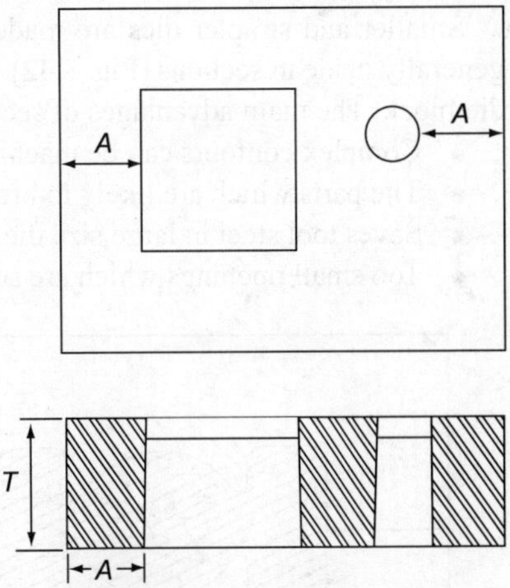

FIG. 8.31 *Die block design*

criterion, the cross sectional area resisting this force is $A \times T$. The impact force should be less than 770 MPa. Otherwise, either A or T should be increased for the safety of the die.

Example 8.7

Design a die block for blanking an 80 mm circle in 1 mm thick C20 cold rolled steel.

From Table 8.2, shear strength = 390 MPa

Shearing load = $390 \times \pi \times 80 \times 1 = 98.018$ kN

Case (i) From Table 8.9,

Die thickness = 22 mm

Die opening to edge = $1.25 \times 22 = 27.5$ mm $\approx$ 30 mm

Hence die block is to be $140 \times 140 \times 22$ mm

Case (ii) From Table 8.10

Perimeter of cut = $\pi \times 80 = 251.327$ mm

Die thickness = 30 mm

Die opening to edge = 30 mm

Hence die block is to be $140 \times 140 \times 30$ mm

Case (iii) From Table 8.11

Die thickness = $0.18 \times 390 = 70.2$ mm

Expansion factor from Table 22.4 is 1.75

Die thickness = $1.75 \times 70.2 = 122.85$ mm $\approx$ 125 mm

Assuming it to be properly supported in die shoe

Die thickness = 65 mm

After adding grinding allowance, die thickness = 70 mm

The die cavity to edge distance = 30 mm

$$\text{Impact pressure} = \frac{98018}{30 \times 70} = 46.68 < 770 \text{ MPa}$$

Hence the die is too safe and therefore the thickness can be reduced to say 25 mm.

$$\text{Impact pressure} = \frac{98018}{25 \times 70} = 130.7 < 770 \text{ MPa}$$

Die block size is therefore $140 \times 140 \times 25$ mm

Smaller and simpler dies are made generally in a single block. But complex shapes or larger dies are generally made in sections (Fig. 8.32). A number of die sections are assembled together to form the complete die block. The main advantages of sectional dies are:

- Complex contours can be machined easily in parts,
- The parts which are likely to break in service can be easily replaced,
- Saves tool steel in large size dies, and
- Too small openings which are sometimes difficult to machine internally, can be partitioned.

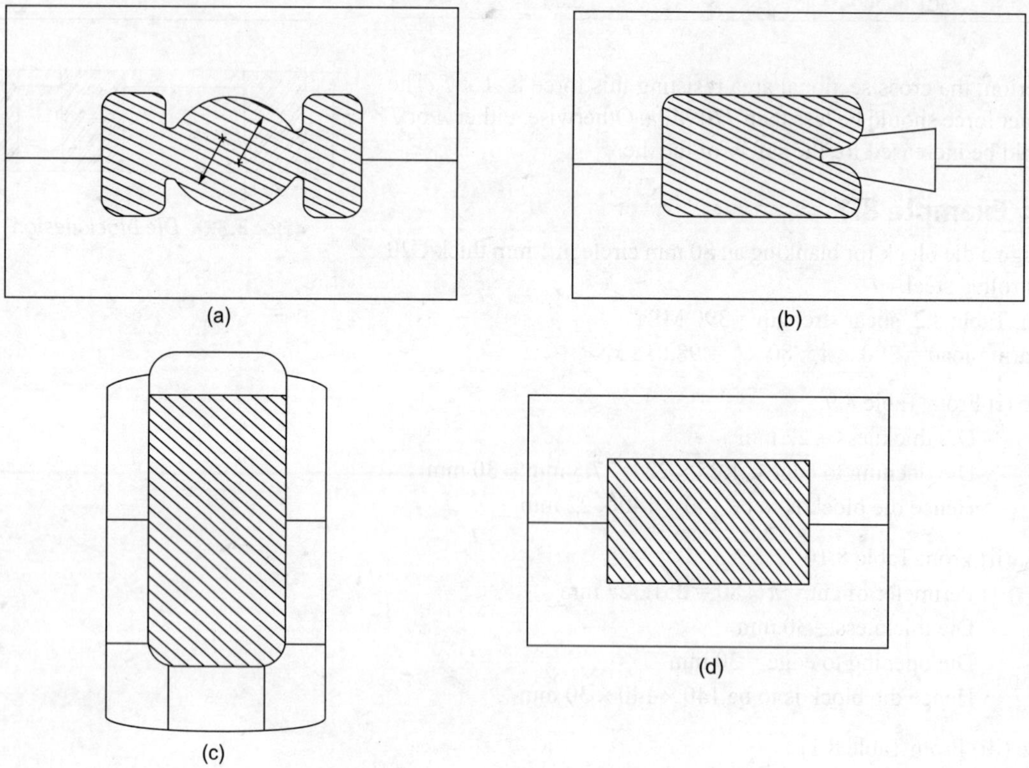

FIG. 8.32 *Sectional die construction*

The die shown in Fig. 8.32(a), has the shape of a cavity which is very difficult to produce as a single die cavity, whereas as shown in two identical pieces, it is easily produced as external cavities. The die cavity is shown hatched in the figure. In the die cavity shown in Fig. 8.32(b), the projecting portion into the cavity is likely to be broken during the operation, can be replaced as an insert very easily. The tapering in the shank of the insert is to ensure the proper location of the insert during the operation of the die. The die cavity shown in Fig. 8.32(c) is very large and therefore machining it from solid tool steel would mean large expense, hence it is made of a number of sections reducing the cost of the die material as well as the machining cost. The die cavity in Fig. 8.32(d) has an internal radius which is very small and cannot be ground internally. In the sectional die construction, the grinding of this internal radius is simplified as shown.

The components of a sectional die are held together by means of dowels and screws to a die holder or a die shoe, as shown in Fig. 8.33(a). The components should be butted against each other and should be strong enough to withstand the shearing forces on them. When sections do not have enough space for dowels and screws, they may be held in the die holder by means of a tapered wedge which is held to the die holder by means of screws as shown in Fig. 8.33(b).

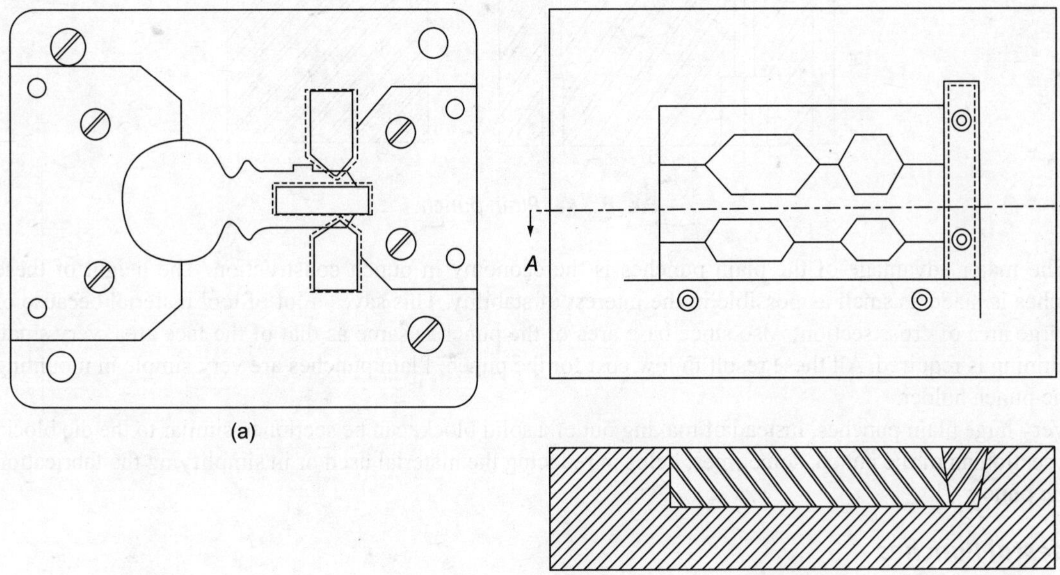

(a)

Section *A* – *A*

(b)

Fig. 8.33 *Sectional die construction*

8.10.3 Punch Design

The choice of the type of punch and its design depends on the shape and size of the pierced or blanked contour and the work material. For example, large cutting perimeters require large punches which are inherently rigid and can be mounted directly. However, smaller size holes require punches which may have to be supported during the operation, and therefore need to have other mechanisms to join it to the punch holder. The punches can be broadly classified as:

- Plain punches,
- Pedestal punches, and
- Punches mounted in punch plate.

Plain Punches

By far, these are the simplest of all types of punches. These are made of solid tool steel block and are directly mounted to the punch holder. These punches are joined to the holder by means of dowels and screws, as shown in Fig. 8.34. These must be large enough to provide the necessary space for dowels and screws as well as the necessary strength to withstand the punching force. The length and width of these punches should be greater than the height of the punch. Sometimes, it may become necessary to have the height of the punch greater than either length or width. This is when the punch is excessively long and the work performed is not heavy or unbalanced.

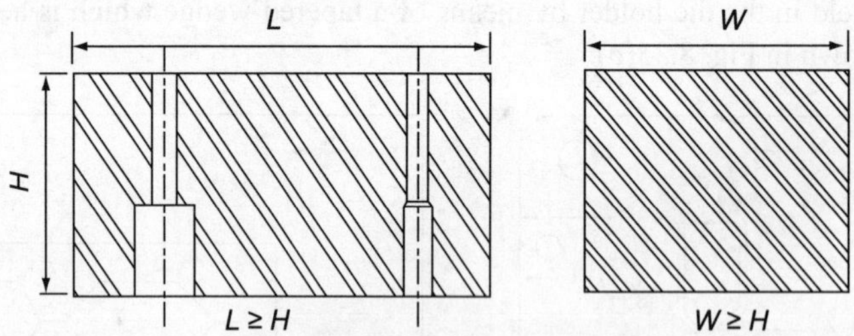

FIG. 8.34 Plain punch

The major advantage of the plain punches is the economy in punch construction. The height of these punches is made as small as possible in the interest of stability. This saves a lot of tool material because of its large area of cross section. Also since base area of the punch is same as that of the face area, very small machining is required. All these result in low cost for the punch. Plain punches are very simple in mounting to the punch holder.

Very large plain punches, instead of making out of a solid block, can be sectioned similar to the die block. The sectioning of the punch, sometimes, helps in reducing the material used or in simplifying the fabrication of the punch.

Pedestal Punches

Also called flanged punches, or shoulder punches, the pedestal punches are characterised by the large base surface compared to the cutting face. The flanged portion which is an integral portion of the punch, as shown in Fig. 8.35, offers excellent stability to the punch. All the cutting force exerted is dispersed by the broad base area and thus pedestal punches are advantageous for heavier work.

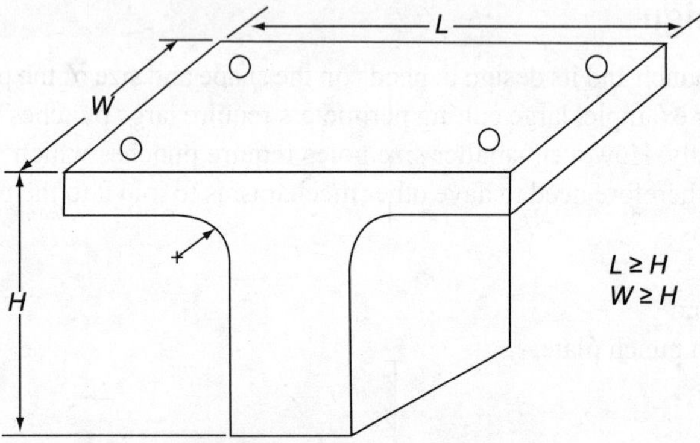

FIG. 8.35 Pedestal punch

The method of mounting is similar to that of the plain punches. In fact, the flanged portion of the punch is easier to mounting. The proportions of the pedestal punches are also similar to that of the plain punches. The length and width of the base should be larger than or equal to the height of the punch. The flange thickness and the fillet radius are to be liberally provided to withstand the large forces coming on the punch.

The fillet radius provided in a pedestal punch reduces the stress concentration, but to strengthen the joining of the punch to the base, an angular fillet as shown in Fig. 8.36(a), would be useful. This is particularly so for narrow punches. The ends of the angular portion should be smoothly blended to the punch side walls as well

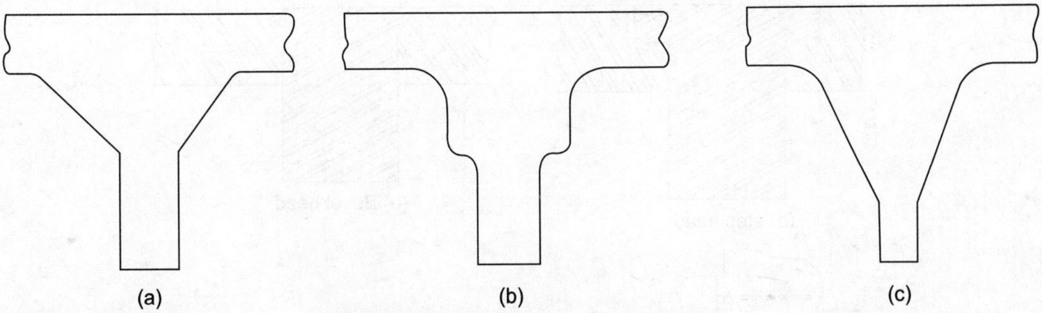

FIG. 8.36 *Pedestal punch construction*

as the flange portion. Other ways of strengthening pedestal punches, when the cutting face area is small and weak, are shown in Fig. 8.36(b) and (c).

Punches Mounted in Punch Plates

A punch plate is used generally to locate and hold the punch in position. This is a useful way of mounting, especially for small punches. There are a number of ways in which a punch is mounted in a punch plate, the choice depending on the size of the punch.

A simple method of assembling a plain punch in the punch plate is shown in Fig. 8.37. Here, the punch has uniform cross section throughout and is fastened to the punch holder by means of the screws through the punch itself. The punch plate has the necessary holes for locating the punch properly. Clearance holes are provided in the punch plate for the positioning of the squared or sharp cornered punches.

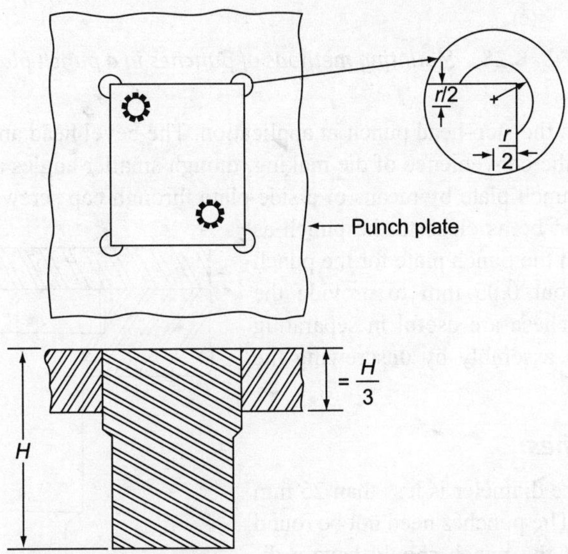

FIG. 8.37 *Mounting of a plain punch in a punch plate*

The other methods of mounting punches in a punch plate are presented in Fig. 8.38. In Fig. 8.38(a) is shown a step-head punch which is positioned as well as retained by the punch plate. The width and height of the step head should be made as small as possible, of the order of 1 to 3 mm. The bevel head punch shown

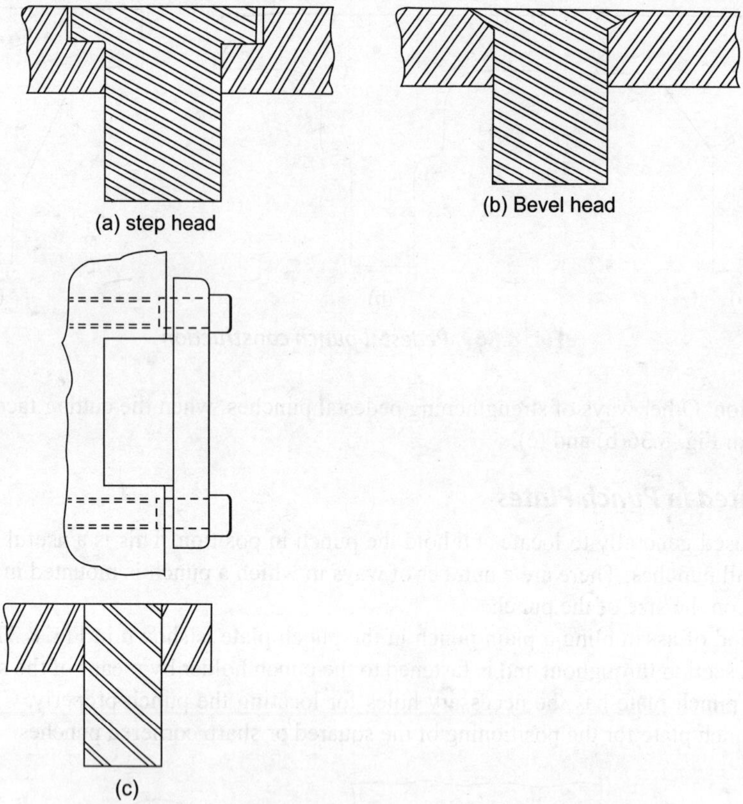

Fig. 8.38 *Mounting methods of punches in a punch plate*

in Fig. 8.38(b) is similar to the step-head punch in application. The bevel head angle is of the order of 30 to 45 degrees depending on the convenience of die making, though smaller angles are preferable for strength. A punch clamped in the punch plate by means of a side plate through cap screws is shown in Fig. 8.38(c). The clamping screws must be as close to the punch as possible. Also, the depth in the punch plate for the punch should be smaller by about 0.05 mm to provide the necessary locking force. These are useful in separating the punch alone from the assembly by unscrewing the two clamping screws.

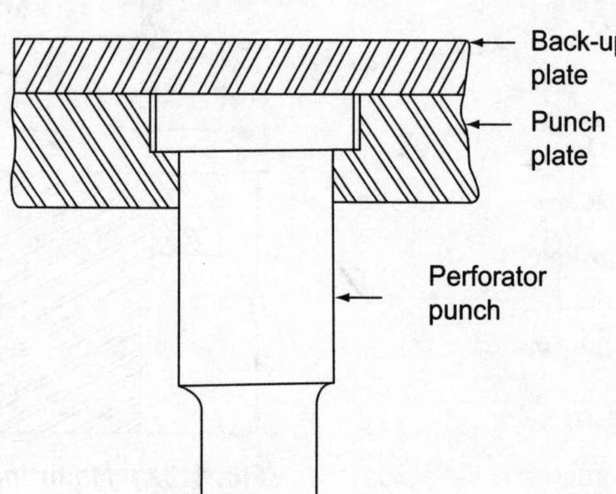

Fig. 8.39 *Perforator punch*

Perforator Type Punches

Punches whose cutting face diameter is less than 25 mm are termed as perforators. The punches need not be round but the inscribed circle of the punch should have a diameter less than 25 mm. As a rule, all the perforators are mounted in a punch plate.

The simplest and the most common perforator is the step-head type shown in Fig. 8.39, mounted in a punch

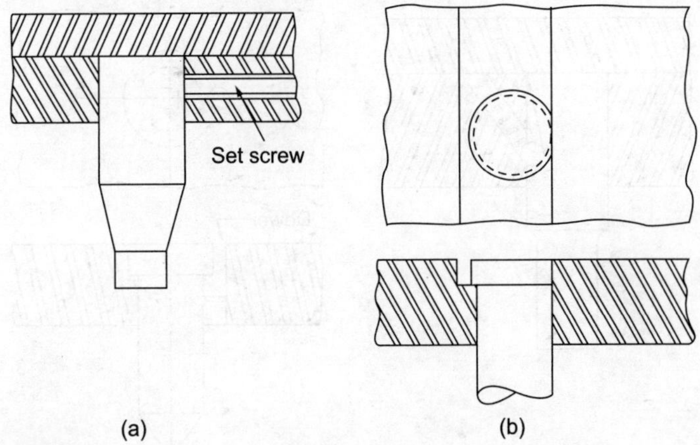

FIG. 8.40 *Methods of mounting perforator punches in punch plates*

plate by means of the step head. The proportions of these punches are standardised and are commercially available in various sizes.

If the cutting face is round, then assembling the punch in any orientation, as in Fig. 8.39, is possible. But for punches with contours other than round, some means for preventing the rotation in the punch plate is necessary. Some typical arrangements are shown in Fig. 8.40 and 8.41. Figure 8.40(a) shows the arrangement wherein the punch is kept in position by a set screw through the punch plate against the punch shank. This would be useful for small production runs and frequent changes. The punch head is ground flat on one side and corresponding slot is made in the punch plate, as shown in Fig. 8.40(b), to keep the punch from rotating. Making the slot and grinding the flat on the punch is fairly an easy job and hence this arrangement is generally used.

Other popular methods of preventing punch rotation are by means of the spring loaded ball as shown in Fig. 8.41(a), and the use of a dowel slot of the size of 3 mm in the head of the punch as shown in (b).

Quill Punches

For piercing very small holes, less than 6 mm, it is desirable to provide extra support to the punch shank by means of a closely fitting quill, as shown in Fig. 8.42. Quilled punches are more expensive if made individually because of the close fitting required between the punch and the quill sizes. Therefore, they are mass produced in various standard sizes.

Back-up Plate

Hardened back-up plates are normally required to be kept between smaller perforator punches and the punch holder. Because of the smaller area of the perforators, they have the tendency to dig into the softer punch holder in the absence of a backup plate. The backup plate is about 6 mm thick.

Slug Ejection

Sometimes, the slug that is punched clings to the punch face and comes along with it during the return stroke. Normally, the die is supposed to restrain the slug from moving along with the punch because of the spring back. But in slugs of very small sizes such as those from thin sheets, the spring back being too small, the slug is likely to be drawn along by the punch. Hence, means are to be provided for the ejection of the slug, particularly in small perforator punches.

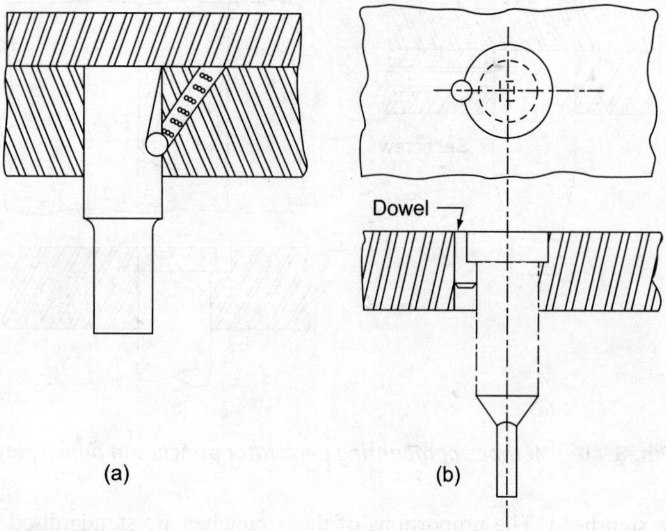

FIG. 8.41 *Methods of preventing rotation of punches in a punch plate*

The most common method employed for slug ejection is the spring loaded pin shown in Fig. 8.43(a). Another method is the blowing of the slug by the compressed air through the punch as shown in Fig. 8.43(b). The compressed air can be supplied continuously or in intervals depending on the punching frequency. The arrangement of compressed air being more complex is generally not used and only the pin ejector is more commonly used. It is not possible to provide the ejector pin in a punch whose diameter is less than about 1 to 1.5 mm. Hence in these cases, the cutting face may be provided with a concave shear or a notch, so that the slug would be distorted and would not cling to the punch face.

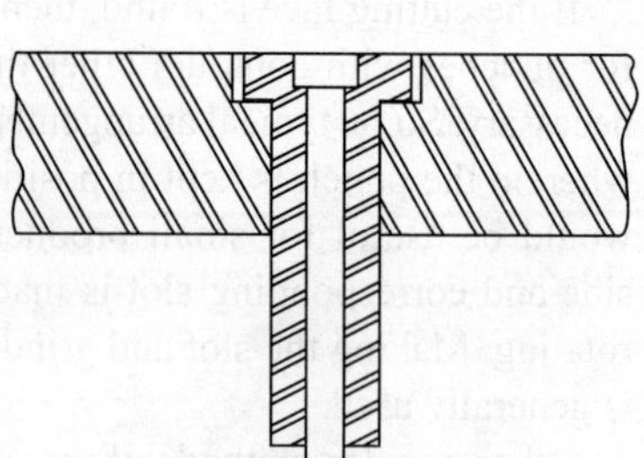

FIG. 8.42 *Quill punch*

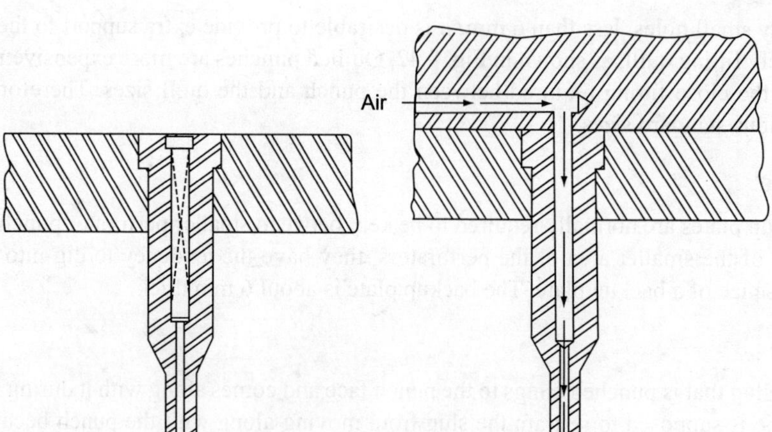

FIG. 8.43 *Slug ejectors*

8.10.4 Pilots

The pilots in progressive dies are used in order to bring the stock into the correct position for the succeeding blanking or piercing operations. The action of a pilot is schematically shown in Fig. 8.44. The stock when fed manually is slightly over fed against a stop, as in Fig. 8.44(a), provided by the pilot moving through the punched hole. The pilot when moves further down brings the sheet properly in position as shown in Fig. 8.44(b).

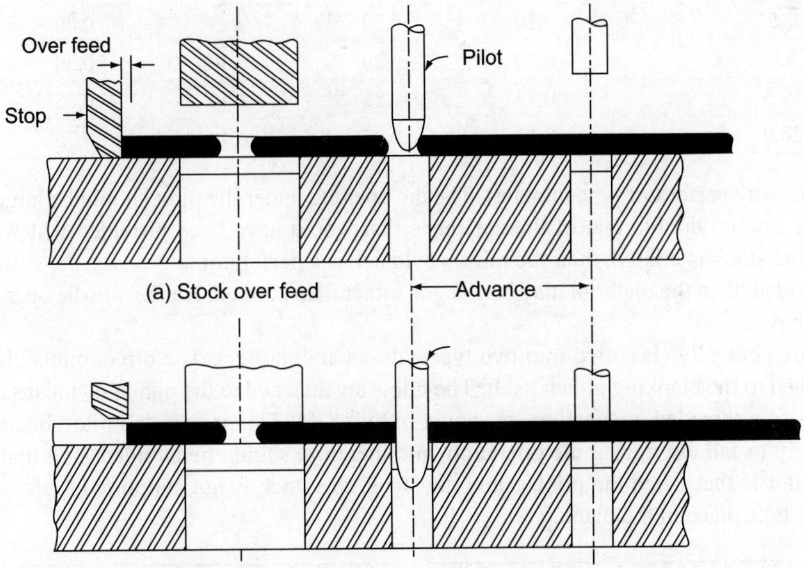

Fig. 8.44 *Piloting action in manually fed stock*

The stock which is fed mechanically is slightly under fed, and the pilot pulls the stock for the correct registering. The main reason for this is that most mechanical feeders have provision for stock movement in only one direction.

The fit between the pilot size and the pierced hole determines the accuracy of the component produced. Too tight a fit between the pilot and pierced hole results in friction which would spoil the component as also, there would be excessive pilot wear. Depending on the type of work, the pilot size may deviate from the pierced hole size as follows:

0.050 to 0.100 mm	average work,
0.025 to 0.050 mm	close work, and
0.013 to 0.018 mm	high precision work

The amount of over feed permissible is presented in Table 8.13.

The length of the pilot should be such that it registers in minimum amount of time. Too short a pilot may not be able to register properly. Too long a pilot cause excessive friction which may be undesirable. The correct size of the pilot should enter the pierced hole fully, before any of the punches come into contact with the stock. The extra length of the pilot beyond the punch face may be approximately of the order of the sheet thickness or 1.5 mm, whichever is greater.

TABLE 8.13 Permissible over feed of stock (mm)

Stock Thickness, (mm)	0.130	0.400	0.750	1.500	3.000
Pilot diameter, mm					
3.0	0.05	0.08	0.13	–	–
4.5	0.08	0.13	0.20	0.25	–
6.0	0.10	0.20	0.25	0.40	–
7.5	0.10	0.20	0.25	0.40	–
9.0	0.10	0.20	0.30	0.50	0.75
12.5	0.13	0.25	0.40	0.75	0.75
20.0	0.13	0.25	0.40	0.75	1.00

Pilots are in a way perforator type punches. The die opening under the pilot should be larger than the pilot diameter by an amount double that of the clearance that would have been provided if it was a perforator. Sometimes, if the stock is misfed, then the pilot would act as a perforator and pierce the stock. If a large die opening is provided, then the material may be flanged rather than pierced. Hence, the die opening is restricted to the size as above.

The pilots are generally classified into two types: direct and indirect. The direct pilots also called punch pilots are attached to the blanking punch itself. The pilots are attached to the punch by means of either a press fit or by means of a threaded connection, as shown in Fig. 8.45. The press fitted pilot, being not positively retained, is likely to fall out during the operation, and therefore should be avoided. The main disadvantage of the punch pilot is that when the pilot enters the stock, the stock is not supported underneath and hence distortions may take place in the blank.

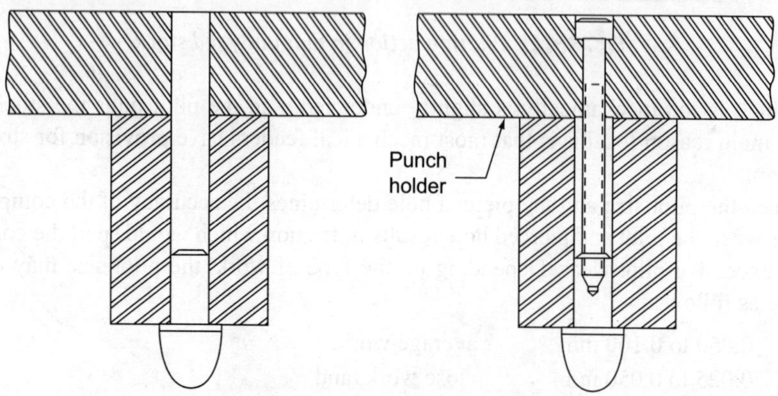

FIG. 8.45 *Punch pilots*

The indirect pilot is not attached to the punch and is on its own. It enters a previously pierced hole, as shown in Fig. 8.46. The indirect pilot should be positioned at some distance away from the blanking punch to provide for the necessary stock support during the entry of the pilot. Solid pilots are sometimes harmful as in case of misfed stocks whose thickness is above 1.5 mm. The pilot in such a case would break. Though, with thinner sheets, it may simply pierce the sheet with no damage to the die. Hence for thicker stocks, it is desirable to use spring loaded pilots as shown in Fig. 8.46.

8.10.5 Stripper and Stock Guide

The stripper removes the stock from the punch after a piercing or blanking operation. The function of a stock guide is as the name implies, guide the stock through the various stations. The strippers are classified into two types: channel or box stripper and spring operated stripper or pressure pad.

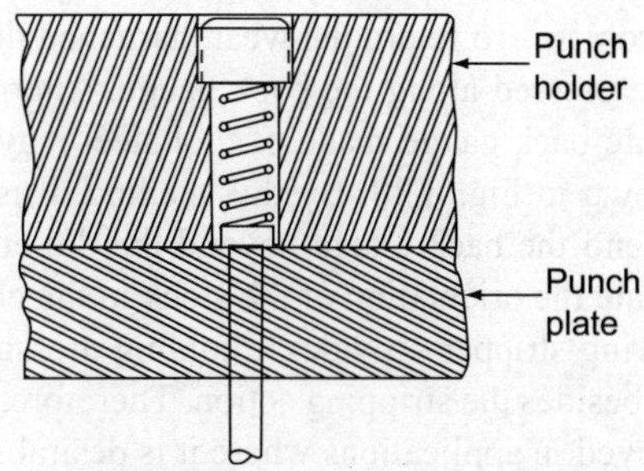

FIG. 8.46 *Spring loaded pilot*

Channel Stripper

The channel stripper is simpler and easier to make. Fewer components are required for its construction and consequently, and is economical. Also it is very rigid and therefore useful where large stripping forces are required. A channel stripper generally consists of a solid rectangular block of the same width as that of the die block for convenience. A slot or channel, slightly longer than the width and thickness of the stock is milled on one side of the block through which the stock is fed, as shown in Fig. 8.47. It is joined to the die block by means of dowels and cap screws as that of the other components.

The height of the channel, H in Fig. 8.47 must be at least 1.5 times the stock thickness. For cases where the stock is to be lifted over a pin stop, the height may suitably be increased. The width, W of the channel should accommodate the stock width and any of its variation over the length of the die. Approximately 0.25% of the die length would be able to take care of variations in most of the materials. If the stock is badly cambered, then the width may accordingly be altered. The clearance hole in stripper for the punch should be such that it is more than the size of the punch but not more than half the sheet thickness. More clearances should be provided for punching harder materials. The thickness of the channel stripper should be able to provide the necessary stripping force. Also, it is necessary that the thickness should be sufficient to accommodate the socket head cap screws used for fastening it to the die block. The latter criterion when satisfied will automatically take care of the former.

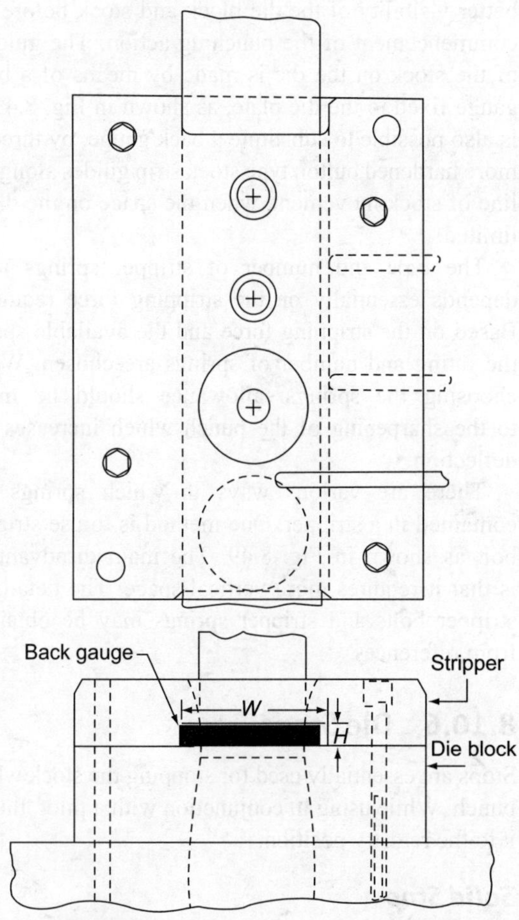

FIG. 8.47 *Channel stripper construction*

The materials generally used for making the channel stripper are low carbon steels. They are normally not heat treated and left in the soft condition only. Occasionally, the stripper may be made of tool steel, if its size is small. One side of the channel, called back gauge, acts as a guide for the stock movement. A stock pusher forces the stock against this back gauge for locating the stock. This is particularly important for close tolerance work with soft stripper where the wear of back gauge is a problem. To retard the wear, hardened dowel pins can be pressed along the back gauge. Alternatively a separate back gauge made of tool steel may be used, as shown in Fig. 8.48. By this method, it is possible to extend the back gauge beyond the die entrance to facilitate the orientation of the stock in the channel.

Spring stripper provides a holding pressure on the stock besides the stripping action. Therefore, these are employed in applications where it is desirable to keep the stock flat during punching or blanking operation. Other advantages of the spring stripper are the better visibility of the die block and stock before the commencement of the punching action. The guiding of the stock on the die is made by means of a back gauge fixed to the die plate, as shown in Fig. 8.49. It is also possible to substitute a back gauge, by three or more hardened button type stock strip guides along the line of stock movement, when the space on the die is limited.

The type and number of stripper springs used depends essentially on the stripping force required. Based on the stripping force and the available space, the rating and number of springs are chosen. While choosing the springs, allowance should be made to the sharpening of the punch which increases the deflection.

There are various ways in which springs are contained in a stripper. One method is to use stripper bolt as shown in Fig. 8.49. The main disadvantage is that it requires more vertical space. The details of stripper bolts and stripper springs may be obtained from references.

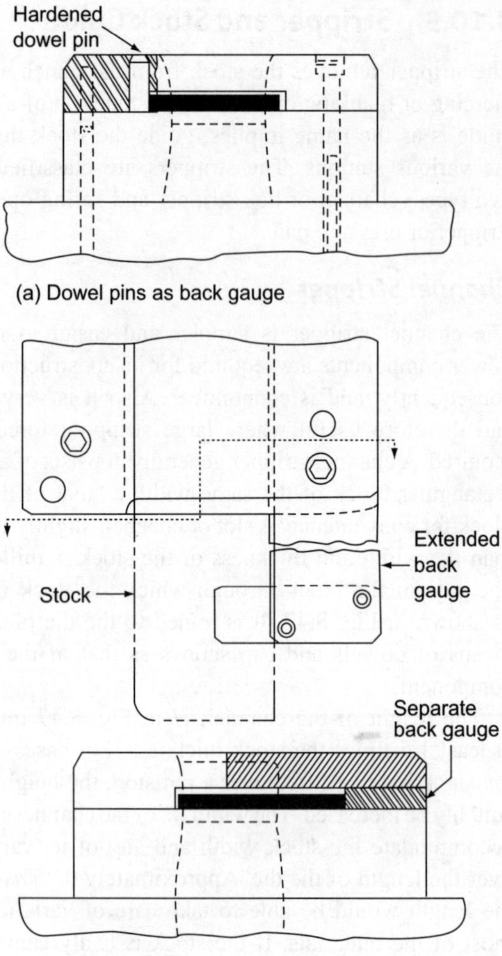

Fig. 8.48 *Separate tool steel back gauge*

8.10.6 Die Stops

Stops are essentially used for stopping the stock while feeding, so as to register the correct position under the punch. While using in conjunction with a pilot, the strip is stopped at a position from which, the pilot brings it to the registry position.

Solid Stop

Figure 8.50 shows a solid stop block which is used as a heel block for the final cut-off punch. Here, the actual registry position is specified by the position of the stop, since it is not using any pilot. The operator needs to be extremely careful to feed the strip up to the stop block and hold it there during punching. Otherwise the punch dimensions may be altered.

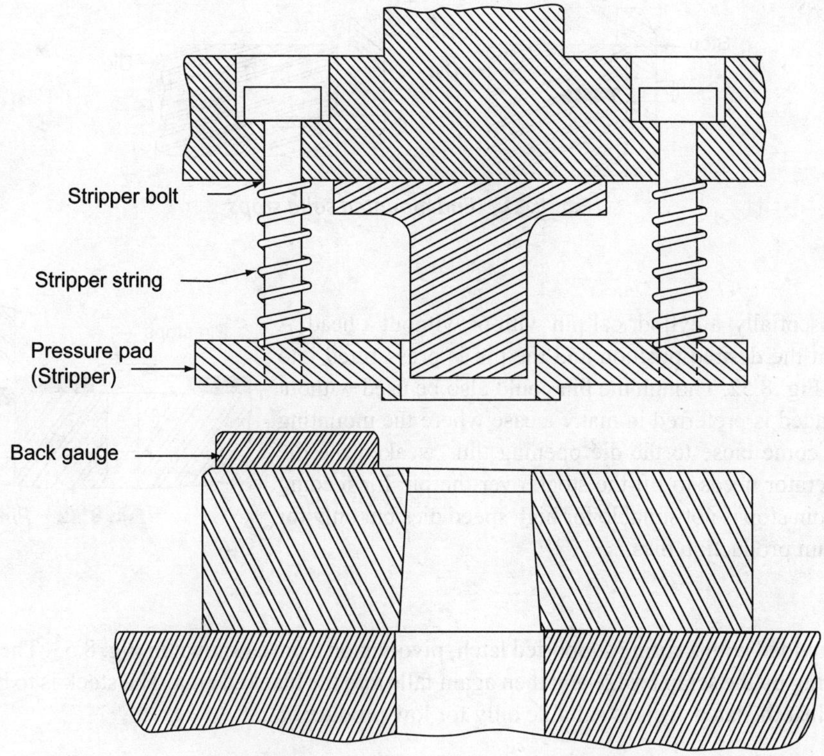

Fig. 8.49 *Spring loaded pressure pad*

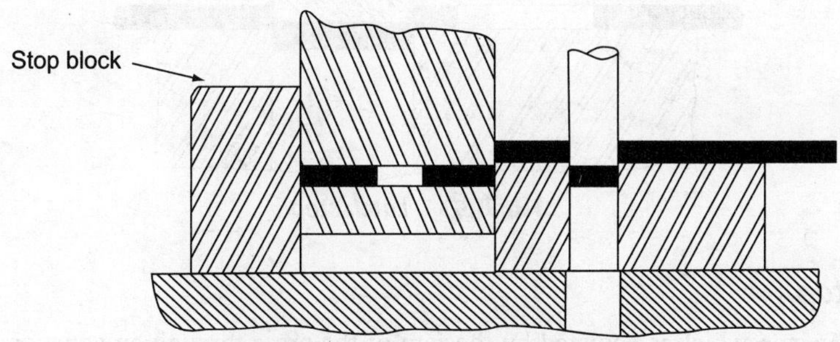

Fig. 8.50 *Solid stop without pilot*

To reduce the operator error, the pilot as shown in Fig. 8.44, is used in conjunction with the stops. The solid stop is made of tool steel and hardened to minimise the wearing because of the repeated strikings of the stock. A simple and more generally useful concept is the end mounting of the solid stop as shown in Fig. 8.51. Since the die block is sharpened, the mounting hole of the end mounted stop is elongated as shown in Fig. 8.51(a), so that no reworking on the stop is required. Alternatively, the design shown in Fig. 8.51(b) would also take care of the die sharpening.

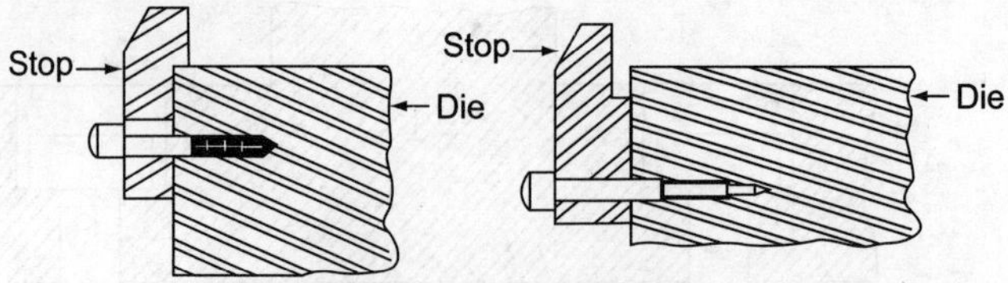

Fig. 8.51 *End mounted solid stops*

Pin Stop

This stop is essentially, a cylindrical pin with or without a head. A hole is made at the desired position and the pin is press fitted into it as shown in Fig. 8.52. Though the pin could also be used without a head, the headed is preferred in many a case where the mounting hole does not come close to the die opening thus weaken the die block. The operator needs to lift the stock over the pin for feeding and hence the pin stop is not suitable for high speed dies but only for low and medium production dies.

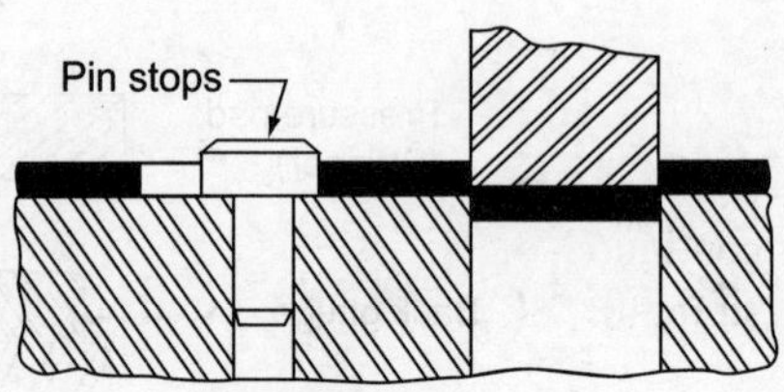

Fig. 8.52 *Pin stop*

Latch Stop

This stop is provided with a spring supported latch, pivoting on a pin as shown in Fig. 8.53. The latch is lifted against the spring by the scrap bridge and then again falls into the blanked area. The stock is to be pulled back to register against the latch. This is suitable only for low production.

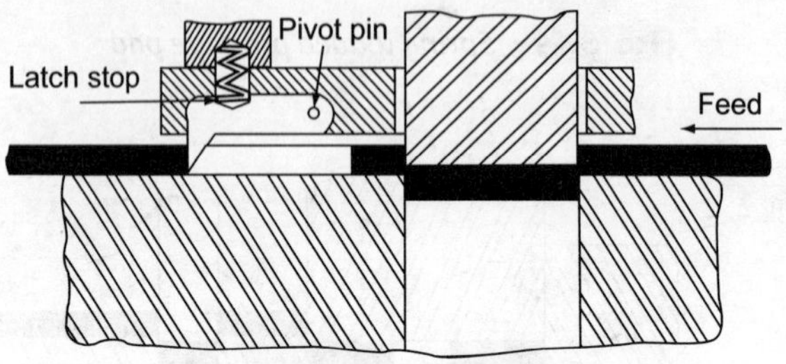

Fig. 8.53 *Latch stop*

Pivoted Auto Stop

This is an automatic stop which is actuated by the ram of the press through an actuator, which sometimes called the trigger, as shown in Fig. 8.54. It consists of a pivoted lever which is connected to the die shoe by a compression spring (not shown in the figure). During the ascent of the ram, the lever is tripped by the actuator so that it comes out of the blanked portion of the stock so that the stock may be fed. When the scrap bridge clears the stop lever, the actuator leaves the stop lever, thus allowing it to fall back into the blanked portion of the stock as shown in Fig. 8.54. A number of variants of this are available for use in different situations.

8.10.7 Stock Strip Layout

Since, the components are to be ultimately blanked out of a stock strip, hence , precaution is to be taken while designing the dies for utilising as much of stock as possible. It is also necessary in progressive dies, to ensure

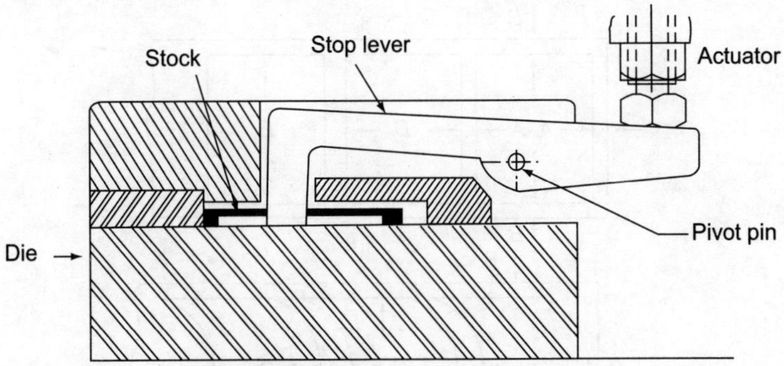

Fig. 8.54 *Pivoted automatic stop*

continuous handling of the scrap on the die block, which means that the scrap strip should have sufficient strength. Some of the terms that are generally used in relation to scrap strip are presented in Fig. 8.55.

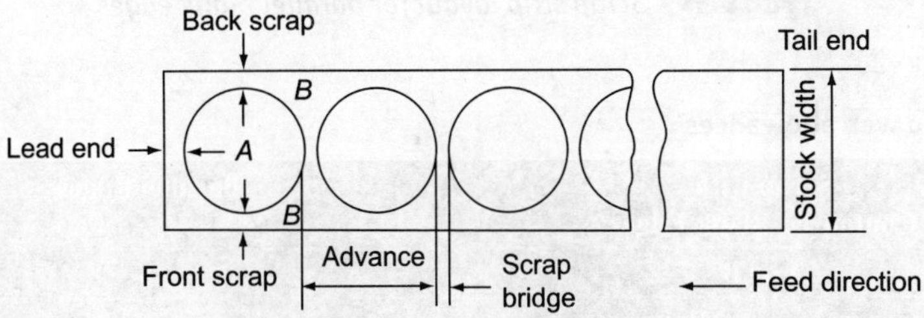

Fig. 8.55 *Scrap strip terminology*

The web of the stock to be left on the scrap strip depends on the thickness and width of the strip, the contour of the blanked shape and the type of blanking done. The parts with contours such as the ones shown in Fig. 8.56, generally provide for an extra material and therefore less scrap needs to be left on the stock strip. Those parts which have consecutive parallel blank edges, as shown in Fig. 8.57, require slightly higher amount of scraps. The scrap values to be left on the stock for these two types of layouts are presented in Table 8.14. These values are for the progressive dies with single pass layouts. For single station dies, the scrap need not be as strong as for progressive dies and hence these values can be reduced by about 25%.

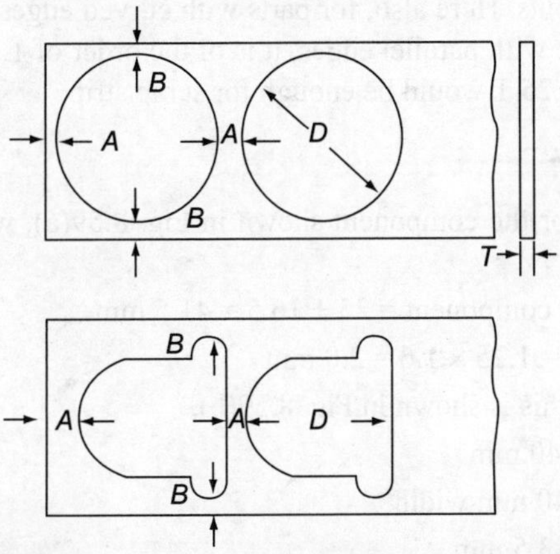

Fig. 8.56 *Scrap strip layout for contoured blanks*

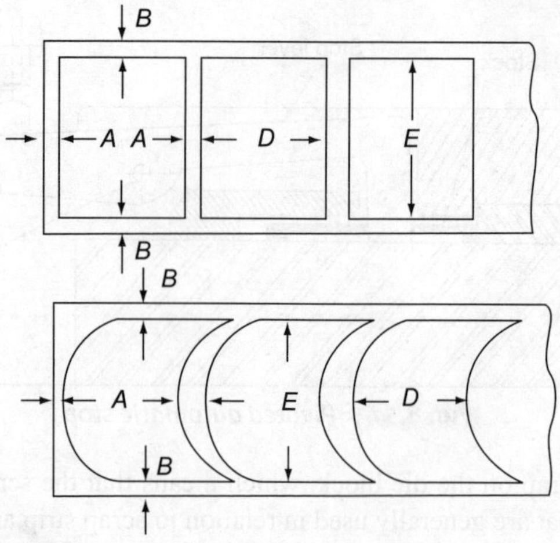

Fig. 8.57 *Scrap strip layout for parallel blank edges*

Table 8.14 Scrap web allowances

Max Dimension (mm)	Scrap with Contoured Blank Edges		Parallel Edges			
	A or B		A		B	
	General	Minimum (mm)	General	Minimum (mm)	General	Minimum (mm)
Up to 25	T	1.20	1.25 T	1.50	1.50 T	1.50
26 to 75	1.25 T	1.60	1.25 T	1.50	1.50 T	1.50
76 to 150	1.50 T	2.40	1.50 T	2.00	1.50 T	2.40
151 to 250	1.50 T	3.00	1.75 T	2.40	1.75 T	3.00
251 to 400	1.75 T	4.00	2.00 T	3.00	2.00 T	4.75

In the case of two-pass layouts such as the ones shown in Fig. 8.58, the scrap to be left can be somewhat less than that of single pass layouts. Here also, for parts with curved edges such as in Fig. 8.58(a), the scrap can be 1.25 T, whereas for those with parallel edges, it is of the order of 1.50 T. If the part has partly curved and partly parallel edges, then 1.25 T would be enough for scrap strip.

Example 8.8

Prepare the scrap strip layout for the component shown in Fig. 8.59(a), which is to be obtained in a brass sheet of thickness 1.6 mm.

The maximum dimension of the component = 25 + 16.5 = 41.5 mm

From Table 8.14, the scrap web = 1.25 × 1.6 = 2.0 mm

A simple single pass layout for this is shown in Fig. 8.59(b).

Strip width = 33 + 2 + 2 = 37 ≈ 40 mm

We can take a standard strip of 40 mm width.

Advance distance = 41.5 + 2 = 43.5 mm

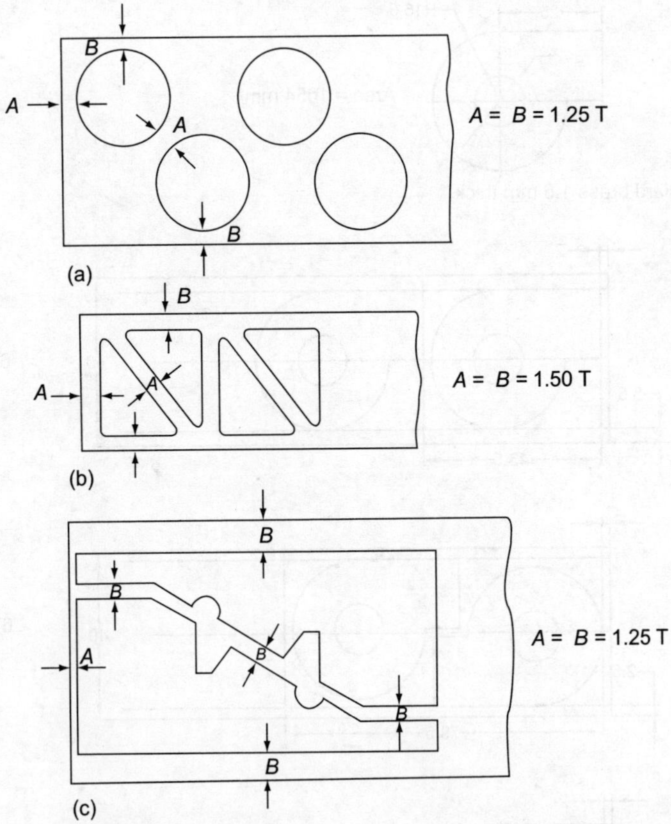

Fig. 8.58 *Scrap strip arrangement for two pass layout*

Area of the cross-section of one component

$$= 25 \times 16.5 + 0.75 \times (16.5)^2$$

$$= 1053.97395 \text{ mm}^2$$

$$\text{Percentage utilisation of stock} = \frac{1053.97 \times 100}{43.5 \times 40} = 60.57\%11$$

A somewhat better two pass layout is shown in Fig 8.59(c).

Strip width required = 33 + 2 + 2 + 2 = 39 ≈ 40 mm

Advance distance = 33 + 2 + 41.5 + 2 = 78.5 mm

$$\text{Percentage utilisation of stock} = \frac{2 \times 1053.97 \times 100}{78.5 \times 40} 12 = 67.13\%$$

A third possible layout with component in vertical position is shown in Fig. 8.59(d).

Advance distance = 33 + 2 = 35 mm

Strip width = 41.5 + 2 + 2 = 45.5 ≈ 46 mm

$$\text{Percentage utilisation of stock} = \frac{1053.97 \times 100}{35 \times 46} = 65.46\% 4$$

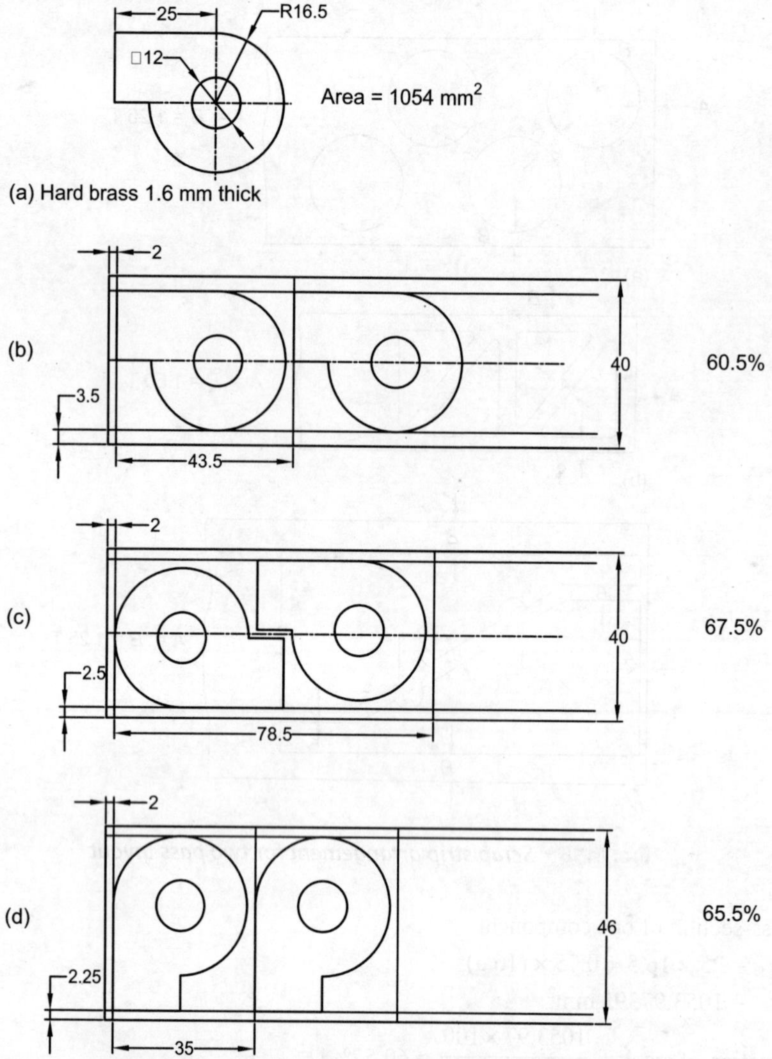

Fig. 8.59 *Possible stock strip layouts for a component*

Though the layout shown in Fig. 8.59(c) gives the highest utilisation, it requires a two pass layout whereas the single pass layout shown in Fig. 8.59(d) is the best with 65.46% utilisation.

8.10.8 Component Design for Blanking

Though every effort is made while preparing the stock strip layout, to maximise the utilisation of the material, sometimes it may be desirable to redesign the shape of the blank to get better stock utilisation. Modifications are needed from the standpoint of reducing as much scrap as possible.

The external contour of the component sometimes may not be conducive to a good layout and as a result, it may become necessary to leave more material on the strip as scrap. For example, when the component, shown in Fig. 8.60(a), is produced by a single pass layout as shown in Fig. 8.60(b), makes for a material

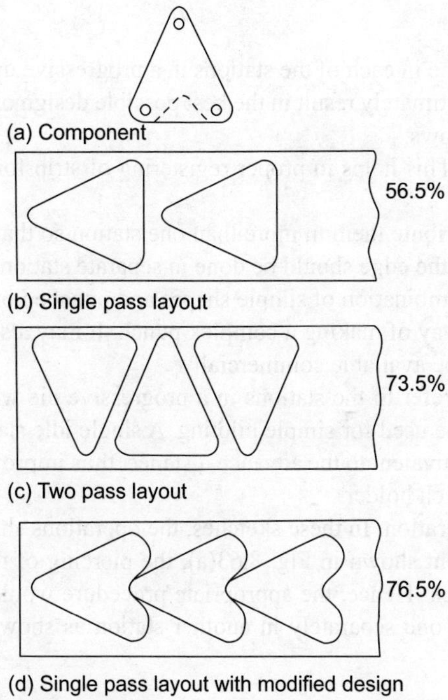

Fig. 8.60 *Improving stock utilisation*

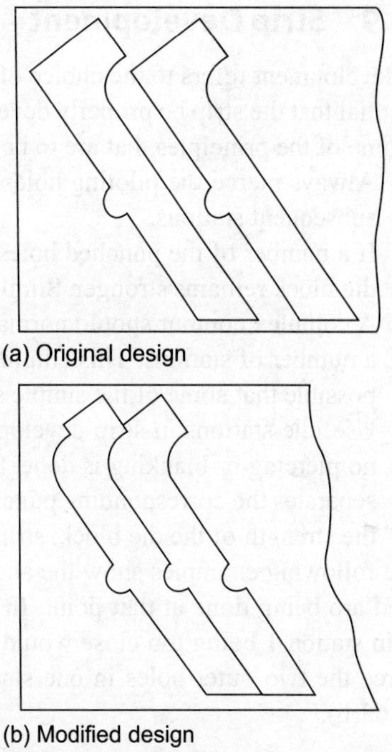

Fig. 8.61 *Redesign of blanks to improve stock utilisation*

utilisation of only 56.5 percent. An improved layout is therefore the two-pass layout with either that the two components are manufactured in a single press stroke or the strip is fed into the die twice. This practice improves the stock utilisation to 73.5%. But feeding stock twice into the die is cumbersome and not convenient. If a modification to the component as shown dotted in Fig. 8.60(a) is made, then the single pass layout shown in Fig. 8.60(d) can be achieved. This redesign besides simplifying the process, also improves stock utilisation to 76%. Any further improvement cannot be possible because of the nature strip of the component.

To get the best material utilisation, the edges of the component in the direction of the stock movement should be as similar as possible. For example, in Fig. 8.60 a small portion of the edge was made similar and a large improvement in stock utilisation was obtained. Figure 8.61(a) shows an example where a slight misfit between the two edges, cause a large amount of scrap. But a careful modification as in Fig. 8.61(b) to make the two parallel edges complimentary makes the scrap almost non-existent.

Sometimes, the extra edge kept on a component because of its angular position may cause unnecessary increase in the stock width, as in Fig. 8.62. In such cases, trimming a small amount of the offending angular projection, as shown by dotted lines in the figure, helps in decreasing the width of the stock without unduly affecting the functionality. This helps in saving a large amount of stock as indicated in Fig. 8.62.

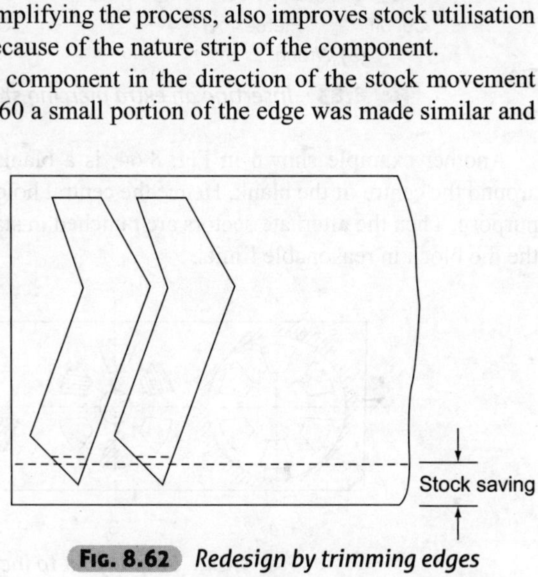

Fig. 8.62 *Redesign by trimming edges*

8.10.9 Strip Development

Strip development refers to the choice of operations to be done in each of the stations in a progressive die. It is essential that the strip be properly developed since it will ultimately result in the best possible design of the die. Some of the principles that are to be followed are as follows.

1. Always pierce the piloting holes in the first station. This helps in proper registering of strip for the subsequent stations.
2. If a number of the punched holes are very close, distribute them in more than one station so that the die block remains stronger. Similarly, holes nearer to the edge should be done in separate stations.
3. A complex contour should normally be split into a combination of simple shapes and punched out at a number of stations. This eliminates the expensive way of making a complex punch. It may also be possible that some of the simple shapes may readily be available commercially.
4. Use idle stations in strip development. Idle stations refer to the stations in a progressive die where no piercing or blanking is done. Some of them may be used for simple piloting. A single idle station separates the corresponding punches by a distance equivalent to the advance distance, thus improving the strength of the die block, stripper and also the punch holder.

The following examples show the above principles in operation. In these sketches, the operations shown hatched are being done at that point. In the strip development shown in Fig. 8.63(a), the piercing of three holes in station 1 being too close would weaken the die block. Hence, the appropriate procedure would be to pierce the two outer holes in one station and the middle one separately in another station as shown in Fig. 8.63(b).

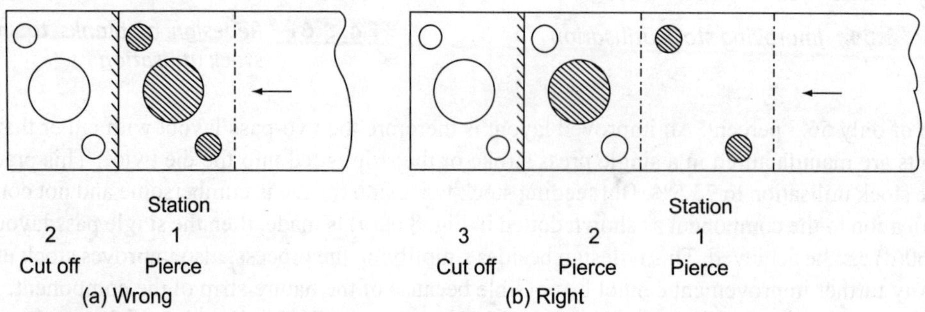

Fig. 8.63 *Inserting an extra piercing station to increase the strength of the die block*

Another example shown in Fig. 8.64, is a blank with six sectors which are very close to one another around the centre of the blank. Hence the central hole is punched in the first station which is used for piloting purpose. Then the alternate sectors are punched in stations 2 and 3 as shown thus maintaining the strength of the die block in reasonable limits.

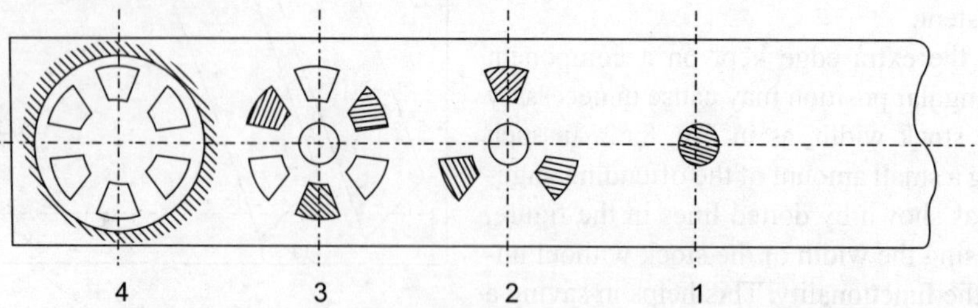

Fig. 8.64 *Strip layout to increase the strength of the die block*

In the strip layout shown in Fig. 8.65(a), the notching and piercing are done in station 1 and cut-off in station 2. But the die block becomes very small because of the cut-off being very close to the other punches. Hence it would be desirable to add an additional idle station 2, as in Fig. 8.65(b) which does the piloting of the strip.

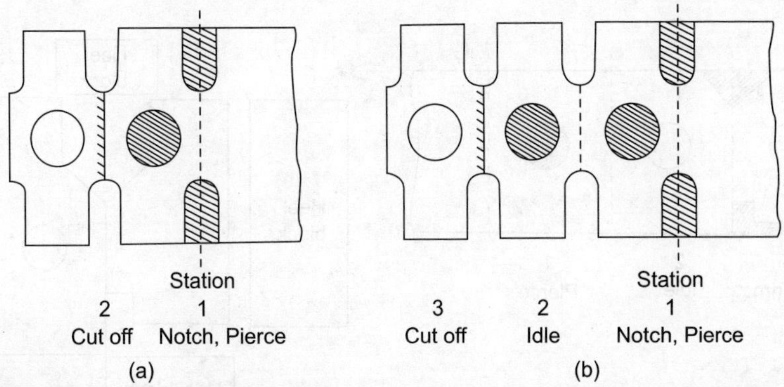

Fig. 8.65 *Incorporating an idle station to improve the strength of the die block*

Two possible strip layouts for a simple component are shown in Fig. 8.66. In the layout shown in Fig. 8.66(a), two separate notches are to be made using heeled punches. But with the layout in Fig. 8.66(b) it is possible to use a simple commercial oblong punch to achieve the notches in the part.

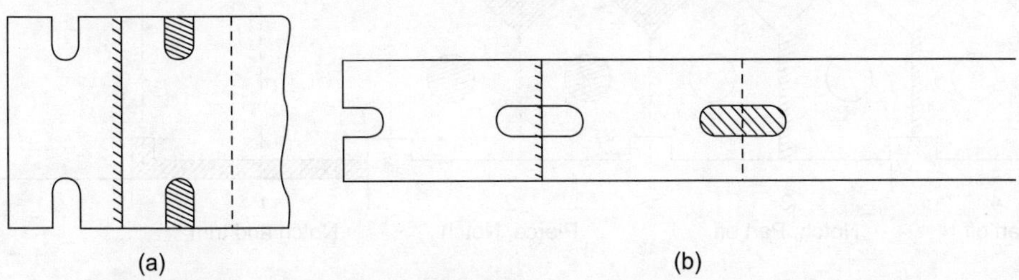

Fig. 8.66 *Change in orientation of a blank to get simple punch shape*

Figure 8.67 shows the strip layout for making a circlip. The holes in the circlip are not useful for piloting and hence the two extra pilot holes are punched in the scrap portion in the scrap portion in the first station as shown in the same figure.

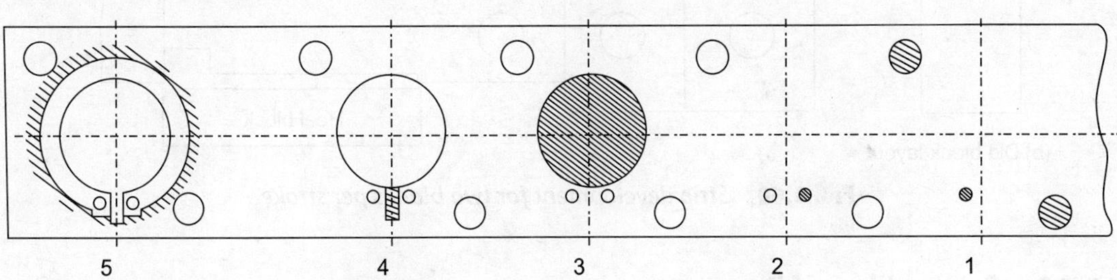

Fig. 8.67 *Strip layout for making a circlip*

The strip layout gets affected by the number of components produced per stroke. For example a strip layout and its die layout are shown in Fig. 8.68 for a simple component which is to be produced at the rate of one per stroke. But when two components are required per stroke, the chosen layout which is different from the previous one is shown in Fig. 8.69.

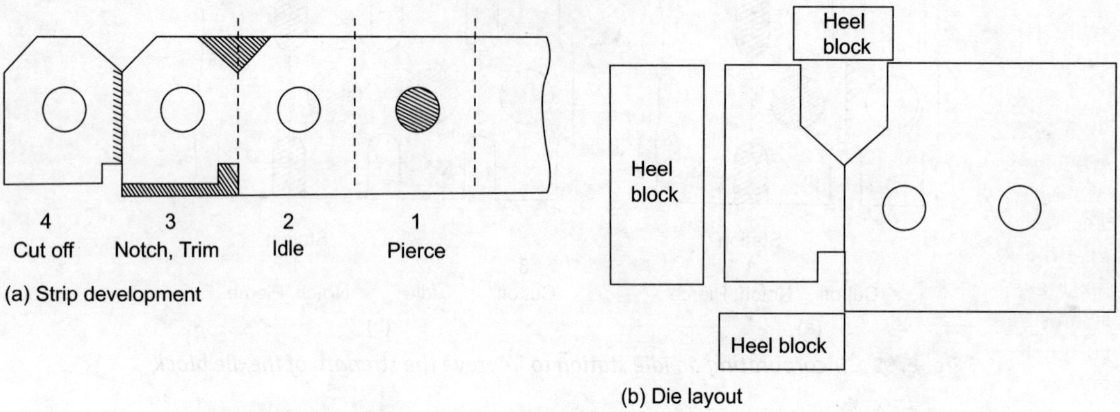

Fig. 8.68 *Strip development and the corresponding die block layout*

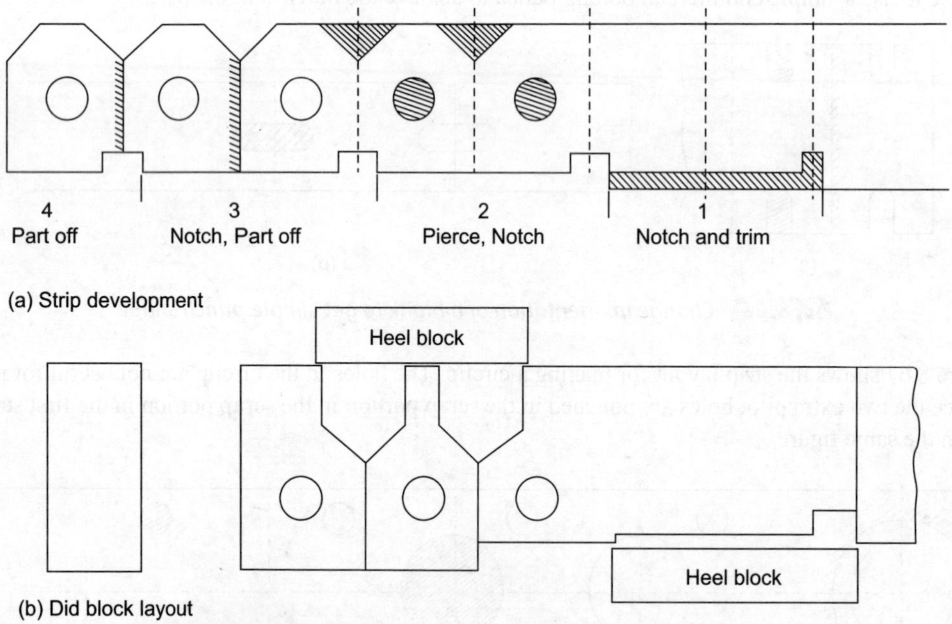

Fig. 8.69 *Strip development for two blanks per stroke*

8.10.10 Centre Line of Pressure

If the shearing to be done is not symmetrical about the axis of the ram, then undesirable side thrust acts on the ram which cause misalignment. It is therefore desirable that the shearing areas are uniformly distributed around the axis of the ram so that no side thrust occurs. To do this, the point on the die block is to be found out, through which the resultant of all the shearing forces is acting and this is coincided with the ram axis. This point is called the centre line of pressure. This in fact, is the centre of gravity of the line (which is proportional to the shearing force) and not the area.

To obtain the centre line of pressure, one has to divide the total blanking contour into known geometrical segments as shown in Fig. 8.70. Then the centre of gravity of each line segment of the total contour is found out. The weighted average of all these line segments would give the centre line of pressure of the total contour. In the Fig. 8.70, the centre of gravity of each of the line segments is shown as x and y distances from a convenient co-ordinate axis system. Then the centre of pressure is given by

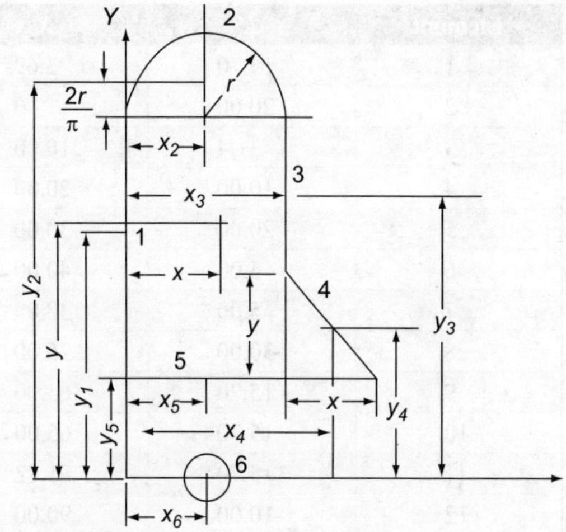

FIG. 8.70 *Calculating the centre of pressure*

$$X = \frac{l_1 x_1 + l_2 x_2 + \cdots + l_6 x_6}{l_1 + l_2 + \cdots + l_6}$$

$$Y = \frac{l_1 y_1 + l_2 y_2 + \cdots + l_6 y_6}{l_1 + l_2 + \cdots + l_6}$$

Where, l1 to l6 = length of each of the line segments.

Example 8.9

Find out the centre of pressure for the die layout shown in Fig. 8.71.

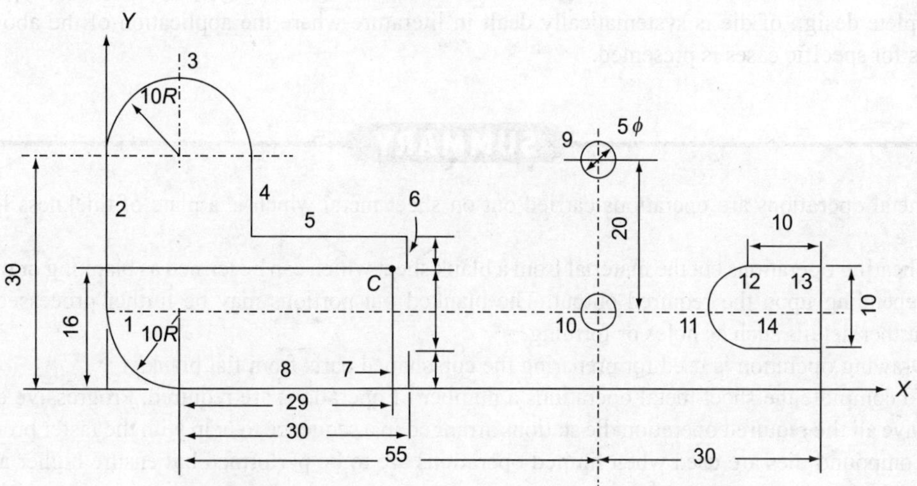

FIG. 8.71 *Example for calculating the centre of pressure*

Element	L	X	y	lx	ly
1	15.70	3.69	3.69	57.93	57.93
2	20.00	0	20.00	0	400.00
3	31.4	10.00	36.37	314.00	1142.00
4	10.00	20.00	25.00	200.00	250.00
5	20.00	30.00	20.00	600.00	400.00
6	5.00	40.00	17.50	200.00	87.50
7	5.00	40.00	2.50	200.00	12.50
8	30.00	25.00	0	750.00	0
9	15.70	65.00	10.00	1020.50	157.00
10	15.70	65.00	30.00	1020.50	471.00
11	15.70	81.82	10.00	1284.60	157.00
12	10.00	90.00	15.00	900.00	150.00
13	10.00	95.00	10.00	950.00	100.00
14	10.00	90.00	5.00	900.00	50.00
	214.20			8397.53	3434.93

$$X = \frac{8397.53}{214.2} = 39.204 \text{ mm}$$

$$Y = \frac{3434.93}{214.2} = 16.036 \text{ mm}$$

The centre of pressure as calculated above is designated as *C* in Fig. 8.71. It is necessary at this stage to see that this point coincides with the centre line of the press ram. For this purpose the die designer has to adjust the die block within the bolster area of the press that was selected for the operation.

So far we have discussed various elements that go into the design of progressive and other types of dies. The complete design of die is systematically dealt in literature where the application of the above design principles for specific cases is presented.

SUMMARY

Sheet metal operations are operations carried out on sheet metal which is a plate of thickness less than 5 mm.

- Shearing operations cut the material from a blank sheet which can be termed as blanking or punching depending upon the required output. The blanked out portions may be further processed to add further details such as holes or forming.
- Drawing operation is used for preparing the cup shaped parts from flat blank.
- To complete the sheet metal operations a number of operations are required. Progressive dies will have all the required operation die stations arranged in a sequence to help with the faster production. Compound dies are used when limited operations are to be performed but ensure higher accuracy for the parts.
- A large number of devices are used in progressive dies to register and automate the movement of stock on the die to the required accuracy.

Questions

8.1 Distinguish between spinning and cup drawing with reference to the processes and components produced.

8.2 Draw a sketch of a punch and a die set used for punching operation. Indicate its various parts.

8.3 Sketch a deep drawing set up, labelling the various important parts.

8.4 Show by a neat sketch, the edge bending operation with a die punch set in position, naming all the important elements of the setup.

8.5 Show by schematic sketches the excessive, correct and insufficient clearances in sheet metal shearing process, indicating the fracture propagation in each case.

8.6 Differentiate between coining and embossing.

8.7 How will you specify a press required for sheet metal operations?

8.8 List various operations generally performed in a sheet metal shop.

8.9 Distinguish between bending and drawing in sheet metal operations.

8.10 Why is it necessary to provide proper clearance between the punch and die in a shearing operation? Give reasons.

8.11 For proper cutting operation in blanking or piercing, on what factors does the clearance depend?

8.12 Washers of 10 mm inside diameter and 20 mm outside diameter are to be obtained from 1 mm thick sheet metal. Sketch the tools required.

8.13 Why is blank holding necessary in a sheet metal drawing operation?

8.14 A sheet which has already been bent in a cold state, offers great resistance to further bending. Explain the reason.

8.15 Why is a shear angle provided in a shearing operation? How do you provide shear angle in the case of punching and blanking operations? Explain with sketches.

8.16 Explain the following die shearing operations: Trimming and shaving.

8.17 What is ironing? Explain its necessity.

8.18 How does the side wall thickness of a drawn component vary? Explain the causes with a sketch.

8.19 Why are the punch radius and die radius essential in a drawing operation?

8.20 What is redrawing? When is this process used?

8.21 What is metal spinning? What types of components are generally manufactured by this process?

8.22 Give typical product specifications where the following processes find their applications:
(a) Stretch forming (b) Coining (c) Embossing (d) Spinning

8.23 How does the grain direction in sheet metal affect the design of:
(a) Bending dies (b) Blanking dies

8.24 Generally, the die opening is straight up to a certain length and tapered thereafter. Explain the reason for such a shape.

8.25 In die shearing operations, die size is the hole size and punch size is the blank size. Explain the validity of the above statement with supporting reasons.

8.26 Where are the following used?
(a) Rigid blank holder (b) Spring loaded blank holder

8.27 Explain the influence of the following parameters on the component produced.
(a) Drawing speed (b) Draw die radius

8.28 What are the criteria which regulate the draw die radius and punch radius in a deep drawing operation?

8.29 Explain what happens when the clearance between the punch and the draw die in the case of the deep drawing operation, is equal to the thickness of the blank.

8.30 What are the requirements of blank holding in shallow drawing?

8.31 Describe the functioning, with sketch, of any one automatic stock stop arrangement.

8.32 With reference to press tools explain the following:
(a) Sectional die construction (b) Fool proofing of die blocks

8.33 How is the grain direction in the sheet metal affect the design of blanking dies?

8.34 Mention the situations when progressive dies are used.

8.35 Specify the basic principles to be followed in the strip development for shearing dies.

8.36 What are the factors that control the design of the springs for spring strippers?

8.37 Mention the advantages of spring stripper over that of channel stripper.

8.38 Explain the advantages and disadvantages of compound dies over progressive dies.

8.39 Explain the function of the backup plate for perforator punches.

8.40 Explain the function of the following components with reference to sheet metal dies:
(a) Pilots (b) Back gauge

8.41 Will the feeding distance be different from advance distance in the case of progressive dies for
(a) Manual feeding (b) Mechanical feeding
Explain your answer.

8.42 For a washer, it is proposed to have the burrs obtained in the two operations, viz blanking and piercing to be on the same side. Suggest a suitable arrangement of the die giving reasons for your choice.

8.43 Explain the use of dowel pins and cap screws in sheet metal die construction.

8.44 What are the methods that can be used for saving material in construction of die blocks? Give any one method showing how the saving is affected.

Problems

8.1 A hole of 10 mm × 25 mm is to be cut in a 3 mm thick sheet. The shear strength of the material is 80 MPa. Estimate the press load required.

8.2 A 60 degree bend is required on a sheet metal component. Should the die angle be equal to, more, or less than 60 degrees? Give reasons in support of your answer.

8.3 Estimate the diameter of the blank required to draw a cup of 10 mm diameter and 10 mm height.

8.4 For a punching operation, a cast steel sheet of thickness 1.5 mm is used. The diameter of the punch used in this operation is 30 mm. The shear strength of this material is known to be 250 MPa. Calculate the following:
(a) Shear area
(b) Force acting on the punch

8.5 Calculate the size of the blank for a tumbler to be deep drawn from a stainless steel sheet of 1 mm thickness. The diameter of the tumbler is 75 mm and height is 150 mm. Assume that the thickness of the blank and the cylindrical portion of the tumbler are not the same. Neglect the effect of the die and punch radii. No ironing is allowed.

8.6 In a particular piercing operation, 20 mm holes are to be punched in a brass sheet of 3 mm thick. The maximum shear strength of the sheet may be taken as 185 MPa. Design the punch and the die sizes as well as the required punch force.

8.7 A cylindrical cup, 100 mm diameter and 85 mm deep, is to be drawn from a steel sheet of deep drawing quality, 1 mm thick. Determine the blank diameter and the other drawing variables for the dies of all draws. Also, estimate the drawing load if the tensile strength is 415 MPa.

8.8 What is the largest diameter that can be pierced in a 1.5 mm thick steel plate of 310 MPa shear strength on a 250 kN press? Design the punch and the die dimensions for this operation.

8.9 Two holes of 50 mm diameter are to be punched simultaneously in a C30 steel stock of 4 mm thick. The shear strength of the material is about 470 MPa. The only available punch press in the shop is of the capacity 200 kN. Examine the possibility, whether the operation could be done on this press. If not suggest any alternative procedure with justification.

8.10 What is the procedure to be adapted for punching a 75 mm diameter hole in a 4 mm thick C60 steel stock with a 250 kN press? The maximum shear strength for the material can be taken as 710 MPa. Make the necessary calculations in support of your answer.

8.11 Calculate the punching loads on the components shown in Fig. 8.72. Also calculate the clearances that are to be provided together with the punch and the die dimensions.

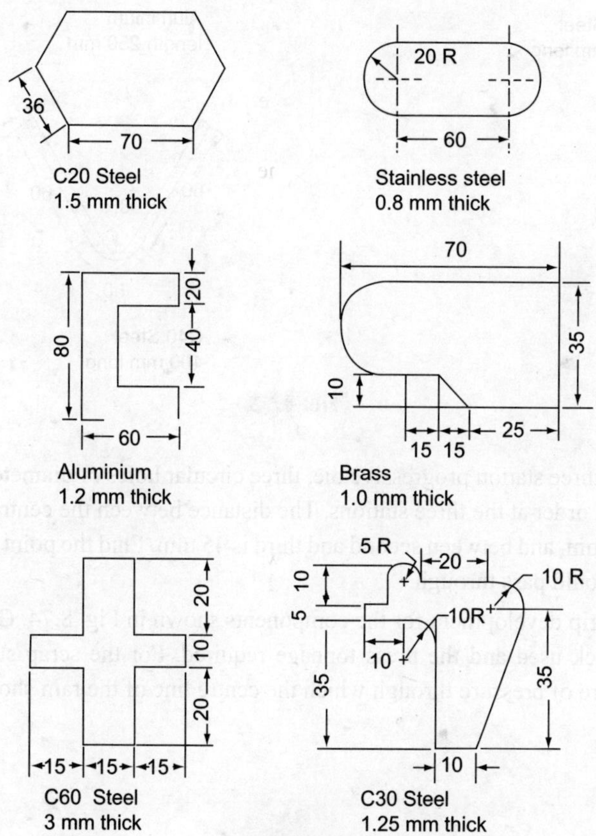

Fig. 8.72

8.12 Calculate the developed lengths for the components shown in Fig. 8.73. Calculate the bending load in each case after deciding on the type of die to be used. Sketch the die sets.

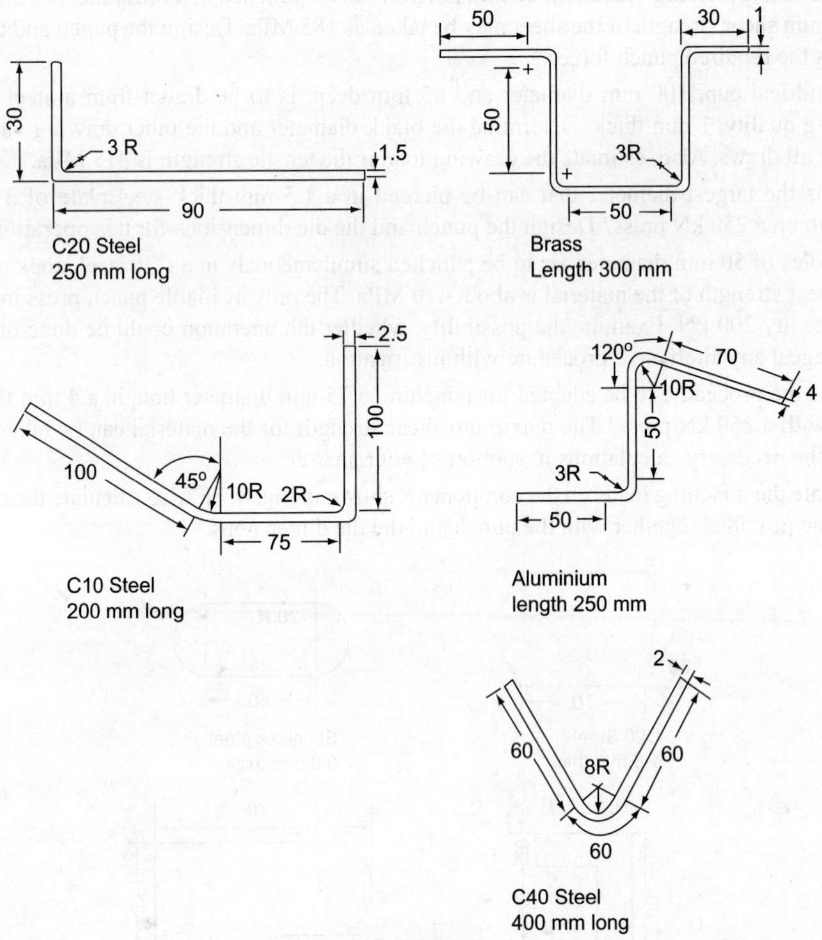

Fig. 8.73

8.13 In a hypothetical three station progressive die, three circular holes of diameters 10, 15 and 30 mm are to be made in that order at the three stations. The distance between the centre of the first hole and the second one is 50 mm, and between second and third is 45 mm. Find the point through which the centre line of the ram should pass through.

8.14 Make the scrap strip development for the components shown in Fig. 8.74. Calculate the approximate percentage of stock used and the press tonnage required. For the scrap strip development shown, calculate the centre of pressure through which the centre line of the ram should pass.

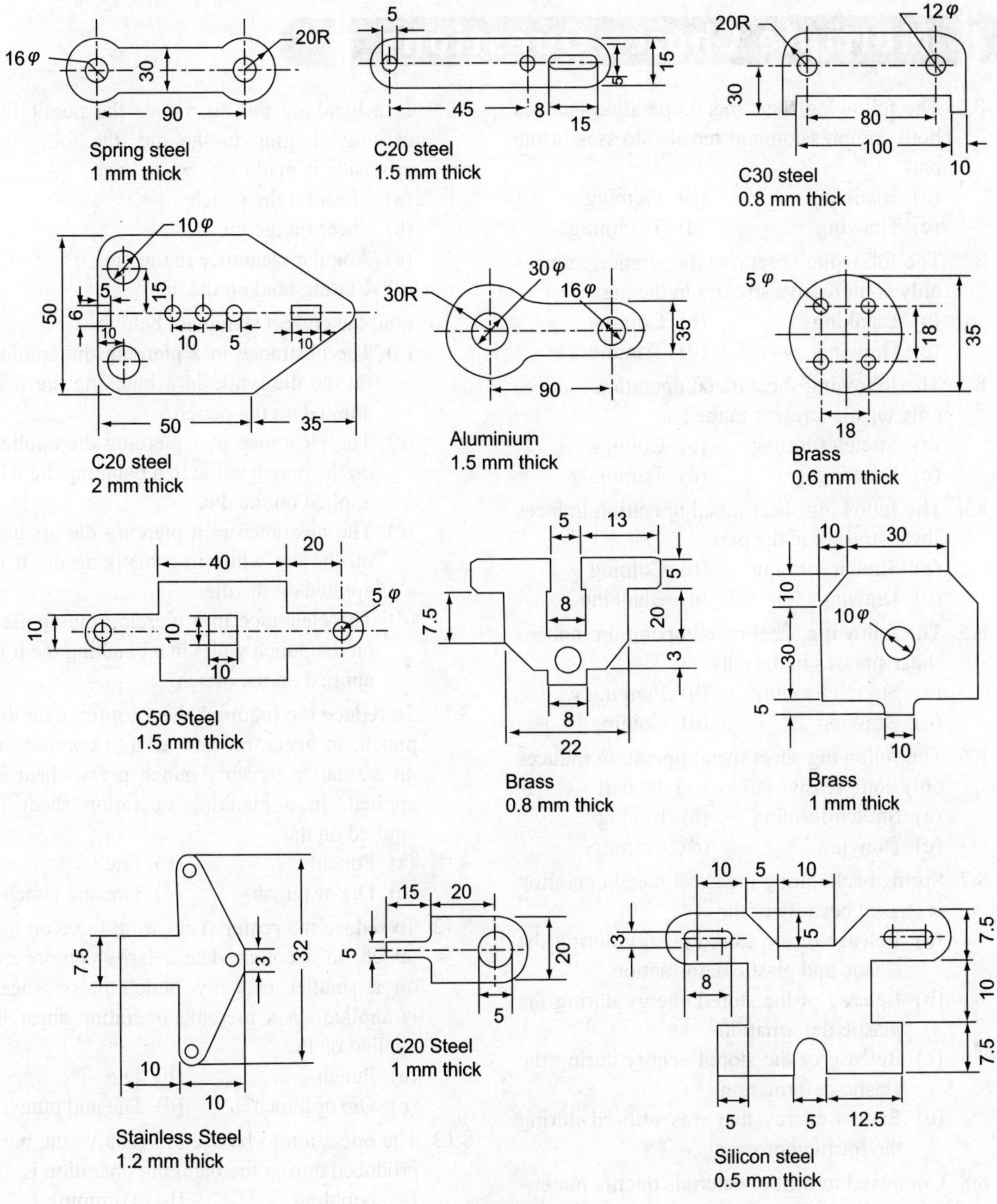
Spring steel
1 mm thick
C20 steel
1.5 mm thick
C30 steel
0.8 mm thick
C20 steel
2 mm thick
Aluminium
1.5 mm thick
Brass
0.6 mm thick
C50 Steel
1.5 mm thick
Brass
0.8 mm thick
Brass
1 mm thick
Stainless Steel
1.2 mm thick
C20 Steel
1 mm thick
Silicon steel
0.5 mm thick

Fig. 8.74

Multiple Choice Questions

8.1 The following sheet metal operation induces both compression and tensile stresses in the part
(a) Blanking (b) Piercing
(c) Drawing (d) Trimming

8.2 The following sheet metal operation induces only compressive stresses in the part
(a) Blanking (b) Coining
(c) Drawing (d) Trimming

8.3 The following sheet metal operation induces only tensile stresses in the part
(a) Stretch forming (b) Coining
(c) Drawing (d) Trimming

8.4 The following sheet metal operation induces shear stresses in the part
(a) Stretch forming (b) Coining
(c) Drawing (d) Blanking

8.5 The following sheet metal operation induces shear stresses in the part
(a) Stretch forming (b) Piercing
(c) Drawing (d) Coining

8.6 The following sheet metal operation induces only compressive stresses in the part
(a) Stretch forming (b) Ironing
(c) Drawing (d) Coining

8.7 Spring back during the sheet metal operation is caused because of the
(a) Release of the stored energy during the elastic and plastic deformation
(b) Release of the stored energy during the plastic deformation
(c) Release of the stored energy during the elastic deformation
(d) Excess energy that was utilised during the forming process

8.8 Compared to brittle materials ductile materials require for shearing operation
(a) Clearance based on material thickness only
(b) Lower clearances
(c) Higher clearances
(d) Same clearance

8.9 In a blanking die, to reduce the possibility of slug clinging to the die, the following provision is made
(a) Shear on the punch
(b) Shear on the die
(c) Angular clearance in the die
(d) Cutting land on the die

8.10 Find the correct statement below:
(a) The clearance in a piercing die applied on the die while in a blanking die it is applied on the punch.
(b) The clearance in a piercing die applied on the punch while in a blanking die it is applied on the die.
(c) The clearance in a piercing die applied on the die while in a blanking die it is applied on the die.
(d) The clearance in a piercing die applied on the punch while in a blanking die it is applied on the punch.

8.11 To reduce the required shearing force on the punch, to accommodate a large component on a smaller capacity punch press, shear is applied. In a blanking operation shear is applied on the
(a) Punch (b) Die
(c) Die or punch (d) Die and punch

8.12 To reduce the required shearing force on the punch, to accommodate a large component on a smaller capacity punch press, shear is applied. In a piercing operation shear is applied on the
(a) Punch (b) Die
(c) Die or punch (d) Die and punch

8.13 The operation to be used to remove the burr produced during the blanking operation is
(a) Nibbling (b) Trimming
(c) Shaving (d) Notching

8.14 Sheet metal drawing operation is used to make
(a) Wire
(b) Cup shaped parts

(c) Tubes
(d) Rods

8.15 Blank holding is normally required in
(a) Blanking
(b) Shallow drawing
(c) Deep drawing
(d) Piercing

8.16 Only axisymmetric cup shapes are produced by this process
(a) Blanking
(b) Spinning
(c) Deep drawing
(d) Shallow drawing

8.17 Bending produces the following types of stresses in the parts
(a) Tensile
(b) Compressive
(c) Tensile and compressive
(d) Shear

8.18 For making a bend of 90 degrees, the part should be bent
(a) More than 90 degrees
(b) Less than 90 degrees
(c) Exactly 90 degrees
(d) None of the above

8.19 The following sheet metal manufacturing process do not experience any spring back
(a) Blanking
(b) Shallow drawing
(c) Deep drawing
(d) Stretch forming

8.20 The following sheet metal process requires a large press
(a) Embossing
(b) Coining
(c) Trimming
(d) Blanking

Answers to MCQs

8.1 (c)	8.2 (b)	8.3 (a)	8.4 (d)	8.5 (b)
8.6 (b)	8.7 (c)	8.8 (c)	8.9 (c)	8.10 (b)
8.11 (b)	8.12 (a)	8.13 (c)	8.14 (b)	8.15 (c)
8.16 (b)	8.17 (c)	8.18 (b)	8.19 (d)	8.20 (b)

Welding Processes

CHAPTER 9

Objectives

After completing this chapter, the reader will be able to

- Understand the varieties of fabrication methods used in manufacturing
- Utilize gas welding processes for low volume and repair work
- Select different arc welding processes for large volume manufacture
- Utilize resistance welding processes for sheet metal joints
- Design welded joints to produce defect free weldments

9.1 FABRICATION METHODS

Joining two or more elements to make a single part is termed as a fabrication process. A fairly large number of industrial components are made by fabrication processes. Common examples are aircraft and ship bodies, bridges, building trusses, welded machine frames, sheet metal parts, etc. The fabrication is often, the most economical method and relies on raw material obtained from one of the primary manufacturing processes such as rolling and extrusion. Hence, it may be called a secondary manufacturing process. The various fabrication processes can be classified as follows:

- Mechanical joining by means of bolts, screws and rivets.
- Adhesive bonding by employing synthetic glues such as epoxy resins.
- Welding, brazing and soldering.

The choice of a particular fabrication method depends on a number of factors pertaining to the joint such as

- Type of assembly permanent, semi-permanent or temporary,
- Materials being joined, steels, cast irons, aluminium, similar or dissimilar metals,
- Economy achieved, and
- The type of service required, such as assembly subjected to heavy loading, impact loading, high temperatures, etc.

Joining obtained by bolts and screws is temporary in nature and can be disassembled whenever necessary. Rivets are semi permanent fastening devices and the joint can be separated only by destroying the rivet without harming the parent elements. These types of mechanical fastening involve making holes in the mating parts which are to be properly accounted for, at the design stage.

Adhesive bonding does not disfigure the joining parts, but would generally have less strength than the mechanical fasteners. But adhesive bonding helps in joining awkward shaped parts or thin sheets which may not lend themselves to mechanical fastening. Similarly, metals and non-metals can be best joined by adhesive bonding as in the case of automobile brake shoe linings. Also, the sandwich structure construction for damping structural vibration relies on the damping capability of the adhesive.

Adhesives are capable of providing the necessary strength to withstand the applied loads for some applications. The most commonly used adhesives can be classified as:

- Thermosetting resins,
- Thermoplastic resins,
- Silicone resins, and
- Elastomers

Out of these, the elastomers which comprise synthetic and natural rubbers are elastic in nature and lack necessary rigidity, cannot be used for structural applications. However, they can be mixed with the thermosetting resins to reduce the brittleness and provide the toughness for shock resistance.

The joint for adhesive application should be properly chosen to provide the necessary cleanliness and the required surface roughness. The type of adhesive chosen should take into account the geometry of the joint, the type of load application and the working conditions such as temperature and humidity. The adhesives are normally applied by either brushing or spraying. The adhesive joint should be cured after assembling by simple air drying or more commonly used baking process. The temperature and time control during the curing process is vital for the development of proper strength.

Welding by contrast to the above fabrication techniques is a metallurgical fusion process. Here, the interface of the two parts to be joined is brought to a temperature above the melting point and then allowed to solidify so that a permanent joining takes place. Because of the permanent nature of the joint and strength being equal to or sometimes greater than that of the parent metal makes welding one of the most extensively used fabrication method.

Welding is not only used for making structures but also for repair work such as the joining of broken castings. Products obtained by the process of welding are called 'weldments'.

Based on the type of joint and the source of the heat input, the welding processes are classified as shown in Fig. 9.1. The abbreviations shown are those approved by the American Welding Society (AWS). The processes shown are only a fraction of the total spectrum of welding processes. In the subsequent chapters, the details of the more commonly used welding processes are covered.

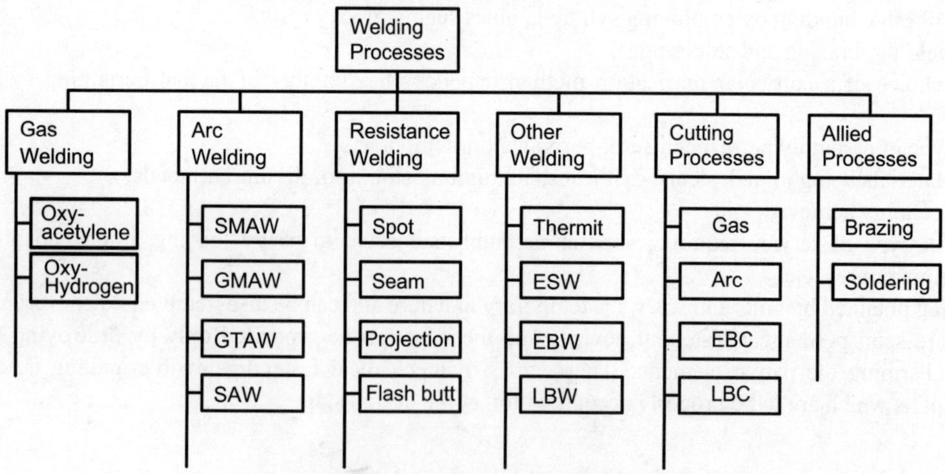

FIG. 9.1 *Welding classification*

9.1.1 General Considerations

Before proceeding to study the details of the various welding processes, some aspects that have relevance to all the welding processes are covered in the following.

Types of Joints

Different types of welding joints are classified as butt, lap, corner, tee and edge joints. These are shown in Fig. 9.2. The choice of the type of joint depends on the weldment being made and the sheet thickness. More details of these will be dealt in the respective welding processes. Each of the above joints can have any of the four different welding positions: flat, horizontal, vertical and overhead as shown in Fig. 9.3 for the butt joint.

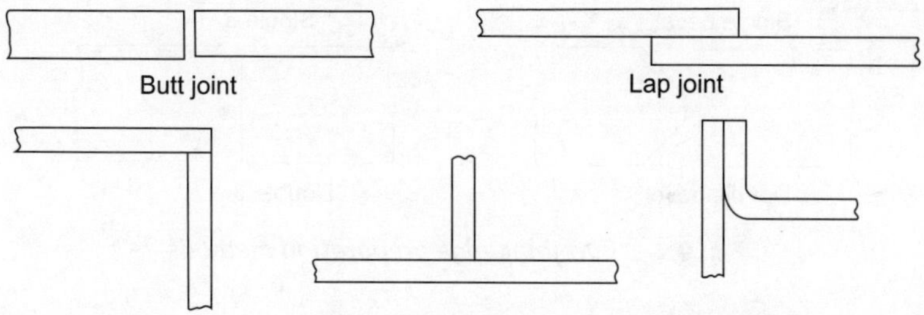

FIG. 9.2 *Types of welded joints*

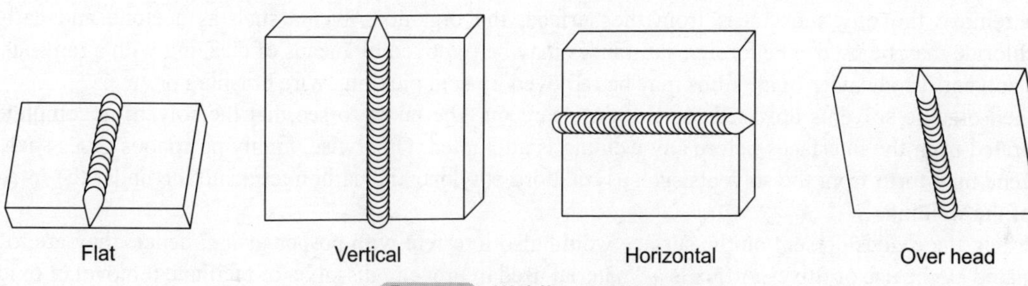

FIG. 9.3 *Welding positions*

The edges of the joint in Fig. 9.2 are shown to be straight. This type of edge preparation is used when the thickness of the two pieces to be joined is small, so that heat of welding penetrates the full depth of the joint. However, when the thickness increases it becomes necessary to prepare the edge in such a way that heat would be able to penetrate the entire depth. To facilitate this, the joint is widened as shown in Fig. 9.4. For very thick plates, the welding needs to be done from both sides. To provide the necessary access into the joint, it could be made as a *V* or *U*, as shown in Fig. 9.4. The V-joint is easier to make but the amount of extra metal to be filled in the joint increases greatly with an increase in the thickness. From this account an *U*-joint is preferable, since the amount of extra metal to be added to fill the joint is generally less beyond a certain plate thickness. However, machining a *U*-joint is more difficult compared to a *V*-joint. The double *V* and double *U* edge preparations are used when welding is to be carried from both sides.

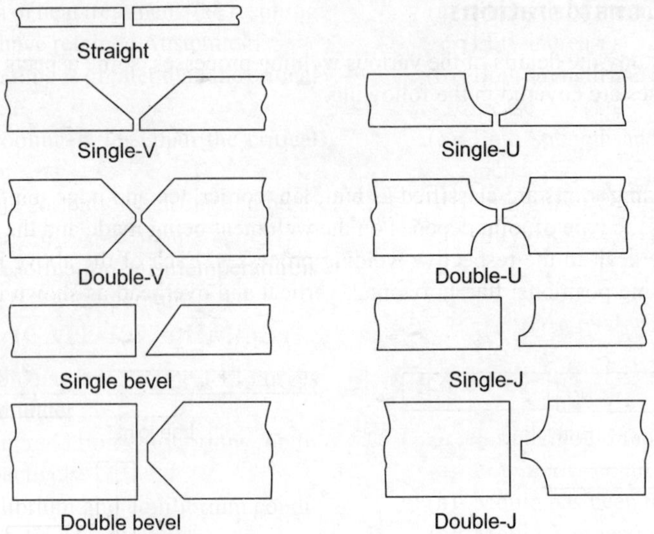

FIG. 9.4 *Butt joint edge preparation methods*

By virtue of the metal being melted at the interface in a welded joint it is necessary that the interfaces are very clean. If the interfaces are not clean, with any oil, dirt, paint or grease residue left would interfere with the proper fusing of the metal and thus weaken the joint. Hence it is essential that the joint surfaces are thoroughly cleaned before the welding is attempted.

To remove the oily substances from the surface, the organic solvents such as acetone and carbon tetrachloride may be used. The foreign substances may be removed by means of cleaning with a rag soaked in the solvent. The heavier oxide films may be removed by acid pickling, wire brushing or emery.

When organic solvents are used for cleaning, care may be taken to see that the solvent is completely evaporated from the interfaces before any welding is attempted. Otherwise, highly poisonous gases such as phosgene may form from the solvents such as trichloro ethylene and carbon tetrachloride under the intense heat of the welding.

Further, the oxides present on the surface would also interfere with proper fusing; hence, they are to be eliminated by the use of fluxes. A flux is a "material used to prevent, dissolve, or facilitate removal of oxides and other undesirable surface substances" as defined by the American Welding Society. The flux is expected to react with the oxides present and form low density slag which would float on top of the molten metal pool protecting it from further oxidation. The types of fluxes used depends on the operation and the parent metal that being welded. Though mild steel may not require any flux since iron oxide has got a very low melting point, non-ferrous materials such as copper and aluminium require fluxes.

Another requirement of welding is a filler metal. Except for the resistance welding process, all the other processes require a filler metal to fill the gap between the parts to be joined. Ideally the composition of the filler metal should be same as that of the base metals which are being joined for proper mixing of the filler metal with base metal. However sometimes it is possible to use additional alloying elements to strengthen the joint.

The following are definitions of some of the welding terms that are generally used.

Backing It is the material support provided at the root side of a weld to aid in the control of penetration.

Base metal The metal to be joined or cut is termed as the base metal.

Bead or weld bead Bead is the metal added during a single pass of welding. The bead appears as a separate material from the base metal.

Crater In arc welding, a crater is the depression in the weld metal pool at the point where the arc strikes the base metal plate.

Deposition rate The rate at which the weld metal is deposited per unit time is the deposition rate and is normally expressed as kg/h.

Fillet weld The metal fused into the corner of a joint made of two pieces placed at approximately 90 degrees to each other (Fig. 9.5), is termed fillet weld.

Penetration It is the depth up to which the weld metal combines with the base metal as measured from the top surface of the joint.

Puddle The portion of the weld joint that melted by the heat of welding is called puddle.

Root It is the point at which the two pieces to be joined by welding is nearest.

Tack weld A small weld, generally used to temporarily hold the two pieces together during actual welding, is the tack weld.

Toe of weld It is the junction between the weld face and the base metal.

Torch In gas welding, the torch mixes the fuel and oxygen and controls its delivery to get the desired flame.

Weld face It is the exposed surface of the weld, as shown in Fig. 9.5.

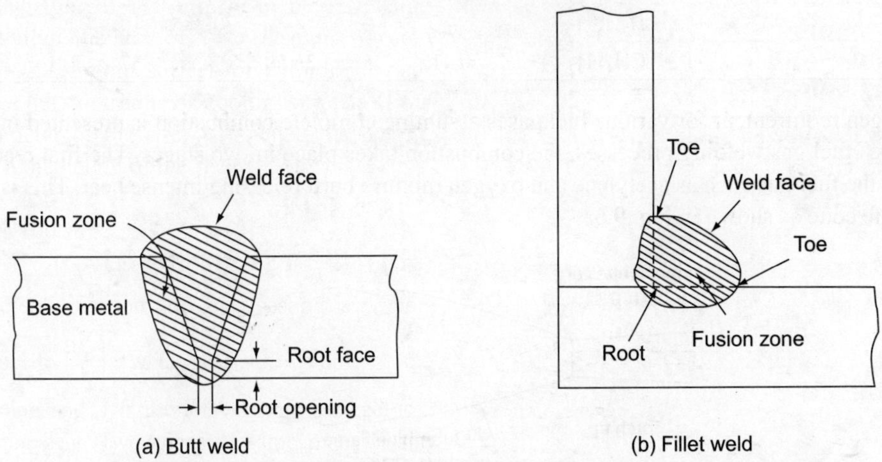

Fig. 9.5 *Terminology to be used with various welded joints*

Weld metal The metal that is solidified in the joint is called weld metal. It may be only base metal or a mixture of base metal and filler metal.

Weld pass A single movement of the welding torch or electrode along the length of the joint which results in a bead is a weld pass.

9.2 GAS WELDING

As the name implies, gas welding also called as oxy-fuel gas welding (OFW), derives the heat from the combustion of a fuel gas such as acetylene in combination with oxygen. The process is a fusion welding process wherein the joint is completely melted to obtain the fusion. The heat produced by the combustion of gas is sufficient to melt any metal and as such is universally applicable.

9.2.1 Fuel Gases

The fuel gas generally used is acetylene because of the high temperature generated in the flame. This process is called oxy-acetylene welding (OAW). The useful fuel gases for gas welding are shown in Table 9.1 with their calorific values and flame temperatures. All of the fuel gases used contain carbon and hydrogen in different amounts as shown by their chemical formulae, which burn in combination with oxygen and release the required heat for melting the metal.

TABLE 9.1 Characteristics of fuel gases

Gas	Chemical Formula	Heat Content, MJ/m^3			Flame Temperature, °C
		Primary	Secondary	Total	
Acetylene	C_2H_2	18.97	36.03	55	3100
Propylene	C_3H_6	16.38	71.62	88	2500
Propane	C_3H_8	9.38	83.62	93	2450
Methyl acetylene propadiene	C_3H_4	21.00	70.00	91.00	2927
Hydrogen	H_2	–	–	10	2390
Natural Gas	CH_4+H_2	0.41	36.59	37	2350

The oxygen requirement for various fuel gases assuming complete combustion is presented in Table 9.2. In all the oxy-fuel gas welding processes, the combustion takes place in two stages. The first reaction takes place when the fuel gas such as acetylene and oxygen mixture burn releasing intense heat. This is present as a small white cone as shown in Fig. 9.6.

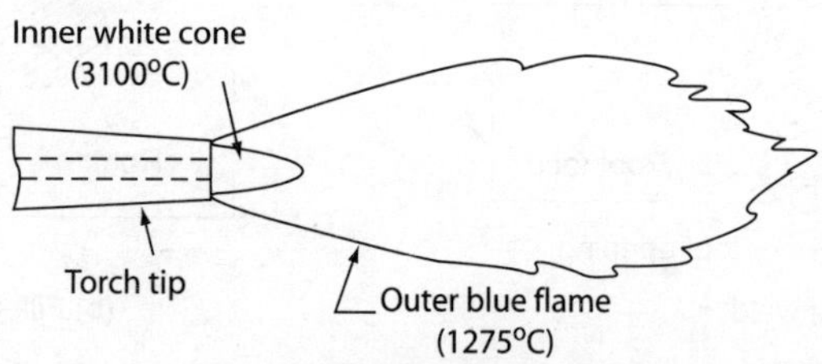

FIG. 9.6 *Schematic of an oxy acetylene gas welding flame (neutral flame)*

For the oxy-acetylene welding, the following reaction takes place in this zone.

$$C_2H_2 + O_2 \rightarrow 2CO + H_2 + 448 \text{ kJ/mol } (18.75 \text{ MJ/m}^3 \text{ of acetylene})$$

The carbon monoxide (CO) and hydrogen produced in the first stage further combine with the atmospheric oxygen and give rise to the outer bluish flame, with the following reaction.

$$4CO + 2H_2 + 3O_2 \rightarrow 4CO_2 + 2H_2O + 812 \text{ kJ/mol } (35.77 \text{ MJ/m}^3)$$

TABLE 9.2 Oxygen consumption for fuel gases (for neutral flame) with complete combustion

Gas	Oxygen Supplied, m^3/m^3		Oxygen Supplied, m^3/kg of Fuel
	By Torch	Total	
Acetylene	1.3	2.5	1.18
MAPP*	2.5	4.0	1.38
Propylene	3.5	4.5	1.94
Propane	4.3	5.0	2.32
Natural Gas	1.9	2.0	2.80

* Methyl acetylene propadiene (stabilised)

Though higher amount of heat is produced in the second stage, since it is distributed over a larger area, the temperature achieved is small (of the order of 1200 to 2000°C) in the flame, which may be used for preheating the steels. The inner white cone temperature is of the order of 3100°C which is used for directly melting the steel joint.

As is shown in Table 9.2, a certain amount of oxygen is required for complete combustion of fuel gases. When the oxygen supply varies, the flame appearance obtained would also vary. Figure 9.6 shows the ideal condition, i.e., the complete combustion, called neutral flame. In neutral flame all the acetylene present is completely burned and thus all the available heat in the acetylene is released. Thus, this is the most desirable flame to be used in oxy-acetylene welding.

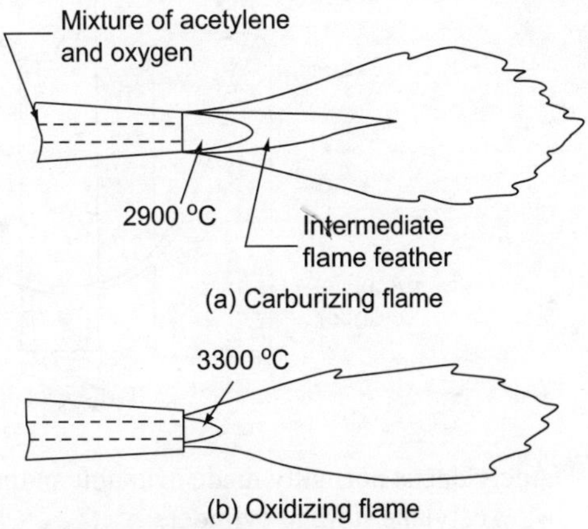

FIG. 9.7 *Different types of oxy-acetylene welding flames*

However, it is also possible to provide more or less oxygen than that theoretically required for complete combustion. When less oxygen is provided, part of the combustible matter is left as it is and it results in a reducing or carburising flame as shown in Fig. 9.7. This flame is similar to the neutral flame, only with the addition of a third phase in between the outer blue flame and the inner white cone as shown in Fig. 9.7(a). It is called 'intermediate flame feather' which is reddish in colour. The length of the flame feather is an indication of the excess acetylene present. Since unburned carbon present, goes into the weld metal pool, the metal appears to boil. This excess carbon causes the steel to become extremely hard and brittle.

The carburising flame is not suggested for general use. However, since this flame provides a strong reducing atmosphere in the welding zone, it is useful for those materials which are readily oxidised, for example, oxygen free copper alloys. It is also used for high carbon steels, cast irons and hard surfacing with high speed steel and cemented carbides. Sometimes a reducing flame with only atmospheric oxygen supply is used to deposit carbon on mould cavity faces to act as an insulator.

When oxygen is in excess, it is called the oxidising flame whose appearance is shown in Fig. 9.7(b). The flame is similar to the neutral flame with the exception that the inner white cone is somewhat small, giving rise to higher tip temperatures (3300°C). There is an excess amount of oxygen present in the flame which badly oxidises the weld metal. Because of the burning of the metal, the weld pool foams and sparks. This also produces a loud noise. This flame would be useful for welding some non-ferrous alloys such as copper base alloys and zinc base alloys. Some alloys of iron such as cast iron and manganese steel are also better welded by an oxidising flame. The presence of excess oxygen in the oxidising flame causes an oxide film to form quickly which provides a protective cover over the base metal pool.

9.2.2 Oxy-Acetylene Welding Equipment

A typical oxy-acetylene welding outfit is shown in Fig. 9.8. It contains the supply units for oxygen and acetylene with associated regulators and the torch which mixes the two gases before they are ignited. The oxygen is normally stored in strong cylinders at a pressure ranging from 13.8 MPa to 18.2 MPa.

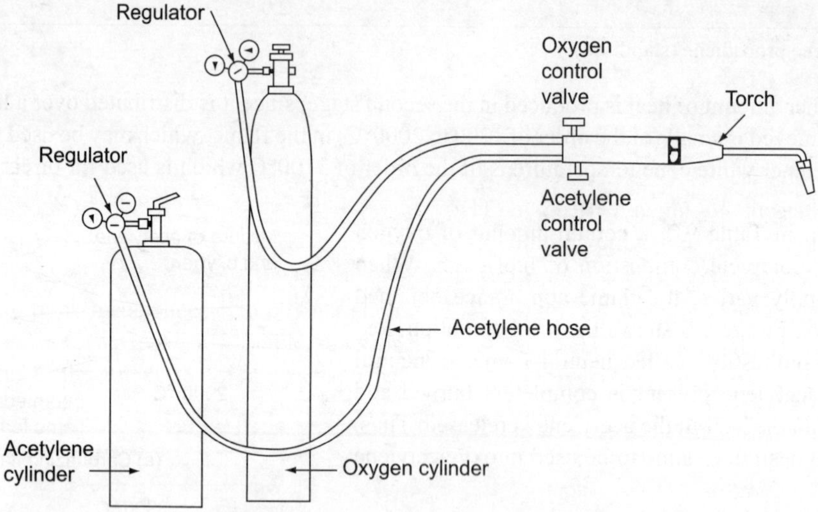

FIG. 9.8 *Oxy-acetylene welding outfit*

Acetylene is normally made available in the following two forms:

- Acetylene storage cylinder
- Acetylene generator

Free acetylene is highly explosive, if stored at a pressure more than 200 kPa, where it becomes very unstable and likely to explode. Hence, acetylene needs to be carefully stored in a strong cylinder, filled with 80 to 85% porous material such as calcium silicate and then filled with acetone which can absorb up to 420 times its volume of acetylene at a pressure of 1.75 MPa. It is expected that the acetylene molecules fit in between the acetone molecules. This helps in storing acetylene at a much higher pressure than permitted when it is in free form.

Acetylene would be released from acetone at a slow rate and thus would not form any pockets of high pressure acetylene. The rate of release depends on the temperature of the gas. Hence the rate of consumption of acetylene should be under the rate of release, which is normally about one-seventh of the capacity of the cylinder per hour. However, if acetylene is drawn at a rapid rate, acetone may also come out along with the acetylene. The presence of acetone in the flame would give it a purple colour. It is not desirable since it reduces the flame temperature.

It is also possible to have an acetylene generator in the place of an acetylene cylinder. Acetylene is normally produced by a reaction between calcium carbide and water which is instantaneous, as shown below:

$$CaC_2 + 2H_2O \rightarrow C_2H_2 + Ca(OH)_2$$

A schematic cutaway view of the acetylene generator is shown in Fig. 9.9. It consists of a cylinder, which is partially filled by water. The calcium carbide is stored in a hopper near the top of the generator. A pressure regulated valve controls the flow of calcium carbide into water, depending on the pressure of the acetylene in the generator. The acetylene is taken out through a gas pipe as shown in Fig. 9.9. The generator would be permitted to produce acetylene to a safe pressure of 100 kPa.

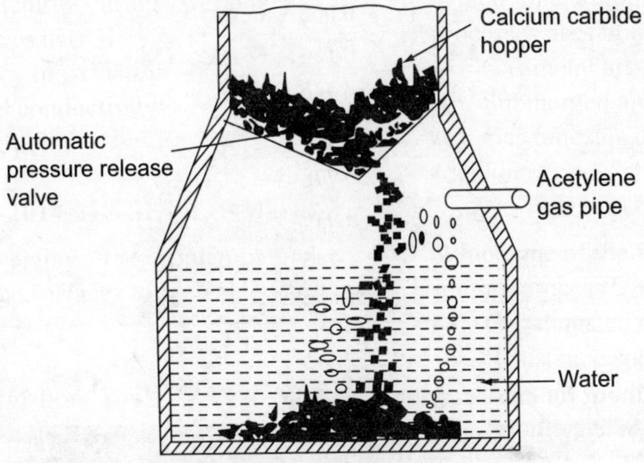

Fig. 9.9 *Acetylene generator*

The oxygen and acetylene from the two cylinders are brought through separate hose pipes to the welding torch as shown in Fig. 9.10. In the torch the two gases are mixed and then flowed out through the nozzle at the torch tip.

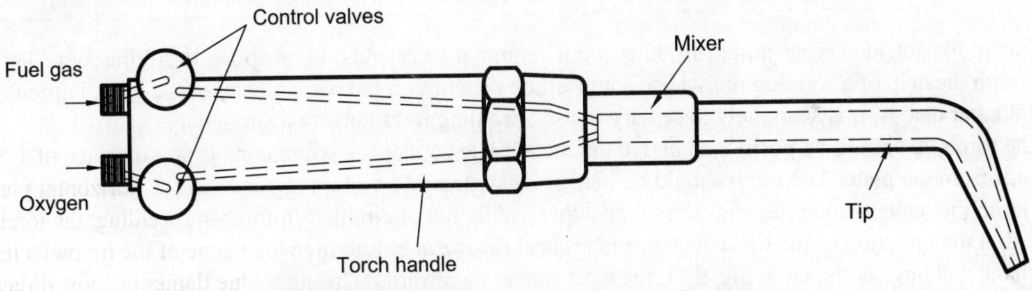

Fig. 9.10 *Oxy-acetylene welding torch details*

9.2.3 Oxy-Acetylene Welding Technique

To light the flame, the acetylene valve on the torch is opened slightly and lighted with the help of a friction spark lighter. The flame draws the oxygen from the atmospheric air and thus results in a reducing flame. Then the acetylene valve is opened to get the required flow of acetylene. The oxygen valve is then slowly opened till the intermediate flame feather of the reducing flame recedes into the inner white cone. The actual adjustment of the flame depends on the type of material to be joined. Table 9.3 provides the details of flames generally used for oxyacetylene welding.

The choice of the torch size depends on the thickness of the metal to be joined. Larger torch tip sizes causes higher amount of oxygen and fuel to flow out causing the release of more heat. Thus for thicker metals, larger tip radii are to be used as shown in Table 9.4.

Table 9.3 Some of the welding parameters for gas welding carbon steels

Metal Thickness mm	Welding Tip Orifice mm	Welding Rod Diameter mm	Oxygen Consumption (m^3/hr)	Acetylene		Welding Speed (m/hr)
				Pressure (kPa)	Consumption (m^3/hr)	
0.8	0.57	1.6	0.031	7	0.0283	
1.6	0.75	1.6	0.062	7	0.0566	
2.4	0.92	1.6 or 2.4	0.156	13.8	0.1420	6.10
3.2	1.09	2.4 or 3.2	0.280	20.7	0.2550	4.88
4.8	1.32	3.2	0.498	27.6	0.4530	4.27
6.4	1.61	3.2 or 4.8	0.778	34.5	0.7080	3.66
8.0	1.85	3.2 or 4.8	0.935	41.3	0.8500	3.05
9.6	2.08	4.8	1.246	48.2	1.1330	2.74
12.7	2.37	4.8	1.869	48.2	1.6990	2.44

Table 9.4 Flame adjustment for oxy-acetylene welding

Metal	Flame	Metal	Flame
Mild steel	Neutral	Aluminium	Slightly carburising
High carbon steel	Reducing	Brass	Slightly oxidising
Grey cast iron		Copper, Bronze	Neutral, Slightly oxidising
	Neutral, slightly oxidising	Nickel alloys	Slightly carburising
Alloy steel	Neutral	Lead	Neutral

Except for outside corner joints, all other joints require a filler metal to be added to fill the joint. This is done with the help of a welding rod whose composition depends on the parent metal of the joint. Particularly with thicker metals, it is absolutely necessary to use a welding rod to obtain a strong weld.

The torch tip should be positioned above the metal plate so that the white cone is at a distance of 1.5 to 3.0 mm from the plate. The torch should be held at an angle of 30 to 45 degrees from the horizontal plane. The torch movement along the joint should be either oscillating or circular. In forehand welding, the torch is moved in the direction of the tip. This tends to preheat the metal before the white cone of the tip melts it. In backhand welding, as shown in Fig. 9.11, the torch moves backwards. The outer blue flames are now directed on the already welded joint.

This allows the joint to be continuously annealed relieving the welding stresses. Also backhand welding allows a better penetration as well as form a bigger weld bead. Backhand welding is generally used for thicker materials.

When the welding rod is used to provide filler material, it is necessary to hold it at a distance of 10 mm from the flame and 1.5 to 3.0 mm from the surface of the puddle (weld metal pool). This way the rod gets preheated and when dipped into the puddle would readily get melted. So as and when filler metal is required, the welder dips the preheated welding rod into the puddle. The welding rod should be swiftly withdrawn after enough filler metal is added.

The preheating of the welding rod should be maintained by keeping it at a proper distance from the flame. Too far a distance makes the rod cooler and would chill the puddle when dipped. Too little a distance makes the tip melt with the result that the molten metal would be blown away by the flame causing uneven bead and poor penetration.

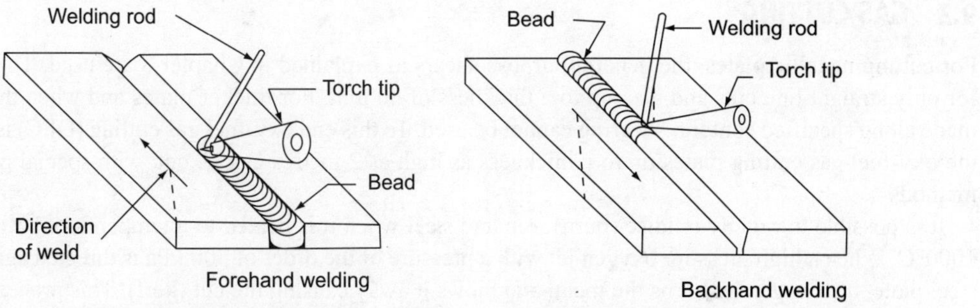

Fig. 9.11 *Forehand and backhand welding techniques*

Oxy-fuel gas welding can be used for all the types of joints and in all positions, though overhead requires additional skill to safeguard the welder. The butt joint edge preparations that required for proper penetration are shown in Fig. 9.12. Thicker plates require more than one pass of the gas torch along the length, to complete the joint. It is called multi-pass welding. A pass is designated as the movement of the torch along the length of the joint once. In multi-pass welding, the first pass (the root pass) is very critical in any welding operation.

The principal advantage of an oxy-acetylene welding setup is the versatility of the equipment. The same equipment with a range of torches would be used for oxygen cutting as well as for brazing and braze welding shown later. Thus, it is useful for general purpose shops particularly smaller ones. Also the source of heat is separate from the filler rod and hence, the filler metal can be properly controlled and heat properly adjusted giving rise to a satisfactory weld. However, this method of welding is somewhat slower and as such cannot compete with the other production welding methods such as electric arc welding discussed later.

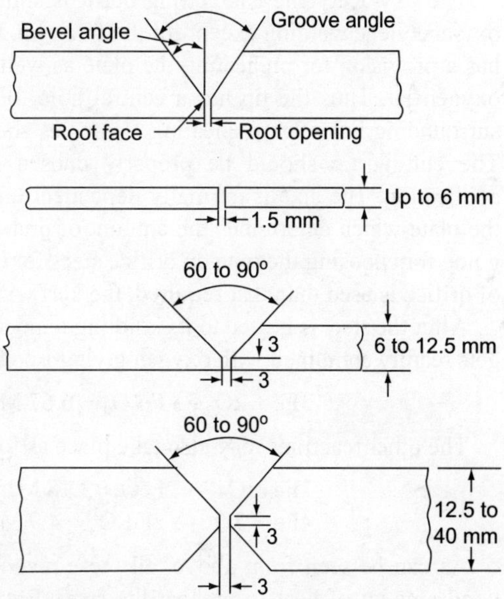

Fig. 9.12 *Suggested butt joint edge preparations for oxy-fuel welding*

9.2.4 Oxy-Hydrogen Welding

In oxy-hydrogen welding (OHW), hydrogen combines with oxygen to generate steam and attains a flame temperature of 2870°C. But the weld pool is not protected from the atmosphere when the oxygen for combustion is completely provided by the torch itself. So generally oxygen in an amount slightly less than that required for complete combustion is provided by the torch whereas the atmospheric oxygen accounts for the burning of the remaining hydrogen. This gives rise to a protective preheating flame that surrounds the main flame. But this reduces the flame temperature to some extent. Because of the lower flame temperature, the oxy-hydrogen welding is generally slow. It is normally used to weld thin sheets of steels and alloys with low melting temperature.

9.3 GAS CUTTING

For cutting metallic plates, the general purpose shears as explained in Chapter 8 are used. These are useful for only straight line cuts and also up to a thickness of 40 mm. For thicker plates and when the cut is to be made along specified contour, shearing cannot be used. To this end oxy-fuel gas cutting (OFC) is useful. With the oxy-fuel gas cutting plates up to a thickness as high as 2 metres can be cut, with special precautions or methods.

It is possible to rapidly oxidise (burn) iron and steel when it is heated to a temperature between 800 and 1000°C. When a high pressure oxygen jet with a pressure of the order of 300 kPa is directed against a heated steel plate, the oxygen jet burns the metal and blows it away causing the cut (kerf). This process invented in 1887 by Thomas Fletcher is extensively used for cutting steel plates of various thicknesses mainly because, the equipment required is simple and can be carried anywhere without handling the heavy steel plates.

The oxy-acetylene gas cutting outfit is similar to that of the oxy-acetylene welding except for the torch tip. Here the torch tip has a provision for preheating the plate as well as providing the oxygen jet. Thus the tip has a central hole for oxygen jet with surrounding holes for preheating flames as shown in Fig. 9.13. The cutting tip should be properly chosen for the intended application. The size is normally dependent on the thickness of the plate which determines the amount of preheating as well as the oxygen jet flow required for cutting. A guide for choosing the oxygen orifice sizes for cutting carbon steels is presented in Table 9.5. If a larger size of orifice is used than that required, the kerf width is wider and larger volumes of oxygen are consumed.

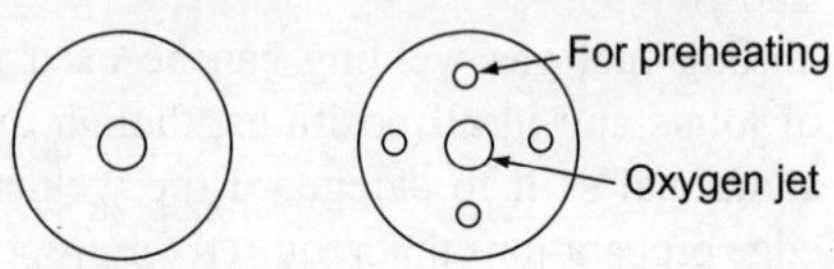

Fig. 9.13 *Differences in torch tips for gas welding and gas cutting*

After the steel is heated to its kindling temperature, which is about 870°C (when it is cherry red colour), it gets readily combined with oxygen giving iron oxide with the following reaction:

$$3Fe + 2O_2 \rightarrow Fe_3O_4 + 6.67 \text{ MJ/kg of iron}$$

The other reactions may also take place as follows

$$2Fe + O_2 \rightarrow 2FeO + 3.18 \text{ MJ/kg of iron}$$
$$4Fe + 3O_2 \rightarrow 2Fe_2O_3 + 4.9 \text{ MJ/kg of iron}$$

As can be seen from above, all these reactions are exothermic in nature and as such would provide a good amount of heat to preheat the steel. But this heating may not be sufficient to bring the steel to its kindling temperature, and hence preheating flames may have to be continued at somewhat lower rate. The heat generated causes the metal to melt and get blown away by the oxygen pressure. In fact, about 30 to 40% of the metal in the kerf is simply blown away, while the rest is oxidised.

Table 9.5 **Tip sizes for cutting carbon steel**

Plate Thickness, mm	Oxygen Orifice Diameter, mm
up to 3	0.65
3 to 6	0.90
6 to 25	1.25
25 to 50	1.60
50 to 100	2.25
100 to 200	3.00
200 to 300	4.25
300 to 400	5.00
400 to 500	6.00

For complete oxidation of steel, about 0.287 m^3/kg of iron is required, but since a part of the unoxidised metal is blown away, the actual requirement would be much less.

To start the cutting operation, the tip must be clean and the preheating flames adjusted properly to neutral flame. The torch tip can be held vertically or slightly inclined in the direction of travel as shown in Fig. 9.13. The torch should be positioned above the metal at a distance of about 1.5 to 3 mm.

The cutting can start at the edge or in the middle of a plate (piercing start). The piercing start may leave ragged cut surface or imperfect start and hence is not generally preferred. Even when the cut is to start at the middle of a plate, a 'lead-in' cut from the edge may be preferable to leave the imperfections away from the cut surface.

After the plate has reached the kindling temperature, the operator should release the oxygen jet to start the cutting, moving the torch in the forehand direction to achieve the desired cut. The kerf (cut edge) appears as shown in Fig. 9.14. The drag lines in the figure show the characteristics of the movement of oxygen stream through the metal.

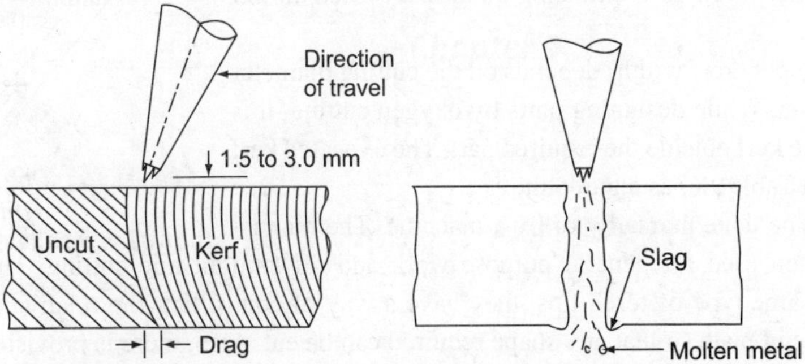

Fig. 9.14 *Positioning of cutting torch in oxy-fuel gas cutting*

Drag is the amount by which the lower edge of the drag line trails from the top edge. A good cut is characterised by very small or negligible drag. To this end, the forward cutting speed of the torch is very important. When the torch is moved too rapidly, the metal at the bottom does not get sufficient heat to get oxidised and cut and hence there is a large drag. Larger drags tend to create turbulence conditions for the oxygen jet which gives rise to very rough and irregular shaped cut edges.

When the torch is moved slowly, all the preheated metal is burnt away by the oxygen jet and a large amount of slag is generated. Since the material is not preheated, the oxygen supply has to be cut-off to continue the cutting. The jet is started only after the material reaches the kindling temperature. This starting off and on of the oxygen jet gives rise to an irregular cut. Some examples of parameters to be used for good oxy-acetylene cutting are presented in Table 9.6.

Table 9.6 **Various process parameters for oxy-acetylene cutting operation**

Metal Thickness, mm	Preheat Orifice Size, mm	Cutting Orifice Size, mm	Oxygen Pressure, kPa	Acetylene Pressure, kPa	Cutting Speed, m/min	Kerf Width, mm
3.2 to 9.5	0.71	0.81	138 to 207	21	0.356 to 0.457	1.5 to 2.5
9.5 to 19.1	1.07	0.97	207 to 276	34	0.305 to 0.381	2.5 to 2.8
19.1 to 38.1	1.09	1.40	276 to 310	34	0.254 to 0.305	2.8 to 3.3
38.1 to 50.8	0.79	1.70	310 to 345	34	0.229 to 0.254	3.3

Though gas cutting is more useful with thick plates, thin sheets (less than 3 mm thick) can also be cut by oxy-acetylene cutting, taking special precautions. Tip size chosen should be as small as possible. If small tips are not available then the tip is inclined at an angle of 15 to 20 degrees, as shown in Fig. 9.15.

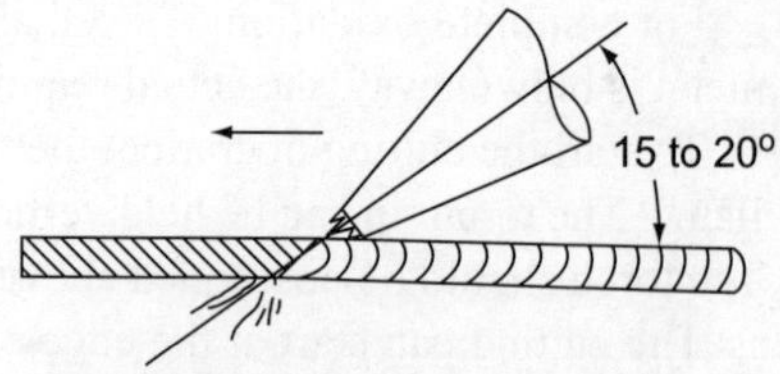

FIG. 9.15 *Recommended torch position for cutting thin steel*

Another important application of gas cutting is in the making of bevels for edges to be used for subsequent welding purposes. For this purpose, the tip is kept at the required angle of the bevel and cut through, as shown in Fig. 9.16. Normally, smaller cutting speeds are employed when thicker metals are to be cut. Also since the cut is at an angle, the actual cut depth is more than the thickness of the plate, which should be borne in mind while attempting this operation. Sometimes even low oxidising flames are used to get quicker preheating.

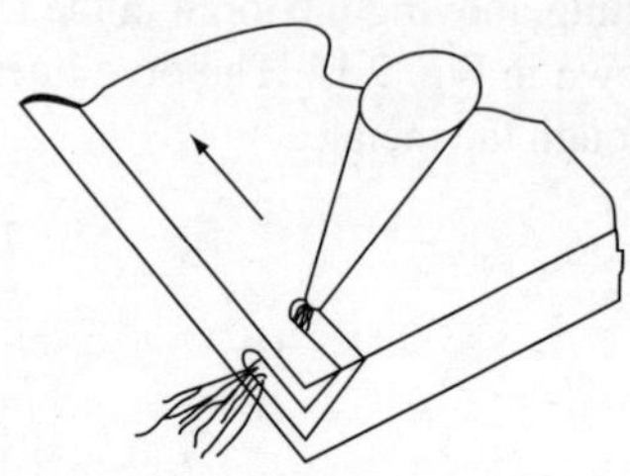

FIG. 9.16 *Obtaining bevel edge in a thin plate*

The width of the cut (kerf width) depends on the cutting diameter of the oxygen orifice. While designing parts by oxygen cutting, it is customary to put the kerf outside the required part. The expected kerf widths are given in Table 9.6, as a guideline.

Gas cutting can be done manually or by a machine. The manual cutting is more often used for general purpose work and for straight line cutting. Though flame cutting machines use the same type of torch tips, they have a way of controlling the relative motion between the torch tip and the metal plate so that any shape required can be cut. Also, there is provision in the machine to hold more than one torch so that larger number of identical pieces can be cut at the same time. The torch is mounted on a rail and both rail and the torch can move simultaneously along the two mutually perpendicular axes in the horizontal plane with the help of servo motors. The information for the movement in these axes is either read optically from a master drawing or traced from a master. Alternatively, microprocessor controlled flame cutting machines are also available, wherein the co-ordinates of the contour required can be directly fed instead of making a master and tracing it.

Oxygen cutting would be useful only for those materials which readily get oxidised and the oxides have lower melting points than the metals. Thus, it is most widely used for ferrous materials. But, it cannot be used for materials such as aluminium, bronze, stainless steel and like metals since they resist oxidation.

When the cutting is done, a large amount of heat is liberated, part of which is utilised in raising the temperature of the metal adjacent to the cut. This metal temperature goes beyond the lower critical temperature and thus, structural transformations occur in these regions. The final microstructure of the metal depends on the cooling rate. Thus, steels with less than 0.3% carbon would not pose any problem since no hardening takes place at the cutting edge. But for steels with higher carbon percentages, the material around the cut should be preheated normally to a temperature of about 250 to 300°C to reduce the cooling rate and avoid the associated brittleness and crack formation at the cut edge. For high carbon steels, it may be necessary to post heat the metal after cutting to further reduce the cooling rate.

Cutting cast iron is more difficult since its melting temperature is lower than that of the iron oxide. Further, if the preheating and post heating are not properly done, formation of white cast iron and the resultant hardness at the edges would be unavoidable. The alloying elements such as chromium and nickel when present in ferrous alloys interfere with the oxidation and are therefore difficult to cut without special provisions.

9.3.1 Powder Cutting

This is the cutting method specially created for the cutting of corrosion resistant steels. The main principle in this process is the injection of iron powder into the oxygen jet while the cut is proceeding. This iron powder is preheated by the time it enters the oxygen jet flowing through the preheating flames of the cutting torch. Instantaneously, this iron gets oxidised by the oxygen jet and release enormous amount of additional heat thus obviating the need for the preheating of the metal and facilitating the cutting.

9.4 ELECTRIC ARC WELDING

In welding, generation of heat by an electric arc is one of the most efficient methods. Approximately, 50% of the energy is liberated in the form of heat. The electric arc welding process makes use of the heat produced by the electric arc to fusion weld metallic pieces. This is one of the most widely used welding process, mainly because of the ease of use and high production rates that can be achieved economically.

9.4.1 Principle of Arc

An arc is generated between two conductors of electricity, cathode and anode (considering direct current, DC), when they are touched to establish the flow of current and then separated by a small distance. An arc is a sustained electric discharge through the ionised gas column called plasma between the two electrodes.

It is generally believed that electrons liberated from the cathode move towards the anode and are accelerated in their movement. When they strike the anode at high velocity, large amount of heat is generated. Also when the electrons are moving through the air gap between the electrodes, also called the arc column, they collide with the ions in the ionised gas column between the electrodes. The positively charged ions, moving from the anode and would be impinging on the cathode, thus liberating heat. About 65 to 75% of the total heat is liberated at the anode by the striking electrons. A temperature of the order of 6000°C is generated at the anode.

In order to produce the arc, the potential difference between the two electrodes (voltage) should be sufficient to allow them to move across the air gap. The larger air gap requires higher potential differences. If the air gap becomes too large for the voltage, the arc may be extinguished. Here, we may make use of an analogy of a person walking on the road. Suddenly, when a deep trench comes in his way, the person would try to jump across it, if it is a short one. But if it is a broader one, then he may move back a little; come running towards the trench and try to jump over it. If it is too broad he may abandon the idea of jumping across it. The energy spent in jumping is much more than what is spent while normal walking. In the case of the arc, the extra energy spent crossing the air gap is liberated as heat.

For convenience of explanation, we have chosen a direct current arc for the above description. But even with an arc of the alternating current (AC), would be similar, with the main difference that the cathode and anode would change continuously and as a result, the temperature across the arc would be more uniform compared to a DC arc.

9.4.2 Arc Welding Equipment

The main requirement in an arc welding set up is the source of electric power. They are essentially of two types:

(a) Alternating current (AC) machines
   (i) Transformer, or
   (ii) Motor or engine driven alternator
(b) Direct current (DC) machines
   (i) Transformer with DC rectifier, or
   (ii) Motor or engine driven generator

The transformer does not have any moving part and as a result operates with less maintenance cost and also has higher efficiency. The power used is also less expensive and practically there is no noise in operation of the welding transformer. In AC welding, normally only transformers are used.

In DC arc welding, a rectifier or a generator can be used to supply the required DC power. In the rectifier type, the power supply is first stepped down by means of a transformer to the required voltage and then silicon controlled rectifiers (SCR) are used to convert AC to DC. These rectifiers are very compact, highly reliable and have high efficiency. The other type is a DC generator which is driven by either an induction motor running on AC or an oil engine. This combination is less efficient, more expensive and noisy in operation.

The welding machines can also be divided into two types, based on the characteristics. The first one is the constant current welding machines or droop curve machines or simply droopers, which vary the welding voltage to account for the change in the arc gap thus maintaining the arc current. As a result of this, the characteristics of the machine (a plot between output volts and output amperes) are a drooping one, as shown in Fig. 9.17. Figure 9.17(a) shows the setting of open circuit voltage (no welding) at 80 volts whereas Fig. 9.17(b) for the same machine shows open circuit voltage is set at 50 volts. It can be seen that for a large change in output voltage, the corresponding change in current is so small that the quality of the weld can be maintained. This is very essential for manual arc welding processes since the maintenance of constant arc is nearly impossible by a human welder.

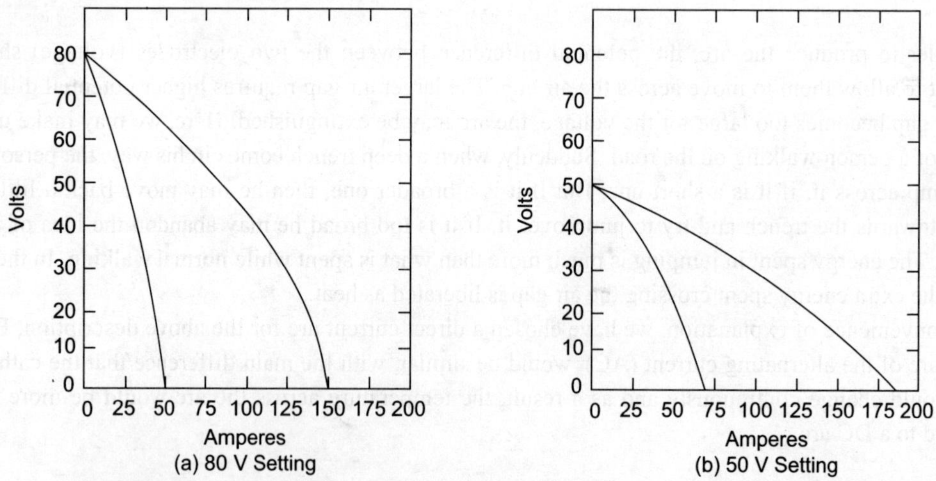

Fig. 9.17 *Characteristic curves of a constant current arc welding machine with different settings*

In contrast to the above, the constant voltage welding machines have a more flat characteristic curve, as shown in Fig. 9.18. The slope of the curve is so flat, that any small change in voltage makes for an extremely large change in the output currents. These systems are generally preferred in the automatic machines since they become self-corrective. For example when the electrode comes a bit closer to the work, the arc voltage drops raising the output current to a very high value. This current instantly melts the electrode and thus maintains the arc gap.

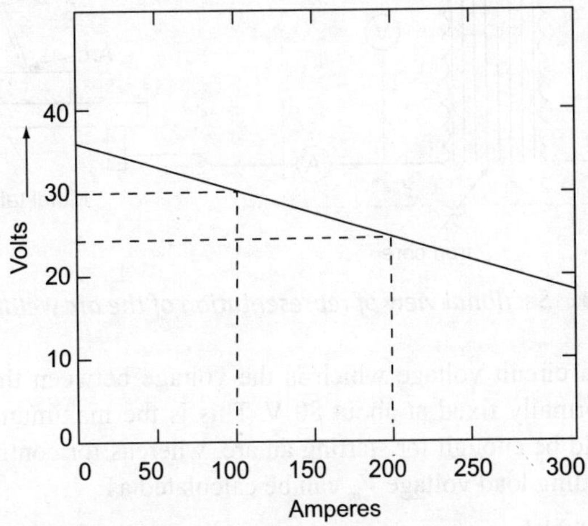

Fig. 9.18 *Characteristic curve of a constant voltage arc welding machine*

Though DC arc welding is more expensive than AC welding, it is generally preferred because of the control of heat input offered by it. As explained earlier, about 70% of the heat is liberated near the anode in DC arc welding. Hence if more heat is required at the work piece side, such as for thicker sheets or for the work materials which have higher thermal conductivity such as aluminium and copper, the work piece can be made as anode, liberating large heat near it. This is termed as 'straight polarity' or DCEN (direct current electrode negative). This gives rise to a higher penetration as shown in Fig. 9.19(a). However for thinner materials, where less heat is required in the weld zone, the polarity could be reversed by making the work piece as negative. This is termed as 'reversed polarity' or DCEP (direct current electrode positive). In reversed polarity, the penetration is small, as in Fig. 9.19(b). In the case of AC welding, the bead obtained is somewhere in between the above two types and is as shown in Fig. 9.19(c). DC arc welding is generally preferred for difficult tasks such as overhead welding, since it can maintain a stable arc.

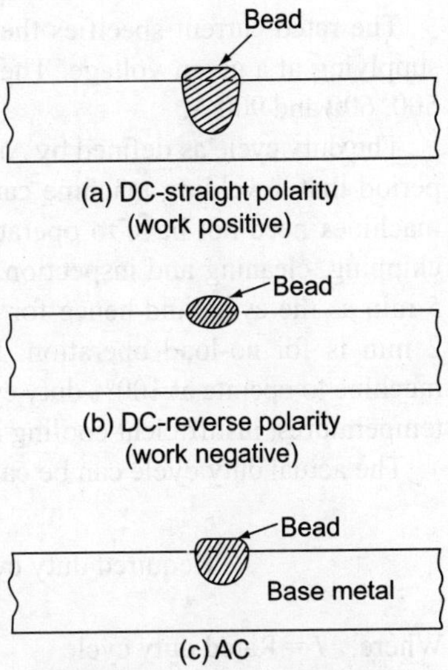

Fig. 9.19 *Weld penetration as affected by the polarity of the*

A typical AC arc welding setup using the transformer is shown in Fig. 9.20. The work piece is kept on a metallic table to which the ground lead of the secondary windings of the welding transformer is connected. The other lead of the secondary is connected to an electrode holder into which the electrode is gripped. When the electrode is brought into contact with the work, the welding takes place.

The arc welding machines are normally specified by means of

- Maximum rated open circuit voltage,
- Rated current in amperes, and
- Duty cycle

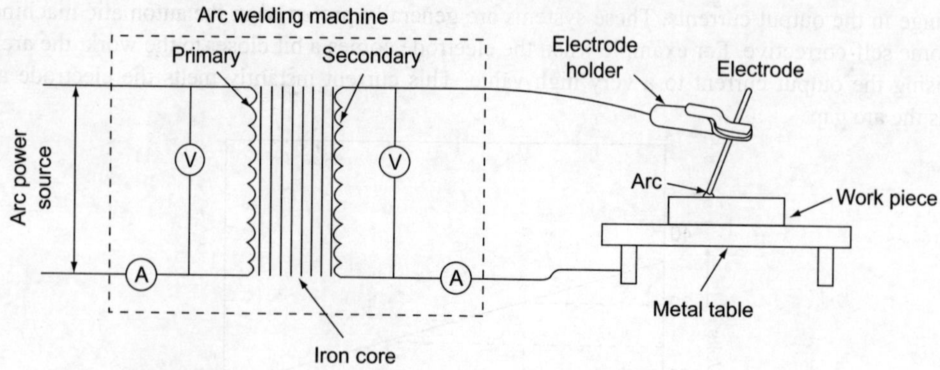

FIG. 9.20 *Sectional view of representation of the arc welding set up*

The maximum rated open circuit voltage which is the voltage between the output terminals when no welding is being done, is normally fixed at about 80 V. This is the maximum and normally, a voltage of the order of 40 to 50 V should be enough for starting an arc, whereas for continuous welding 20 to 30 V is sufficient. The minimum welding load voltage V_m can be calculated as

$$V_m = 20 + 0.04\, I$$

Where, I is the load current in amperes.

The rated current specifies the maximum current in amperes (A) that a welding machine is capable of supplying at a given voltage. The preferred current ratings as per BIS:1851-1966, are 150, 200, 300, 400, 500, 600 and 900 A.

The duty cycle as defined by American Welding Society (AWS) is "The percentage of time in a 10 minute period that a welding machine can be used at its rated output without over loading". Most of the welding machines need not have to operate the full time, since a good length of time is spent in setting up, metal chipping, cleaning and inspection. Normally, a 60% duty cycle is suggested. The Indian standard specifies 5 min as the cycle and hence for every 5 min of welding transformer operation, 3 min is for welding and 2 min is for no-load operation. But continuous, automatic welding machines may require the welding machine to operate at 100% duty cycle. Sometimes the suggested duty cycle can be lowered for high ambient temperatures, insufficient cooling air quality and low line voltage.

The actual duty cycle can be calculated as follows:

$$\text{Required duty cycle, } T_a = \left(\frac{I}{I_a}\right)^2 T$$

Where, T = Rated duty cycle

I = rated current at the rated duty cycle

I_a = Maximum current at the rated duty cycle

9.4.3 Electrodes

The electrodes used for providing heat input in arc welding are of two types, the consumable and the non-consumable electrodes.

When the arc is obtained with a consumable electrode, the weld metal under the arc melts as also the tip of the electrode. The molten metal from the electrode and that obtained from the base metal gets intimately mixed under the arc and provides the necessary joint after solidification. So in this process, once the arc is initiated, the electrode is continuously consumed and hence, the electrode should be moved continuously towards the work piece to maintain the constant arc length. Since the electrode continuously melts, it also acts as the filler rod to provide the filler metal into the joint. Thus, the functions of providing a filler metal and heat are both built into a single electrode. This, however, does not mean that the welding speed should be lowered for want of proper penetration. Wide ranges of electrodes and welding machines are available which can be used to get the proper heat input as well as the metal deposition as required. Consumable electrodes are made of various materials depending on the purpose and chemical composition of the metals to be welded. Thus, they may be made of steel, cast iron, copper, brass, bronze or aluminium.

It is also possible to use non-consumable electrodes made of carbon, graphite or tungsten. The carbon and graphite electrodes are used only in DC welding, whereas tungsten electrodes are used for both AC and DC welding. The filler metal required has to be deposited through a separate filler rod. Thus, in this welding method it is possible to properly control the heat input as well as the amount of filler metal deposited, since both are separately controlled. When consumable electrodes are used, the welding process is called metal arc welding whereas when non-consumable electrodes are used they are termed by the electrode material used, for example, carbon arc welding or tungsten arc welding.

A consumable electrode, used in welding, can be either bare or coated. The coated electrode also called stick electrode, is used for the manual arc welding process. The coating on the electrodes serves a number of purposes which are detailed as follows with reference to Fig. 9.21.

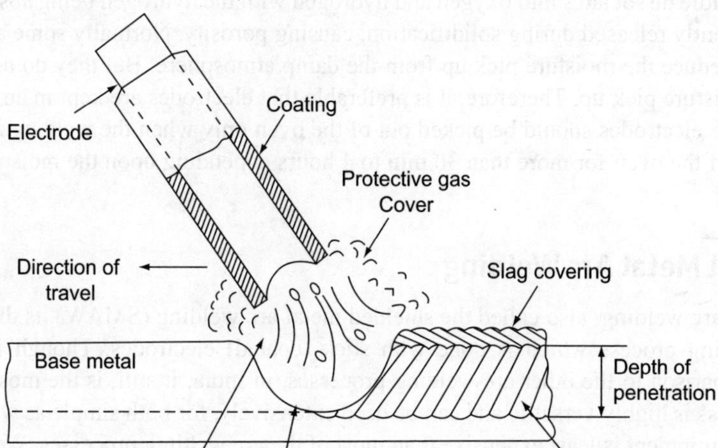

Fig. 9.21 *Cut away view of the arc welding with a coated electrode*

1. The coatings give off inert gases such as carbon dioxide under the arc heat, which shields the molten metal pool and protects it from the atmospheric oxygen, hydrogen and nitrogen pick up thus, reducing contamination of the weld metal.
2. The coatings provide flux to the molten metal pool, which mixed with the oxides and other impurities present in the puddle, forms a slag. The slag being lighter, floats on the top of the puddle and protects it against the surrounding air during the weld bead solidification. The slag covering also helps the metal to cool slowly preventing the formation of a brittle weld. When the weld is sufficiently cooled, the slag can be removed exposing the shiny weld underneath.

3. Some elements that are required for stabilisation of the arc are also added in these coatings. The coatings are different for AC welding and DC welding.
4. Special alloying elements can be introduced through these coatings to improve the strength and physical properties of the weld metal.
5. Since the electrode coating is consumed at a slower rate compared to the core (filler metal), the coating would be extended beyond the electrode as shown in Fig. 9.21. This helps in concentrating the arc and directing the filler metal in the joint. Also the thermal losses to the atmosphere from the electrode tip are reduced. This would increase the available heat for melting the electrode and thus help in improving the metal deposition rate.
6. The coatings are normally insulators of electricity and thus, permit the electrode to be used in narrow grooves, and other difficult locations without causing any short circuiting problems.
7. It is possible to include iron powder in the coating in large amounts so that the electrode can be kept in contact with the work piece, which may be necessary for welding in overhead and other positions. Also this increases the penetration and metal deposition rate.
8. The coatings also contain materials which can control the slag to be viscous or fluid. Viscous slag would be useful for making welds in vertical position to cover the metal puddle for a longer time.

The stick electrodes are normally available in diameters 3.2, 4, 5, 6, 8 and 9 mm and the length is 350 or 450 mm.

One of the major concerns with the coated electrodes is the moisture pick up by the coating. This moisture when enters the puddle dissociates into oxygen and hydrogen with the hydrogen being absorbed by the liquid metal and subsequently released during solidification, causing porosity. Normally some additives are added to the coatings to reduce the moisture pick up from the damp atmosphere. But they do not ensure complete safety from the moisture pick up. Therefore, it is preferable that electrodes are kept in an oven to keep them thoroughly dry. The electrodes should be picked out of the oven only when the welding is to be done. They should not be out of the oven for more than 30 min to 4 hours depending upon the moisture picking time of a particular coating.

9.4.4 Manual Metal Arc Welding

The manual metal arc welding, also called the shielded metal arc welding (SMAW), is the most extensively used manual welding process which is done with stick (coated) electrodes. Though in USA, its use is decreasing in comparison to the other arc welding processes, in India, it still, is the most used arc welding process. This process is highly versatile and can be used extensively, for both simple as well as sophisticated jobs. Further, the equipment is least expensive than most of the arc welding processes. Welds by this process can be made in any position.

Job of any thickness can be welded by shielded metal arc welding. But very small thicknesses below 3 mm may give rise to difficulty in welding because of their lack of rigidity. Similarly very large thicknesses above 20 mm may take a long time for filling up the joint groove.

The shielded metal arc welding can be done with either AC or DC power source. The typical range of the current usage may vary from 50 to 500 A with voltages from 20 to 40 V.

The main disadvantage of the shielded metal arc welding process is the slow speed. The typical weld metal deposition rates may be in the range of 1 to 8 kg/hr in the flat position. This reduces substantially for the vertical and overhead positions. Further, a lot of electrode material is wasted in the form of unused end, slag and gas. There are more chances of slag inclusions in the bead. Also special precautions are needed to reduce moisture pick up so that it would not interfere with the welding.

The typical shielded metal arc welding setup with AC power source is shown in Fig. 9.20. The electrodes for the welding operation should be selected properly depending on the requirements of the welding. The main points to be considered are

1. The composition of the base metal, which determines the electrode composition.
2. The tensile strength of the required joint.
3. The thickness of the base metal. For thinner metals the current setting should be lower.
4. The required metal deposition rate.
5. The type of arc welding equipment used. DC arc welding equipment would be preferable for overhead welding.
6. The weld position, flat, horizontal, vertical or overhead. A flat position can accommodate a larger size electrode. Also to increase metal deposition rate, coatings with iron powder can be used. In the case of vertical and overhead positions, it is necessary that the weld pool be smaller for better control. This would necessitate a small size electrode.

It is generally preferable to use the largest size electrode that is permitted for a given welding situation.

The covered electrodes to be used for manual metal arc welding are standardised for various base materials. The coding is used for marking the electrodes as per the relevant Indian Standards. For example, BIS: 815-1966 describes the coding of the electrodes to be used for mild steel and low alloy high tensile steel.

The electrodes are marked with a 6 digit numeral associated by a prefix and a suffix. The meaning of these and the various values it can take is shown schematically in Fig. 9.22.

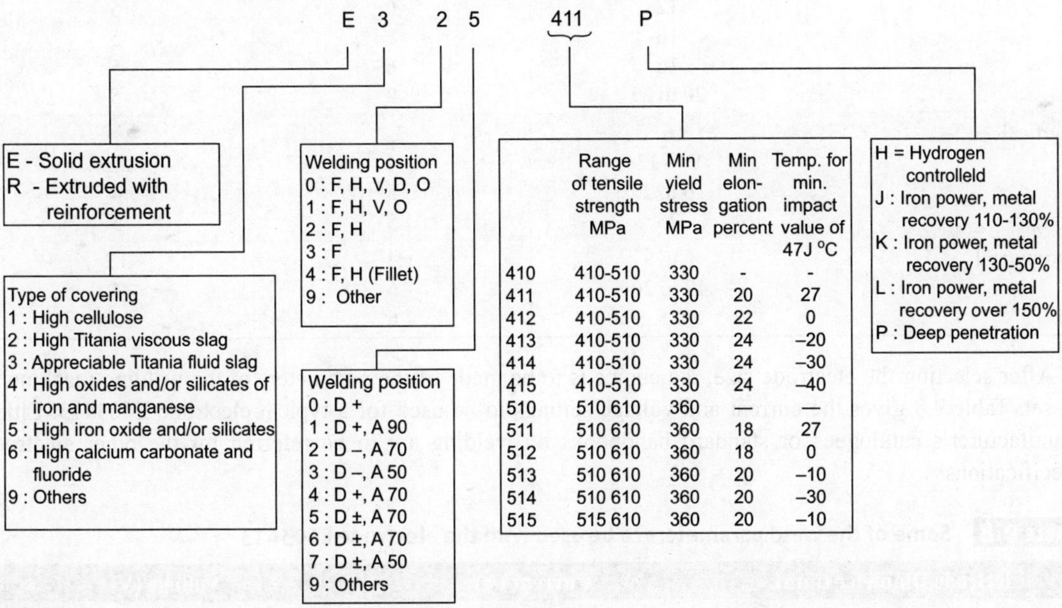

FIG. 9.22 *Coated electrode designation for manual metal arc welding of mild steel using coated electrodes as per BIS: 815-1966*

The high cellulose type of electrode coating gives rise to a voluminous gas shield but also burns off quickly. Thus, the deep penetration can be achieved but the weld finish is somewhat coarser. When the content of titania (as rutile, titanium-white or ilmenite) is high, smooth arc is produced giving rise to voluminous slag and as such should normally be used for the flat position only. But the weld finish is smooth. Iron oxide when present in the coatings gives rise to higher metal deposition rates. The other materials that may be present as fluxes in coatings are manganese oxide, calcium carbonate and fluoride. The slag becomes fairly fluid when appreciable amounts of calcium carbonate and fluoride are present. These are used for high tensile structural steels.

The welding current category D+ refers to direct current with electrode positive (reverse polarity) and A90 refers to the open circuit voltage not less than 90 V for a reference electrode size of 4 or 5 mm. When the electrode size varies, the voltage also changes accordingly. For example, for 2.5 to 8 mm electrodes, it may vary from 100 to 70 V in case of A90.

The size of the electrode to be chosen is based on the thickness of the plate to be welded and the weld position. Table 9.7 gives suggested electrode sizes based on these parameters. The root electrode size is to be used for the root pass in the case of a multipass welding.

TABLE 9.7 Electrode sizes based on the metal thickness to be welded.

Weld Position	Metal Thickness (mm)	Electrode Diameter (mm)	Root Electrode Diameter (mm)
All	3 to 6	3.2	–
Down hand	6	4	–
	9	6	4
	12	8	4
	12	6	5
	16	8	5
	16	6	5
	20 to 25	9	5
Vertical	6	3.2	–
	9 to 12	4	3.2
	16 to 20	5	4
	25	5	5
Horizontal	6 to 9	4	3.2
	12 to 20	6	4
	25	6	5

After selecting the electrode size, the choice is to be made of the parameters of the welding machine to be set. Table 9.8 gives the current and voltage settings to be used for a typical electrode E603413. Either manufacturer's catalogues or standard handbooks on welding are to be referred for the other electrode specifications.

TABLE 9.8 Some of the weld parameters to be used with the electrode E603413

Electrode Diameter (mm)	Current (A)	Voltage (V)
3.2	100 to 150	18 to 22
4	140 to 200	20 to 24
5	200 to 220	21 to 25
6	275 to 350	23 to 27
8	375 to 475	24 to 28

The type of edge preparation required for butt welds in manual metal arc welding process is presented in Fig. 9.23 with the necessary dimensional details.

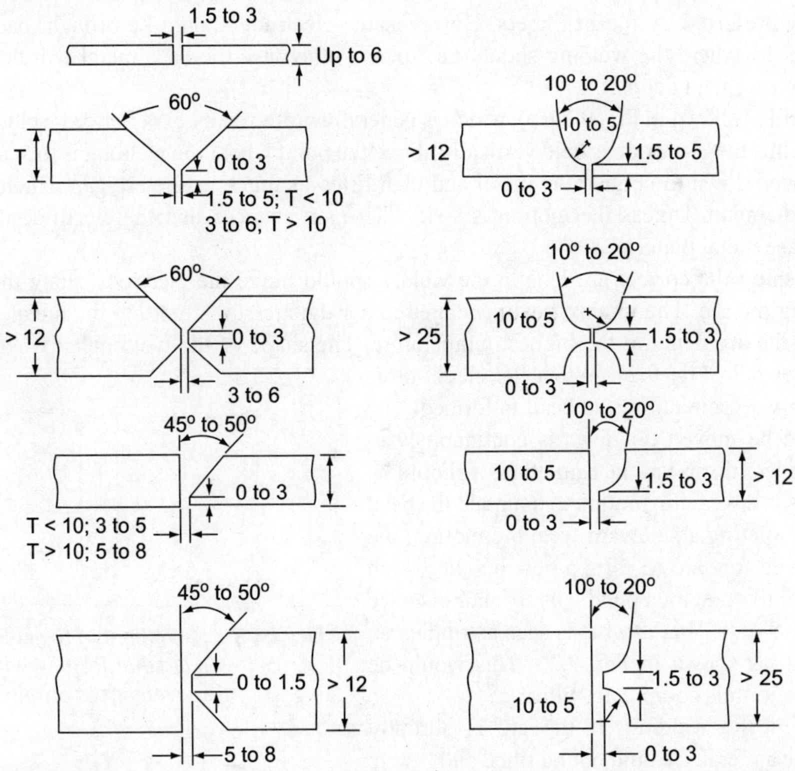

Fig. 9.23 *Butt edge weld preparation details for manual metal arc welding*

Having chosen the electrode, the welding machine is set and the edge is made ready for welding. To start the arc, first the welder has to make a contact between the electrode and the work piece so that current flow is established. Then the electrode should be moved away from the work piece by a very small amount so that the arc is established. To accomplish this generally two different methods are employed, which are shown in Fig. 9.24.

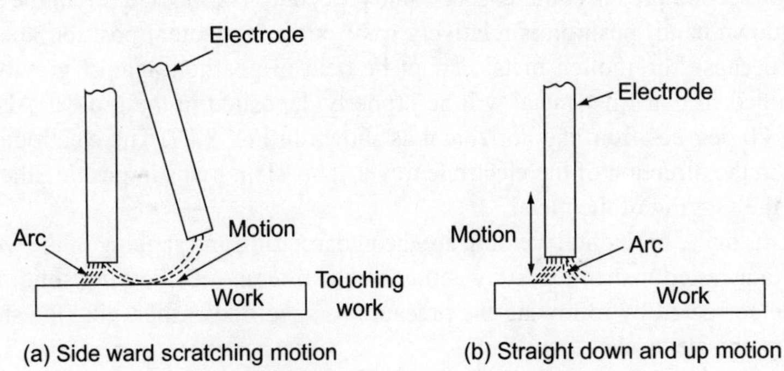

Fig. 9.24 *Method of striking an arc*

In the first method shown in Fig. 9.24(a), the electrode is moved in an arc so that it will scratch the work metal and thus establish the current flow. The normal distance by which the electrode is separated from the work metal is same as that of the diameter of the electrode wire. The scratching method is an easier method and is generally preferred by the beginners. However, the electrode should be brought back immediately to the point of start where the welding should take place, otherwise the base metal will unnecessarily get defaced by the weld metal deposit.

The other method shown in Fig. 9.24(b), which is generally preferred by experienced welders, is called the tapping start. In this the electrode is held vertically above the point where the welding is to start and in a swift motion it is moved down to contact the metal and then lifted as much as the arc gap which is same as the electrode wire diameter. Unless the motion is swift, there are chances that the electrode itself would get welded to the base metal plate.

After establishing the correct arc length, the welder should move the electrode along the length of the joint maintaining the arc. The intense heat generated under the arc starts melting the metal, with the metal at the centre of the arc being at the highest temperature. The shape of the weld metal when the arc crater is formed is shown in Fig. 9.21. When the electrode is moved in the forward direction, the bead is formed. The electrode should be moved downwards continuously to maintain the arc length, and at the same time it should be moved sideways in a weaving motion to maintain the bead width. After completing a sideward weaving motion, the electrode is moved forward to form a new puddle which is separated from the previous puddle by a small distance of the order of 1.5 mm. This is what is seen as ripples on the welded joint, as shown in Fig. 9.25. This would be continued till the joint is completely filled.

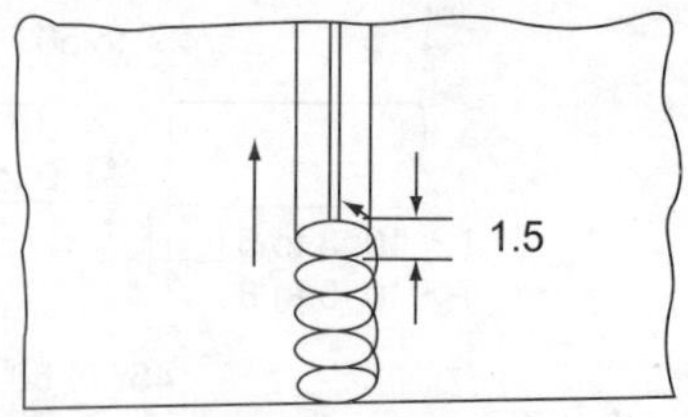

Fig. 9.25 *Movement of the electrode in a circular fashion with forward movement to obtain the bead*

At the end of the welding, if the arc is abruptly extinguished, the arc crater would not be filled and hence, a depression would be left in the joint. Therefore, the arc has to be slowly extinguished by the gradual decrease of the welding current, which ensures a complete filling of the arc crater.

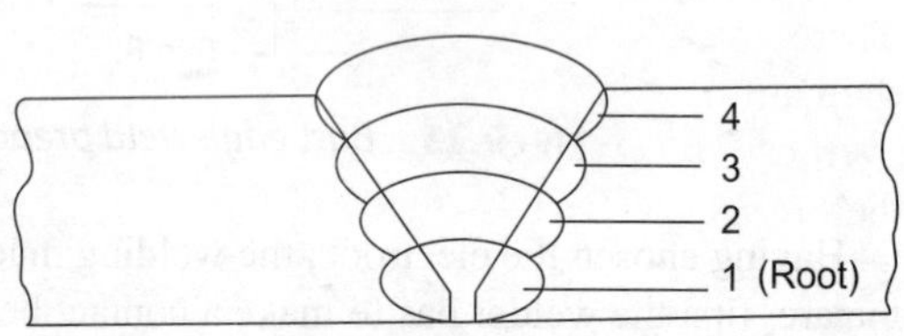

Fig. 9.26 *A typical multipass bead formed by manual metal arc welding*

In multipass welding, as shown in Fig. 9.26, the brittle slag coating present on the bead after the root pass is made is chipped off and then the area is cleaned with a wire brush, before the second pass is commenced. Same procedure is followed for all the subsequent passes.

Welding in flat (down hand) position is relatively easy, but in horizontal position, the welding becomes somewhat difficult because the molten metal cannot be held in position against gravity. Generally, short arc length is maintained so that filler metal will be properly deposited in the groove. Also, the electrode is maintained at about 20 degrees from the horizontal as shown in Fig. 9.27. The electrode is also positioned at about 20 degrees in the direction of the electrode travel. This slant helps in pushing the molten metal into shape and reducing the sagging of the metal.

The major problems in the horizontal welding are the undercutting and sagging of the weld bead, as shown in Fig. 9.28 in an exaggerated manner. Gravity acting on the molten metal during solidification causes this sagging. This can be corrected by following the precautions specified earlier, viz., the short arc length and angular positioning of the electrode.

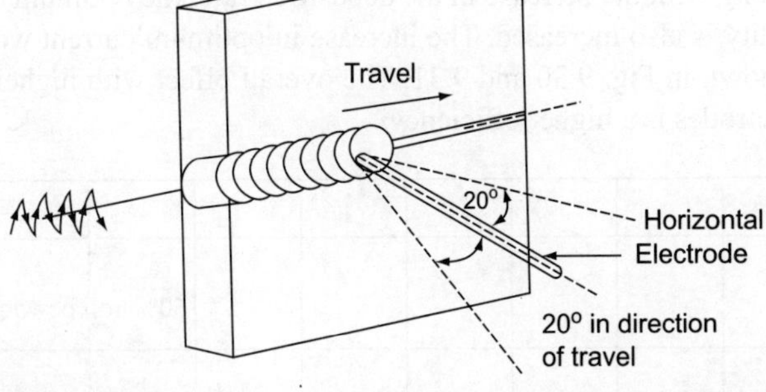

Fig. 9.27 *The position of electrode for welding in horizontal position*

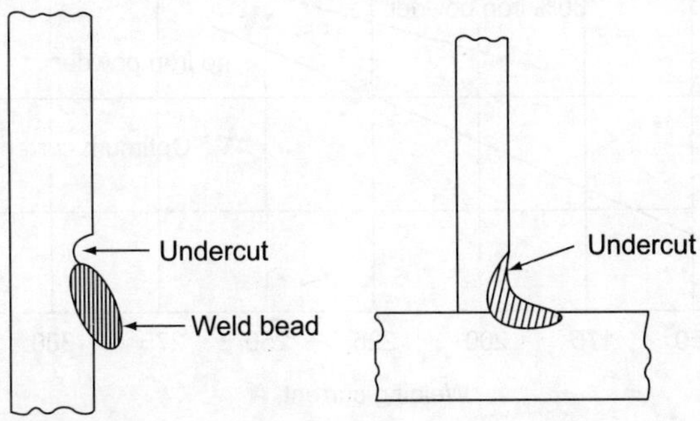

Fig. 9.28 *Undercut that is likely in horizontal welding*

There are two positions in the vertical, the vertically upward and vertically downward directions. Out of these, the vertically upward direction is most preferred because of the strong weld obtained. In the downward welding, the slag is likely to move down due to the gravity and mix with the weld metal. Therefore, generally only vertically upward welding is preferred. Here also to fight the gravity, the electrode is held at an angle of about 20 degrees from the vertical plane as in Fig. 9.29, with a short arc length so that the filler metal will be easily moved into the groove. Further, the electrode is moved in and out of the puddle so that a proper bead is formed. In the vertical position, it is necessary to provide the welder with the necessary protective clothing because of the possibility of the spill over of the liquid metal and sparks on him.

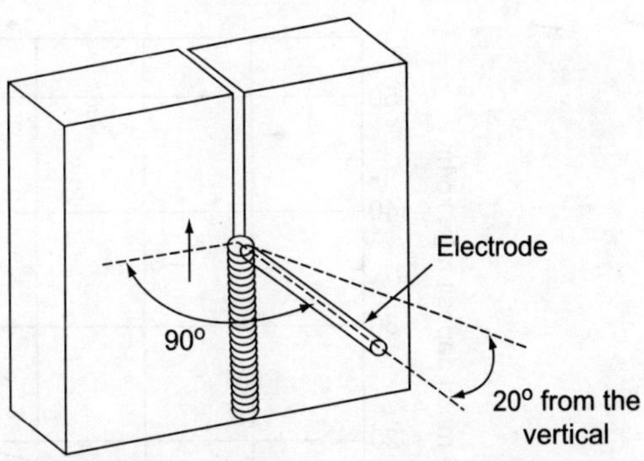

Fig. 9.29 *Positioning of electrode for welding in vertical upward position*

The overhead welding is very similar to the flat position. Because of the greater possibility of the weld metal falling, the electrode sizes used are small with a very short arc length.

Though welding in any position is feasible, the flat position is the best since higher weld metal deposition rates as well as good bead quality can be achieved. Because of this in many a situation, special fixtures are used to rotate the jobs in such a way that the weld joint comes into the flat position.

It has been observed that adding iron powder to the coating of the electrodes causes an increase in the metal deposition rates. Along with the increase in the deposition rate, the optimum current setting to be used for the good welding quality is also increased. The increase in optimum current would be higher for higher iron powder content as shown in Fig. 9.30 and 9.31. The overall effect with higher current and higher iron powder content in the electrodes is a higher efficiency.

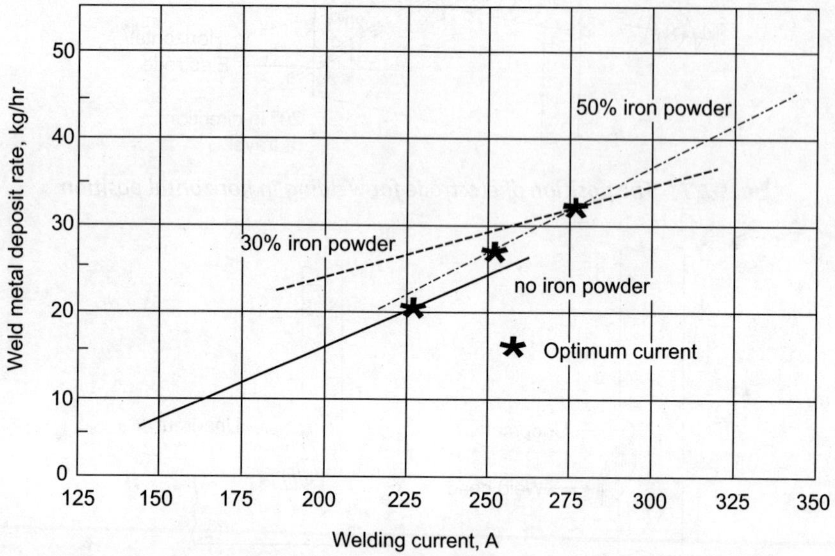

Fig. 9.30 *Effect of iron powder in the electrode coating of shielded metal arc welding process (AC)*

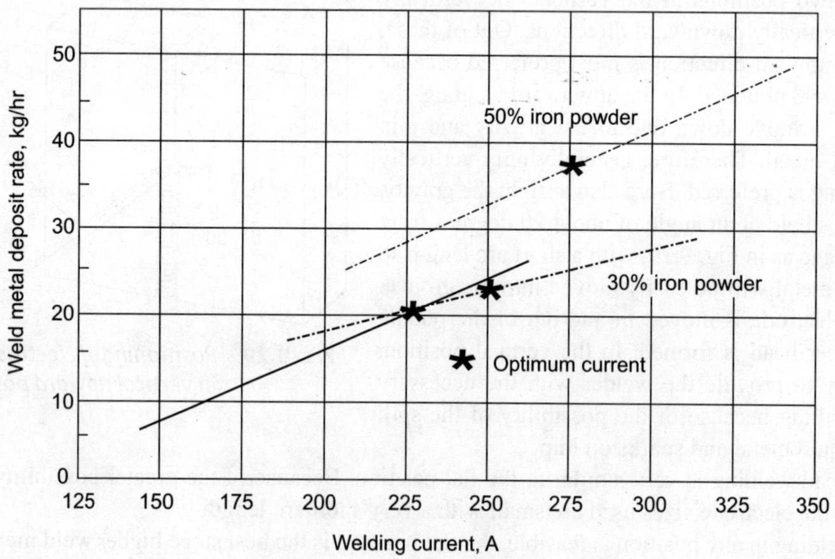

Fig. 9.31 *Effect of iron powder in the electrode coating of shielded metal arc welding process (DC electrode positive)*

Arc Blow

An AC arc once started is more stable, but it is not so with the DC arc. The predominant problem faced with the DC arc is the "arc blow", the deflection of the arc by means of the magnetic fields setup due to the flow of the welding current. All electrical conductors are surrounded by magnetic flux lines around them as shown in Fig. 9.32(a). These magnetic flux lines move easily in metal but not in air.

When the electrode moves forward, these flux lines move with the electrode. But when the weld comes to the edge of the plate, as shown in Fig. 9.32(b) and Fig. 9.32(c), or takes a turn then the flux lines move out of the base metal which is not possible. This causes high magnetic pull on the arc which results in a backward arc blow as shown in Fig. 9.32(b). The result of an arc blow is an excessive spatter (throwing out the tiny droplets of weld metal out of the joint on to the base metal which appear as tiny dots on the metal plates), as also incomplete fusion and reduced welding speed. When a large slag is produced, the arc blow melts the slag causing still more excessive spatter.

The problem of arc blow gets multiplied when welding is done on strongly magnetic materials such as nickel alloys because of the strong magnetic fields set up by these metals.

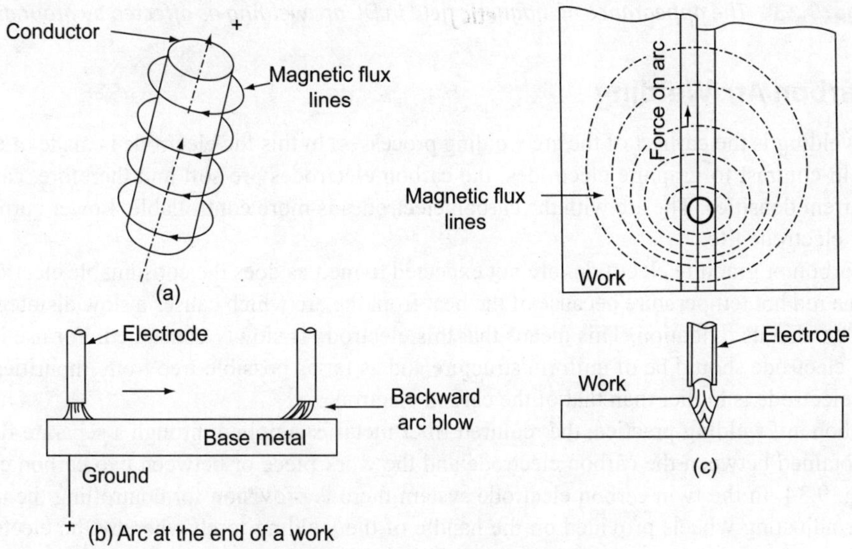

Fig. 9.32 *Arc blow in DC arc welding*

Some of the methods that are used to reduce the severity of the arc blow problem are to

1. Change to AC welding because of the continuous change in the polarity, the effect of magnetic field is nullified
2. Reduce the current used so that the strength of the magnetic field gets reduced
3. Use a short arc length so that the filler metal would not be deflected but carried easily to the arc crater
4. Put steel blocks near the end of the plate in contact with the base metal so that the magnetic flux lines would flow through them and reduce the arc blow
5. Place more than one ground lead from the base metal (preferably one each from the ends of the base metal plate as in Fig. 9.33)

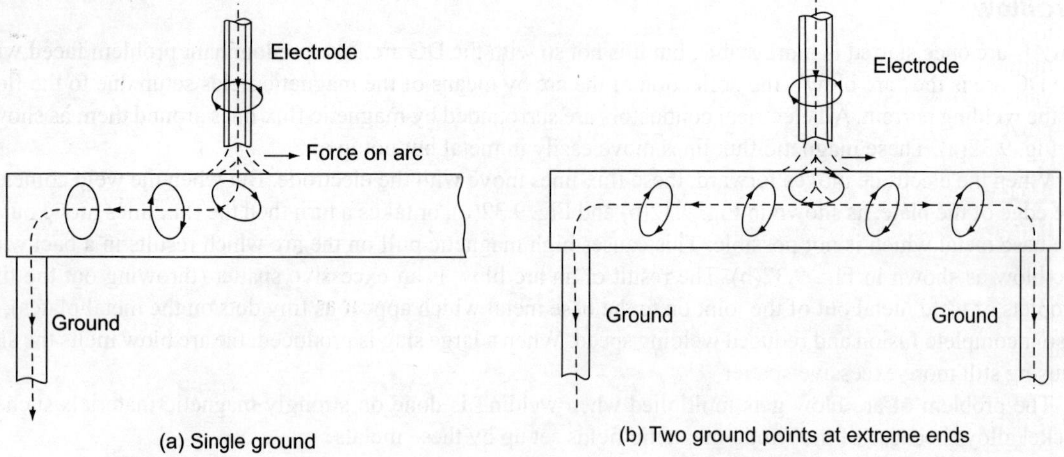

Fig. 9.33 *The appearance of magnetic field in DC arc welding as affected by grounding*

9.4.5 Carbon Arc Welding

Carbon arc welding is the earliest of the arc welding processes. In this the electrode is made of either carbon or graphite. In contrast to graphite electrodes, the carbon electrodes are soft and therefore, cannot take up very high current densities. The arc with the carbon electrodes is more controllable. Lower currents also add to the higher electrode life.

Though carbon or graphite electrodes are not expected to melt as does the consumable electrodes, they do get heated to a red hot temperature because of the heat from the arc which causes a slow disintegration of the electrode tip as also its oxidation. This means that this electrode is slowly consumed. For use in carbon arc welding, the electrode should be of uniform structure and as far as possible free from impurities. The life of the graphite electrode is higher than that of the carbon electrode.

In the carbon arc welding practice, the required filler metal is supplied through a separate filler rod. The arc can be obtained between the carbon electrode and the work piece or between two carbon electrodes, as shown in Fig. 9.34. In the twin carbon electrode system there is provision for controlling the arc length by means of the adjusting wheels provided on the handle of the welding torch whereby the electrodes can be either brought in contact with one another or taken apart.

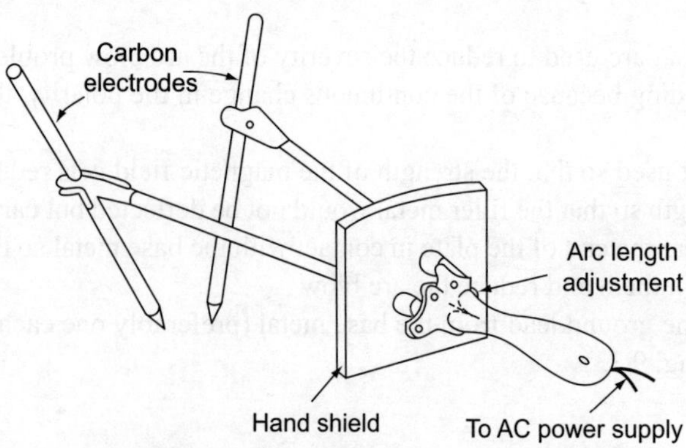

Fig. 9.34 *Twin carbon arc welding torch*

Generally, DC power supply with electrode negative is used for the single electrode carbon arc welding to minimise the heat generation near the electrode side so that the wear (consumption) of the electrode is maintained at a minimum rate. The tip of the electrode is made conical. The typical current settings for the various electrode sizes are shown in Table 9.9 for guidance. For the twin carbon arc welding process, AC power source is normally used since no special advantage is derived from using DC supply.

TABLE 9.9 Current requirements for carbon electrodes

Electrode Diameter, mm	Welding Current, A		Maximum Current Density (A/mm^2)	Metal Deposited (kg/h)
	Min	Max		
3.2	0	35	4.48	–
4.5	25	60	3.41	–
6	50	90	2.88	–
7.5	80	125	2.56	–
10	120	185	2.34	–
11	140	210	2.20	0.680
12.5	170	260	2.08	1.134
16	230	370	1.89	2.040
20	290	490	1.74	2.722
22	350	615	1.60	–
25	400	750	1.50	–

Because of the separation of the heat source from the filler metal, better control of heat input is possible in the case of carbon arc welding. It is possible to weld thicker plate to thinner plate using twin carbon electrode welding by providing additional heat to the thicker plate before providing the filler metal and heating the joint. The major problem is the blow holes that are caused because of the turbulence associated with the DC power source (due to the magnetic arc blow). Though carbon arc welding is not suitable to overhead or vertical welding positions, very high mechanised welding speeds could be obtained by the process in the flat position.

9.4.6 Inert Gas Shielded Arc Welding

The endeavour of the welder is always to obtain a joint which is as strong as the base metal and at the same time the joint is as homogeneous as possible. To this end, the complete exclusion of oxygen and other gases which interfere with the weld pool to the detriment of the weld quality is very essential. In manual metal arc welding, the use of stick electrodes does this job to some extent but not fully. In inert gas shielded arc welding processes, a high pressure inert gas flowing around the electrode while welding would physically displace all the atmospheric gases around the weld metal to fully protect it.

It all started in 1890 when a patent was granted for welding with an electrode which is provided with a shielding of flowing carbon dioxide gas. But in 1930, the first use of helium and argon as shielding gases for arc welding with non-consumable electrodes was demonstrated for commercial use. Since then rapid strides have taken place in gas shielded arc welding and it is now used extensively.

The shielding gases most commonly used are argon, helium, carbon dioxide and mixtures of them. Argon and helium are completely inert and therefore they provide a complete inert atmosphere around the puddle when used at sufficient pressure. But when these gases are used, they should be of high purity (99.95% purity). Any contamination in these gases would decrease the weld quality. Hydrogen, if present or generated because of the dissociation of moisture, would give rise to weld porosity.

Argon is normally preferred over helium because of a number of specific advantages. It requires a lower arc voltage, allows for easier arc starting and provides a smooth arc action. A longer arc can be maintained with argon, since arc voltage does not vary appreciably with arc length. This helps particularly in manual TIG welding accounting for the human variability. It is more economical in operation. Also it is the heaviest of the shielding gases used as such it generally requires a lower flow rate for good shielding action. However, this may not be true when welding is done in overhead position because of the tendency of the lighter gases moving upwards and ending up near the weld zone. Argon is particularly useful for welding thin sheets and for out of position (vertical, horizontal and overhead) welding.

The main advantage of helium is that it can withstand the higher arc voltages. As a result, it is used in the welding where higher heat input is required such as for thick sheets or for higher thermal conductive materials such as copper or aluminium.

Carbon dioxide is the most economical of all the shielding gases. Under the arc, the carbon dioxide decomposes to carbon monoxide (CO) and oxygen which subsequently combines back to form carbon dioxide when they cool. It requires slightly higher currents which cause more agitation in the weld puddle resulting in the trapped gases to rise and thus, reduces the weld porosity.

Both argon and helium can be used with AC as well as DC welding power sources. However, carbon dioxide is normally used with only DC with electrode positive. Carbon dioxide with negative electrode tends to cause large electrode spatter and an unstable arc. In such situations, the electrodes are treated with caesium and sodium to stabilise the arc.

9.4.7 Tungsten Inert Gas Welding (TIG)

Tungsten inert gas (TIG) welding or gas tungsten arc welding (GTAW) is an inert gas shielded arc welding process using non-consumable electrode. The electrodes may also contain 1 to 2% thoria (thorium oxide) mixed along with the core tungsten or tungsten with 0.15 to 0.40% zirconia (zirconium oxide). The pure tungsten electrodes are less expensive but will carry less current. The thoriated tungsten electrodes carry high currents and are more desirable because they can strike and maintain a stable arc with relative ease. The zirconia added tungsten electrodes are better than pure tungsten but inferior to thoriated tungsten electrodes.

A typical tungsten inert gas welding setup is shown in Fig. 9.35. It consists of a welding torch at the centre of which is the tungsten electrode. The inert gas is supplied to the welding zone through the annular

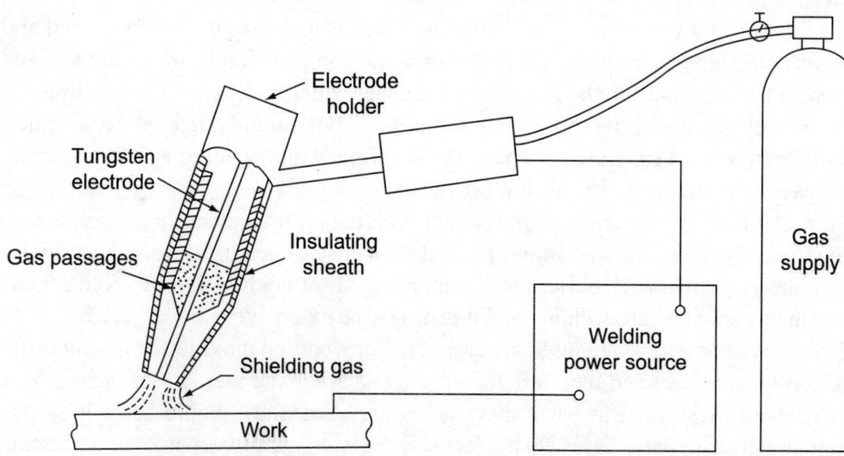

FIG. 9.35 *A typical tungsten inert gas (TIG) welding set up*

path surrounding the tungsten electrode to effectively displace the atmosphere around the weld puddle. The smaller weld torches may not be provided with any cooling devices for the electrodes, as in Fig. 9.34, but larger ones are provided with circulating cooling water.

The TIG welding process can be used for the joining of a number of materials though the most common ones are aluminium, magnesium and stainless steel. The typical combination of TIG set ups to be used with these and other metals are presented in Table 9.10.

TABLE 9.10 The metals that can be commonly welded by TIG welding and their choice of setups.

Material	Electrodes	Power Supply Used	Preferred Shielding Gas
Stainless steel	Thoriated tungsten	DCEN	Argon
Aluminium	All types	AC	Argon
Magnesium	Tungsten	AC	Argon
Deoxidised copper			
Monel	Thoriated tungsten	DCEN	Argon
High carbon steel			
Cast iron	Thoriated tungsten	AC or DCEN	Argon

Power Sources

The power sources used are always the constant current type. Both direct current (DC) and alternating current (AC) power supplies can be used for TIG welding. When DC is used, the electrode can be negative (DCEN) or positive (DCEP). With DCEN more heat is generated near the work piece and consequently the electrode does not get heated to a great extent. But when DCEP is used, a large amount of heat is liberated at the electrode itself thereby limiting the maximum current that can be carried by an electrode. Roughly, the current carrying capacity of a DCEN electrode is about 10 times as high as that of a DCEP electrode.

The DCEP is sometimes utilised to break down the oxides on the surface of the metals such as aluminium. The electrons from the oxide layer move towards the positive electrode weakening the surface layer. The positively charged ions from the electrode would then be able to easily break the surface layer and thus would help in obtaining the fusion. Similarly, when AC is used, the half cycle during which the electrode is positive, the electrons from the oxide layer would be moving towards the electrode, whereas in the other half the electrons from the electrode would be able to easily break the oxide layer on the work piece surface. Thus, an AC arc welding is likely to give rise to a higher penetration than that of DCEP. Hence, DCEP is normally used for welding thin metals whereas for deep penetration welds DCEN is used. DCEP also causes larger heat affected zones and more weld distortion than DCEN.

The DC power supply used for TIG can be either a steady one or more often a step pulsed one. In the case of the step pulsed current machine, the current level is maintained at two levels, as shown in Fig. 9.36(a). The low level is called the background current which is used for cooling the weld metal. The other is the peak current used when the actual melting (welding) takes place. During the background current period the arc is maintained but very small heat input goes to the weld and, as a result, the arc crater cools. This type of step pulsed DC source is, particularly, useful for welding in out of positions (other than flat position) since it allows for the controlled heating and cooling. Otherwise, the electrode is to be flipped away slightly from the arc crater to allow for the cooling of the puddle before it is moved forward again. But the pulsed DC arc welding provides for proper solidification during the background current period when the torch is moved forward for forming the next spot (bead), as shown in Fig. 9.36(b).

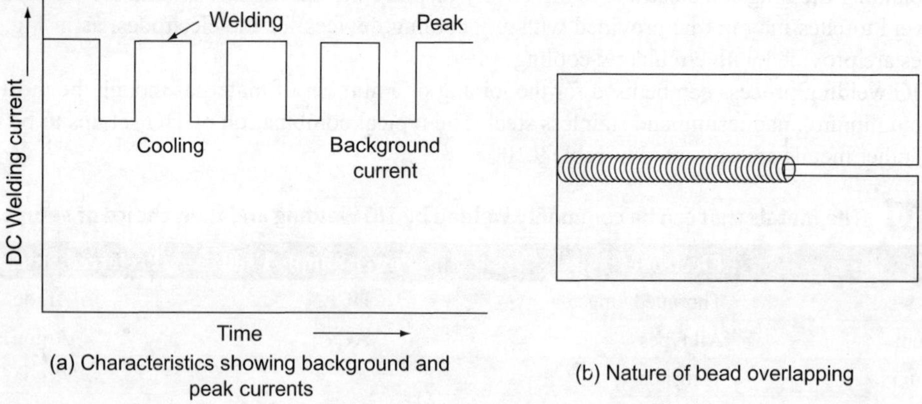

FIG. 9.36 *Step pulsed DC supply (a) Characteristics showing background and peak currents (b) Nature of bead overlapping*

When alternating current (AC) is used for TIG welding, the current continuously changes its direction. It changes its direction 50 times every second (in 50 Hz power supply) such that half the time it is operating as DCEN and rest, as DCEP. A typical AC wave form is shown in Fig. 9.37(a) which is termed as balanced wave since the positive side and the negative side are identical in magnitude. But the TIG welding machine would not behave as normal AC. During the period when the electrode is positive, the electrons move from the flat work piece surface to the small sized tip of the electrode restricting the flow of electrons. This is termed as 'rectification' and is responsible for the reduced current flow during DCEP portion of the AC wave as shown in Fig. 9.37(b). This is known as unbalanced wave.

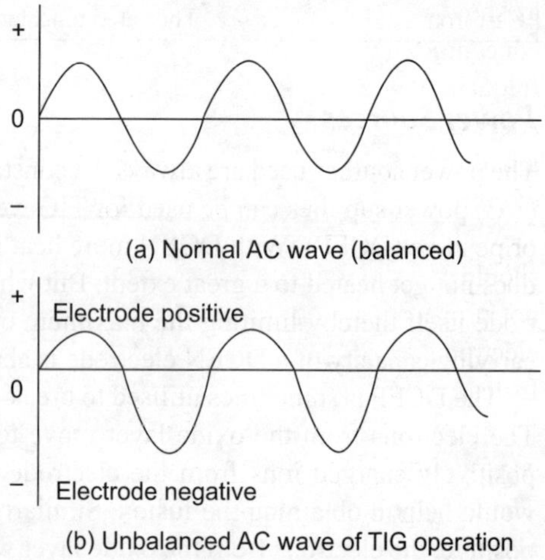

FIG. 9.37 *The nature of alternating current (a) normal AC wave (balanced) (b) unbalanced AC wave of TIG operation*

This rectification in an AC cycle during the time when the electrode becomes positive will make the AC arc a highly unstable one. To maintain a steady arc in an unbalanced AC welding machine, a very high voltage, very high frequency (several MHz) and low current (less than one ampere) power supply is superimposed on the unbalanced wave. This maintains the shielding gas ionised during the period when the electrode is positive and thus maintains the arc continuously. There are quite a few advantages of an unbalanced AC arc welding machine compared to a balanced wave machine. It is less expensive. Since less current flows when the electrode is positive, less heat is liberated near the electrode. This permits a higher current carrying capacity for the electrodes which results in better penetration.

It is possible to provide a balanced wave, as shown in Fig. 9.37(a), by incorporating a large number of capacitors in series to provide the necessary current discharge during the time when the electrode is positive. These capacitors get charged during the time when the electrode is negative. A balanced wave maintains a steady arc and therefore is preferred for better weld quality. Since during the positive electrode cycle more current flows than in a balanced wave, better removal of the oxide layer is possible. However, these machines are more expensive compared to the unbalanced type.

Electrodes

The tungsten electrodes used for welding should be clean and completely free from any kind of contamination such as molten filler metal.

If the arc is started by first touching the base metal and withdrawn, the electrode tip may pick up the base metal which causes the subsequent sputtering and loss of metal in the electrode tip. Also, the electrode may get consumed quickly if it is allowed to get oxidised, since tungsten oxide has a lower melting temperature. The oxidation occurs when the electrode is allowed to cool in the atmosphere after welding. Hence, the shielding gas flow should be maintained for some time after extinguishing the arc so that the electrode gets sufficiently cooled in a protective atmosphere rather than in the oxidising normal atmosphere.

The electrode tip should be prepared for proper weld penetration. The typical shapes that can be used are shown in Fig. 9.38. Though it is possible to use these electrodes without any tip preparation, it would be better to prepare the tip since it enhances the weld quality. For AC welding with high frequency (AC-HF) unbalanced machines, the tip should be pencil pointed as shown in Fig. 9.38(e), so that the HF current gets concentrated and the arc is easily initiated (high frequency current tend to flow through the surface). Also, once the arc is initiated the tip gets heated and a ball of tungsten is formed which reduces the effect of current rectification and thus, stabilises the AC arc.

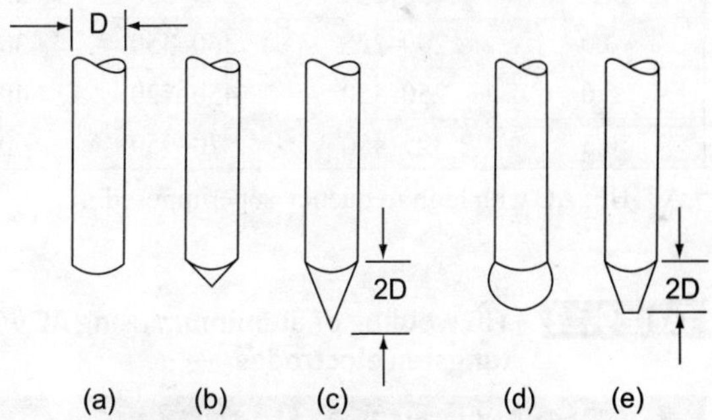

Fig. 9.38 *Typical electrode tip shapes*

With DCEN, the electrode would be made conical as in Fig. 9.38(c). While grinding, the tip concentricity should be maintained, otherwise the gas flow would become uneven making some part of the puddle not properly shielded and thereby, causing contamination of the weld joint in that portion. Pure tungsten electrodes are never made into conical point since the end is likely to melt and contaminate the weld metal. Instead, it is better to make full round ball at the tip, as shown in Fig. 9.38(d).

Welding Technique

The welding technique used for TIG is essentially similar to that of the gas welding. The edge preparation is also similar to that of gas welding. Backing of the joint is sometimes preferable to provide good appearance and uniformity of the weld. The metallic backing plates used are provided with a small groove of a depth of the order of 0.4 mm near the root, with the width being about 3 to 4 times the depth. The backing plate is removed after the welding is over.

The current settings to be used depend on the type of power supply and the electrode used. The typical ranges of these values used for various electrode sizes are presented in Table 9.11.

Sometimes filler metals may have to be used depending on the base metal. The filler metal for TIG (GTAW) welding is generally a bare wire. The size of the filler metal depends on the base metal thickness. The sizes of the filler rods are shown in Table 9.12, for various metal thicknesses for aluminium welding.

The nozzle or shield size (the diameter of the opening of the shroud around the electrode) to be chosen depends on the shape of the groove to be welded as well as the required gas flow rate. The gas flow rate depends on the position of the weld as well as its size. All these parameters for TIG welding of aluminium are summarised in Table 9.12. Too high a gas consumption would give rise to turbulence of the weld metal pool and consequently porous welds.

TABLE 9.11 The suggested current range in amperes for TIG welding with argon gas and different types of electrodes.

Electrode Diameter, mm	Pure Tungsten		2% Thoriated Tungsten			Zirconium Tungsten AC-HF A
	AC-HF* A	DCEN A	AC-HF A	AC A	DCEN A	
1.0	10–60	15–80	20–80	20–60	25–85	20–80
1.6	50–100	20–150	50–150	60–120	50–160	50–150
2.5	100–160	125–225	130–250	100–180	135–235	130–250
3.15	150–210	225–360	225–360	160–250	250–400	225–360
4.0	200–275	360–450	300–450	200–320	400–500	300–450
5.0	250–350	450–720	400–550	290–390	500–750	400–550
6.3	325–450	720–950	600–800	340–525	750–980	600–800

* AC-HF: AC with high frequency superimposed.

TABLE 9.12 TIG welding of aluminium using AC with high frequency welding machine and argon gas with tungsten electrodes.

Metal Thickness (mm)	Filler Rod Size (mm)	Electrode Diameter (mm)	Current (A)	Shield Diameter (A)	Arc Travel Speed (cm/min)	Edge Preparation	Gas Flow Rate (L/h)
1.6	1.6	1.6	60–85	10	32	SCB	425
3.15	2.5	2.5–3.15	125–150	10–12.5	27.5	SCB	560
5.0	3.15	3.15–4.0	180–225	12.5	20	SSV	720
6.3	5.0	4.0–5.0	240–280	12.5	20	SSV	850

SCB - Square close butt
SSV - Single side V with 60 or 70 degrees included angle and root opening of 0 to 2.5 mm.

Because of the use of shielding gases, no fluxes are required to be used in inert gas shielded arc welding. However for thicker sections, it may be desirable to protect the root side of the joint by providing a flux or preferably a shroud of inert gas.

9.4.8 Gas Metal Arc Welding (GMAW)

Metal inert gas arc welding (MIG) or more appropriately called as gas metal arc welding (GMAW) utilises a consumable electrode and hence, the term 'metal' appears in the title. There are other gas shielded arc welding processes utilising the consumable electrodes, such as flux cored arc welding (FCAW) all of which can be termed under MIG. Though gas tungsten arc welding (GTAW or TIG) can be used to weld all types of metals, it is more suitable for thin sheets. When thicker sheets are to be welded, the filler metal requirement makes GTAW difficult to use. In this situation, the GMAW comes handy.

The typical set up for GMAW (or MIG) process is shown in Fig. 9.39. The consumable electrode is in the form of a wire reel, which is fed at a constant rate, through the feed rollers. The welding torch is connected to the gas supply cylinder which provides the necessary inert gas. The electrode and the work piece are connected to the welding power supply. The power supplies are always of the constant voltage type only. The current from the welding machine is changed by the rate of feeding of the electrode wire.

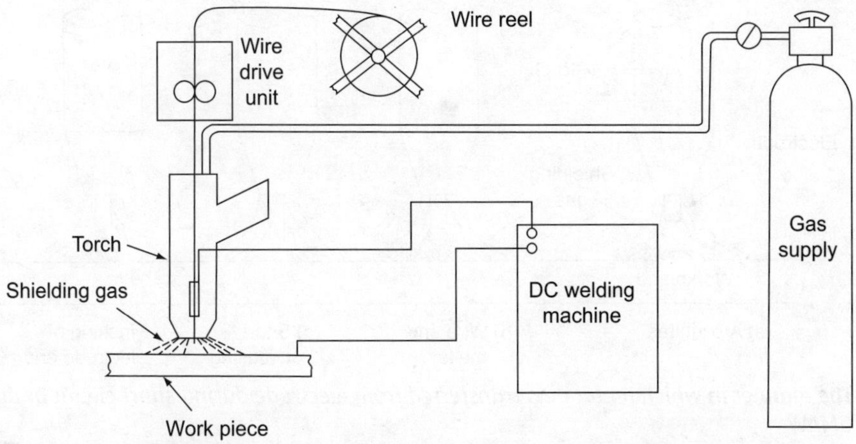

Fig. 9.39 *Schematic of a gas metal arc welding set up*

Normally DC arc welding machines are used for GMAW with electrode positive (DCEP). The DCEP increases the metal deposition rate and also provides for a stable arc and smooth electrode metal transfer. With DCEN, the arc becomes highly unstable and also results in a large spatter. But special electrodes having calcium and titanium oxide mixtures as coatings are found to be good for welding steel with DCEN.

Metal Transfer

In the GMAW process, the filler metal is transferred from the electrode to the joint. Depending on the current and voltage used for a given electrode, the metal transfer is done in different ways. They are:

- Short circuit or dip transfer
- Globular transfer
- Spray transfer
- Pulsed spray transfer
- Rotating spray transfer

The above are found to occur with varying rates of currents, the lowest rate giving the short circuit transfer with the rest arranged in the increased order of currents. Consequently, the metal deposition rates also increase in the order as shown in Table 9.13, except for pulsed spray transfer.

Table 9.13 **Deposition rates with different types of metal transfers.**

Metal Transfer Method	Deposit Rate, (kg/h)
Short circuit	0.9 to 2.7
Globular	1.8 to 3.2
Spray	2.7 to 5.4
Pulsed spray	0.9 to 2.7
Rotating spray	6.3 to 13.5

The short circuit metal transfer occurs with relatively low current settings of the order of 75 to 175 A for an electrode diameter of 0.9 mm. Figure 9.40 shows the manner in which the sequence of metal transfer takes place in short circuit or dip transfer method.

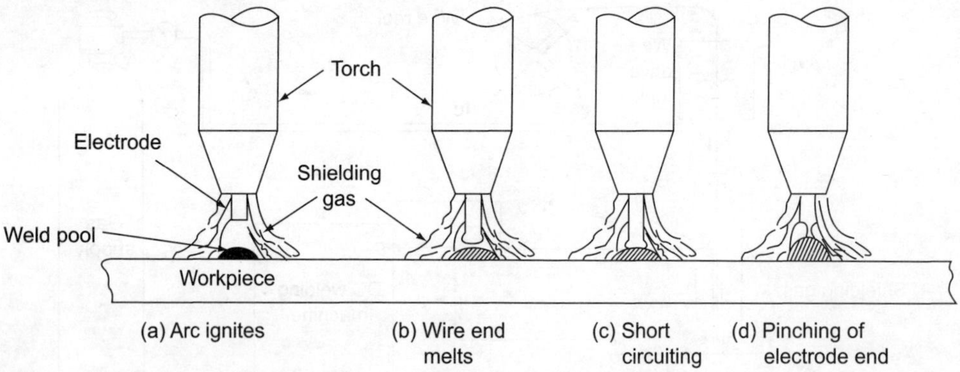

FIG. 9.40 *The manner in which metal gets transferred from electrode during short circuit or dip transfer in GMAW*

Figure 9.40(a) shows the initiation of arc. Under the intense heat of the arc, the electrode tip melts and forms a globule of molten metal at the tip as shown in Fig. 9.40(b). As the electrode wire is fed towards the work piece, the molten tip touches the weld metal pool and thus short circuits the electrode to the work piece as in Fig. 9.40(c). This reduces the voltage across the arc. The metal tip then gets pinched as shown in Fig. 9.40(d) by the surface tension of the weld metal pool as well as the magnetic force due to the current flow. Finally, the metal is pinched away and the arc gets ignited again, and the cycle is repeated all over again.

The number of times that the pinching takes place depends on the inductance of the welding machine used and the parameters set. The inductance of the welding machine controls the rate at which the short circuiting current increases. Too low an inductance gives rise to a very high short circuiting current and consequently high pinching rate. With high inductance, the short circuit current becomes low and results in somewhat lower pinching force. The effect of inductance, measured in terms of the response rate (raising rate of current due to short circuiting, kA/s), is given in Table 9.14 for the commonly used electrode diameters. It can be observed that the optimum response rate depends on the electrode wire size. Hence, the welding machines to be used for short circuiting transfer make use of a variable choke system which can be tuned for different wire sizes. The frequency of the metal transfer may be of the order of 50 to 200 per second. This rate also depends on the open circuit voltage and the wire feed rate employed. Too high an open circuit voltage reduces the frequency of short circuiting. The low wire feed rates also makes for lower frequencies. The optimum values of these parameters are to be obtained for any welding situation either through experience or by trial and error method, with some starting point from the welding machine manual.

TABLE 9.14 **GMAW (CO_2) welding parameters for short circuiting transfer of metal**

Wire Size (mm)	Wire Feed Rate (m/min)	Arc Voltage (V)	Average Current (A)	Response Rate (kA/s)
0.8	5	18	100	50 to 150
1.2	2.5	19	120	40 to 130
1.6	1.7	20	160	20 to 75

The main advantage of using the short circuit metal transfer is the lower penetration, which can be effectively used for welding thin sheets. Poorly fit joints as well as all difficult to reach positions (such as overhead) can also be successfully welded. Practically, there is no spatter since the metal is transferred only when the electrode touches the work piece. Also less amount of metal is affected around the weld joint because of the smaller heat input.

Globular or drop transfer of metal occurs at slightly higher currents than those used in short circuit transfer. Different possibilities of globular transfer are shown in Fig. 9.41. The higher currents used in globular transfer melt the metal at a higher rate thus forming, a globule of molten metal at the tip of the electrode. This metal drop size is considerably larger than the electrode diameter. It stays with the electrode as long as the force due to surface tension of the liquid metal is more than the gravity. However, when the size increases, the gravity pulls it off from the tip and the metal droplet gets deposited on the work piece. If the size is too big, then the droplet may in fact short circuit the electrode as shown in Fig. 9.41(b). The droplet may also get disturbed by the magnetic pull due to the current flow and cause spatter as in Fig. 9.41(c). To reduce the possibility of spatter, it is desirable to maintain a short arc with a deep crater as in Fig. 9.41(d). This deep crater accommodates all the metal coming in the form of droplets and the spatter could be practically eliminated. This is termed as submerged or buried arc.

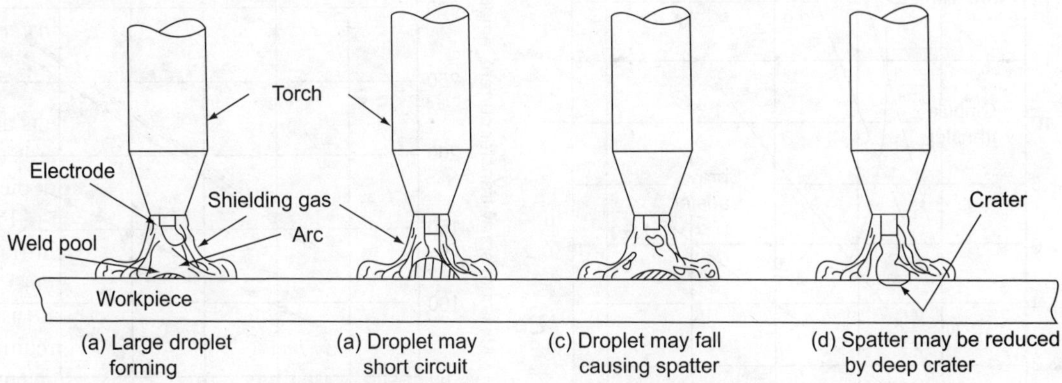

Fig. 9.41 *Globular metal transfer in GMAW*

Since the metal is transferred by gravity, this method cannot be used in out of position welding and can only be used for flat (down hand) welding position. Globular transfer occurs when carbon dioxide is used as the shielding gas.

Spray transfer of metal occurs when the current is increased beyond that of the globular transfer current. As the current is increased, the magnetic pull on the molten metal increases and subsequently the molten metal is detached from the electrode tip by this pull, irrespective of the gravity force. Since gravity has no effect on the metal transfer, spray transfer welding can be done in any position, horizontal, vertical or overhead. As the arc heat melts the electrode, as shown in Fig. 9.42, the metal is continuously attracted towards the work by the magnetic force like a spray from a water jet.

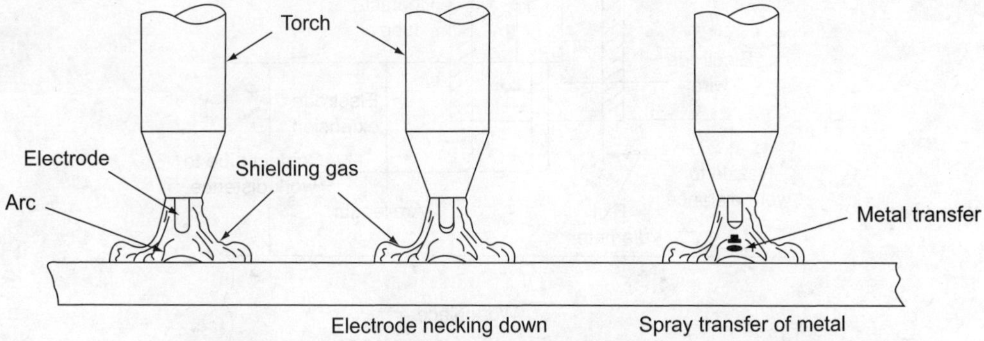

Fig. 9.42 *Spray transfer of metal from the electrode in GMAW*

The current at which the metal transfer changes from globular to spray type is termed as the globular to spray transition current. This transition current is different for different electrode materials. Aluminium requires much lower currents to achieve the spray transfer, as shown in Fig. 9.43. The transition current increases directly with the electrode diameter. Hence, current density is not critical. This effect is shown in Fig. 9.44. The transition current decreases with an increase in the electrode extension. The electrode extension is the amount by which the electrode wire is out of the contact tube, as shown in Fig. 9.45. As shown, with still higher currents the spray drops become finer.

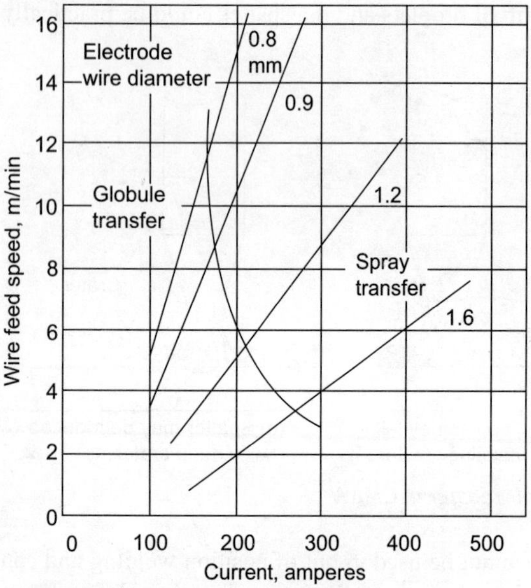

Fig. 9.43 *Effect of electrode diameter on globule-spray transfer current for various wire feed speeds for mild steel electrode*

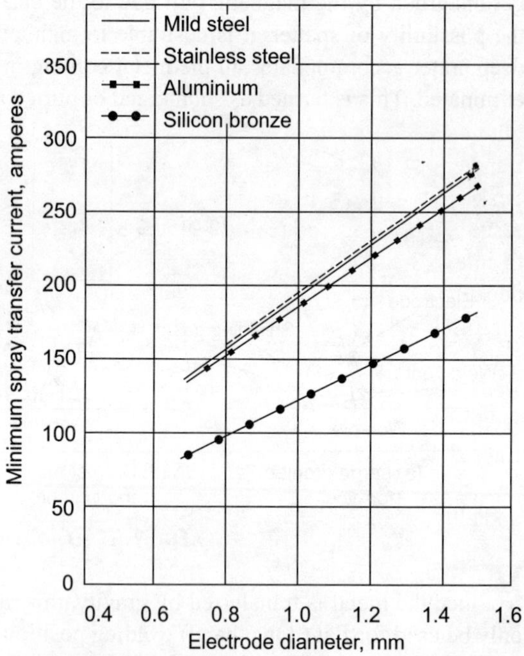

Fig. 9.44 *Effect of electrode diameter on the globular to spray transfer current for different electrode materials*

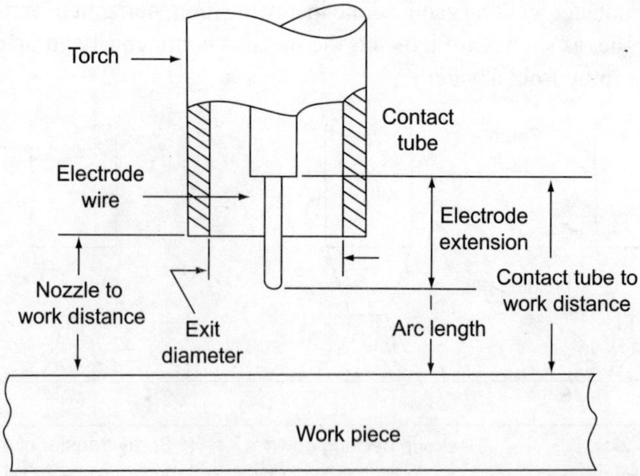

Fig. 9.45 *Terminology used in gas metal arc welding process*

The spray transfer of metal occurs only when the shielding gas contains at least 90% argon. It will not occur when carbon dioxide is used as the shielding gas. The arc in spray transfer can be properly directed. Also it provides a deep penetration since the metal spray pattern is concentrated and stable. When the electrode extension is made longer, it gets preheated, which increases the burn off rate of the electrode resulting in less penetration. This is the most normal method of metal transfer method used in GMAW predominantly with argon as the shielding gas.

Because of the larger heat inputs and deeper penetration, the spray transfer is not suitable for thin materials which are to be welded in out of positions. For such applications, the pulsed spray transfer is generally used. It is similar to the spray transfer except that the current required for melting the electrode tip is given in regular pulses rather than continuously. The welding current termed as the background current, which is lower than that required for spray transfer, is maintained throughout. During this time, the arc is maintained but no metal transfer takes place. This current is increased at regular frequency, to a level required for spray transfer, called the peak current. As a result, the spray transfer of metal only occurs in periods when the peak current is maintained. Pulsing of peak current reduces the total heat input to the base metal and thus, results in lower metal deposition rate, as can be seen from Table 9.13. This helps in welding of thin sections or in overhead welding. But the equipment for the spray metal GMAW process is expensive because of the additional pulsing power supply required.

As the welding current is increased beyond what is required for the spray transfer, the rotating spray transfer results. In this, the end of the electrode wire melts and rotates in a spiralling or helical pattern. Because of this rotation, the weld pool is extended over a larger area, resulting in poor penetration. This method is useful for depositing large amount of metal. Similar to drop to spray transfer, the rotating spray transition current increases directly with the electrode diameter and decreases with the electrode extension.

Typically used voltage and current settings for GMAW power sources are shown in Table 9.15 for steels and aluminium alloys.

TABLE 9.15 The machine settings used for GMAW process

Electrode Diameter (mm)	Mild Steel and Low Alloy Steels				Aluminium Alloys			
	Short Circuit Transfer		Spray Transfer		Short Circuit Transfer		Spray Transfer	
	Voltage (V)	Current (A)	Voltage (V)	Current (A)	Voltage (V)	Current (A)	Voltage (V)	Current (A)
0.8	15–21	70–130	24–28	150–265	15–18	45–120	22–28	90–150
0.9	16–22	80–190	24–28	175–290	17–19	50–150	22–28	100–175
1.2	17–22	100–225	24–30	200–315	16–20	60–175	22–28	120–210
1.6	–	–	24–32	275–300	–	–	24–30	160–300
2.4	–	–	24–33	350–600	–	–	24–32	220–450

Shielding Gases

The gases that can be used for GMAW are argon, helium, nitrogen, oxygen, carbon dioxide and a mixture of the above gases in various proportions. Each of these has their effect on the formation of the bead and the penetration.

Argon reduces the spatter and concentrates the arc. As a result it gives deep penetration welds and is the most widely used of all the shielding gases. Also, argon ionises easily requiring smaller arc voltages. It also has lower thermal conductivity and hence conducts heat very slowly from the arc to the weld zone. Thus, it is good for welding thin sheets. It is also good for out of position welding because of the lower voltages employed. A large percentage of argon is essential, if spray transfer of metal is desired.

Helium is the most expensive of all the shielding gases. Helium has a better thermal conductivity and hence, is useful for thicker sheets as well as for metals having higher thermal conductivity such as copper and aluminium. The filler metal deposition rate by helium is much higher compared to argon because of the higher current carrying capacity.

The arc in carbon dioxide shielding gas is unstable and therefore a very short arc is to be used to reduce the metal spatter. This is the least expensive of all the shielding gases. Since about 7.5% of the carbon dioxide decomposes into carbon monoxide and oxygen in the arc, deoxidisers such as aluminium and silicon are to be used while using carbon dioxide. It is a heavy gas and therefore covers the weld zone very well. The metal transfer is only globular with the carbon dioxide shielding gas.

The mixture of gases such as argon-helium, argon-carbon dioxide and argon-oxygen are used for special applications where these constitutional gases provide the different capabilities that are desired. Table 9.16 gives the typical shielding gas combinations that are found useful for welding various materials.

TABLE 9.16 Shielding gases that can be used for different materials in welding by GMAW process.

Metals	Shielding Gas
Aluminium	Argon Argon + 50% Helium Helium
Magnesium	Argon Argon + 75% Helium
Copper (Deoxidised)	Argon Argon + 75% Helium
Carbon steel	Carbon dioxide Argon + 2% carbon dioxide
Low alloy steel	Argon + 2% carbon dioxide Argon + 5% carbon dioxide Argon + 2% oxygen Argon + 5% oxygen
High alloy steel	Argon + 25% carbon dioxide
Stainless steel	Argon + 1% Oxygen Helium + 7.5% Argon + 2.5% Carbon dioxide
Nickel	Argon Argon + 75% Helium

Shielding gases should always flow in a laminar manner without causing any turbulence. The turbulence causes weld contamination. When too small a gas flow is allowed, the weld area would not be sufficiently protected since all the atmospheric air around it could not be displaced by the small gas flow. The gas flow rates to be used depend upon the thickness of the sheet being welded, the position of the weld as well as the base material.

Electrodes

The electrode wire comes generally in the form of coils. Various wire compositions are available depending on the base metal composition. Wires of very small diameter are more expensive. The normal sizes may be of the order of 0.5 to 3.2 mm. The electrode wires to be used for steels generally have the deoxidisers added to it. The deoxidisers help in the reduction of oxidation of the weld metal as well as the porosity. But the deoxidisers increase the cost of the electrode wire also.

The electrode wire is produced by wire drawing where lubrication is essential. Any lubricant film left on the wire interferes with the welding quality in the form of unstable arc and porosity of the weld, and therefore should be properly cleaned before they are used. Also, the electrode wire should be of proper diameter throughout. When the diameter of the electrode wire varies, the arc is likely to wander.

Welding Technique

The variables that affect the weld quality are the electrode stick out (extension), the travel speed and the welding method, besides the electrode diameter and the welding machine parameters.

A minimum electrode stick out of the order of 10 mm is to be maintained to safeguard the contact tube from burning. Larger electrode extensions decrease the melting rates and increase the penetration. The effect of electrode extension on the melting rate also depends on the size of the electrode, as shown in Fig. 9.46. It is possible to give the melting rate (MR) in terms of the various parameters as follows

$$MR = aI + bLI^2$$

Where, MR = melting rate, kg/h

I = the welding current, A

L = electrode extension, mm

a, b = constants whose values are given below for 1.6 mm wire size with DCEP

	a	b
Aluminium	5.4×10^{-3}	4.4×10^{-6}
Mild steel	8.6×10^{-3}	2.5×10^{-5}

As can be seen from the above, the effect of the electrode extension is minimal for aluminium compared to mild steel. This is because of the higher electrical conductivity of the aluminium compared to the mild steel. This results in lesser resistive heating and consequently, the melting rate is not greatly affected by the electrode extension for aluminium.

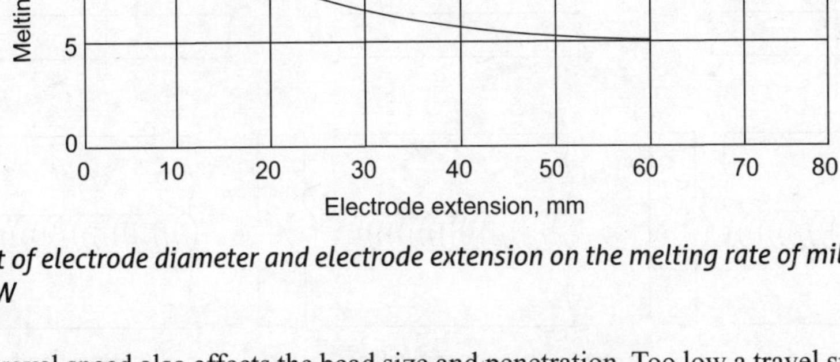

Fig. 9.46 *Effect of electrode diameter and electrode extension on the melting rate of mild steel electrode in GMAW*

The electrode travel speed also affects the bead size and penetration. Too low a travel speed causes a large deposition of the filler metal whereas too high a speed does not give enough time for melting the base metal and consequently, results in less deposition. Different electrodes require different optimum travel speeds which are normally provided by the electrode manufacturer. Table 9.17 gives some specifications for welding mild steel with short circuit metal transfer using carbon dioxide shielding gas.

TABLE 9.17 Typical welding conditions for butt joints in vertical position with a root gap of 1.6 mm.

Material Thickness (mm)	Wire Diameter (mm)	Welding Current (A)	Arc Voltage (V)	Wire Feed Rate (m/min)	Welding Speed (cm/min)	Joint Preparation	Number of Passes
1.0	0.8	60	18	2.5	45	SCB	1D
1.6	0.8	80	19	3.4	60	SCB	1D
3.2	0.8	120	20	5.5	45	SEOB	1D
6.5	1.2	150	20	3.25	25	SSV	2D
9.5	1.2	150	20	3.25	15	SSV	3U
12.5	1.2	150	20	3.25	15	SSV	4U
	1.6	160	21	1.87	15	DSV	4U
19	1.2	150	20	3.25	15	SSV	4U
	1.6	160	21	1.87	15	DSV	4U
25	1.2	150	20	3.25	15	DSV	5U
	1.6	160	21	1.87	15	DSV	5

SCB - Square close butt; SEOB - Square edge open butt
SSV - single sided V with 60 degree included angle
DSV - double sided V with 60 degree included angle

The welding method determines the positioning of the gun with respect to the base metal. The three possible ways of positioning the welding gun, shown in Fig. 9.47, are called forehand, perpendicular or backhand.

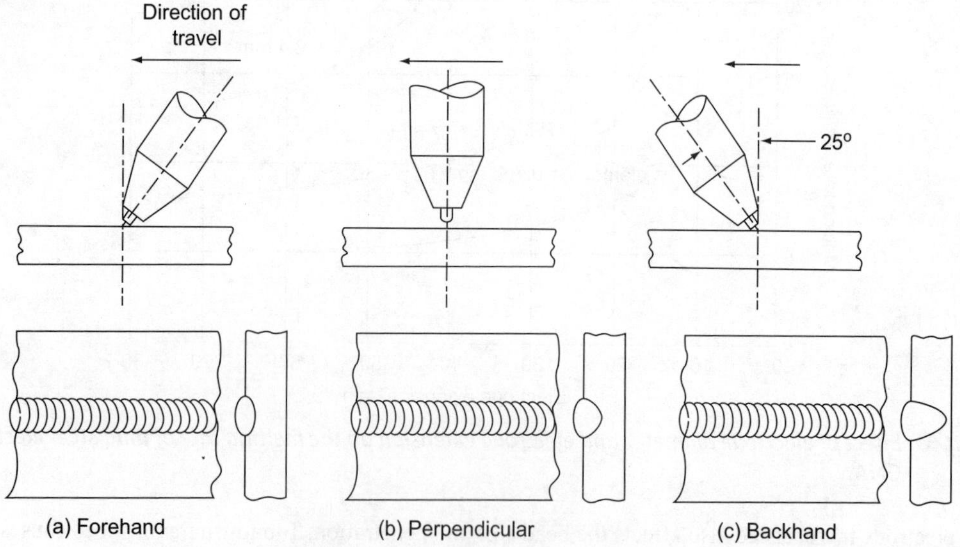

FIG. 9.47 *Positioning of electrode gun with respect to the base metal plate*

In forehand welding, the torch points in the direction of the electrode travel with a slant backwards from the vertical, as shown in Fig. 9.47(a). Here, the arc heat is directed forward and as a result the penetration would be the least. In the backhand technique, the gun is slanted in the direction of the travel by an angle of about 25 degrees as in Fig. 9.47(c). In this position the arc heat is directed on the weld bead to give the best penetration. The perpendicular position results in a penetration, midway between the above two techniques. The backhand technique is generally preferred because it gives a more stable arc and lesser spatter.

The rest of the welding techniques and the precautions to be taken are similar to the other arc welding processes.

Flux Cored Arc Welding

The flux cored arc welding (FCAW) process is a modification of the GMAW where the solid electrode wire is replaced by a tubular electrode containing flux at the centre of the electrode throughout its length. First introduced in 1960, this process has been finding wide usage and replacing other arc welding processes. Because of the availability of flux in the welding zone, the weld metal of any required composition can be closely controlled as well as obtain a smooth weld bead.

FCAW is normally used for the welding of structural steels and alloy steels though, is not limited to them. This process has the advantages of the stick electrodes in the form of flux availability to remove the oxides and other contaminants in the form of slag, which covers the weld bead during solidification and cooling and protects it for longer periods. Alloying elements and deoxidisers can be added to the joint. This process gives rise to a deeper penetration and as a result many a times the edge preparation can be eliminated.

The equipment used for flux cored arc welding is similar to that used for GMAW. Since the electrode is not a solid wire, the feeding units should be properly designed so as not to put excess pressure on the electrode resulting in its flattening.

9.4.9 Submerged Arc Welding (SAW)

The submerged arc welding (SAW) is used for doing faster welding jobs. It is possible to use larger welding electrodes (12 mm) as well as very high currents (4000 A) so that very high metal deposition rates of the order of 20 kg/h or more can be achieved with this process. Also, very high welding speeds (5 m/min) are possible in SAW. Some submerged arc welding machines are able to weld plates of thickness as high as 75 mm in butt joints in a single pass. Though submerged arc welding can be used even for very small thicknesses, of the order of 1 mm, it is more economical for larger welds only.

The schematic representation of a typical submerged arc welding process is presented in Fig. 9.48. The arc is produced while the consumable electrode wire which is continuously fed into the weld zone as in GMAW. The welding zone is completely covered by means of a large amount of granulated flux which is delivered ahead of the welding electrode by means of a welding flux feed tube. The arc occurring between the electrode and the work piece is completely submerged under the flux and not visible from outside. A part of the flux melts and forms the slag, which covers the weld metal, as shown in Fig. 9.48. The unused flux is collected and reused.

The power source used with submerged arc welding can be either AC or DC. Both constant voltage and constant current type machines can be effectively used. However, for larger electrodes, a constant current power supply is preferred. The current ratings of the SAW machines are, in general, two to three times higher than that of the similar GMAW machines. Though there is no special advantage, the AC machines may sometimes be preferred when arc blow becomes a severe problem.

With a single wire SAW, arc blow is not encountered with AC supply. But sometimes two wires may be used to deposit larger amounts of metal. These two electrodes being connected to two separate power

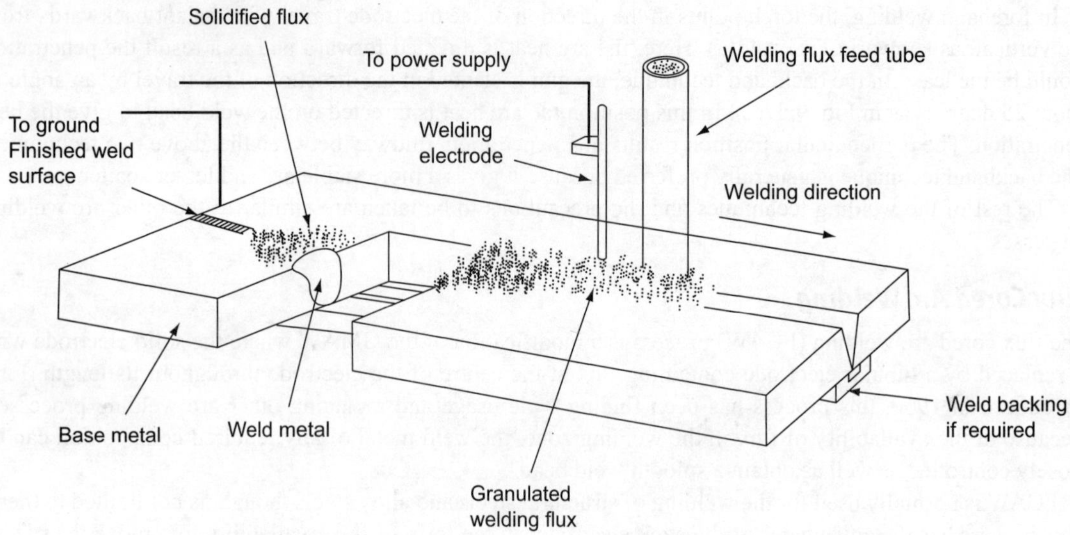

Fig. 9.48 *Schematic view of a submerged arc welding in progress*

sources, the arc blow is likely to occur because of the interference of the two magnetic fields surrounding the two electrodes, if the two currents are in phase. To reduce this problem, generally, the two power supplies are so adjusted that when one of the supplies is in peak, the other would be set to zero current so that no opposing magnetic fields would be set. One of the electrodes is called leading and the other trailing. The two electrodes are physically separated, in the joint, one positioned slightly ahead of the other.

Since the arc is completely submerged in the flux, there is no spatter of the molten metal. Since this process uses loose granulated flux to cover the joint, it is not possible to carry out in any position other than the flat or down hand position. The out of position welds are difficult to carry, also because of the large metal pools that are generated in the SAW process.

The electrode wires normally used are of sizes 1.6, 2, 2.5, 3.15, 4, 5, 6.3, and 8 mm. The wires should be smooth with no surface imperfections or contaminants. Since the wire feed rate is normally very high, it is not possible to manually feed the wire into the joint.

Being a large volume process, the submerged arc welding produces large amount of molten weld metal which takes some time for solidification. Hence, it is imperative in SAW, to provide some way of containing this molten metal. The weld metal backing is normally used. The backing plates can be with or without grooves. The most commonly employed are the copper plates which can be cooled with internal running water, when necessary. Normally for thin plates, plain copper backing plates without any cooling water would be enough. Pure copper, because of its high thermal conductivity, would remove heat quickly from the molten weld pool and allow it solidify fast. For welding plates that are less than 3 mm thick, copper backing plates without grooves are used. For thicker plates, grooves with a depth of 0.5 to 1.5 mm and width of 6 to 20 mm are made for providing the necessary support to the molten metal.

The type of edge preparation and joint design used with submerged arc welding is slightly different from the other arc welding processes because of the larger deposition rates and deeper penetrations involved. Some typical joint preparations are shown in Fig. 9.49. The complete edge preparation details can be obtained from handbooks or from Indian standards such as BIS:823-1964 and BIS:4353-1967.

Some of the variables that affect the weld quality in submerged arc welding are: the electrode diameter, voltage and current settings, welding speed and the width, and the depth of the flux covering used.

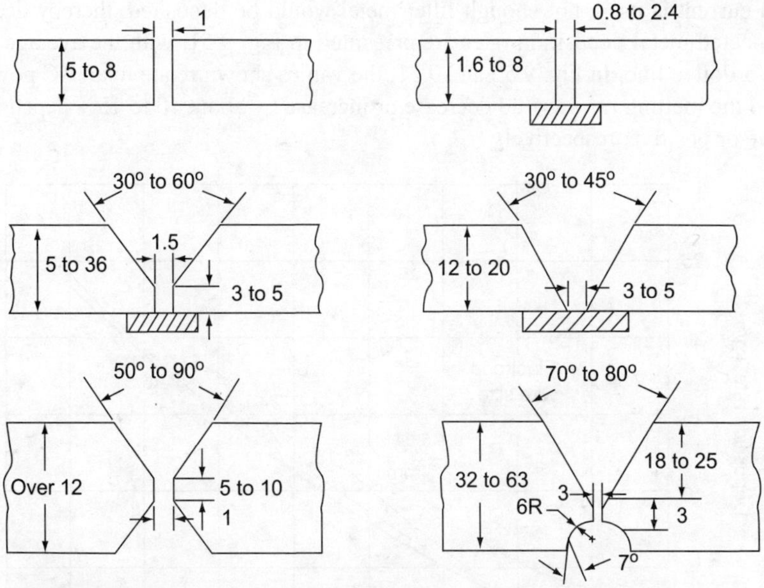

Fig. 9.49 *Some typical edge preparations for submerged arc welding*

The amount of current setting used determines the wire burn off rate for different electrodes as shown in Fig. 9.50. If higher welding current is used than required, additional heat and filler metal would be input into the weld joint. This enlarges the heat affected zone as well as causes unnecessary reinforcement of the weld joint which is an unnecessary exercise.

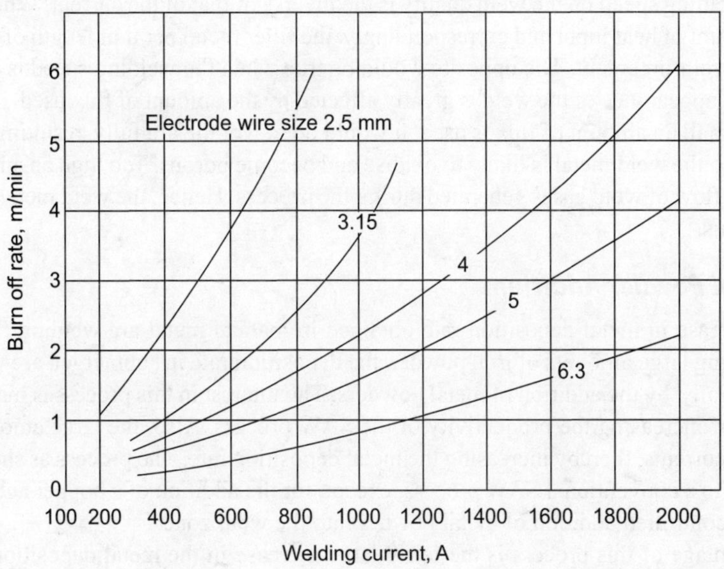

Fig. 9.50 *Electrode burn off rates for various electrode sizes and currents in AC welding (SAW)*

If too small a current is used, not enough filler metal would be deposited, thereby decreasing the joint strength. The expected metal deposition rates are presented in Fig. 9.51, with the average rate of 13.5 kg/h per kA shown as a dotted line. In Fig. 9.50 and 9.51, the values shown relate to an AC power supply. When DC power is used the melting rates would decrease or increase by about 10 to 15% depending on the use of electrode negative or positive, respectively.

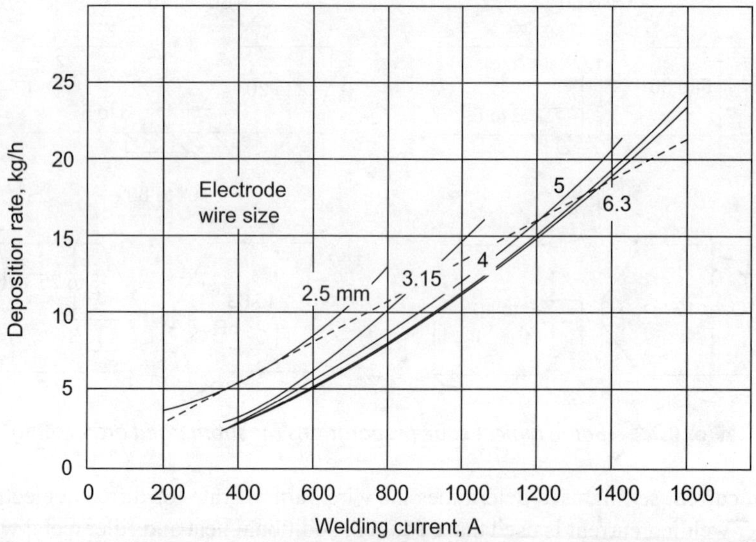

Fig. 9.51 *Metal deposition rate in AC submerged arc welding as affected by welding current*

The effect of welding speed on the weld quality is the inverse of that of the current. When welding speed is increased, the amount of heat input and correspondingly, the filler metal per unit length of the joint decreases, so also is the joint reinforcement. The opposite would be true when the welding speed is increased.

The quality and appearance of the weld is greatly affected by the amount of flux used, in terms of its width and depth. If too small an amount of flux is used, it would not cover the arc fully, resulting in the arc flashing and spattering. Also the weld metal is likely to oxidise and become porous. Too high an amount of flux on the arc deters the easy flow of weld gases generated during the process. Hence, the weld metal would be distorted and becomes porous.

SAW with Metal Powder Additions

Similar to the increase in metal deposition rate obtained in manual metal arc welding (SMAW) with stick electrodes containing large amounts of iron powder, the deposition rate in submerged arc welding can also be increased substantially, by the addition of metal powders. The interest in this process is mainly because of the inexpensive way of increasing the productivity of the SAW process. Also, the large amount of iron powder helps increase the currents, thereby increasing the metal deposition rate. The process as shown in Fig. 9.52, is essentially similar to a conventional SAW process, except, for the addition of a hopper and a metering device to provide for the controlled addition of metal powder into the weld zone.

The main advantage of this process is the fairly large increase in the metal deposition rate, as shown in Fig. 9.53. The metal powder used to improve the joint mechanical properties can have any composition. It is also possible to use this process in even multi-wire submerged arc welding, whereby very large deposit rates of as high as 90 kg/h can be realised. Since the welding is done in a short time, the weld distortions are minimised.

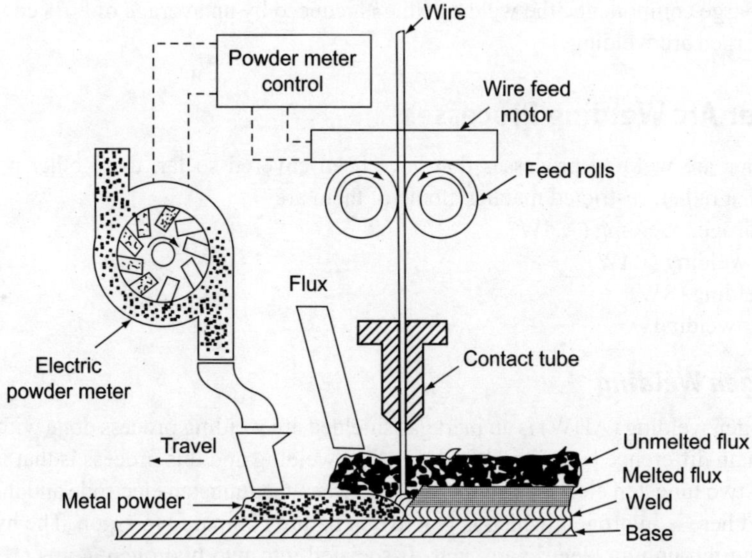

FIG. 9.52 *Submerged arc welding with metal powder addition*

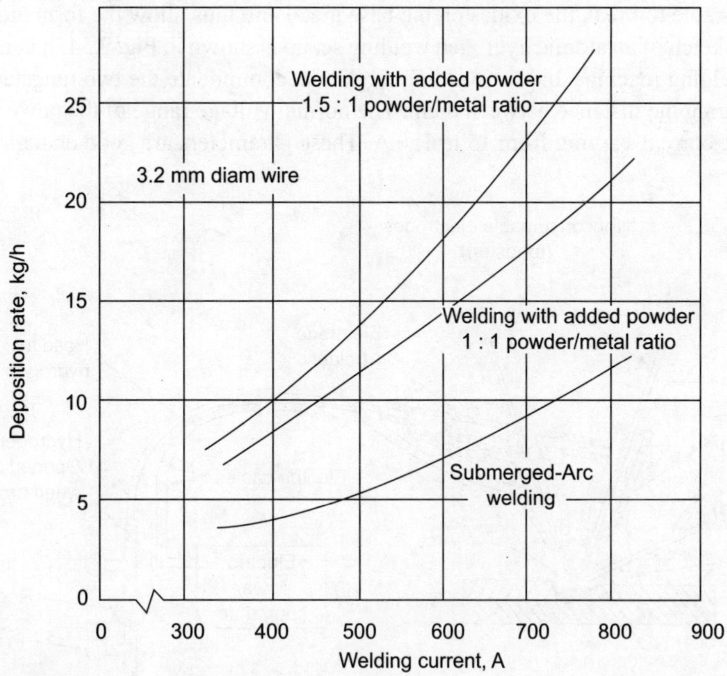

FIG. 9.53 *The metal deposition rates in SAW with metal powder addition*

It is also possible to use a continuous strip as an electrode in the place of the solid wire in a process which is termed as submerged arc strip welding. Typical widths may range from 25 mm to 125 mm. Because of the larger width of the electrode, wide joints can be made. This is used for cladding or overlaying of large components. For large components, the welding time is reduced by an average of 80% compared to that of a single wire submerged arc welding.

9.4.10 Other Arc Welding Processes

Besides the various arc welding processes that have been covered so far, there other processes available which are used in, a rather, restricted manner. Some of them are

- Atomic hydrogen welding (AHW)
- Plasma arc welding (PAW)
- Stud arc welding (SW)
- Fire cracker welding

Atomic Hydrogen Welding

The atomic hydrogen welding (AHW) is an inert gas shielded arc welding process done with non-consumable electrodes. The main difference between TIG (or GTAW) welding and this process is that in AHW, the arc is obtained between two tungsten electrodes rather than between the tungsten electrode and the work piece. The shielding gas used here is hydrogen, which is reactive in nature compared to argon. The hydrogen molecule (H_2), when passing through an electric arc, gets dissociated into two hydrogen atoms ($H+$). The hydrogen atoms are highly reactive. They form hydrogen molecule and combine with oxygen, if present, to form water vapour and thus release intense heat for the necessary melting of the joint. Because of its reactivity, the atomic hydrogen is able to break the oxides on the base metal and thus allow the formation of a clean weld.

The schematic sketch of an atomic hydrogen welding set up is shown in Fig. 9.54. It consists of a hydrogen cylinder, an AC welding machine and the welding torch to accommodate the two tungsten electrodes with a provision for changing the distance between them. The normal voltage range of the power supply is between 50 to 75 V with the current varying from 15 to 150 A. These parameters are good enough for electrode sizes of 1 to 5 mm.

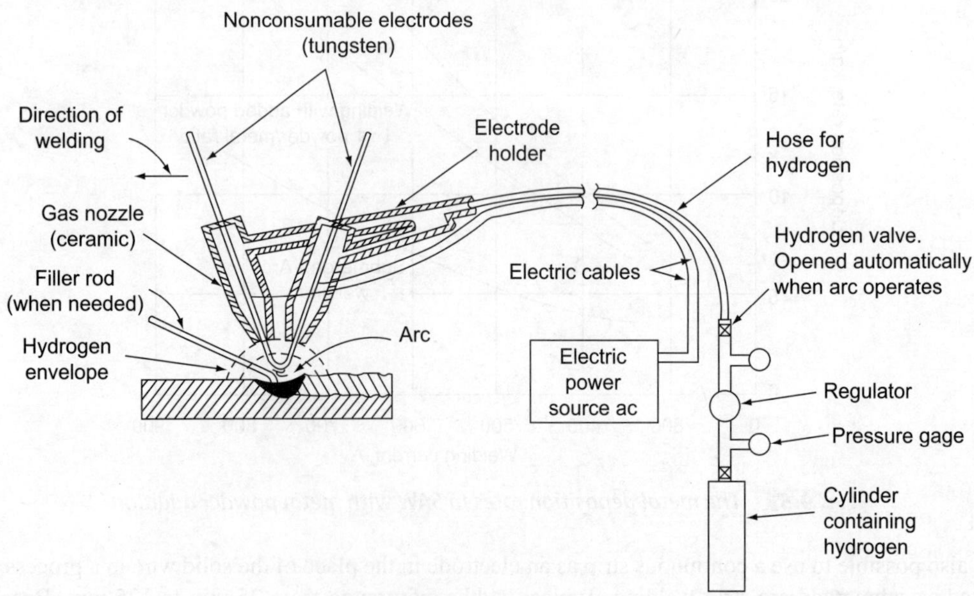

FIG. 9.54 *Set up of atomic hydrogen welding (courtesy Campbell op cit, p 398, McGraw-Hill, New York)*

The path of electron travel (arc) between the two electrodes is not a straight path as in other arc welding processes. Instead, they travel in the form of a fan. This is because the hydrogen atoms formed by the arc causes a downward force because of which the electron flow is slightly deviated. This fan shape can be changed by altering the distance between the electrodes and the current level. The DC machines could also be used in atomic hydrogen welding, but because of the electron flow being only in one direction, the wear of the electrode is particularly high and as a result only AC power supply is used.

When hydrogen atoms recombine near the work piece surface, they generate a temperature of the order of 300°C. Because of this heat, the molten metal becomes highly fluid and therefore, atomic hydrogen welding is used for the flat positions only. Filler metal when needed is melted intermittently in the arc fan for fusing with the base metal.

When properly performed, the atomic hydrogen welding gives an extremely clean weld with excellent quality. It is generally used for welding of tool steels containing tungsten, nickel and molybdenum as also for hard surfacing and repairing of moulds, dies and tools. Though it can be used for any job, its high cost prohibits its general usage.

Plasma Arc Welding

Plasma is a state of the matter when part of the gas is ionised making it a conductor of electric current. It is the state of the matter present in between the electrodes in any arc. The plasma arc welding (PAW) closely resembles the TIG process in which it also uses a non-consumable tungsten electrode and a shielding gas such as argon. The main difference is in the construction of the torch. In plasma arc welding, the plasma arc is tightly constrained as shown in Fig. 9.55. A small amount of pure argon gas flow is allowed through the inner orifice surrounding the tungsten electrode to form the plasma gas. Because of the squeezing action of the constraining nozzle, the arc in PAW is concentrated and straight. This constriction increases the heat contained per unit volume of the arc plasma. Thus, arc temperatures of the order of 11 000°C are not unusual in PAW. The filler metal if required is fed into the arc as in GTAW process.

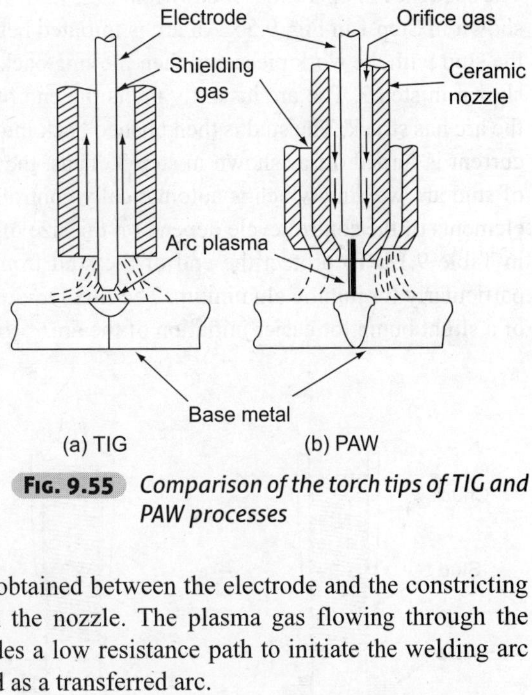

Fig. 9.55 *Comparison of the torch tips of TIG and PAW processes*

To initiate the arc in PAW, a low current pilot arc is obtained between the electrode and the constricting nozzle, which ionises the plasma gas flowing through the nozzle. The plasma gas flowing through the constriction reaches a very high temperature and provides a low resistance path to initiate the welding arc between the electrode and the work piece. This is termed as a transferred arc.

The plasma gas itself is not sufficient to protect the weld metal and therefore, a large volume of inert shielding gas is allowed to flow through an outer gas nozzle surrounding the inner nozzle, as shown in Fig. 9.55. The shielding gases that can be used are argon, helium or a mixture of the above with that of hydrogen.

The power sources used are DC with electrode negative for better electrode life. These are normally the constant current or drooper type of power supplies. The ampere ratings are generally higher than that of the corresponding GTAW machines.

The actual welding by PAW is done by means of a technique called as 'key hole'. Because of the high temperature reached in the PAW process, the base metal beneath the arc is completely melted and a thin key hole is formed. As the torch advances, the metal melted ahead, flows back into the key hole. Thus, the key hole continuously travels along the direction of the torch. Because of the likely key hole formation, this process is limited to the down hand welding position only.

The heat input is properly controlled because of its concentration as well as uniformly deep penetration is possible. Greater lengths between the electrode tip and the work piece facilitate easy application of the filler metal rod, without the contamination of the electrode. The metal deposit rates are higher than that of GTAW. The heat affected zone around the weld metal is small due to the concentration of the arc. The disadvantages are however, that the welding equipment is expensive and also the nozzle surrounding the electrode needs a frequent replacement.

Stud Arc Welding

The stud arc welding (SW) is a process for faster joining of the studs to the work pieces such as machine assemblies. The equipment consists of a gun, similar to a GMAW torch, which holds the stud to be welded. The sequence of operations is shown in Fig. 9.56. Weld gun holding the stud is positioned for the assembly as shown in Step 1 in Fig. 9.56. An arc is initiated between the stud and the metal plate by first short circuiting the stud with the work piece and then moving back the stud slightly, by means of a motor inside the torch as shown in step 2. The arc instantly melts the end of the stud as well as the portion of the work piece where the arc has struck. The stud is then pushed back into the metal pool in the work piece and simultaneously, the current is turned off as shown in step 3. Thus, the stud gets welded to the plate. This is the complete cycle of stud arc welding which is automatically controlled. The current to be used and the timing of the various elements of the control cycle depend on the size of the stud to be used. Typical system parameters are given in Table 9.18. To protect the end of the stud from oxidation, sometimes inert gas shielding may be used particularly for joining aluminium. The flux required is added to the tip of the stud to be welded, in the form of a slight bump for easier initiation of the arc.

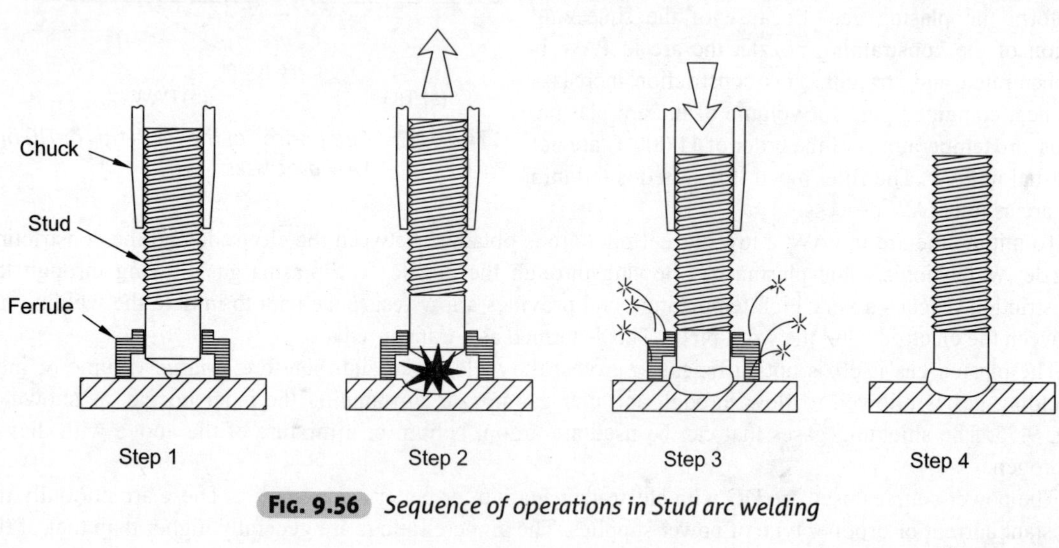

Fig. 9.56 *Sequence of operations in Stud arc welding*

TABLE 9.18 Recommended operating parameters for stud arc welding process

Stud Size, mm	Current, A	Time, seconds
6	350	0.25
8	420	0.31
10	580	0.43
12	720	0.50
14	880	0.59
16	1040	0.66
20	1440	0.78
24	1900	0.85

Fire Cracker Welding

Fire cracker welding is a variation of the manual metal arc welding process (SMAW). The joint to be welded is filled with a stick electrode which is kept in place by means of a glass fibre tape or a copper retaining bar as shown in Fig. 9.57. An arc is initiated by short circuiting the electrode to the work piece. Once initiated, it is an automatic process till the electrode is completely burned off. Flux is provided by the coating on the electrode as well as the shielding gas.

By this method it is possible to weld in flat position only. The weld quality obtained is similar to the shielded metal arc welding process. Because of its automatic nature, very little skill is required in its operation. It has found application in ship building industry.

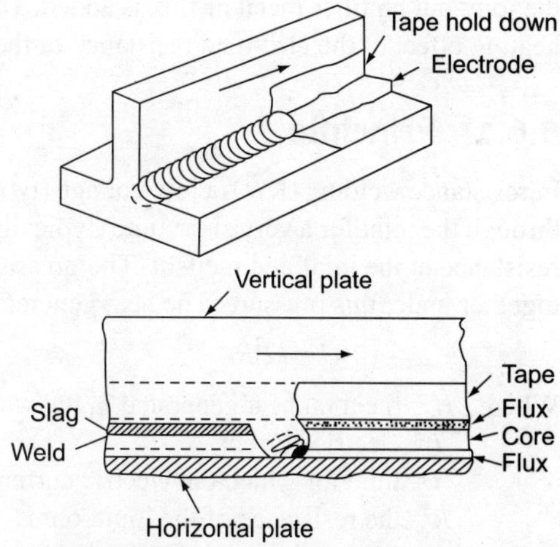

FIG. 9.57 *Fire cracker welding process*

9.5 ARC CUTTING

In arc cutting, the metal is simply melted by the intense heat of the arc and is then blown away by the force of arc itself or by other gases such as air or shielding gases. It is also possible to burn the gases by sending a jet of oxygen. Depending on the source of heat input, there are various processes such as carbon arc cutting (CAC), air carbon arc cutting (AAC), oxygen arc cutting (AOC), shielded metal arc cutting (SMAC), gas metal arc cutting (GMAC), gas tungsten arc cutting (GTAC) and plasma arc cutting (PAC).

In all these processes, the equipment used is similar to that used for the corresponding welding process with the exception of the torch which is different. The torch holds the electrode as also has the provision for the supply of high pressure gas wherever needed.

The carbon arc cutting is the simplest process which uses a carbon electrode to obtain the required arc. The metal that is cut is blown away by the arc force and gravity. This produces very rough edge and therefore finds very little application except in scrap cutting.

In the air carbon arc cutting (AAC), the arc is normally obtained between a copper coated graphite or carbon electrode and the work piece, with the molten metal being forced out by means of a compressed air at a pressure of 550 to 690 kPa. It may be possible to use a very low pressure of the order of 280 kPa in some manual torches for field application but is not generally recommended. The air consumption is in the range of 85 to 1400 L/min depending on the thickness of the metal being cut. The copper coating is used to reduce the oxidation of the electrodes and to help to cool the electrode.

In oxygen arc cutting (AOC), a hollow tubular electrode is used to obtain the arc. Compressed oxygen is forced through the hollow portion so that the metal is oxidised and blown in a fashion similar to oxy-fuel gas cutting (OFC).

The shielded gas cutting processes (GMAC and GTAC) are used for cutting those materials, which have oxidation problems, such as aluminium, stainless steel or nickel alloys. In plasma arc cutting (PAC) very high temperatures of the order of 14 000°C are generated. Therefore, any metal can easily be melted and blown away by this process.

9.6 RESISTANCE WELDING

The welding processes covered so far are fusion welding processes where only heat is applied in the joint. In contrast, resistance welding process is a fusion welding process where both heat and pressure are applied on the joint but no filler metal or flux is added. The heat necessary for the melting of the joint is obtained by the heating effect of the electrical resistance of the joint and hence, the name resistance welding.

9.6.1 Principle

In resistance welding (RW), a low voltage (typically 1 V) and very high current (typically 15 000 A) is passed through the joint for a very short time (typically 0.25 s). This high amperage heats the joint, due to the contact resistance at the joint and melts it. The pressure on the joint is continuously maintained and the metal fuses together under this pressure. The heat generated in resistance welding can be expressed as

$$H = kI^2Rt$$

Where, H = the total heat generated in the work, J
I = electric current, A
t = time for which the electric current is passing through the joint, s
R = the resistance of the joint, ohms
and k = a constant to account for the heat losses from the welded joint.

The resistance of the joint, R is a complex factor to know because it is composed of

(a) The resistance of the electrodes,
(b) The contact resistançe between the electrode and the work piece,
(c) The contact resistance between the two work piece plates, and
(d) The resistance of the work piece plates

The amount of heat released is directly proportional to the resistance. It is likely to be released at all of the above mentioned points, but the only place where large amount of heat is to be generated to have an effective fusion is at the interface between the two work piece plates. Therefore, the rest of the component resistances should be made as small as possible, since the heat released at those places would not aid in the welding.

Because of the squaring in the above equation, the current, 'I' needs to be precisely controlled for any proper joint.

The schematic representation of the resistance welding process is shown in Fig. 9.58. The main requirement of the process is the low voltage and high current power supply. This is obtained by means of a step down transformer with a provision to have different tappings on the primary side, as required for different materials. The secondary windings are connected to the electrodes which are made of copper to reduce their electrical resistance. The time of the electric supply needs to be closely controlled so that the heat released is just enough to melt the joint and the subsequent fusion takes place due to the force (forge welding) on the joint. The force required can be provided either mechanically, hydraulically or pneumatically, as shown in Fig. 9.58. To precisely control the time, sophisticated electronic timers are available.

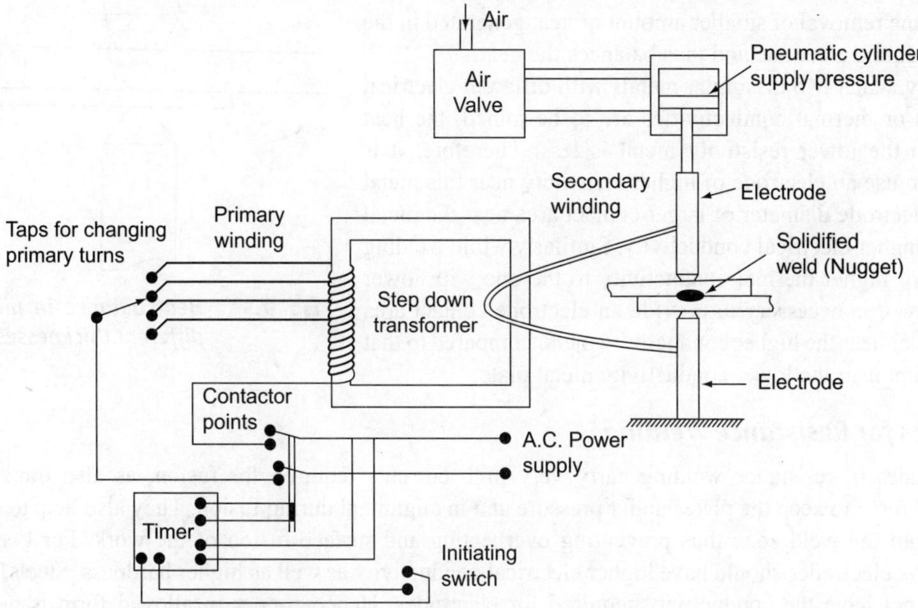

Fig. 9.58 *Schematic of a resistance welding set up*

The critical variable in a resistance welding process is the contact resistance between the two work piece plates and their resistances themselves. The contact resistance is affected by the surface finish on the plates, since the rougher surfaces have higher contact resistance. The contact resistance also will be affected by the cleanliness of the surface. Oxides or other contaminants, if present, should be removed before attempting resistance welding.

The lower resistance of the joint requires very high currents to provide enough heat to melt it. The average resistance may be of the order of 100 micro ohms, as a result, the current required would be of the order of tens of thousands of amperes. With a 10 000 A current passing for 0.1 s, the heat liberated is

$$H = (10\,000)^2\,(0.0001)\,(0.1) = 1000 \text{ J}$$

This is typical for the welding of 1 mm thick sheets. The actual heat required for melting (assuming the weld area as a cylinder of 5 mm diameter and 1.5 mm height) would be of the order of 339 J. The rest of the heat is actually utilised in heating the surrounding areas and lost at other points.

The welding force used has the effect of decreasing the contact resistance and consequently, an increase in the welding current for the proper fusion.

Heat Balance

One of the very important characteristics of the resistance welding process is the transfer of heat to the two parts being joined differently so that proper fusion is obtained, even when the plates are dissimilar from the stand point of material or thickness. For example, consider the case of two metal pieces of the same composition and different thicknesses which are to be joined in a lap joint as shown in Fig. 9.59. The proper fusion can be obtained only if there is proper 'heat balance'. This is taken care of by providing an electrode with a smaller contact area at the thinner sheet and a thicker electrode at the thicker sheet together with very high current densities for short times. This eliminates the removal of smaller amount of heat generated in the sheet through the electrode and thus balances the heat.

Similarly, when two dissimilar metals with different electrical resistivities or thermal conductivities are to be joined, the heat liberated in the lower resistivity metal is less. Therefore, it is necessary to use an electrode of higher resistivity near this metal or use an electrode diameter of larger contact area near the metal which has higher electrical conductivity. Similarly while welding a material of higher thermal conductivity to the one with lower conductivity, it is necessary to provide an electrode contact area that is smaller near the higher conductivity metal compared to that electrode kept near the lower conductivity metal plate.

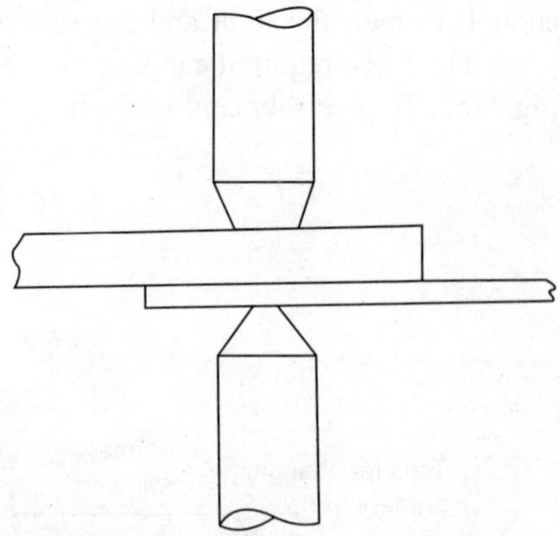

FIG. 9.59 *Heat balance in metals of different thicknesses*

Electrodes for Resistance Welding

The electrodes in resistance welding carry very high currents required for fusion, as also transmit the mechanical force to keep the plates under pressure and in alignment during fusion. They also help to remove the heat from the weld zone thus preventing overheating and surface fusion of the work. For both these purposes, the electrodes should have higher electrical conductivity as well as higher hardness. Steels, though strong, do not have the conductivity required for electrodes. Hence copper in alloyed form is generally used for making electrodes. Though pure copper has high electrical and thermal conductivities, it is poor in mechanical properties.

Copper cadmium (0.5 to 1.0%) alloys have the highest electrical conductivity with moderate strengths and therefore are used for welding non-ferrous materials such as aluminium and magnesium alloys. Copper chromium (0.5 to 0.8 %) alloys have slightly lower electrical conductivities than above but better mechanical strength. These are used for resistance welding of low strength steels such as mild steel and low alloy steels. When cobalt and beryllium are added to copper, its conductivity is decreased to a great extent but the strength is increased. Hence, these are used for welding higher heat resisting alloys such as stainless steels and steels with tungsten and other such alloying elements.

The electrode tips are available in a number of different configurations as shown in Fig. 9.58, to allow for different welding situations. Since a large amount of heat is liberated in the electrodes, it is necessary to cool the electrodes to maintain their strength. To this end, cooling water is circulated through the electrodes as shown in Fig. 9.60.

Some of the very important advantages of the resistance welding process are:

1. Very little skill is required to operate the resistance welding machine. Mostly the machines are semi-automatic or automatic and hence properly set for a particular operation. Anyone with a little training would be able to operate the machine.
2. These are very well suited for mass production, as they give a high production rate.
3. There are no consumables used in this process except for the electrical power and a relatively smaller electrode wear. As a result, it is a very economical process.
4. Heating of the work piece is confined to a very small part, which results in less distortion.
5. It is possible to weld dissimilar metals as well as metal plates of different thicknesses.

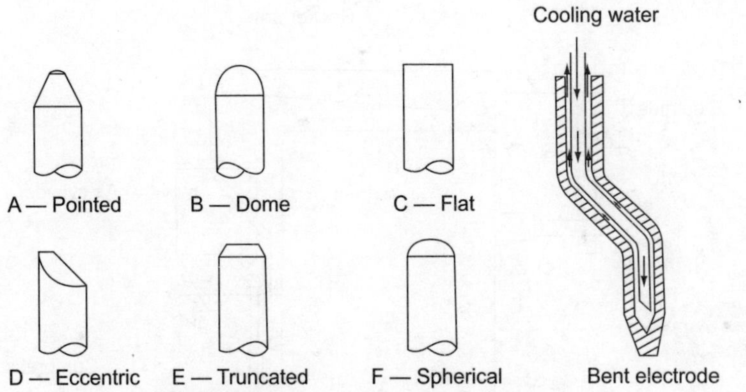

FIG. 9.60 *Different electrode tip shapes*

In spite of these advantages, there are certain limitations and disadvantages of the resistance welding process. They are

1. The resistance welding machine is highly complex with various elements such as a heavy transformer, electrodes and heavy conductors for carrying the high currents, the electrode force applying mechanism such as a pneumatic cylinder and its supply, the heavy machine structure to support the large forces and an expensive timing arrangement. All these make the resistance welding machine highly expensive, unless it is used for large scale production.
2. Certain resistance welding processes are limited only to lap joints. This limits its use to sheet metal whose thickness is less than 3 mm. The lap joints have an inherent crevice between the sheets, which is responsible for stress concentration and a loss of fatigue life. Also for materials that are prone to corrosion, the lap joint may be a source of trouble.

The various resistance welding processes of interest are:

- Resistance spot welding
- Resistance seam welding
- Projection welding
- Upset welding
- Flash welding
- Percussion welding

9.6.2 Resistance Spot Welding

The process description made so far is called the resistance spot welding (RSW) or simply spot welding. This is the most common resistance welding process. This is essentially done to join two sheet metal jobs in lap joint forming a small nugget at the interface of the two plates, as shown in Fig. 9.61.

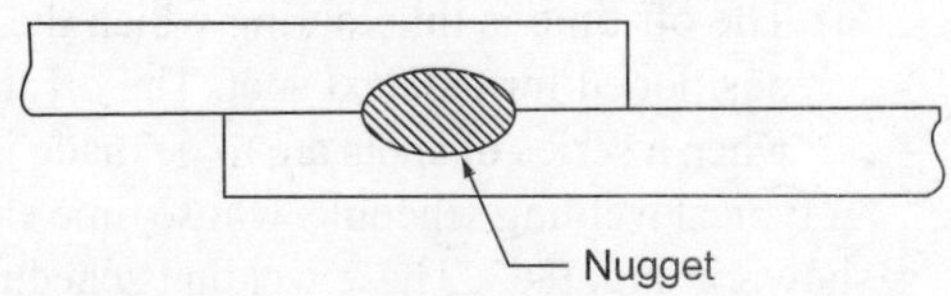

FIG. 9.61 *Formation of a nugget in spot welding*

A typical resistance spot welding machine is shown in Fig. 9.62. It essentially consists of two electrodes, out of which one is fixed. The other electrode is fixed to a rocker arm (to provide mechanical advantage) for transmitting the mechanical force from a pneumatic cylinder. This is the simplest type of arrangement. The other possibility is that of a pneumatic or hydraulic cylinder being directly connected to the electrode without any rocker arm.

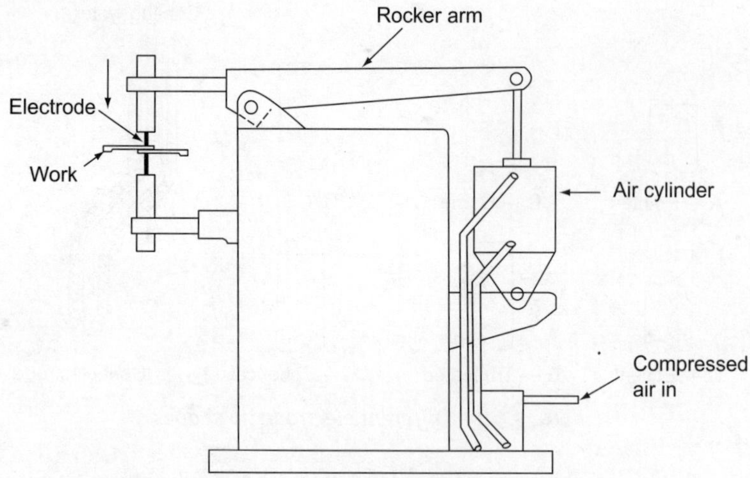

Fig. 9.62 *Rocker arm type resistance spot welding machine*

For welding large assemblies such as car bodies, portable spot welding machines are used. Here the electrode holders and the pneumatic pressurising system are present in the form of a portable assembly which is taken to the place, where the spot is to be made. The electric current, compressed air and the cooling water needed for the electrodes is supplied through cables and hoses from the main welding machine to the portable unit.

In spot welding, a satisfactory weld is obtained when a proper current density (A/sq mm) is maintained. The current density depends on the contact area between the electrode and the work piece. With continuous use, if the tip becomes upset and the contact area increases, the current density will be lowered and consequently the weld is obtained over a large area. This would not be able to melt the metal and hence there would be no proper fusion.

A resistance welding schedule is the sequence of events that normally take place in each of the welds. The events are:

- The squeeze time is the time required for the electrodes to align and clamp the two work pieces together under them and provide the necessary electrical contact.
- The weld time is the time of the current flow through the work pieces till they are heated to the melting temperature.
- The hold time is the time when the pressure is the maintained on the molten metal without the electric current. During this time, the pieces are expected to be forge welded.
- The off time is time during which the pressure on the electrode is taken off so that the plates can be positioned for the next spot. The off time is not normally specified for simple spot welding, but only when a series of spots are to be made in a predetermined pitch.

A typical welding schedule, whose time elements are measured in terms of the time unit for a cycle of AC, is shown in Fig. 9.63. These welding schedules can be precisely controlled with the modern programmable electronic controllers.

The forging pressure is applied to ensure that during the solidification, the nugget does not develop any porosity or crack. Porosity in spot welds may also develop if the forging pressure is released prematurely. The welding pressure used depends on the thickness of the metal and the geometry of the weld. It is generally preferable to have the pressure increased by two to three times during solidification phase, compared to that during the melting.

The use of high welding current for short periods causes excessive heat generation in the weld area. If the pressure is not properly applied immediately after welding or if the pressure is insufficient, then the porosity may develop at the centre of the nugget or cracks may form.

The metals expand due to the temperature. But the metal under the electrode tip cannot expand in the axial direction of the electrode because of the restraining force. However, while cooling, the metal under this area would reduce in size under the absence of any such restraining force. This gives rise to a visible concave depression under the electrode tip. Though this is very small, it cannot be totally eliminated. The only way in which this can be reduced is to use short duration currents such that heat from the welding zone is not to be conducted far beyond.

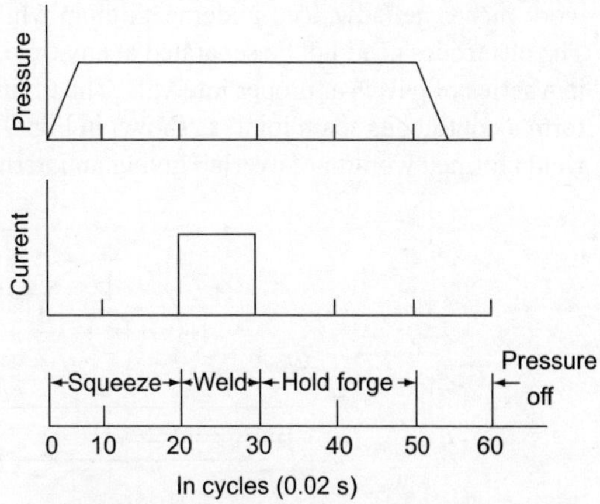

FIG. 9.63 *Simple electric resistance welding schedule*

Surface indentation is also possible sometimes due the expulsion of the molten metal from the weld zone. The main reason for this is the generation of excessive weld heat by a smaller electrode contact area, improper alignment of the work and the electrodes, too long a time for the current flow, too large an electrode force and improper mating surfaces. The expulsion of the molten metal may takes place, if the spot is done very near the edge.

The depth of penetration of the weld would normally vary from 0.3 to 0.8 times the thickness of the joining members. A depth of penetration less than this would not give rise to enough strength whereas a very high penetration nearing the thickness would bring the nugget outside and disfigure the appearance of the sheet. This also reduces the life of the electrode tip. Also, the penetration should be equal from both the sides, even while welding thicker plate to a thinner plate.

The surface melting of the plate may also be caused by poor electrode cooling. Dirty or oxide surfaces on the work piece may interfere with the heat flow into the electrode and thus may allow the top surface to melt.

The spacing of spots affects the current required for fusion. If the spacing becomes small, some current would be shunted through a weld already made, and as a result, higher currents may be required to obtain proper melting. Therefore it is necessary that the spots are not made closer than required.

Designing of spot weld parameters are essentially based on the sheet thicknesses being welded. The typical equations for welding mild steel up to a thickness of 3.2 mm are given below.

$$\text{Electrode tip diameter} = 2.54 + (t_1 + t_2) \text{ mm}$$

$$\text{Weld time} = 2.36 \times (t_1 + t_2) \text{ cycles}$$

$$\text{Current} = 3937 \times (t_1 + t_2) \text{ A}$$

$$\text{Electrode force} = 876 \times (t_1 + t_2) \text{ N}$$

Where, t_1 = thickness of the first plate, mm
t_2 = thickness of the second plate, mm

9.6.3 Resistance Seam Welding

The resistance seam welding (RSEW) is a specialised process of spot welding. Here the cylindrical electrodes are replaced by disc electrodes as shown in Fig. 9.64. The disc electrodes are continuously rotated so that the work pieces get advanced underneath them while at the same time the pressure on the joint is maintained. The electrodes need not be separated at any time. The current is applied through the heavy copper electrodes in a series of pulses at proper intervals. The timing is adjusted so that the pulses overlap each other and thus, form a continuous seam joint, as shown in Fig. 9.64. If the timing of the pulses is increased, then each of the welds formed would not overlap giving an intermittent seam.

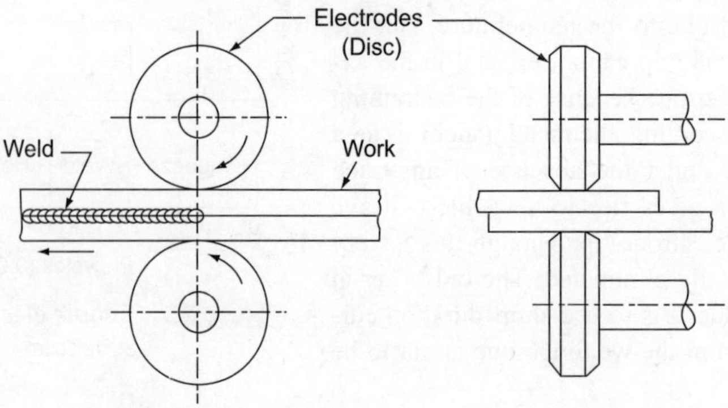

Fig. 9.64 *Resistance seam welding process*

To achieve the pressure tightness, the seam welds should at least overlap 10 to 25% of each nugget.

9.6.4 Projection Welding

Projection welding (RPW) is another variation of spot welding, where one of the sheets to be joined, is provided with a number of projections to help localise the current at a predetermined spot, as shown in Fig. 9.65. The projections are generally very small, of the order of 0.8 mm and are obtained by means of embossing. Because of the localisation being achieved by the use of projections, it is no more necessary to use a small sized electrode. As the welding current passes through these projections, they soften, get melted and a fusion joint is made under the pressure applied from the electrode.

There are a number of advantages in using projection welding. They are:

1. It is possible to weld more than one spot at a given time. The number of welds depends on the number of projections that can come under the tip of the electrode. Up to about 6 welds per cycle can be made, which substantially reduces the cost of manufacture.
2. The welds may be placed closer than possible in spot welding.
3. Because of the large sized electrodes used, their life is much longer than that of the spot welding electrodes. The larger contact area makes for very limited deformation of the welding tip.
4. The electrodes can be made to take the shape of the assembly fixtures, as in Fig. 9.65(c). They can help in holding the work pieces together in proper alignment during the welding process.
5. Proper heat balance can be easily obtained in projection welding by making the projections in thicker plates while welding sheets of different thicknesses. For welding dissimilar metals, the projections are to be made on the material having higher electrical conductivity to provide the proper heat balance.
6. The uniformity and appearance of the weld is much better compared to spot welding.

A disadvantage of this process is that a prior operation of embossing is required before welding. Also, the electrodes are often to be specially made for a particular operation. This can only be justified when production rates and volumes are large.

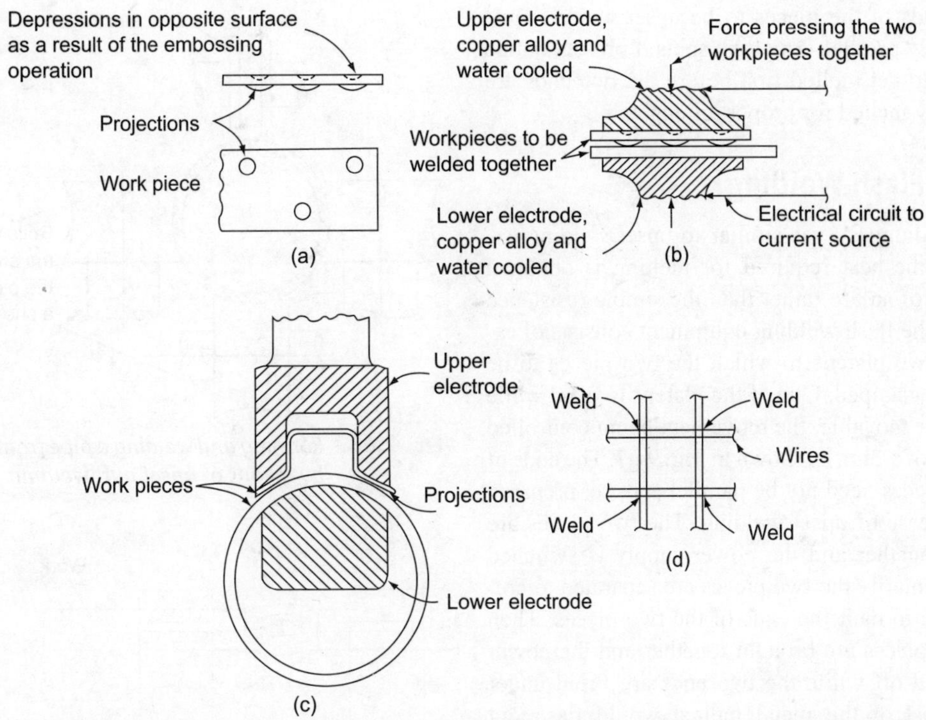

Fig. 9.65 *Features of projection welding (a) Projections on the work piece, (b) Projection welding process, (c) Special electrodes to serve as assembly fixtures, (d) Cross-wire welding (courtesy Campbell op cit, p 410, McGraw-Hill, New York)*

The projections can be of any shape, though circular are easier to make and which also provide uniform strength. The other shapes used are elongated circle in one direction, rounded top or annular projection around a hole. The only requirement is that these projections should be stiff enough to withstand the electrode force before the current is passed. Joining of wires is a natural projection welding method, as shown in Fig. 9.65(d). The side of a wire has a limited contact with the wire, therefore, welding can be easily carried out without any pre-operations.

9.6.5 Upset Welding

In upset welding (UW), the pieces to be joined are brought together to mate with each other in butt joint compared to the lap joint that is used so far. The two pieces are held tightly together and current is applied, so that the heat is generated through the contact area between the two plates. Because of the joint being under pressure, the ends of the two pieces gets slightly upset and hence its name. This is useful for joining the two ends of rods or similar pieces.

A variant of the upset welding, where a continuous seam weld is obtained, is shown in Fig. 9.66. This is the process used for making electric resistance welded (ERW) pipes starting from a metal plate of suitable thickness. The plate is first formed into the shape of the pipe with the help of the three roll set as shown in Fig. 9.66. The ends of the plate would then be forming the butt joint. The two rotating copper disc electrodes are made to contact the two ends of the plate through which the current is passed. The ends get heated and then forge welded under the pressure of the rolls.

The ends of the pieces to be upset welded must be perfectly parallel. Any high spots if present on the ends would get melted first before the two ends are completely melted for proper joining.

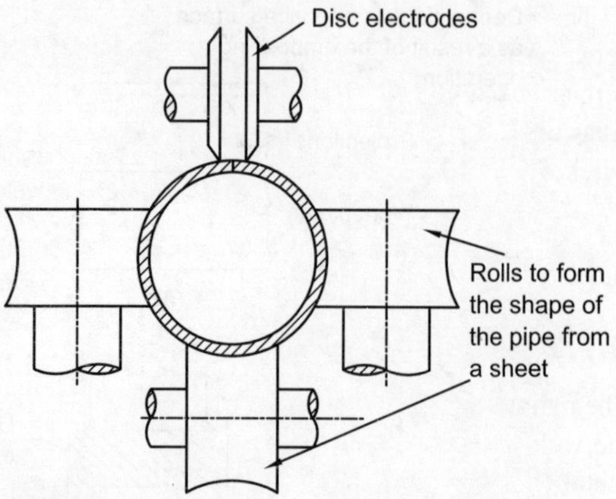

Fig. 9.66 *Forming and welding a pipe from a sheet by means of upset butt welding process*

9.6.6 Flash Welding

Flash welding (FW) is similar to upset welding except that the heat required for melting is obtained by means of an arc rather than the simple resistance heating. The flash welding equipment consists of essentially two platens to which the two pieces to be joined are clamped. One of the platens is fixed while the other is movable, the movement being controlled by means of a cam, as shown in Fig. 9.67. The ends of the two pieces need not be parallel and, so, prepared as in the case of upset welding. The two pieces are brought together and the power supply is switched on. Momentarily the two pieces are separated to create the arc to melt the ends of the two pieces. Then again the pieces are brought together and the power is switched off while the two ends are fused under force. Most of the metal melted would flash out through the joint and forms like a fin around the joint. It is generally a faster operation compared to that of upset welding and would be automatically controlled by a cam arrangement.

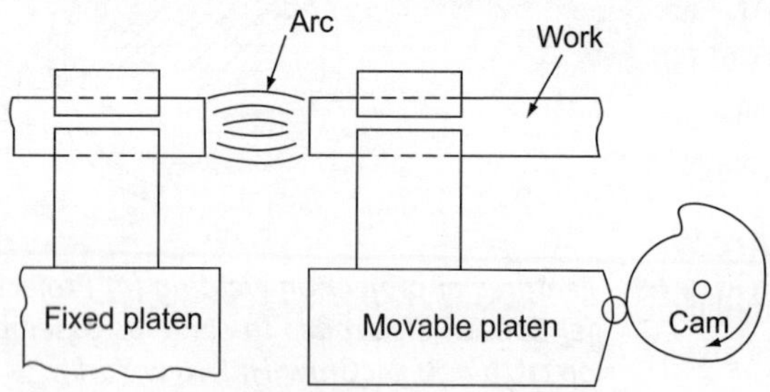

Fig. 9.67 *Flash welding*

9.6.7 Percussion Welding

Percussion welding is similar to upset welding where the welding is completed in a short time with a high intensity arc between the two parts to be joined. The molten surfaces are then squeezed by the impacting of the work pieces forcing part of the liquid metal from the joint to go outside. The equipment used for percussion welding has one part fixed while the other one is clamped in a spring loaded slide for fast movement. This movement is achieved by the quick release of energy from the spring when the trigger is released, as shown in Fig. 9.68.

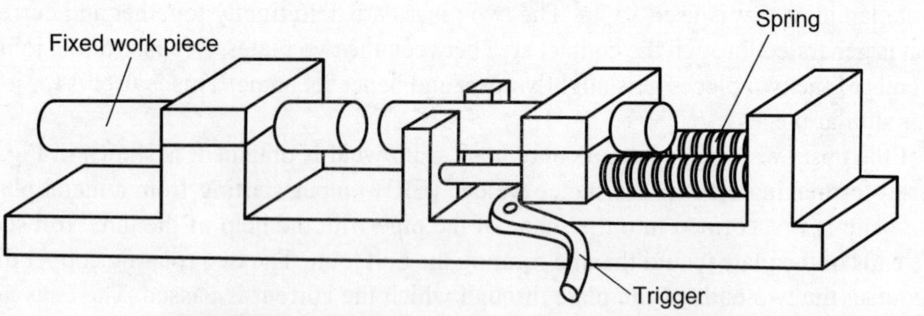

Fig. 9.68 *Schematic of percussion welding setup*

The two pieces to be joined are connected to a DC power supply. When released, an arc will be stuck between the two work surfaces thereby melting the joint. Further movement of the movable part applies percussive force in the joint and makes it to solidify under pressure. This process is limited to small surface areas only. It is possible to join 1 mm diameter wires within a duration of less than 100 micro-seconds. The wires would be travelling with a velocity of 1 m/s. Generally used for joining small rods or tubes.

9.7 WELDING DESIGN

9.7.1 Heat Input

The main function of the heat that would be input into the welding joint is to melt the base metal and filler metal (if used), so that proper fusion of the joint is achieved. The characteristic weld joint as obtained by arc welding which is called bead-on-plate is shown in Fig. 9.69. In this, there are two distinct areas identified;

Ar representing the area of the filler metal melted
Am representing the area of the base metal melted

The main energy input is dependent upon the heat source used.

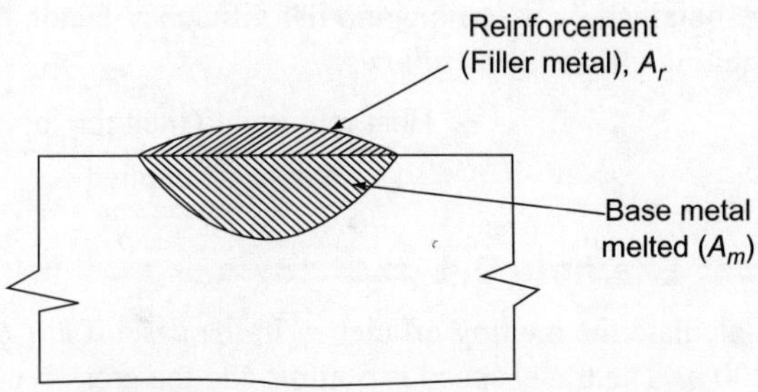

Fig. 9.69 *Cross-section of a bead on plate*

Arc Welding

In the case of arc welding, the heat input, *P* in watts, is by the direct conversion of the electrical energy and is given by

$$P = VI$$

Where, *V* is the potential of the power source in Volts,
I is the current in Amperes

For the melting of the weld joint, it is necessary to know the exact amount of heat that is entering the joint. Since the electrode will be moving at a given velocity, *v*, the actual heat available for melting the joint is given by

$$H = \frac{P}{v} \text{ J/s}$$

Where, *v* is the travel speed of the electrode, mm/s

This is the actual heat generated at the tip of the electrode and ideally should be available for melting the joint. However, the actual heat utilised by the joint depends upon how this heat is transferred from the tip of the electrode to the joint. Hence a factor of heat transfer efficiency, f_1 is to be defined to take into account of any losses in between such that the net heat available at the weld joint, H_{net} is to be given by

$$H_{net} = \frac{f_1 VI}{v} \text{ J/s}$$

When the heat source is concentrated such as an electric arc, the value of f1 would be greater than 0.8 and would almost approach 1.0. The following are some expected values for the various processes:

GTAW	0.21 to 0.48
SMAW and GMAW	0.66 to 0.85
SAW	0.90 to 0.99

Though the net heat as calculated above reaches the weld joint, all of it cannot be utilised for melting since part of it would be conducted away from the joint by the base metal as reflected in the heat affected zone as shown in Fig. 9.70. The actual heat distributed into the surrounding metal would depend upon the welding process as well as the process parameters including the joint design.

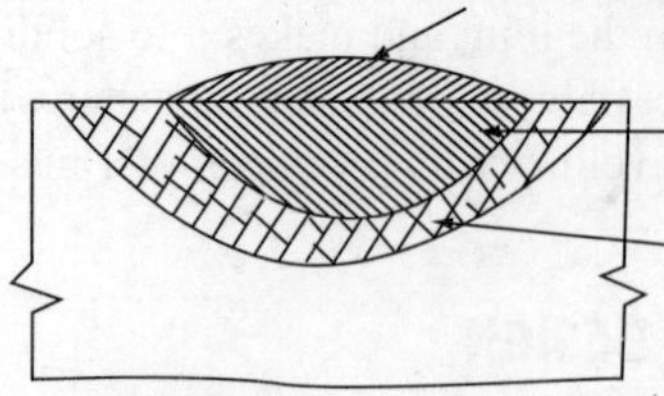

Fig. 9.70 *Cross-section of a bead on plate showing the heat affected zone*

The net heat that is actually utilised for melting can be obtained by assuming another efficiency factor, f_2 as melting efficiency as follows:

$$f_2 = \frac{\text{Heat required to melt the joint}}{\text{Net heat supplied}}$$

Example 9.1

Calculate the melting efficiency in the case of arc welding of steel with a potential of 20 V and current of 200 A. The travel speed is 5 mm/s and the cross-sectional area of the joint is 20 mm^2. Heat required to melt steel may be taken as 10 J/mm^3 and the heat transfer efficiency as 0.85.

Net heat supplied = 0.85 × 20 × 200 = 3400 W

Volume of base metal melted = 20 × 5 = 100 mm^3/s

Heat required for melting = 100 × 10 = 1000 J/s

$$\text{Melting efficiency} = \frac{1000}{3400} = 0.2941$$

Resistance Welding

In the case of resistance welding, the heat input is given by

$$H = I^2Rt$$

Where, R is the resistance of the joint in ohms, and
t is the time in seconds for which the current is flowing through the joint.

Example 9.2

Two steel sheets of 1.0 mm thick are resistance welded in a lap joint with a current of 10 000 A for 0.1 second. The effective resistance of the joint can be taken as 100 micro ohms. The joint can be considered as a cylinder of 5 mm diameter and 1.5 mm height. Density of steel is 0.00786 g/mm^3 and heat required for melting steel be taken as 10 J/mm^3.

Heat supplied = $10\,000^2 \times 100 \times 10^{-6} \times 0.1$ 1 = 1000 J

$$\text{Volume of the joint} = \frac{\pi \times 5^2 \times 1.5}{4} = 29.452 \text{ mm}^3$$

Heat required for melting = 29.452 × 10 = 294.52 J ≈ 295 J

Heat lost to surroundings = 1000 – 295 = 705 J = 70.5%

In comparison to spot welding, projection welding with a short cycle time would be more efficient as shown in the following example.

Example 9.3

Two steel sheets of 1.0 mm thick are resistance welded in a projection welding with a current of 30 000 A for 0.005 second. The effective resistance of the joint can be taken as 100 micro ohms. The joint can be considered as a cylinder of 5 mm diameter and 1.5 mm height. Density of steel is 0.00786 g/mm^3 and heat required for melting steel be taken as 10 J/mm^3.

Heat supplied = $30000^2 \times 100 \times 10^{-6} \times 0.005 = 450$ J

$$\text{Volume of the joint} = \frac{\pi \times 5^2 \times 1.5}{4} = 29.452 \text{ mm}^3$$

Heat required for melting = $29.452 \times 10 = 294.52 \text{ J} \approx 295$ J

Heat lost to surroundings = $450 - 295 = 155 \text{ J} = 34.4\%$

The higher efficiency achieved is because of the short time the current is flowing such that the heat losses are minimised.

9.7.2 Heat Flow

Heat transfer in a welded joint is a complex phenomenon involving three dimensional movement of a heat source. Heat from the weld zone is transferred more to the other parts of the base metal by means of conduction. Similarly heat is also lost to surroundings by convection from the surface, with radiation component being relatively small except near the weld pool. Thus the analytical treatment of the weld zone is extremely difficult.

Hence certain assumptions are made such that analytical equations could be developed for important parameters such that they could be utilised in practical situations. Since the metallurgical structure of the weld zone is controlled by the (critical) cooling rate, it would be of great interest in the case of steels particularly those that are heat treatable.

For relatively thick plates, the cooling rate is given by

$$R = \frac{2\pi k (T_c - T_0)^2}{H_{net}}$$

Where, T_0 = Initial plate temperature, °C

k = thermal conductivity of base metal, J/mm . s . °C

R = Cooling rate at the weld centre line, °C/s

T_c = Temperature at which the cooling rate is calculated, °C

Though this cooling rate is at the centre line, even at the weld boundary it may be a little lower, but would be sufficiently accurate in predicting the properties of the heat affected zone. The temperature, T_c at which the cooling rate is being calculated should be based on the metallurgical considerations of the base metal. For most of the steels a temperature of 550°C is satisfactory.

If the plates are relatively thin requiring less than four passes, the following equation can be used:

$$R = 2\pi k \rho c \left(\frac{h}{H_{net}}\right)^2 (T_c - T_0)^3$$

Where, h = thickness of the base metal, mm
ρ = density of base metal, g/mm^3
c = specific heat of the base metal, J/g . °C

However, the distinction between thin and thick plates becomes arbitrary without a quantitative evaluation. Thus a 'relative plate thickness factor', τ is defined as follows:

$$\tau = h\sqrt{\frac{\rho c(T_c - T_0)}{H_{\text{net}}}}$$

$\tau \le 0.75$ thin plate equation is valid
$\tau \ge 0.75$ thick plate equation is valid

The heat flow condition under various plate thicknesses are schematically shown in Fig. 9.71. Typical cooling rate of 6°C/s has been generally found to be the maximum limit for most of the steels to provide satisfactory metallurgical quality without martensitic transformations.

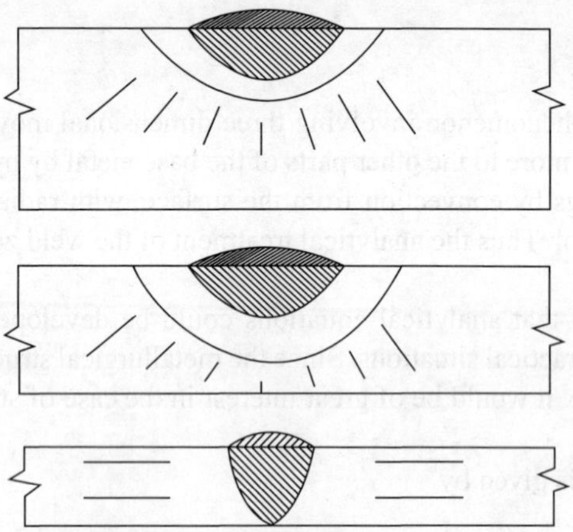

FIG. 9.71 *Defining the plate thickness for cooling rate calculations*

Example 9.4

Find the best welding speed to be used for the welding of 6 mm steel plates with an ambient temperature of 30°C with the welding transformer set at 25 V and current passing is 300 A. The arc efficiency is 0.9 and the possible travel speeds are 6 to 9 mm/s. The limiting cooling rate for satisfactory performance is 6°C/s at a temperature of 550°C.

Given $T_0 = 30°C$
$k = 0.028$ J/mm.s.°C
$R = 6°C/s$
$T_c = 550°C$
$V = 25$ V
$I = 300$ A
$h = 6$ mm
$f_1 = 0.9$
$\rho c = 0.0044$ J/mm^3 °C

Assume the travel speed, $v = 9$ mm/s

$$\text{Heat input, } H_{net} = \frac{f_1 VI}{v} = \frac{0.9 \times 25 \times 300}{9} = 750 \text{ J/mm}$$

To check whether it is a thin or thick plate,

$$\tau = h\sqrt{\frac{\rho c(T_c - T_0)}{H_{net}}} = 6\sqrt{\frac{0.0044(550-30)}{750}} = 0.3314$$

This being less than 0.6, it is a thin plate. Calculate the cooling rate using the thin plate equation

$$R = 2\pi k \rho c\left(\frac{h}{H_{net}}\right)^2 (T_c - T_0)^3$$

$$= 2\pi 0.028 \times 0.0044\left(\frac{6}{750}\right)^2 (550-30)^3$$

$$= 6.9659 \text{ °C/s}$$

This cooling rate being a little higher than the critical cooling rate required, we may reduce the travel speed to 8 mm/s and recalculate.

Assume the travel speed, $v = 8$ mm/s

$$\text{Heat input, } H_{net} = \frac{0.9 \times 25 \times 300}{8} = 843.75 \text{ J/mm}$$

To check whether it is a thin or thick plate,

$$\tau = h\sqrt{\frac{\rho c(T_c - T_0)}{H_{net}}} = 6\sqrt{\frac{0.0044(550-30)}{843.75}} = 0.312$$

This being less than 0.6, it is still a thin plate. Calculate the cooling rate using the thin plate equation

$$R = 2\pi k \rho c\left(\frac{h}{H_{net}}\right)^2 (T_c - T_0)^3 = 2\pi\, 0.028 \times 0.0044\left(\frac{6}{843.75}\right)^2 (550-30)^3 = 5.504 \text{ °C/s}$$

This being a satisfactory cooling rate, the welding speed can be finalised at 8 mm/s.

The same set of equations could also be utilised to calculate the preheat temperature required to avoid martensitic transformation in the weld zone.

Another important parameter that needs to be calculated is the peak temperature reached at any point in the material. This would help in identifying what type of metallurgical transformations are likely to takes place in the heat affected zone (HAZ). Peak temperatures can be calculated [28.1] using the following equations:

$$\frac{1}{T_p - T_0} = \frac{\sqrt{2\pi e}\ \rho c h y}{H_{net}} + \frac{1}{T_m - T_0}$$

Where, T_p = temperature reached at a distance of y mm from the fusion boundary. At the fusion boundary, $T_p = T_m$

T_m = melting temperature of the base metal, °C

e = base of natural logarithm = 2.718 218 28

This equation can be used to calculate the extent of heat affected zone as well as the type of microstructure present in the heat affected zone. This would be particularly true in the case of heat treatable steels.

Example 9.5

Calculate the heat affected zone for example 9.4 with a welding speed of 7 mm/s. Melting temperature of steel can be taken as 1510°C.

Making use of the peak temperature equation, let us calculate the temperatures reached at intervals of 1 mm each from the fusion zone:

Heat input, $$H_{net} = \frac{f_1 VI}{v}$$

$$= \frac{0.9 \times 25 \times 300}{7} = 964.29 \text{ J/mm}$$

At y = 1 mm

$$\frac{1}{T_p - T_0} = \frac{\sqrt{2\pi e}\ \rho chy}{H_{net}} + \frac{1}{T_m - T_0}$$

$$= \frac{\sqrt{2\pi e} \times 0.0044 \times 6 \times 1}{964.29} + \frac{1}{1510 - 30}$$

or T_p = 1298°C

At y = 2 mm T_p = 1139°C

At y = 3 mm T_p = 1015°C

At y = 4 mm T_p = 916.3°C

At y = 6 mm T_p = 768.3°C

At y = 8 mm T_p = 662.6°C

At y = 10 mm T_p = 583.4°C

At y = 12 mm T_p = 521.8°C

At y = 14 mm T_p = 472.5°C

At y = 16 mm T_p = 432.3°C

The same are plotted in Fig. 9.72 to show the extent of transformation taking place.

It is possible to see how the peak temperatures reached in the heat affected zone will be affected by a change in the welding speed.

Let v = 8 mm/s

Heat input, $$\frac{0.9 \times 25 \times 300}{8} = 843.75 \text{ J/mm}$$

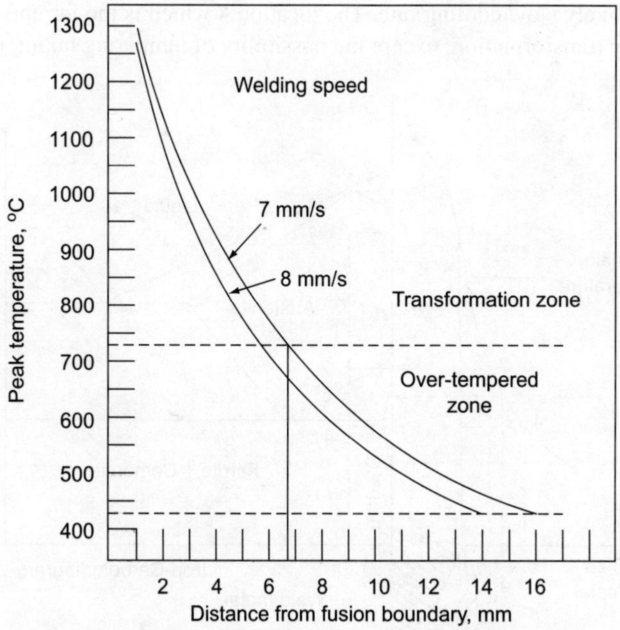

FIG. 9.72 *Peak temperatures reached in the base metal as varied by the welding speed*

Recalculating the peak temperatures with this new heat input gives the following set of values:

At $y = 1$ mm $T_p = 1272°C$
At $y = 2$ mm $T_p = 1100°C$
At $y = 3$ mm $T_p = 970°C$
At $y = 4$ mm $T_p = 868°C$
At $y = 6$ mm $T_p = 719°C$
At $y = 8$ mm $T_p = 615°C$
At $y = 10$ mm $T_p = 538°C$
At $y = 12$ mm $T_p = 479°C$
At $y = 14$ mm $T_p = 432°C$

Heat affected zone can be defined for steels as about 730°C below which no transformation occurs as shown in the iron-carbon equilibrium diagram. However, below that temperature some steels are likely to be tempered. Hence in such cases the heat affected zone to include the tempering transformation can be taken as 430°C.

The peak temperature as calculated above is shown in comparison with the iron carbon equilibrium diagram to obtain the relative metallurgical structure in the heat affected zone in Fig. 9.73. As can be seen from the figure in zone 1, the metal is in the austenitic region, and is likely to be transformed into martensite if the cooling rate is sufficiently high. This is the zone where there is a likelihood of high residual stresses and crack formation. In zone 2 it is still austenite and there is a possibility of transformation exists depending upon the cooling rate for this location and the alloying elements present in the steel. In location 3, there is part austenite and part ferrite (since it is a hypo eutectoid steel) and is likely to form pearlitic structure upon cooling in view of the likely slow cooling rate. The location 4 which is the far end of the heat affected zone is not likely to have any transformation, except the possibility of tempering taking place.

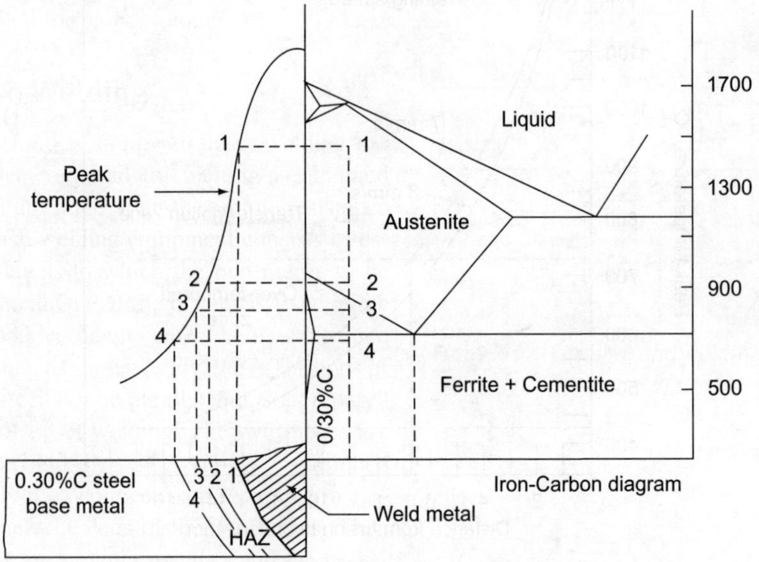

FIG. 9.73 *Iron-carbon equilibrium diagram alongside the peak temperatures reached in the heat affected zone*

The above equation is derived for single pass operation and thin plates. Though this can be applied to multi-pass operations, the T_0 value would change in view of the heat received during the prior passes. If that temperature can be accurately predicted, then this can be applied up to 4 passes without loss of much accuracy in the prediction.

9.7.3 Distortions

One of the major problems found with weldments is the distortion. Distortion is caused mainly because of the shrinkages that take place in weldments. The shrinkage taking place in a weldment depends upon the geometry and type of the weld. There are three types of distortions possible in weldments.

(a) Transverse shrinkage occurring perpendicular to the weld line.

(b) Longitudinal shrinkage occurring parallel to the weld line, which is very small of the order of about 0.1% of the weld length and hence can be neglected.

(c) Angular change as a rotation about the weld line.

Transverse Shrinkage in Butt Joints

The transverse shrinkage in a butt joint in a single pass weld, as shown in Fig. 9.74, is a result of contraction of the base metal, which was expanded during welding. The transverse shrinkage can be estimated by using the following equation for carbon and low alloy steels:

$$s = 5.08 \times \frac{A_w}{t} + 1.27 \times d$$

Where, s = transverse shrinkage, mm

t = plate thickness, mm

d = root opening, mm

Aw = cross-sectional area of weld, mm^2

During a multi-pass welding, the total shrinkage is a sum total of the individual shrinkages in each of the passes. However, the major contribution comes from the first pass. It is observed experimentally that the total transverse shrinkage is proportional to the logarithm of the total weight of weld metal deposited. The relation is as follows:

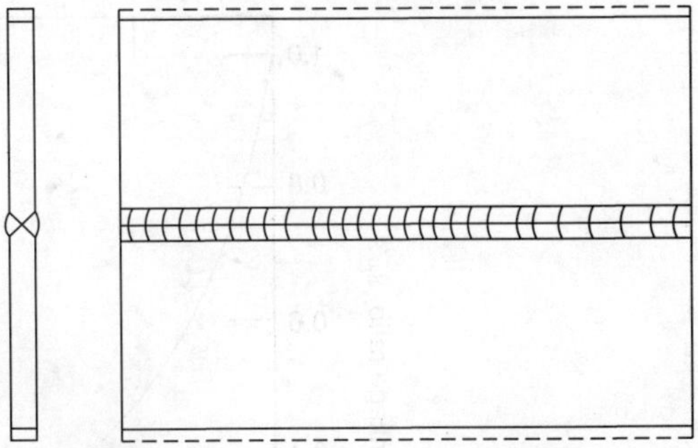

FIG. 9.74 *Transverse shrinkage in butt joints*

$$st = s_0 + b \times (\log w - \log w_0)$$

Where, st = total transverse shrinkage

s_0 = transverse shrinkage in the first pass

w = total weight of the weld metal deposited

w_0 = weight of the weld metal deposited in the first pass

b = a constant

Based on this relationship, the methods for reducing the transverse shrinkage are:

(a) Decrease the total weight of the weld metal, w, and/or

(b) Increase the metal deposited in the first pass.

The other factors which influence the transverse shrinkage are given in Table 9.19.

TABLE 9.19 **Variables affecting the transverse shrinkage in butt joints**

Process Variable	Effect
Root opening	Shrinkage increases with root opening
Joint design	Single V produces more shrinkage compared to double V
Electrode diameter	Larger sized electrodes reduces shrinkage
Degree of constraint	Shrinkage decreases with an increase in the degree of constraint

Restraining the work pieces to expansion is one of the ways to reduce the transverse shrinkage. Constraining the work piece as shown in Fig. 9.75 produces a reaction stress. The amount of reaction stress for a unit transverse shrinkage is defined as the degree of restraint, *ks*. The experimental observations of the effect of degree of restraint on the shrinkage are shown in Fig. 9.76. The experimental results can be fitted into an equation of the form as follows:

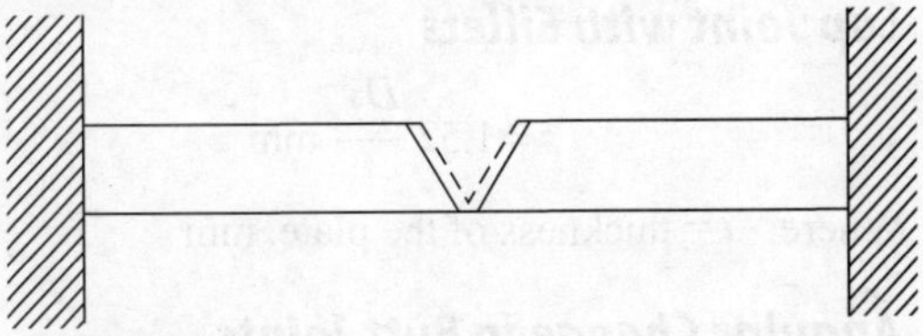

FIG. 9.75 *Effect of restraining the work pieces in butt joint*

$$\frac{s_r}{s} = \frac{1}{1 + 0.086 \times k_s^{0.87}}$$

Where, s_r = transverse shrinkage of a restrained joint

s = transverse shrinkage of an unrestrained joint

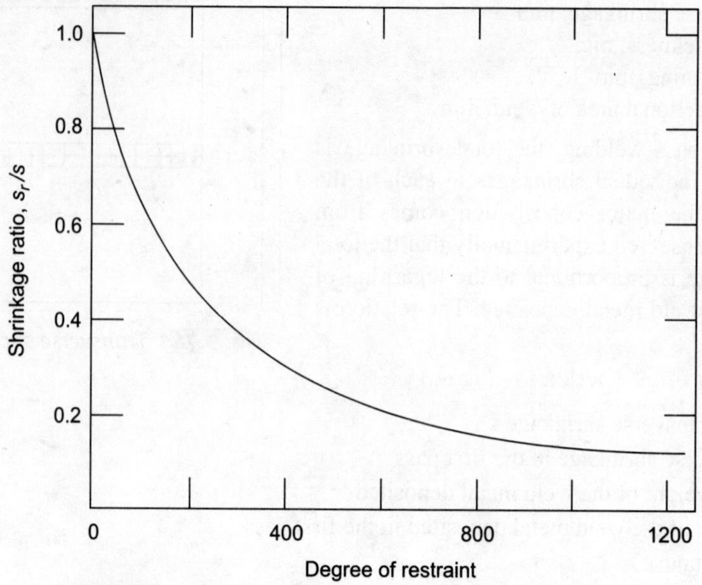

FIG. 9.76 *Variation of shrinkage with degree of restraint*

In the case of *T*-joints and fillet welds, the transverse shrinkage is much smaller and can be easily estimated using the following relations:

T-joint with Two Fillets

$$s = 1.02 \frac{D_f}{t_b} \text{ mm}$$

Where, s = transverse shrinkage, mm
D_f = length of fillet leg, mm
t_b = thickness of bottom plate, mm

Lap Joint with Fillets

$$s = 1.52 \frac{D_f}{t} \text{ mm}$$

Where, t = thickness of the plate, mm

Angular Change in Butt Joints

Angular change, as shown in Fig. 9.77, occurs due to the non-uniform transverse shrinkage in the thickness direction. Experimentally it was observed in the case of double V-groove that the distortion was minimum when the following condition is satisfied:

$$\frac{t_1 + 0.5 \times t_3}{t_1 + t_2 + t_3} = 0.62$$

Where, t_1 = depth of the first *V*-groove as shown in Fig. 9.78.
t_2 = depth of the last *V*-groove as shown in Fig. 9.78.
t_3 = depth of root face, mm

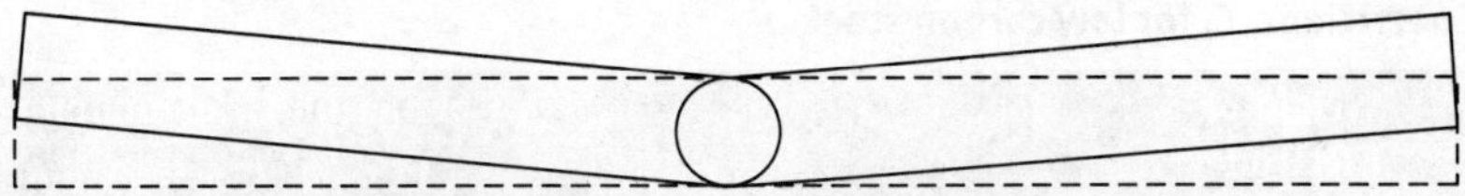

Fig. 9.77 *Angular distortion in butt joints*

Angular Distortion of Fillet Welds

Fillet welds in a structure are affected by the way the structure is designed and the type of restraint provided. For example, in a frame structure as shown in Fig. 9.79, when there is no restraint the angular distortion would be as shown in (a). However, if the stiffeners are restrained as shown in (b), then a wavy distortion would be appearing. The evaluation of such a distortion has to follow the design principles of framed structures. The distortion at any point x, will be given by

$$\frac{\delta}{L} = 0.25 \cdot \varphi - \left[\frac{x}{L} - 0.5\right]^2 \cdot \varphi$$

Where, δ = distortion at x

L = length of span

ϕ = angular change

x = distance from the centre line of the frame

The maximum value of delta is at $x = 0.5$ L, when

$$\delta = 0.25\ \varphi \cdot L$$

The angular change ϕ is related to the angular change ϕ_0 of an unrestrained welding under the same welding conditions as follows:

$$\varphi = \frac{\varphi_0}{1 + \dfrac{2R}{LC}}$$

Where, C = constant whose values are given in Table 9.20

R = rigidity given by the following relation

$$R = \frac{Et^3}{12 + (1 - v^2)}$$

Where, v = Poisson's ratio

E = Modulus of elasticity, MPa

t = thickness of the flange plate, mm

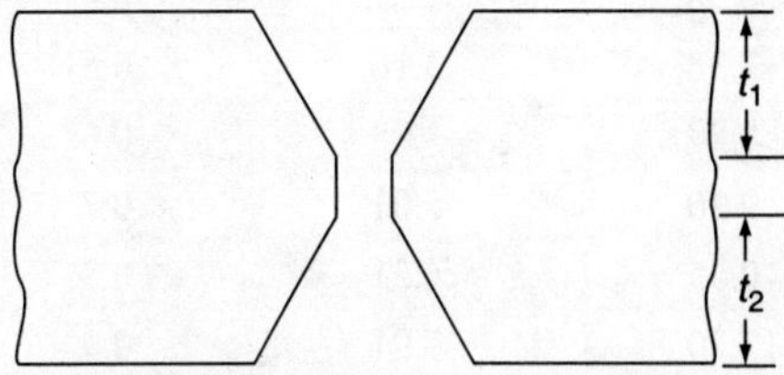

Fig. 9.78 *Double V-joint configuration*

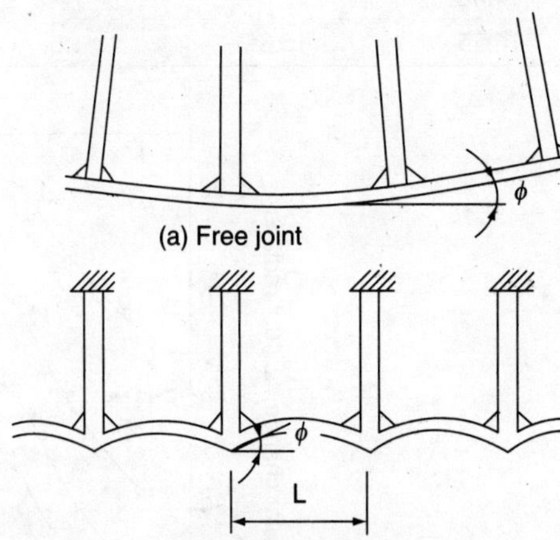

Fig. 9.79 *Angular distortion of fillet welds with and without restraint*

The values of ϕ_0 are given in Fig. 9.80. In this Fig. 9.80, the parameter, w is given by

$$w = \frac{12 D_f^3\, \rho}{f_1}$$

TABLE 9.20 **Coefficient, *C*, for low carbon steel**

Fillet Weld Size, mm	Weight of Electrode Consumed, g/cm	$\log_{10} w$	Coefficient, C, kg mm/mm			
			t = 10 mm	t = 13 mm	t = 18 mm	t = 25.4 mm
6.58	2.51	0.4	5400	19 900	76 100	170 100
7.38	3.16	0.5	4700	18 000	65 200	142 400
8.29	3.98	0.6	4100	16 300	56 100	130 200
9.30	5.01	0.7	3800	15 000	48 800	125 000
10.45	6.31	0.8	3500	13 600	43 000	116 800
12.20	7.91	0.9	3300	12 200	38 900	112 000
13.15	10.00	1.0	3100	11 000	36 100	108 200
14.80	12.60	1.1	3000	9 800	35 200	105 000
16.55	15.85	1.2	2950	8 800	34 800	102 000

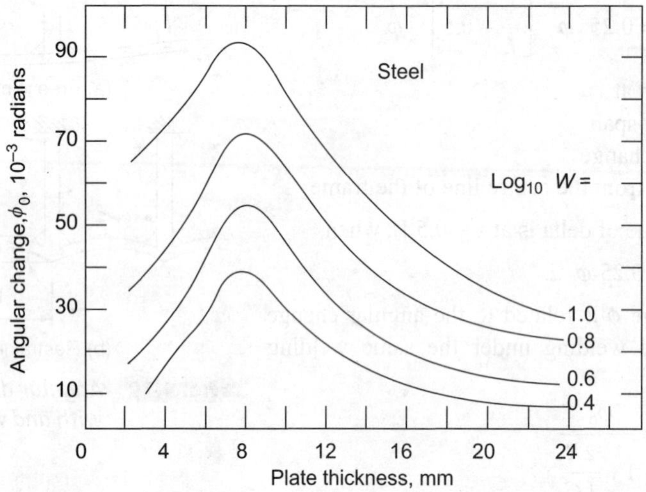

FIG. 9.80 *Angular distortions in unrestrained fillet welds in steel*

Where, ρ = density of the weld metal = 7.85 g/cm^3

D_f = length of the fillet leg, mm

f_1 = Deposition efficiency = 0.657

$w = 0.058\ D_f^2$

Control of Distortion

As described previously, distortions are inevitable in a welding operation. It is therefore necessary to find ways by which these can be minimised to prepare a satisfactory weldment.

One of the important ways to control the distortions is a good design of the product with minimum number of joints. Also a proper design of the joint helps in reducing the magnitude of the problem. In spite of this, there still will be distortion and therefore during the assembly process care should be taken to see that the distortions are controlled. There are two possibilities:

- Estimate the likely amount of distortion as detailed above, and preset the members to compensate for this distortion.
- Assemble the parts correctly and then apply a proper restraint to minimise the distortion during the welding process.

Though the first method looks attractive, is a bit difficult to apply except for simpler assemblies. A typical example is shown in Fig. 9.81.

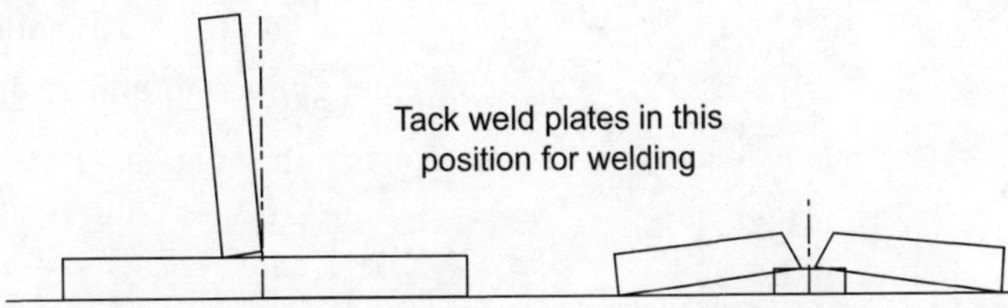

Fig. 9.81 *Pre-setting the work pieces to compensate the distortion for simple joints*

The most generally preferred method in the industry is using restraint. There are many ways used for restraining such as clamps, fixtures and even tack welds. Though restraining reduces the distortions, it would cause high residual stresses which may lead to cracking. Hence it is necessary to carefully apply restraint without causing too high a magnitude of harmful residual stresses.

Another method available for reducing the distortions is the preheating of the members of the weldment such that the heat of welding would be properly balanced. The experimental results obtained in the case of a *T*-joint are shown in Fig. 9.82. The distortion is reduced by almost 50% by preheating the bottom plate. In Fig. 9.82, the parameter, *Z* is defined as follows to account for all the parameters of the welding process:

$$Z = \frac{I}{t\sqrt{vt}}$$

Where, I = welding current, A

t = thickness of bottom plate, mm

v = welding speed, mm/s

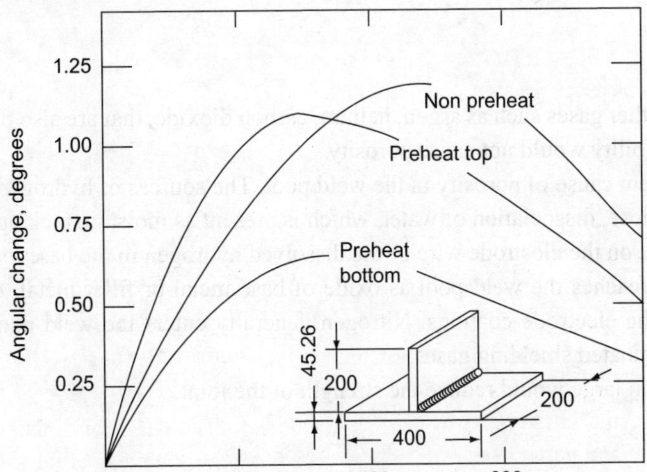

Fig. 9.82 *Effect of preheat on the resultant distortion for various weld parameters*

9.7.4 Defects in Welding

In view of the severe thermal regime through which the welding process proceeds, the weldments are likely to be affected and if proper care is not taken, likely to end up with certain defects. Distortions have been discussed in greater detail earlier, and we will see the other defects here. The defects likely are:

- Undercut
- Incomplete fusion
- Porosity
- Slag inclusion
- Hot cracking
- Cold cracking
- Lamellar tearing

Undercut

This appears like a small notch in the weld interface. This is generally attributed to the improper welding technique or excessive welding current. This is mainly caused by the incorrect manipulation of the electrode while depositing the bead, particularly, in horizontal and vertical welding.

Incomplete Fusion

This will be seen as a discontinuity in the weld zone. The main causes for this defect are improper penetration of the joint, wrong design of the joint or incorrect welding technique including the wrong choice of the welding parameters. The main parameter that controls is the welding current, if it is lower than required, it would not sufficiently heat all the faces of the joint to promote proper fusion. Also the improper cleaning of the joint hinders the fusion of the metal in the joint.

Porosity

Porosity in welding is caused by the presence of gases which get entrapped during the solidification process. The main gases that cause porosity are:

- Hydrogen,
- Oxygen, and
- Nitrogen

Though there are other gases such as argon, helium, carbon dioxide, that are also present in the weld pool, in view of their insolubility would not cause porosity.

Hydrogen is the main cause of porosity in the weld pool. The sources of hydrogen could be the electrode coatings such as cellulose, dissociation of water, which is present as moisture pick up of electrodes, leftover wire drawing lubricant on the electrode wire or the dissolved hydrogen in the base metal and filler metal.

Oxygen generally reaches the weld pool as oxide of base metal or filler metal, or in some compounds of fluxes present in the electrode coatings. Nitrogen generally enters the weld pool through atmospheric nitrogen or the contaminated shielding gases.

Porosity if present in large would reduce the strength of the joint.

Slag Inclusion

Slag is formed by the reaction with the fluxes and is generally lighter. In view of its low density, it will float on top of the weld pool and would be chipped off after solidification. However, the stirring action of the high intensity arc would force the slag to go into the weld pool and if there is not enough time for it to float, may get solidified inside the fusion zone and ends up as slag inclusion. Also in multi-pass welding, the slag solidified in the previous pass is not cleaned before depositing the next bead may cause slag inclusion. Some of the factors that cause slag inclusion are:

- High viscosity of weld metal
- Rapid solidification
- Insufficient welding heat
- Improper manipulation of the electrode
- Undercut on previous pass

Slag inclusion like porosity, weakens the metal by providing discontinuities.

Hot Cracking

Generally occurs at high temperature and the size can be very small to visible. The crack is, in most parts, intergranular and its magnitude depends upon the strains involved in solidification. They are more likely to form during the root pass when the mass of the base metal is very large compared to the weld metal deposited. It can be prevented by preheating the base metal, increase the cross-sectional area of the root bead, or by changing the contour or composition of the weld bead.

Cold Cracking

Cold cracking generally occurs at room temperature after the weld is completely cooled. This can be generally seen in the heat affected zone. The causes are:

- Excessive restraint of the joint which induces very high residual stresses
- Martensitic transformations making the metal very hard as a result of rapid cooling

Stress relieving the weldment immediately would help in relieving the residual stresses and the potential for cracking gets reduced. Also pre and post heating of the weldment helps in reducing the cooling rates and the consequent locking of the stresses.

Lamellar Tearing

It is generally seen at the edge of the heat affected zone. It appears as a long and continuous visual separation line between the base metal and the heat affected zone. This is caused by the presence of the elongated inclusions such as Mn, Fe and S in the base metal. It can also be caused by the weld configuration which gives rise to high residual tensile stresses in the transverse direction.

9.7.5 Weld Testing

As described above, there is likelihood that various defects are to be present in the weldments. Therefore it is necessary to test the welds for soundness before they can be used. These can be broadly divided into two types:

- Destructive tests
- Non-destructive tests

Destructive Weld Tests

Mechanical testing of weldments involve standard material testing procedures explained in Chapter 2 such as hardness test, tensile test, impact strength test etc., can be used. In addition to these standard tests, specific weldment testing procedures such as bend test that is used for testing the ductility of the weld joint are also used. The purpose of destructive weld testing as practiced in industry is to establish the welding procedures, welders, and welding operators. It can also be less frequently used for quality control for which non-destructive testing is used.

Non-destructive Testing of Weldments

The traditional non-destructive test techniques used are visual inspection, eddy current, magnetic particle testing, liquid penetrant and ultrasonic testing and X-Ray. The details of some of the non-destructive test methods are given in chapter 5. Same test could be utilized for the purpose of weld testing. Eddy current and ultrasonic methods are becoming increasingly popular for sub-surface defect checking. These require a careful probe positioning as wrong placement could lead to false positives or no detection of a real defect.

SUMMARY

Fabrication processes are used for joining multiple parts to form working assemblies.

- Gas welding uses a fuel gas such as acetylene together with oxygen to generate sufficient heat to melt the weld joint. It is used for low volume and repair work. It is economical and versatile.
- Heat produced by an electric arc is used to melt and join thick materials in arc welding. It is a very versatile process and a large number of variants are available in this process. For engineering applications requiring very high quality welds generally GTAW and GMAW processes are used. The SAW process is used for joining very thick plates in a single pass.
- Electric resistance welding creates a joint by heating the joint without any filler metal. The heating takes place due to the resistance offered the plates to be joined. This is a low cost process for mass production.
- Designing a weld joint requires the complete understanding of the physics of the process involved. Since most of the welding processes utilize heat, it is important to know the flow rates of heat through joint to estimate the possible distortion, and heat affected zone.

Questions

9.1 Give a brief classification of the various fabrication processes.

9.2 Enunciate the factors based on which the type of fabrication method is chosen for a given application.

9.3 What are the kinds of temporary fabrication methods you know of?

9.4 What are the specific advantages of adhesive bonding?

9.5 Give the typical applications for adhesive bonding.

9.6 Mention the applications you know of where the different kinds of fabrication methods are employed.

9.7 How is the kind of adhesive chosen for bonding?

9.8 Why is welding extensively used?

9.9 What are the kinds of joints that are normally employed for welding purposes? Give their sketches.

9.10 Discuss the need for edge preparation in welding.

9.11 What is meant by 'penetration'? Explain its relevance to welding.

9.12 Discuss the need for cleaning the surfaces to be welded.

9.13 How do you classify the different weld positions?

9.14 What is flux? Why is it essential to use it in some welding situations?

9.15 What is filler metal? Explain its importance in welding, giving its composition.

9.16 'Two plates were welded together and then the strength of the joint was tested. It was found that the weld was stronger than either of the plate'. Do you think that the above statement is incorrect? Comment, giving valid reasons.

9.17 Describe the principle of an oxy-fuel gas welding process.

9.18 List the various fuel gases that are used for the oxy-fuel gas welding processes, describing the applications of each of them.

9.19 Describe the types of flames obtained in an oxy-acetylene gas welding process giving the applications.

9.20 Why is the neutral flame extensively used in oxy-acetylene welding?

9.21 Describe the oxidising flame with a neat sketch. Mention the materials in welding of which this flame is used.

9.22 Briefly describe the oxyacetylene welding equipment.

9.23 How is the acetylene required for gas welding, is stored?

9.24 Describe the oxy-acetylene gas welding technique.

9.25 Explain the advantages and applications of oxy-acetylene welding.

9.26 Write short note on oxy-hydrogen welding.

9.27 Describe the principle of oxy-fuel gas cutting.

9.28 What do you understand by the term 'kerf' in gas cutting? Explain its relevance.

9.29 What are the parameters that decide the drag in oxy-fuel gas cutting? Explain how a good cut is achieved.

9.30 Explain the effects of the torch speed on the cut in gas cutting.

9.31 Explain the problems posed by steels with different carbon percentages, in oxygen cutting and how they are overcome.

9.32 Explain various methods available for oxy-acetylene gas cutting in industrial practice.

9.33 Show the cross section of a gas cutting operation labelling important details.

9.34 How is a gas cutting torch different from welding torch?

9.35 Distinguish between arc and gas welding processes from the point of view of heat concentration, temperature, ease of operation and running cost.

9.36 What are the parameters that control the weld quality in manual metal arc welding (also called as SMAW)?

9.37 Why is the tungsten inert gas welding preferred for welding aluminium plates? Give reasons.

9.38 The possibility of distortion is very low in the case of submerged arc welding. Comment giving reasons.

9.39 Why is it normally necessary to use filler material in welding with tungsten arc? Give reasons.

9.40 You are familiar with the arc welding with coated electrodes. From the point of ease of operation, this method is said to be more convenient than the tungsten inert gas welding process. Comment giving reasons.

9.41 How do you specify an electric arc welding power source?

9.42 How is an arc obtained in arc welding?

9.43 What are the advantages of AC equipment over DC equipment in arc welding?

9.44 How is polarity defined in the case of a DC welding source? How is it advantageously used?

9.45 What is a bead? Explain with a sketch.

9.46 State the important functions of flux coatings of electrodes used in manual metal arc welding process.

9.47 Explain the TIG and MIG systems of arc welding. Give the applications of each.

9.48 Explain the characteristics of arc welding machines, viz., constant current and constant voltage. Mention the applications of each of them.

9.49 Explain the reasons why DC arc welding is more used than AC arc welding for specialised applications.

9.50 What are the problems encountered with the use of coated electrodes? Explains how these can be taken care of.

9.51 Explain briefly the procedure of manual metal arc welding process.

9.52 Briefly explain the coding method used for the electrodes used in SMAW.

9.53 Explain the method of obtaining a weld in horizontal position by SMAW.

9.54 What are the problems faced in SMA welding in vertical and horizontal welding positions.

9.55 What is an arc blow? Explain with a neat sketch the causes of the arc blow, its effects on welding and the methods of reducing the arc blow problem.

9.56 How is the arc blow problem in AC welding taken care of?

9.57 Briefly explain the process of carbon arc welding. What material are the electrodes used in carbon arc welding, made of?

9.58 Why is generally DC power supply used in carbon arc welding?

9.59 Explain the necessity of inert gas shielded arc welding process.

9.60 What are the shielding gases that are most commonly used in the inert gas shielded arc welding process? Answer along with their applications.

9.61 What are the typical electrode materials used in tungsten inert gas welding?

9.62 Why an AC power supply is not normally used in TIG welding process?

9.63 Describe the features of the power supply used in TIG welding process.

9.64 Describe briefly the technique of welding in TIG process.

9.65 Describe the flux materials used in TIG welding process.

9.66 What are the differences between TIG and MIG (or GMAW) welding processes?

9.67 Explain the process of metal transfer in GMAW process with neat sketches.

9.68 Explain the relationship between the shielding gas used and the type of metal transfer, occurring in GMAW process.

9.69 Explain the GMA welding technique.

9.70 Briefly explain the flux cored arc welding process mentioning its advantages.

9.71 What are the applications of the submerged arc welding and how is it different from the inert gas shielded metal arc welding process?

9.72 Explain how the welding quality in submerged arc welding is affected by some of the welding parameters.

9.73 What are the advantages that are gained by adding metal powders in SAW process?

9.74 Briefly explain the process of atomic hydrogen welding.

9.75 What are the differences between plasma arc welding and gas tungsten arc welding?

9.76 Write a short note on the plasma arc welding process.

9.77 Explain the process of stud arc welding with a neat sketch and explain how it is different from other arc welding processes?

9.78 Briefly explain the process of fire cracker welding with its applications.

9.79 How is arc cutting different from oxy-fuel cutting process?

9.80 Briefly explain any one of the arc cutting processes.

9.81 What are the various methods available for arc cutting of metals? Compare them with reference to the applications, ease of use and quality of the cut obtained.

9.82 Explain the process of air carbon arc cutting.

9.83 Explain the resistance welding process giving the equipment, parameters controlled and the applications.

9.84 What are the parameters to be controlled in the resistance welding process?

9.85 Distinguish giving reasons, between joining of sheets in resistance welding and arc welding, from the point of current required and duration of the welding.

9.86 Show by a schematic sketch, a resistance welding operation labelling the important features.

9.87 Mention the various types of resistance welding operations, you know of.

9.88 Can dissimilar metals be welded by resistance welding? If so, give the necessary precautions.

9.89 State the advantages of projection welding.

9.90 Explain how a tube can be manufactured from sheet by a suitable welding process.

9.91 Explain the principle underlying the resistance welding process. Give names of products wherein the following processes are used:
(a) Spot welding (b) Seam welding (c) Flash welding

9.92 Briefly explain the projection welding process.

9.93 Briefly explain the percussion welding process.

9.94 What do you understand by the term 'heat balance'? Explain its relevance.

9.95 What materials are used for making the resistance welding electrodes? Explain their applications.

9.96 What are the specific advantages and disadvantages of the resistance welding process?

9.97 Explain the relevance of welding schedules.

9.98 Write a brief note on flash welding.

9.99 Explain how the actual heat input in the case of any arc welding can be calculated.

9.100 Compare the melting efficiencies of arc welding and resistance welding. Explain the reasons for the differences found.

9.101 Discuss the heat flow and its consequences in the case of an arc welded joint.

9.102 Describe the effect of various welding parameters that control the metallurgical structure of the heat affected zone in C50 steel.

9.103 Describe the major forms of distortions that are likely to occur in an arc welded butt joints.

9.104 Describe the major forms of distortions that are likely to occur in an arc welded tee joints.

9.105 Explain the effect of welding process parameters on the extent and nature of distortions found.

9.106 What is the effect of restraining a joint? Explain how it could be advantageously utilised in minimising the distortions.

9.107 What are the methods available for controlling the distortions in welded assembly structures? Describe their relative effects and applications.

9.108 What are the defects that are generally found in welding? Describe their cause and remedies.

9.109 Briefly explain the weld testing methods followed for welding quality control.

Problems

9.1 In a given arc welding operation, the power source is at 20 V and current at 300 A. If the electrode travel speed is 6 mm/s, calculate the cross-sectional area of the joint. The heat transfer efficiency is taken as 0.80 and melting efficiency as 0.30. Heat required to melt the steel is 10 J/mm^3.

9.2 In a resistance welding of a lap joint or two mild steel sheets of 1.5 mm thick, a current of 10 000 A is passed for a period of 0.1 seconds. The effective resistance of the joint is 120 micro ohms. Density of steel is 0.00786 g/mm^3 and heat required to melt is 1381 J/g. The joint can be considered as a cylinder of 5 mm diameter and 2.25 mm in height. Calculate the percentage of heat distributed to the surroundings.

9.3 A projection welding of 3 spots is obtained with a current of 35 000 A for a period of 0.01 second. If the effective resistance of the joint is taken as 120 micro ohms. If the joint can be considered as a cylinder of 6 mm diameter and 2.5 mm in height, calculate the heat distributed to the surroundings.

9.4 The arc welding of 6 mm steel plates in butt welding is being carried out at a welding speed of 9 mm/s. The welding transformer is set at 25 V and the current flowing is 300 A. Taking the arc efficiency as 0.85, calculate the temperature to which the plates should be preheated such that the critical cooling rate is kept below 6 °C/s at a temperature of 550°C for satisfactory metallurgical quality.

9.5 Determine the appropriate welding speed to be used to weld 8 mm C50 steel plates when the power source is at 30 V and current at 325 A. The arc efficiency is 0.80 while the maximum cooling rate allowed is 6°C/s at a temperature of 550°C. Possible welding speeds are 6, 7, 8 and 9 mm/s.

9.6 Calculate the size of the heat affected zone with the same parameters as in problem 9.4, and discuss the likely metallurgical structures in HAZ.

9.7 Calculate the size of the heat affected zone with the same parameters as in problem 9.5, and discuss the likely metallurgical structures in HAZ.

9.8 It is necessary to design the weldment using the Shielded Metal Arc Welding (SMAW) process for the following plates (i) 15 mm and (ii) 25 mm. The following parameters are obtained from previous experience:

Voltage = 25 V
Current = 250 A
Heat transfer efficiency = 0.85
Electrode travel speed = 8 mm/s
Cooling rate = 6°C/s
Temperature for cooling rate calculation = 550°C

Calculate the necessary preheating temperature of the base metal to reduce the residual stresses. Comment on the results.

Multiple Choice Questions

9.1 Welding is a
(a) Permanent joint
(b) Semi-permanent joint
(c) Temporary joint
(d) Temporary or Semi-permanent joint

9.2 Mechanical fastening is a
(a) Permanent joint
(b) Fusion joint
(c) Temporary joint
(d) Temporary or Semi-permanent joint

9.3 Soldering is a
(a) Permanent joint
(b) Fusion joint
(c) Leak proof joint
(d) Temporary or Semi-permanent joint

9.4 For very thick plates the following edge preparation is preferable
(a) Single V (b) Single U
(c) Double U (d) Double V

9.5 The fuel gas that has highest flame temperature (heat capability) in gas welding is
(a) Propylene (b) Propane
(c) Acetylene (d) Hydrogen

9.6 The flame type that is normally used to weld cast iron in oxy-acetylene welding method is
(a) Carburising flame
(b) Neutral flame
(c) Oxidising flame
(d) Natural flame

9.7 The flame type that is normally used to weld high carbon steels in oxy-acetylene welding method is
(a) Natural flame (b) Neutral flame
(c) Oxidising flame (d) Reducing flame

9.8 Filler metal is not required in the following type of welding process
(a) Oxy-acetylene welding
(b) Arc welding
(c) Resistance welding
(d) Oxy-hydrogen welding

9.9 Drag in oxy-acetylene cutting can be reduced by
(a) Increasing the travel speed of the torch
(b) Decreasing the travel speed of the torch
(c) Using a larger hole in the torch for the oxy-acetylene jet
(d) Using a larger hole in the torch for the oxygen jet

9.10 Straight polarity in arc welding is obtained with
(a) Alternating current electrode with electrode being positive
(b) Direct current electrode with electrode being positive

(c) Direct current electrode with electrode being negative
(d) Alternating current electrode with electrode being negative

9.11 Consumable electrode for manual metal arc welding of steel is made of
(a) Tungsten
(b) Steel
(c) Copper
(d) Cadmium copper

9.12 Electrode for gas metal arc welding of steel is made of
(a) Tungsten
(b) Steel
(c) Copper
(d) Cadmium copper

9.13 For welding thin work pieces of aluminium sheet the following welding process is preferred
(a) Manual Metal Arc Welding
(b) Gas Metal Arc Welding
(c) Gas Tungsten Arc Welding
(d) Submerged Arc Welding

9.14 Advantage of coated electrode for shielded metal arc welding process is
(a) Gives off inert gases and protect the molten weld pool
(b) Provides arc stabilizing compounds
(c) Provides flux to remove oxide from the weld joint
(d) All of the above

9.15 Arc blow in manual metal arc welding is caused by
(a) The use of AC welding with non-consumable electrodes
(b) The use of DC welding with consumable electrodes
(c) The use of AC welding with consumable electrodes
(d) None of the above

9.16 Argon is the most widely used inert gas in gas tungsten arc welding because
(a) It requires a lower arc voltage, allows for easier arc starting and provides a smooth arc action.
(b) It provides a short arc
(c) It is lighter compared to other inert gases
(d) It can withstand higher arc voltage

9.17 Filler rod used in welding mild steel plates using gas metal arc welding is made of
(a) Mild steel
(b) Low carbon steel
(c) Medium carbon steel
(d) No filler rod is used

9.18 For overhead welding of thin plates in gas metal arc welding, the preferred metal transfer should be
(a) Short circuit or dip transfer
(b) Globular transfer
(c) Spray transfer
(d) Pulsed spray transfer

9.19 The welding process preferred to weld ship hulls is
(a) Manual Metal Arc Welding
(b) Gas Metal Arc Welding
(c) Submerged Arc Welding
(d) Flux Cored Arc Welding

9.20 In resistance spot welding, the heat required to melt the joint is mainly because of the resistance of the
(a) Electrodes
(b) Work piece joint
(c) Contact resistance of the electrode with work piece
(d) All of the above

Answers to MCQs

9.1 (a)	9.2 (d)	9.3 (c)	9.4 (c)	9.5 (c)
9.6 (a)	9.7 (d)	9.8 (c)	9.9 (b)	9.10 (c)
9.11 (b)	9.12 (b)	9.13 (c)	9.14 (d)	9.15 (b)
9.16 (a)	9.17 (d)	9.18 (d)	9.19 (c)	9.20 (d)

Case Study

METAL ARC WELDING – SKIN CANCER

It is well known that exposure to UV radiation from the Sun can lead to skin cancer. However, it is not limited to the radiation from the Sun alone and can be caused due to extended exposure to any UV source. Arc welding produces strong UV radiation that can be classified as UVA (400–315 nm), UVB (315–280) and UVC (280–100 nm). UVA penetrates deeper into the skin but does not cause DNA damage, UVA leads to maximum DNA damage and UVC is usually absorbed into the atmosphere but is equally dangerous as UVB.

Welders are constantly exposed to direct radiation and it has been seen that personal safety equipment are not effective in stopping the UV radiation from reaching the skin. This case study presents the results of a research that was carried out to see if metal arc welding can cause different skin cancers like basal cell carcinoma (BCC), squamous cell carcinoma (SCC), actinic keratosis (AK) and cutaneous malignant melanoma (CMM). The study was carried out with over 4000 participating male welders for a period of one year, selected from a welder population of over 10,000. The study was carried out at a Danish welding company. Health histories were gathered before the study to rule out any pre-existing conditions.

A high risk for BCC and AK located at the neck was found among the welders where the exposure is high compared to other locations. This increased with number of years of welding. Other areas of exposure like face and head were not at high risk. Risk of other forms of cancer did not increase with number of years in welding. Interestingly, the risk increased over 130% in case of workers who were working outdoors, possibly due to the combined effect of radiation from the welding arc and Sun. Also workers who were paid more had better chances to go on vacations leading to higher exposure to Sun and hence higher chances of skin cancer. In addition, it was noted that higher socio-economic group had more opportunity to have regular medical check-ups and hence earlier and better detection of skin cancer.

Hence, metal arc welding has significant effect on skin cancer around the neck and the occurrences increase with person's solar exposure. It has been recommended to use safety industrial clothing that cover the neck and apply sunscreen to prevent the exposure, decreasing the chances of getting skin cancer. The study provided no evidence for skin cancer at other locations due to metal arc welding.

It may be remembered that skin cancer is not the only occupational health hazard posed by welding. IARC (International Agency for Research on Cancer) identifies ocular melanoma (a cancerous condition developed in and around eye in the cells that produces pigment) as a major health issue in welders.

CHAPTER 10

Other Fabrication Processes

Objectives

After completing this chapter, the reader will be able to

- Understand the different types of welding processes used for special fabrication applications
- Learn brazing and soldering applications

10.1 THERMIT WELDING

Thermit Welding (TW) is a process which was traditionally used for the welding of very thick plates. Though this was used for welding large sections such as locomotive rails, ship hulls and broken large castings, its use has decreased to a great extent in the present time, because of the availability of other simpler methods such as submerged arc welding.

The heat source utilised for fusion in thermit welding is the exothermic reaction of the thermit mixtures. A typical thermit mixture for welding steels is aluminium and iron oxide. When the intimately mixed thermit powder is brought to its ignition temperature of 1200°C, the thermit reaction starts. Aluminium has greater affinity towards oxygen, and as a result, it reduces the ferric oxide to liberate iron and in the process, releases heat. The following reaction takes place.

$$3\ Fe_3O_4 + 8\ Al \rightarrow 9\ Fe + 4\ Al_2O_3 + 3.01\ \text{MJ/mol}$$

The temperature reached is of the order of 3000ºC. The enormous amount of heat liberated melts both the iron and aluminium oxide to a very fluid state. Because of the large differential in the densities, aluminium oxide would be floating on the top with the molten steel settling below. Once started, the reaction continues till all the thermit mixture in the reaction vessel or ladle is completely reduced.

Though the theoretical temperature is 3088°C, in practice, the heat losses from the crucible make it about 2500°C. The addition of fluxes or other mild steel and alloying additions reduce the temperature still below, to ultimately a super heat of the order of 400 to 550°C.

The heat released during the operation is of the order of 35 kJ/kg of thermit mixture. The complete reaction takes place in a total of 1 min, irrespective of the amount of thermit mixture present in the crucible. For example, about 1 or 2 tons of molten steel with a super heat of about 500°C can be generated in less than a minute. Though such a large amount of heat is generated, the safety of the thermit mixture is assured because the reaction can only start when the mixture is brought to its ignition temperature.

There are a large number of other thermit mixtures available for welding different materials. For example, for welding of power cables made of copper the following thermit mixture is used.

$$3\,CuO + 2\,Al \rightarrow 3\,Cu + Al_2O_3 + 1.152\ MJ/mol\ (4866°C)$$

$$3\,Cu_2O + 2\,Al \rightarrow 6\,Cu + Al_2O_3 + 1.089\ MJ/mol\ (3138°C)$$

The thermit welding process is essentially a casting process where the molten metal obtained by the thermit reaction is poured into the refractory cavity made around the joint. The two pieces to be joined are properly cleaned and the edge is prepared. Then wax is poured into the joint so that a pattern is formed where the weld is to be obtained. A moulding flask is kept around the joint and sand is rammed carefully around the wax pattern, as shown in Fig. 10.1 providing the necessary pouring basin, sprue and risers. A bottom opening is provided to run off the molten wax. The wax is melted through this opening which is also used to preheat the joint and make it ready for welding.

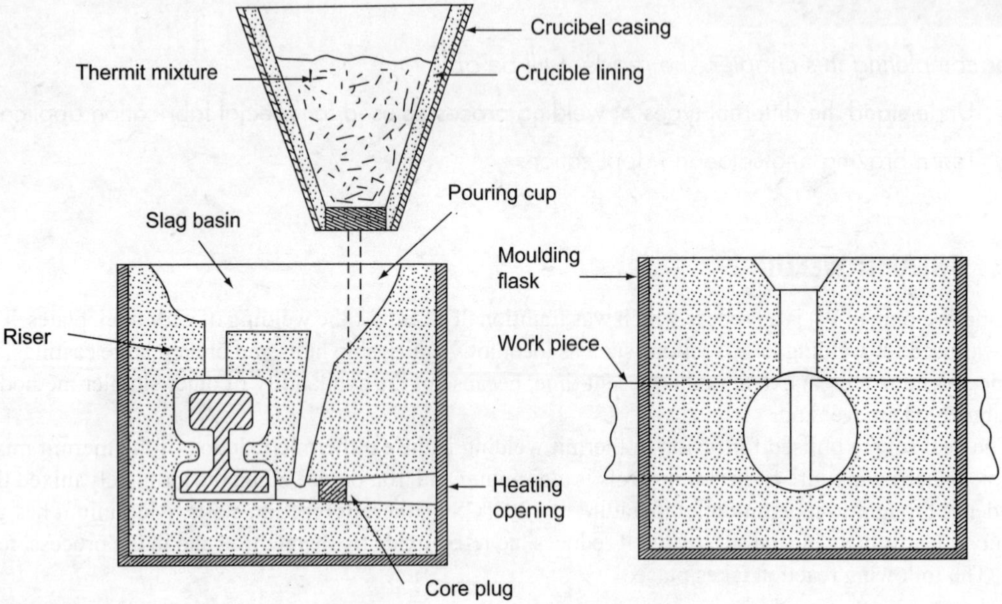

Fig. 10.1 *Set-up for thermit welding*

The thermit mixture which is mixed with fluxes is filled into a ladle through a bottom opening. The opening is initially closed. The igniting mixture which is normally barium peroxide or magnesium is placed at the top of the thermit mixture. The igniting mixture is lighted by means of a heated metal rod, whereby the complete reaction takes place and molten metal is produced. The bottom plug of the ladle is opened and the metal is allowed to flow into the mould prepared. The weld joint is allowed to cool slowly.

Besides making a fast weld, thermit welding also provides a reasonably strong weld. The strength of thermit welded joint reaches that of a forged metal without any defects.

10.2 ELECTRO-SLAG WELDING

The Electro-Slag Welding (ESW) process is developed essentially to weld very large plates without any edge preparation. This is essentially a single-pass process using a consumable electrode for filling the gap between the two heavy plates. The heat required for melting the plates and the electrode is obtained initially by means of an arc, so that flux will form the molten slag. Once the molten slag is formed, the arc is extinguished and the heat of welding is obtained by the resistance heating of the slag itself.

The typical electro-slag welding set-up is shown in Fig. 10.2. For effective welding, it is necessary to maintain a continuous slag pool and, therefore, the best way to maintain it is to weld vertically. The slag pool is contained in the groove with the help of two water-cooled copper dam plates which move along with the weld, as shown. The size and type of electrodes chosen for a given operation depends on the width of the joint. In Fig. 10.2, two electrodes for feeding through the feed rollers (not shown) into the weld zone are shown.

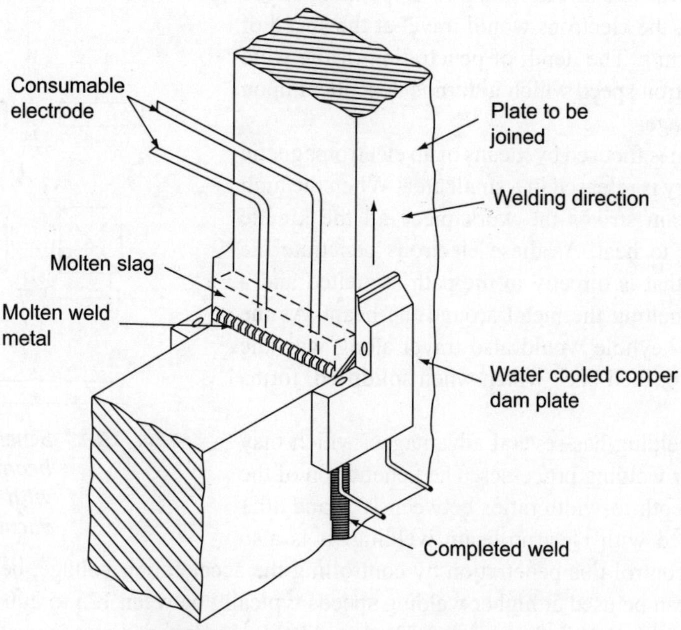

Fig. 10.2 *Electro-slag welding process*

The flux required to maintain a satisfactory amount of slag is fairly small, of the order of 0.2 to 0.3 kg per metre of weld length irrespective of the plate thickness. Thus, the heat utilised for melting the slag is much less. Most of the heat supplied in electro-slag welding, thus, melts the joint. By this process, a plate of 200 mm thickness can easily be welded in a single pass. Better welds can be done because of vertical welding as any gas present easily bubbles out through the slag. The heating and cooling of the edge is more gradual. The slag floating at the top would be preheating the joint before the actual melting by the heat liberated from the electrode. Whatever be the thickness of the plate, no edge preparation is required.

Electro-slag welding is useful for welding very thick plates; the work pieces whose thicknesses are less than 50 mm may not be economical for welding by this process unlike submerged arc welding.

Some typical applications of this welding process are in the fabrication of high-pressure vessels, frames of heavy mechanical and hydraulic presses, rolling mill frames, ship hulls, locomotive frames, etc.

10.3 ELECTRON BEAM WELDING

The heat source in Electron-Beam Welding (EBW) for melting the joint is a focused beam of high velocity electrons. The electron beam upon impinging the work piece releases the necessary heat by converting its kinetic energy. A schematic sketch of electron-beam welding is shown in Fig. 10.3.

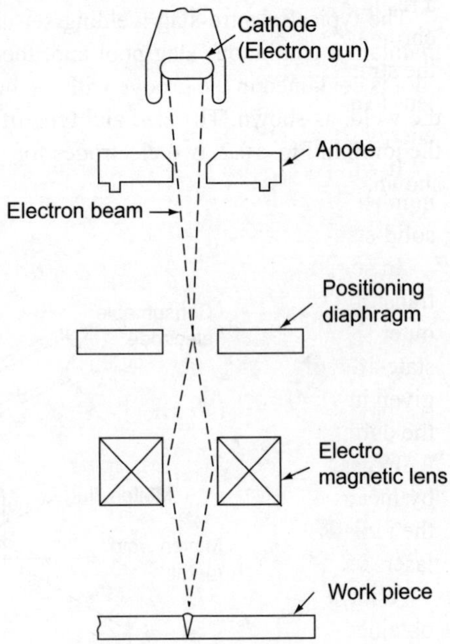

Fig. 10.3 *Schematic of electron-beam welding set-up with work piece in non vacuum conditions*

The cathode (heated filament) within the electron gun is the source of a stream of electrons. These electrons are accelerated towards the anode because of the large potential difference that exists between them. The potential differences that are used are of the order of 30 kV to 175 kV. The higher the potential difference, higher would be the acceleration. The current levels are low ranging, between 50 mA to 1000 mA. Depending on the accelerating voltage, the electrons would travel at the speed of 50,000 to 200,000 km/s. The depth of penetration of the weld depends on this electron speed which in turn is dependent upon the accelerating voltage.

The electron beam is focused by means of an electromagnetic lens so that the energy is released in a small area. When the high velocity electron beam strikes the work piece all the kinetic energy is converted to heat. As these electrons penetrate the metal, the material that is directly in the path is melted and a keyhole is formed melting the metal around the beam. As the beam traverses, the keyhole would also travel along with the molten metal being pushed back which when solidified, forms the joint.

Electron-beam welding has several advantages, which may not be found in other welding processes. The penetration of the beam is high. The depth-to-width ratios between 10:1 and 30:1 can be easily realised with electron-beam welding. It is also possible to closely control this penetration by controlling the accelerating voltage, beam current and beam focus. The process can be used at higher welding speeds typically between 125 to 200 mm/s. Filler metal or flux are not needed to be used in this process. The heat liberated is low and also is in a narrow zone, thus the heat affected zone is minimal as well as weld distortions are virtually eliminated.

It is possible to carry out the electron-beam welding with work pieces in open atmosphere (a pressure of 100 kPa), as shown in Fig. 10.3, in partial vacuum (0.13 to 13 Pa) or in a hard vacuum (0.13 to 133 MPa). For welding in vacuum, the work piece is enclosed in a vacuum created box. When an electron beam moves in the normal atmosphere, the electrons would be impinging with the gas molecules in the atmosphere and would thus be scattered. This scattering increases the spot size of the electron beam and consequently, there is lower penetration. Penetration increases as the vacuum increases due to the fact that increased vacuum decreases the scattering of the electron beam. The other advantage of using vacuum is that the weld metal is not contaminated.

The vacuum electron-beam welding methods have been well established with large amount of energy provided for. But the maintenance of a vacuum chamber with the work piece is an inconvenience. Therefore, the non-vacuum welding process has been finding general use, particularly in the automotive industry. The further development made in the process is introduction of computer numerical control for the purpose of controlling the weld geometry.

10.4 LASER-BEAM WELDING

The LASER (Light Amplification by Stimulated Emission of Radiation) is a concentrated beam of coherent monochromatic radiation. The normal white light consists of a number of colours and a number of waves. As a result, it is not possible to collimate ordinary light without losing its intensity. It is possible to use a monochromatic source as laser provided all the waves are of single phase (coherent). This is achieved by means of the stimulation. Because of the coherency, it is possible to concentrate the laser beam by means of an optical lens to a spot of any desired size without appreciably losing any of its intensity. Thus, the laser beam is a high energy source of heat to melt (even evaporate) the joint for fusion welding in Laser-Beam Welding (LBW).

It will be beyond the scope of this book to discuss the various details of lasers, which can be found in a number of modern physics books. In laser beam welding, two types of lasers are generally used. They are solid-state lasers and gas lasers.

In solid-state lasers, the light is emitted from a glass or a single crystal (such as ruby) that is doped with transition elements (such as chromium for ruby). When a normal white light impinges on the crystal, the outer shell electrons of the dope elements go to a higher energy metastable state. They return to the normal state after emitting the extra energy in the form of a photon. All the photons that are stimulated to emit at a given instant so that they will form a coherent radiation, which can be concentrated by optical lenses. Thus, the output would be normally in pulses. The power ratings of such units may be up to 2 kW.

In gas lasers, the gas (such as carbon dioxide) molecules are excited to the higher vibrational energy level by means of an electric discharge. The transition from this high energy level to the normal level generates the radiation, which is coherent and gets focused by means of the usual optical lenses. Continuous wave gas lasers using carbon dioxide gas with powers upto 20 kW are used for laser-beam welding.

With low power lasers typically less than 1 kW, the penetration would not be much and the weld is obtained by means of complete welding of the joint near the surface. But as the power is increased, the large heat density obtained would cause the metal at the centre of the laser jet to be vaporised with a key hole being formed similar to that of the electron-beam welding. Temperatures within this keyhole can reach as high as 25,000°C. This gives rise to a larger penetration. Instead of heat being conducted mainly downward from the surface, it is conducted radially outward from the keyhole, forming a molten region surrounding the vapour. As the laser beam moves along the work piece, the molten metal fills in behind the keyhole and solidifies to form the weld. This technique permits large welding speeds depending on laser size.

There are two types of lasers that are being used for welding operation: CO_2 and Nd:YAG. The Nd:YAG laser light is absorbed quite well by conductive materials, with a typical reflectance of about 20 to 30 percent for most metals. Using standard optics it is possible to achieve focused spot sizes as small as 0.025 mm diameter. However, the CO_2 laser has an initial reflectance of about 80 percent to 90 percent for most metals and requires special optics to focus the beam to a minimum spot size of 0.075 to 0.100 mm diameter. However, whereas Nd:YAG lasers might produce power outputs up to 500 watts, CO_2 systems can easily supply 10 kW and greater.

Metals with low boiling points produce a large amount of metal vapour, initiating the breakdown of gas and plasma generation in the region of high beam intensity just above the metal surface. This plasma by absorbing the laser energy can block the beam passes, forming bubbles at the root of the weld. If the viscosity is high, these bubbles do not escape before the molten metal solidifies. Although the melting point of metals does not have a significant effect on laser weldability, it must be reached during the initial absorption of energy. Thus, low-melting-point materials are easier to weld with a laser than high melting point metals.

Laser-beam welding can be used for obtaining deeper penetration welds even with dissimilar metals. They would give rise to very small heat affected zones. The laser-beam welding process would reduce the roughness of the welded surface. These can also be used for joining multilayer materials with differing thermal properties also.

The advantages of laser beam welding are:

- Faster welding rates
- Low distortion
- No flux or filler metal required

- Single-pass two-side welding
- Non-contact, eliminating any debris build-up
- Can reach otherwise inaccessible locations
- Shorter cycles and higher uptimes
- Welds magnetic materials

In all the welding methods that have been discussed so far, the metal in the joint is melted under the heat source and the fusion is obtained with or without the application of pressure. There are other classes of welding processes that are discussed here, where the metal is only brought into the plastic state without the actual melting and the joining is obtained by the application of large forces.

10.5 FORGE WELDING

This is the oldest of the metal-joining processes known to humankind. In this process the ends of the parts to be joined are heated to a temperature slightly below the solidus temperature and a pressure is applied so that a fusion joint is obtained. This is the familiar method used by village blacksmiths. The force can be applied in repeated blows manually or by a machine, or continuously by rotating rolls.

In forge welding of ferrous alloys, a thin coating of iron oxide is readily formed when the heated metal is exposed to atmospheric oxygen. Unless this oxide is removed from the weld zone, it would interfere with the fusion of the joint. Hence, it is necessary to apply fluxes so that the iron oxide becomes more fluid and under the pressure flows out of the joint, thus, aiding a proper weld. The fluxes that are normally used are,

- Borax ($Na_2B_4O_7 . 10H_2O$), and
- Silica sand.

Fine silica sand is cheaper than borax and may be used for sprinkling on low-carbon steels. Wrought iron and low-carbon steels are easily forge welded. The use of fluxes for these materials can be avoided because their oxides would melt at much lower temperatures than the forge-welded temperatures.

The grain size in the joint is likely to increase in size because of the heating. But the subsequent metal working done on the joint properly controls the grain size and gives rise to good mechanical properties.

The forge-welding operation is a very slow and labour-intensive process. The quality of the weld depends to a great extent on the skill of the welder. As a result, it is more expensive and, therefore, is hardly used in industrial practice.

10.6 FRICTION WELDING

The heat required for welding in this process is obtained by the friction between the ends of the two parts to be joined. One of the parts to be joined is rotated at a high speed around 3000 revolutions per minute, and the other part is axially aligned with the second one and pressed tightly against it, as shown in Fig. 10.4. The friction between the two parts raises the temperature of both the ends. Then the rotation of the part is stopped abruptly and the pressure on the fixed part is increased so that the joining takes place. This process is termed Friction Welding (FRW).

The machine for friction welding is similar to a centre lathe. Though a centre lathe could be used for smaller size jobs, the bigger ones require a special welding machine. This is necessary because in a lathe, the power available would not be sufficient. The power requirements of friction welding may be between 25 kVA to 175 kVA, which is far beyond that of the many general-purpose centre lathes. Further, the quick disengagement and the instantaneous braking of the rotating part would not be possible in a lathe.

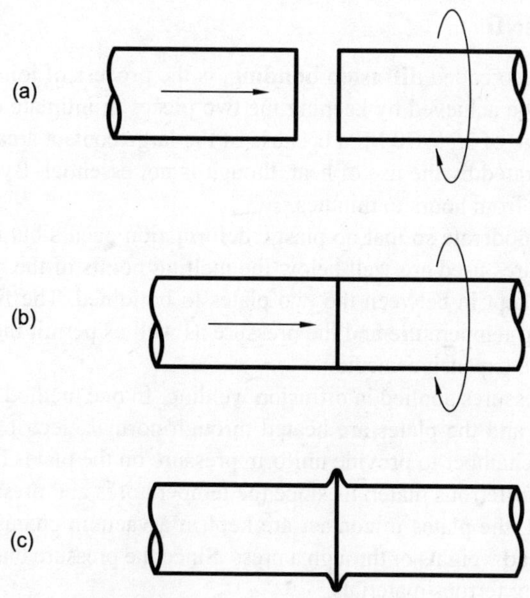

Fig. 10.4 *Friction-welding process*

The major parameters in friction welding are the rotational speed and the axial pressure applied. The axial pressure applied depends on the strength and hardness of the metals being joined. The pressure may range from 40 MPa for low-carbon steels to as high as 450 MPa for alloy steels. The rotational speed used may also change the requirement of the pressure. High speeds reduce the pressure required for joining. The rotational speeds may be of the order of 1500 to 3000 rpm. The other variable that needs to be closely controlled is the time of contact between the two parts. The total welding time that is taken in the friction welding is between 2 to 30 seconds.

The main advantage of friction welding is the ease with which the joining can take place. Edge cleaning is not a problem, since the oxides and contaminants present would easily be removed during the initial rubbing. Since the joining takes place by diffusion rather than by actual melting, even dissimilar metals can be joined. Also, the heat generated being small well below the melting temperature, there will be no distortion and warping. The only distortion apparent at the joint is the upset portion of the metal which can be removed by machining later. The quality of the weld achieved is very high and it is economical in operation. Any unskilled operator can work on a friction-welding machine since it is completely automatic in operation.

Because of the high quality of the weld obtained, friction welding has been widely accepted in the aerospace industry as well as the automobile industry for welding of critical parts.

Another variant of friction welding is **inertia welding**. In this process, the energy required for welding is stored in a flywheel, which is accelerated to the required speed. Once the flywheel has reached the operational speed, it is disengaged from the drive motor by the clutch and the axial force is applied on the rotational part by the fixed part. This gradually decreases the rotational speed of the flywheel which finally stops after some time and in the meantime the joint is heated to the forging temperature. Then the pressure on the joint is increased so that a weld is made.

10.7 DIFFUSION WELDING

Diffusion Welding (DFW), also called **diffusion bonding**, is the process of joining two parts purely by the diffusion. The diffusion can be achieved by keeping the two pieces in intimate contact under pressure. The pressures used are in the range of 35 to 70 MPa because of the large contact areas used. The diffusion being a rate process, can be accelerated by the use of heat, though is not essential. By the application of heat, the bonding time can be reduced from hours to minutes.

The pressures should be moderate so that no plastic deformation occurs but a complete intimate contact is maintained. The temperatures used are well below the melting points of the metals being joined. A filler metal is generally used and kept in between the two plates to be joined. The filler metal generally aids in the reduction of the operating temperature and the pressure as well as permit the use of normal atmosphere around the joint instead of the expensive vacuum.

There are two types of pressures applied in diffusion welding. In one method, the plates are kept together in contact inside a chamber and the plates are heated through normal electric resistance. During heating, inert gas is pumped into the chamber to provide uniform pressure on the plates being joined. This method is normally used for joining non-ferrous materials since the temperatures and pressures required are relatively small. In the second method, the plates in contact are kept in a vacuum chamber and the force is applied mechanically by means of dead weights or through a press. Since the pressure that can be applied is high, this method is suitable for welding ferrous materials.

When a filler material is used, it is expected that the temperature required for bonding is reduced by the formation of a eutectic alloy. Thus, the filler metal diffuses into the two metal plates and forms a eutectic joint. The filler metal thickness may be of the order of 5 to 25 microns.

Diffusion welding can be used for joining metals to metals as well as metals to non-metals. Because of diffusion being responsible for the joint, metallurgically the joint would be sound with no likely defects. The weld is very neat requiring no further processing on the joint. Because of the joint covering a large area, the process is very useful for making laminates of the exotic materials. Diffusion welding is more expensive and, therefore, can only be justified for closer tolerance work or for expensive materials.

10.8 EXPLOSION WELDING

In Explosion Welding (EXW), detonation of explosives is used to accelerate a part to move towards the other plate at a fast rate, so that the impact creates the joint as shown in Fig. 10.5. As the plate moves at high velocity and meets the other plate with a massive impact, very high stress waves (the order of thousands of MPa) created between the plates, which clears all the oxide and scales present in the interface and making a clean joint.

Generally, low-detonation-velocity (2.4 to 3.6 km/s) explosives are used for explosion welding. These produce sound welds without any shock effects on the plates joined. The detonation velocity changes with the thickness of the plate being welded. The thicker plates require higher velocities. High-velocity explosives are more difficult to use and are also likely to damage the work pieces causing spalling along the edges and fissuring at the interface.

Use of explosives is dangerous. Therefore, they can only be used after obtaining the necessary permission from the concerned government department. There are safe methods available for the use of explosives.

The largest application of explosion welding is the cladding of metals for the purpose of corrosion prevention. With explosion welding, very large plates can be cladded. The joining of dissimilar metals, such as titanium to steel, aluminium to steel, and aluminium to copper can be successfully carried out. Joining of tube to tube sheets of large heat exchangers has been successfully done using explosion welding.

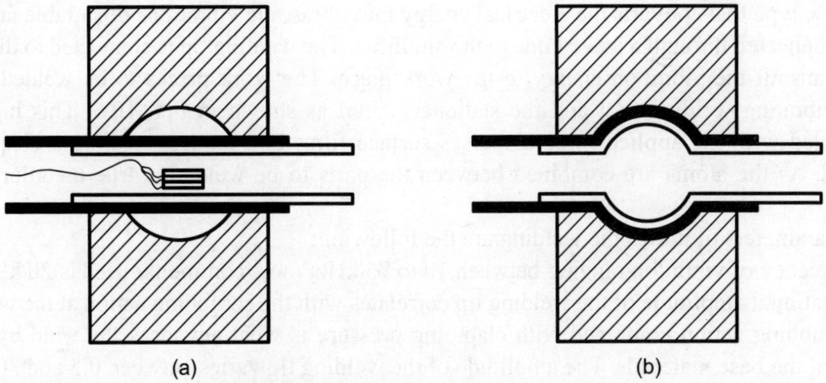

Fig. 10.5 *Explosion-welding process*

10.9 ULTRASONIC METAL WELDING

Ultrasonic Metal Welding (USW) is a solid state welding process that produces coalescence through the simultaneous application of localised high frequency vibratory energy. Ultrasonic metal welding can be applied for joining both similar and dissimilar metal work pieces or welding a metal to a metallised substrate, e.g. ceramics or glass. The coalescence is caused by the application of external pressure and energy from the ultrasonic vibrations. In an ultrasonic welding machine, ultrasonic vibration is generated by one or more transducers that convert the high frequency electrical energy generated by the power supply into mechanical vibratory energy of the same frequency. This vibration is transmitted into the work pieces with tip vibration in a plane parallel to the plane of the weld interface.

A typical ultrasonic metal welding equipment is shown schematically in Fig. 10.6. It consists of an electrical power supply unit with a sine-wave generator, a timer and a frequency-adjusting unit. The electrical power from this unit is input to the electromechanical transducer which can be either a magnetostrictive or

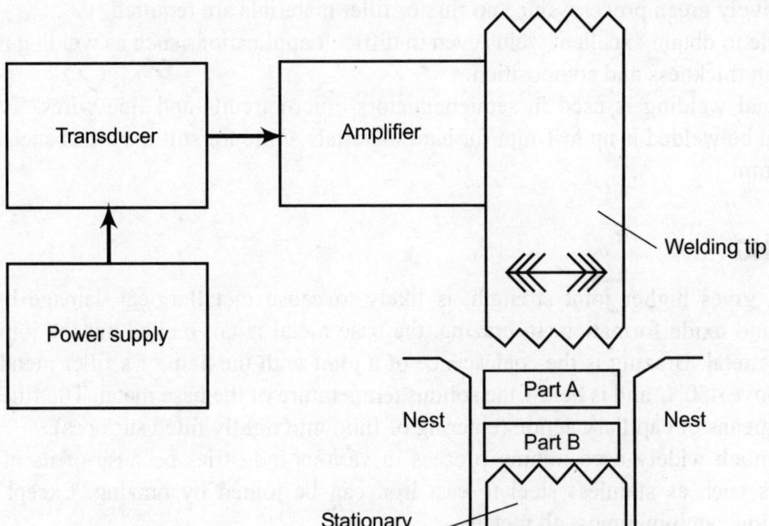

Fig. 10.6 *Typical arrangement of ultrasonic metal welding*

the piezoelectric type that converts the electrical energy into ultrasonic vibrations of suitable amplitude. The transducer is connected through a waveguide to the amplifier. The welding tip is connected to the waveguide/amplifier to transmit the vibration energy to the work piece. The work pieces to be welded are clamped between the vibrating welding tool and the stationary anvil as shown in Fig. 10.6. This high frequency vibration coupled with the applied force disperses surface films and oxides, creating a clean surface for diffusion weld. As the atoms are combined between the parts to be welded, a true metallurgical bond is produced.

The main parameters of ultrasonic welding are the following:

- The frequency of vibration ranging between 10 to 75 kHz; most commonly used is 20 kHz.
- The vibrational amplitude of the welding tip correlates with the scrubbing action at the weld interface. This scrubbing action combined with clamping pressure is what advances the weld by diffusing or mixing of the base materials. The amplitude of the welding tip varies between 0.5 and 30 μm.
- The weld time depends on the energy required for joining. Its interaction with the amplitude and clamping force provide the requisite energy for joining. The duration of the welding operation ranges from 10 ms to several seconds.
- The clamping pressure depends on the hardness of the work material and would generally be lower than about 0.30 times its hardness.

Advantages

- Ultrasonic metal welding provides excellent electrical, mechanical and thermal connections between similar and dissimilar metals. This process is ideal for bonding conductive materials such as copper, aluminium, brass, gold and silver.
- It requires relatively less energy to join compared to other processes. A typical weld uses under 2 kW and is completed in less than one-half second.
- The temperature rise during the welding process is low and as a result, the metallurgical structure of the weldment is not affected.
- Welding tip is not affected during the welding operation and requires no maintenance. It, therefore, lasts for several hundred thousand cycles. Consequently, the tooling cost is very low.
- It is a relatively green process, since no flux or filler materials are required.
- It is possible to obtain excellent welds even in difficult applications such as welding materials that are dissimilar in thickness and composition.

Ultrasonic metal welding is used in semiconductors, microcircuits and fine wires. Typical maximum thickness that can be welded is up to 1 mm for hard materials while for soft materials such as aluminium, it can go up to 2.5 mm.

10.10 BRAZING

Welding, though gives higher joint strength, is likely to cause metallurgical damage by way of phase transformations and oxide formation. In brazing, the base metal is not melted but the joint is obtained by means of a filler metal. Brazing is the coalescence of a joint with the help of a filler metal whose liquidus temperature is above 450°C and is below the solidus temperature of the base metal. The filler metal is drawn into the joint by means of capillary action (entering of fluid into tightly fitted surfaces).

Brazing is a much widely used joining process in various industries because of its many advantages. Dissimilar metals such as stainless steel to cast iron can be joined by brazing. Except aluminium and magnesium, brazing can join almost all metals.

Because of the lower temperatures used, there is less distortion in brazed joints. Also, in many cases, the original heat treatment of the plates being joined is not affected by the brazing heat. The joint can be quickly finished without much skill. Because of the simplicity of the process, it is often an economical joining method with reasonable joint strength. The brazed joints are reasonably stronger, depending on the strength of the filler metal used.

But the brazed joint is generally not useful for high temperature service because of the low melting temperature of the filler metal. The colour of the filler metal in the brazed joint also may not match with that of the base metal.

Because the filler metal reaches the joint by capillary action, it is essential that the joint is designed properly. The clearance between the two parts to be joined should be critically controlled. Another important factor to be considered is the temperature at which the filler metal is entering the joint. While designing a brazed joint, care is to be taken to see that the differences in the coefficients of thermal expansion of the two pieces to be joined are properly accounted for. If there is too much of clearance, the capillary forces may not be sufficient to draw the filler metal into the joint, whereas insufficient clearance may have too small an amount of filler metal to give rise to any effective strength. The main types of brazing joints used are shown in Fig. 10.7.

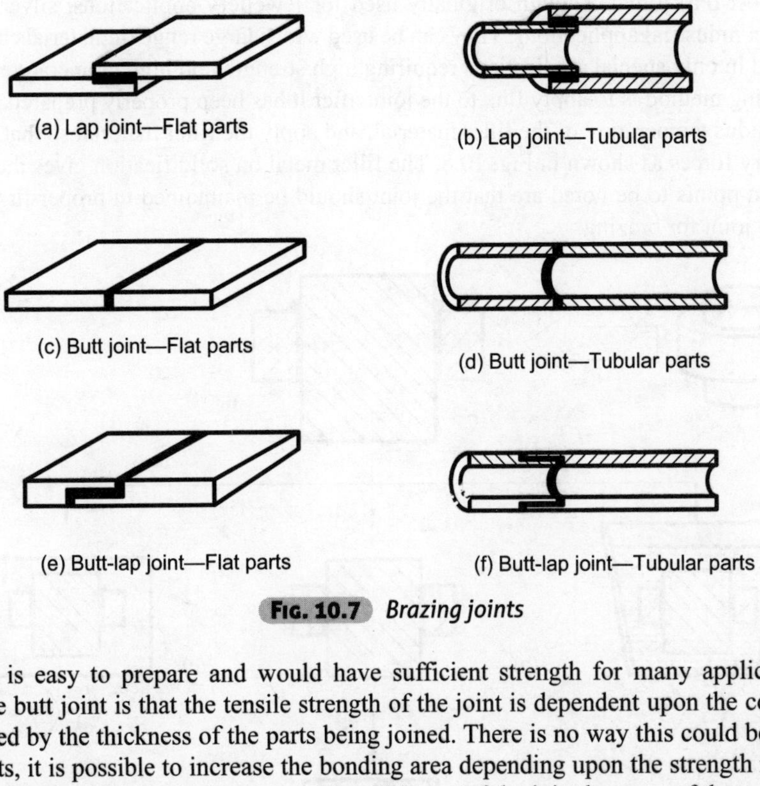

(a) Lap joint—Flat parts

(b) Lap joint—Tubular parts

(c) Butt joint—Flat parts

(d) Butt joint—Tubular parts

(e) Butt-lap joint—Flat parts

(f) Butt-lap joint—Tubular parts

Fig. 10.7 *Brazing joints*

A butt joint is easy to prepare and would have sufficient strength for many applications. The main limitation of the butt joint is that the tensile strength of the joint is dependent upon the contact area, which in turn is dictated by the thickness of the parts being joined. There is no way this could be increased. In the case of lap joints, it is possible to increase the bonding area depending upon the strength requirements. The main problem of the lap joint is the increase in the thickness of the joint because of the overlap of the parts. This can be used where it is not objectionable. To combine the advantages of the two types, the butt-lap joint is used. Though it requires higher joint preparation effort, it can be compensated by increased strength of the joint along with a uniform thickness for a given application.

In brazing, joints need to be extremely clean. Any grease or oil present in the joint prevents the flow of filler metal. Hence, the joint should be thoroughly cleaned using proper solvents. Oxides and scales present are removed by acid pickling. Fluxes are added into the brazed joint to remove any of the oxides present or prevent the formation of the oxides so that the base metal and the filler metal remain pure during the joining. The fluxes generally used are combinations of borax, boric acid, chlorides, fluorides and tetraborates, and other wetting agents. The fluxes used for ferrous materials are mixtures of borax and boric acid in a paste form. A popular composition is 75% borax and 25% boric acid. Alkaline biflourides are used for brazing of stainless steel, aluminium or beryllium copper alloys. A special flux containing sodium cyanide is used in brazing tungsten to copper. The fumes from all these fluxes are harmful and therefore proper ventilation and care in use are essential.

Enough flux should be applied at the joint so that it lasts throughout the brazing session. The method of application of the flux is by spraying, brushing or with the help of a pressurised applicator.

Depending on the type of base metals brazed, a number of filler metals are available. The joint obtained in brazing is by means of the diffusion of the filler metal into the base material, associated with the surface alloying. Copper based materials are generally used for brazing ferrous materials. Copper alloys having high zinc (70%) content are not extensively used because of their brittleness, but those with less zinc content are used for brazing of various steels. Aluminium-silicon filler materials are used for brazing aluminium.

Silver brazing makes use of a silver based filler metal. Silver brazing is used to give high strength (tensile strength up to 900 MPa) joints. Though originally used for jewellery applications, silver brazing is now extensively used in industrial applications. They can be used with a large range of materials but because of its high cost it is used in only special applications requiring high strength and high temperature service.

A general brazing method is to apply flux to the joint after it has been properly prepared, heat the joint to bring it to the liquidus temperature of the filler material, and apply the filler material so that it flows into the crevice by capillary forces as shown in Fig. 10.8. The filler metal on solidification gives the necessary joint strength. The main points to be noted are that the joint should be maintained in proper fit with cleanliness before heating the joint for brazing.

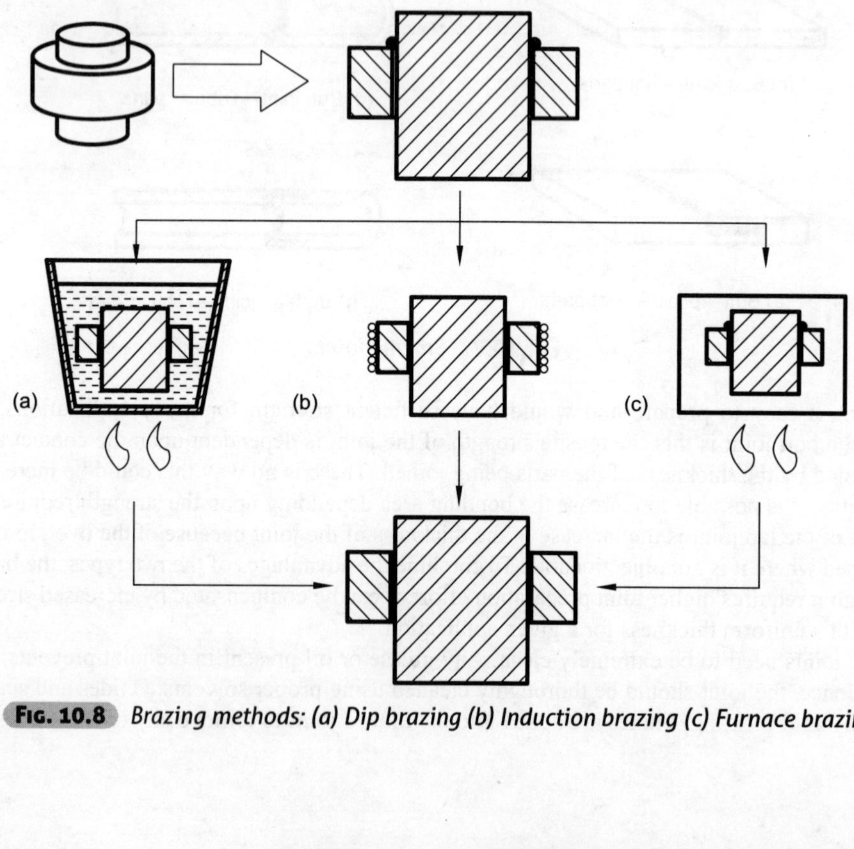

FIG. 10.8 *Brazing methods: (a) Dip brazing (b) Induction brazing (c) Furnace brazing*

Heat sources that are used for brazing are a molten bath of brazing filler metal, oxy-acetylene torch, controlled atmosphere furnace, electrical resistance heating and induction heating. In torch brazing, a reducing flame (oxy-acetylene or propane) is generally used to inhibit oxidation. Torch brazing is generally a manual operation with the operator having considerable skill in judging the operating conditions of the brazing. It is also possible to mechanize torch brazing by employing multiple torches with a conveyor line that can bring the brazing joints to the torches.

In dip brazing, a molten salt bath or metal bath is used to supply heat to the base metal as shown in Fig. 10.8(a). A brazing joint is prepared and dipped into the bath for quick melting of the brazing alloy. When the assembly is taken out of the bath, the joint cools and joint strength is obtained. It is a faster method for preparing brazing joints. In induction brazing Fig. 10.8(b), heat is obtained from the high frequency current induced in the work by means of the electrical coils. Electrical resistance heating set up is similar to the electrical resistance welding, except that brazing requires less power.

In furnace brazing, a furnace with a controlled atmosphere, which is either neutral or reducing, is used. The assemblies to be joined enter the furnace on a conveyor and will pass through the various heating and cooling portions of the furnace to make the joint.

10.11 BRAZE WELDING

Braze welding is similar to brazing, where the joint is obtained by means of a filler metal, whose liquidus temperature is above 450°C and below the solidus temperature of the base metal. But in braze welding the filler metal reaches the joint without the capillary action since the joint gap is more. As a result, the types of joints that can be used in braze welding are much more varied.

The joint design for braze welding is similar to that used for oxy-acetylene gas welding. Thick layers of filler metals are deposited in braze welding since filler metals enter the joint by gravity. The cleaning and flux requirement for braze welding is similar to that of brazing.

Typical filler metals used in braze welding are brasses with zinc content up to 40%. Here again, the joint is obtained by diffusion of the filler metal into the base metal and by surface alloy formation. These are responsible for good strength achieved by braze welding. Silver based alloys are not used in braze welding because of the large amounts of filler metals required which would increase the cost of the joint.

10.12 SOLDERING

Soldering is a method of joining similar or dissimilar metals by means of a filler metal whose liquidus temperature is below 450°C.

Though soldering obtains a good joint between the two plates, the strength of the joint is limited by the strength of the filler metal used. Soldering is normally used for obtaining a neat leak proof joint or a low resistance electrical joint. The soldered joints are not suitable for high temperature service because of the low melting temperatures of the filler metals used.

The joint design (Fig. 10.9) used for soldering is similar to that of brazing since the filler metal enters the soldered joint by capillary action. A soldered joint is weaker compared to that of the brazed joint. As a result, other means of mechanical fastening is used in addition to soldering to provide the necessary joint strength.

The soldering joints also need to be cleaned meticulously to provide chemically clean surfaces to obtain a proper bond. Solvent cleaning, acid pickling and even mechanical cleaning are applied before soldering.

To remove the oxides from joint surfaces and to prevent the filler metal from oxidising, fluxes are generally used in soldering. Rosin and rosin plus alcohol based fluxes are least active type and are generally used for electrical soldering work. Because of the content of acids, these are corrosive at soldering temperature. They can be easily cleaned after the soldering.

The organic fluxes such as zinc chloride and ammonium chloride are quick acting and produce efficient joints. But because of their corrosive nature, the joint should be thoroughly cleaned of the entire flux residue from the joint. These are to be used for only non-electrical soldering work.

Fluxes are normally available in the form of powder, paste, liquid or in the form of core in the solder metal. It is necessary that the flux should remain in the liquid form at the soldering temperature and be reactive to be of proper use.

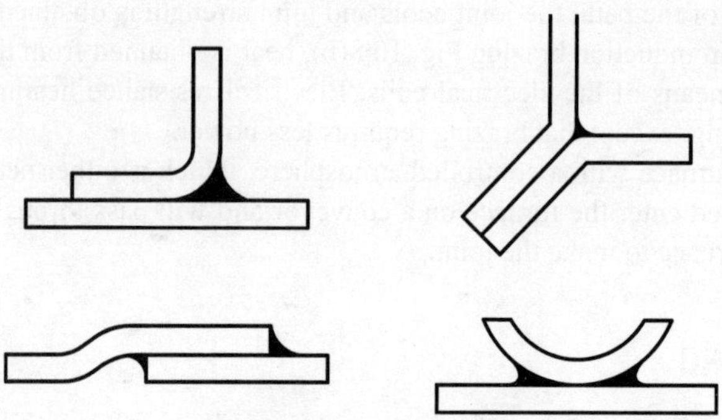

Fig. 10.9 *Soldering joints*

The filler metals used are normally called as solders which are essentially alloys of lead and tin. The alloys of lead and tin have lower liquidus temperatures as shown in Fig. 2.6. The eutectic alloy (62% Tin + 38% Lead) has the lowest at 183°C. Some of the soldering filler metals that are used are presented in Table 10.1 along with their solidus and liquidus temperatures. To improve the mechanical properties and temperature resistance, solders are added to other alloying elements such as zinc, cadmium and silver in various proportions as shown in Table 10.1.

Table 10.1 **Composition and melting temperatures of some filler metals used in soldering.**

S. No.	Weight Percentage of Composition					Temperature	
	Tin	Lead	Silver	Cadmium	Zinc	Solidus °C	Liquidus °C
1	10	90	—	—	—	268	301
2	25	75	—	—	—	183	267
3	40	60	—	—	—	183	235
4	62	38	—	—	—	183	183
5	70	30	—	—	—	183	192
6	96	—	4	—	—	221	221
7	62	36	2	—	—	180	190
8	2.5	97	0.5	—	—	303	310
9	91	—	—	—	9	199	199
10	70	—	—	—	30	199	311
11	—	—	5	95	—	338	393
12	—	—	—	40	60	265	335
13	—	—	—	10	70	265	399
14	—	—	—	82.5	17.5	265	265

The most commonly used soldering methods are,

- With soldering iron (flame or electrically heated)
- Dip soldering
- Wave soldering

A soldering iron is a copper rod with a thin tip which can be used for flattening the soldering material. The soldering iron can be heated by keeping in a furnace or by means of an internal electrical resistance whose power rating may range from 15 W for the electronic applications to 200 W for sheet metal joining. This is the most convenient method of soldering but somewhat slower compared to the other methods.

In dip soldering, a large amount of solder is melted in a tank which is closed. The parts that are to be soldered are first cleaned properly and dipped in a flux bath as per the requirement. These are then dipped into the molten solder pool and lifted with the soldering complete. Wave soldering is a variant of this method wherein the part to be soldered (e.g., an electronic printed circuit board, PCB) is not dipped into the solder tank, but a wave is generated in the tank so that the solder comes up and makes the necessary joint. This is normally a continuous process with the PCB's being continuously moving on top of the solder tank and the waves become continuously generated. It is used for mass production of the electronic equipment.

There are also other methods available for soldering, such as torch soldering, oven soldering, resistance soldering, induction soldering and infrared soldering.

SUMMARY

There are a number of special welding processes that are used for unique applications that cannot be handled by the conventional welding processes such as arc welding.

- Thermit welding utilises a thermit mixture to quickly melt a large volume of metal to be poured into the welding joint.
- Electro-slag welding is used to weld large plates by the heat generated from the electric heating of the slag produced in the weld joint.
- Electron-beam and laser-beam welding are used to weld large-aspect ratio parts using the heating produced by the electron and laser beams respectively.
- Solid state welding ensures that the weld is created without the direct addition of heat for melting the metal.
- Brazing produces a joint without actually melting the parent metal. This ensures less distortion while maintaining good strength.
- Soldering is used mainly to provide continuity and leak proof joints.

Questions

10.1 For welding heavy rail sections, thermit welding is often used. Explain how the heat necessary for the joining process is obtained.

10.2 How can the shape of the joint be effectively controlled in a thermit welding process?

10.3 What are the specific advantages and applications of thermit welding?

10.4 What are the specific situations in which electro-slag welding is used?

10.5 Describe the electro-slag welding process.

10.6 Compare electro-slag welding process with that of submerged-arc welding from the standpoint of heat liberated, joint preparation and welding position.
10.7 Describe the electron-beam welding process.
10.8 Explain how the atmosphere around the work piece affects the weld obtained in electron-beam welding.
10.9 What do you understand by laser-beam welding?
10.10 Explain various types of lasers used in the laser-beam welding process.
10.11 Write a short note on laser-beam welding detailing the applications.
10.12 What are the advantages claimed of friction welding?
10.13 Describe briefly about forge welding.
10.14 Explain the reasons why forge-welding process is not commercially used.
10.15 Explain the process of friction welding, giving the applications.
10.16 Is it possible to use a centre lathe for friction welding? Support your answer with reasons.
10.17 What applications would require diffusion welding?
10.18 Explain the process of diffusion welding.
10.19 Write a brief note on explosion welding.
10.20 How is brazing different from welding?
10.21 Why is brazing more extensively used in industrial practice?
10.22 What filler metals are generally used in brazing?
10.23 Explain what you understand by silver soldering or silver brazing.
10.24 What is the requirement of fluxes in brazing? Give details of some of the fluxes used in brazing with their applications.
10.25 Explain briefly how brazing is carried out.
10.26 What do you understand by braze welding?
10.27 Compare brazing and braze welding.
10.28 What are the typical applications of soldering?
10.29 Distinguish between brazing and soldering from the point of view of the filler metals used, applications and the strength of the joint obtained.
10.30 What fluxes are generally used in soldering?
10.31 Give the typical compositions of the filler metals used in soldering with their applications.

Multiple Choice Questions

10.1 Thermit welding is used for welding
(a) thin plates in a single pass
(b) very thick plates in a single pass
(c) very thick plates in two passes
(d) very thick plates in multiple passes

10.2 Thermit welding utilises the heat liberated from the
(a) chemical reaction between two compounds
(b) burning of a solid fuel along with the filler material in the joint
(c) burning of a liquid fuel along with the filler material in the joint
(d) electric induction of the joint

10.3 Thermit mixtures used in thermit welding normally use the following metal:
(a) Manganese (b) Magnesium
(c) Aluminium (d) Alum

10.4 Electro-slag welding utilises the heat liberated from the
(a) chemical reaction between two compounds
(b) resistance heating of the slag in the joint
(c) electric arc between the electrode and the filler material in the joint
(d) electric induction of the joint

10.5 Electro-slag welding is used for welding
(a) rolling mill frames
(b) ship hulls
(c) frames of heavy mechanical and hydraulic presses
(d) all of the above

10.6 Electron-beam welding is used for welding
(a) high depth-to-width ratios
(b) low depth-to-width ratios
(c) depth-to-width ratios equal to one
(d) with large heat affected zone due to large heat generated

10.7 Filler metal or flux is not required for the following process:
(a) Thermit welding
(b) Electron-beam welding
(c) Electro-slag welding
(d) Arc welding

10.8 Welding process that requires no direct heat application in the joint:
(a) Thermit welding
(b) Electron-beam welding
(c) Friction welding
(d) Laser-beam welding

10.9 Welding process that requires no direct heat application in the joint:
(a) Thermit welding
(b) Diffusion bonding
(c) Friction welding
(d) Laser-beam welding

10.10 Filler metal or flux is not required for the following process:
(a) Thermit welding
(b) Laser-beam welding
(c) Electro-slag welding
(d) Arc welding

Answers to MCQs

10.1 (b)	10.2 (a)	10.3 (c)	10.4 (b)	10.5 (d)
10.6 (a)	10.7 (b)	10.8 (c)	10.9 (b)	10.10 (b)

CHAPTER 11

Powder Metallurgy

Objectives

After completing this chapter, the reader will be able to

- Understand the basics of powder metallurgy process
- Learn about various metal powder production methods
- Know the different steps involved in powder metallurgy part preparation
- Understand the advantages and limitations of powder metallurgy
- Learn other developments in the compaction process.
- Design parts for powder metallurgy process

11.1 INTRODUCTION

Powder metallurgy is the name given to a process in which metallic powders are heated below their melting temperatures to achieve the bonding. In reality the powder metallurgy (P/M) process involves compacting of metal or alloy powders into the desired shape after blending and then heated in a controlled atmosphere at a temperature below the melting point in order to achieve the bonding of the particles to get the desired properties. The powder metallurgy process enables to produce parts in their final shape eliminating the need for any additional machining. Raw material is not wasted during the processing while unusual materials or mixtures can be utilized. It is possible to get parts with unique properties not possible by any other manufacturing process. Most of the powder metallurgy parts are in the size range of less than 2 kg, though parts as large as 20 kg were made. Large parts require very expensive tooling and as such are not widely made by powder metallurgy.

The limit to the size of a PM component is based on the final density and the available press size. Typical values are the projected area (area perpendicular to the pressing direction) usually between 4 and 16,000 mm^2, with length between 0.8 to 150 mm, although 75 mm is the practical maximum. There are parts weighing above 10 kg in production today.

Though powder metallurgy was appeared to have been used by Egyptians around 3000 BC, the modern developments have started in the mid or late nineteenth century. The early interest was in the self lubricated bearings. Later the invention of incandescent light required the filament to be made by P/M process. The invention of tungsten carbide in 1920's utilized the P/M process. Later the automobile industry had utilized the P/M process to a great extent and even today it accounts for a large volume of its usage. Many of the enhancements in various powder metallurgy technologies have taken place in the last 70 years.

The powder metallurgy process requires a number of processes that need to be carried out as shown in Fig. 11.1. It first starts with the production of metal or alloy powders of the requisite size and distribution. The metal or alloy powders need to be blended with suitable additives and lubricants. The thorough blending of powders and additives ensures that the additives are uniformly distributed which would facilitate the compaction process later. This blended powder is placed in the die and then pressed or compacted by a punch. There are a number of compacting methods that are used and can be broadly classified as cold, warm and hot compacting. In each of these varieties there are more processes that are possible such as die compacting, pressure-less sintering, isostatic extrusion, injection moulding, rolling, slip casting, cold forming, etc., some of which will be discussed later. After compacting, the material is termed as 'green compact'. This has the overall shape of the part required but does not have enough strength as a working part since the powders are not bonded together. To achieve the bonding, the green compact is kept in a furnace with the requisite atmosphere and heated for a finite time. During this process the lubricants in the compact gets evaporated while the bonding takes place. This is termed as sintering. After the sintering, the P/M part can be optionally done with other manufacturing operations such as repressing, coining, sizing, resintering, forging, rerolling, or metal infiltration or finishing operations such as machining, heat treating, steam treating, plating, tumbling, shot peening or oil impregnation.

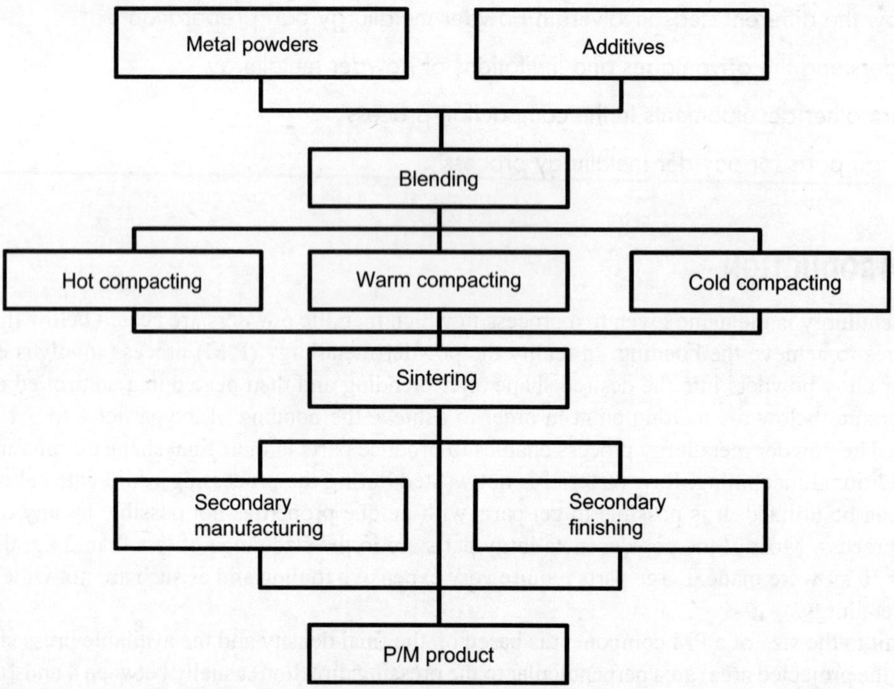

Fig. 11.1 *Simplified flow diagram indicating the various operations in powder metallurgy processing*

11.2 PRODUCTION OF METALLIC POWDER

The first step in the overall PM process is making metal powders. The final properties of a P/M product depend upon the properties of the metal or alloy powders that were used in its production. The important characteristics of powders include the particle shape, size, and the size distribution. There are a number of processes used for the manufacture of metal/alloy powders:

- Solid state reduction,
- Atomisation,
- Chemical, and
- Electrolysis

Solid-State Reduction

This process is generally used for producing iron powder. In this process the selected metal/alloy is crushed, mixed with carbon and passed through a continuous furnace where a reaction takes place, which leaves a cake of sponge metal. This sponge metal is then crushed after separating from all non-metallic material. Then it is sieved to produce powder. The purity of the powder is dependent on the purity of the raw materials. The powder particles are irregular and sponge-like which can be readily compressed to give good green strength.

Atomisation

Atomisation breaks molten metal into small droplets by rapidly freezing before the droplets come into contact with each other or with a solid surface. The atomisation is achieved by bringing the thin molten metal stream in contact with the impact of high-energy jets of gas or liquid. Air, nitrogen and argon are commonly used gases and water is the liquid most widely used. In atomisation the particle shape is determined largely by the rate of solidification and varies from spherical, if a low heat capacity gas is employed, to highly irregular if water is used. By varying the design and configurations of the jets, pressure and volume of the atomising fluid, thickness of the stream of metal etc., it is possible to control the particle size distribution over a wide range.

This technique is applicable to all metals that can be melted and is used commercially for the production of iron, copper, alloy steels, brass, bronze, aluminium, tin, lead, zinc and cadmium. It can also be used in selected instances for high melting point materials such as tungsten, titanium and rhenium. As shown in Fig. 11.2, the liquid metal is siphoned by the high velocity jet of gas expanding through the nozzle. The liquid metal will be atomised and sprayed into the collection chamber.

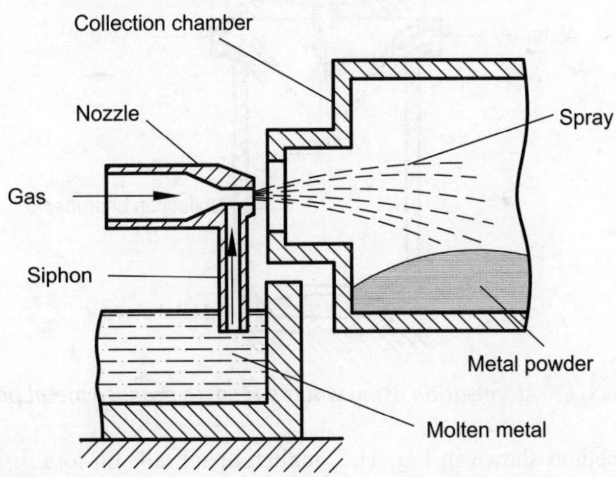

Fig. 11.2 *Atomisation using a gas stream to produce metal powder*

Another variation of the gas atomisation process is shown in Fig. 11.3. The molten metal flows by gravity into a thin stream which is immediately atomised by the high pressure gas jets coming from both sides thereby forming spherical particles, finally collected in the collection chamber.

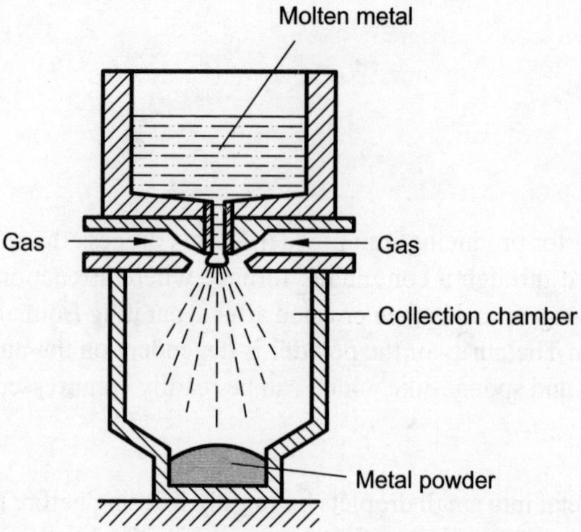

FIG. 11.3 *Another method of atomisation using a gas stream to produce metal powder*

Figure 11.4 shows the atomisation process similar to Fig. 11.3, except high pressure water jet is used in place of gas. Water provides higher cooling rate, however the particles produced are not spherical, but are irregular in shape. One disadvantage of using water is that the surface of the powder particles gets oxidized. This can be taken care of by using synthetic oils in place of water.

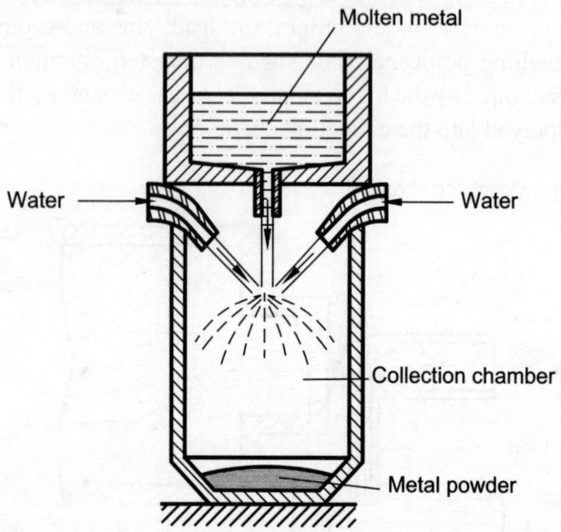

FIG. 11.4 *Atomisation using water stream to produce metal powder*

In the rotating disk method shown in Fig. 11.5, molten metal falls on to a disk that is rotating at high speeds. The liquid metal that is impinging on the disk will be thrown out rapidly into small droplets by the disk which are solidified and collected in the collection chamber.

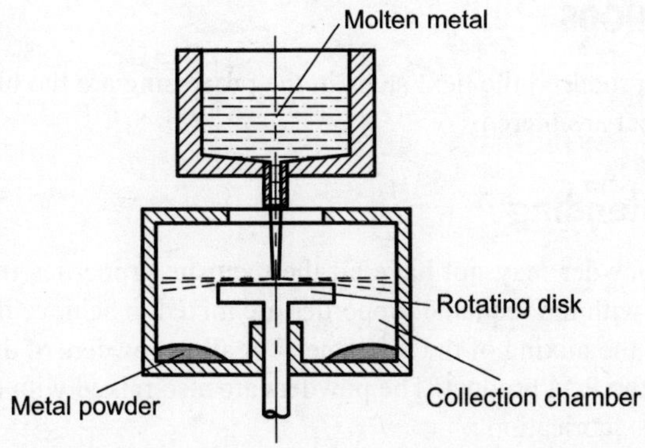

Fig. 11.5 *Atomisation using a rotating disk to produce metal powder*

Another method of atomisation is the rotating consumable electrode method as shown in Fig. 11.6. Here the metal is melted by the arc between the rotating consumable electrode and the stationary electrode. Since the electrode is rotating, the molten metal is atomised by the centrifugal force and collected in the chamber in which inert gas is filled.

Electrolysis

The desired metal made as anode in an electrolytic cell, such that it is dissolved by the electrolyte in the cell and then transported and deposited on the cathode in a spongy or powdery form. The deposit is removed, washed and dried to get the metal powder. Copper is the primary metal produced by electrolysis but iron, chromium and magnesium powders are also produced using this process.

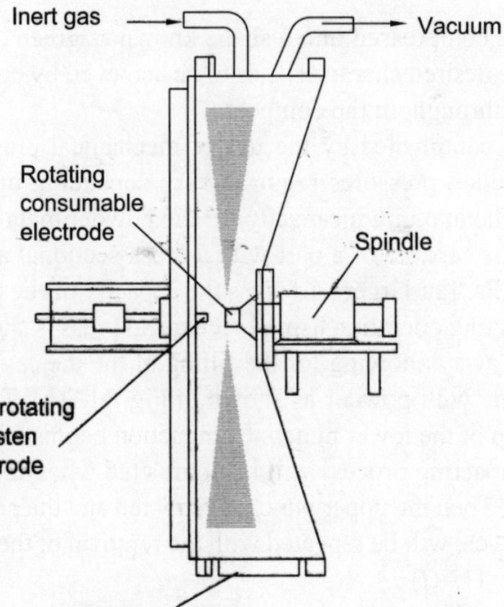

Fig. 11.6 *Atomisation using a rotating consumable electrode to produce metal powder*

11.3 PROCESSING METHODS

After the metallic powder is produced the next steps in the processing are the blending of the powders and compacting before the compact is sintered.

11.3.1 Mixing and Blending

Generally a single metallic powder may not have all the requisite properties required for a part. Hence a number of different powders with the requisite properties are mixed to achieve the necessary balance of the properties. Blending refers to the mixing of the same metal or alloy powders of different size distributions to reduce the porosity levels in the P/M product. The powders are also mixed with other additives to help with the alloying process as well as lubrication.

The main function of the lubricant is to reduce the friction between the powder and the die walls, core rods, etc., where the powder slides during the compaction process. This ensures that the desired uniformity of density from top to bottom of the compact. The lubricant also helps in reducing the friction for easy ejection of the compact and minimises the tendency to form cracks. Popular lubricants are stearic acid, stearin, metallic stearates, especially zinc stearate, and increasingly, other organic compounds of a waxy nature.

Blending and mixing is normally done by using mechanical processes. Typically the mixing machines will be rotating drums in which the powder is loosely filled to the extent of about 20 to 40% of the volume. It may have internal baffles to help throwing the powder away so that there will be thorough mixing. Over-mixing should be avoided, as it will increase the apparent density of the mix. Also over-mixing usually reduces the green strength of the subsequent compacts probably by completely coating the whole surface of the particles, thereby reducing the area of metal-to-metal contact on which the green strength depends. This will also reduce the sintered strength.

11.3.2 Compacting

In compacting loose powder is compressed into a shape known as green compact which is a very important step in powder metallurgy. The desired characteristics to be achieved by compacting are high product density and uniformity of that density throughout the compact.

Compacting is generally accomplished by the use of mechanical presses and rigid tools but hydraulic presses are also used. Compaction pressures required vary depending upon the type of material used and range from 40 to 1650 MPa. The maximum capacity for the powder metallurgy press may be of the order of 1 MN or less. That translates as capable of a part with a cross-sectional area of 2500 mm^2 with an average compacting pressure of 400 MPa. For larger part sizes the capacity of the press has to be accordingly higher.

Typical sequence of compacting operation using mechanical press is shown in Fig. 11.7. At the start of the cycle, the upper punch moves away allowing for the filling of the die cavity, as shown in Fig. 11.7(a). Feed shoe completely charges the die with powder as shown in Fig. 11.7(b). The amount of powder filled in the die is controlled by the location of the lower punch. Compaction begins after the withdrawal of the feed shoe as shown in Fig. 11.7(c). Compacting process will be completed when the punches complete their intended travel as shown in Fig. 11.7(d). Then the upper punch is retracted and the part is ejected using the lower punch as shown in Fig. 11.7(e). The cycle will be repeated with the removal of the green compact and recharging for the next cycle as shown in Fig. 11.7(f).

It is necessary that in order to achieve consistent quality in the final parts, the amount of powder delivered to the die and the movement of the punches in the die set remains the same for all the components in the batch. This depends upon powder particle shape which affects the bulk flow rate which in turn affects the amount of powder delivered to the tooling. Also the shape of the particle affects the final density of the component by controlling the compressibility.

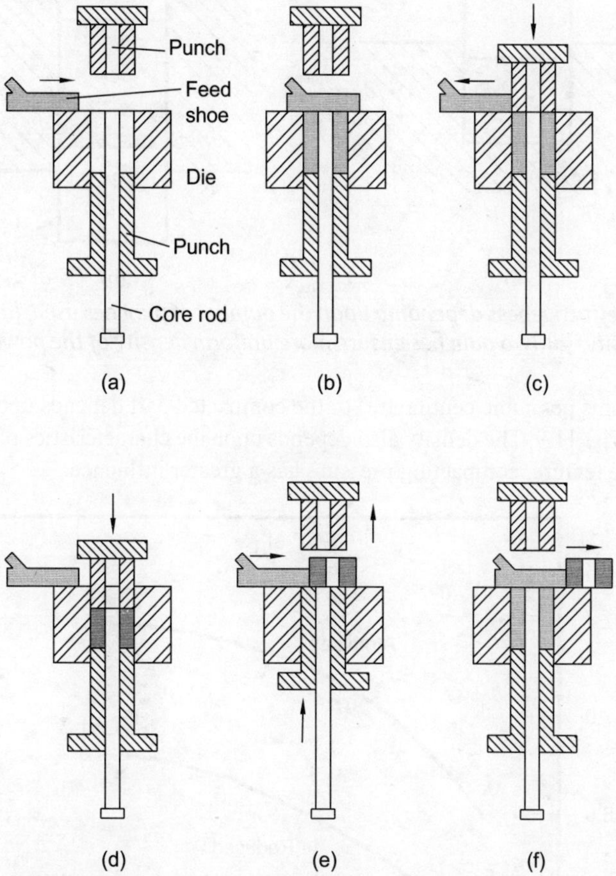

Fig. 11.7 *Compacting cycle for a single level component*

The powder particles move primarily in the direction of the applied force. The powder simply compresses until an equal and opposite force is created. The mechanics of compaction are largely governed by friction between the die and the powder and between the powder particles. The goal during the compaction cycle is to impart uniform density throughout. During the early stage of compaction, the contact between particles increases with less micro-porosity. As the compaction increases, plastic deformation of the particles occurs before the final stage of compaction when cold welding and interlocking of the particles occur, imparting sufficient strength to the green part for ejection and further handling.

Use of a single punch will not guarantee uniform compaction. With a single punch as shown in Fig. 11.8(a), the powder nearer to the punch gets compressed more. The compacting force will not travel all along the depth of the part and as a result the powder near the bottom of the die gets very little compacting. For thin parts a single punch may be sufficient. However for thicker parts as shown in Fig. 11.8(b), two punches in a double acting press will be used. The use of two punches ensures that the density is more uniform throughout the part.

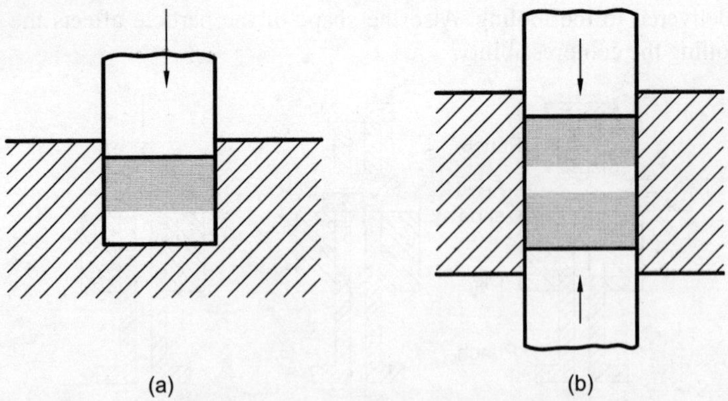

FIG. 11.8 ***Compacting effectiveness depending upon the number of punches used; (a) Single punch with non uniform density; (b) Two punches ensure more uniform density of the powder compacting***

The green density (grams per cubic centimetre) of the compacted part depends upon the pressure used for compacting as shown in Fig. 11.9. The density also depends upon the characteristics of the metal powder such as size, shape and surface texture, compacting pressure has a greater influence.

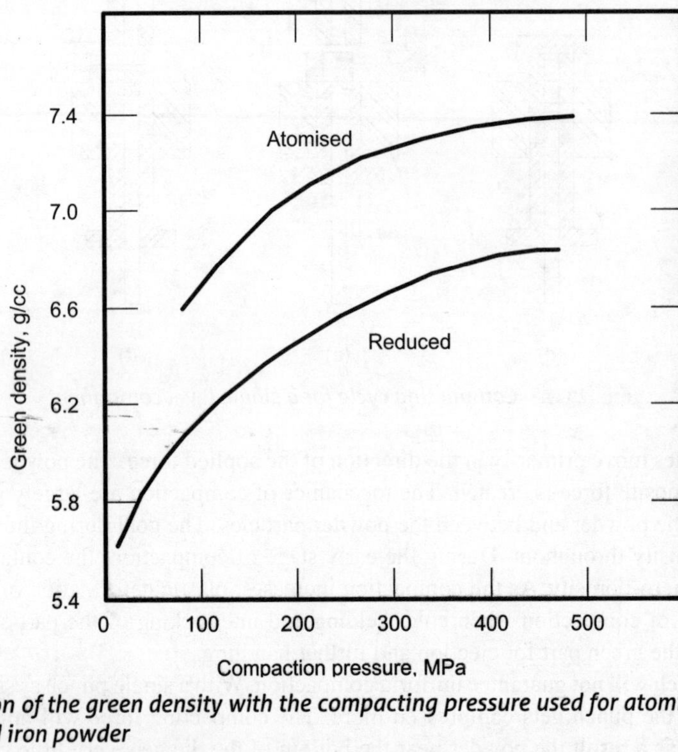

FIG. 11.9 ***Variation of the green density with the compacting pressure used for atomised and conventional reduced iron powder***

Degree of compaction experienced by the metal powder depends upon the displacement that can be achieved during the compaction process. For example as shown in Fig. 11.10(a), for a part with two levels is compacted by a single punch, then the density of the powder in left side as in Fig. 11.10(b) gets less compacted compared to the right portion. As a result the density of the material on the left will have lower density compared to the metal on the right which will have a higher density. This is undesirable from the part strength point of view.

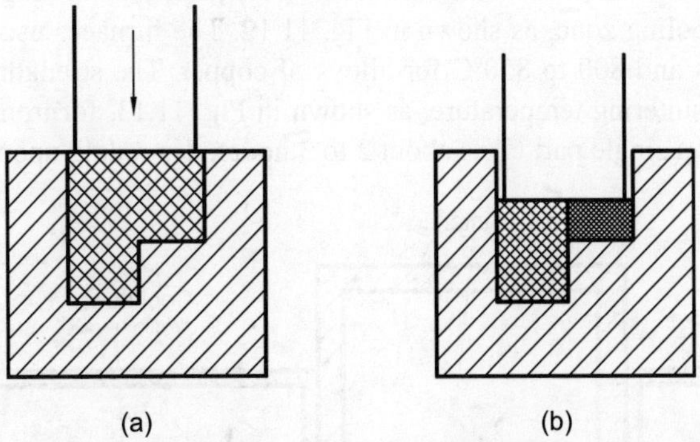

FIG. 11.10 *Compaction of a two level part, (a) Initial condition; (b) part with different densities after compaction*

Similarly it is impossible to produce uniform compaction in different thicknesses of parts with a single punch as shown in Fig. 11.11(a). Hence in such cases it is necessary to provide a separate punch for each thickness of the part as shown in Fig. 11.11(b).

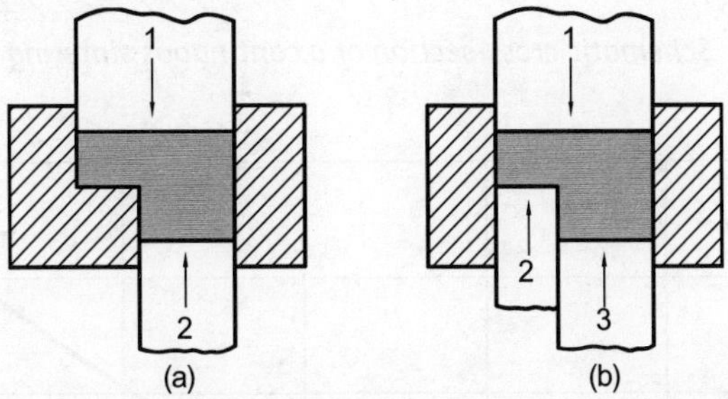

FIG. 11.11 *Compaction of a two level part, (a) Initial condition; (b) part with different densities after compaction*

Tooling is a very critical element in the powder metallurgy process. The tooling must be robust enough to last from several hundred thousand cycles without appreciable wear and damage. The tools are subjected to high compaction pressures, often as high as 690 MPa for some steel components, as well as the abrasive characteristics of the powders. Tooling is made with hardened tool steel while for higher volume production involving abrasive powders tungsten carbide is used.

The die set must be able to withstand radial pressure during compaction and hold tolerance in the horizontal direction of the component. The die set experiences the radial pressure during compaction and wears along the walls due to the ejection motion of the green part. The punches must possess high compressive yield strength, toughness and fatigue strength. Core rods are used to shape through holes in the component. Like the dies, core rods should be made from cemented carbides.

11.3.3 Sintering

Sintering refers to the heating of the green compact in an oven. The heat is supposed to join the various grains (metal powder) into a single mass, thus developing the necessary strength. The strength obtained in the process depends on the temperature and time the powder compact is supposed to be in the oven.

The traditional furnace found in the PM industry is a mesh belt furnace with three operating zones; a preheat or de-lube zone where the lubricants and binders are burned off, a sintering zone where the necessary strength develops, and a cooling zone, as shown in Fig. 11.12. The furnaces usually operate between 1100 to 1200ºC for ferrous parts and 800 to 850ºC for alloys of copper. The strength achieved by the sintering process depends upon the sintering temperature, as shown in Fig. 11.13, for iron with 1.25% graphite. The trip through the furnace for a single part takes about 2 to 3 hours, depending upon the size of the part.

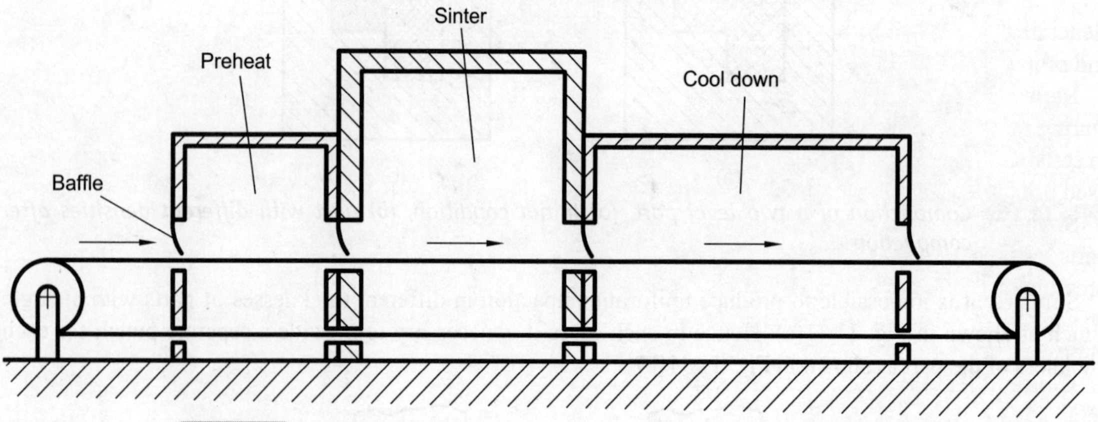

Fig. 11.12 *Schematic cross-section of a continuous sintering belt furnace*

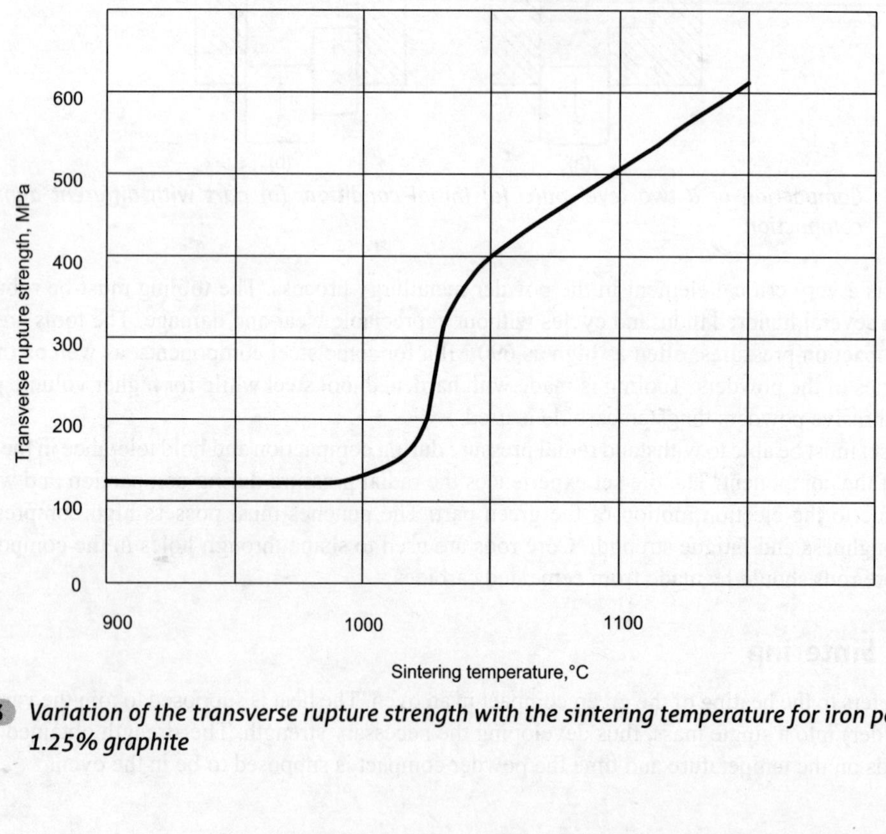

Fig. 11.13 *Variation of the transverse rupture strength with the sintering temperature for iron powder with 1.25% graphite*

Some components are sintered at a temperature 150°C higher, a process called high temperature sintering, an effort made to enhance mechanical properties. Another variation on the process is called sinter-hardening, accomplished by using a controlled cooling rate in the cooling section of the belt furnace, transforming the steel matrix of a ferrous part to martensite, thereby eliminating the need for a secondary hardening step.

11.3.4 Secondary Operations

Generally some optional secondary operations are performed on sintered part to achieve the final dimensions and properties of the part.

Repressing is performed on PM components to increase the density and improve the mechanical properties. During re-pressing the density of the part is generally increased, especially if the as-sintered density is low. In certain cases where strength and other mechanical properties are required to be at maximum, re-pressing is used principally to achieve such densification. Further improvement is achieved by re-sintering.

Sizing is performed to modify the surface shape and provide stricter dimensional control. Generally for parts 'as-sintered' to be accurate to a tolerance of –0.050 mm per mm, in the direction at right angles to the pressing direction, and 0.10 mm per mm parallel to the pressing direction. Sizing may be done in the die that was used for compacting the powder in those cases where the dimensional change on sintering is controlled at or very near to zero, but commonly separate sizing tools are used.

Coining is used to imprint or emboss the face(s) of the component in contact with the punch that has the requisite details.

Heat Treatment PM components can be heat treated to improve strength and hardness, can be steam treated to make the surface hard and more wear resistant as well as contributing to improved corrosion resistance and the sealing of porosity. During the quenching process, because of the porosity inherent in sintered parts, they should not be immersed in corrosive liquids such as salt baths, water, or brine and only oil cooling is preferred. Heating should also be done in a gas atmosphere.

Powder metallurgy parts have inherent porosity built into them because of the structure of the part and the compaction pressure used. Normally these pores are harmful for the mechanical strength of the part. However these pores can be filled with materials such as metals, oils or plastics to achieve special properties that cannot be obtained by any other manufacturing process.

Infiltration The P/M component is dipped into a low melting temperature alloy liquid, such that the liquid would flow into the voids simply by capillary action, thereby decreasing the porosity and improving the strength of the component. The process is used quite extensively with ferrous parts using copper as infiltrant but to avoid erosion, an alloy of copper containing iron and manganese, is often used.

Impregnation Impregnation is similar to the infiltration, except that in this case the powder metallurgy component is kept in an oil bath. The oil will penetrate into the voids by capillary forces and remains there. The oil will be used for lubrication of the component when necessary. During the actual service conditions, the oil will be released slowly to provide the necessary lubrication. The components can absorb between 12 and 30 % oil by volume. It is used on PM self-lubricating bearing components since the late 1920's.

Some times resin (thermo-setting or other plastic materials) impregnation can be used to enhance machinability and improve ductility and prepare parts for plating. The benefits to be obtained include some increase in mechanical properties, sealing of the pores which may provide pressure tightness and will also prevent the entry of potentially corrosive electrolyte during a subsequent plating operation. Additionally the machinability of sintered parts is improved.

Finishing Operations

A sintered component can be finished or treated just like any other metal component. A number of finishing operations such as deburring, plating, coating, etc. are performed depending upon the part requirements.

11.4 ADVANTAGES

There are many advantages of using Powder Metallurgy. They are:

1. The P/M parts can be produced to the neat net-shape requiring very little finishing operations. For higher dimensional accuracy coining or sizing operations can be used. It is possible to achieve a surface finish between 0.80 – 1.20 micron (Ra) from a die with a surface finish of the order of 0.50 – 0.80 microns.
2. Powder metallurgy process will not cause any waste products during the processing and as such is very economical.
3. Reasonably complex shapes can be produced by powder metallurgy which cannot be economically machined or cast.
4. It is possible to produce parts with a combination of materials that is not possible by other processes. For example, metals and ceramics can be intimately mixed.
5. Automation of the P/M process can be easily accomplished reducing the labour requirements and increasing the quality.
6. Certain metals that cannot be manufactured by other processes can be done by powder metallurgy. Case in point is that of tungsten and tungsten carbide.
7. Provides controlled porosity for self-lubrication by infiltration of lubricating oils.

However in spite of such advantages, there are certain limitations for the process, which need to be considered carefully before selecting the process.

1. The tooling cost is generally high, and therefore can only be justified for mass production.
2. Raw material cost is high. However it can be justified because most of the material is utilised without any wastage during the processing.
3. Because of the presence of residual porosity the mechanical properties are inferior compared to the corresponding cast or machined part.
4. With complex part geometries, the flow of metal powder into deep cavities and corners is a problem, and as a result, there will be variation in the density in the part at different locations, which translates into different properties.

11.5 OTHER COMPACTION METHODS

In the conventional compaction process the powder gets compacted in the direction of force, while in the direction perpendicular to the force the compaction is limited. Isostatic pressing ensures the density remains the same throughout the part.

Cold Isostatic Pressing (CIP) In Cold Isostatic Pressing (CIP), the powder is contained in a flexible mould made of rubber or other elastomer material as shown in Fig. 11.14(a). The flexible mould is then pressurized by means of a high-pressure water or oil as shown in Fig. 11.14(b). Thus the powder is compacted with the same pressure in all directions, and, since no lubricant is needed, high and uniform density can be achieved. Then the pressure is released and the finished part is ejected from the mould. The process removes many of the constraints that limit the geometry of parts compacted unidirectional in rigid dies. Long thin-walled cylinders and parts with undercuts present no problem.

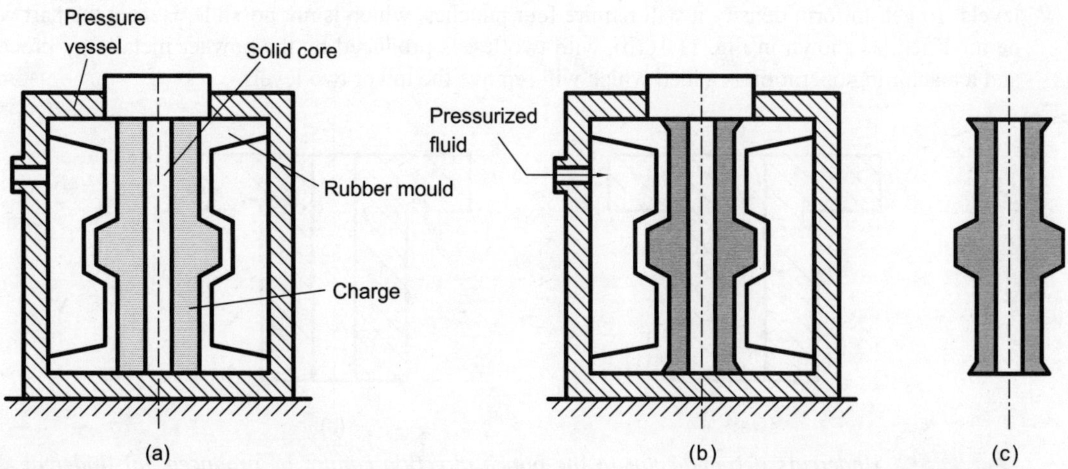

Fig. 11.14 *Cold isostatic pressing, (a) Charging of the rubber mould, (b) Hydrostatic pressure applied to the mould, (c) Finished part*

Hot Isostatic Pressing (HIP) Hot Isostatic Pressing (HIP) is carried out at high temperature and pressure using a gas such as argon. Because of the high temperature the flexible mould is made of sheet metal. In HIP, compaction and sintering are completed simultaneously. The process is used in the production of billets of super alloys, high speed steels, titanium, ceramics etc., where the integrity of the materials is a prime consideration.

11.6 DESIGNING FOR P/M

Because of the nature of the powder metallurgy process it is important that the parts should be properly designed to take advantage of the process characteristics. The following are some guidelines that should be kept in mind while designing the parts to be produced by powder metallurgy process.

- The geometry of the part should be such that the part can be easily ejected from the die. No draft is required. Perpendicular sidewalls and holes and recesses parallel to punch travel are required. Holes can be any shape including cylindrical as long as they remain parallel to the axis of the punch travel.
- Choosing the wall thickness needs careful consideration. Long thin walls are difficult to produce. Long walls with a length to thickness ratio exceeding 8 to 1 should be avoided. Thin walls result in fragile tools.
- Undercuts on the horizontal plane (perpendicular to the pressing direction) as shown in Fig. 11.15(a) cannot be produced. If required it needs to be produced without the undercut and then machined as a secondary operation, as it inhibits part ejection. If an undercut is required at the junction as shown in Fig. 11.15(a), the alternative is to press a semicircular groove at the horizontal portion of the junction as shown in Fig. 11.15(b).
- Abrupt changes in section should be avoided since they introduce stress raisers which may lead to crack formation as a result of the stresses induced by the elastic expansion or spring back that takes place as the compact is ejected from the die.
- Ideally the part should have a uniform cross-section. However when parts have different cross-sections, achieving uniform density is a problem. Hence, maximum number of levels in the part should be limited depending upon the type of equipment used. For example in Fig. 11.16(a), the part has four

levels. To get uniform density it will require four punches, which is not possible. Hence, the part will be modified, as shown in Fig. 11.16(b), with two levels produced by the powder metallurgy process and a machining operation is added which will remove the lower two levels.

Fig. 11.15 *Undercuts perpendicular to the punch direction cannot be produced, (a) Undercut needs to be machined, (b) undercut parallel to the punch movement can be produced*

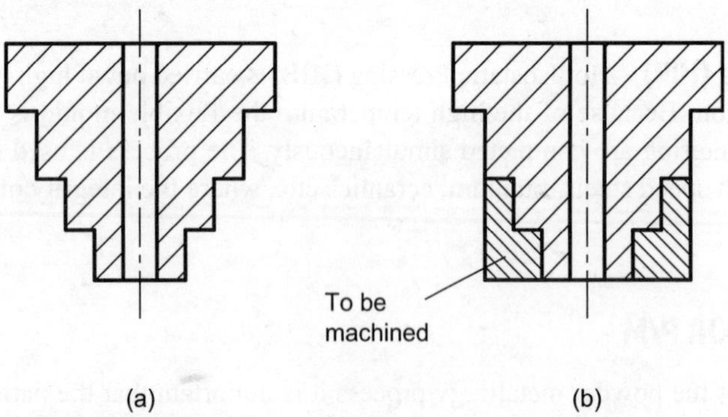

Fig. 11.16 *The number of levels in a part should be limited to maintain uniform density, (a) Part with four levels requires four lower punches to get uniform density, (b) Part is modified so that two levels are produced by P/M and the other two levels are produced by machining as shown.*

- Tolerances that can be achieved by powder metallurgy are comparable to other metal working processes as shown in Table 11.1.

Table 11.1 **Tolerances achievable after pressing and sintering**

	Practical, mm	Possible, mm
Length (±)	0.190	0.130
Inside diameter (±)	0.100	0.050
Outside diameter (±)	0.100	0.050
Concentricity (±)	0.150	0.100
Flatness on ends (±)	0.130	0.100
Parallelism of ends (±)	0.130	0.100

- Screw threads cannot be produced and hence will be machined after the powder metallurgy processing.
- Tooling with generous fillets tends to last longer and result in higher component manufacturing speeds and greater tool integrity.

11.7 METAL INJECTION MOULDING (MIM)

P/M parts can also be made by Metal Injection Molding (MIM) which was developed in the 1970s. In this process, the binder used is a thermoplastic resin is mixed with metal powder which is then injected into the die using the injection moulding machine that is used for making thermoplastic parts. The part obtained from the injection moulding machine is the green compact which then goes through the sintering process to achieve the necessary strength. Before sintering, the binder that was used for forming the shape is removed by heating or using a solvent. It is possible to get a higher density product with MIM that would increase its strength. Schematic representation of MIM in comparison to the conventional powder metallurgy is shown in Fig. 11.17.

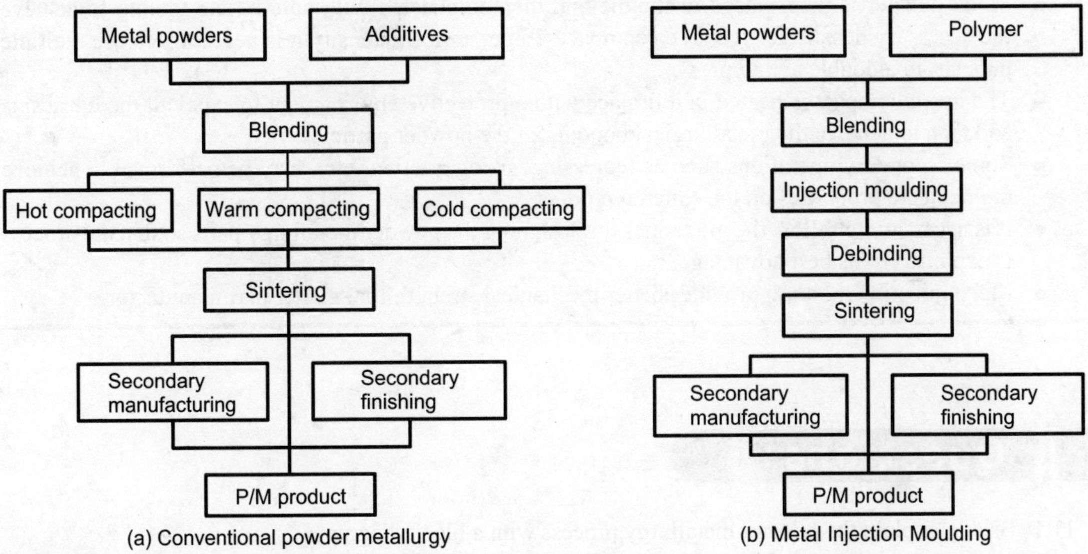

Fig. 11.17 *Different steps in the processing of powder metallurgy parts, (a) Conventional powder metallurgy, (b) Metal Injection Moulding (MIM)*

The MIM process utilizes all types of metallic powders similar to the conventional process though stainless steel is widely used. The particle sizes used are with a mean diameter of 2 to 10 μm. It is generally preferred to use fine grains. The metallic powders are compounded with wax and a polymeric binder. The binder used should be able to mix a high volume of powders, typically 60% by volume that can be injection moulded at suitable elevated temperature. Further, it should be easily removed by de-bonding in a reasonably short and environmentally friendly process.

Early binders used in MIM were mixtures of wax and stearic acid with a polymer such as polyethylene or polypropylene. The removal of this type of binder requires a very careful heating in a thermal process lasting 24 or more hours. Alternatively, a polyacetal binder system developed by BASF is based on polyoxymethylene (POM). De-bonding of this is to be done in a gaseous acid environment, i.e., highly concentrated nitric or oxalic acid that can be completed in a shorter time frame of about 3 hours.

Rest of the processing is similar to the conventional powder metallurgy operation. Some of the application where this is used is in dental and medical application such as Orthodontic brackets, gears used in surgical stapling units, drug delivery devices, joint replacement parts, etc. It is also used in the automobile sector such as rocker arms for variable valve stroke engines, shift levers, turbo charger vanes, etc. Other applications are for mobile phones, consumer electronics, fiber optic parts, etc.

SUMMARY

Powder metallurgy is unique in providing exceptional properties that cannot be achieved by other manufacturing processes.

- The starting point of powder metallurgy is the metal powder which is generally obtained by atomisation from liquid metal. There are a number of methods that are possible for this.
- Different metal/alloy powders will be mixed with additives and then blended to ensure uniform properties and appropriate lubrication.
- Metal powder is then placed in the die and then compacted with appropriate tooling to achieve the necessary density for the green compact. To get uniform density it is necessary to use multiple punches in a double acting press.
- The green compact is heated in a furnace with a protective environment to expel all the lubricants and then to achieve the metallurgical bonding of the powder particles.
- Some secondary operations such as repressing, sizing, plating, etc., may be performed to achieve the requisite properties on the sintered product.
- It is necessary to follow the guidelines for designing the powder metallurgy parts so that the process is exploited to its best advantage.
- Metal injection molding provides better mechanical strength for smaller part manufacture.

Questions

11.1 Briefly explain the powder metallurgy process with a block diagram.

11.2 What are the various methods available for making the metal powder?

11.3 Explain any one of the atomisation process used for preparing the metallic powder.

11.4 What do you understand by mixing and blending with reference to powder metallurgy?

11.5 Describe the movement of powder particles during compaction in powder metallurgy process. What features are responsible for the fact that powder does not flow and transmit pressure like a liquid?

11.6 Explain sintering process in connection with the powder metallurgy.

11.7 Explain the following terms with reference to the powder metallurgy process: Sintering, Infiltration, and Impregnation.

11.8 Give the advantages of powder metallurgy parts.

11.9 Give a brief account of cold isostatic pressing.

11.10 Give a brief account of metal injection moulding.

Multiple Choice Questions

11.1 Powder metallurgy process involves the following operations in sequence
(a) Powder mixing, sintering, compacting, and finishing
(b) Powder mixing, sintering and finishing
(c) Powder mixing, compacting, sintering and finishing
(d) Powder mixing, compacting, and sintering

11.2 Metal powder particle size is reduced during atomisation by
(a) Decreasing gas velocity
(b) Increasing gas velocity
(c) Decreasing gas pressure
(d) Decreasing metal volume

11.3 Identify the correct statement among the following:
(a) Blending and mixing refer to the same process of mixing different types of metal or alloy powders of different size distributions
(b) Blending is mixing the different metal or alloy powders of different size distributions while mixing is mixing same types of metal or alloy powders
(c) Blending is mixing the same metal or alloy powders of different size distributions while mixing is mixing different types of metal or alloy powders
(d) Blending and mixing refer to the same process of mixing same types of metal or alloy powders of different size distributions

11.4 Identify the correct statement among the following:
(a) Single punch is sufficient to achieve uniform compacting of powder metallurgy parts
(b) Multiple punches are required to achieve uniform compacting of powder metallurgy parts
(c) Any type (single or multiple) of punch will be able to achieve good compacting of the powder metallurgy parts
(d) Punching with variable pressure will be able to achieve good compacting of the powder metallurgy parts

11.5 Green density of the powder metallurgy part will be increased by
(a) Increasing the sintering temperature
(b) Decreasing the sintering temperature
(c) Increasing the compacting pressure
(d) Decreasing the compacting pressure

11.6 Final strength of a powder metallurgy part can be increased by
(a) Increasing the sintering temperature
(b) Decreasing the sintering temperature
(c) Increasing the lubricant in the powder mix
(d) Decreasing the compacting pressure

11.7 To improve the mechanical properties of a powder metallurgy part, the following finishing operation is used
(a) Repressing (b) Sizing
(c) Coining (d) Impregnation

11.8 To improve the dimensional accuracy of a powder metallurgy part, the following finishing operation is used
(a) Repressing (b) Sizing
(c) Coining (d) Impregnation

11.9 To improve the self lubricating capacity of a powder metallurgy part, the following finishing operation is used
(a) Repressing (b) Sizing
(c) Infiltration (d) Impregnation

11.10 To improve the strength of a powder metallurgy part, the following finishing operation is used
(a) Coining (b) Sizing
(c) Infiltration (d) Impregnation

Answers to MCQs

11.1 (c)	11.2 (b)	11.3 (c)	11.4 (b)	11.5 (c)
11.6 (a)	11.7 (a)	11.8 (b)	11.9 (d)	11.10 (c)

CHAPTER 12

Plastic Processing

Objectives

After completing this chapter, the reader will be able to

- Understand the basics and history of plastic materials for engineering applications
- Learn about different plastic materials and their properties for engineering application
- Know the different extrusion processes used in making plastic parts
- Understand the injection moulding process and other variants of the same
- Know the various blow moulding operations used in making plastic parts
- Learn thermoforming as a method to make low-cost plastic parts with simple equipment
- Understand the moulding methods used for thermosetting materials
- Design plastic parts taking the process and material requirements into account

12.1 INTRODUCTION

Polymers are new materials that have been extensively used in engineering starting from the nineteenth century. A monomer is the single building block used in creating the polymer by the process of polymerisation. Polymer is basically a linked monomer with a number of them being linked based on the type of polymer. Most polymers are based on carbon and to that extent they are called organic materials. However, there are certain inorganic polymers also in existence.

Though engineering use of polymers started with 1800's, they are in existence from a long time. Most of them are of biological nature and are used as a source of food. However, engineering polymers are of interest to us in this chapter. There were significant developments in the invention of rubber vulcanisation and phenolic plastics during the nineteenth century. The real use of engineering plastics exploded after the Second World War in the twentieth century.

Plastics are formed when small organic molecules are stitched together to form a long chain. This process is called **polymerisation**. Those organic molecules suitable for polymerisation are called monomers. It is necessary for a monomer to be at least bifunctional (capable of forming two covalent bonds) to polymerise. The bifunctional monomer will attach other monomers at the front and back end to form a chain. Some monomers are polyfunctional; capable of forming three are more bonds. These can form chains which are three-dimensional arrangements of chains called a **network**.

Polymerisation is generally of two types; addition and condensation. The **addition polymerisation** consists of breaking the double carbon bond (C=C) in bifunctional polymers, so that the chain can be formed. Some of the monomers that are commonly used in addition polymerisation to produce thermoplastic materials are shown in Table 12.1. For example, ethylene when polymerised will be called polyethylene.

TABLE 12.1 Typical plastic materials made by addition polymerisation

Plastic Monomer	Chemical Formula
Styrene	$CH_2=CH—C_6H_5$
Vinyl chloride	$CH_2=CHCl$
Propylene	$CH_3CH=CH_2$
Methyl methacrylate	$CH_2=CC_3O_2H_4$
Vinyl acetate	$CH_3COOCH=CH_2$
Ethylene	$CH_2=CH_2$

In **condensation polymerisation**, two different organic molecules react to form a plastic molecule. The reaction generally results in the separation of a small molecule such as H_2O as a by-product. Plastics made via condensation polymerisation, can degrade when exposed to water and high temperatures. Under these conditions, depolymerisation occurs, severing the polymer chains. Some typical plastic materials formed by condensation polymerisation are shown in Table 12.2.

TABLE 12.2 Typical plastic materials made by condensation polymerisation

Plastic	Condensation Polymerisation Reactants
Epoxy (araldite)	Epichlorohydrin + bisphenol A
Phenol-formaldehyde (Bakelite)	Phenol + formaldehyde
Polyamide (Nylon)	Hexamethylene diamine + adipic acid
Urethane	Isocyanate + alcohol

Resin is the term generally used for uncompounded ingredients or monomers that are mixed but not yet polymerised. Examples are thermosetting resin. Sometimes it is also used as a synonymous term for plastics, for example **thermoplastic resin** instead of thermoplastic material.

As the polymerisation proceeds the number of carbon chains will intermingle to form large volumes. Depending on the orderliness of these chains, the polymers will be termed **amorphous (disorderly)** or **crystalline (orderly)**. Crystallinity, or orderliness, of the chains controls the properties of the plastics. Amorphous polymers are less dense compared to crystalline plastics. Crystallinity in plastics increases the strength and toughness.

12.2 PROPERTIES OF PLASTICS

12.2.1 Density

Plastic materials have relatively low density compared to metallic materials. Also, their density depends upon the processing method used to prepare the final structure. For example, expanded polystyrene foam used in packaging may have a density of 10 kg/m^3 compared to its usual density of 500 kg/m^3. The low weight is often an advantage for some applications.

12.2.2 Strength

Plastic materials show both elastic and plastic behaviour depending upon the temperature. They are generally good in compression while under tensile loading they do not exhibit sufficient strength. So the designers of plastic products that experience tensile loading have to develop geometries with high stiffness. The strength of plastics in tensile loading can be greatly increased by reinforcing with appropriate fibre in the required directions.

12.2.3 Toughness

Toughness of plastic materials has a wide range with acrylic and polystyrene being brittle, while other plastics are tough. Some plastic materials such as polycarbonate are virtually unbreakable. Plastics become brittle if they are subjected to subzero temperatures and/or contain severe stress concentrations, such as sharp corners and notches. To improve the impact strength of brittle plastics, such as polystyrene and rubber fillers are added at the cost of reducing the strength and stiffness.

12.2.4 Viscoelasticity and Creep

Plastics under short-term loading with low deflections and small loads at room temperatures behave elastically like a spring, returning to their original shape after the load is removed. No energy is lost or dissipated during this purely elastic behaviour. They also exhibit a viscous behaviour under long-term heavy loads or elevated temperatures. While still solid, plastics will deform and flow similarly to a very high-viscosity liquid. This time-and-temperature-dependent behaviour occurs because the polymer chains in the part slip and do not return to their original position when the load is removed. This dual behaviour is termed **viscoelasticity**.

Creep is a consequence of plastics' viscoelastic behaviour and is the deformation that occurs over time when a material is subjected to constant stress at elevated temperature. Under these conditions, the polymer chains slowly slip past one another. Because some of this slippage is permanent, only a portion of the creep deformation can be recovered when the load is removed.

12.3 ADDITIVES IN PLASTICS

A number of additives are normally used with plastic materials to modify their behaviour, improve properties, or reduce the overall cost, thereby increasing the range of application of plastics. Additives often determine the success or failure of a resin or system in a particular application.

Plasticizers are mixed with plastics to improve their flow characteristics and decrease the brittleness. These are liquid organic compounds with fairly large molecular weight. The glass-transition temperature is the temperature at which an amorphous polymer becomes brittle on cooling, or soft on heating. Plasticizers lower the glass transition temperature thus making the plastic softer and more flexible at the room temperature. For example, polyvinyl chloride (PVC) is brittle at the room temperature. However, when a plasticizer such as di-iso-octyl is mixed, PVC becomes flexible, and is used as wrapping tape for electrical insulation.

Fillers are inexpensive materials that are added to plastics to reduce their cost. They are generally inert and will not react with plastics. However, many of the filler materials increase the hardness and impact strength of plastics. Common fillers are wood flour, quartz, glass spheres, talc, calcium carbonate and alumina trihydrate.

Flame retardants are added to plastics to reduce the flammability of plastics by preventing oxygen reaction and improving charring. Elements such as boron, nitrogen, chlorine, antimony and phosphorous are added for this purpose. These sometimes can cause problems by reducing flexibility, tear resistance, tensile strength and heat deflection. Generally required for electrical and medical-housing applications, flame retardants and their amounts vary with the inherent flammability of the base polymer.

Reinforcing agents are the materials that are specifically added to plastic materials to raise the mechanical properties. Reinforcing can affect practically all of the properties. Some examples of the change in mechanical properties by the addition of glass fibre to the common engineering plastics are given in Table 12.3. Some of these agents bond with the plastic structure imparting regularity.

TABLE 12.3 **Effect of glass-fibre reinforcements on the mechanical properties of plastic materials**

Plastic	Increase %		
	Tensile Strength	Hardness	Impact Strength
ABS	250	Nr	–50
Epoxy	200	20	500
Nylon	200	Nr	20
Polycarbonate	225	15	–800
Polystyrene	150	25	100

Nr – Not reported

Stabilisers are used to stabilize the properties of the plastic throughout its useful life. They resist heat, strength deterioration and reduce the effect of radiation on bonds within the chains.

Colourants are used to give colour to the plastics. Both organic dyes and inorganic pigments are used to add colour to plastic materials. Pigments disperse rather than dissolve in plastics thereby reducing transparency of the material. They hide flaws such as air bubbles, making it difficult to judge quality.

12.4 PLASTIC MATERIALS

Polymeric materials of interest to engineers can be broadly classified into plastics and elastomers as shown in Fig. 12.1. Plastic materials are further classified into thermoplastic and thermosetting materials based on the processing methods.

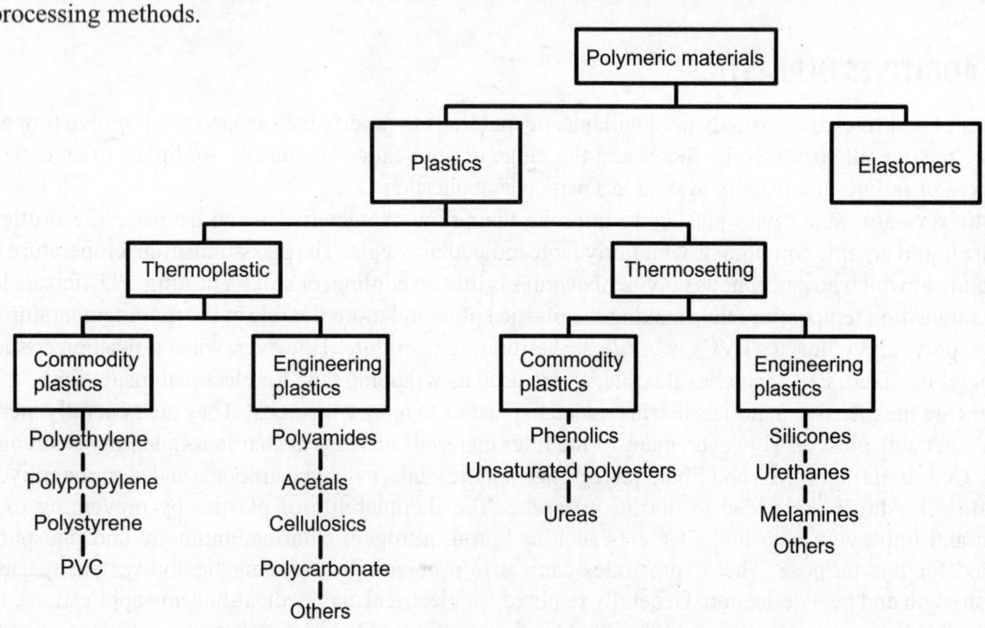

FIG 12.1 *Classification of the different polymeric materials*

12.4.1 Thermoplastics

These are the plastics that can be softened and melted by heat and can then be formed into required shape when it is hot. These materials can be melted a number of times. That means it is possible to recycle thermoplastics. However, in industry frequent remelting is avoided since some chemical degradation occurs during remelting. Thermoplastic materials tend to consists of long polymer chains with little breadth, akin to a two-dimensional structure. The fabrication processes used for thermoplastics such as blow moulding and injection moulding are less expensive compared to those methods used for thermosetting materials.

Polyethylene

The basic monomer for this is ethylene (C_2H_4) with a carbon-to-carbon double bond ($H_2C{=}CH_2$). It is generally made from petroleum products. There are a variety of polyethylenes: High-Density PolyEthylene (HDPE), Low Density PolyEthylene (LDPE), and Linear Low Density PolyEthylene (LLDPE). Low density grades have lower strength and lower melting point. High density grades are more crystalline. Low density grades are used for packaging films because of their clarity while the high density grades are used for injection-moulded consumer items and blow-moulded parts. Major advantage of polyethylene is its resistance against many acids and bases and its flexibility. Limitation of this material is its low strength and low heat resistance compared to other engineering polymers. Also its mechanical properties degrade when exposed to certain environments such as UV (ultraviolet) radiation from sunlight.

Polypropylene

The monomer (C_2H_3-CH_3) for this is the substitution of a methyl group (CH_3) for one H atom in ethylene (C_2H_4). It requires special processing utilising a catalyst to achieve the polymer chain with the CH_3 molecules are all on one side. It is much harder, stronger and tougher compared to HDPE. As a result, it is more suitable moulded parts than HDPE. Also its chemical resistance is better than polyethylene. The density of polypropylene is low of the order of 0.90 to 0.915 g/cm^3. It is used for luggage, battery cases and toolboxes because of its low weight. Carpets and ropes are other major uses of this material.

Polyvinyl Chloride (PVC)

It is generally called PVC and is the most used in terms of volume. The monomer is ethylene (C_2H_4) with one H atom replaced by one chlorine atom (C_2H_3Cl) and is made from acetylene's (C_2H_2) reaction with hydrochloric acid in the presence of a catalyst. Plasticized PVC containing plasticizers such as phthalate ester have low strength and are generally used for decorative laminates, upholstery and other consumer products. PVC is considered non-inflammable and therefore, in its plasticized form is used for electrical insulation. Rigid PVC without any plasticizer has sufficient strength and is almost considered as engineering plastic. Most widely used application is for piping. Also it is resistant to most acids and bases. The main limitation of this material is its low toughness and notch sensitivity.

Polystyrene

This is one of the largest used thermoplastic almost amounting to 20% of the total plastic usage. It is made from ethyl benzene which is ethylene with one H atom replaced by a benzene ring (C_6H_6). The benzene rings are located randomly along the polymer chain. The presence of benzene ring inhibits deformation and provides hardness and brittleness. It has poor impact strength and as a result is not used in engineering plastics. Because of its low cost the widest use of it is in throwaway items such as the plastic tableware and food containers.

Acrylonitrile Butadiene Styrene (ABS)

It is a terpolymer (consisting of three polymers) of Acrylonitrile, Butadiene and Styrene called ABS. The butadiene chain in this terpolymer is grafted to the side of the acrylonitrile-styrene chain. This is a very important engineering thermoplastic and finds wide use in industry. It has excellent toughness and very good formability. Because of this reason, this material is suitable to most of the plastic processing methods such as thermoforming and moulding.

Polymethyl Methacrylate (PMMA)

These are more commonly called **acrylics** with tradenames such as Perspex and Plexiglas. These are made by reacting methylacrylic acid (CH_2=C-CH_3-$COOCH_3$) with an alcohol. From the ethylene molecule, one H is replaced by CH_3 and other H is replaced by $COOCH_3$ group. It is clear and rigid and is used for sight glasses, machine guards, windows, low-cost lenses and safety glasses. They can be readily injection moulded.

Fluorocarbons

These are a class of polymers with a fluorine atom substituted for the H atom. More familiar plastic in this family is the tradename Teflon which is polytetrafluoroethylene (PTFE). The monomer for this is the ethylene with all four H atoms replaced by fluorine atoms (C_4F_4). PTFE is one of the most chemically inert material known and, therefore, used for seals, tubing and small containers used for handling very aggressive chemicals. It cannot be used for structural materials because of its low strength. Also, it is very expensive. Its processing is different from other plastics since it does not melt and flow. It is compacted from powder and then sintered similar to the powder-metallurgy methods studied earlier.

Polyamides

It is a family of products obtained by condensation of an acid (RCOOH) with amine (NH_2R). These are commonly referred to as nylons with variants such as Nylon 6, Nylon 11 and Nylon 12. The suffixes refer to the number of carbon atoms in each of the building blocks. For example, Nylon 6 is $C_6H_{11}ON$. Though nylon is a tradename, in view of its widespread use over a long period (since 1930s), polyamide is frequently referred to as nylon. Nylon is crystalline thermoplastic with very high mechanical properties compared to other thermoplastics. One major problem with nylon is the moisture absorption (up to 10% by weight). With moisture absorption, the tensile strength is reduced and the dimensions increase. Nylon is a widely used engineering plastic for parts such as gears, cams, and slides for machinery.

Polyacetals

These are different from other polymers since they have a hetero chain with C and O compared to carbon chain in other thermoplastics. The basic monomer used is formaldehyde. The chain is composed of the polyoxymethylene (POM) where methylene (CH_4) with one H atom replaced by CH_2OH, while the other is replaced by O, which in turn is connected to CH_2OH. These are similar to nylon and are more widely used engineering plastics. One major problem is the high shrinkage of acetal during injection moulding.

Cellulosics

It is the major component of all plant material. The cellulose molecule [$C_4H_5O_2(OH)_2CH_2OH$] is very complex and the bonding is obtained by the oxygen (O) linkage. The common cellulose polymers are cellulose acetate, cellulose acetate butyrate and cellulose acetate propionate. Cellulose acetate is often used for films and fibres. A few common cellulose polymers are rayon and cellophane. These are generally of low cost with good mouldability and weatherability. Consumer application is the major domain where cellulosics are widely used.

Polyesters

These are usually the polyethylene terephthalate (PET) material and used for a good number of engineering and consumer applications. It has been used for auto parts, gears, cams and water bottles. PETG is the glycol modified PET and is also used for similar applications. A major application is the clothing and auto-tyre re-inforcements. These generally have higher strength compared to other engineering plastics.

Polycarbonates

It is also a polyester made by the combination of carbonic acid and an aromatic bisphenol. These are amorphous polyesters with good transparency, excellent mouldability and high-impact strength. They are more expensive but preferred in applications where high impact strength is desired. Other mechanical properties are comparable to that of ABS and nylon.

Some of the common thermoplastic materials are given in Table 12.4 with their mechanical characteristics.

TABLE 12.4 Thermoplastic materials

Plastic	Famous Trade Name	Relative Tensile Strength	Relative Impact Strength	Behaviour Under Load
Acetal	Delrin	High	Medium	Strong, stiff and good toughness
ABS (Acrylonitrile butadiene-styrene)		Medium	High	Strong, tough and resilient
Acrylic (polymethyl methacrylate)	Perspex, Plexiglas	High	Low	Rigid and strong
Nylon (Polyamide)	Zytel	Medium	High	Tough and resilient
Polycarbonate		High	High	Strong and tough
Polyethylene		Low	High	Tough
Polypropylene		Low	Medium	Tough
Polystyrene		Medium	Low	Hard and rigid
Polyvinyl chloride (PVC)		High	Low	Strong and rigid
PVC—acetate copolymer		Low	High	Flexible

12.4.2 Thermosetting Materials

These are plastics that cannot be melted once they are solidified. The raw materials for thermosetting materials are usually called resins, which are mixed and placed in the mould and heated and compressed, during this process the materials achieves the strength and hardness. Polymerisation occurs by strong network bonds (cross linking) with the application of heat, pressure and/or time. The cross links between neighbouring polymer molecules limit chain movement. These materials are characterised by a three-dimensional network of molecules. These materials could not be recycled until recently. When heated, these materials burn and char. Thermosetting materials degrade rather than soften when exposed to excessive heat. Some of the common thermosetting plastic materials are given in Table 12.5 with their mechanical characteristics. The manufacturing processes used are more expensive compared to thermoplastic materials.

TABLE 12.5 Thermosetting materials

Plastic	Famous Trade Name	Relative Tensile Strength	Relative Impact Strength	Behaviour Under Load
Epoxy	Araldite	High	Low	Hard, strong, elastic
Melamine		High	Low	Hard, strong, elastic
Phenolic (Phenol-formaldehyde)	Bakelite	High	Low	Hard, strong, elastic
Urethane		Medium	High	Tough and strong

Phenolics

This is the oldest thermosetting plastic. These plastics have the basic structure in the form of a ring similar to benzene ring, with one H atom replaced by OH, called phenol. The moulding resin is obtained by mixing phenol with formaldehyde (CH_2O) to obtain the monomer phenol-formaldehyde. Phenol-formaldehyde forms a network of structure that is hard and very rigid. Polymerisation of this is accomplished by introducing cross-linking it into a three-dimensional structure by inputting energy in the form of heat. The tradename **Bakelite** for phenol-formaldehyde is widely used to produce moulded electrical parts such as plugs, switches, electric sockets, etc. These are by far the largest volume in usage among the thermosetting plastics. Another major usage is as adhesive for plywood. They have excellent mechanical strength and high toughness. These are also the most economical plastic material and hence used extensively.

Alkyds

This is a derived name from alcohol and acid, and is a family of thermosetting plastics. These are essentially polyesters but for all practical purpose kept separate from them in view of the processing used. The common acids used are phthalic and isophthalic acids with glycerine and ethylene glycol as the common alcohols. These are widely used in paints.

Melamine Formaldehyde

Melamine is another very rigid thermosetting plastic. This can be easily moulded and is low-cost, rigid, non-toxic and has good toughness. This is widely used in food handling.

Typical mechanical properties of engineering plastics (thermoplastic as well as thermosetting) are given in Table 12.6.

12.5 EXTRUSION OF PLASTICS

As explained in Chapter 7, extrusion is the process of confining the material in a closed cavity and then allowing it to flow from only one opening so that the metal will take the shape of the opening. The operation is identical to the squeezing of toothpaste out of the toothpaste tube.

Extrusion can be used to process most thermoplastics such as polyethylene, polypropylene, polyurethane, polystyrene, polyamide, polyester and flexible polyvinyl chloride. A characteristic that often differentiates extruded from injection-moulded plastics is the viscosity of the plastic at normal processing temperatures. Extruded plastics often have a higher melt viscosity that allows the extrudate to retain the shape imparted to it by the die while the extrudate is in the cooling stages.

TABLE 12.6 Mechanical properties of plastic Materials

Plastic	Molecular Packing	Mechanical Properties at Room Temperature			
		Tensile Strength, MPa	Percent Elongation	Compressive Strength, MPa	Izod Impact Strength, Nm @24.4°C
Polystyrene	Amorphous	52	2	97	0.4
High-impact polystyrene	Amorphous	34	10	52	0.8 to 13.6
Acrylics	Amorphous	69	6	103	0.5
Polycarbonate	Amorphous	62	100	69	20.3
ABS	Amorphous	41	30	55	8.1
Acetal	Crystalline	69	5	124	2.7
Nylon	Crystalline	83	400	69	3.4
Polypropylene	Crystalline	31	500	48	1.4
Polyethylene (High density)	Crystalline	28	600	21	13.6
Polyethylene (Medium density)	Crystalline with some amorphous region	17	600	21	10.9
Polyethylene (Low density)	Semi-Crystalline	10	700	21	No break
Epoxy	Cross-linked network	69	3	138	1.1
Phenolic	Cross-linked network	48	2	69	0.5

It is possible to combine a variety of resins to gain special physical, biological or chemical properties. Also, as explained earlier, additives that include lubricants, thermal stabilisers, antioxidants and colourants, can be used during the extrusion process to enhance processing characteristics of the polymer or to alter product properties.

The plastic that is generally in the form of pellets or granules is fed into the extruding machine through a hopper as shown in Fig. 12.2. This machine is known as an **extruder**. Pellets move into the barrel by gravity. The barrel is heated by electric heaters which soften the pellets in the barrel. The screw moves the pellets continuously and the mechanical working and the friction present generate more heat to soften the plastic. The heating of the barrel may be reduced once sufficient heat is supplied. The rotating screw has three main sections as shown in Fig. 12.2. In the feed section, the pellets are moved from the hopper and preheated. In the compression section, the plastic is softened sufficiently so that it will flow smoothly like a liquid. Finally in the metering section the plastic is completely homogenised and sufficient pressure is developed so that the material will be forced through the die. The feeding capacity of the screw is determined by its geometry and speed of rotation.

There is a breaker plate (a plate with several small holes) between the die and the barrel which improves the mixing of the plastic before entering the die. The extruded product called extrudate is cooled by blowing air or by passing through a water-filled channel.

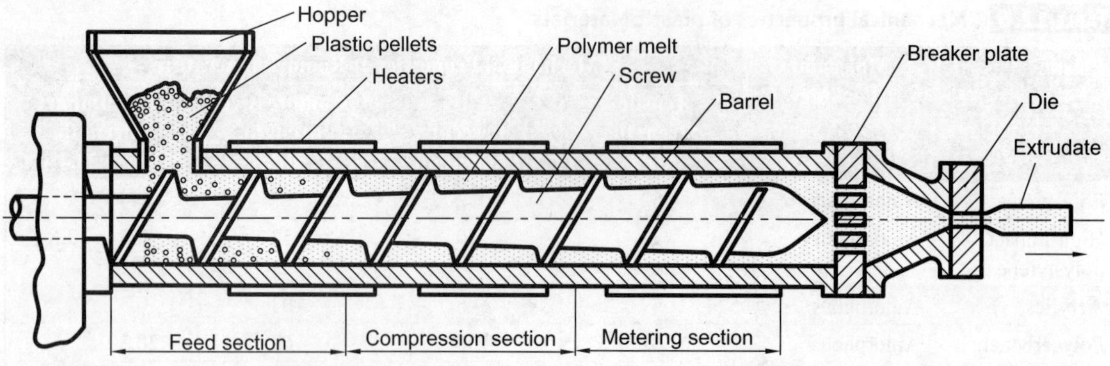

FIG. 12.2 *Constructional features of a plastic extruder*

The die is a metal plate placed at the end of the extruder with a cut out of the shape of the cross section of the product being manufactured. Any complex external contour can be extruded with relatively simple and inexpensive tooling. An example of the die cross section and the product produced by it are shown in Fig. 12.3. A die, which is a simple plate with the shape of the part cut through the die, with no transition guiding the material, is a low-cost design used for low-volume production. However, a streamlined die provides a better flow of the plastic with minimum number of defects. In this type of die, channels to flow the material evenly throughout the die are made. This design is most expensive and ideal for high-volume jobs.

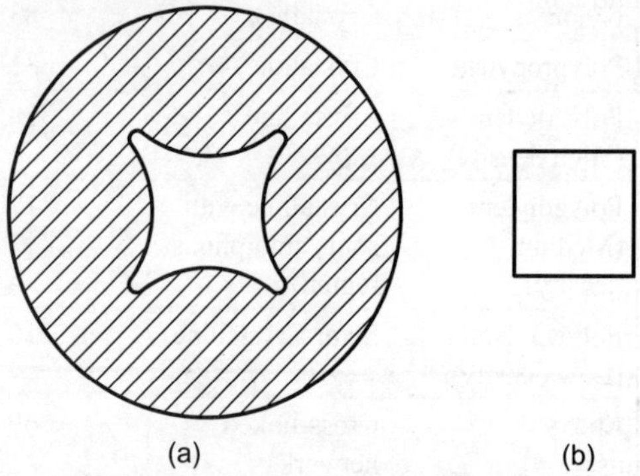

FIG 12.3 *Typical shape of a final product (b) obtained from the die cross section (a)*

The parts for extrusion should have consistent wall thickness as it allows for an even flow of material through the die that produces more controlled parts with a lower tooling cost. As far as possible, avoid designing profiles with hollow sections, since they add significantly to the cost of both the part and the tooling.

There are a number of plastic-extrusion processes used in the industry.

Solid Extrusion

This is the process that is just described earlier. This produces a part with a uniform cross section.

Hollow Extrusion

This process is used to extrude parts that have hollow cross sections such as pipes and tubes. The dies used for this operation need to have special spider sections to form the hollow part. The dies also need to have the streamlining sections built into the die to allow for the plastic to flow smoothly. The dies are more expensive compared to those used for solid extrusion.

Co-extrusion

This is the process of extruding two or more materials through a single die with two or more orifices arranged so that the extrudates from the multiple openings merge and weld together into a laminar structure before chilling.

Extrusion Advantages

- Equipment is simple and relatively inexpensive
- Short lead times since it uses simple dies which can be very quickly produced
- Relatively low tooling costs
- Overall cost of the parts produced by extrusion is low
- Practically, any cross section can be easily produced

12.6 INJECTION MOULDING

Injection moulding is similar to pressure die casting discussed in Chapter 6. In this process, plastic material in a highly softened state is forced to flow at high pressure through a nozzle into the mould cavity. The plastic solidifies in the die and then ejected by opening the die. The shape of the component is almost in its final form and can be produced in an extremely fast rate. Typical cycle times may be of the order of 10 to 30 seconds. Unlike molten metal in diecasting, plastic melts have a high viscosity and have to be injected with a large force into the hollow mould cavity. More melt must also be packed into the mould during solidification to avoid shrinkage in the mould. Injection moulding pressures usually range from 70 MPa to 200 MPa.

Injection moulding is the most widely used plastic-processing method. It can be used to produce a wide variety of products. Very complex parts can be produced whose sizes may range from very small (50 g) to very large (25 kg) with excellent control of tolerances. Most polymers may be injection moulded, including thermoplastics, fibre reinforced thermoplastics, thermosetting plastics and elastomers. Structural injection moulding is also possible in which a core and skin may be made of different polymers. Reaction injection moulding and liquid injection moulding, which differ in the manner of mixing ingredients, involve the injection of liquid polyurethane systems that polymerise within the mould. Moulds with moving and unscrewing mandrels are not unusual, and allow the moulding of parts with multiple cavities and internal and external threads. Metallic inserts such as screws, pins and strips can also be placed in the mould cavity to become an integral part of the injection-moulded product.

The schematic of injection-moulding machine construction is shown in Fig. 12.4 and 12.5. The machine in Fig. 12.4 shows a machine that has reciprocating and rotating screw that is used for pressurising the softened plastic. This is the common pressurising system found with most modern plastic-injection-moulding machines. The other machine, shown in Fig. 12.5, shows a reciprocating plunger with a torpedo for pressurising. The spreader or torpedo is used to produce uniform flow around machine component peripheries and to produce desired flow-induced molecular structure. This is less common in the current industrial machines.

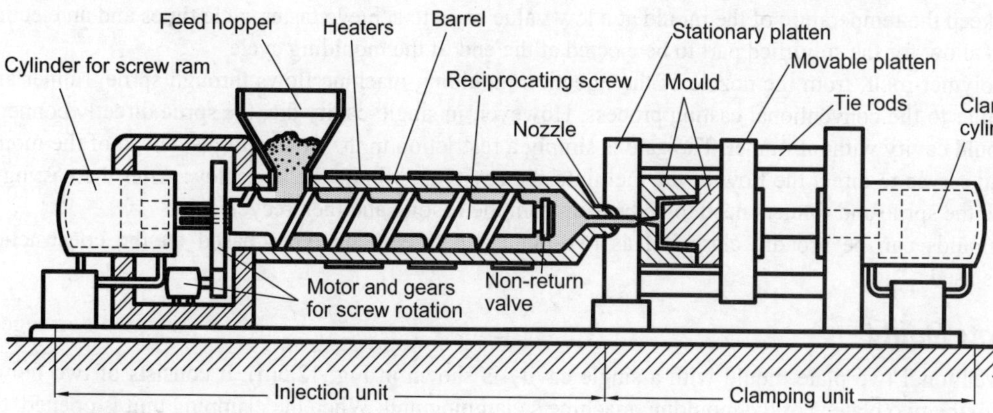

Fig 12.4 *Constructional features of an injection-moulding machine with a reciprocating screw*

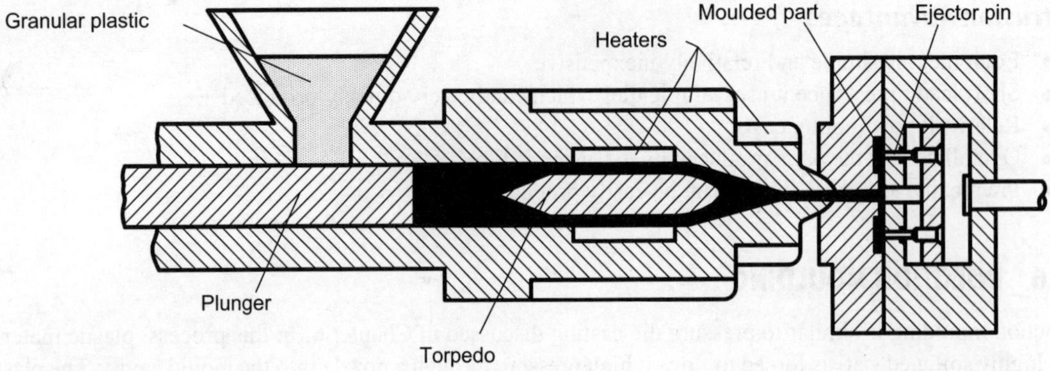

Fig. 12.5 *Schematic of an injection-moulding machine with a reciprocating plunger and torpedo*

The machine can be broadly broken into two parts, an injection unit and a clamping unit. The injection unit of the reciprocating screw type machine is very similar to the system described for extrusion machine. The function of the screw is to soften and pressurise the melt so that it will be injected into the mould through the nozzle. The plastic granules or pellets will be fed by gravity from the hopper into the barrel. The rotating screw will move them through the plunger, while receiving heat from the electric resistance heaters around the barrel. As the softened plastic reaches the end of the barrel, the screw acts like a ram and then pushes the plastic through the nozzle. A non-return valve at the end of the screw ensures that plastic will not flow along the threads of the screw. After the injection is over, the screw returns to its normal position to continue with the next moulding cycle.

The clamping unit takes care of operating the mould. It holds the two halves of the mould securely in the mould-closed position for the purpose of injection process and opens the mould after the cooling of the plastic inside the mould and ejects the casting from the mould. The structure of the clamping unit is very rigid to provide the necessary locking force, to take care of the injection pressure depending upon the size of the part being moulded. Later the mould is closed to start the next moulding cycle.

12.6.1 Injection Moulds

The quality of the part will be dictated by the methods used in the design and manufacture of the mould. It is, therefore, necessary to understand the various elements of a mould to get the best out of it. The mould consists of the mould cavity, passage way to transport the plastic melt into the mould cavity, a cooling system that will keep the temperature of the mould at a low value so as to achieve faster cycle times and an ejection system to allow for the solidified part to be ejected at the end of the moulding cycle.

The polymer-melt, from the nozzle of the injection moulding-machine flows through sprue, runner and gates similar to the conventional casting process. However, in single-cavity die, the sprue directly connects to the mould cavity without runner. The gate is simply a restriction in the flow path just ahead of the mould cavity and serves to direct the flow of the melt into the cavity and limits the back flow. After the casting is solidified, the sprue and runners need to be broken from the casting and then recycled.

The moulds can be broadly classified as two-plate and three-plate types based on the construction principles used.

Two-Plate Mould

The conventional two-plate mould with a single cavity is shown in Fig. 12.6(a). It consists of two halves fastened to the two platens of the moulding machine's clamping unit. When the clamping unit is opened, the two mould halves open, as shown in Fig. 12.6(b) for ejecting the solidified part. The parting surface is where the mould opens to remove the part. This is also called a **single daylight mould**. Daylight refers to the space between the mould halves when it is open.

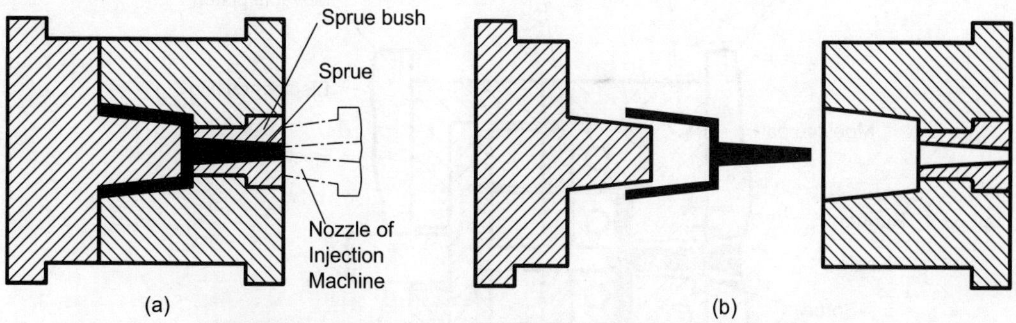

Fig. 12.6 *Schematic of a two-plate mould with a single cavity*

It is also possible to have a two-plate mould with multiple cavities to produce more than one part in a single shot, as shown in Fig 12.7(a). In this figure, there are two cavities to make two parts simultaneously. The parting surface is similar to Fig. 12.6 where the mould opens to remove the parts. It may be noticed in this case that the plastic melt flows from the nozzle of the injection barrel into the mould cavity through a sprue, runner and gates similar to sand casting. The **gates** constrict the flow of plastic melt into the mould cavity. There are one or more gates for each cavity in the mould depending upon the number of cavities. The two-plate mould is the most common mould in injection moulding.

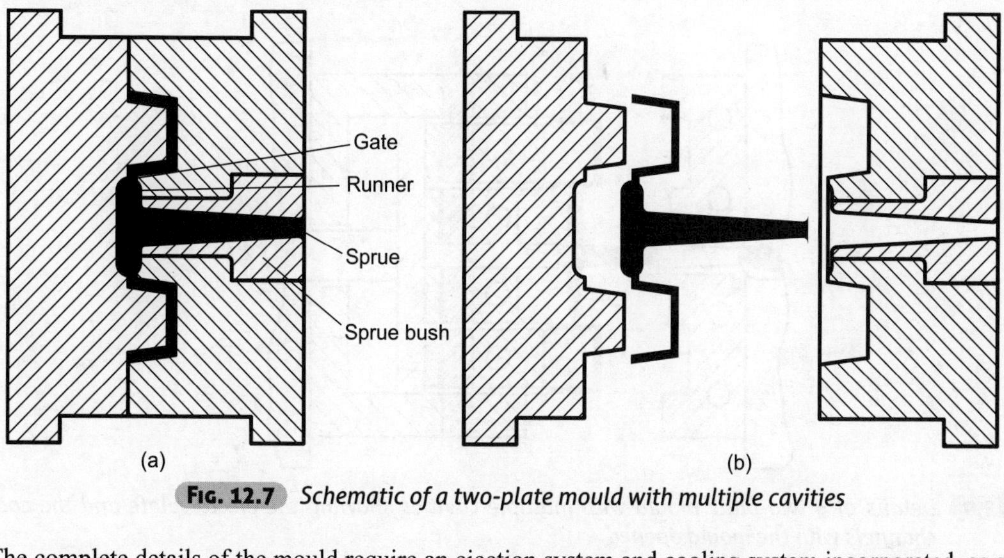

Fig. 12.7 *Schematic of a two-plate mould with multiple cavities*

The complete details of the mould require an ejection system and cooling system incorporated, as shown in Fig. 12.8. The ejection system is needed to eject the moulded part from the cavity at the end of the moulding cycle. Ejector pins are built into the moving half in the ejector-plate housing, as shown in Fig. 12.8. The natural shrinkage of the part allows the solidified part to stick to the moving half of the mould. At the end of the moulding cycle when the mould opens, part will move with the moving half. Then the part is ejected by the ejector pins a shown in Fig. 12.9. The parts and runners will fall by gravity into a container that is placed underneath the mould, thus automating the mould cycle.

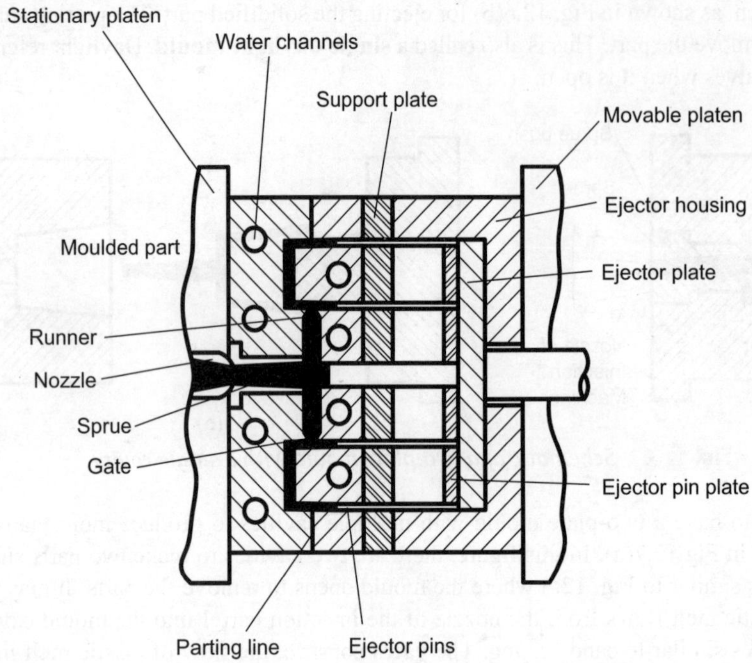

Fig. 12.8 *Details of a two-plate mould with multiple cavities showing the ejector plate and the cooling channels*

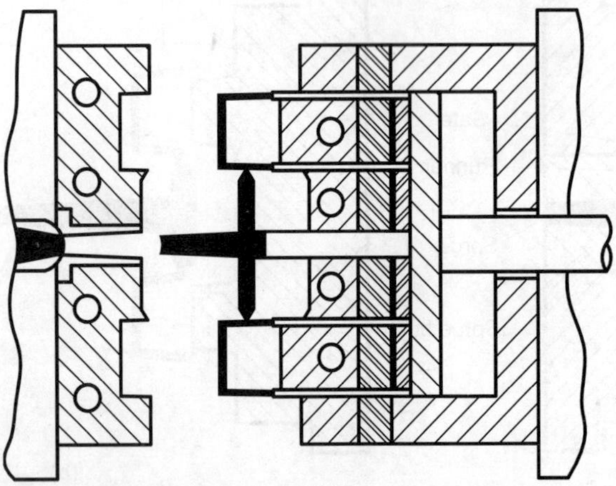

Fig. 12.9 *Details of a two-plate mould with multiple cavities showing the ejector plate and the cooling channels with the mould opened*

As the moulding cycle is repeated, heat gets accumulated in the mould, thereby raising its temperature. A provision needs to be made in the moulds to extract this heat so that the moulding cycle remains short. This is accomplished by providing a cooling system in the mould. Cooling channels are made in the mould around the mould cavity such that cooling water can be circulated through them as shown in Fig. 12.8. An external water pump is connected to these passageways in the mould continuously to remove the heat from the hot plastic. It is also necessary to remove the air from the mould cavity as the plastic rushes in. Though much of the air passes through the small ejector-pin clearances in the mould, it might be necessary to machine narrow air vents into the parting surface to permit the escape of air present in the mould cavity. The size of the air vents is 0.03 mm deep and 12 to 25 mm wide which is large enough to permit the air to escape to the outside but are too small for the viscous plastic melt to flow through.

Three-plate Mould

An improvement of the two-plate mould where the complete automation of the moulding cycle can be done is shown in Fig. 12.10. It consists of a feed plate that is attached to the stationary platen of the machine. The flow of plastic melt is through a gate located at the base of the part allowing more even distribution of melt into the sides of the part. In the two-plate mould the plastic melt enters through the side possibly creating a weakness at the weld line. The second plate is the floating cavity plate while the third plate is the ejector plate housing connected to the movable platen of the injection-moulding machine. When all the plates are moved as required, there are two spaces between the mould plates as shown in Fig. 12.11, and hence it is called a **double-daylight mould**. As shown in Fig. 12.11 the runner system and part gets separated during the ejection process without operator intervention.

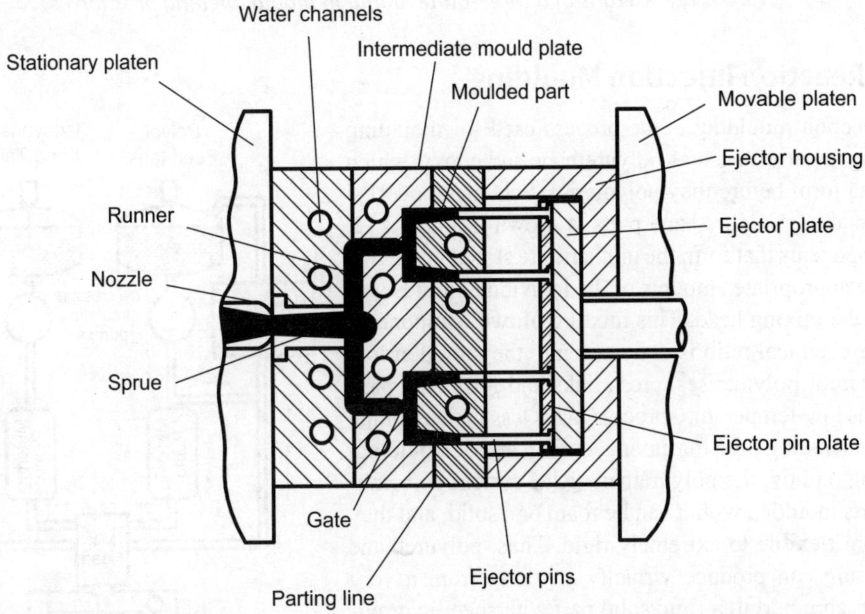

Fig. 12.10 *Details of a three-plate mould with multiple cavities showing the ejector plate and the cooling channels*

In the moulds described so far, the plastic melt in the sprue and runners is allowed to cool and solidify. These are separated from the part and recycled. However, the energy spent in the recycling is a waste and also the material deteriorates as the number of remelts of the material takes place. To reduce this hot runner, moulds are used. It contains a heated runner manifold block within the mould. While the plastic in the mould cavity solidifies, the material in the sprue and runner channels remains molten, ready to be injected into the cavity in the next mould cycle. Hot runner moulds reduce the overall cost by reducing the recycling of sprues and runners.

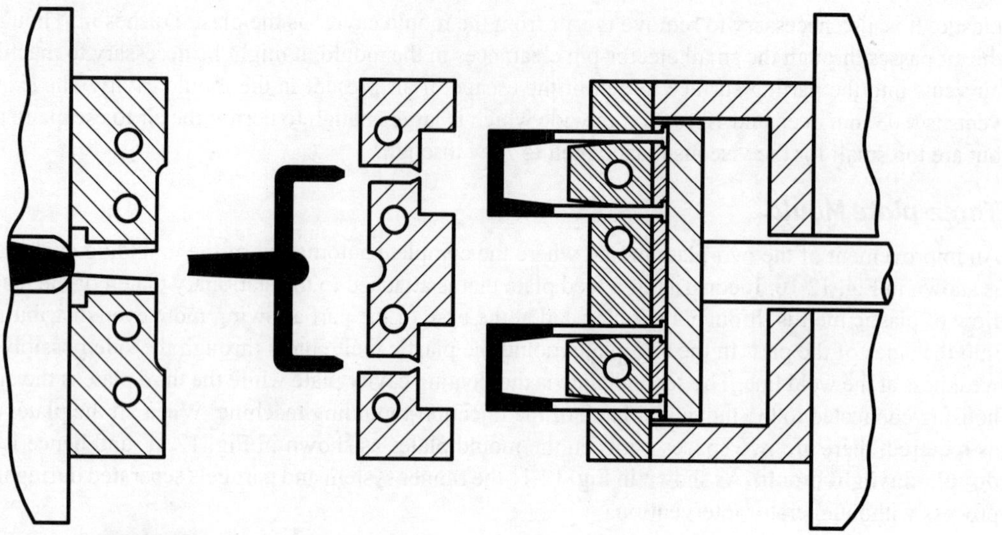

Fig. 12.11 *Details of a three-plate mould in mould opening position*

12.6.2 Reaction Injection Moulding

Reaction injection moulding is the process used for moulding thermosetting materials such as polyurethane and epoxy, which exist in liquid form before they polymerise. Schematic of RIM process for making polyurethane parts is shown in Fig. 12.12. The two components that form the urethane are stored in the two tanks and the appropriate amounts of the individual liquids will be mixed in the mixing head. This mixture of two components in the proper chemical ratio is delivered into the mould at low pressure, where it polymerises, cross-links and forms the part. Typically, this low-temperature process takes less than a minute to complete, including time for mixing, curing, and demoulding.

Depending on how the polyurethane RIM system is formulated, the parts moulded with it can be foam or a solid, and they can vary from flexible to extremely rigid. Thus, polyurethane RIM processing can produce virtually anything from a very flexible foam-core part to a rigid solid part with specific gravities ranging from 0.2 to 1.6. The process was originally developed to mould very large automobile parts such as bumpers, interior trim panels and spoilers. Because of the low pressures used, the cost of the mould is much less compared to the conventional injection moulding.

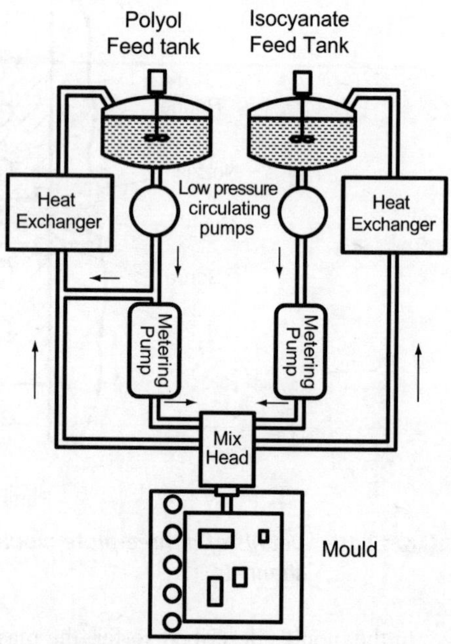

Fig. 12.12 *Schematic of a reaction injection-mould system*

12.6.3 Liquid Injection Moulding

Liquid injection moulding is used for moulding silicone products. In this process pumping, systems deliver a two-part liquid silicone (catalyst and cross-linker) directly into a mixer for homogenisation. This mix is then injected directly into heated mould cavities in as little as 3 to 10 seconds using a relatively low pressure. Moulding and vulcanisation (curing) occur inside the mould cavities within 10 to 90 seconds due to the high mould temperature. Entire processing takes place within a completely closed system ensuring cleanliness. Main advantages of this process are

- Automated process with completely integrated processing (this ensures cleanliness with minimised contamination risk)
- The injection pressures used are low which means it requires less clamping force for the moulds
- Faster cycle times with consistent part quality

12.6.4 Co-Injection Moulding

This is a process that uses two materials to mould a part. The two materials have different properties with one being hard that forms the skin, while the softer one forms the core through injection-moulding process. It requires two injection units for the two different plastics being used. The skin material is injected first into the mould cavity and is immediately followed by a core material. As the skin material flows into the cavity, the material next to the cavity walls freezes and when the core material enters, it displaces the skin material in the centre of the channel by pushing the skin ahead. As the skin material flows ahead, it continues to freeze on the walls producing the skin layer.

Typical applications of this process are the following:

- The core can be formed to reduce the weight of the part and at the same time absorbs the noise.
- Cores can be glass filled to get improved mechanical properties.
- Environment friendly since recycled material can be embedded as core without worrying about the appearance of the part. Glass-filled cores for improved physical properties.
- High gloss skin material over structural core material for combination of aesthetic and structural properties.

12.7 BLOW MOULDING

Blow moulding is the process of inflating a hot, hollow, thermoplastic preform or parison inside a closed mould so that its shape conforms to that of the mould cavity. A wide variety of hollow parts, including plastic bottles can be produced from many different thermoplastic materials using this process.

The cost of blow-moulded parts is higher than that of injection-moulded parts but lower than rotational-moulded parts. Tooling costs are lower compared to injection moulding. The cycle times are relatively higher compared to injection moulding because of the process involved. Typical parts made are bottles, toys, air ducts for automobiles, chemical and gasoline tanks, and a number of household goods.

12.7.1 Extrusion Blow Moulding

An extrusion blow-moulding machine consists of an extruder similar to that used with plastic extrusion, which softens the plastic and forms it into a tube (called a parison or preform) through a conventional-type die and a split-body mould as shown in Fig. 12.13. The die closes around the parison, sealing both ends as shown in Fig. 12.13(b). A blow pin is inserted into the parison to inflate it, causing it to expand and confirm the shape of the mould cavity Fig. 12.13(c). Again, the mould is cooled and once the part has solidified, the mould opens and the part is removed Fig. 12.13(d). Extrusion blow-moulding is a continuous process that is used to mostly to manufacture small, thin-walled parts but can produce parts such as large drums.

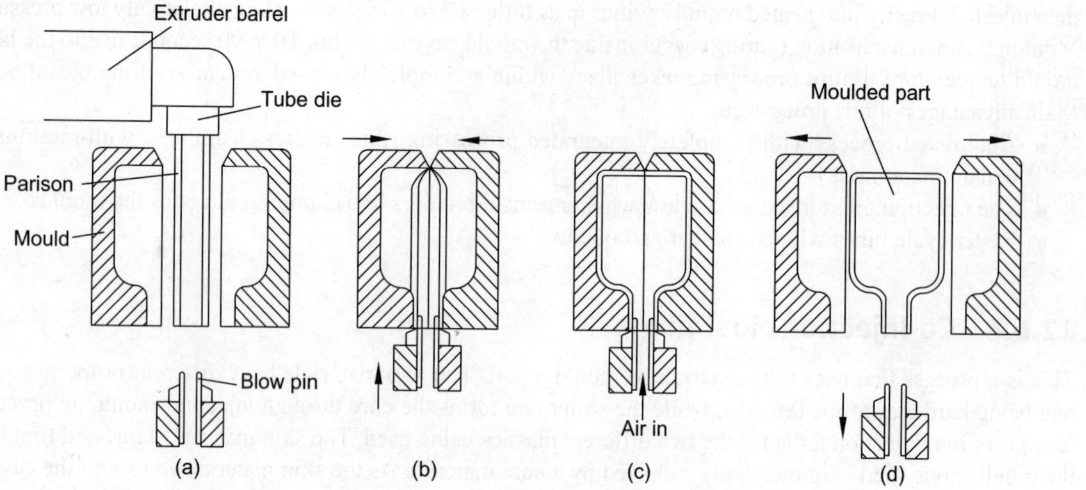

FIG. 12.13 *Operating steps in a blow-moulding operation: (a) Hollow parison is formed with extruder (b) The mould closes around the parison and blow pin inserted (c) Air is blown into the parison thus expanding it to confirm to the contours of the mould (d) The mould opens and the finished part is removed*

Extrusion blow moulding can be used for a wide variety of container shapes, sizes and neck openings. These machines are very fast and often can produce 300 to 350 bottles per hour. Extrusion blow moulds are generally much less expensive than injection blow moulds and can be produced in a much shorter period of time. This process can use many different plastics, including HDPE, PVC, PC and PP. The capital investment required for extrusion blow moulding is low and is suitable for small production runs. Some disadvantages usually include a high scrap rate, a limited control over wall thickness, and some difficulty of trimming away excess plastic.

In the blow moulds, coolant flow channels are provided to accelerate part cooling, thereby reducing the cycle time. Also, there are raised regions on the die face to pinch off and seal the parison before blowing. Some times recessed regions are provided for flash to flow into, thereby minimising the potential for mould separation due to flashing. Vents may also be provided which are small channels at the mould wall end, to allow air to escape from between the part wall and mould surface.

Mould material strength is not an important criterion as blow-moulding pressures are relatively low compared to other moulding operations. Majority of moulds are made from high strength aluminium alloys. However, mould wear may become a problem. Plated steel and beryllium-copper are alternative materials for moulds or these more wear-resistant materials can be used for various components of aluminium moulds, e.g. inserts and pinch-offs.

Extrusion blow moulding is usually used to make items of weight greater than 350 g such as containers for food, laundry or waste.

12.7.2 Injection Blow Moulding

Injection blow moulding is a two-stage process with the parison being produced in the first stage which needs to be transferred to the blow mould. Thus the first operation is identical to the injection moulding described earlier. The plastic melt is injected into a heated preform mould around a hollow mandrel blow tube or core rod as shown in Fig. 12.14(a). The work piece for the second stage in the blow-moulding process is the parison and the blow-rod assembly. This assembly is then transferred to the blow mould and the mould is closed around it as shown in Fig. 12.14(b). Between the preform production and the blow-moulding processes, the heated preform may be held in a temperature-conditioning stage or a cooled preform is re-heated. If the preform mould is provided with cooling channels, the coolant may be heated to a temperature lower than the melt temperature but high enough so that the preform can be directly transferred to the blowing station with no, or little, temperature conditioning. Air is introduced through the mandrel to inflate the part to conform to the internal cavity of the mould as shown in Fig. 12.14(c). The air is injected into the plastic at a pressure between 0.50 to 1.0 MPa. Once cooled, the mould opens, and the part is indexed to the ejection part of the machine where the finished part is removed from the mandrel as shown in Fig. 12.14(d).

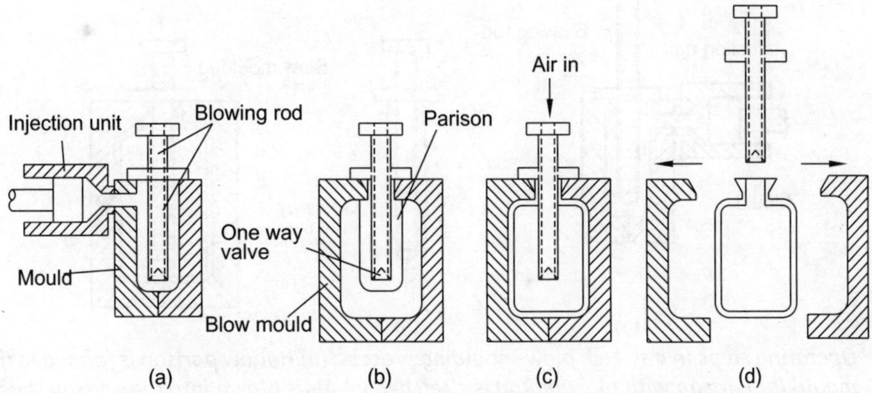

FIG. 12.14 *Operating steps in a injection blow moulding process: (a) Hollow parison is formed in the injection mould (b) Parison with blowing rod is transferred to the blow mould (c) Air is blown into the parison thus expanding it to confirm to the contours of the mould (d) The mould opens and the finished part is ejected*

Injection blow moulding is used to achieve very accurate wall thickness, high-quality neck finish, and wide mouth openings and to process polymers that cannot be extruded. Generally, an injection blow-moulded container's material is distributed evenly throughout and do not need any trimming or reaming. It is suitable for smaller containers. Injection-blown containers usually have a set gram weight which cannot be changed unless a whole new set of blow stems are built. Usual applications include pharmaceutical, cosmetic single-serving liquor bottles that weigh less than 350 g.

12.7.3 Stretch Blow Moulding

Stretch blow moulding is best known for producing PET bottles commonly used for water, juice and a variety of other products. It produces a part with the required properties for the work material by producing desirable molecular orientation. In this process, a preform or parison is elongated mechanically in the mould and then expanded radially in the blowing process. As a result, the desirable resulting molecular orientation yields a material with increased strength. This biaxial stretching of material increases the tensile strength, barrier properties, drop impact, clarity and top load in the container. With these increases, it is usually possible to reduce the overall weight in a container by 10 to 15 percent less than when producing a container by simpler blow-moulding techniques. Stretch blow moulding is only used for difficult to blow crystalline and crystallisable polymers such as polypropylene and polyethylene terephthalate (PET).

In order to produce and retain the desired structure and specified properties, the stretching and blowing processes need to be carried out at temperature lower than in other blow-moulding processes. Thus, the allowable temperature range for this will be smaller and hence is more difficult to control. A temperature-conditioning station in-line is required, or a reheating operation needed for preforms. This increases the difficulty of process design and operation, and material specification since plastic properties depend on temperature history.

Stretch blow moulding starts the production of a preform similar to the injection blow moulding process. The parison and the blow-rod assembly is then transferred to the blow mould, and the mould is closed around it as shown in Fig. 12.15(b). The preform is reheated before it is placed in the bottle mould. The softened parison is stretched to about twice its original length as shown in Fig. 12.15(b). Compressed air is then blown into the stretched parison to expand to the contours of the mould, as shown in Fig. 12.15(c). Once the bottle is cooled, the mould is opened and the finished bottle is removed from the mould cavity.

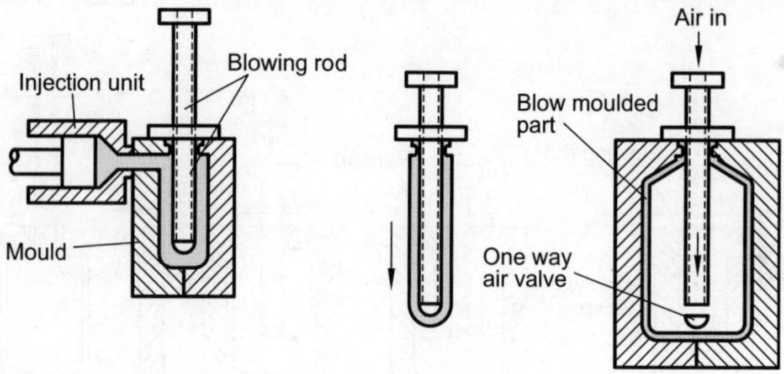

Fig. 12.15 *Operating steps in a stretch blow-moulding process: (a) Hollow parison is formed in the injection mould (b) Parison with blowing rod is stretched (c) Air is blown into the parison thus expanding it to confirm to the contours of the mould*

A comparative evaluation of the different blow-moulding techniques is given in Table 12.7.

Table 12.7 Comparison of the different blow-moulding techniques

Operation	Advantages	Disadvantages
Extrusion blow moulding	High production rate Low tooling cost Wide selection of equipment	Large amount of scrap Uses recycled scrap Limitation on wall thickness Trimming facilities required
Injection blow moulding	No scrap Excellent thickness control Accurate neck finish Outstanding surfaces Can produce low volumes	High tooling cost Large objects not possible
Stretch blow moulding	Economical Improved properties Accurate control of wall thickness Reduced weight allowed	

12.8 THERMOFORMING

In this process, a thermoplastic sheet can be formed into a three-dimensional shape by the application of heat and differential pressures. First, the plastic sheet is clamped to a frame and uniformly heated to make it soft and flowable. Then a differential pressure (either vacuum, pressure or both) is applied to make the sheet conform to the shape of a mould or die positioned below the frame. There are a number of variants in this process.

It is possible to use most of the thermoplastic materials. The starting material is plastic sheet of uniform thickness. The tooling used is generally made with machined aluminium. However, for short run parts, even wood can be used as the mould material. Thus, tooling costs are generally low. It is a relatively simple process and is used for making such parts as covers, displays, blister packaging, trays, drinking cups and food packaging.

12.8.1 Vacuum Thermoforming

Vacuum forming is a thermoforming process that forms thermoplastic sheets into three-dimensional shapes through the application of heat and vacuum. During vacuum-forming process, plastic material is heated (170 to 200°C) until it becomes pliable, and then is placed over a mould of the requisite shape and drawn in by a vacuum until it takes on the desired shape as shown in Fig. 12.16. The application of a vacuum draws out the air between the mould and the sheet so that the plastic conforms to the mould exactly. This is accomplished through venting holes in the mould that are joined to vacuum lines. The mould also has an integrated water cooling system that brings the temperature of the plastic to the set temperature needed. Once the curing temperature is reached after the part is formed, air flows back into the mould and separates the new part from the mould.

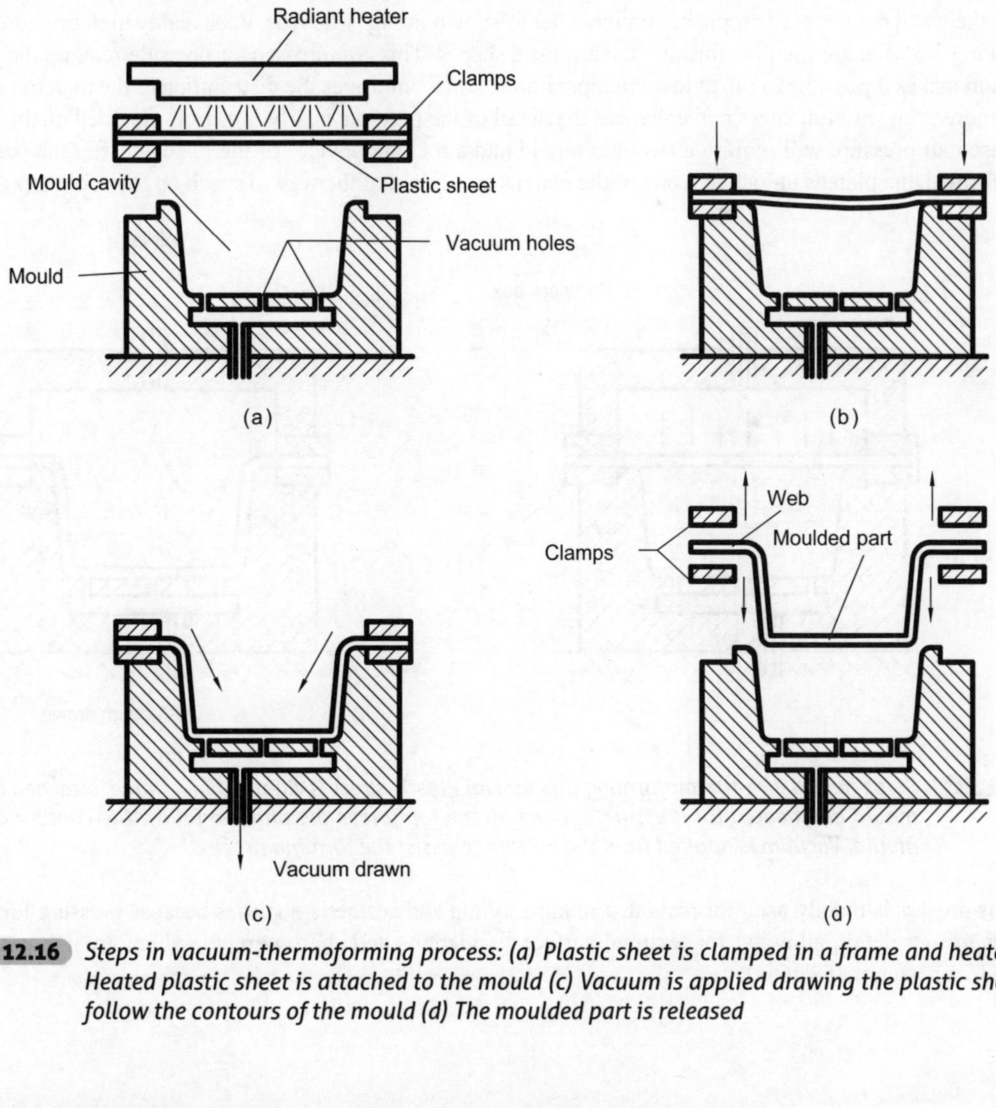

FIG. 12.16 *Steps in vacuum-thermoforming process: (a) Plastic sheet is clamped in a frame and heated (b) Heated plastic sheet is attached to the mould (c) Vacuum is applied drawing the plastic sheet to follow the contours of the mould (d) The moulded part is released*

The heating time depends on sheet thickness, surface finish, material colour, heater temperature, and the type of heaters used. Infrared heating elements are the most commonly used heating source. Generally, excessive sheet heating leads to colour shift, discolouration, surface blistering, de-lamination, and loss of both physical and electrical properties. Cooling begins the instant the sheet is transferred from the oven to the forming station. The transfer time should be only a few seconds. A good vacuum system with the capacity to quickly evacuate the volume of the mould is essential. Typical values for the vacuum developed by the vacuum pump should be about 35 Torr.

Tooling for vacuum thermoforming is less expensive compared to injection moulding; as a result it is more cost-effective. Normally, it is an economical choice for small and medium production runs. Time of production is generally short. Sharp, precise details are available for many products, which makes vacuum formed plastics an attractive alternative to other moulding processes.

12.8.2 Pressure Thermoforming

Pressure thermoforming is an improvement over vacuum forming as in that it utilises both vacuum and compressed air to force the plastic sheet against the mould. As the platens are closed, the vacuum pulls on one side of the sheet and compressed air pushes on the other as shown in Fig. 12.17(b). Reasonably high pressures approaching 3.5 MPa are used for forming the requisite shapes. This compressed air pressure reduces the cycle time and makes it possible to run at lower temperatures. It also improves the distribution of the material creating a more even wall thickness and enhances the detail of the part to a nearly injection-moulded quality. The increased air pressure will require a stronger mould and a locking device for the platens. After the part has been formed, the platens unlock and one of the platens moves out of the way to speed up the cooling process.

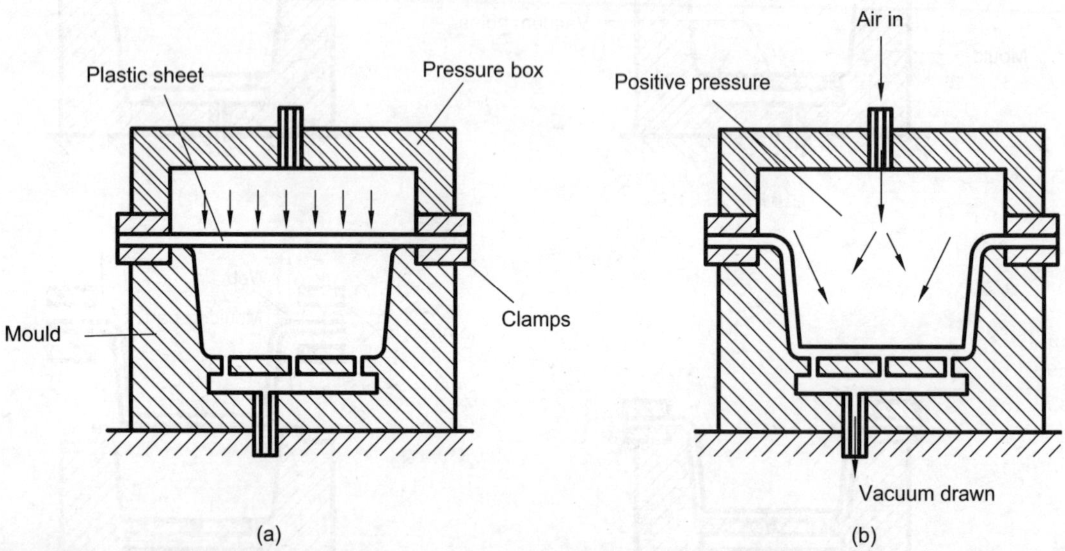

FIG. 12.17 *Steps in pressure-thermoforming process: (a) Plastic sheet is clamped in a frame, attached to the mould and heated (b) Pressure applied on the top forces the plate to follow the contour of the mould. Vacuum is applied from the bottom to assist the forming process*

This process is mainly used for parts that require styling and aesthetic qualities because pressure forming creates greater detail, allowing for textured surfaces, undercuts and sharp corners, which are not as easily created with vacuum forming.

12.8.3 Drape Forming

The moulds shown in the two processes are called as negative (female) moulds. Since the part is drawn towards the negative mould, the exterior of the part will have the exact surface detail that is present on the mould surface. The interior side will be an approximation of the same. The problem with the negative moulds is that forcing the material to follow the negative contours is more difficult. Figure 12.18 shows a thermo-forming operation with a positive mould (male mould). It is termed drape forming.

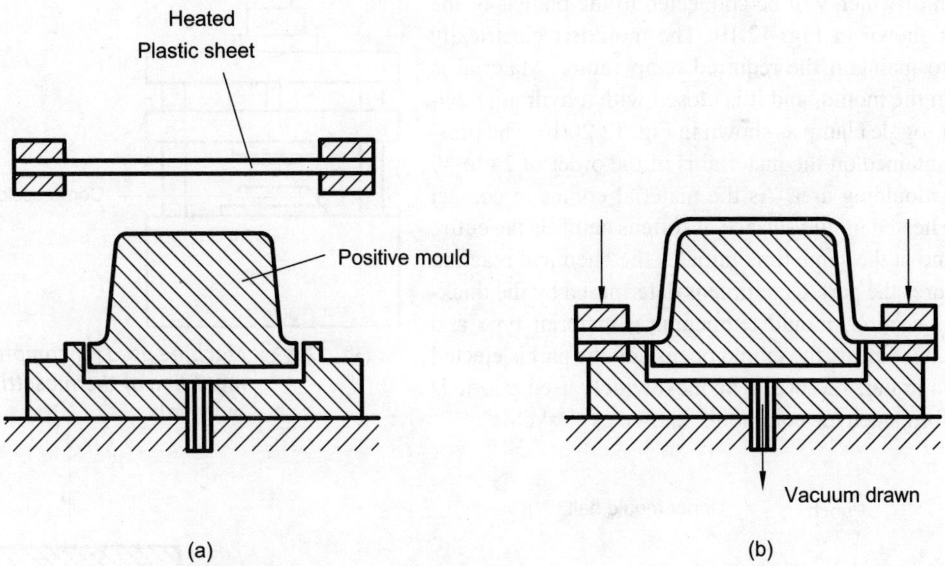

FIG. 12.18 *Steps in thermoforming with a positive mould*

In drape forming, the sheet is framed and heated, it is mechanically stretched, and a pressure differential is then applied to form the sheet over a positive mould. In this case, however, the sheet touching the mould remains close to its original thickness similar to cup forming of sheet metal as explained in Chapter 8. It is possible to drape form items with a depth-to-diameter ratio of approximately 4 to 1.

Positive moulds are easier to build and generally cost less than the negative moulds. However, positive moulds can be easily damaged. Drape forming can also be used with gravitational force alone. For multi-cavity forming, such as tote trays, female moulds are preferred because they do not require as much spacing as male moulds.

There are many variants of thermoforming processes such as plug-assist forming, vacuum-snap back, billow forming, stretch forming, matched die forming and twin-sheet forming.

12.9 THERMOSETTING MATERIALS

Unlike the thermoplastic materials, thermosetting materials cannot be remelted once polymerisation takes place. Hence, resin compounds of the thermosetting materials need to be formed before the polymerisation process. As a result, the processes used for thermosetting materials are different from that of the thermoplastic materials that were discussed so far. The mould used for these materials should be able to provide the conditions for polymerisation within the moulding machine. Generally, heat and pressure are the two variables that are permitted to vary during the polymerisation process. RIM discussed earlier for thermosetting materials (epoxy and urethane) do not require heat or high pressure for polymerisation.

12.9.1 Compression Moulding

Compression moulding is the oldest plastic-processing method. A compression mould is made of two halves with one each of which will be connected to the platens of the press as shown in Fig. 12.19. The mould is electrically heated to maintain the required temperature. Material is placed in the mould, and it is closed with a hydraulic cylinder, or toggle clamp as shown in Fig. 12.20(b). The pressure maintained on the material is of the order of 14 to 40 MPa of moulding area. As the material comes in contact with the heated mould surface, it softens and fills the entire cavity and at the same time initiates the chemical reaction which cures the part. Cure time is determined by the thickest cross section, mould temperature, material type and grade. After curing, the mould opens and the part is ejected as shown in Fig. 12.20(c). The most widely used plastic is phenol-formaldehyde commonly known as 'Bakelite'.

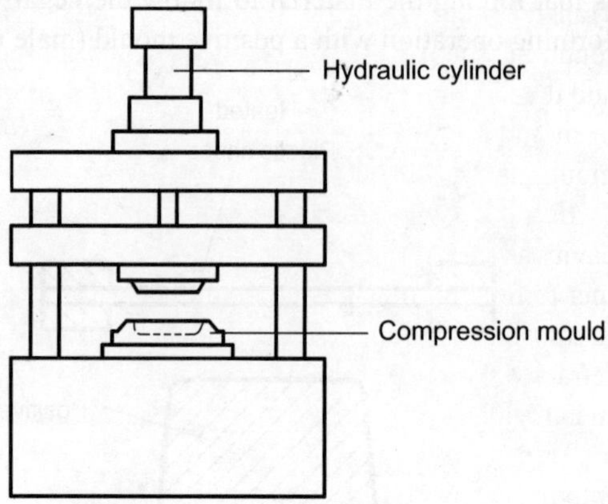

Fig. 12.19 *Machine used for compression moulding of thermosetting materials*

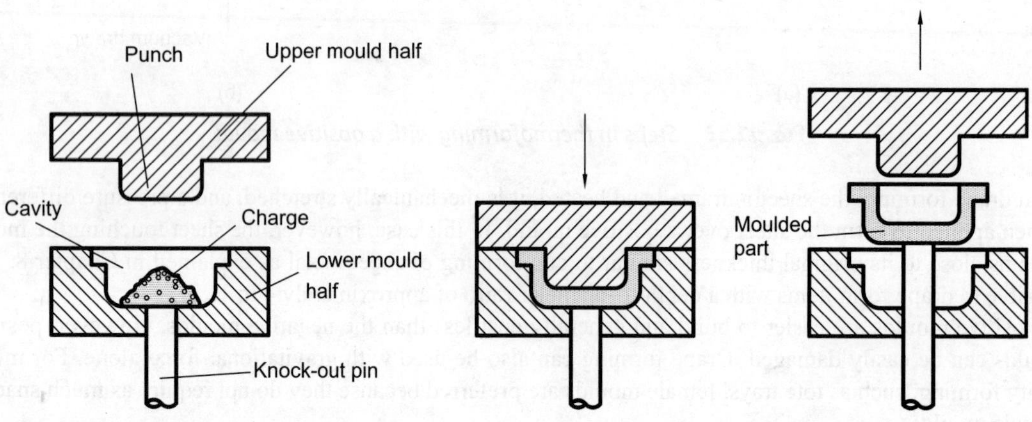

Fig. 12.20 *Steps in compression moulding for thermosetting materials*

The compression moulds are simple and the tooling costs are relatively low. Most of the material in the mould is utilised with very little waste products. The stability and the dimensional accuracy that can be achieved are good. Since the moulding pressures required are low, it is possible to mould large parts with relatively small presses. The process is good for large parts. Shrinkage is minimised and closely reproducible.

In spite of these many advantages of the processes, there are certain limitations which need to be considered while planning this process. The charge placed in the mould need to be tightly controlled. The depth of the part is limited because of the necessity to maintain the rigidity of the punch. Often it is limited to 2 to 3 times the diameter of the punch. Mould design gets tougher for complex part geometries, particularly with uneven parting lines.

12.9.2 Transfer Moulding

Transfer moulding is very similar to compression moulding and is developed to avoid the disadvantages found in that process. In this method, thermosetting charge is heated and compressed in a separate chamber and then injected in to the closed mould where it is allowed to cool and solidify. Transfer moulding is capable of moulding part shapes that are more intricate than compression moulding but not as intricate as injection moulding.

In transfer moulding, the thermosetting charge is loaded into transfer pot immediately above the mould cavity as shown in Fig. 12.21(a). Here, the charge is heated and pressure is applied to force the softened polymer to flow into the heated mould where curing takes place, as shown in Fig. 12.21(b). Curing time required in transfer moulding is less than that required for compression moulding. After the curing, the transfer ram is retracted while the moulded part is ejected from the mould as shown in Fig. 12.21(c). Sprue and the leftover metal with the ram called cull are scrap and need to be discarded since they cannot be recycled.

Transfer moulding also lends itself to moulding with inserts, in which a metal or ceramic insert is placed into the cavity prior to injection and the heated plastic bonds to the insert during moulding. Since the mould is closed before the plastic is injected, the time taken for injection takes less time. It is possible to have longer core pins since they can be supported at both ends. Tool maintenance is relatively low because of the smaller pressures involved. However, it generates waste material in the form of cull and sprue. Compared to compression moulding, high moulding pressures are required for the transfer process, so fewer cavities can be put into a press of the same capacity.

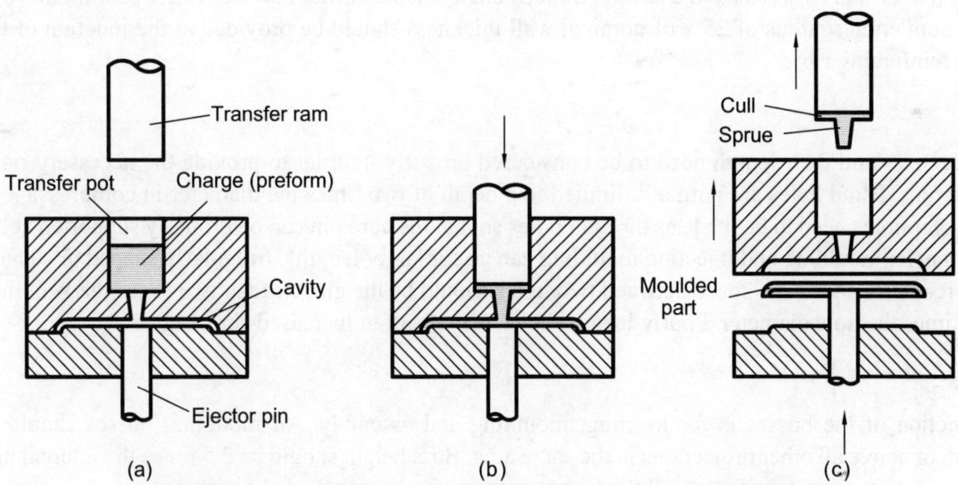

FIG. 12.21 *Steps in transfer moulding for thermosetting materials*

12.10 PLASTIC PRODUCT DESIGN

The designer should first determine the most desirable moulding process that is to be used for the manufacture. Then the designer should design the part taking the best aspects of that process into account. A few of the general guidelines for designing the plastic parts are given here.

Wall Thickness

Walls of plastic parts should be maintained as uniform as practically possible. Non-uniform walls will result in warpage, high stress concentrations and production difficulties. Also, parts with uniform walls will fill properly and will fit together because variable shrinkage is minimised. Heavy sections should be cored-out to maintain uniform wall thickness.

Minimum wall thickness to be used depends upon the process. For injection moulding, preferable wall thickness range is 1.8 to 3.0 mm, while anything lower than 0.40 mm will be extremely difficult to produce. For compression moulding, minimum wall thickness is 1.5 mm.

Draft

The part needs to be released and ejected from the mould. For this purpose, draft or taper needs to be provided on all surfaces that are normal to the parting plane. The shrinkage of part will allow it to be released from the external walls easily while it sticks tightly to the cores. General taper can vary depending upon the surface between 1 and 5°. It is also important to consider the characteristics of the plastic material before deciding the draft angle to be used. Large parts call for more draft than small parts. Thin-walled parts that undergo high-pressure injection moulding need more draft than parts that are subjected to lower-pressure moulding.

Radii and Fillets

It is necessary to provide generous radii and fillets from the strength point of view as well as the flow of plastic melt through the mould cavity. Typically, the inside corner radius should be 50% of the wall thickness while that of the outside should be 150% of the wall thickness. This ensures a smooth flow without any abrupt change in flow direction. Also, both sides of the corner will display equal amounts of shrinkage preventing the sink marks. This also leads to a strong, dimensionally stable corner that will resist post-mould warpage. A minimum corner radius of 25% of nominal wall thickness should be provided at the junction of the wall and the reinforcing rib.

Holes

The hole sizes and their length need to be considered properly in order to provide the necessary rigidity to the core pins. Blind holes are normally limited to a depth of two times the diameter in compression moulding. Longer holes need to be made as through holes so that the core pin can be properly supported at both the ends. Transfer moulding and injection moulding can go for larger lengths. Irregularly shaped or square holes are not recommended. The space between two holes should be the greater than two times the wall thickness or two times the hole diameter. Poorly located holes may result in increased shrinkage.

Bosses

The function of the bosses is for locating, mounting and assembly. All mounting bosses should project 0.40 mm or above all other projections in the same area. Boss height should be 2.5 times the internal diameter of the boss. To maintain uniform wall thickness, separate the bosses from the walls.

Ribs

Ribs are often the practical way to increase the stiffness of large surfaces. They will also help as feeders to allow the material flow for isolated bosses. Rib thickness should be smaller than the wall thickness. Ribs should be provided with general draft. Ribs should not be thicker than 2/3 the nominal wall thickness. Thicker ribs will create sink marks. The distance between two ribs should not be less than two times nominal wall thickness. Very thick ribs can cause internal voids as well as sink marks on the part surface opposite from where they are attached. Ribs that are too tall can create mould filling and venting problems.

SUMMARY

Plastics have been used extensively from the nineteenth century for engineering applications.

- Polymerisation of the plastic molecules is an important process in ensuring the properties that are required. The type of polymerisation depends upon the type of monomer and accordingly the final mechanical properties will depend on it.
- Polymer materials can be divided into thermoplastic, thermosetting and elastomers. Each of these types has specific properties that lend themselves to differing engineering and consumer applications.
- Extrusion is commonly used for constant cross section parts both solid and hollow. This is a low-cost process, though hollow parts require expensive dies.
- Injection moulding is the most widely used process for thermoplastic materials. The plastic material is heated and compressed in a barrel and then injected into the mould. Upon solidification, the part is ejected from the mould. It is a fast processing method and is good for really complex parts.
- Injection moulds are broadly divided into two-plate and three-plate moulds. Two-plate moulds are more common. Hot runner moulds help in the complete automation of the process without the requirement for any operator.
- Reaction injection moulding is used for thermosetting materials whose resins are in liquid form.
- Blow moulding is used for making large and small hollow parts. It is fast and economical for large parts. Heated tubular preform is blown by compressed air in a mould where the preform expands to confirm to the contours of the mould.
- Injection blow moulding is used for parts that require extra precision. In the stretch blow moulding, parts acquire higher mechanical strength because of the longitudinal stretching and radial blowing.
- Thermoforming is a relatively simpler process for making parts from heated plastic sheet. Vacuum and pressure are applied in different processes to ensure that the part follows the contours of the mould.
- Compression moulding is used for making parts with thermosetting materials such as phenol-formaldehyde. Transfer moulding is an improvement over compression moulding. The thermosetting material is heated and compressed in a separate chamber and then injected into the mould.
- Care has to be taken to design parts to be made of plastic materials. Certain attention needs to be paid to the wall thickness, draft, fillets, corner radii, bosses, ribs and holes to take advantage of the capability of different processing methods.

Questions

12.1 What are the two types of polymerisation methods in plastics?

12.2 What are the common additives used in plastics?

12.3 How do you classify the polymeric materials? Show with a simple flow diagram.

12.4 Explain the extrusion process for thermoplastic materials.

12.5 Explain the various methods (only the principle) available for injection moulding of plastics. Give the comparative advantages and applications for the same.

12.6 What are the different types of moulds used in injection moulding? Give their relative applications.

12.7 Explain reaction injection-moulding process.

12.8 Explain co-injection moulding process.

12.9 Explain the various methods (only the principle) available for blow moulding of thermoplastics giving their relative applications.

12.10 Name the processes used for making of (a) plastic bottles such as those used for storing 1 litre shampoo, (b) plastic top cover of a plain paper copier (photostat machine), and (c) plastic buckets of 20 litre capacity.

12.11 Explain injection blow-moulding process.

12.12 Explain stretch blow-moulding process with applications.

12.13 Describe the thermoforming process. What are its applications?

12.14 Give a brief account of some salient points to be considered for designing parts for plastic processing.

Multiple Choice Questions

12.1 Polyvinyl chloride is a
(a) Thermoplastic material
(b) Thermosetting material
(c) Elastomeric material
(d) Epoxy resin

12.2 Phenol-formaldehyde is a
(a) Thermoplastic material
(b) Thermosetting material
(c) Elastomeric material
(d) Epoxy resin

12.3 Identify the material produced by addition polymerisation
(a) Epoxy
(b) Phenol-formaldehyde
(c) Polyamide (Nylon)
(d) Polyvinyl acetate

12.4 Identify the material produced by condensation polymerisation:
(a) Polystyrene
(b) Polyamide (Nylon)
(c) Polypropylene
(d) Polyvinyl chloride

12.5 Extrusion is generally used for
(a) Thermosetting materials
(b) Thermoplastic materials
(c) Elastomers
(d) Phenol-formaldehyde

12.6 The advantage of using plastic-extrusion process is
(a) Equipment is simple and relatively inexpensive
(b) Short lead times
(c) Relatively low tooling costs
(d) All of the above

12.7 The most widely used plastic-processing method in the industry is
(a) Extrusion
(b) Injection moulding
(c) Blow moulding
(d) Thermo forming

12.8 The mould that provides highest productivity in injection moulding is
(a) Two-plate mould
(b) Three-plate mould
(c) Hot-runner mould
(d) Single-plate mould

12.9 To make complex parts in thermosetting materials the following process is used:
(a) Extrusion
(b) Injection moulding
(c) Reaction injection moulding
(d) Blow moulding

12.10 For making hollow parts with thin sections the following process is used:
(a) Extrusion
(b) Injection moulding
(c) Reaction injection moulding
(d) Blow moulding

Answers to MCQs

12.1 (a)	12.2 (b)	12.3 (d)	12.4 (b)	12.5 (b)
12.6 (d)	12.7 (b)	12.8 (c)	12.9 (c)	12.10 (d)

CASE STUDY

ENERGY CONSUMPTION IN BULK FORMING

Due to the increasing population of the world, the energy demand is severe. Further, energy generation leads to a significant carbon footprint that affects the environment. Nearly about 20% of the energy is wasted in the industrial sector in the United States. Hence, it is the need of the hour to choose the processes and improve them to make them more sustainable. In this case study, the energy consumption of bulk forming was compared with additive and subtractive manufacturing. Bulk forming processes for plastics include injection moulding, thermoforming and compression moulding, where as in metals, the methods include die casting, sand casting, investment casting and forging.

In the present case study, injection moulding, milling (subtractive) and fused deposition modeling (FDM-Additive) are considered. A model of 30 × 30 × 5 mm with two holes and slots was selected. Specific Energy Consumption (SEC) is used for the purpose of comparing the different manufacturing processes. The energy consumed in the production of a component part is defined as SEC. It is defined as 1 joule per unit volume of material for subtractive processes and 1 joule per unit mass of material for additive processes.

Injection Moulding

The mould was machined using a CNC machine. The prepared aluminium mould was used for injection moulding. The warmup of the moulds took 27 minutes. The average power consumption was 534 W. The total energy consumption for cutting the mould and the injection moulding process was 832.4 Wh/part. However, the specific energy consumption (SEC) drastically decreased with the number of parts as the actual processing energy consumption was only 37 Wh/part.

Milling

In this study a CNC machine was used to manufacture the part. The process consisted of milling drilling and slotting. The total processing time was 1123 s including 15 min preparation time. The total energy consumption was 40.8 Wh.

Additive Process

A small FDM machine was used in the study. The thickness of each layer in the process was 0.254 mm. and the setup time was 31 minutes. The actual process time was 18 minutes. Before the process, the machine needed to be warmed up for 25 minutes before the fabrication. The total energy consumption was 717.1 Wh/part. In this energy consumption, the actual process consumed only 75.1 kWh kg^{-1}. When more than three parts were made, the cost increased with wage cost.

For fewer parts, FDM was economical, but for larger number of parts, injection moulding and milling proved to be better options.

Yoon, H. S., Lee, J. Y., Kim, H. S., Kim, M. S., Kim, E. S., Shin, Y. J., Chu, W.S. & Ahn, S. H. (2014). A comparison of energy consumption in bulk forming, subtractive, and additive processes: Review and case study. *International Journal of Precision Engineering and Manufacturing-Green Technology*, *1*(3), 261-279.

CHAPTER 13

Ceramics and Glasses

Objectives

Ceramics have been utilized as an engineering material in a number of industries thanks to the major developments that have taken place in its processing methods. After completing this chapter, the reader will be able to

- Understand the importance of ceramic as an engineering material
- Identify the differences between ceramics and other engineering materials
- Understand different types of ceramics used in traditional environments and engineering situations
- Understand different processing methods used in ceramic manufacture

13.1 INTRODUCTION

Ceramics have been extensively used in the human history. Archaeologists regularly unearth a number of manmade ceramic samples dating as far back as 25,000 BC. Early ceramic materials are simple burned clay while the later examples provide a lot of embellishments to demonstrate the artistic nature of the people involved in making them. Ceramics are essentially inorganic and non-metallic materials. Most of the ceramics are typically crystalline in nature and are compounds formed between metallic and nonmetallic elements. Typical examples are aluminum and oxygen (aluminium oxide - Al_2O_3), silicon and carbide (silicon carbide - SiC), silicon and nitrogen (silicon nitride - Si_3N_4) calcium and oxygen (calcium oxide - CaO), and zirconium and oxygen (zirconium oxide - ZrO_2). Ceramics are generally processed at a high temperature during their manufacture. The general characteristics of ceramic material is high hardness, brittleness and high temperature resistance. Greek term keramos, meaning "a potter" or "pottery" is the source for English ceramic term. Early Greeks have used the term keramos to indicate "burned stuff" or "burned earth" indicating the nature of processing to manufacture them from earth.

Ceramics have distinct properties that are different from other engineering materials such as metals and polymers. The property variation is given qualitatively in Table 13.1.

TABLE 13.1 Comparison of Metals, polymers and ceramics

Property	Metal and Alloys	Polymers	Ceramics
Density	High	Very low	Low
Hardness	High	Very low	Very high
Wear resistance	Low	Low	High
Ductility	High	High	Low
Elastic modulus	High	Low	Very high
Temperature resistance	High	Low	Very high
Thermal conductivity	High	Low	Depends on material
Electrical conductivity	High	Low	Depends on material
Thermal expansion	Low	Very low	Depends on material

Ceramics can be broadly categorized into various types as shown in Table 13.2.

TABLE 13.2 Classification of ceramics

<table>
<tr><td rowspan="12">Ceramic materials</td><td rowspan="2">Traditional Ceramics</td><td>Structural Clay Products</td><td>Bricks, Tiles, Pipes, and Sanitary Ware.</td></tr>
<tr><td>White wares</td><td>Pottery, dinner ware, porcelain and terracotta.</td></tr>
<tr><td rowspan="5">Engineering ceramics</td><td>Fireclay refractories</td><td></td></tr>
<tr><td>Silica refractories</td><td></td></tr>
<tr><td>Basic refractories</td><td></td></tr>
<tr><td>Electro-ceramics</td><td>Insulators, semiconductors, metallic, superconductors, ionic conductors, etc.</td></tr>
<tr><td>Special</td><td>Hard surface coatings such as TiN, WC, and diamond.</td></tr>
<tr><td rowspan="2">Glasses</td><td>Glasses</td><td></td></tr>
<tr><td>Glass ceramics</td><td></td></tr>
<tr><td>Abrasives</td><td></td><td>Alumina, Silicon carbide, Cubic boron nitride</td></tr>
<tr><td>Cements</td><td></td><td></td></tr>
<tr><td>Advanced ceramics</td><td>Bio-ceramics, cutting tool materials, etc.</td><td></td></tr>
</table>

Compared to metals, modern ceramics have a large range of advantages that make them as indispensable engineering material used in specified applications.

- Ceramic materials have excellent temperature resistance, often up to 1100°C and above. As a result these materials can be used in situations that reach high temperatures without affecting the performance.
- Some of the ceramic materials are hard and because of their ability to maintain the hardness even at high temperatures make them as excellent cutting tool materials. Examples are silicon carbide, boron carbide, cubic boron nitride and diamond.
- These have very low friction coupled with hardness and abrasion resistance to qualify them to be excellent for special class of bearing applications without the need for any lubrication.
- Being lighter than metals, ceramics are excellent for aerospace applications.
- Engineering ceramics are less expensive than super alloys providing an opportunity to replace them where possible.

- Another advantage with replacing super alloys is that ceramics do not require expensive alloying elements such as cobalt, chromium, nickel and tungsten.

A comparison of the properties of ceramics with some common engineering metals is given in Table 13.3.

TABLE 13.3 Comparison of some ceramic materials with common metals

Material	Melting Point, °C	Density, kg/m^3	Hardness, Moh Scale	Modulus of Elasticity, GPa	Coefficient of Thermal Expansion, K^{-1}
Al_2O_3	2050	3690	9	300	8.1×10^{-6}
SiC	2800	3200	9	140	4.3×10^{-6}
ZrO_2	2660	5600	8	240	6.6×10^{-6}
BeO	2550	3000	9	400	10.4×10^{-6}
Mild steel	1370	7900	5	170	15×10^{-6}
Aluminium	660	2700	3	70	24×10^{-6}

13.2 CERAMIC MATERIALS

The properties of ceramic materials are dependent on the types of atoms present, the types of bonding between the atoms and the way the atoms are packed together. The mechanical properties of ceramics depends in addition to the microstructure, on atomic defects and pores present inside the material.

One of the oldest and widely used raw materials in ceramic making is the clay. One example is kaolinite which is aluminum silicate, $Al_2Si_2O_5(OH)_4$. Rocks that are rich in kaolinite are called China clay. When added with water, kaolinite forms a soft and plastic material that is well suited for forming. Other raw materials used are quartz, flint and feldspar. Industrial ceramic raw materials are alumina (Al_2O_3), and zirconia (ZrO_2). In addition to these materials, carbides and nitrides of materials such as silicon (SiC), tungsten (WC), titanium (TiN) are also used.

Traditional Ceramics

Traditional ceramics with which most people are familiar and will come across during their daily activities. In recent years, the processing technology of these materials has advanced substantially with improved products and reduced costs.

Clay based ceramic ware of medium porosity (4 to 20%) is called earthenware. These can be glazed or unglazed in their finished form. Example products that are made in this type are the kitchen ware, table ware and art ware. Depending upon the application, they will have different porosities and fired at different temperatures.

Stoneware is another class of ceramic that can be either vitreous or semi-vitreous. The materials used in them are non-refractory fireclays. Some of them will also have barium-containing as well as large amounts of iron oxide. Vitreous stoneware has low porosity levels (0 to 5%). Fully vitreous (no porosity) ceramic is called China and porcelain synonymously and contains flint and feldspar in addition to clay. They are glazed at high temperatures, and exhibit strength, hardness and chemical durability.

Engineering Ceramics

Engineering ceramics are used in engineering applications that have a wider range of materials and properties.

Refractories are the main ceramic products that are extensively used in engineering to withstand elevated temperatures. They have to withstand thermal cycling, thermal shock and mechanical fatigue, at these elevated-temperature environments.

Fireclay refractories are the most widely used refractory and find application in many industries that have ingredients such as kaolin, alumina and aluminosilicates. Kaolin based Fireclay refractories have high refractoriness and high load resistance. The performance of refractories increase with an increase in alumina content. Alumina Refractories have a large percentage of aluminium oxide ranging from 80 to 99%. These have higher refractoriness and higher load bearing capacity. Silica refractories are used in glassmaking industry because of their low degree of spalling due to low thermal expansion of silica. Their main ingredient is silica with additions of lime, iron oxide and alumina. Basic refractories on the other hand contain magnesite (MgO), calcia (CaO) and chrome (Cr_2O_3) for use in metallurgical industries. These are not suitable for acidic environments.

13.3 CERAMIC PROCESSING

Processing methods for ceramics depends upon the final application. Various processes available are similar to the conventional metal processes. Powder metallurgy process can be used to compact the ceramic powder using isostatic compacting followed by sintering. Similar to plastic processing, injection moulding and extrusion can be used to shape the ceramic parts in solid state. Additive manufacturing methods can also be used for ceramic processing such as fused deposition modelling and 3D printing. In liquid form ceramics can be processed by slip casting.

Slip Casting

Slip casting is the method used to make thin and uniform walled ceramic components. Slip is a fluid with ceramic particles in a liquid that is water with some chemicals that help the powder to disperse throughout the fluid. The mould is generally made of gypsum with sufficient porosity to help the removal of liquid from the slip. The steps in slip casting are shown schematically in Fig. 13.1.

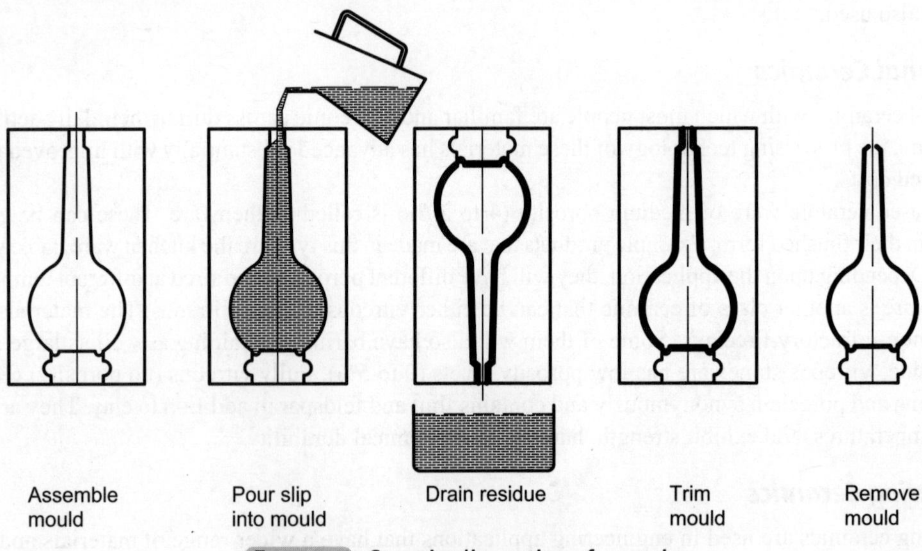

FIG. 13.1 *Steps in slip casting of ceramics*

To start the process the two halves of the mould are assembled to form the shape of the part to be produced. Then slip will be poured into the mould cavity to completely fill the cavity. The liquid in the slip is removed gradually by the mould wall through the capillary action. This allows the ceramic particles to form the shape around the cavity with uniform thickness. The ceramic particles will not have sufficient strength at this stage. Then the fluid from the mould will be drained to leave the shape of the part in the mould. When the mould is disassembled, the green body of the part is seen. This is the shape that need to be made. The green body is carefully removed and coated with a suitable glaze to provide a smooth surface where required. Then it is transferred to a furnace that provides the heat to sinter the part around 1400°C to get the necessary strength.

13.4 GLASSES

Glass is known to man for a very long time. Archeologists have found glass artefacts as far back as 4000 BC in the form of glazing on stone beads. Glass is amorphous material that has been rapidly cooled from liquid state in such a way that the atoms remain in a randomly arranged atomic state. Glass does not have a long range periodic crystalline structure. A schematic representation of glass is shown in Fig. 13.2.

Glass manufacturing starts by mixing a number of raw material powders. The main ingredients are sand (SiO_2), limestone ($CaCO_3$) and soda ash (Na_2CO_3). Each ingredient in the mixture provides a specific function for the final use of glass. Some of the functions provided by these include network formers, fluxing agents, property modifiers and colorants. These raw materials are heated such that a homogeneous viscous liquid of glass is formed. Then this liquid will be manipulated to obtain the desired shape.

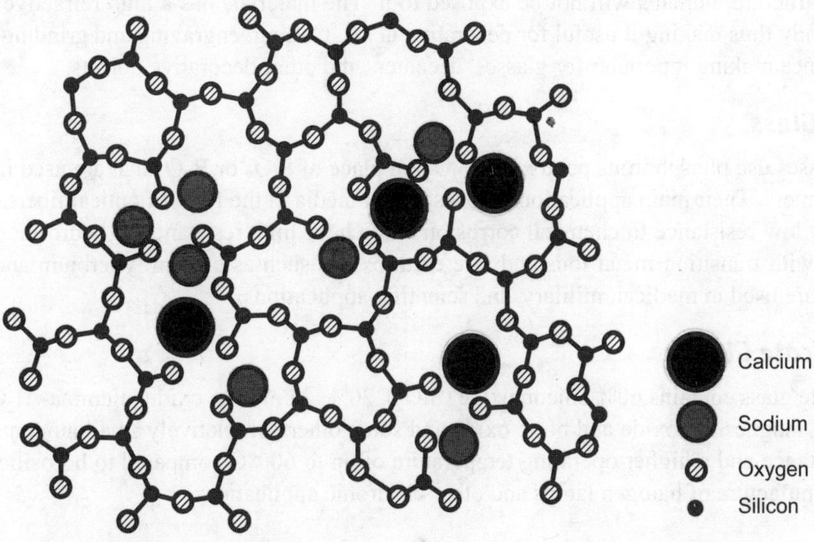

FIG. 13.2 *Schematic 2D representation of a soda-lime silicate glass*

13.4.1 Types of Glasses

There are a large variety of glasses available to suit the different applications.

Soda Glass or Soda-lime Glass

It is the most common variety of glass that is used in tableware, window glass, bottles, etc. It is one of the most refined, optimized and automated processes, that is relatively inexpensive. Typical composition is Silica (SiO_2) - 70%–74%; sodium oxide (Na_2O) - 12%–16%; calcium oxide (CaO) - 5%–11%; magnesium oxide (MgO) - 1%–3%; and aluminium oxide (Al_2O_3) - 1%–3%. The through-put of this type of glass is high during mass production. It is not durable as some of the other glasses. Soda glass is normally colourless but some chemicals can be added to give it different colours such as green, blue or brown.

Borosilicate Glass

Borosilicate Glass is the glass that can withstand heat and has good thermal shock resistance. It is the material used for manufacturing ovenware with the trade name Pyrex. It has low coefficient of thermal expansion as a result will develop fewer internal stresses during heating and cooling cycles. Borosilicate glass consists of silica (70–80%) and boric oxide (7–13%) with smaller amounts of the alkalis (sodium and potassium oxides) and aluminium oxide. Apart from the home baking use, it is extensively used in the chemical and pharmaceutical industries.

Lead Glass

Crystal glass with the sparkle is also known as lead glass. The main application of this to make a wide variety of decorative objects. In the ingredients calcium oxide is replaced by lead oxide, and potassium oxide instead of sodium oxide. Any glass with less than 24% lead oxide is called crystal glass. Since lead is locked inside the chemical structure, humans will not be exposed to it. The materials has a high refractive index making it sparkle brightly thus making it useful for decorative items. Cutting, engraving and grinding highlights the crystals brilliance making it popular for glasses, decanters and other decorative objects.

Phosphate Glass

Phosphate glasses use phosphorous pentoxide (P_2O_5) in place of SiO_2 or B_2O_3 that are used in other glasses as network formers. Their main application is as laser gain media in the form of optical fibers. These glasses have relatively low resistance to chemical corrosion while have high resistance to hydrofluoric acid. These can be doped with transition metal ions and rare earth oxides such as erbium, ytterbium and neodymium. These glasses are used in medical, military, and scientific applications.

Aluminosilicate Glass

Aluminosilicate glass contains 60% silicon oxide (SiO_2), 20% aluminium oxide (alumina-Al_2O_3) along with calcium oxide, magnesium oxide and boric oxide and some others in relatively small amounts. It has better chemical resistance and a higher operating temperature of up to 600°C compared to borosilicate glass. It is used in the manufacture of halogen lamps and other electronic applications.

SUMMARY

Ceramics have properties that are not available from traditional engineering materials. Though ceramics are known for long, their utilization in engineering has increased in the last century with developments in processing technologies. It is used in aerospace applications where it has been able to reduce the use of expensive metals usage. Fireclay refractories are the most used engineering ceramics for its ability to withstand high temperature. Ceramic processing following traditional manufacturing methods such as isostatic compression and sintering, injection moulding etc., together with slip casting that is specific to ceramic manufacturing. Glass is an amorphous solid that is quickly cooled. There are a large variety of glasses used in industry and continuously new developments are coming particularly with the miniaturization of electronics.

Questions

13.1 Give a comparison of ceramic properties with that of the conventional engineering materials.
13.2 Give the advantages to be gained by using ceramic materials.
13.3 What are the raw materials that are generally used in ceramics?
13.4 Give a brief account of traditional ceramics.
13.5 Give a brief account of engineering ceramics.
13.6 Explain the slip casting process for making ceramic vessel.
13.7 Give a brief method of glass manufacturing with the ingredients used in its manufacture.
13.8 Give a brief account of different glasses.

CHAPTER 14

Composite Materials

Objectives

After completing this chapter, the reader will be able to

- Understand the basics of composite materials
- Learn about classification of composites
- Understand the properties and processing methods of Polymer Matrix Composites
- Learn about other composites

14.1 INTRODUCTION

The heterogeneous combination of two or more materials is called a composite material. The constituent materials differ in form or composition on a macroscale. The materials that are added to the parent material may be reinforcing elements, fillers and/or binders. The composite material will be designed in such as a way that it maximizes specific performance properties. The added materials will not dissolve or merge completely and therefore normally exhibit an interface between one another. Some examples are wood (cellulose fibers in a matrix of lignin and hemicellulose), concrete (sand, cement, and stone) and fibre reinforced plastic (polymer with glass or carbon fibers).

Nature has been using composites quite extensively. Bone is a good example that is made of hydroxyapatite (mostly calcium phosphate) which is hard but brittle and collagen (protein) which is a soft and flexible material. The combination of the two gives the ability for the bone to support the body. Humans have also been making composites for thousands of years. Mud brick used as the building material is a good example of a composite. Mud has been found to be a good building material because of its good compressive strength. However it lacks any tensile strength. To compensate for that, straw has been added which has good tensile strength but poor compressive strength. By combining the two, early humans have been making excellent building blocks for residence structures.

Typically composites have two phases. Matrix is the primary phase that is continuous in character and is generally more ductile and less hard. It should have low density with good shear properties. It keeps the second phase dispersed throughout and transmits and shares the load with it. The matrix should be chemically compatible with fibres over the working range and protect their fibre surface from damage.

The second phase that generally provides the necessary strength to the matrix is in a discontinuous form. It should have high strength, high stiffness and low density. The fibres should be stable and maintain their strength through the rigors of handling of fabrication. The fibres should have mechanical property variation across the whole size minimal and their arrangement within the composite as regular as possible. Because it provides the strength it is sometimes called the reinforcing phase.

Some typical composite designs are shown schematically in Fig. 14.1. These are explained in detail later.

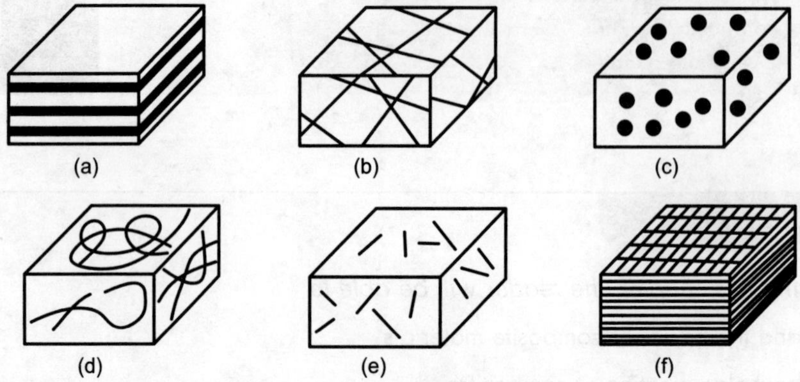

Fig. 14.1 *Different types of composite materials; (a) Laminate; (b) Reinforcement with long fibres; (c) Particle reinforcement; (d) Continuous fibres arranged randomly; (e) Short fibres; (f) Woven fabric mats.*

Advantages of composites

1. The biggest advantage of composite materials is that they are light as well as strong. This is normally expressed as strength-to weight and stiffness-to-weight ratios. Composite materials have been found to have one of the highest strength-to-weight ratios (strength divided by density). When used as structural material in automobiles or aero planes composites provide higher performance coupled with substantial fuel savings.
2. It is possible to tailor make the properties for a given application by carefully choosing appropriate combination of matrix and reinforcement material and its manufacturing method.
3. Composites provide design flexibility as many of the composite manufacturing methods allow for moulding into complex shapes as required. For example it is possible to achieve complex double-curvature parts with a smooth surface finish in one manufacturing operation.
4. It is possible to design and fabricate complex geometry in a single operation, thereby reducing the part count.
5. It is possible to reduce the production cost for specific product as there are wide ranges of composite manufacturing processes available, many of them are more efficient.
6. Generally composites offer excellent resistance to heat, corrosion, chemical attack and outdoor weathering.
7. Composites are less likely to breakup under stress. If a crack is formed it will not spread like in a piece of metal because it will be blocked by the fibres.
8. A lot of waste materials can be recycled during the composite manufacturing process.

Disadvantages of composites

1. Many of the constituents of composite materials are expensive and a careful selection need to be made to control the costs involved.

2. Repairing of composite is not an easier task particularly for doing on site. Depending upon the composite and specific processing involved in its original manufacture, not all facilities such as hot curing may be readily available on site.

Composites have been used in a number of domains of activities. Modern aviation both civil and military is extensive user of composites. It is also the main sector that drove the innovation in composites. A large number of composite parts can be found in modern aircraft such as wing, body panels and tail sections, propellers and rotor blades, internal structure and fittings. The use of composites in aircrafts is continuously increasing. In the 1950's, Boeing 707 had just 2% composite materials in the structure. In contrast, Boeing 787 dreamliner aircraft utilizes 50% advanced composites in its airframe structure. Similarly, Airbus A380 accounts for about 25% of the total airframe. Some of the components that use composites in aircraft are landing gear doors, floor beams and floor boards, leading and trailing edge panels on the wing and stabilizer, etc. Helicopters and military vehicles, missiles as well as space applications utilize composites to a great extent.

14.2 CLASSIFICATION OF COMPOSITES

Composites can be classified based on matrix material used as follows:

Polymer Matrix Composites (PMCs) These are composed of a matrix from either thermosetting resins such as epoxy or unsaturated polyester, or thermoplastic such as polycarbonate, polyvinyl chloride, nylon and polystyrene. The dispersed material is the fibres such as glass, carbon, steel or Kevlar. These are the most commonly used type of composites.

Metal Matrix Composites (MMCs) These are composed of a metallic matrix such as aluminium, magnesium, iron, etc. with a dispersed fibre that could be either metal or ceramic. The metallic fibre examples are tungsten, molybdenum, lead, etc. The ceramic reinforcements are oxides and carbides such as cemented carbides and other cermets.

Ceramic Matrix Composites (CMCs) These are composed of ceramic matrix such as Al_2O_3 and SiC embedded with fibres. These are specifically used for high temperature applications. These are less commonly used compared to other composites.

Composites can also be classified into different types based on the nature of the dispersed material is shown in Fig. 14.2.

Structural

Structural composites are further categorised into laminates and sandwich panels, as described next.

Laminates In this laminae are stacked and bonded in proper arrangement such that the orientation of fibre direction varies with each successive layer allowing it to withstand the loads much better. This provides a balanced in-plane stiffness. Each layer called ply is generally thin about 0.2 mm with reinforcing fibres that are in a specific orientation in a thermosetting polymer such as uncured epoxy. The fabricator then lays out each these layers with each successive layer in layup sequence in different orientation to achieve the required properties Fig. 14.1(a).

Sandwich Panels The construction of sandwich panel is such that two strong faces (steel or aluminium) are provided with a lower density strong core in between. The core is generally of honeycomb structure to provide large bending stiffness. The faces and the core are bonded properly so that load can be transferred between them.

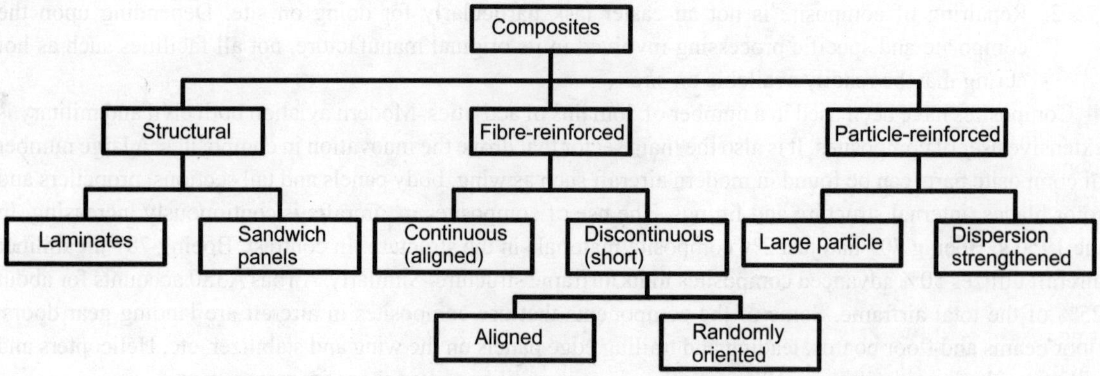

FIG. 14.2 *Classification of composites based on dispersed material*

Fibre-reinforced Composites

The reinforcement is in the form of a fibre. Fibres used are generally glass, Kevlar (DuPont's trade name for aramid fibre) or carbon/graphite. Fibres being strong in tension provides a large improvement in strength for fibre-reinforces composites materials. An example of widely used material is the fibre-glass that is continuous glass fibres in a polymer matrix. The other type of fibres used are whiskers which are the thin single crystals that have large length to diameter ratios. These have very high strength but are very expensive and difficult to manufacture. Examples are graphite, silicon carbide and silicon nitride.

Continuous fibre reinforced Composites These will have the long fibres aligned in preferred orientation Fig. 14.1(b).

Short-fibre reinforced composites These utilize short fibres, typically length less than 100 times the diameter in the form of discontinuous fibres Fig. 14.1(e). The fibres can be oriented in preferred direction or can be randomly oriented.

Particle-reinforced Composites

These are relatively simpler composites because of their production and forming. The dispersed phase is in the form of particles that are distributed in the matrix providing the reinforcement Fig. 14.1(c). These are less effective compared to the fibre-reinforced composites. They increase strength and stiffness and are normally utilized where high wear resistance is required. The main advantage of particulate-reinforced composites is its low cost. An example is the concrete paved roads for low maintenance roads.

Large Particle-reinforced Composites The particle size used in this composite is relatively large. The reinforcement capability is indicated by the bonding of the matrix and particle. Though the particles can have a variety of geometries, it would be preferred to have small size and even distribution throughout the matrix Fig. 14.1(c). Some examples are concrete (cement, sand and gravel) and plastics with fillers added to modify the properties.

Dispersion-strengthened Particle-reinforced Composites In this composite the particles are very small in size (less than 0.1 μm in diameter). By uniformly dispersing the particulate the strength is increased. The materials used can be metallic or non-metallic oxides. The strength increases with the increase in volume fraction of the particulates.

14.3 POLYMER MATRIX COMPOSITES (PMC)

As defined earlier polymer matrix composites have organic polymer matrix with strengthening fibres in the matrix. Matrix holds and protects the fibres in place while transmitting the load to them. Advanced composites are a class of Polymer matrix composites that have high mechanical properties (strength and stiffness) compared to the normal reinforced plastics and are used in aerospace applications. Reinforced plastics are relatively inexpensive and are widely used.

Matrix

The matrix phase used in PMC can be classified as thermosetting resins or thermoplastic resins. Thermosetting resins are the most commonly used resins in commercial PMC. These include epoxies, and polyamides, though epoxies are the most widely used resins. These materials during the curing phase crosslinks the polymer chain thus providing a three dimensional structure and stability to the polymer. These generally have better dimensional stability, and good resistance to high temperature. Thermoplastic resins used include polyesters, polyphenylene sulfide, polyamide imide, and polyether ether ketone (PEEK). Though these are not as good as thermosetting resins in terms of their high temperature resistance, these are easier from the manufacturing point of view. Another advantage of thermoplastic resins is their ability to be reused by forming into a different shape by heating.

Fibres

Fibres are the strong reinforcement with their strength depending upon the diameter. The various materials used are glass, graphite/carbon, aramid, nylon, metal and whiskers (single crystal). The strengthening of the material by the fibre depends upon the mechanical properties of the fibre. However, it also depends upon the amount and orientation of the fibre along with the surface interaction between the fibre and the matrix.

A large variety of fibres are used in composites:

Glass fibre is the most common fibre used for reinforced plastics. These are generally two types, S-glass which is silica-magnesium glass used in high strength composites and E-glass which is borosilicate glass used in electrical insulation. These are relatively easy to produce by drawing from the glass melt. As a result these are less expensive compared to other fibres and are widely used.

Carbon (predominantly amorphous) and graphite (predominantly crystalline) fibres are much stronger compared to glass with rigidity ten times and density half. This has got higher fatigue resistance compared to glass. However, this is more expensive than glass fibre.

Polymeric fibres have low density, low stiffness with medium to high strength. Examples are aramid (Kevlar®) and UHMWPE (Ultra high molecular weight polyethylene).

Ceramic fibers have high strength and low density. Examples are alumina, silicon carbide, etc. One of the main characteristics of the ceramic filters is their high thermal resistance and stability.

Metallic fibres are heavy and are relatively less expensive. Because of their higher density they are not much preferred.

A comparison of some of the properties of fibres used in composites is given in Table 14.1.

TABLE 14.1 Properties of some fibres used in FRP

	Density, g/cm^3	Elastic Modulus, *E* GPa	Elastic Modulus Density
Glass (E type)	2.55 to 2.6	72 to 85	30
Glass (S type)	2.485 to 2.495	86 to 93	36
Carbon T1000GB (PAN based)	1.80	291	162
Carbon K13D (Pitch based)	2.20	940	427
Aramid (Kevlar® 29)	1.44	70.5	49
Aramid (Kevlar® 49)	1.44	112.4	78
Polyethylene (Spectra® 75)	0.97	133	137
Polyethylene (UHMWPE Dyneema SK99)	0.970	155	160
Alumina, Al_2O_3 (Fibre FP)	3.95	379	96
Silicon Carbide, SiC (Nicalon®)	2.55	220	86
Steel	7.8	207	27

Minimum length of the fibre that is necessary to get the required strength is termed as Critical length. It is given by

$$\text{Critical fibre length, } l = \frac{\sigma d}{2\tau} \tag{14.1}$$

Where, σ = Tensile strength of the fibre

d = fibre diameter

τ = fibre-matrix bonding strength (or shear yield strength of the matrix, whichever is smaller)

For continuous and aligned fibre composites, the elastic modulus of the composite in the longitudinal direction, E_{cl} is given by,

$$E_{cl} = E_m V_m + E_f V_f \tag{14.2}$$

Where, E_m = Elastic modulus of the matrix

E_f = Elastic modulus of the fibre

V_m = Volume fraction of the matrix

V_f = Volume fraction of the fibre

The elastic modulus of the composite in the transverse direction, E_{ct} is given by,

$$E_{ct} = \frac{E_m E_f}{V_m E_f + V_f E_m} \tag{14.3}$$

The longitudinal tensile strength of the composite, σ_{cl} is given by,

$$\sigma_{cl} = \sigma_m(1 - V_f) + \sigma_f V_f \tag{14.4}$$

As explained above, there are a wide range of matrix materials and fibre materials that result in various manufacturing methods that are used for the fabrication of composites. The basic steps are:

- Impregnation of the fiber with the resin,
- Forming of the structure,
- Curing (thermoset matrices) or thermal processing (thermoplastic matrices), and
- Finishing.

Some of the manufacturing processes used for composites are:

- Injection molding for fast and high volume work with thermoplastics that have very short fibres
- Sheet moulding for automotive body panels
- Resin transfer molding for complex parts with control in fibre orientation
- Filament winding with moderate speed for complex geometries generally used for aircraft parts such as fuselage
- Pultrusion for continuous cross section parts
- Prepreg tape laying is slow, laborious and expensive with lot of automation used for aerospace structures, Prepreg is in the form of a tape or cloth that is the matrix that has already been impregnated with the fibre.

14.4 METAL MATRIX COMPOSITES

Metal Matrix Composites are generally used to provide specific properties to the composite such as improving creep performance, resistance to higher temperature, improve dimensional stability, increase abrasion resistance or decrease the density. About 75% of the MMC are particulate composites while other are fibrous composites. A few examples of Metal Matrix Composites are

- Aluminium based composites with reinforcing materials such as Al_2O_3, Al_4C_3, and SiC. These are generally used for aeronautical and automotive applications.
- Copper based composites with reinforcements such as Al_2O_3, BeO, ThO_2, SiO_2, and TiO_2. Improve thermal stability and higher heat resistance for use in electrical applications because of their higher electrical conductivity.
- Silver based composites with reinforcements such as Al_2O_3, CdO, SnO_2, and others. These are primarily used for electrical contacts of different relays, switches and circuit breakers.
- Nickel based composites with reinforcements such as ZrO_2, HfO_2, and ThO_2. It can withstand up to 1650°C and used in aviation and space technology.

The processing methods can be broadly classified into two types:

- Solid state processing
  - Powder metallurgy
  - Diffusion bonding
  - Physical vapor deposition
- Liquid state processing
  - Squeeze casting
  - Stir casting
  - Infiltration

14.5 CERAMIC MATRIX COMPOSITES

Ceramic Matrix Composites improve toughness of the conventional ceramic materials which are brittle by adding carbides and oxides in the dispersed phase. The reinforcements added act as barriers to crack propagation. The fibres used are carbon, silicon carbide, oxides and SiC whiskers. The fibres are typically in the size range of about 10 μm while the whiskers (single crystal fibres) are about 1 to 5 μm. Some typical CMCs are Al_2O_3 matrix with SiC whisker, Si_3N_4 matrix with SiC whisker, Al_2O_3 matrix with ZrO_2, and Si_3N_4 matrix with SiC particulate.

SUMMARY

Composites are the heterogeneous combination of multiple materials to get better properties than possible with any single material. The matrix and the reinforcements can be either metal, non-metal or ceramics thereby providing a large range of materials for different applications. Since the introduction in early 1950's composites have found extensive usage in a large range of industries. The largest volume composites used are the polymer matrix composites with the example of fibreglass. Large variety of fibres have been developed to provide special properties for different products based on the requirements. It is possible to theoretically estimate the composite properties based on the matrix and reinforcement mechanical properties. Traditional manufacturing processes have been generally modified to use for the fabrication of the composites.

Questions

14.1 How do you define a composite material?
14.2 Are composite materials required? Compare them with traditional materials.
14.3 Give some examples of composite materials that you are familiar with.
14.4 What are the advantages and disadvantages of composites?
14.5 What is the function served by the matrix in a composite material?
14.6 What is the function served by the fibres in a composite material?
14.7 What is the difference between fibre and particle reinforced composites? Give an example.
14.8 Give a classification of composites based on the matrix material used.
14.9 Give a classification of composites based on the reinforcement material used.
14.10 What are the different types of fibres used in composites? How are their properties compared?
14.11 Write a short note on metal matrix composites.

GATE Previous Years Questions

Chapter 2

2.1 The process of reheating the martensitic steel to reduce its brittleness without any significant loss in its hardness is

(a) Normalizing (b) Annealing
(c) Quenching (d) Tempering

(GATE-2014-ME-SET-1, 1-Mark)

2.2 During normalizing process of steel, the specimen is heated

(a) Between the upper and lower critical temperature and cooled in still air
(b) Above the upper critical temperature and cooled in furnace
(c) Above the upper critical temperature and cooled in still air
(d) Between the upper and lower critical temperature and cooled in furnace

(GATE-2012-ME-1-Mark)

2.3 If a particular Fe-C alloy contains less than 0.83% carbon, it is called

(a) High speed steel
(b) Hypoeutectoid steel
(c) Hypereutectoid steel
(d) Cast iron

(GATE-2007-ME-1-Mark)

2.4 The percentage of carbon in gray cast iron is in the range of

(a) 0.25 to 0.75 percent
(b) 1.25 to 1.75 percent
(c) 3 to 4 percent
(d) 8 to 10 percent

(GATE-2004-ME-1-Mark)

2.5 From the lists given below, choose the most appropriate set of heat treatment process and the corresponding process characteristics

Process	Characteristics
P. Tempering	1. Austenite is converted into bainite
Q. Austempering	2. Austenite is converted into martensite
R. Martempering	3. Cementite is converted into globular structure
	4. Both hardness and brittleness are reduced
	5. Carbon is absorbed into the metal

(a) P-3, Q-1, R-5 (b) P-4, Q-3, R-2
(c) P-4, Q-1, R-2 (d) P-1, Q-5, R-4

(GATE-2004-ME-2-Marks)

2.6 Hardness of steel greatly improves with

(a) Annealing (b) Cyaniding
(c) Normalizing (d) Tempering

(GATE-2003-ME-2-Marks)

2.7 The ductility of a material with work hardening

(a) Increases
(b) Decreases
(c) Remains unaffected
(d) Unpredictable

(GATE-2002-ME-2-Marks)

2.8 Cast steel crankshaft surface is hardened by

(a) nitriding
(b) normalizing
(c) carburising
(d) induction heating

(GATE-2000-ME-1-Mark)

2.9 The alloying element mainly used to improve the endurance strength of steel materials is

(a) Nickel (b) Vanadium
(c) Molybdenum (d) Tungston

(GATE-1997-ME-1-Mark)

2.10 On completion of heat treatment, the resulting structure will have retained Austenite if
(a) Rate of cooling is greater than the critical cooling rate
(b) Rate of cooling is less than the critical cooling rate
(c) Marten site formation starting temperature is above the room temperature
(d) Martensite formation finish temperature is below the room temperature

(GATE-1997-ME-1-Mark)

2.11 The iron-carbon diagram and the TTT curves are determined under
(a) Equilibrium and non-equilibrium conditions respectively
(b) Non-equilibrium and equilibrium conditions respectively
(c) Equilibrium conditions for both
(d) Non-equilibrium conditions for both

(GATE-1996-ME-1-Mark)

2.12 Match the terms used in connection with heat-treatment of steel with the micro structural/physical characteristics: Terms Characteristics

List - I	List - II
A. Pearlite	P. Extremely hard and brittle phase
B. Martensite	Q. Cementite is finely dispersed in ferrite
C. Austenite	R. Alternate layers of cementite and ferrite
D. Eutectoid	S. Can exist only above 723°C
	T. Pertaining to state of equilibrium between three solid phases
	U. Pertaining to state of equilibrium between one liquid and two solid Phase

(GATE-1992-ME-2-Marks)

2.13 When the temperature of a solid metal increases.
(a) Strength of the metal decreases but ductility increases
(b) Both strength and ductility of the metal decrease
(c) Both strength and ductility of the metal increase
(d) Strength of the metal increases but ductility decreases

(GATE-2005-ME-1-Mark)

CHAPTER 3

3.1 Green sand mould indicates that
(a) Polymeric mould has been cured
(b) Mould has been totally dried
(c) Mould is green in colour
(d) Mould contains moisture

(GATE-2011-ME-1-Mark)

3.2 Gray cast iron blocks 200 × 100 × 10 mm are to be cast in sand moulds. Shrinkage allowance for pattern making is 1%, the ratio of the volume of pattern to that of the casting will be
(a) 0.97 (b) 0.99
(c) 1.01 (d) 1.03

(GATE-2004-ME-2-Marks)

3.3 Hardness of green sand mould increases with
(a) Increase in moisture content beyond 6 percent
(b) Increase in permeability
(c) Decrease in permeability
(d) Increase in both moisture content and permeability

(GATE-2003-ME-1-Mark)

3.4 Shrinkage allowance on pattern is provided to compensate for shrinkage when
(a) The temperature of liquid metal drops from pouring to freezing temperature
(b) The metal changes from liquid to solid state at freezing temperature
(c) The temperature of solid phase drops from freezing to room temperature
(d) The temperature of metal drops from pouring to room temperature

(GATE-2003-ME-2-Marks)

3.5 Disposable patterns are made of
(a) Wood (b) Rubber
(c) Metal (d) Polystyrene
(GATE-2000-ME-1-Mark)

3.6 Which of the following materials requires the largest shrinkage allowance, while making a pattern for casting?
(a) Aluminium (b) Brass
(c) Cast Iron (d) Plain Carbon Steel
(GATE-1999-ME-1-Mark)

CHAPTER 4

4.1 The part of a gating system which regulates the rate of pouring of molten metal is.
(a) Pouring basin (b) Runner
(c) Choke (d) In gate
(GATE-2016-ME-SET-1, 1-Mark)

4.2 A cylindrical job with diameter of 200 mm and height of 100 mm is to be cast using modulus method of riser design. Assume that the bottom surface of cylindrical riser does not contribute as cooling surface. If the diameter of the riser is equal to its height, then the height of the riser (in mm) is
(a) 150 (b) 200
(c) 100 (d) 125
(GATE-2016-ME-SET-1, 2-Marks)

4.3 The solidification time of a casting is proportional to $\left(\frac{V}{A}\right)^2$, where V is the volume of the casting and A is the total casting surface area losing heat. Two cubes of same material and size are cast using sand casting process. The top face of one of the cubes is completely insulated. The ratio of the solidification time for the cube with top face insulated to that of the other cube is
(a) $\frac{25}{36}$ (b) $\frac{36}{25}$
(c) 1 (d) $\frac{6}{5}$
(GATE-2015-ME-SET-1, 2-Marks)

4.4 A cube and a sphere made of cast iron (each of volume 1000 cm^3) were cast under identical conditions. The time taken for solidifying the cube was 4s. The solidification time (in s) for the sphere is ________.
(GATE-2015-ME-SET-2, 2-Marks)

4.5 Ratio of solidification time of a cylindrical casting (height = diameter) to the cubic casting of side two times the height of cylindrical casting is ________.
(GATE-2015-ME-SET-3, 2-Marks)

4.6 An aluminium alloy (density 2600 kg/m^3) casting is to be produced. A cylindrical hole of 100 mm diameter and 100 mm length is made in the casting using sand core (density 1600 kg/m^3). The net buoyancy force (in Newton) acting on the core is ________.
(GATE-2014-ME-SET-1, 2-Marks)

4.7 A cube shaped casting solidifies in 5 minutes. The solidification time in minutes for a cube of the same material, which is 8 times heavier than the original casting will be
(a) 10 (b) 20
(c) 24 (d) 40
(GATE-2013-ME-1-Mark)

4.8 In a gating system, the ratio 1:2:4 represents
(a) Sprue base area : runner area : ingate area
(b) Pouring basin area : ingate area: runner area
(c) Sprue base area : ingate area: casting area
(d) Runner area : ingate area : casting area
(GATE-2010-ME-1-Mark)

4.9 The primary purpose of a sprue in a casting mould is to
(a) Feed the casting at a rate consistent with the rate of solidification
(b) Act as a reservoir for molten metal
(c) Feed molten metal from the pouring basin to the gate
(d) Help feed the casting until all solidification takes place
(GATE-2002-ME-1-Mark)

4.10 Chills are used in moulds to
(a) achieve directional solidification
(b) reduce the possibility of blowholes
(c) reduce freezing time
(d) smoothen metal flow for reducing splatter
(GATE-1998-ME-1-Mark)

CHAPTER 5

5.1 The hot tearing in a metal casting is due to
(a) High fluidity
(b) High melt temperature
(c) Wide range of solidification temperature
(d) Low coefficient of thermal expansion
(GATE-2014-ME-SET-3, 1-Mark)

5.2 Two streams of liquid metal, which are not hot enough to fuse properly, result into a casting defect known as
(a) Cold shut (b) Swell
(c) Sand wash (d) Scab
(GATE-2009-ME-1-Mark)

5.3 Misrun is a casting defect which occurs due to
(a) Very high pouring temperature of the metal
(b) Insufficient fluidity of the molten metal
(c) Absorption of gases by the liquid metal
(d) Improper alignment of the mould flasks
(GATE-2004-ME-1-Mark)

5.4 Light impurities in the molten metal are prevented from reaching the mould cavity by providing a
(a) Strainer (b) Button well
(c) Skim bob (d) All of the above
(GATE-1996-ME-1-Mark)

5.5 In a green-sand moulding process, uniform ramming leads to
(a) Less chance of gas porosity
(b) Uniform flow of molten metal into the mould cavity
(c) Greater dimensional stability of the casting
(d) Less sand expansion type of casting defect
(GATE-1992-ME-1-Mark)

CHAPTER 6

6.1 Which of the following engineering materials is the most suitable candidate for hot chamber die casting?
(a) Low carbon steel
(b) Titanium
(c) Copper
(d) Tin
(GATE-2007-ME-1-Mark)

6.2 An expendable pattern is used in
(a) Slush casting
(b) Squeeze casting
(c) Centrifugal casting
(d) Investment casting
(GATE-2006-ME-1-Mark)

6.3 Match the following:

List - I	List - II
(A) Sand casting	(1) Symmetrical and circular shapes only
(B) Plaster mould casting	(2) Parts have hardened skins and soft interior
(C) Shell mould casting	(3) Minimum post-casting processing
(D) Investment casting	(4) Parts have a tendency to warp
	(5) Parts have soft skin and hard interior
	(6) Suitable only for non-ferrous metals

(GATE-1998-ME-2-Marks)

6.4 Centrifugally cast products have
(a) Large grain structure with high porosity
(b) Fine grain structure with high density
(c) Fine grain structure with low density
(d) Segregation of slug towards the outer skin of the casting
(GATE-1993-ME-2-Marks)

CHAPTER 7

7.1 In a single pass rolling process using 410 mm diameter steel rollers, a strip of width 140 mm and thickness 8 mm undergoes 10% reduction of thickness. The angle of bite in radians is
(a) 0.006 (b) 0.031
(c) 0.062 (d) 0.600
(GATE-2012-ME-2-Marks)

7.2 The maximum possible draft in cold rolling of sheet increases with the
(a) Increase in coefficient of friction
(b) Decrease in coefficient of friction
(c) Decrease in roll radius
(d) Increase in roll velocity
(GATE-2011-ME-1-Mark)

7.3 Cold working of steel is defined as working
(a) At its re-crystallization temperature
(b) Above its re-crystallization temperature
(c) Below its re-crystallization temperature
(d) At two thirds of the melting temperature of the metal
(GATE-2003-ME-1-Mark)

7.4 Hot rolling of mild steel is carried out
(a) At re-crystallization temperature
(b) Between 100°C to 150°C
(c) Between re-crystallization temperature
(d) Above re-crystallization temperature
(GATE-2002-ME-1-Mark)

7.5 Thickness of a metallic sheet is reduced from an initial value of 16 mm to a final value of 10 mm in one single pass rolling with a pair of cylindrical rollers each of diameter of 400 mm. The bite angle in degree will be
(a) 5.936 (b) 7.936
(c) 8.936 (d) 9.936
(GATE-2007-ME-2-Marks)

CHAPTER 8

8.1 Match the following metal forming processes with their associated stresses in the work piece.

List - I		List - II	
P	Coining	1	Tensile
Q	Wire Drawing	2	Shear
R	Blanking	3	Tensile and compressive
S	Deep drawing	4	Compressive

(a) P-4, Q-1, R-2, S-3
(b) P-4, Q-1, R-3, S-2
(c) P-1, Q-2, R-4, S-3
(d) P-1, Q-3, R-2, S-4
(GATE-2012-ME-1-Mark)

8.2 Calculate the punch size in mm, for a circular blanking operation for which details are given below:

Size of the blank	25 mm
Thickness of the sheet	2 mm
Radial clearance between punch and die	0.06 mm
Die allowance	0.05 mm

(a) 24.83 (b) 24.89
(c) 25.01 (d) 25.17
(GATE-2012-ME-2-Marks)

8.3 The shear strength of a sheet metal is 300 MPa. The blanking force required to produce a blank of 100 mm diameter from a 1.5 mm thick sheet is close to
(a) 45 kN (b) 70 kN
(c) 141 kN (d) 3500 kN
(GATE-2011-ME-2-Marks)

8.4 In the deep drawing of cups, blanks show a tendency to wrinkle up around the periphery (flange). The most likely cause and remedy of the phenomenon are, respectively
(a) Buckling due to circumferential compression; Increase blank holder pressure
(b) High blank holder pressure and high friction; Reduce blank holder pressure and apply lubricant
(c) High temperature causing increase in circumferential length; Apply coolant to blank
(d) Buckling due to circumferential compression, decrease blank holder pressure
(GATE-2008-ME-2-Marks)

8.5 Match the correct combination for following metal working processes.

Processes	Associated State of Stress
P. Blanking	1. Tension
Q. Stretch Forming	2. Compression
R. Coining	3. Shear
S. Deep Drawing	4. Tension and Compression
	5. Tension and Shear

(a) P-2, Q-1, R-3, S-4
(b) P-3, Q-4, R-1, S-5
(c) P-5, Q-4, R-3, S-1
(d) P-3, Q-1, R-2, S-4

(GATE-2007-ME-2-Marks)

8.6 A 2 mm thick metal sheet is to be bent at an angle of one radian with a bend radius of 100 mm. If the stretch factor is 0.5, the bend allowance is

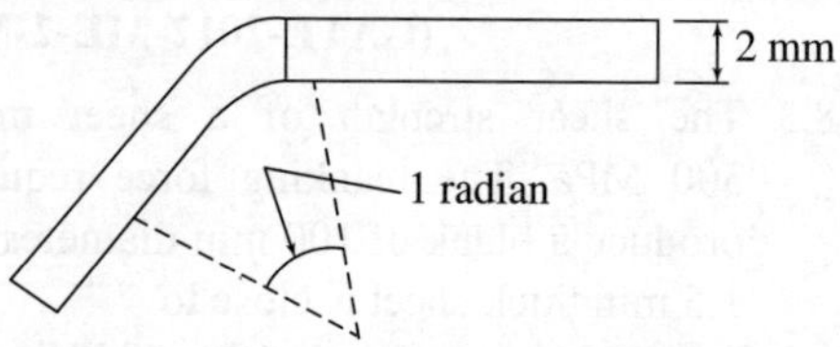

(a) 99 mm (b) 100 mm
(c) 101 mm (d) 102 mm

(GATE-2005-ME-2-Marks)

8.7 A shell of 100 mm diameter and 100 mm height with the corner radius of 0.4 mm is to be produced by cup drawing. The required blank diameter is

(a) 118 mm (b) 161 mm
(c) 224 mm (d) 312 mm

(GATE-2003-ME-2-Marks)

8.8 A metal disc of 20 mm diameter is to be punched from a sheet of 2 mm thickness. The punch and the die clearance is 3%. The required punch diameter is

(a) 19.88 mm (b) 19.94 mm
(c) 20.06 mm (d) 20.12 mm

(GATE-2003-ME-2-Marks)

8.9 In a blanking operation, the clearance is provided on

(a) The die
(b) Both the die and the punch equally
(c) The punch
(d) Neither the punch nor the die

(GATE-2002-ME-2-Marks)

8.10 The cutting force in punching and blanking operations mainly depends on

(a) The modulus of elasticity of metal
(b) The shear strength of metal
(c) The bulk modulus of metal
(d) The yield strength of metal

(GATE-2003-ME-2-Marks)

8.11 In deep drawing of sheets, the values of limiting draw ratio depends on

(a) Percentage elongation of sheet metal
(b) Yield strength of sheet metal
(c) Type of press used
(d) Thickness of sheet

(GATE-1994-ME-1-Mark)

CHAPTER 9

9.1 Under optimal conditions of the process the temperatures experienced by a copper work piece in fusion welding, brazing and soldering are such that

(a) $T_{\text{welding}} > T_{\text{soldering}} > T_{\text{brazing}}$
(b) $T_{\text{soldering}} > T_{\text{welding}} > T_{\text{brazing}}$
(c) $T_{\text{brazing}} > T_{\text{welding}} > T_{\text{soldering}}$
(d) $T_{\text{welding}} > T_{\text{brazing}} > T_{\text{soldering}}$

(GATE-2016-ME-SET-1, 1-Mark)

9.2 The welding process which uses a blanket of fusible granular flux is

(a) Tungsten inert gas welding
(b) Submerged arc welding
(c) Electro slag welding
(d) Thermit welding

(GATE-2016-ME-SET-2, 1-Mark)

9.3 In a linear arc welding process, the heat input per unit length is inversely proportional to

(a) Welding current
(b) Welding voltage
(c) Welding speed
(d) Duty cycle of the power source

(GATE-2015-ME-SET-1, 1-Mark)

9.4 The major difficulty during welding of aluminium is due to its
(a) High tendency of oxidation
(b) High thermal conductivity
(c) Low melting point
(d) Low density

(GATE-2014-ME-SET-1, 1-Mark)

9.5 In solid-state welding, the contamination layers between the surfaces to be welded are removed by
(a) Alcohol
(b) Plastic deformation
(c) Water jet
(d) Sand blasting

(GATE-2014-SET-1, 1-Mark)

9.6 Within the Heat Affected Zone (HAZ) in a fusion welding process, the work material undergoes
(a) Micro structural changes but does not melt
(b) Neither melting nor micro structural changes
(c) Both melting and micro structural changes after solidification
(d) Melting and retains the original microstructure after solidification

(GATE-2014-SET-4, 1-Mark)

9.7 Match the CORRECT pairs.

Processes		Characteristics/ Application	
P.	Friction Welding	1.	Non-consumable electrode
Q.	Gas Metal Arc Welding	2.	Joining of thick plates
R.	Tungsten Inert Gas Welding	3.	Consumable electrode wire
S.	Electroslag Welding	4.	Joining of cylindrical dissimilar materials

(a) P-4; Q-3; R-1; S-2
(b) P-4; Q-2; R-3; S-1
(c) P-2; Q-3; R-4; S-1
(d) P-2; Q-4; R-1; S-3

(GATE-2013-ME-1-Mark)

9.8 Which one among the following welding processes uses non-consumable electrode?
(a) Gas metal arc welding
(b) Submerged arc welding
(c) Gas tungsten arc welding
(d) Flux coated arc welding

(GATE-2011-ME-1-Mark)

9.9 Which one of the following is a solid state joining process?
(a) Gas tungsten arc welding
(b) Resistance spot welding
(c) Friction welding
(d) Submerged arc welding

(GATE-2007-ME-1-Mark)

9.10 The strength of a brazed joint
(a) Decreases with increase in gap between the two joining surfaces
(b) Increases with increase in gap between the two joining surfaces
(c) Decreases up to certain gap between the two joining surfaces beyond which it increases
(d) Increases up to certain gap between the two joining surfaces beyond which it decreases

(GATE-2005-ME-1-Mark)

9.11 Two 1 mm thick steel sheets are to be spot welded at a current of 5000 A. Assuming effective resistance to be 200 micro-ohms and current flow time of 0.2 second, heat generated during the process will be
(a) 0.2 Joule (b) 1 Joule
(c) 5 Joules (d) 1000 Joules

(GATE-2004-ME-1-Mark)

9.12 In Oxy-acetylene gas welding, temperature at the inner core of the flame is around
(a) 3500°C (b) 3200°C
(c) 2900°C (d) 2550°C

(GATE-2003-ME-1-Mark)

9.13 Match the following:

Work Material	Type of Joining
(P) Aluminium	(1) Submerged Arc Welding
(Q) Die Steel	(2) Soldering
(R) Copper Wire	(3) Thermit Welding
(S) Titanium Sheet	(4) Atomic Hydrogen Welding
	(5) Gas Tungsten Arc Welding
	(6) Laser Beam Welding

(a) P-2, Q-1, R-2, S-6
(b) P-3, Q-1, R-5, S-2
(c) P-1, Q-3, R-1, S-5
(d) P-3, Q-4, R-2, S-6

(GATE-2003-ME-2-Marks)

9.14 Which of the following arc welding processes does not use consumable electrodes?
(a) GMAW
(b) GTAW
(c) Submerged Arc Welding
(d) None of these

(GATE-2002-ME-1-Mark)

9.15 The temperature of a carburizing flame in gas welding is ___________ that of a neutral or an oxidizing flame.
(a) Lower than (b) Higher than
(c) Equal to (d) Unrelated to

(GATE-2002-ME-2-Marks)

9.16 Preheating before welding is done to
(a) Make the steel softer
(b) Burn away oil, grease, etc, from the plate surface
(c) Prevent cold cracks
(d) Prevent plate distortion

(GATE-1996-ME-1-Mark)

9.17 The electrodes used in arc welding are coated. This coating is not expected to
(a) Provide protective atmosphere to weld
(b) Stabilize the arc
(c) Add alloying elements
(d) Prevents electrode from contamination

(GATE-1994-ME-1-Mark)

9.18 In d.c. welding, the straight polarity (electrode negative) results in
(a) Lower penetration
(b) Lower deposition rate
(c) Less heating of work piece
(d) Smaller weld pad

(GATE-1993-ME-2-Marks)

Chapter 11

11.1 The operation in which oil is permeated into the pores of a powder metallurgy product is known as
(a) Mixing (b) Sintering
(c) Impregnation (d) Infiltration

(GATE-2011-ME-1-Mark)

Solutions

Chapter 2

2.1 (d)
Tempering is the process to reduce brittleness of martensitic steel.

2.2 (a)
Normalizing is process where steel is heated to certain temperature and then cooled in room atmosphere. Hence attains a single phase which gives the time to the grains to grow and form large grains. Thus option A is correct.

2.3 (d)

2.4 (c)

2.5 (c)

2.6 (b)

2.7 (b)

2.8 (c)
2.9 (b)
2.10 (d)
2.11 (a)
2.12 A-R, B-P, C-S, D-T
2.13 (a)

Chapter 3

3.1 (d)
3.2 (a)

volume of casting

$$= 200 \times 100 \times 10$$
$$= 2 \times 10^5 \text{ mm}^3$$

volume of pattern

$$= (200 \times 100 \times 10) \times (0.99)^3$$
$$= 2 \times 10^5 \times 0.970$$

Ratio of volume of pattern to casting

$$= \frac{\text{volume of pattern}}{\text{volume of casting}} = 0.97$$

3.3 (c)
3.4 (c)
3.5 (d)
3.6 (a)

Chapter 4

4.1 (c)

Choke has minimum area out of sprue, runner and ingate

4.2 (a)

d_c = 200 mm $d_r = h_r$ c = Casting
h_c = 100 mm h_r = ? r = Rises
$M_r = 1.2\ M_c$

$$\left(\frac{V}{S}\right)_r = 1.2\left(\frac{V}{S}\right)_c$$

$$\frac{\frac{\pi}{4}d_r^2 h_r}{d_r h_r + \frac{\pi}{4}d_r^2} = 1.2\frac{\frac{\pi}{4}d_c^2 h_c}{d_c h_c + \frac{\pi}{4}d_c^2 x^2}$$

$$= \frac{d_r^2 h_r}{4d_r h_r + d_r^2} = \frac{1.2 d_c^2 h_c}{4d_c h_c + 2d_c^2}$$

$$\frac{h^3{}_r}{4h_r^2 + h_r^2} = \frac{1.2 \times (200)^2 \times 100}{4 \times 200 \times 100 + 2 \times 200^2}$$

$$[\because d_r = h_r]$$

$$\frac{h_r}{5} = 1.2 \times \frac{200 \times 10}{4 \times 100 + 2 \times 200}$$

$$h_r = 1.2 \times 5 \times \frac{200 \times 100}{4 \times 200} 125 \times 1.2 = 150$$

4.3 (b)

$$t \propto \left(\frac{V}{A}\right)^2$$

$$A_1 = 5a^2 \quad (1)$$

$$;\ A_2 = 6a^2 \quad (2)$$

$$\because V = \text{constant}$$

$$t \propto \frac{1}{A^2}$$

$$\frac{t_1}{t_2} = \left(\frac{A_2}{A_1}\right)^2$$

$$\frac{t_1}{t_2} = \left(\frac{6a^2}{5a^2}\right)^2 = \frac{36}{25}$$

4.4 (6.15)

$$\text{solidification time} = k\left(\frac{V}{A}\right)^2$$

$$\text{for cube, } t = 4 = k.\left(\frac{a^3}{6a^2}\right)^2 = k.\left(\frac{a}{6}\right)^2$$

$$k.\frac{a^2}{36} = 4$$

$$ka^2 = 36 \times 4 \quad (1)$$

vol. of cube = vol. of sphere

$$\Rightarrow \frac{4}{3}\pi r^3 = a^3$$

$$\Rightarrow r = \left(\frac{3}{4\pi}\right)^{1/3} a$$

solidification time for sphere,

$$t' = k.\left(\frac{\frac{4}{3}\pi r^3}{4\pi r^2}\right)^2 = \left(\frac{r}{3}\right)^2$$

$$\Rightarrow k\left(\left(\frac{3}{4\pi}\right)^{1/3}.a/3\right)^2$$

$$\Rightarrow \left(k.a^2 \times \left(\frac{3}{4\pi}\right)^{2/3}\right)\Big/9$$

$$= 6.1573\,\text{sec}$$

4.5 (0.25)

$$t_s = k\left(\frac{V}{A}\right)^2$$

for cylindrical cavity (1)

$$t_1 = k\left(\frac{\frac{\pi}{4}d^2h}{\pi dh + \frac{\pi}{2}d^2}\right)^2 = k\left(\frac{d}{6}\right)^2 \; (d = h)$$

for cubic casting (2)

$$t_2 = k\left(\frac{a^3}{6a^2}\right)^2 = k\left(\frac{a}{6}\right)^2 = k\left(\frac{d}{3}\right)^2 \; (a = 2d)$$

$$\therefore \frac{t_1}{t_2} = \frac{\left(\frac{d}{6}\right)^2}{\left(\frac{d}{3}\right)^2} = \left(\frac{1}{2}\right)^2 = 0.25$$

4.6 (7 to 8)

$d = 1600\ \text{kg/m}^3$

$\rho = 2600\ \text{kg/m}^3$

Net buoyancy force

= weight of liquid displaced – weight of solid body

$$= v \times \rho \times g - v \times d \times g = vg(\rho - d)$$

$$= \frac{\pi}{4}d^2L \times g(\rho - d)$$

$$= \frac{\pi}{4}0.1^2 \times 0.1 \times 9.81(2600 - 1600)$$

$= 7.7\text{N}.$

4.7 (b)

$$\text{since } t = k\left(\frac{\text{volume}}{\text{Surface Area}}\right)^2$$

for a cube of side 'x'

$$\frac{v}{A} = \frac{x^3}{6.x^2} = \frac{x}{6}$$

$$\therefore \frac{t_1}{t_2} = \frac{x_1^2}{x_2^2} \Rightarrow \frac{5}{t_2} = \frac{x_1^2}{x_2^2}$$

$$\Rightarrow t_2 = 5\left(\frac{x_2}{x_1}\right)^2 \quad (1)$$

for same material which is 8 times heavier

$v_2 = 8V_1$ as densities are same

$$x_2^3 = 8.x_1^3 \quad (2)$$

$$\therefore x_2 = 2x_1$$

From equation (1) and (2),

$$t_2 = 5.\left(\frac{2x_1}{x_1}\right)^2 = 20\ \text{min}$$

4.8 (a)

4.9 (c)

4.10 (a)

Chapter 5

5.1 (c)

5.2 (a)

5.3 (b)

5.4 (d)

5.5 (c)

Due to uniform ramming, the density of the moulding sand will be same throughout the flask. Due to the uniform density, it will acquire greater dimensional stability

Chapter 6

6.1 (d)

6.2 (d)

6.3 A-4, B-6, C-5, D-3

6.4 (b)

Casting is made by the principle of centrifugal force phenomena. Material flashes on the outer surface of the flask with a high force leading to fine grain structure and high density.

Chapter 7

7.1 (c)

$$\Delta H = D(1-\cos\alpha)$$

$$\cos\alpha = 1-\frac{\Delta H}{D} = 1-\frac{0.1\times 8}{410} \Rightarrow \alpha = 3.57^\circ$$

$$\alpha = 3.57\times\frac{\pi}{180} \text{ radians} \Rightarrow \alpha = 0.062 \text{ radians}$$

7.2 (a)

7.3 (c)

7.4 (d)

7.5 (d)

h_1 = initial thickness = 16 mm,

h_2 = final thickness

= 10 mm

Roll diameter = D = 400 mm,

$$\tan\theta = \sqrt{\frac{h_1-h_2}{D/2}},$$

where θ = Angle of bite $\therefore \theta = 9.826$

Chapter 8

8.1 (d)

8.2 (a)

Diameter of punch = Diameter of Blank – 2 × radial clearance – die allowance

$= 25 - 2\times 0.06 - 0.05 = 24.83$ mm

8.3 (c)

Blanking force = $\tau . A_s = 300\times\pi dt$

$= 300\times\pi\times 100\times 1.5 = 141\ kN$

8.4 (b)

8.5 (d)

8.6 (c)

Bend allowance

$= \alpha(r+kt) = 1\times(100+(0.5\times 2)) = 101$ mm

8.7 (c)

Blank diameter,

$$D = \sqrt{d^2+4dh} = 223.60 \text{ mm}$$

8.8 (a)

Required Punch diameter

$$= 20 - \frac{2\times 6}{100} = 19.88 \text{ mm}$$

8.9 (c)

8.10 (b)

8.11 (d)

Chapter 9

9.1 (d)

9.2 (b)

Submerged arc welding uses a blanket of fusible granular flux

9.3 (c)

9.4 (a)

9.5 (b)

9.6 (a)

9.7 (a)

9.8 (c)

9.9 (a)

9.10 (d)

9.11 (d)

Heat generated = $I^2RT = 1000$ J

9.12 (b)

9.13 (d)

9.14 (b)

9.15 (b)

9.16 (c)

Preheating before welding is done to slow the cooling rate of the weld and the base material, resulting in softer weld metal and heat affected zone.

9.17 (d)

The important functions of electrode coatings are:

1. To stabilize the arc.

2. To generate gases to act as a shield against the surrounding atmosphere
3. To control the rate at which the electrode melts.
4. To act as a flux to protect weld against formation of oxides, nitrides and other inclusions.
5. To add alloying elements to the weld zone to enhance the properties of the joint

9.18 (B and D)

In general, DCSP results in:

- Higher Penetration rate
- More heating of work piece (around 70% of heat is generated at work piece
- Lower deposition rate

Chapter 11

11.1 (c)

Appendix

INTRODUCTION TO SI UNITS

Throughout this book, only S I (international system) units are used since they have been universally adopted.

The base units related to this book are presented in Table A1 along with their standard symbols.

TABLE A1 SI Units

Quantity	Unit	Unit Symbol
Length	metre	m
Mass	kilogram	kg
Time	second	s
Electric current	ampere	A
Temperature	kelvin	K
Amount of substance	mole	mol

When the value of any quantity goes beyond a certain limit, it is customary in SI units to provide prefixes to indicate the multiples and submultiples of the SI base units. The standard prefixes with their meanings are presented in Table A2. While specifying the prefixes only one is to be used, e.g.

$$3.2 \times 10^{-12}\ \text{m} = 3.2\ \text{pm}$$

It is customary to represent units with multiples and submultiples of 10 power raised to the power of multiples of 3 so that the actual value is between 1000 and 0.001.

TABLE A2 Unit prefixes

Multiplication Factor	Expressed as	Prefix	Symbol
1 000 000 000 000 000 000	10^{18}	exa	E
1 000 000 000 000 000	10^{15}	peta	P
1 000 000 000 000	10^{12}	tera	T
1 000 000 000	10^{9}	giga	G
1 000 000	10^{6}	mega	M
1 000	10^{3}	kilo	k
100	10^{2}	hecto	h*
10	10^{1}	deka	da*
0.1	10^{-1}	deci	d*

(Contd.)

0.01	10^{-2}	centi	c
0.001	10^{-3}	milli	m
0.000 001	10^{-6}	micro	μ
0.000 000 001	10^{-9}	nano	n
0.000 000 000 001	10^{-12}	pico	p
0.000 000 000 000 001	10^{-15}	femto	f
0.000 000 000 000 000 001	10^{-18}	atto	a

* refers to the prefixes which are not generally preferred.

In writing numbers either in the text or in tables, it is customary not to use comma (,) for separating the digits of 1000 and above but to use a blank space instead. For example, 213,012 should be written as 213 012. However, when the number is only of four digits, the number need not be spaced in text. But in tables, if there are other numbers which are of more than four digits in the same column, then the spacing is used.

Besides these base units, there are a number of other units which are derived from them to indicate the specific properties such as stress, power etc. The details of these are given in Table A3. Besides these there are some units which are not part of the SI units but they continue to be used because of the wide spread familiarity. These units have been mentioned in Table A4.

TABLE A3 Derived units

Quantity	Name	Symbol	Formula
Frequency	hertz	Hz	s^{-1}
Force	newton	N	$m.kg.s^{-2}$
Energy, work	joule	J	N.m
Power	watt	W	$j.s^{-1}$
Electric potential	volt	V	$W.A^{-1}$
Electric resistance	ohm	Ω	$V.A^{-1}$
Pressure or stress	pascal	Pa	$N.m^{-2}$
Linear velocity	—	—	$m.s^{-1}$
Density	—	—	$kg.m^{-3}$

TABLE A4 Some of the derived units whose use is predominant though not part of SI units

Quantity	Name	Symbol	Formula
Volume	litre	L	1000 mm^3
Time	minute	min	60 s
Time	hour	h	3.6 ks

Some of the conversion factors that are used in the book, are presented in Table A5.

TABLE A5 Conversion factors

English to Metric Units			Metric to English Units		
From	**To**	**Multiply**	**From**	**To**	**Multiply**
Length					
mil	millimeter (mm)	0.0254	millimeter (mm)	mil	39.37
inch (in)	mm	25.4	mm	in	0.0394
in	centimeter (cm)	2.54	cm	in	0.3937
foot (ft)	centimeter	30.48	cm	ft	0.0328
ft	meter (m)	0.3048	m	ft	3.2808
yard (yd)	m	0.9144	m	yd	1.0936
Area					
in^2	mm^2	645.16	mm^2	in^2	0.0016
in^2	cm^2	6.4516	cm^2	in^2	0.155
ft^2	cm^2	929.03	cm^2	ft^2	0.0011
ft^2	m^2	0.0929	m^2	ft^2	10.7639
yd^2	m^2	0.8361	m^2	yd^2	1.196
Volume					
in^3	cm^3	16.3871	cm^3	in^3	0.061
fluid ounce	cm^3	29.5735	cm^3	fluid ounce	0.0338
ft^3	dm^3	28.3169	dm^3	ft^3	0.0353
ft^3	m^3	0.0283	m^3	ft^3	35.3147
yd^3	m^3	0.7646	m^3	yd^3	1.3079
in^3/lb	m^3/kg	3.6E-05	m^3/kg	in^3/lb	27,680
ft^3/lb	m^3/kg	0.0624	m^3/kg	ft^3/lb	16.018
Mass					
pound (lb)	gram	453.5924	gram	pound	0.0022
pound	kilogram (kg)	0.4536	kg	pound	2.2046
pound	metric ton	0.00045	metric ton	pound	2204.6
U.S. ton (short)	metric ton	0.9072	metric ton	U.S. ton (short)	1.1023
Force					
lbf	N	4.448	N	lbf	0.225
Density					
lb/in^3	kg/m^3	27.680	kg/m^3	lb/in^3	3.6E-05
lb/in^3	g/cm^3	27.68	g/cm^3	lb/in^3	0.03613
lb/ft^3	g/cm^3	0.016	g/cm^3	lb/ft^3	62.43
lb/ft^3	kg/m^3	16.0185	kg/m^3	lb/ft^3	0.0624

(*Contd.*)

Temperature					
in/(in °F)	m/(M °C)	1.8	m/(M °C)	in/(in °F)	0.556
°F	°C	(F–32)/(1.8)	°C	°F	(1.8 C)+32
°F	°K	(F+459.67)/(1.8)	°K	°F	(1.8 K)–459.67
Pressure					
lb/in^2	kPa	6.8948	kPa	lb/in^2	0.145
lb/in^2	MPa	0.00689	MPa	lb/in^2	145
lb/in^2	GPa	6.9E-06	GPa	lb/in^2	145,038
lb/in^2	bar	0.0689	bar	lb/in^2	14.51
Energy and Power					
ft lbf	J	1.3558	J	ft lbf	0.7376
in lbf	J	0.113	J	in lbf	8.850
ft lbf/in	J/m	53.4	J/m	ft lbf/in	0.0187
ft lbf/in	J/cm	0.534	J/cm	ft lbf/in	1.87
ft lbf/in^2	kJ/m^2	2.103	kJ/m^3	ft lbf/in^3	0.4755
kW	metric horsepower	1.3596	metric horsepower	kW	0.7355
U.S. horsepower	kW	0.7457	k W	U.S. horsepower	1.3419
Btu	J	1055.1	J	Btu	0.00095
Btu	W h	0.2931	W h	Btu	3.412
Btu in/(h ft^2 °F)	W/(m °K)	0.1442	W/(m °K)	Btu in/(h ft^2 °F)	6.933
Btu/lb	kJ/kg	2.326	kJ/kg	Btu/lb	0.4299
Btu/(lb °F)	J/(kg °C)	4187	J/(kg °C)	Btu/(lb °F)	0.000239
V/mil	MV/m	0.0394	MV/m	V/mil	25.4
Output					
lb/min	g/s	7.560	g/s	lb/min	0.1323
lb/h	kg/h	0.4536	kg/h	lb/h	2.2046
Velocity					
in/min	cm/s	0.0423	cm/s	in/min	23.6220
ft/min	m/min	0.3048	m/min	ft/min	3.2808
ft/s	m/s	0.3048	m/s	ft/s	3.2808

Bibliography

1. Abramov, G.: *Foundry Practice for young workers,* Mir Publishers, Moscow, 1986.
2. Adams, C.M. Jr., *Cooling Rates and Peak Temperatures in Fusion Welding,* Welding Journal Research Supplement, Vol. 37, No. 5, 1958, pp 210 - 215.
3. AFS.: *Foundry Sand Handbook,* American Foundrymen's Society, Des Plaines, 1963.
4. AFS: American Foundry Society, *Casting Defects Handbook: Iron & Steel,* George Goodrich, Technical Editor, American Foundry Society, Schaumburg, IL, 2008.
5. AFS: *Cupola and its Operation,* American Foundrymen's Society, Des Plaines, 1965.
6. AFS: *Patternmaker's Manual,* American Foundrymen's Society, Des Plaines, 1960.
7. Alexander, J.M. and R.C.Brewer: *Manufacturing Properties of materials,* Van Nostrand, London, 1963.
8. Altan, T., Soo-lk Oh and H.L. Gegel: *Metal Forming, Fundamentals and Applications,* American Society of Metals, Metals Park, Ohio, 1983.
9. Althouse, A.D., C.H. Turnquist, W.A. Bowditch and K.E. Bowditch: *Modern Welding,* The Goodheart-Wilcox, South Holland, 1997.
10. ASM: *ASM Metals Handbook Vol. 15, "Casting"*, American Society of Metals, Metals Park, 2008.
11. ASM: *Forging Design Handbook,* American Society of Metals, Metals Park, 1965.
12. ASM: *Metals Handbook Vol. 1 Properties and Selection: Iron and Steels,* American Society of Metals, Metals Park, 1990.
13. ASM: *Metals Handbook, Vol. 8 Mechanical testing,* American Society of Metals, Metals Park, OH, 2000.
14. ASM: *Source book on Electron Beam and Laser Welding,* American Society of Metals, Metals Park, 1981.
15. Beeley, P. R.: *Foundry Technology,* Butterworths, London, 2001.
16. Bishop, H.F. E.T. Myskowski and W.S. Pellini: A Simplified Method for Determining Riser Dimensions, *Trans AFS, Vol 63, 1955, pp 271 - 281.*
17. Blazynski,T.Z.: *Design of Tools for Deformation Processes,* Elsevier/Applied Science, London, 1986.
18. Bradley, E.F.: *High Performance Castings - A Technical Guide,* ASM International, Metals Park, 1989.
19. Brick, R.M., A.W. Pense and R.B. Gordon: *Structure and Properties of Engineering Materials,* McGraw-Hill, New York, 1977.
20. Brown, J.R. ed.: *Foseco Foundryman's handbook,* 10th Ed., Butterworth Heinemann, Oxford, 1994.
21. Caine,J.B.: *Design of Ferrous Castings,* American Foundry men's Society, Des Plaines, 1963.
22. Campbell, J.S.: *Principles of Manufacturing Materials and Processes,* McGraw-Hill, New York, 1961.
23. Chvorinov,N.: Theory of Solidification of Castings, *Geisserei, Vol 27, 1940, pp 177 - 225.*
24. Clark, D.S. and W.R.Varney, *Physical Metallurgy for Engineers*, Litton Educational, Belmont, 1962.

25. Connor, L.P. (ed.), *Welding Handbook, Vol. 1, Welding Technology,* 8th Edition, American Welding Society, Miami, 1987.
26. Cook, N.H.: *Manufacturing Analysis,* Addison-Wesley, Reading, 1966.
27. Dalls, D.B. (ed): *SME Tool and Manufacturing Engineers Handbook*, McGraw-Hill, New York, 1976.
28. Dave Smith: *Welding Skills and Technology,* Gregg Division, McGraw-Hill, New York, 1984.
29. Davidson, A. (ed): *Handbook of Precision Engineering, Vol 10 Forming Processes,* Macmillan, London, 1974.
30. Davies, D.J. and L.A. Oelmann: *The Structure, Properties and Heat Treatment of Metals,* Pitman Books, London, 1983.
31. Dieter,G.E.: *Mechanical Metallurgy,* 3rd ed., McGraw-Hill, New York, 1986.
32. Donaldson, C., G.H.LeCain and V.C.Goold: *Tool Design,* McGraw Hill, New York, 1973.
33. Easterling, K., *Introduction to the Physical Metallurgy of Welding,* Butterworths, London, 1983.
34. Evans,R.M., R.P.Meister and W.C.Brayton: Fire Cracker Welding for ship yard Applications, *Welding Journal, Vol. 55, no 7, 1976, pp 555-565.*
35. Feldmann,H.D.: *Cold Forging of Steel,* Hutchinson, London, 1961.
36. Flinn, R.A.: *Fundamentals of Metal Casting,* Addison-Wesley, Reading, 1963.
37. Hayden,H.W., W.G.Moffatt and J.Wulff: *The Structure and Properties of Material Vol. III Mechanical Behaviour,* John Wiley, New York, 1965.
38. Heine, R.W., C.R. Loper and P.C. Rosenthal: *Principles of Metal Casting,* McGraw Hill, New York, 1967.
39. ISO/TR 10108:1989: *Steel - Conversion of hardness values to tensile strength values,* International Standards Organisation, 1989.
40. Jaggi, O.P.: *Dawn of Indian Technology,* Atma Ram, New Delhi, 1969.
41. Jain, P.L.: *Principles of Foundry Technology*, 5th ed., Tata McGraw-Hill, New Delhi, 2009.
42. Jhaveri, P., Moffatt, W.G. and Adams, C.M. Jr., The effect of Plate Thickness and Radiation on Heat Flow in Welding and Cutting, *Welding Journal, Vol. 41, No. 1, 1962, pp 12 - 16.*
43. Jutz, H. and E. Scharkus (ed): *Westermann Tables for the Metal Trade,* Wiley Eastern, New Delhi, 1976.
44. Kaye, A. and A. Street: *Diecasting Metallurgy,* Butterworth Scientific, London, 1982.
45. Koshal, D. (ed.): Manufacturing Engineers Reference Book, Butterworth Heinemann, Oxford, 1993.
46. Krauss George, *Principles of Heat Treatment of Steel,* American Society for Metals, Metals Park, 1980.
47. Maleka, A.H. (ed): *Electron Beam Welding,* McGraw-Hill, London, 1971.
48. Manko, H.H.: *Solders and soldering,* McGraw-Hill, New York, 1964.
49. Marek, C.T.: *Fundamentals in the Production and Design of Castings,* John Wiley, New York, 1963.
50. Masterov, V. and V. Berkovsky: *Theory of Plastic Deformation and Metal Working,* Mir Publishers, Moscow, 1975.
51. Matousek, R.: *Engineering Design,* Blackie, Glasgow, 1963.
52. Mohan, J., R.M. Krishnan, B.R. Nijhawan, P.K. Gupte and B.V. Somayajulu: *Foundry Moulding Sands of India,* Council of Scientific & Industrial Research, New Delhi, 1961.
53. Morris,J.L.: *Welding Processes and Procedures,* Prentice-Hall, Englewood Cliffs, NJ, 1958.
54. Mukherjee, P.C.: *Fundamentals of Metal Casting Technology,* 2nd ed., Oxford & IBH Publishing, New Delhi, 1988.
55. Niebel, B.W.and A.B. Draper: *Product Design and Process Engineering*, McGraw-Hill, New York, 1974.

56. Orlov, P.: *Fundamentals of Machine Design Vol. 3*, Mir Publishers, Moscow, 1977.
57. Ostergaard, D.E.: *Advanced Die Making,* McGraw-Hill, New York, 1967.
58. Ostergaard, D.E.: *Basic Die Making,* McGraw-Hill, New York, 1967.
59. Patan, B. (ed): *Electric Slag Welding,* Mir Publishers, Moscow, 1967.
60. Patte, H.E.: *Technological Advances in Welding and other Joining Processes,* Battelle Press, Columbus, 1982.
61. Pearson, C.E. and R.N.Parkins: *The Extrusion of Metals*, Chapman & Hall, London, 1961.
62. Polukhin, P., N. Fedosov, A. Korolyov and Y. Matveyev: *Rolling Mill Practice*, Peace Publishers, Moscow, 1968.
63. Porter, L.F. and P.C. Rosenthal: Fluidity Testing of Grey Cast Iron, *Trans AFS, Vol. 60, 1952, p 725.*
64. Rajan, T.V., C.P. Sharma and Ashok Sharma: *Heat Treatment Principles and Techniques*, Prentice Hall, New Delhi, 1988.
65. Rossi, B.E.: *Welding Engineering,* McGraw-Hill, New York, 1954.
66. Ruddle,R.W.: *The Running and Gating of Sand Castings,* The Institute of Metals, London, 1956.
67. Ruddle, R.W.: *The Solidification of Castings*, The Institution of Metals, London, 1957.
68. Sabroff, A.M. et al: *Forging Materials and Practices*, Reinhold, New York, 1968.
69. Sachs,R.J.: *Theory and Practice of Arc Welding*, Van Nostrand Reinhold, New York, 1960.
70. Singer, C. E.J. Holmyard and A.R. Hall ed: *A History of Technology, Vol. I,* Oxford at Clarendon Press, Oxford, 1954.
71. SME: *Die Design Handbook*, McGraw Hill Book, New York, 1965.
72. SME: *Fundamentals of Tool Design,* 6th Edition, Society of Manufacturing Engineers, Dearborn, MI, 2010.
73. Street,A.C. (ed): *Diecasting Book,* Port cullis Press, London, 1977.
74. Suryanarayana, A.V.K.: *Testing of Metallic Materials,* Prentice Hall, New Delhi, 1985.
75. Titov, N.D. and Yu.A. Stepanov: *Foundry Practice*, Mir Publishers, Moscow, 1981.
76. Tselikov,A.: *Stress and Strain in Metal Rolling,* Mir Publishers, Moscow, 1967.
77. Tselikov, A.I. and V.V.Smirnov: *Rolling Mills,* Pergamon Press, Oxford, 1965.
78. Upton, B.: *Pressure Diecasting Part 1 Metals, Machines, Furnaces,* Pergamon Press, Oxford, 1982
79. Wallace, J.F. and E.B. Evans: Risering of Grey Iron Castings, *Trans AFS, Vol 67, 1959.*
80. Wallace, J.F., and E.B. Evans: Gating of Grey Iron Castings, *Trans AFS , Vol 65, 1957, pp 267 - 275.*
81. Weisman, C. (ed): *Welding Handbook, vol 1, Fundamentals of Welding,* American Welding Society, Florida, 1976.
82. Wlodawer, R.: *The Directional Solidification of Steel Castings,* Pergamon Press, Oxford, 1966.
83. Woods, P.F.: *Production Welding,* McGraw-Hill, London, 1971.
84. Wusatowski, Z.: *Fundamentals of Rolling,* Pergamon Press, Oxford, 1969.

Index

E

F

G

H

I

J

K

L